T0344969

Precision Cosmology
The First Half Million Years

Cosmology seeks to characterise our Universe in terms of models based on well-understood and tested physics. Today we know our Universe with a precision that once would have been unthinkable. This book develops the entire mathematical, physical and statistical framework within which this has been achieved.

It tells the story of how we arrive at our profound conclusions, starting from the early 20th century and following developments up to the latest data analysis of big astronomical datasets. It provides an enlightening description of the mathematical, physical and statistical basis for understanding and interpreting the results of key space-based and ground-based data. Subjects covered include general relativity, cosmological models, the inhomogeneous Universe, physics of the cosmic background radiation, and methods and results of data analysis.

Extensive on-line supplementary notes, exercises, teaching materials, and exercises in Python make this the perfect companion for researchers, teachers and students in physics, mathematics, and astrophysics.

Bernard J. T. Jones is Emeritus Professor at the Kapteyn Astronomical Institute of the University of Groningen. His research has covered many areas of astrophysics, both theoretical and observational, with a strong emphasis on cosmology, where he is widely published and cited.

Precision Cosmology

The First Half Million Years

BERNARD J. T. JONES

Kapteyn Astronomical Institute
University of Groningen

CAMBRIDGE
UNIVERSITY PRESS

Shaftesbury Road, Cambridge CB2 8EA, United Kingdom

One Liberty Plaza, 20th Floor, New York, NY 10006, USA

477 Williamstown Road, Port Melbourne, VIC 3207, Australia

314–321, 3rd Floor, Plot 3, Splendor Forum, Jasola District Centre, New Delhi – 110025, India

103 Penang Road, #05–06/07, Visioncrest Commercial, Singapore 238467

Cambridge University Press is part of Cambridge University Press & Assessment, a department of the University of Cambridge.

We share the University's mission to contribute to society through the pursuit of education, learning and research at the highest international levels of excellence.

www.cambridge.org
Information on this title: www.cambridge.org/9780521554336

DOI: 10.1017/9781139027809

First published 2017

A catalogue record for this publication is available from the British Library

Library of Congress Cataloging-in-Publication data
Names: Jones, B. J. T. (Bernard Jean Trefor), 1946–
Title: Precision cosmology : the first half million years / Bernard J.T.
Jones, Rijksuniversiteit Groningen, The Netherlands.
Description: Cambridge : Cambridge University Press, 2017. | Includes
bibliographical references and indexes.
Identifiers: LCCN 2016058586 | ISBN 9780521554336 (alk. paper)
Subjects: LCSH: Cosmology. | Cosmology–Mathematics. | Cosmological distances.
Classification: LCC QB981 .J664 2017 | DDC 523.1–dc23
LC record available at https://lccn.loc.gov/2016058586

ISBN   978-0-521-55433-6   Hardback

To the memory of my teenage daughter Suzie,
in fulfilment of my promise to her.

Contents

Preface

The subject matter in this book covers the theory and observations relating to the earliest epochs of our evolving Universe, from its singular origin to just beyond the epoch of recombination: the first half million years. This is a fundamental period of the Universe's lifetime, and the period from which the key parameters describing it can be determined. As our understanding of the Universe grew, and new observations were found to rule out the simplest models, new physical entities and concepts had to be introduced to ensure consistency between theory and observation. The subsequent period, from recombination to the present epoch, is the era during which stellar and galactic physics takes place, and is not covered in this volume.

Precision Cosmology is a teaching book and a technical reference book which aims to provide a clear and understandable description of the applied mathematics and physics required to establish the high level of precision which can now be achieved for the study of the Universe – Precision Cosmology. It details the background and mathematical tools needed to enable a student embarking on advanced studies and research to understand what others have done and published.

This book is also an experiment in presentation. The printed text is accompanied by extensive online supplements, which provide a high level of didactic mathematical material too extensive to be included in the main text. These supplements are freely available for download and are intended as a source of teaching material for both teachers and students across a range of topics in physics, mathematics and statistics. They are suitable for all levels of university study, from undergraduate to postdoctoral research. The online supplements and other support material are available from

www.BernardJones.net/PrecisionCosmology.html

Figures in the text marked ⓒ are available on this website and are free for use under the terms of the CC-BY-SA Creative Commons License. Exercises and slides for teaching purposes are also to be found on this site.

In this way, much of the book should be accessible to physics and mathematics undergraduates who have had courses in mechanics, electromagnetism and special relativity. The aim is to take the student beyond this basic knowledge, right up to the level of being able to follow and understand research papers, and thus to understand today's cosmology in its proper context as a branch of physics. In this way the text should be equally useful to those postgraduates wishing to embark on research into cosmology.

Our understanding of the Universe, the science of cosmology, has evolved remarkably over the past century and, in particular, over the years since the discovery in 1965 of the Cosmic Microwave Background Radiation (CMB). That discovery established the basic paradigm of modern cosmology: the Hot Big Bang theory. So, by the turn of the 21st century another revolution was already under way: a revolution based on big science and big data, with data gathering projects consuming considerable resources in terms of telescopes, computers, detectors, etc., and involving big teams of engineers and scientists, numbering in their hundreds or even thousands. Cosmology today is largely about the acquisition, modelling and analysis of data with which we may confront our theories.

This book describes the science that forms the basis for our understanding of the Universe and drives the design of experiments to test that understanding. The main focus is on the methods and tools that are are available to model and analyse the data from these experiments. This is the basis for our profound conclusions about the nature of our Universe.

Good science needs corroboration from alternative directions. However, unlike almost every other branch of physics, we have only one example of the object of our study: our Universe is unique and we cannot study any other for the purposes of corroboration. We cannot just take another sample. Corroboration therefore requires considerable ingenuity and that, to some extent, has come via simulations of the Universe, from analysis of the distribution of galaxies and from gravitational lensing studies. These very different approaches to determine the cosmological parameters yield very similar results. Yet none of these, taken alone, would provide convincing evidence that our interpretation is correct. However, taken collectively, and combining also with the CMB, Baryonic Acoustic Oscillations (BAOs) and supernova data, our conviction that everything is self-consistent is strengthened.

Fifty years before the publication of this book, cosmology was a small but important branch of Einstein's General Theory of Relativity (GR). Observational cosmology at that time was once described as merely being 'a search for two numbers' that would define which of the relativistic models would best describe the Universe. Fifty years before that our Galaxy was believed to be the entire Universe: all the nebulae, clusters and stars were thought to be part of the one super-system, enshrined first as the 'Kapteyn Universe' and a little later as Harlow Shapley's 'meta-Galaxy'. At that time, Einstein's theory had only just achieved its final form.

Today, at the start of the 21st century, we have explored, mapped and modelled our Universe using vast quantities of data obtained and analysed by teams of thousands of engineers and scientists. We have determined the six basic parameters that describe our model with remarkable accuracy, often to two or three significant digits. While this effort has successfully yielded a convincing paradigm, it has revealed a number of important issues such as the mysteries of dark matter and dark energy. There is now strong evidence that 72% of the Universe comprises what is referred to as 'dark energy', and we have no idea what this is. A lot of ingenuity will go into future research to discover its nature.

The Structure of the Book

The text of the book is divided into five fairly independent parts, and each part is then divided into chapters. The five parts are:

I. 100 years of Cosmology
A recap on the highlights of the past century. Much of this could be regarded as introductory material for the parts that follow, especially as regards setting the material in a holistic context.

II. Newtonian Cosmology
Simple Newtonian cosmological models and what they can explain.

III. Relativistic Cosmology
Special and general relativity and the greater precision it gives.

IV. The Physics of Matter and Radiation
The physics of the cosmic background radiation.

V. Precision Tools for Precision Cosmology
Statistical methods, precise observations and data processing techniques.

Bibliographic Material

The bibliography is hardly complete, despite there being some 1000 references – there are today tens of thousands of papers that are relevant to cosmology. The cited papers are mainly the ones on which I have drawn in order to write this book. Generally, such papers are either unique, or they have a nice way of explaining their goals and methods that I will have borrowed for my text. The citation of those latter authors is one way I can acknowledge my gratitude for their clarity of exposition.

I also prefer to cite papers that are generally available over the internet either through an openly accessible archive, or through the generosity of the publisher. I have avoided citing papers that are only available to those having subscriptions to journals or who are willing to pay for downloading copies of the articles. I hold to the belief that the tax payer has already paid generously for the science we do and, as a consequence, should have full and free access to its results.

Use of Internet Resources

Much of the information that is on the periphery of my own area of expertise has been culled from the numerous excellent web pages of astronomers, physicists and statisticians around the world. This material is mostly in the form of conference presentations, lecture notes, diagrams and e-prints. This has obviously played a part in my own education both as an astrophysicist and as a teacher. It would be impossible to even recall where most of the acquired knowledge came from, let alone cite it. I can only thank all those people without specific acknowledgement.

This book would not have been possible without access to that vast and freely available resource: the Internet. Access to the SAO/NASA Astrophysics Data System Digital Library (ADS) has been invaluable, facilitating my trawling through thousands of articles. The help of Edwin Henneken of ADS has been much appreciated. Of course, there are also

Wikipedia, NASA, and the e-print arXiv who make and keep information freely accessible. These are an essential part of scientific research that can only be fully appreciated by those of us who are old enough to have been without them.

Personal Acknowledgements

Much of the material here has been culled from my own research, review articles, scientific papers (published and unpublished), text books, resources on the world wide web, and even from the lecture courses I attended as an undergraduate and graduate student many years ago. Inevitably the clearest methods and explanations I have learned from these sources have found their way into my own lecture courses and hence into this text. In particular I am indebted to my own teachers of long ago whose clarity enthused me about relativity and cosmology through their lectures and their books: C.W. Kilmister, F.A.E. Pirani, H. Bondi, D.W. Sciama, M.J. Rees, G.F.R. Ellis. Their superlative lectures are still burned into my brain. Throughout my career I have been fortunate to have mentored and supervised a large number of talented graduate students. That process has been a key part of my own education both as a teacher and a scientist.

I am indebted to Jan Tauber, Planck Project Scientist, for allowing me to use Planck pictures and edit them for publication in grey tones. Indeed, I would like to thank all those who granted permission to use their figures and those who offered advice.

On a more personal level it is important that I acknowledge the help from my wife, Janet, in editing this book and making it a reality. Without her stringent hands-on help I would still have over 1500 pages of text and no book. She was able, despite protestation, to throw out material until it got down to some 700 pages, just short of the Cambridge University Press 750 page limit. Her corrections to my grammar, compiling lists of errors, and for general encouragement when things got difficult or even seemed impossible, are in retrospect highly appreciated. She made it happen. Indeed, Cambridge University Press (CUP) have been incredibly tolerant of my ever-shifting deadlines, starting in the mid-1970s when this project was first conceived. Cosmology changed rapidly since that time, and evidently faster than I could write. The efforts of my Copy-Editor, Dr Richard Smith, in bringing this volume into shape are particularly appreciated.

Rien van de Weygaert has played a singular role in this long term project, for which I am eternally grateful. He also provided me with a 'home' in Groningen. Rien has acted as my mentor throughout that period, and his help, advice and support have been more appreciated than I can put into words. Most of the book in its present form was written while at the Kapteyn Institute. I am proud that I should be able to put the Kapteyn Institute as my affiliation on the title page of this book.

Last but not least, my entire family is owed a tremendous vote of thanks for their encouragement and support, despite their wondering whether it would ever be possible for me get this book finished. Well it's now finished, and a testament to the value of their support.

Bernard J.T. Jones
Kapteyn Astronomical Institute

Notation and Conventions

Throughout this book I have used 'astronomer's units' on the grounds that these are the units in which most of the astrophysics literature and textbooks are written. So brightness is given in 'magnitudes' and distances of astronomical object in parsecs, and so on. However, there are some instances that are a matter of choice of particular conventions.

Special Relativity

In the chapters on relativity and the conventions I have adopted there are best described as 'MTW' ((Misner *et al.*, 1973)) since that is a commonly used textbook for general relativity. Those conventions also accord with Hawking and Ellis (1975), Wald (1984) and the wonderful *Varenna Lectures* of Ellis (1971) (reprinted as Ellis (2009b)).

Throughout the section on General Relativity we use geometrised units in which $G = 1$ and $c = 1$ unless explicitly shown otherwise. Where particularly relevant, as in discussing observed data, these units will be converted to the dimensional units as are typically used in astronomy (i.e. astronomer's units).

The *signature of the metric* will be taken as $(-+++)$ and space-time coordinates in this system will be described as 4-dimensional vectors as in (x^0, x^1, x^2, x^3), the x^0 coordinate representing the time-like coordinate. This is the space-like $(1, 3)$ format. In that format the Minkowski metric for the line element ds when the coordinates are labelled (t, x, y, z) is

$$ds^2 = -c^2 dt^2 + dx^2 + dy^2 + dz^2 = -c^2 dt^2 + d\mathbf{r}^2,$$

with which the *proper time*, $d\tau$, is defined as

$$d\tau^2 = -ds^2.$$

With this, *time-like* intervals have $c^2 dt^2 > d\mathbf{r}^2$, i.e. $d\tau^2 > 0$ and, conversely, *space-like* intervals have $ds^2 > 0$, i.e. $d\tau^2 < 0$. Accordingly the *proper time* corresponding to a time-like interval separating two events is

$$d\tau = \sqrt{dt^2 - dr^2/c^2},$$

while the space-like separation between two events separated by a space-like interval is

$$d\ell = \sqrt{dr^2 - c^2 dt^2}.$$

General Relativity

Lower case Latin indexes $a, b, c, \ldots$ on vector and tensor quantities, as in u^a, and T^{ab} will take values $0, 1, 2, 3$, while lower case Greek indices $\alpha, \beta, \gamma, \ldots$ will take on the values $1, 2, 3$. The summation convention, implying summation over repeated indices, is used on lower case Latin and lower case Greek indices, but is suppressed when using upper case Latin indices.

We need to define some important symbols and tensors. Firstly, given a metric tensor g_{ab} we define the Christoffel symbols Γ^a_{bc} as

$$\Gamma^a_{bc} = \tfrac{1}{2} g^{ae} [g_{eb,c} + g_{ec,b} - g_{cb,e}].$$

With this we can define the Riemann Tensor as

$$R^a{}_{bcd} = \Gamma^a_{bd,c} - \Gamma^a_{bc,d} + \Gamma^a_{mc} \Gamma^m_{bd} - \Gamma^a_{md} \Gamma^m_{bc}.$$

The Ricci tensor R_{ab} and Ricci scalar R are the following contractions of the Riemann tensor:

$$R_{ab} = R^m{}_{amb} \quad R = R^m{}_m.$$

The Einstein tensor is

$$G_{ab} = R_{ab} - \tfrac{1}{2} g_{ab} R.$$

Covariant derivatives with respect to the metric g_{ab} are denoted by the symbol ∂_a, or, using subscripts:

$$u^a_{;b} \equiv \partial_b u^a = \frac{\partial u^a}{\partial x^b} + \Gamma^a_{mb} u^m = \partial_b u^a + \Gamma^a_{mb} u^m = u^a_{,b} + \Gamma^a_{mb} u^m.$$

Covariant derivatives do not commute:

$$u^a_{;bc} - u^a_{;cb} = -R^a{}_{mbc} u^m.$$

The signature used here is $(- + + +)$, as in MTW. To transform from the equally widely used $(+ - - -)$ signature, change the signs of the following:

$$g_{ab}, \quad \Box = g^{ab} \partial_a \partial_b, \quad R^a{}_{bcd}, \quad R_{ab}, \quad T^a{}_b,$$

but leave these unchanged:

$$R_{abcd}, \quad R_a{}^b, \quad R, \quad T_{ab}.$$

Electromagnetism

The electromagnetic units used are also 'astronomer's units', as opposed to the widely taught SI units. The varieties of electromagnetic units are discussed at some length in Appendix A. Since the goal here is to discuss Maxwell's equations and not to discuss the experimental side of electromagnetism, the units used here follow the system that is the most convenient and concise for this purpose.

PART I

100 YEARS OF COSMOLOGY

Emerging Cosmology

> *The Universe is as it is because it was as it was*
>
> Herman Bondi,
> Lectures at King's College London, 1965

1.1 Introduction

Almost every civilisation throughout history has had a cosmology of some kind. By this we mean a description of the Universe in which they live based on their state of knowledge. The Vikings, for example, had a complex cosmology in which the world and its inhabitants were controlled by a set of Gods, both good and bad. Nature was ruled not by the laws of physics, but by the forces of nature controlled by the whims of these Gods. However bizarre that may seem to us now, at the time this belief-set dominated human behaviour: its social mores and values.

Today we live in a Universe that is described by physical laws. What is remarkable is that these laws have more often than not been discovered on the basis of laboratory experiments, and subsequently found to work on the vastest scales imaginable. That fact leads us to believe that our explanations of the Universe are a valid description of what is actually happening. We do not need to invoke special laws just to explain the cosmos and our position within it.

The physical laws governing the Universe and its constituents were discovered over a period of several centuries. Some might say this path to realisation started with Copernicus putting the Sun at the centre of everything rather than the Earth. Others might argue that this was merely descriptive and that knowledge of the laws started to emerge following on from the work of Kepler, Galileo and Newton. However one sees it, by the beginning of the 20th century, with Einstein's Theory of Relativity, the scene was set to embark on a journey of observational cosmology which 100 years later would lead to most scientists agreeing that we have a self-consistent theory of the Universe based on known laws of physics. Some, no doubt, would go as far as to say that the current view was incontrovertible.

Just as the early map makers measured and marked out our planet, the cosmographers of the 20th century marked out and mapped the Universe. Just as those map makers and those who used the maps showed that the Earth was round, the cosmographers of the 20th century have shown that the Universe we see is, in the large, homogeneous and isotropic, and, most remarkably, is expanding and began a finite time in our past. In addition, the evidence was that the birth of the Universe was phenomenally hot and so the theory of this origin became known as the Hot Big Bang theory.

The fact of the cosmic expansion from a hot singular state a finite time ago in our past must surely be one of the outstanding revelations of 20th century science. Some, myself included, might even argue that it is one of the most fundamental discoveries in all of science.

Acceptance of the finite age expanding model did not come easily. The idea of a finite age Universe had been around for over 50 years when, in 1965, there was a remarkable discovery that effectively set the seal on this notion. This was the discovery by Penzias and Wilson of a cosmic electromagnetic radiation field left over from the initial event. The radiation field is now referred to as the Cosmic Microwave Background Radiation (CMBR).

This radiation field served to establish the physical model of the Universe, and within a few years of the discovery many scientists were working on exploring the consequences of that model. Observation of the Universe went hand in hand with theoretical advances to clarify and test the Hot Big Bang Theory, and now this is one of the paradigms of modern science.

However, as it turned out, not everything was perfect: there were still surprises in store. Observers, while mapping out the furthest reaches of the Universe, discovered that the cosmic expansion was not quite as simple as had been first thought. It appears that although we can construct models of the Universe that are fully consistent with this observational data, that consistency can only be achieved by supposing that most of our Universe is made up of matter of an as yet unknown kind. The material we are familiar with, the atoms that make up the substances of everyday life, are but a small fraction of everything. We are as certain about the existence of this 'dark matter' as we are about the Big Bang itself.

The challenge is to understand the paradox that the dominance of dark matter presents us with. We have studiously avoided moulding the locally determined laws of physics to explain the phenomenon of cosmic expansion, yet we find we are forced to introduce something we did not know about before. Depending on who you are, this is either a disturbing or exciting situation.

In this book I will describe the story that leads to this remarkable conclusion, concentrating on the observations and the theoretical framework within which they are interpreted.

1.2 Pre-20th Century Cosmology

Big advances in science have often been driven by advances in technology, and this is no less true in astronomy and cosmology. Galileo with his telescope opened up the Solar System, Herschel built the largest telescopes of their time (see Figure 1.1) and produced the first great catalogue of galaxies. Hubble had access to the greatest telescopes of his day: first the Crossley $60''$, followed by the Hooker $100''$ and then the Hale $200''$. That story continues right up to the present day, as we shall see in the chapters to come.

Fig. 1.1   The iconic picture of 19th century astronomy: William Herschel's great '40 foot' reflecting telescope. The telescope, designed and built by Herschel, first saw light in 1789 and was last used in 1815. This was the largest telescope ever built for a period of over 50 years when, in 1845, the Earl of Ross built his giant reflector having a 72″ mirror. Both telescopes had mirrors made of a reflective copper-tin alloy.
(From Encyclopædia Britannica, 1797.)

It is worth recounting the story of the people whose collective insight eventually led to an appreciation of one of the most remarkable discoveries made by humankind: that our Universe has a finite age. Scientifically, all the evidence points in this direction. That evidence has been accumulated as part of a century-long process bringing together the capabilities of observational astronomy and the fundamental physics of gravitation. However, the process starts long before that, arguably with Galileo turning his telescope to the heavens, and certainly with William Herschel who systematically studied the heavens beyond the stars with the most powerful telescopes of his day.

Short biography: **William Herschel** (1738–1822) came to England around 1759 and took up a career as a musician. Herschel became interested in astronomy in 1773, and very quickly went on to master the technology of building reflecting telescopes. In 1781 he discovered the planet Uranus and shortly after that became court astronomer to the King of England, George III. Much of his work was done in collaboration with his sister Caroline whose own work was acknowledged in 1828 by the Royal Astronomical Society with the award of the Society's Gold Medal.

The existence of objects in the sky that were neither stars nor planets was only appreciated towards the end of the 18th Century. Halley (1716) had listed six nebulous or diffuse objects, but the first serious catalogues were those of de la Caille (1755) and of Messier (1784). This latter catalogue was said to have been drawn up to help astronomers searching for comets from erroneously picking up known nebulae. It was a compilation of the work of others and has been occasionally added to since its first publication. The Messier catalogue has had a lasting impact on astronomy. Today, Messier's catalogue contains 110 objects of which 40 are galaxies.

The year following the publication of Messier's Catalogue, saw William Herschel's first catalogue of nebulae: it contained over 1000 objects. Over the following years that catalogue grew to list 2500 objects (Herschel, 1789, 1800). William Herschel's son, John Herschel, expanded his father's catalogue to create *The General Catalogue of Nebulae and Clusters* containing over 5000 nebulae (Herschel, 1864). This was soon used as the basis of several important studies of the all-sky galaxy distribution: Abbe (1867), Proctor (1869) and Waters (1873) (see Figure 1.2). Here, for the first time, was a map of the deep sky showing the zone of obscuration by the Milky Way and the clustered distribution of the nebulae (which had, by that time, started to be referred to as 'galaxies').

There is a more detailed overview of the history of this period in the article by Lundmark (1956) and the books by North (1965) and Saslaw (1999).

1.2.1 Observation in the Early 20th Century

By the early 1920s it was apparent that the Universe was mainly populated by galaxies, or, as they were referred to then, 'Island Universes'. Originally, the galaxies were thought by many astronomers to be parts of our own Galaxy of Stars, but early research showed that they were in fact distant stellar systems comparable in scale to our own Galaxy.

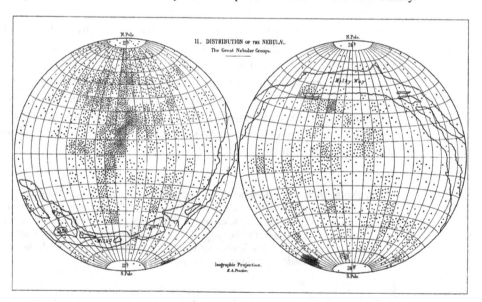

Fig. 1.2   R.A. Proctor's first all sky map showing the Milky Way and the distribution of nebulae (galaxies) over the sky (Proctor, 1869, Fig. 2). The obscuration from the Milky Way is quite evident, as is the clustered distribution of galaxies. The Virgo Cluster of Galaxies around (12^h, $+15°$) is clearly visible. The amateur astronomer Waters (1873) produced a similar map based on Abbe's (1867) refined analysis of the Herschel Catalogue (Herschel, 1864) *(© Royal Astronomical Society (1869).)*

The work of Henrietta Swan Leavitt in which she discovered that the period of Cepheid variable stars was related to their absolute magnitude (Leavitt, 1908; Leavitt and Pickering, 1912) is fundamental in establishing the cosmological distance scale.[1] She had been finding and studying variable stars in the Magellanic clouds and noticed that those which were identified as Cepheid variables having the longest period were the brightest, and since they were all at roughly the same distance they were the intrinsically brightest. Her data was shortly afterwards used by Hertzsprung (1913) to put the distance of the Small Magellanic Cloud (SMC) at 30 000 light years.[2] This makes the SMC the first object to be identified as extra-galactic.

> **Short biography: Henrietta Leavitt** (1868–1921) graduated from what is nowadays known as Radcliffe College in 1892. During her senior year there she took a course in astronomy and went on to take an unpaid job at the Harvard Observatory in 1893 when the observatory was under the directorship of Edward C. Pickering. In 1902 her position became permanent and she was awarded a nominal salary of $0.30 per hour for her work, which consisted mainly of the tedious job of measuring and cataloguing variable stars in the Magellanic clouds. Pickering treated her as a mere lab assistant and did not allow her to follow up on this work. She died of cancer in 1921, after which her work was taken on by Hubble in determining the distances to nearby galaxies.

When Hubble (1926) estimated the distance to the nearby galaxy M33 he did not cite Leavitt's ground-breaking paper, referring instead to Shapley (1918a), citing it and referring to it as 'Shapley's period luminosity curve'. Shapley (1918a,b) in his work on distances makes no reference to Henrietta Leavitt as having discovered the relationship, but merely makes a passing reference: 'Some years ago Miss Leavitt found a similar relation between the apparent photographic brightness and the length of the period for the Cepheid variable stars in the Small Magellanic Cloud'.[3]

In recent years, Leavitt's period-luminosity relationship has played a fundamental role in establishing the extragalactic distance scale. A key goal of the Hubble Space Telescope (HST) was to determine the Hubble constant, H_0, to within $\pm 10\%$ using Cepheid variables as the fundamental calibrators of the secondary distance measures. There were three such

---

[1] The first of these papers is a long-term study of 1777 variable stars in the Magellanic Clouds. She describes her photographic material and procedures in the first paragraphs of the first paper: this was evidently a formidable undertaking both in terms of the photometry and the data handling. The second of these papers starts 'The following statement regarding the periods of 25 variable stars in the Small Magellanic Cloud has been prepared by Miss Leavitt'. She then goes on to report that 'A remarkable relation between the brightness of these variables and the length of their periods will be noticed'.

[2] Hertzspring's paper has a misprint at this point, reporting 3 000 light years for an effective parallax of 0.0001″. It was common in the early years of the 20th century to quote distances in terms of the parallax of the object when seen from the opposite points on the Earth's orbit. That defines the *parsec* unit of distance as D parsecs $= 1/p$ arc seconds (″) for a parallax p: A 1″ parallax corresponds to a distance of 1pc, (hence the name *parsec*, attributed to Herbert Turner in 1913). 1pc = 3.261 ... light years.

[3] The period luminosity data shown in Shapley (1918a, Fig. 1) for periods < 200 days is very similar to the data shown in Leavitt and Pickering (1912, Fig. 2). In both papers the period is displayed using the logarithm of the period. Leavitt, however, simply put a straight line through the points. That was enough to make her point, though on closer examination there is a manifest curvature in the data points (see Figure 1.3). Shapley (1918a, Table XI) provided a table of the curve he had fitted to the data values which would have made Hubble's task a little easier than using Leavitt's plot. It would be fair to say that Shapley's main contribution was the calibration of the Cepheid period luminosity relation using parallaxes and other methods, and that was what Hubble needed.

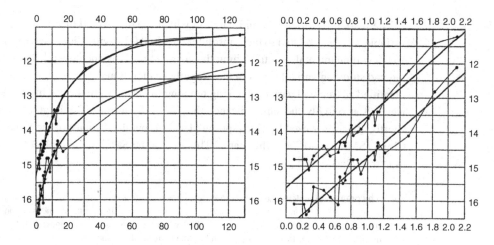

Fig. 1.3   Henrietta Leavitt's Period–Luminosity diagram for 25 Cepheid variable stars in the Small Magellanic Cloud (Leavitt and Pickering, 1912, Figs. 1 & 2). The vertical axis is the apparent magnitude (brightest at the top) and two curves are shown on each panel, one fitting the data for the brightness at maximum light and the other for the brightness at minimum. The horizontal axis is the period in days, linear in the left panel and logarithmic on the right. This is arguably one of the most important plots in the history of astronomy.
(Source: Harvard College Observatory Circular, 1912)

'Key projects' awarded to research groups in 1986. There were many advantages in using HST that would allow the study of Cepheids in galaxies outside of the Local Group and up to ten times the distance that could be achieved from ground-based observatories. The final summary of the achievements, as given in Freedman *et al.* (2001), is $H_0 = 72 \pm 2$ km s^{-1} Mpc^{-1}, with a possible additional systematic error of some ± 7 km s^{-1} Mpc^{-1}. The largest contribution to the systematic error is the absolute zero point of the Cepheid period–luminosity relation. More recent use of the Spitzer Space Telescope has allowed a further increase in accuracy by reducing the systematic errors to give $H_0 = 74.3 \pm 2.1$ km s^{-1} Mpc^{-1} (Freedman *et al.*, 2012).

In those early years of the 20th century the quest for distance estimators for nebulae was intense. Although the Cepheid method could be locally calibrated and could take us to the nearest galaxies, it involved taking and calibrating numerous high resolution photographic plates and searching for Cepheid variable stars: Leavitt's achievement was indeed quite remarkable. There were at the time two alternatives to Cepheids: use of Novae and observations of the brightest stars. Lundmark (1956) has written an eyewitness review on the situation at that time. Lundmark was himself one of the first people to estimate the distances to several nearby galaxies: he provided a detailed overview in his paper of 1925 (Lundmark, 1925).

The year before, Lundmark (1924) had written a paper on distances to galaxies and, using Slipher's velocity data (Slipher, 1915, 1917), had plotted a velocity–distance diagram (*loc. cit.* Figure 5). He comments that 'Plotting the radial velocities against these relative distances (Fig. 5) we find that there may be a relation between the two quantities, although not a very definite one'. The situation was left to Hubble (1926, p.356 *et seq.*) to resolve.

1.2.2 Who Discovered the Cosmic Expansion?

One of the key historical issues is the question of who did what and when? Who discovered the expansion of the Universe? That discovery is attributed to Hubble, but several people, notably Lundmark and Lemaitre, had considered cosmic expansion before his paper was published. Should they be credited with the discovery rather than Hubble?

This all happened in the brief period between around 1900 and 1930 and there are no witnesses left who can testify from experience. There is only the documentary evidence of the papers that were published and the letters and notes that were left. I want to take a slightly different approach in looking at this: I want to put myself as a researcher in the period 1925–1930 and address the questions of who knew what. What was the environment and the thinking that such a researcher would have found?

Two suggestions will emerge from this. Firstly: Hubble's approach used data that was not available to any others, so the others could not have done what he did. The others might have suspected the truth, but, unlike Hubble, they would have been unable to establish their claim convincingly. Secondly, there was an enormous gulf between 'observers', who gathered and analysed data and 'theorists', who at that time were often lost in the maze of complexities brought up by Einstein's theory. Hubble had never really come to terms with the interpretation of his discovery in terms of an expanding Universe. Conversely, few, if any, mathematicians of the time could relate the data to the coordinate systems used in their equations. Eddington in his book had noted, as others had done, the preponderance of positive radial velocities in the data available to him as 'very striking'. Yet he did not question the reason for this, he did not say 'the Universe seems to be flying apart', nor did Hubble in his famous paper of 1929.

1.2.3 General Relativity and Cosmology

It was apparent shortly after the publication of Einstein's General Theory of Relativity that this theory provided the appropriate framework for discussing mathematical models of the Universe.

Hubble (1929) discovered the relationship between redshift and distance that bears his name, but the interpretation of that result was not immediate. Hubble himself, in that important paper, never used either of the words 'expansion' or 'recession', but instead alluded to '. . . the possibility that the velocity–distance relation may represent the de Sitter effect . . .', going on to explain briefly what the de Sitter effect was. The title of Hubble's paper was *A Relation between Distance and Radial Velocity among Extra-Galactic Nebulae*. But, no matter what Hubble's thinking on the interpretation might have been, it was a landmark paper in cosmology: he had found direct evidence relating the redshift of the spectral lines with distance.[4]

---

[4] Given that Hubble did not conclude that the Universe was expanding, there then arises the question of 'Who discovered the expanding Universe?'. There are many interesting discussions on this point, focusing mainly

Soon after Einstein published his General Theory of Relativity (Einstein, 1915), a number of people, notably de Sitter (1917a, 1918) and Einstein (1917a) himself, produced solutions to his field equations that could describe a Universe governed by the influence of the force of gravity. There was little data with which to test these models, and so they remained just that, abstract mathematical models.

Prior to the publication of Hubble's 1929 paper several papers had been published presenting exact homogeneous and isotropic solutions to the Einstein Field Equations, notably by Einstein, by de Sitter, by Friedmann (1922, 1924) and by Lemaître (1927, 1931a).

The solutions of Friedmann[5] and Lemaître were non-static, expanding, solutions that now underpin all of modern cosmology. The work of Friedmann was certainly an important mathematical triumph, but it had absolutely no connection with data or with the real world (see Heller (1985)). Lemaître actually calculated a redshift–distance relationship for his solution and later championed the notion of the 'primeval atom' in which the Universe expanded from a denser state. But it was perhaps Robertson who first wrote about the expanding Universe, even before the publication of Hubble's 1929 paper (Robertson, 1928, 1929).

Four scientists, Friedmann, Lemaître, Robertson and Walker established the theoretical basis for modern cosmology within the framework of Einstein's General Theory of Relativity. They provided what should be referred to as the Friedman–Lemaître–Robertson–Walker equations for a homogeneous and isotropic expanding cosmological model that is derived from Einstein's theory of general relativity. The appropriate abbreviation should be 'FLRW models', though in the literature we variously see this referred to as the 'Friedmann–Lemaître solution' or the 'Robertson–Walker metric'. North (1965) gives the opinion that while Robertson wound up two decades of discussion on homogeneous and isotropic solutions to the Einstein equation, Lemaître set the pattern for the future of cosmology. There is a nice technical overview by Lemaître himself of how he saw his 'primeval atom' model in Lemaître (1958). My own opinion is that this vastly underestimates Robertson's contribution.

Short biography: Georges Lemaître (1894–1966) obtained his first degree in mining engineering at the University of Brussels in 1913. During the first World War he served as an artillery sergeant in the Belgian army, after which in 1919 he went to the University of Louvain to study physics and mathematics. During that period he studied the works of Einstein. He was ordained as a priest in 1923, and then spent a year in each of Cambridge University and the Massachusetts Institute of Technology (MIT), where he was awarded a PhD, before returning to Belgium and becoming Professor of Astrophysics at the Catholic University of Louvain in 1927. 1927 was the year in which he first published his relativistic solution describing an expanding Universe with a cosmological constant.

on the paper of Lemaître (1927) and its English language translation (Lemaître, 1931a). That aspect of the question is discussed in the papers of Kragh and Smith (2003) and of Kragh (2008), and more recently by Livio (2011) and Nussbaumer and Bieri (2011).

[5] There is no consistent transliteration of the spelling of Friedman's name from the Russian. Hence we variously see Fridman, Friedmann, and Friedman. The lunar crater named after him is 'Fridman'. The ADS bibliographic database uses 'Friedmann', so this form will be used here.

1.3 The Expansion Law

The famous Curtis–Shapley debate on whether spiral nebulae were within the galaxy or external systems had taken place in 1920 (Curtis, 1920, 1921; Shapley, 1919, 1921). Not surprisingly, no conclusion was reached, both astronomers stuck to their position. It was not until December 1924 that the astronomical community really appreciated that galaxies, or 'nebulae' as they were referred to at that time, were systems outside of our Galaxy. Hubble had a paper read at a meeting of the American Astronomical Society giving distances for three local galaxies, M31, M33 and NGC6822 using the Cepheid variables as distance indicators.[6] Even conceptually, the notion of an expanding Universe populated by galaxies was far from the then-current thinking.

Several technological advances helped to provide the necessary data. Large reflecting telescopes were built and data could be acquired onto photographic plates. Despite the size of the telescopes, the photographic emulsions were not very sensitive and so acquiring data was a slow process. Moreover, the data could only be acquired by the few who had access to these instruments.

The first published list of galaxy redshifts was that of Slipher (1915) (see also Slipher (1917)). The list had 15 galaxies. Most of the velocities in the list were large and positive and this led astronomers of the time to add a so-called 'K-term' to the solution for the motion of the Sun relative to distant stars and nebulae. Remember, at that time astronomers were not even sure whether these objects were extragalactic, so it would have been difficult to reach any profound conclusions.

Short biography: **Vesto Slipher** (1875–1969) After getting his PhD at Indiana University in 1909, Slipher spent his entire professional career at the Flagstaff Observatory in Arizona. In 1912 he obtained the first spectrum of a galaxy, M31, with an exposure of 6^h50^m using a $24''$ refractor telescope, revealing a systematic shift in the position of the spectral lines towards the blue. The velocity of approach of M31 was $-300\,km\,s^{-1}$, this was, at the time, the highest velocity measured in an astronomical object. His conclusion was that *extension of the work to other objects promises results of fundamental importance*. Despite the difficulty of measuring fainter objects, Slipher had indeed obtained spectra for 15 galaxies by 1915, 11 of which were red-shifted, and added a further six galaxies by 1917, all of which showed red-shifted spectral lines at hitherto unprecedented velocities. The final list he gave to Eddington in 1923 contained 41 galaxies.

His 1913 spectrum of NGC4595 (the *Sombrero* galaxy) revealed, for the first time the tilted spectral lines that were indicative of systemic rotation in galaxies. Later spectra found the same phenomenon in other galaxies. Slipher had also discovered and measured the rotation of galaxies.

Slipher expanded the list, giving measured radial velocities for some 41 galaxies.[7] He, Wirtz, Lundmark and Hubble all noted that the galaxies in this sample, with

---

[6] This is based on the comments of Sandage (1961b, p4.). The articles by Hoskin (1976) and Trimble (1995) are far more detailed, the former giving details and analysis of the talks as they were presented and the latter eloquently describing the important backdrop against which this was played out in 1920.

[7] Strictly speaking Slipher measured the redward shift of the spectral lines. It was natural to interpret this as a velocity since that is how they measured the radial velocities of stars. Thus the galaxies were assigned the velocities that corresponded to the observed redshift as if the redshift did reflect the radial velocity. However,

only a few exceptions, displayed positive radial velocities: they were moving away from us.

The extended data set of 41 velocities was published by Eddington in Section 70 of his book *Mathematical Theory of Relativity* (Eddington, 1923). It is certainly significant that this section of Eddington's book is on 'de Sitter's Spherical World', since this reflects the general scientific reaction to this dataset. If these objects were to be outside our Galaxy, this was at the time the only available explanation of the recession phenomenon.

Eddington noted in his text, as others had done, the preponderance of positive (receding) velocities, but evidently hesitated to draw the correct conclusion. It was in fact Wirtz (1922, 1924) who first suggested that the expansion term describing the velocity field (the 'K-term', Wirtz (1918)) might be a function of distance from the Sun. Wirtz came to this conclusion by noting that the galaxies of smaller diameter tended to have the greater radial velocities. Following up on this, Lundmark (1925) plotted a redshift magnitude diagram for this sample, but found no convincing trend. A reading of Lundmark's paper shows the extent to which there was confusion about the true nature of the 'nebulae': Lundmark used the same radial velocity data as Hubble but analysed the distances in a different manner and so failed to fully recognise what Hubble had seen.

The key data was spectroscopic. Slipher (1915, 1917)[8] had been able to get the spectra of the light from several brighter galaxies, and by 1923 that list had grown to 41 galaxy radial velocities. This sample was a key part of the data that Hubble (1929) used when realising that the fainter and presumably more distant objects were receding from us the fastest. Hubble established the law that now bears his name on the basis of data from only 41 galaxies, none of which he had himself obtained. The sample was slowly built up over the next decades.

By 1925 it was known that the galaxies were systems not unlike our own Galaxy, lying far beyond the limits of our own system. The quest for radial velocities of fainter galaxies was continued by Humason who provided the largest radial velocity known at the time (Humason, 1929). The paper immediately following Humason's in the journal was Hubble's (1929) confirmation of the existence of the redshift–distance effect that now bears his name; see Figure 1.4. Hubble did not however conclude that the effect was due to a general expansion of the system of galaxies, but referred to the phenomenon as the 'de Sitter Effect'.

It should be remembered that during those early years one of the most debated issues was whether the observed radial velocities were to be interpreted as being due to an overall cosmic expansion, or to the de Sitter Effect, or to some unknown agency such as 'tired light'.

in the context of 1920s general relativity and cosmology it would not have been clear that this interpretation was correct. The then-popular de Sitter model of the Universe explained redshifts in terms of a geometrical or gravitational effect, not a velocity. Hubble's own interpretation of the 'velocity–distance relationship' that he had discovered was that he had detected the de Sitter effect.

8   At that time the standard interpretation of the redshifts was the so-called de Sitter effect, see Sandage (1975, *Stars and Stellar Systems vol. IX*). It is perhaps surprising that even as late as 1953, Hubble expressed doubts about the interpretation of the recession velocity being due to the cosmic expansion and not some other phenomenon (see his George Darwin Lecture (Hubble, 1953)).

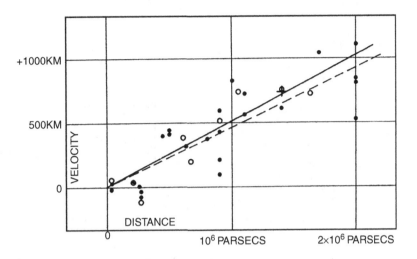

Fig. 1.4 Hubble's original diagram showing the magnitude redshift relationship (Hubble, 1929, Fig. 1). This was the first published evidence for the expansion of the Universe. The galaxies plotted here are from his Table 1. The nearer ones had their distances estimated from Cepheid variables, while the further ones used the brightest visible stars, calibrated via the Cepheid distances of the nearer galaxies.
(© Proc. Nat. Acad. Sci., public domain.)

Short biography: **Edwin Hubble** (1889–1953) obtained a degree in Mathematics, Astronomy and Philosophy at University of Chicago in 1910, and then went on to Oxford as a Rhodes Scholar where he studied Jurisprudence, Literature and Spanish, getting his MA in 1913. Some time after returning to the USA he did a PhD at the Yerkes Observatory in Chicago. The PhD title was 'Photographic investigations of faint nebulae': the scene was set. In 1919 he got a staff position at the Mount Wilson Observatory, hired by George Ellery Hale. In the decade that followed he would make one of the most momentous discoveries of modern science.

Within a year of the publication of Hubble's paper, de Sitter (1930) had recognised an issue with Hubble's work on the expansion: the galaxies plotted were a heterogeneous set of objects and not really a set of standard candles.[9] We see this in Hubble's Table I where (in the last column) there is a spread of four magnitudes in the absolute luminosity ascribed to the galaxies in his sample on the basis of the determined distances and apparent magnitudes. de Sitter addressed this problem by selecting galaxies on the basis of their surface brightness. Hubble (1926)[10] had already identified the mean surface brightness as a useful tool in classifying galaxies and had published the data necessary for de Sitter to produce his version of the velocity–distance diagram.

[9] From the point of view of Hubble's own work in producing the velocity–distance diagram, using standard candles was not particularly important: he could estimate the distances directly. However, de Sitter realised that it was possible to correct for the variation in intrinsic luminosity of the sample by using the surface brightness as an indicator of absolute magnitude. Hubble himself had provided the magnitude and diameter measurements of his galaxies that made this possible (Hubble, 1926).

[10] This paper is Hubble's most cited paper, while the expansion paper is only his second most cited. In this formidable paper, Hubble describes the basic galaxy types, S, SB, E and Irr and their subtypes, and discusses the number–magnitude counts.

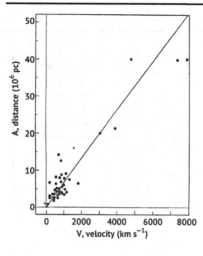
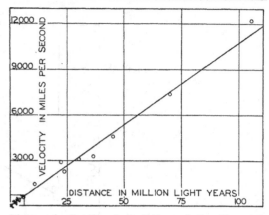

Diagram showing the velocity-distance relation. The small black dots in the lower left-hand corner represent the only available observations up to 1928. The open circles represent recent observations.

Fig. 1.5 Two Hubble expansion diagrams published in 1930 and 1931. Left: de Sitter (1930, Fig. 3) who selected galaxies using a surface brightness criterion, and Right: Humason (1931, Fig.2 p.151) who was starting on his quest to get ever higher redshifts for the Hubble diagram (the original figure caption has been retained). See also Hubble and Humason (1931, Fig. 5 p.165). Notice that de Sitter plots the velocity on the horizontal axis, which is now the current convention. Notice also that the distances quoted are for different values of the Hubble expansion rate. (de Sitter's diagram has been edited for greater clarity, see footnote 11.)
(Diagrams reproduced from the cited papers. Left: © Proc. Nat. Acad. Sci., public domain. Right: © 1931 University of Chicago Press, with permission.)

Importantly, the surface brightness is independent of distance, and so could be calculated without knowing the distance to the galaxy using the apparent magnitude m and the apparent diameter, d. Galaxies having the same mean surface brightness have the same value of $\log d + 0.2m$. de Sitter's version of the Hubble diagram is shown in the left panel of Figure 1.5.[11]

At the time of Hubble's seminal work on the expansion, 43 redshifts were known, most of which had been measured by Slipher. By 1931 Humason (1931) had added a further 46 radial velocities of galaxies in eight clusters (see also Hubble and Humason (1931)). By using the average velocities and distances for the clusters he was able to radically improve the expansion diagram (see the right-hand panel of Figure 1.5). In that paper Humason stated explicitly that the phenomenon was due to the recession of the galaxies, though, in the light of the controversy over the interpretation of the phenomenon, he was guarded in his declaration and referred to 'apparent velocities'.

Short biography: **Milton Humason** (1891–1972) left school at the age of 14, finding work as a mule driver in George Ellery Hale's project to build a telescope on Mount Wilson in the Sierra Nevada mountains. In 1917 he become a janitor at the Mount Wilson observatory, going on to become a highly respected night assistant. In 1919 Hale made him a staff member, despite strong protests on the grounds that Humason

---

[11] The available, scanned, copy of de Sitter's published graph was of rather low quality, so the above figure has been edited for clarity. In the original, de Sitter had distinguished between spiral and elliptical galaxies, but the distinction was poorly indicated and so has been removed in this rendition. The irregulars, denoted by crosses have not been changed.

had received no formal education. Humason was undoubtedly one of the driving forces behind the early study of the cosmic expansion. Humason's work culminated in the great Humason, Mayall and Sandage catalogue of galaxy redshifts published in 1956, shortly after his retirement in 1954.

The doubling of the number of redshifts out to a far greater distance than ever achieved previously was due to the availability at Mount Wilson of a new spectrograph lens that allowed shorter exposure times by a factor of ten or so. Humason (1931) pushed the redshift–magnitude relation to far greater depths, but it was not until after the papers of Humason (1936a,b) that the relationship was referred to as the Hubble Diagram.[12]

Hubble was not working in a vacuum, and he did have access to the largest telescopes in the world. That enabled him to see individual stars in galaxies beyond our local Galaxy neighbourhood and estimate real distances. But he could not have achieved that without the earlier work of Henrietta Swan Leavitt, Knud Lundmark and Vesto Slipher.

In 1932 Shapley and Ames published the very important Shapley–Ames Catalogue (Shapley and Ames, 1932). This was an all-sky catalogue of 1246 galaxies brighter than magnitude 13.2, which is 50% complete down to magnitude 12.7.[13] The catalogue had no redshift data, but it was an important jumping-off point for the galaxy catalogues of the 1950s, 1960s and 1970s. It was the standard listing of nearby bright galaxies for 50 years, until it was revised by de Vaucouleurs and de Vaucouleurs (1964); de Vaucouleurs et al. (1976, 1991) (known respectively as RC1,RC2 and RC3) and by Sandage and Tammann (1981).

Short biography: **Adelaide Ames** (1900–1932) came to astronomy in 1922 when she joined Shapley's graduate program at Harvard College Observatory as an assistant to Harlow Shapley. She had graduated from Vassar that year and originally wanted to pursue a career in journalism. The following year she was joined by Cecelia Payne-Gaposhkin, who had come from England to join Shapley. The two women, both born in the same year, became close friends. Ames died in a boating accident after her canoe capsized the year the catalogue was published.

1.3.1 Interpretation of the Redshifts

Astronomers in the 1920s and 30s were still focused on notions of gravitational redshifts or the so-called 'de Sitter effect' that had been derived as consequences of Einstein's General Relativity. Hubble was evidently unaware of the solutions of Friedmann and Lemaître and in his 1929 paper attributed the relationship to the 'de Sitter effect'. In the years before Hubble's 'discovery', it had been noticed that the fainter galaxies in Slipher's list had the higher velocities and with that there had been two notable suggestions that this could be interpreted as a consequence of a *cosmic* expansion.[14] Lemaître (1927, 1931a, the edited translation into English) had calculated the cosmic expansion rate based on Hubble's previously published distance estimates and came up with a value similar to the value that

---

[12] In the second of these papers (Humason, 1936b) Humason plots velocity against the magnitude of the 5th ranked cluster galaxy, an important choice of distance indicator.

[13] The catalogue was updated posthumously with Adelaide Ames as sole author in 1950 (Ames, 1950).

[14] As opposed to the rapid expansion of a local subsystem of nebulae. Hubble (1925) had already established the distances to M31 and M33 and shown that they were far beyond anything previously known.

would be obtained by Hubble two years later. Robertson (1928) boldly suggested that the redshift–distance relationship reflected the expansion of the Universe and had provided the theoretical explanation for the phenomenon. Given Slipher's velocities, the key was, of course, knowing the distances to the galaxies and Hubble had provided that. So we would certainly be justified in calling the Hubble velocity–distance relationship the 'Hubble–Lemaître–Robertson Law'.

A few years after the publication of Hubble's paper, Hubble collaborated with Richard Tolman on writing an important 35 page paper, Hubble and Tolman (1935). In that paper, they discussed the relationship between observations of galaxies and the issues in interpreting the Hubble diagram in the light of relativistic cosmological models. The numerous issues involved in making quantitative photometric observations of galaxies that are limited by the wavelength sensitivity of the photographic emulsions are discussed at length, and all set within the framework provided by the metric of Friedmann, Lemaître and Robertson.[15] Arguably, the paper of Hubble and Tolman marked the start of observational relativistic cosmology.

By 1956, Humason *et al.* (1956) were able to publish a catalogue of almost 1000 galaxy redshifts confirming not only the Hubble law, but allowing the first determination of the parameters of the cosmic expansion: the rate of expansion and the acceleration of the expansion. With that came the determination of the age of the Universe. This was not, however, without problems. Although the apparent brightness of galaxies was known from measurements, the intrinsic brightness was not known. Hence the distance to the galaxies could not be established, and the scale and age of the Universe were therefore uncertain. The uncertainty was considerable: it took many decades to sort this out.[16]

1.3.2 The Physics of the Big Bang – George Gamow

Not everyone, and notably Hubble himself, were happy with the expanding Universe interpretation of the redshift–distance relation and there were important alternative theories. Notable among these was the 'Steady State Theory' which proposed that, instead of all the matter in the Universe being created at one instant in a cataclysmic event, the material would be created continuously at just the rate required to fill up the space left by the expansion. The Big Bang versus Steady State controversy raged for 20 years.

The definitive evidence about the physical nature of the Universe came in 1965 with the discovery of the Cosmic Microwave Background Radiation ('CMB'). This radiation field was interpreted as natural consequence of a *Hot* Big Bang theory in which the Universe was hotter in the past.[17] The idea that the Big Bang would have been hot enough to synthesise

---

[15] The final sentence of this great paper is 'It also seems desirable to express an open-minded position as to the true cause of the nebular red-shift, and to point out the indications that the spatial curvature may have to play a part in the explanation of the existing nebular data'. This was a statement that the Hubble diagram would eventually provide a measure of the geometric curvature of the Universe.

[16] The discovery of Quasars in the early 1960s provided a flurry of excitement since they were bright objects with higher recession velocity than any hitherto measured galaxy. Schmidt (1965) measured the first redshift in excess of 1.0. However, they were manifestly not standard candles and since there was no way of estimating their intrinsic luminosity they added little to the story of the expansion of the Universe except perhaps to cause a certain amount of controversy, see Section 2.5.

[17] Russian astrophysicists generally, and perhaps more descriptively, referred to this as the 'Relict Radiation'.

the elements came with George Gamow, and with his work came the notion that there would be a radiation field left over from the Big Bang.

Gamow had been the strongest advocate of the Hot Big Bang theory in the decades prior to 1965. As early as 1946, a time when all aspects of nuclear physics were being intensively researched, he was interested in the origin of the chemical elements and he needed a very hot place to create them. The ideal place would be a Hot Big Bang.[18] On the basis of his insight, Gamow (1956) could even attempt to calculate the temperature that would be necessary to make these elements. His guess was quite close, though it was the Soviet physicist Smirnov (1965) who in 1964 presented what was probably the first correct calculation using a limited number of nuclear reactions.

As it turned out, Gamow's notion of the Hot Big Bang, encapsulated in Figure 1.6, is essentially what cosmologists would advocate today.[19] In his early book on cosmology,

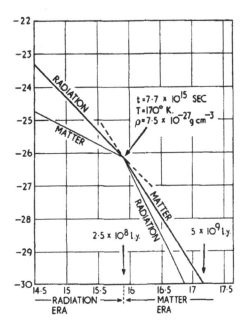

Fig. 1.6   Ten years before the discovery of the cosmic microwave background, George Gamow (1956) made this sketch about the history of a Universe emerging from a Hot Big Bang. The densities, in gm cm^{-3}, of matter and radiation (ordinates) are plotted as a function of time in seconds (abscissae). Gamow notes the transition from the radiation dominated era to the matter dominated era on the plot. The axes are logarithmic and the displayed numbers are based on values for the cosmic matter and radiation densities and for the expansion rate that seemed reasonable at that time.
A version of the diagram was used by Dicke *et al.* (1965) to interpret the discovery of the CMB and is reproduced in Figure 3.3. The modern version of this diagram appears in Figure 7.4.
(© *Elsevier 1956, reproduced with permission.*)

[18] The origin of the Universe provides a place in which deuterium, tritium, helium, lithium and other light elements can be created: it is difficult to make these in the required quantities anywhere else and the Hot Big Bang makes the right amounts (see Section 7.3.2).

[19] It is interesting to note that Gamow's 1956 paper *loc. cit.* was possibly the first time this diagram, describing the *physics* of the cosmic expansion, appeared in print. Gamow had already calculated the numbers with which

Gamow (1952, p.42) derives what is now the standard formula for the temperature of the Universe as a function of time[20] : $T_{\text{Kelvin}} = 1.5 \times 10^{10}/t_{\text{sec}}^{1/2}$ and substitutes the value of the approximate age of the Universe as known at that time, $t \sim 10^{17}$ s to get a present day temperature estimate of $T \sim 50$ K. [21] A few years later he improves on this by considering the matter and radiation together (Gamow, 1956, Eq. 10) and concludes with a value of 6 K. There were many uncertainties at that time, such as the age of the Universe and the current density of matter, which make this no more than a well-motivated approximation.

to make up such a graph in earlier papers (Gamow, 1946, 1948, 1949, 1952). However, those earlier estimates were biased by highly uncertain values for the present age of the Universe (typically taken to be 2×10^9 years) and other physical parameters. During the decade prior to 1956, the papers Gamow wrote with his co-workers, and the papers of those co-workers, were almost entirely dominated by the nuclear physics aspects of the early expansion with the wish to build all known abundances starting only with neutrons.

[20] The equations for the change in temperature of a gas of black-body radiation had already been derived within the framework of General Relativity by Tolman (1934b, Section 171), who showed that the spectrum remains a black-body for which the temperature would vary inversely with a power of the volume (*loc. cit.* Eq. 171.6). This can be traced back to Tolman (1931). Gamow (1948) transformed Tolman's mathematics to the language of physics.

[21] This rather crude estimate requires only a knowledge of the time and temperature at which the radiation field is created. There is no nuclear physics in this (see Chapter 7).

2   The Cosmic Expansion

The inter-war years, 1918–1939, were a period of coming to terms (a) with Einstein's General Relativity and (b) with Hubble's discovery of the redshift–distance relationship. By the end of the period our cosmological framework was understood well enough in terms of an expanding homogeneous and isotropic solution of the Einstein equations and it probably seemed a matter of acquiring redshift in order to settle the parameters of the model. Two parameters would do the job.

It could not have been imagined that by 1955 there would be a heated argument between two camps: Gamow, who said there was a Hot Big Bang, and Hoyle, Bondi and Gold who said there was not. Added to that was another heated, even acrimonious, argument about the interpretation of the counts of recently discovered radio sources made by Ryle in Cambridge, England and Mills in Sydney, Australia. Ten year after that we had the Quasars and Cosmic Microwave Background Radiation (CMB) that, at the time, not everyone believed was cosmic in origin.

This chapter relates some of that story. It is an essential part of explaining how come we are where we are.

2.1 Models of the Cosmic Expansion

Several spatially homogeneous solutions of the equations of the General Theory of Relativity were discovered within the first decade following their publication (Einstein, 1916a). At that time relatively little was known about the Universe: it was still uncertain whether or not the nebulae were merely parts of our own Galaxy, although, through the pioneering work of Slipher, it was known that most of them were rushing away from us. Little or nothing was known about the homogeneity or isotropy of the Universe, but the assumption of homogeneity and isotropy would simplify the largely intractable Einstein equations. During the decade following the publication of the Einstein equations several important cosmological solutions were discovered. These are discussed next.

It is interesting in this context to read and compare the texts of Eddington (1923), published before Edwin Hubble's study of the nebulae and his consequent discovery of the cosmic expansion, and the text of Tolman (1934b) which was published shortly thereafter. The search for understanding the Universe in terms of models is beautifully described in Michael Heller's *Ultimate Explanations of the Universe* (Heller, 2010).

2.1.1 Models of the First Post-GR Decade

Two solutions played an important role early on: the static solution of Einstein (1917a) and the empty but expanding solution of de Sitter (1917b): both solved the Einstein equations with a 'cosmological constant'. Einstein introduced the cosmological constant in order that he could achieve a static solution of his equations where the force of gravity due to matter was balanced by the repulsive force due to the constant. At that time, he had no reason to believe that the Universe was anything but static. This solution is covered in detail in Section 15.6.3.

de Sitter, on the other hand, sought a solution of the Einstein equations with a positive cosmological constant containing no matter. The cosmological constant would drive expansion and this would be a General Relativistic analogue of the Minkowski Space. The remarkable property of this solution was that light rays between a comoving observer and source were red-shifted: this was a manifestation of the curvature of the de Sitter space-time. At the time when Hubble announced the expansion law through his velocity–distance relationship, he thought that he had detected the 'de Sitter effect'.

The only problem was that the Universe was not in fact empty, and so could not be properly modelled by the de Sitter solution. Nonetheless, the de Sitter solution has today become one of the most important solutions of the Einstein equations. It forms a basis for understanding the earliest moments of the cosmic expansion (the proposed inflationary period), and the evidence is that the Universe is currently entering into a phase where a cosmological constant is driving the Universe into a de Sitter expansion.

Providing an expanding model with matter was left up to Friedmann, who in two papers (Friedmann, 1922, 1924) derived the equations for both open and closed universes that are the basis for understanding the dynamics of the cosmic expansion as we now see it.[1] Independently, but a little later, Lemaître (1927) derived the same closed solution, and in addition provided a remarkably detailed physical explanation as to how they would be relevant to the Universe of galaxies. We should perhaps talk of the 'FRW metric' and the 'FLRW Universe'. At that time Hubble had not published his work on the velocity distance relationship, but already Lemaître clearly foresaw what was to come, and even calculated the cosmic expansion rate on the basis of data from Slipher and from Hubble himself. Not surprisingly, he got the same value as Hubble was later to publish. This is one of the great stories of scientific discovery.

In 1925, another homogeneously expanding solution was published by Kasner (1925a,b). The importance of this solution was that the expansion was not isotropic. At the time this was not considered as a potential cosmological solution. However, it was later found to be one of a class of solutions that might represent more general homogeneous but anisotropically expanding cosmological models. These models were seriously studied in the cosmological context during the late 1960s and early 1970s.

---

[1] Michael Heller has written an interesting article about Friedmann, his two famous papers, and his philosophy (Heller, 1985). Also, Belenkiy (2013) has written a fine article about Friedmann, commenting on the relationship of his work to that of Lemaître. I am grateful to John Peacock for discussions and for drawing my attention to Belenkiy's work.

2.1.2 Eddington at the RAS Meeting of 9 May 1930

At the Friday, 9 May 1930 meeting of the Royal Astronomical Society in London, Edding-ton presented a paper, *The Instabililty of Einstein's Spherical World*. He commented that after seeing the work of Lemaître (1927), it was easy to show that the Einstein static solu-tion was unstable, and therefore unacceptable as a model for the Universe. He also rejected the de Sitter model because it was empty. Eddington then concluded that the only way forward was to adopt Lemaître's family of solutions. It was following that meeting that Eddington saw to it that Lemaître's work was published in English in the *Monthly Notices of the Royal Astronomical Society*. The publication of the English language version of Lemaître's paper (Lemaître, 1931a) was a direct consequence of that. Friedmann's work was, by comparison, left out in the cold.

2.1.3 The Einstein de Sitter Model

In 1931, both Einstein and de Sitter were independently studying Friedmann's non-static solution of the Einstein Equations. As it happened, in early 1932 both Einstein and de Sitter met up while visiting Mount Wilson Observatory and came up with a model having zero cosmological constant for which they could calculate the relationship between the expansion factor and the measured redshift of a galaxy. This became known as the Einstein de Sitter model, which played a central role in theoretical cosmology for the rest of the 20th century (Einstein and de Sitter, 1932). With this, Einstein dropped the cosmological term from his equations, but de Sitter, who had always voiced issues over the apparent age of the Universe, generally preferred to keep it in his work since this gave him the freedom to adjust the age.

It was the Einstein de Sitter paper and the paper presented by Eddington to the Royal Astronomical Society that triggered the investigation of non-static models of the Universe.[2]

2.1.4 Friedmann, Lemaître, Robertson and Walker

These four names lie at the very foundation of relativistic cosmology as we know it today. Each wrote one or more papers on the non-static homogeneous and isotropic solu-tion of the Einstein equations that now bears their names. Friedmann (1922, 1924)[3] and Lemaître (1927, 1931a)[4] independently presented homogeneous and isotropic solutions of the Einstein equations, including Einstein's cosmological constant.

---

[2] It was not until 1931 that Einstein acknowledged in print (Einstein, 1931) that the correct interpretation of the redshift data was most probably the cosmic expansion as described by the solutions of Friedmann. Previously, Einstein had criticised and later retracted his criticism of Friedmann's work (Einstein, 1922, 1923). See Heller (1985) for a discussion of Friedmann's work.

[3] Friedmann published his two papers in the well-respected and well-known German language journal *Zeitschrift für Physik* in 1922 (closed models) and 1924 (open models). In the 1922 paper his name appears as 'Friedman', while in the 1924 paper his name appears as 'Friedmann'.

[4] Lemaître (1927) was published in French in an important but less widely read journal, *Annales de la Société Scientifique de Bruxelles*. The English language translation appeared in 1931, though with the omission of an important paragraph (see main text) and the omission of the fourth conclusion to the paper's concluding remarks. The translator also added in a reference to Friedmann (1922), together with some additional references

Lemaître's 1931 paper, which appeared in the *Monthly Notices of the Royal Astronomical Society* at the behest of Eddington, was a slightly edited translation of his 1927 paper. The major edit was the removal of the discussion between the important Equations (23), and Equation (24) (which has itself been abridged). Those deleted French-language paragraphs are Lemaître's quantitative prediction of the expansion law as derived on the basis of his model.

> **Remark:** The missing paragraph in the English language version of Lemaître's analysis of the cosmic expansion (Lemaître, 1931a, p.56) translates as follows:
> *Using the 42 nebulae in the lists of Hubble and Stromberg, and taking account of the Sun's proper motion* ($300 \, \text{km s}^{-1}$ *in the direction* $\alpha = 315°$, $\delta = 62°$*), we find a mean distance of* 0.95 *million parsecs and a radial velocity of* $600 \, \text{km s}^{-1}$*, giving* $625 \, \text{km s}^{-1}$ *at* 10^6 *parsecs.*
> Lemaître adds a footnote remarking that he would have got $575 \, \text{km s}^{-1}$ at 10^6 parsecs if he had not chosen to add a radial weighting function to the data.
> This statement does not appear in the English translation Lemaître (1931a), nor do either the all-important preceding data analysis or the following commentary. The data analysis in arriving at this conclusion would have been a major part of the effort in getting to the results described in the paper.

The solutions of Friedmann and Lemaître were known to Robertson (1929, 1933)[5] and to Walker (1935). What Robertson and Walker achieved was to provide a general solution of the Einstein field equations that satisfied the constraints of being both homogeneous and isotropic to all observers at all times, and to show that all previously known such relativistic solutions were particular cases of their solution.[6] Indeed, their result is even stronger than that: it is a geometric result concerning Riemannian space-times and does not depend specifically on the Einstein equations.

> **Remark:** The mathematical results of Robertson and of Walker tell us that the *only* spatially homogeneous and isotropic solution to the Einstein equations is what we now call the Robertson–Walker metric. This is a result about the geometry of space-time. The Einstein equations serve to determines the arbitrary function that describes the dynamics of the metric: the scale factor.

that post-dated the original French version, but not including the paper of Hubble (1929) announcing the discovery of the cosmic expansion.

[5] Interestingly, Robertson (1929, footnote 11) takes the opportunity of apologising for not citing the earlier work of Lemaître (1925) in his own (Robertson, 1928) paper, but was evidently unaware of the important Lemaître (1927) paper. In this paper (Robertson, 1929, footnote 3), he is somewhat critical of Friedmann in regard to the mathematical assumptions that underlie his solution, a remark that is also noted by Tolman (1934b, footnote p.362).

[6] We never see Walker referred to except in the combination 'Robertson–Walker', yet they never wrote a paper together.

Given the overwhelming evidence we have today for the homogeneity and isotropy of the Universe on the largest scales, this is a result that is fundamental in our understanding of cosmology.

It is entirely appropriate that our basic model of the Universe should honour all four names, with the abbreviation 'FLRW'.[7,8]

2.1.5 Milne's Model

A Newtonian cosmological model is one in which the material in the Universe expands uniformly due to a hypothetical explosion taking place at some instant of time. The fastest moving particles emanating from the explosion move the furthest and in the absence of gravitational forces each particle would maintain a constant velocity giving rise to a linear velocity–distance relationship. In such a model the matter is expanding against a static background space. Newtonian theory makes no statements about light propagation and so, in principle, in such a model the farthest particles could be moving faster than the speed of light.

Milne (1935) sought to make the background against which this expansion took place, not the empty space of Newtonian theory, but the empty space of special relativity: Minkowski space. In doing so, Milne totally ignored any role that gravitation may play in the evolution of the Universe. Like the naïve Newtonian model there is no expansion of the space time itself: the matter merely moves against a static background. Milne's model is often referred to as a 'kinematic model' and is quite contrary to the precepts of General Relativity. Robertson (1935) wrote a superb critique of Milne's point of view in the light of his own work (Robertson, 1933) on cosmology from the point of view of the Theory of General Relativity. Milne (1933) wrote a rebuttal of Robertson's critique.[9]

By contrast, in general relativistic models it is the background geometry itself that is unfolding and that is why a test particle in a vacuum space-time can exhibit a redshift as seen from a stationary observer. It is also the reason why, in general relativistic models, we can have different definitions of 'distance' depending on whether we use intrinsic brightness or size as our means of estimating distance. If we embrace general relativity, then the basic properties of the Milne model are reproduced by the solution to the Friedmann–Lemaître equations for an empty Universe ($\rho = 0$) and zero cosmological constant that is geometrically open ($k = -1$).

---

[7] The annotated bibliography of Robertson (1933) is a fine resource concerning who did what and when prior to that date.

[8] There are a number of articles asserting that one or more of these authors deserves precedence over the others either (a) as the founder of relativistic cosmology, or (b) as the one who provided the equations that best describe our Universe, or (c) as the theorist who foresaw Hubble's discovery of the expansion of the Universe. No one person can be thought of as having addressed all three of these options. Our fundamental general relativistic model of the Universe, now called *The Standard Model*, was truly a joint and largely independent effort by all four.

[9] See also Bondi (2009, Ch. XI) for an incisive discussion of 'kinematic relativity' and the Milne model.

2.2 The Expansion Law

The expansion law is the relationship between the object's spectral redshift and some distance indicator such as its brightness. For objects of the same intrinsic brightness we would expect a tight relationship expressing how the indicated distance depends on redshift. However, 'redshift' often gets translated as 'velocity' and we come up with the admirably simple Hubble law: $v = Hr$. However, confusion can arise in understanding how to interpret this, especially when the velocity v exceeds the velocity of light. Part of this problem is that the assigned recession velocity is not in fact a velocity: it is merely an (incorrect) interpretation of the measured shift in the spectral lines. This shift should be interpreted within the framework of general relativity.

2.2.1 The Concept of Distance is not Simple

In the curved space times of general relativistic cosmology things are not intuitively simple, particularly where the propagation of light is concerned. For example, the apparent size of an object may even increase when its distance from us becomes greater! This is due to the differential gravitational deflection of light beams between object and observer by the curvature of space-time: geodesics in curved space times have non-zero curvature.[10] In a curved space-time we take the shortest distance between two points to be determined by the path taken by a light ray travelling between the points: a 'null geodesic'.

The situation in cosmology is further confused by the fact that as an object becomes more distant from us we are seeing it as it was in the past. When the object was at the age that we see, we were in fact much closer to it: the expansion of the Universe has since that time carried the object away from us. The sense of confusion is made worse by a lack of consistency of terminology.

There is a fine discussion concerning this by Hogg (1999) and Bunn and Hogg (2009).

2.2.2 Redshift is not Only Due to Velocity

Redshift is directly observable but the velocity of recession is not. The redshift is the relative amount by which the spectral lines are shifted from their rest wavelength, relative to the rest wavelength:

$$z = \frac{\lambda_{\text{obs}}}{\lambda_{\text{rest}}} - 1, \tag{2.1}$$

where λ_{rest} is the rest wavelength of the spectral line and λ_{obs} is the wavelength at which it is observed. On the basis of the Doppler effect we might (näively) write the velocity v corresponding to a redshift z as

---

[10] Lines of longitude on the surface of the Earth are geodesics. The measured distance between points having the same latitude on two neighbouring geodesics increases as the points move away from the pole, reaching a maximum at the equator, after which the distance deceases to zero at the opposite pole. It is this decrease that gives rise to this apparent anomalous behaviour.

$$z = \frac{v}{c}, \tag{2.2}$$

where c is the velocity of light. This clearly goes wrong if we interpret values of $z > 1$ in this way, the speed of light is the maximum speed for the transmission of information. So we might use the special relativistic version of this, which is

$$1 + z = \frac{\lambda_{\text{obs}}}{\lambda_{\text{rest}}} = \sqrt{\frac{1 + v/c}{1 - v/c}} \quad \Rightarrow \quad \frac{v}{c} = \frac{(1+z)^2 - 1}{(1+z)^2 + 1} \quad \text{Special Relativity only!} \tag{2.3}$$

At least this has the virtue that z becomes infinitely large as $v \rightarrow c$, but the fact remains that in General Relativity infinite redshift does not imply a velocity equal to c (Davis and Lineweaver, 2001).

Equation (2.3) is technically incorrect in a general relativistic cosmological model simply because it cannot be true in all coordinate systems.[11] Indeed, within the framework of general relativistic models it is not clear that we should even be talking about a 'recession velocity' as there may be a contribution to the redshift due to the curvature of the space-time (see Harrison (1993) and Hogg (1999) for commentary on this).[12]

Notwithstanding this, when a value for 'recession velocity' is quoted, as in 'moving away from us at 98% the speed of light', Equation (2.3) is used to convert the observable redshift into the quotable number representing a velocity.[13]

> **Remark:** The only observable is the redshift z.
> Other quantities relating to the galaxy, such as distance, recession velocity or the look-back time to the epoch when we are seeing the galaxy, must be derived within the framework of some precisely defined *general relativistic* context.

This reflects a fundamental failing of Newtonian cosmological models: there is no mechanism for handling the propagation of light correctly, not even Special Relativity. In some circumstances, we can extend Newtonian cosmology by grafting on to it some results from general relativity (Bondi, 2009, Section 9.5). We shall do this in Section 5.10.

2.2.3 Local Relative Velocities

We see statements that a galaxy has a peculiar velocity of $300\,\text{km}\,\text{s}^{-1}$, by which we mean its velocity relative to its surroundings, or that radio jets are seen expanding at ten times the speed of light. These are local phenomena where cosmology plays no part except insofar as the objects concerned may be at great distances.

Peculiar, non-cosmic, components of velocity are generated through gravitational interactions with other structures in the Universe and so serve as an important probe of that

---

[11] This was first discussed within the framework of the de Sitter solution by Robertson (1928, Section 4, p.843).

[12] Notably, de Sitter (1931) remarked that *The theory of [general] relativity brought the insight that space and time are not merely the stage on which the piece is produced, but are themselves actors playing an essential part in the plot.*

[13] Feynman (1965, p.181) makes the astute comment that *It makes no sense to worry about the possibility of galaxies receding from us faster than light, whatever that means, since they would never be observable by hypothesis.* See Kiang (1997, 2003), Davis and Lineweaver (2001, 2004); Lineweaver and Davis (2005) for this and other words of wisdom on the subject of superluminal cosmic expansion.

structure. The only observable we have with which we can define a peculiar velocity v_{pec} is the observed redshift z_{obs} of the object and the redshift of some (close-by) point of reference, z_{cos}. When $|z_{obs} - z_{cos}| \ll 1$ this provides a value for the peculiar velocity via the equation

$$v_{pec} = \frac{z_{obs} - z_{cos}}{1 + z_{cos}} c, \qquad v_{pec} \ll c \tag{2.4}$$

z_{cos} is, in effect, the redshift the object would have had at that position were there no external influences driving a non-cosmic component of velocity. There is no direct way of determining z_{cos}. In the simplest case where a galaxy is a member of a large cluster of galaxies we might venture to simply average the redshifts of all member galaxies and use that as z_{cos}.

2.2.4 Measuring Velocities Faster than Light

One of the most common misconceptions in cosmology arises in relation to the upper limit imposed on measured velocities by the speed of light, c. It is important to recognise that we can and do measure velocities greater than the speed of light, there is nothing counter-relativistic in that. Considerable superluminal velocities have, for example, been measured for radio jets observed over periods of decades using VLBI techniques.[14] Some stars, such as classical novae and supernovae undergoing explosive outbursts, show evidence of super-luminal expansion in their light echoes. The light echoes are from the illumination of surrounding medium or interstellar gas and dust on the line of sight.

2.3 Emergence of Cosmology 1945–1965

In the aftermath of the Second World War there was a surge of development in three areas: nuclear physics, which came from work on the atomic bomb; radio astronomy, an outgrowth of research into radar; and computing, which was developed largely in relation to cryptography. All these had existed prior to 1939 in some embryonic form, and the

---

[14] The expansion velocity of the jet, projected on the sky, is derived simply by observing the change in size of the jet with time. If the distance to the source is known from the redshift of the parent object, we have a direct measurement of the length of the jet as projected on the sky. If the angle the jet makes with the sky is θ, then is is easy to show that the velocity measured in this way is

$$v_\perp = v \frac{v \sin\theta}{1 - \beta \cos\theta}, \quad \beta = \frac{v}{c}. \tag{2.5}$$

This has a maximum value

$$v_{\perp,max} = \gamma \beta c, \quad \gamma = (1 - \beta^2)^{-1/2} \tag{2.6}$$

(γ is the usual Lorentz factor). This maximum is achieved for values of β close to 1 and angles such that the jet is moving close to the line of sight. A jet moving at $\beta = 0.9$ having an angle of $25°$ with the line of sight appears to be expanding at $\sim 2c$, while a jet with $\beta = 0.99$ and having an angle of $10°$ with the line of sight is seen to expand at $\sim 7c$. The famous jet of the galaxy M87 has a measured expansion velocity of $\sim 6c$ using data taken over the period 1994–1998 by the Hubble Space Telescope.

advances benefited from the needs of war. This was a period of uncertainty in cosmology highlighted by the conflict between the Steady State Theory and the Hot Big Bang Theories of the Universe, whether redshifts were really cosmological or intrinsic to the sources, and uncertainty of a factor two as to the value of the Hubble constant (was it 50 km s^{-1} Mpc^{-1} or 100 km s^{-1} Mpc^{-1}?). New types of extragalactic object were being discovered, notably quasars, and cosmic structure was being revealed.[15]

The opening of the Hale 200$''$ telescope and the 48$''$ Schmidt telescope on Mount Palomar in 1948 was symbolic of a drive to push astronomy to the limits of technological capabilty. Both telescopes, equipped with modern detectors, are still operational today. It was not until the 1970s that European astronomers had access to comparable facilities.[16]

Thus, from the point of view of cosmology, the scene was set for a rapid development in understanding the origin of the chemical elements and stellar evolution, for the studies of ever more distant galaxies, and for the discovery of new classes of object: radio galaxies and quasars.

2.3.1 Two Cosmological Paradigms

Cosmology prior to 1939 was mainly focused on establishing the general relativistic framework for the apparent cosmological expansion. During the period 1939–1946 there had been substantial advances in technology that would allow cosmology to proceed in other directions at a greater pace. In particular there was the rise in our understanding of nuclear physics, triggered by the earlier discovery of the neutron, and the possibility of mapping the sky at radio wavelengths, triggered by developments in radar technology. Added to that was the advent of the 200$''$ Hale telescope which would enable the study of galaxies at even greater distances.

On the theoretical side, two theories of cosmology emerged at about the same time: the Hot Big Bang theory, pioneered by George Gamow, and the Steady State Theory, pioneered by Herman Bondi, Tommy Gold and Fred Hoyle. We shall cover various aspects of these theories in more detail in the following chapters, but given the intensity of the debate between followers of either of these two views during this period, it would be remiss not to explain why this was for a period of time one of the great schisms of 20th century science.[17]

---

[15] The status of cosmology in the 1960s is nicely encapsulated in the volume of review articles edited by Sandage *et al.* (1975, *Galaxies and the Universe*). Although published in 1975, the book was commissioned in 1962, and the articles were received from authors over the period 1964–1973. So by the time the volume appeared some of the articles were somewhat outdated. The discovery of the CMB, for example, came at the start of this period and was briefly incorporated into the fine article by Peter Scheuer (*Radio Astronomy and Cosmology*, *loc. cit.* Ch.18.). However, the articles in the volume span the issues that were considered central to cosmology towards the end of the 1960s.

[16] Anglo Australian Telescope 3.9 m, 1974; ESO 3.6 m, 1977 and CFHT 3.6 m, 1979.

[17] These were far from being the only proposed cosmologies. During the 1960s there were arguments about a number of alternatives, in particular the matter anti-matter symmetric cosmologies of Alfvén and Klein (1962), and the post-CMB version of Omnès (1969). These were countered by, for example, Jones and Jones (1970) and the debate continued actively for over a decade with thrust (Aldrovandi *et al.*, 1973) and counter-thrust (Steigman, 1976, 2008). The debate continues to the present time (Prokhorov, 2015; Baur *et al.*, 2015).

2.3.2 Gamow's Hot Big Bang

Gamow had been at the University of Leningrad during the period 1923–1929, studying under Alexander Friedmann up until 1925 when Friedmann died. The years that followed found him working in the arena of nuclear phsyics at Göttingen, Copenhagen and Cambridge, but it was not until 1933 that he defected from the Soviet Union and went to the United States. There he worked on the source of energy in stars (Gamow, 1935; Chandrasekhar *et al.*, 1938) and on galaxy formation within the framework of the Friedmann Lemaître expanding Universe, with Edward Teller (Gamow and Teller, 1938, 1939a,b).

It is not surprising that after the war he took up working on the details of the early phases of what became dubbed as the Hot Big Bang Universe. He assembled a small team with Alpher and Herman and together they established a model that is very close to what has now become our standard model.[18] Gamow's team had written many papers about the radiation field that would be left over from that first event and they had approached several groups at various times asking whether this could be detected. The range of uncertainty in the expected radiation temperature was largely a consequence of uncertain physics combined with systematic errors in establishing the timescale of the cosmic expansion. It is perhaps not surprising that they could not encourage others to make a try at detecting this radiation. As it happens, such a detection would have been feasible even in the 1950s. However, there were many other interesting experiments to do with more certain outcomes and which were more mainstream. Nobody seriously considered making this temperature measurement until Dicke and his group at Princeton took up the challenge in the early to mid-1960s.

2.3.3 The Steady State Cosmology

The problem with the Big Bang theory was that it required that the Universe 'start' at some finite time in the past when there was an instant of creation of everything we see. This concept of an origin for the Universe had been enshrined both in Gamow's work and in Lemaître's notion of the *Primeval Atom*. It was hardly surprising that, at that time and right up until the end of the 1960s, not everyone felt confident in supporting the notion of such a cataclysmic initial event.

In 1948, three young scientists at Cambridge decided that continual creation of matter would be far more acceptable than creating the entire Universe at a single instant. They conceived of an expanding Universe in which the material evacuated by the expansion would be replaced at a steady rate by newly created material. This would produce a model Universe that was not only homogeneous and isotropic, but one that was also the same at all times. It would have neither a beginning nor an end.[19]

[18]   As close, in fact, as the then-current ignorance of nuclear physics and of the basic cosmological parameters would allow. There would later on be claims as to who did what and the inevitable discussion of assigning credit to individuals within the group. The clear point here is that Gamow had been the driving force behind the Hot Big Bang theory from before the war, and continued to be almost until the time of his death.

[19]   Einstein had, around 1931, attempted to produce a steady state model Universe containing a Λ-term and matter: in other words a generalisation of his empty static Universe. The four page manuscript of an unpublished paper, *Zum kosmologischen Problem*, *c*1931, was discovered in the Einstein Archive. The manuscript is

Two papers presenting the Steady State model from somewhat different points of view were published in the same issue of the *Monthly Notices*.[20] The first was by Bondi and Gold (1948), who took an approach based simply on a generalisation of the (Copernican) Cosmological Principle. They invoked a *Perfect Cosmological Principle* that the Universe looked the same from all places and *at all times*. The other version, by Hoyle (1948), approached the theory via a generalisation of the Einstein field equations that would allow a steady state expanding solution.[21] No specific mechanism for this creation of material was suggested in either paper, but nor had any mechanism ever been suggested by the adherents of the Hot Big Bang model whereby the Universe might suddenly come into existence.

The issue of the creation of material, ostensibly out of nothing, was of course central to the great debate.[22] It was beyond what conventional physics could address. Yet this critique was never seen as an issue for the Big Bang alternative, which pushed back the origin of all material into the unknown arena of quantum gravity, the physics of the so-called Planck era, when the Universe as we know it should suddenly spring into existence.

Hoyle had postulated a field, called the *C-field*, as the source of created matter. As the Universe expanded energy would be extracted from the C-field to produce normal matter, but because of the expansion its density ρ would remain constant, as it must for the Universe to be the same at all times.

William McCrea was perhaps the most prolific contributor to the Steady State theory, and, in particular, in elucidating the physical principles underlying the theory. In one of his

translated and discussed in fascinating detail by O'Raifeartaigh and McCann (2014) and O'Raifeartaigh *et al.* (2014), who concluded that Einstein perhaps did not publish this because it contained an unrecoverable error.

[20] It is occasionally remarked that the radical difference in their presentations of the Steady State Theory by Bondi and Gold and by Hoyle reflects a difference in the two versions of the theory. This is incorrect: Bondi and Gold had shown that their Perfect Cosmological Principle is by itself enough to determine the metric, a reiteration of an earlier result of Robertson (1929) concerning stationary solutions of the Einstein equations. See also footnote 21.

[21] Hoyle (1948, Equation 22) had rewritten the Einstein equations in the form having the creation term on the left-hand (geometric) side of the equations,

$$R_{ab} - \frac{1}{2} g_{ab} R + C_{ab} = -\kappa T_{ab}, \tag{2.7}$$

and wrote the matter content on the right-hand side as $T_{00} = \rho c^4$ (in his units), with all other components being zero. There is a small irony here in that Einstein, in an attempt to derive a non-vacuum steady state theory in 1931 (O'Raifeartaigh *et al.*, 2014; O'Raifeartaigh and McCann, 2014), made the same assumption about the energy momentum tensor and found he could not derive a solution (see footnote 19). Hoyle remarks that 'The C_{ab} term plays a rôle similar to that of the cosmical constant in the de Sitter model, with the important difference, however, that there is no contribution from the C_{00} component. As we shall see, this difference enables a Universe, formally similar to the de Sitter model, to be obtained, but in which ρ is non-zero.' A little later in the paper (*loc. cit.* eq. 26) he arrives at the exponentially expanding de Sitter solution in which the mass density ρ is constant.

Bondi and Gold (1948, p. 260) had arrived at the same de Sitter solution on the basis of their Perfect Cosmological Principle and without writing any equations, simply referring to the earlier work of Robertson (1935) and Walker (1935).

Later, McCrea (1951) was to move the C_{ab} term to the right hand side and regard it as a field. Then finally, Hoyle and Narlikar (1962, Eq. 7) derived the creation term from an action principle suggested to them by M. H. L. Pryce. The discussion following this last paper is of considerable interest in the light of earlier criticisms of the theory.

[22] In terms of the present cosmological parameters, the creation rate required would have been on the order of one hydrogen atom per cubic meter every 1 Gyr.

earliest papers on this subject, McCrea (1951) suggested a way of reconciling the Steady State theory with General Relativity (GR). Hoyle in 1948 had modified GR by adding an extra *Creation Term*, the *C-field*, to the Einstein equations (see Equation 2.7). This was not unlike Einstein's own modification of his equations when adding the Λ-term in order to achieve a static model Universe, and Hoyle's modification had seemed no less arbitrary. However, McCrea suggested that this term might be simply viewed as a part of the energy momentum tensor on the right-hand side of the field equations, rather than as an intrinsic part of the Einstein equations. In doing this, McCrea showed that Hoyle's C-field could be viewed as material with the equation of state $p + \rho c^2 = 0$, arguing that the consequent 'negative pressure' should be viewed as a *zero-point stress*, and emphasising that the C-field should be viewed as a field that pervades the intergalactic medium. The idea was subsequently developed further by Davidson (1959). (See also the extensive discussion of North (1965, Ch.10, Sections 3,4).) This interpretation changed somewhat in a later version of the Steady State theory by Hoyle and Narlikar (1962, 1963, 1964)[21] where the C-field had still had $p + \rho c^2 = 0$, but with negative mass density and positive pressure. See McCrea (1964) for a deeper discussion of this and also Harrison (1995) for some interesting remarks.

Because of the strict balance between creation and expansion, the Steady State theory made a number of rather precise predictions. Most importantly was the fact that the Steady State Universe was geometrically flat: the expansion rate, i.e. the Hubble constant, would be determined only by the total material density. Bondi and Gold showed (*loc. cit.* figure 1) how the counts of cosmic objects, such as galaxies, would grow with distance. The slope of the Hubble relationship was fixed as was the observed age distribution of galaxies. Everything would be determined in terms of the expansion rate, i.e. the Hubble constant. This lack of freedom implied that the theory could easily be tested by observation. This was in many ways the perfect example of a cosmological theory. All that we can say now is that it is a pity that the Universe is not like that: it was a beautiful theory.[23]

The irony today is that the C-field pervading the Universe as introduced by Bondi, Gold and Hoyle was little different from the field that we currently attribute to the as-yet not understood dark energy field, Λ, that pervades the Universe today and that makes up 70% of the total energy density of the Universe.

---

[23] It has been suggested by some historians (see for example, Kragh (2007, Section 4.2)) that the Steady State theory was merely an aberration on the part of a few people that others outside of their particular community cared little about. That was simply not the case: European science had played a key role in the earlier, pre-1939, development of cosmology and after 1945, the UK, where the debate raged most strongly, became an important centre for developments in theoretical cosmology. After 1945, apart from Gamow's group, cosmology research in the USA, with its unique observational facilities, was concerned mostly with data acquisition. Neither Tolman nor Robertson, the pre-war heroes of relativistic cosmology in the USA, nor Milne and Dingle who had been active pre-war in the UK, published any research papers after 1945. Of the 'old pre-war brigade' only McVittie continued publishing papers on general relativistic cosmology, and then from a largely observational point of view. General relativity moved in other directions.

It is somewhat remarkable that Hoyle's 1948 paper has achieved most of its citations in papers written since the year 2000, largely because of the parallel with the dark energy Λ-field.

2.4 Radio Astronomy Source Counts

The first catalogue of radio sources was the First Cambridge catalogue ('1C') of 1950, containing some 50 sources. By 1951 about 100 radio sources were known.[24] Positional accuracy was at first rather low, generally no better than 1 degree, but by 1952 extragalactic objects had been identified with two of these sources. Later, CygA was identified by Baade and Minkowski (1954) as a collision between two distant galaxies ($v \simeq 17\,000\,\mathrm{km\,s^{-1}}$) galaxies.

The following years saw a number of major radio astronomical facilities come on line in the UK (1951, the Jodrell Bank MKI), the USA (Greenbank 1959) Australia (the Mills Cross in 1954 and the Parkes telescope in 1961) and the Netherlands (Dwingeloo in 1956 followed by Westerbork in 1960). These were capable of finding the radio stars and determining positions with sufficient accuracy to warrant a search for the optical counterpart of the source using a large optical telescope (usually the 200″ on Mount Palomar).[25]

The review by Jauncey (1975) covers the surveys made during this period in great detail. See Condon *et al.* (2012) for a more recent take on this subject.

2.4.1 The $\log N$–$\log S$ Plot

A central tool in the statistical analysis of the source catalogues generated by radio surveys was the $\log N$–$\log S$ plot, describing the number of sources detected down to a flux level S. There would be more sources at smaller values of S. For a homogeneous distribution of sources all having the same luminosity, or power, P, the distance out to which we find sources brighter than S varies as $R \propto S^{-1/2}$. This is a trivial consequence of the inverse square law that the observed brightness S varies as $1/R^2$. The volume included by a radius R is $V \propto R^3 \propto S^{-3/2}$, and so, if the source density is uniform, the number of sources observed that are brighter than S is $N(> S) \propto V \propto S^{-3/2}$. We can write this as:

---

[24] At that time and during most of the 1950s, they were referred to by the Cambridge group as *radio stars*, a name which was in use for over a decade before being replaced by the term *radio sources*.

[25] In 1951, Ewen and Purcell (1951) detected the 21 cm line of neutral hydrogen from the galactic plane using a purpose built radiometer. The existence of the line had been predicted by van de Hulst (1945). Two other groups announced the detection of the 21 cm line at the same time: Muller and Oort (1951) from Leiden which contained an addendum by Pawsey (1951) announcing that Christiansen and Hindman had also detected it.

Edward Purcell went on to share the 1952 Nobel Prize in Physics with Felix Bloch for their work on nuclear magnetic resonance. There was no mention of Purcell's work on the 21 cm line in the official citation, despite which many authors have implied or stated that he won the prize for the discovery of the 21 cm line. The official Nobel biography only states that 'He has made some contribution to the subject of radioastronomy'.

The discovery of the 21 cm line was the birth of 21 cm astronomy. There followed a period of intense research and development and by 1958, Oort *et al.* (1958, Fig. 4) were able to publish the first full map of the galactic plane. These were exciting times for radio astronomy, seeing world-wide cooperation and commitments to funding large projects.

Many of these early papers published prior to 1955 are available in the book by Sullivan (1982, *Classics in Radio Astronomy*).

$\log N$–$\log S$ relationship for sources uniformly distributed in Euclidean space

$$\log N(> S) = -\tfrac{3}{2} \log S + \text{const.} \tag{2.8}$$

where the constant reflects the intrinsic luminosity, P, of the sources: the bigger is P, the bigger is $N(> S)$ because we are including additional sources that are further away. If the sources do not have the same luminosity, but are drawn from the same luminosity function at all distances (i.e. the sample is statistically homogeneous) then it is easy to show that Equation (2.8) still holds: the slope of the $\log N$–$\log S$ relationship is -1.5 for a homogeneous sample. If either the density or the luminosity function were dependent on the distance R, the slope might no longer be -1.5.

The *differential source counts* are the counts per unit interval of S:

$$\Delta N(S) \propto S^{-5/2} \Delta S. \tag{2.9}$$

Plotting $\Delta N(S)$ displays the counts in statistically independent bins, which is more appropriate for model fitting than the integrated counts $N(S)$. Since we are interested in how the counts compare to the case of a uniform distribution of statistically identical sources, it is convenient to make logarithmic plots of the quantity $N(S)S^{3/2}$, which would be a horizontal line independent of S for the uniform model. An example of data analysed by Wall and Cooke (1975) is shown in Figure 2.1 where source count data obtained at different frequencies is shown.[26]

So the $\log N$–$\log S$ plot for a catalogue is a relatively simple way of ascertaining whether a sample of objects was drawn from a homogeneous distribution. However, interpreting a slope that is different from -1.5 is not straightforward as deviations could be due to either source density or luminosity evolution (or both). Longair (1966) gave a detailed interpretation of the source counts within the framework of the Robertson–Walker metric. The slope might be different for other reasons, notably experimental systematics. The fainter sources near the limit of the telescope sensitivity might be confused with noise, or, if there are too many sources at the fainter levels there might be more than one source per resolution element (the beam-width) resulting in a systematic counting error.

One of the key issues which arose in interpreting the source count data occurred when comparing surveys working at very different frequencies: if the rate of evolution of the sources was different at different frequencies, then the two surveys would give different values for the slope of the $\log N$–$\log S$ plot. So it is possible that one survey, yielding a slope of -1.5 might give the impression that there was no evolution, while another survey would yield a dramatically different conclusion. This is in fact what happened when comparing source counts from Cambridge and from Sydney in the 1950s.

---

[26] The shaded regions of the diagram are the source counts inferred from what is referred to as a $P(D)$ analysis of the data. The $P(D)$ method deals only with the signal amplitude in the raw data and, in principle, enables a source count that is independent of assumptions concerning source identification and hence independent of assumptions about the instrumentation. It makes no assumptions about Gaussianity of the signal and is free of confusion effects. It can count sources that would be below the nominal threshold for detection using source separation algorithms. The only downside is that it is statistical and does not produce a source catalogue. See, for example, Ryle and Scheuer (1955) and Scheuer (1957).

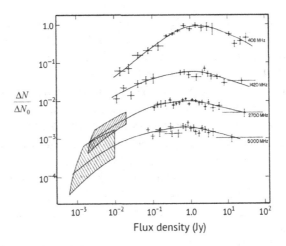

Fig. 2.1 Differential source counts, ΔN, at four different frequencies, normalised to the counts, ΔN_0, for a uniformly distributed population of sources (Wall and Cooke, 1975, Fig. 8). Thus a uniformly distributed non-evolving source distribution in a spatially flat cosmology would be a horizontal line.

The curves have been shifted vertically by an arbitrary amount for clarity.

In this way we see the rise in the counts relative to a uniform distribution at high fluxes, S, and falling off towards lower fluxes after reaching a maximum. The curves are manifestly frequency dependent, the high frequency counts are flatter than the low frequency counts.

The shaded area is the extended counts determined from $P(D)$ analysis.

(© Royal Astronomical Society, with permission.)

What is clear is that, in a perfect experiment, if the Steady State Theory is correct, the slope should be -1.5 at all frequencies: the hypothesis underlying the Steady State Theory is that the Universe is homogeneous and the same at all times.

2.4.2 The Cambridge 2C Survey

The first major survey was the Cambridge 2C survey (3.7 m wavelength) of Shakeshaft *et al.* (1955) which listed over 1900 radio sources, most of which were isotropically distributed sources of small angular diameter and some 100 of which could be identified with NGC objects. The data was analysed in the seminal paper of Ryle and Scheuer (1955) who showed that, while the sources were isotropically distributed, they did not represent a spatially homogeneous population, there being far more faint objects than would have been expected had this been a uniform distribution of similar sources. This paper would have important future ramifications and undoubtedly marked the entry of radio astronomy into the arena of cosmology.

Ryle and Scheuer asserted that *Attempts to explain the observations according to steady-state theories offer little hope of success, but there seems every reason to suppose that they might be explained in terms of an evolutionary theory,* and by 1957, Scheuer (1957) using a refined method of analysis was able to assert that *the observations are therefore inconsistent with a uniform distribution of radio sources in space.*

This of course worried Hoyle who wrote to Mills in Australia. It was 1955, Mills' *Cross* telescope had just come on-line the previous year and was taking data. What was available certainly indicated a discrepancy of some sort Sullivan (1984, p.153, Mills' article), and so Mills was able to reassure Hoyle. The problem was displayed in the detailed 33 page paper of Mills and Slee (1957) which undertook a detail source by source comparison of an area common to the Cambridge and Sydney surveys. The conclusion they reached (*loc. cit.* p.169) was that *It is obvious that the Cambridge sources, with an average density of about two per beamwidth, are likely to be affected by confusion, while the effect might be expected to be trivial in the Sydney catalogue averaging one source per 17 beamwidths.* This was strong stuff, and the riposte came with Scheuer's $P(D)$ analysis ((Scheuer, 1957). A rift had been created between the two groups and the resolution would not be found until there existed deeper and higher resolution surveys with redshifts for a large sample of the detected sources.

Which is what happened. Cambridge proceeded to the important 3C survey in 1959 (Slee, 1959) and its subsequent revision, the 3CR catalogue (Bennett, 1962). Meanwhile, the Australians completed their survey over the period 1958–1961 (Mills *et al.*, 1958, 1960, 1961). The slope of the Cambridge counts decreased from -3.0 to -1.8, but the Australian counts remained flatter and closer to the magic -1.5. Somewhat later Pooley and Ryle (1968) published the analysis of an even deeper Cambridge survey revealing a bright end slope of -1.85, flattening out as the sources got fainter.

Following that, further deeper and higher resolution surveys were undertaken, but the goal lost some of its impetus because of two important discoveries: the Quasars and the cosmic microwave background radiation (CMB).

The quasars were extreme extragalactic point-like radio sources, and the increase in their number with redshift could not be doubted. This signalled the end of the Steady State theory unless the redshifts of those objects had some other cause than the cosmic expansion. The argument continued into the 1970s. But the discovery of the CMB was decisive. Although the debate on Big Bang versus Steady State did continue after that discovery, this was the end of a fine theory. The Steady State theory would be rejected, but not for any of the reasons that could have been anticipated in 1950, except perhaps in the mind of George Gamow.

2.5 Quasars

By the early 1960s, when more accurate source positions became available from the Cambridge 3C survey catalogue, a number of radio sources came to be identified with point-like objects. These point-like sources became known as 'Quasi-stellar sources', or *Quasars*, or *QSOs* for short. The first identification of a radio source, 3C48, with a star-like object came in 1960 in an announcement at the AAS meeting of December 1960, when it was described as possibly being 'a nearby star'. Identification and photometric study of stellar objects identified with 3C48, 3C196 and 3C286 were published by Matthews and Sandage (1963), but while the paper was being written a very accurate position for 3C273 was obtained from measuring a lunar occultation of the source by Hazard *et al.* (1963).

That enabled an unambiguous identification of the optical counterpart with which Schmidt (1963) obtained a spectrum with a number of lines that he was able to identify as being Balmer lines of hydrogen, yielding what for the time was a high redshift of $z = 0.16$. With this information Greenstein (1963) was able to assign a redshift of $z = 0.37$ to a spectrum of 3C48.

There then followed 3C47 at $z = 0.425$ and 3C147 at $z = 0.545$. By 1965, Schmidt (1965) was able to publish a list of far ultraviolet emission lines that were being shifted into the visible region of the spectrum, together with a list of redshifts for nine quasars, the largest of which was 3C9 at $z = 2.01$.[27]

Since those early days, the number of quasars having known redshifts has increased by a factor of ~ 10 every decade. In 2004, Croom *et al.* (2004, Fig. 3), published a QSO $N(z)$ diagram based on the 23 338 QSOs in the 2QZ survey. By January 2015 Flesch (2015) compiled a listing of 510 764 comprising 424 748 QSOs plus other active objects (Seyfert galaxies, AGNs BL Lac objects and so on). Much of this progress has been due to successive releases of the Sloan Digital Sky Survey (SDSS).[28]

2.5.1 Evidence for Cosmic Evolution with Redshift

The list of redshifts quickly grew ever larger and by 1970 Schmidt (1970)[29] was able to show that, for QSOs identified in the 3CR catalogue, the density evolution with redshift gave a number density per comoving volume varying as $(1 + z)^6$ for a $q_0 = 1$ cosmological model. This corresponded to a slope of -1.85 at the bright end of the $\log N$–$\log S$ curve, flattening out slightly towards fainter magnitudes. This was therefore entirely consistent with the $\log N$–$\log S$ curve derived for the 3CR source counts by Pooley and Ryle (1968). Schmidt's 1970 paper, with its detailed modelling of the evolutionary properties of the radio and optical QSO luminosity function also established a number of properties of QSO evolution that would be important in interpreting the radio source counts.

This was direct evidence that the quasar population had evolved since a redshift of $z \sim 2$ and that would pose a severe problem for the classical Steady State theory. From that point on, mainstream cosmology accepted this as *prima facie* evidence for an evolutionary cosmology that would be amply confirmed by future QSO surveys and by the discovery of the cosmic background radiation two years later.

2.5.2 The Redshift Controversy

However, not everyone was convinced by the conclusion from QSO evolution models. It still remained to be shown that the cosmological redshifts were in fact the consequence of

---

[27] At that time the spectra were taken on photographic plates at the prime focus of the $200''$ telescope using a low dispersion (400 Å per mm.). Later developments in detector technology would speed up the process of getting spectra of faint objects by a huge factor.

[28] Pâris and Petitjean (2014)
https://www.sdss3.org/dr10/algorithms/qso_catalog.php

[29] This is the paper where the editor of the *Astrophysical Journal*, S. Chandrasekhar, inserted a footnote to the paper's title saying that it was 'with regret' that the Journal would 'concede' to recognise the term 'quasar'.

the cosmological expansion. This gave rise to another of the great debates of astronomy, the so-called *redshift controversy*. If these redshifts were due to the cosmic recession, these objects would be 10−30 times brighter than the brightest galaxies hitherto known. More-over, the emitted radio power would be formidable. There was good reason to question the by-now standard interpretation of the observed redshift. This important debate spanned the period from 1965 to around 1985.[30]

The argument against the cosmological origin of QSO redshifts came from two sides. On the one hand, Geoffrey Burbidge argued that there was a pile-up of redshifts at around $z \simeq$ 1.95 and further that there was some evidence for a periodicity in the redshift distribution (Burbidge, 1967; Burbidge and Burbidge, 1967). At the time, there were only seven objects known to have emission line redshifts in excess of 1.9, of which only five had more than one visible absorption line. Burbidge, in the first of these papers, noticed that these five objects all showed good evidence for an absorption line redshift of $z_{abs} = 1.95$. The second of these papers extended the discussion to ten objects, out of which eight had shown the $z_{abs} \simeq 1.95$ absorption lines. Thereafter, as the list of QSOs grew there came a succession of papers discussing the status of the $z \sim 1.95$ peak, adding in claims of periodicities in the redshift distribution. A few years later Burbidge and O'dell (1972) were able to analyse some 249 QSOs, together with 57 active compact objects having emission lines redshifts, again providing evidence of the same phenomena.

Not surprisingly the claimed periodicity was hotly contested. Knight *et al.* (1976), for example, simulated the emission line distribution and gave a reassessment of the likelihood of there being any periodicity.

By 1980, Hewitt and Burbidge (1980) had generated a redshift catalogue of 1549 QSOs, including redshifts from objective prism and grism surveys. These were emission line redshifts with the distribution shown in Figure 2.2. Even when excluding the objective prism/grism objects, the redshift distribution showed a peak around a redshift of $z \simeq 2$. The authors concluded that this distribution *shows the peaks and possible periodicities in the redshift distribution which have been described for more than a decade.*

These conclusions were generally criticised on the grounds that this effect was most probably due to selection effects in QSO samples, while the veracity of the claimed periodicity was frequently criticised, Box and Roeder (1984); Scott (1991).

The final version of the Hewitt–Burbidge catalogue appeared in 1993 (Hewitt and Bur-bidge, 1993) and contained 7315 objects, more than half of which had been obtained using objective prism/grism techniques. The redshift histogram also appeared (as Figure 3) and showed the broad-peaked distribution around $z \sim 2$.[31]

On the other hand, Halton Arp argued that QSOs were too frequently close in the sky to nearby galaxies, concluding that the QSO had to be physically associated with the nearby galaxy. This supported a number of theoretical papers that had proposed that QSOs had

---

[30] The discussion did not stop at that point, papers on this subject continued appearing even into the 21st century (Burbidge and Napier, 2001).

[31] Despite its heterogeneity, the 1993 Hewitt–Burbidge catalogue played a significant role in QSO research for almost a decade after its publication. It also showed clearly decline in the QSO density at redshifts $z \gtrsim 2$. See Richards and Strauss (2006, Fig. 20) using 15343 QSOs from the Sloan DR3 release.

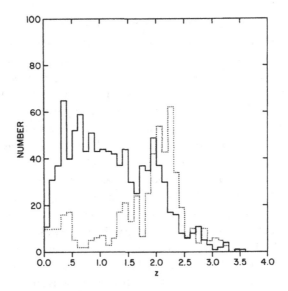

Fig. 2.2   Redshift distribution $N(z)$, of QSOs in the Hewitt and Burbidge (1980) quasar redshift catalogue. The solid line histogram is the redshift distribution of QSOs discovered by positional methods or by colour techniques. The hump around $z \sim 2$ is apparent. The dotted line histogram shows the redshift distribution of of QSOs discovered on objective prism/grism plates. The distribution peaks around $z = 2-2.5$ because objects having this range of redshifts have prominent emission lines visible in their photographic spectra.
(From Hewitt and Burbidge (1980, Fig. 3). Reproduced by permission of the AAS.)

been ejected from our galaxy. Both Arp and Burbidge were noted astronomers who had made major contributions in other fields.

George Field, who was at that time the chair of the Astronomy section of the American Association for the Advancement of Science, proposed a Symposium on the redshift controversy centred around a debate in which the arguments for and against the cosmological interpretation of the redshifts would be put in two 45 minute presentations. Halton Arp would argue the non-cosmological case while the eminent John Bahcall would argue the conventional cosmological interpretation. Each speaker would have 45 minutes to present their case with opportunities for rebuttal and questions from the floor. The meeting took place on 30 December 1972 in Washington DC and resulted in a published proceedings, Field *et al.* (1973, *The Redshift Controversy*).

This was not to be the final word on the subject, but the mainstream thinking at the time was, and still is, overwhelmingly in favour of the simple cosmological origin for the redshifts.

2.6 Nuclear Physics

The great advances in nuclear physics before and during the Second World War had a highly significant dual impact on cosmology. Firstly, they set the scene for research into stellar evolution, the product of which was an understanding of the chemical abundances of stars of diverse types and an understanding of their evolution. Secondly, there was the quest to understand, in terms of cosmology, the origin of all the chemical elements. There were two lines of thought: the elements were built from a primordial soup of neutrons during an early hot phase of the cosmic expansion, or the elements were built in stars. These two lines would in fact converge, making the very lightest elements in the early Universe, and the rest in stars, but that would take almost two decades to realise fully.

The idea of building up the chemical elements from hydrogen or from protons through a sequence of nuclear reactions goes back to the 1930s with the relatively unknown work of Walke (1935) and a series of papers and discussions in the late 1920s and 1930s by Gamow (Gamow, 1928, for example) To put a historical perspective on this, it is worth noting that the element deuterium was first produced in 1931 by Urey *et al.* (1932) and the neutron was discovered in 1932 by Chadwick (1932). The discovery of the neutron provided nuclear mechanisms through which stars could synthesise elements, as was soon after shown by Bethe and Critchfield (1938) and Bethe (1939).

Gamow was later to postulate that a Hot Big Bang might be an ideal place to synthesise all of the chemical elements: this part of the story is covered in Chapter 7. In particular, the synthesis of the element helium was perceived to be a central issue since it could not be made in stars, they were not hot enough.[32]

The confirmation that stars do in fact generate heavy elements came in 1952 with the discovery of technetium (Tc) lines in the spectra of S-type stars. No isotope of Tc has a half-life longer than 4.2×10^6 yr (the half-life of ^{98}Tc) and so it is not seen to occur naturally on Earth. The Tc lifetime is orders of magnitude shorter than the lifetime of these stars and so the element must have been synthesised deep in the stars and dredged up to the surface.

2.6.1 Ages of Oldest Objects

By 1948 the issue of the age of the Universe had become a central issue. This is beautifully reviewed by McCrea (1984) from the point of view of his own recollections. Hubble had determined the value of the Hubble constant at 560 km s^{-1} Mpc^{-1} which, within the framework of the FRWL models with zero cosmological constant, put an upper bound

[32] The helium which we handle must have been put together at some time and some place. We do not argue with the critic who urges that the stars are not hot enough for this process; we tell him to go and find a hotter place. Eddington (Section 209, p.301. 1926, *The Internal Constitution of the Stars*). Eddington never pursued this because he did not believe hydrogen was the major constituent of stars, despite Cecelia Payne-Gaposhkin's seminal work of 1925 (Payne, 1925a,b) showing that stars were mostly hydrogen. The 'hotter place' is, of course, the Big Bang.

In his obituary to Cecelia Payne-Gaposhkin, Gingerich (1982) writes 'One implication of her study was astonishing beyond belief: the overwhelmingly high abundances of hydrogen and helium, which on the urging of Henry Norris Russell she rejected as astrophysically absurd'. To quote from the 1925 paper where she assembles in Table 5 her determination of the abundances of the elements, *Hydrogen and helium are omitted from the table. The stellar abundance deduced in these elements is improbably high, and is almost certainly not real. Russell and Compton have suggested that the astrophsyical behaviour of the Balmer lines . . .* (Payne, 1925a, p.197). It was some years before it was realised that she was correct. McCrea (1928, 1929) built stellar structure models in which he showed, possibly for the first time, that the models would only work if the abundance of hydrogen and helium was indeed high, but he did not cite the work of Payne.

on the age of the Universe of 1.8×10^9 yr.[33] This was lower than estimates of the age of the Earth based on radioactive decay evidence from old rocks, but the feeling was that this problem would eventually right itself. However, Gamow (1949) was concerned enough about the discrepancy that he discusses the notion of having a cosmological constant $\Lambda \sim 2.9 \times 10^{-17}\,\mathrm{s}^{-1}$, a value exceeding that of the Hubble constant ($1.8 \times 10^{-17}\,\mathrm{s}^{-1}$ in the same units).

As McCrea *loc. cit.* points out, the age dilemma was not the reason for dreaming up the Steady State theory, and indeed poses a problem for the Steady State theory because we would expect to find really old objects for which we even now have no evidence, McCrea (1950).[34] The Steady State theory was simply introduced to avoid the need for postulating a cosmos of finite age that somehow 'started' some ten billion years ago. Continuous creation of material seemed no less plausible than the creation of all matter at one instant, and indeed, McCrea's re-interpretation of the original C-field puts that process on the same footing as the Λ-term that we now accept.

If the age of the Universe were finite, we should find no objects that are older than the presumed age. One of the early problems was that the distances to extragalactic objects were underestimated by a considerable factor. This meant that the estimated age of the Universe would be considerably less than the actual age: there appeared to be an age dilemma. This dilemma persisted into the 1990s. In 1995 (see Figure 2.3), several cosmologists gave opinions as to the likely values of the various cosmological parameters, we see a row where the age of the Universe is one of the parameters. This reflected the fact that, at that time, the age estimates of the oldest globular clusters could be older than the age of the Universe for certain choices of the other parameters. Age imposed a constraint.

2.6.2 Cosmic Nucleosynthesis

The synthesis of the elements, and in particular helium, had been one of the prime motivations for the development of a Big Bang theory with a hot singular origin. Shortly after the discovery of the cosmic microwave background radiation, nucleosynthesis was pursued with renewed vigour and, in those early days, played a key role in constraining the values of the cosmological density parameter. Importantly, it also provided a constraint on the number of neutrino species and in this sense heralded the birth of astro-particle physics.

In the pre-1965 era the notion that the Big Bang had been a hot enough place to manufacture elements other than hydrogen was driven by one man: George Gamow. Prior to World War II, Gamow had been working on nuclear energy sources and stellar evolution (Gamow, 1938; Gamow and Teller, 1938) and had written two papers on galaxy formation (Gamow and Teller, 1939b,a). After the war, Gamow proposed his first Hot Big Bang models in the hope that such models could produce all the elements, starting with basic building blocks of neutrons and protons and a radiation field. The difficulty in getting past

---

[33] To put this in perspective, the analysis of the Planck CMB anisotropy data yields an age of $13.8 \times 10^9 \pm 0.04$ yr.

[34] Gamow (1954a), following a discussion with Baade, who had commented that the spread in the colours of elliptical galaxies was rather narrow, remarked that 'it would be very strange indeed' if their ages were significantly different. He acknowledged that a complete theory of stellar evolution would be needed to establish this.

Fig. 2.3   A cosmology school in Leiden, 1995, in which the lecturers gave their opinions as to the values of the cosmic parameter. The poll was not entirely serious: it was intended to show the diversity of opinion as to the best guess for any of the parameters rather than the values for any particular model. The prognosticators were from left to right: Alain Blanchard, Bernard Jones (standing), John Peacock, Peter Coles, Vincent Icke, Rien van de Weygaert and Peter Katgert (standing). They were asked their opinions (guesses at that time) about the Hubble constant, H_0, the Ω_0 parameter, the value of Λ, the curvature, k (whether the Universe was open, closed or flat), and the age of the Universe. The variety of the answers illustrates the uncertain situation in 1995, however, the prescience of RvdW is noteworthy.

See also Figure 2.5.

atomic numbers 5 and 8 made it clear that only the light elements could be synthesised in a Hot Big Bang and that the rest would have to be made in stars. However, it was equally clear that the Hot Big Bang would produce a substantial quantity of helium, perhaps as much as 25% of the mass of the Universe. Pre-1961, in the absence of any data to the contrary, this was regarded as a weak point of Gamow's theory (see footnote 6).

The culmination of the programme initiated by Gamow after World War II was possibly the paper by Alpher *et al.* (1953), who presented detailed calculations of the early thermal history of the Hot Big Bang model. Significantly, they concluded as follows:

> **Concluding remark from Alpher *et al.* (1953)**
>
> Finally, we should like to point out that all of the results presented in this paper follow uniquely from general relativity, relativistic quantum statistics and β-decay theory without the introduction of any free parameters, so long as the density of matter is very small compared with the density of radiation.

Gamow's work on the synthesis of elements, and on cosmology in general, was the first real step in promoting cosmology from its state as a solution to the Einstein equations into the world of physical theory. If the solutions of Friedmann and Lemaître were the first revolution in cosmology, then Gamow's discussion of the physical processes that would take place in such cosmologies was the second. Perhaps he was somewhat ahead of the curve in that the technology of the time did not allow much in the way of verification for his ideas: it would have been necessary to have both measurements of the present day helium abundance and the detection of a thermal relict radiation to have established his ideas beyond serious doubt. It would take more than another 15 years to do this and he would live only 2 years beyond the discovery of the radiation field he had foreseen.

2.7 HMS – the First Redshift Catalogue

The Humason, Mayall and Sandage catalogue of galaxies (Humason *et al.*, 1956, simply referred to as 'the HMS catalogue') provided redshifts and photometry for 800 galaxies' redshifts acquired at the Mount Wilson and Palomar Observatories and at the Lick Observatory. There were also redshifts for 26 galaxy clusters. This was a formidable task given the technology of the time. A galaxy at a redshift of 12 418 $km\,s^{-1}$ required an exposure of 6.7 hours, while one at a redshift of 27 128 $km\,s^{-1}$ required exposures of 8 hours and 9 hours. While these examples were extreme, typical exposures were generally a few hours.

The bottom line of the HMS paper was to confirm the linearity of the redshift magnitude relationship first established by Hubble and extended by Humason. The cluster data did show tentative evidence for a deviation from linearity that would indicate a deceleration of the cosmic expansion.[35]

The 74 page HMS paper is truly remarkable. Not only did it push the boundaries of knowledge far beyond anything that had gone before, it also set the standards for attacking the problem of deriving cosmological information from redshift surveys, discussing every conceivable effect and presenting the results in a model of scientific writing (see Figure 2.4).

2.7.1 Status of the Hubble Diagram 1975–1977

Twenty years after the publication of the HMS paper, Gunn and Tinsley (1975) published a paper entitled *An accelerating Universe*, assessing what the Hubble diagram at that time would suggest about the cosmological parameters (see Figure 2.5). Their analysis is a clear summary of the situation faced at that time by attempts to determine cosmological parameters.

---

[35] Interestingly, the authors add the qualifying sentence in the *Abstract* of the paper that 'This non-linearity indicates deceleration of the cosmic expansion if interpretation is made by theoretical equations due to Robertson'. (See p.147 *et seq* of the paper for the full discussion.) The cosmic expansion was still, in 1956, not something that would be talked of with confidence.

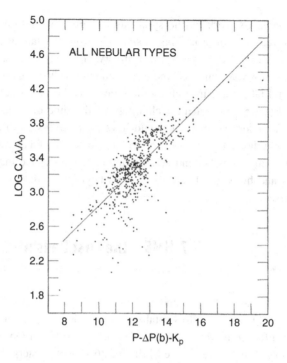

Fig. 2.4 The Humason, Mayall and Sandage Hubble diagram for 474 field galaxies. Note that they plot magnitudes and do not try to interpret magnitudes as distances. Hubble's original diagram (see Figure 1.4) had no galaxies with velocities greater than 1200 km s^{-1}. It is interesting to reflect that this diagram would show nothing of any significance were it restricted to those low velocity galaxies!
(From Humason et al. (1956, Fig. 10.). Reproduced by permission of the AAS.)

In the intervening years Gunn had enhanced the Hubble diagram using the brightest cluster galaxies as standard candles (Gunn and Oke, 1975), and Tinsley had been working on the effects of galaxy evolution over the look-back time relevant to the Hubble diagram (Tinsley, 1970, 1972, 1975). Gunn and Oke had put limits on the value of the deceleration parameter $q_0 = -0.15 \pm 0.57(1\sigma)$, which led to a true value 2σ below zero after applying Tinsley's correction for stellar evolution effects. Hence they concluded that, on the basis of these evolutionary models, the data would indicate an accelerated expansion.

At that time Gunn and Tinsley were able to assert, in the abstract to that paper, that on the basis of their models for the Hubble diagram:

From Gunn and Tinsley (1975, 'An accelerating Universe')
New data on the Hubble diagram, combined with constraints on the density of the Universe and the ages of galaxies, suggest that the most plausible cosmological models have a positive cosmological constant, are closed, too dense to make deuterium in the Big Bang, and will expand for ever.

The conclusion of the paper stated that given reasonable assumptions:
one cannot escape the conclusion that Λ is non-zero and positive.

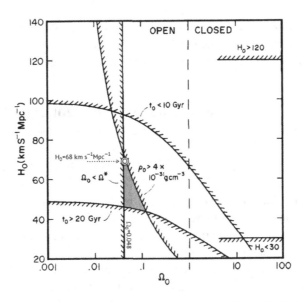

Fig. 2.5 Bounds on the cosmological parameters according to Gunn and Tinsley (1975). The labelled hashed lines show the excluded regions leaving only the small shaded triangle of allowable possibilities for the present Hubble constant, H_0, and the present density of baryonic matter, Ω_0. A key role is played by three lines. 1: $\Omega_0 = \Omega^*$, which states that Ω_0 must be bigger than the material we see. 2: $\rho_0 > 4 \times 10^{31}$ gr cm^{-3} so that we can obtain cosmological deuterium. 3: $t_0 = 20$ Gyr is the line for a cosmic age of 20 Gyr, it is asserted that the Universe cannot be that old. The values of Ω_0 and H_0 from the Planck experiment are indicated on the diagram. A non-zero Λ merely changes the ages line and the position of the Open/Closed line. See also Figure 2.3 *(From Tinsley (1977b, Fig. 2 p.326). © (1977) John Wiley & Son, with permission.)*

This was perhaps the first statement to explicitly assert that using the Hubble diagram, a case could be made in favour of a positive cosmological constant. They did qualify this with the final important remark that . . . *the situation is, as is usual in observational cosmology, that one awaits new and better data.*

Ostriker and Tremaine (1975) pointed out that the central cluster galaxies could accrete neighbouring galaxies by the process of dynamical friction. If that were to happen, this might compensate for, or even dominate, the effects of stellar evolution such as those which Gunn and Tinsley had invoked in their galaxy evolution corrections to the Hubble diagram.[36] Tinsley (1977a,c) revisited the problem and concluded that it would not be possible to determine the deceleration parameter q_0 from the Hubble diagram alone. Indeed, she concluded that, at best, the Hubble diagram (using galaxies) would provide more information about galaxy evolution than about the expansion of the Universe.

Galaxy evolution is a complex process that even now is not well understood, though much has changed in the past four decades. Cosmological computer simulations describe complex dynamical processes such as galaxy and cluster building, while star formation and evolution is handled by what are referred to as semi-analytic modelling (Benson, 2010).

---

[36] On the basis of data gathered in the ENACS survey of rich galaxy clusters, Jones and Mazure (1996) suggested (a) galaxy luminosity functions were not universal, but that the range spanned by the 10th–20th ranked galaxies was stable, giving estimates of the distance with an error of only 10%−15%.

To determine the nature of our Universe would require something radically different, and that was to come in the shape of the Supernova Hubble diagram and in observations of the power spectrum of the cosmic background radiation.

2.8 Lambda – the Spanner in the Works

2.8.1 Λ, Einstein and Lemaître

From the earliest days of general relativity, both Einstein and Lemaître considered gravitational field equations having a so-called 'cosmological constant', Λ. Einstein introduced it to provide a solution for a homogeneous and isotropic Universe in which the force of gravity that tended to make everything collapse in upon itself would be counter-balanced by a repulsive force due to this hypothetical Λ-field. This 'Einstein Universe' would be static and non-expanding, though distant objects would manifest the redshift phenomenon.

Later Lemaître (1927, 1931a, Eq. (2) and (3)) presented the equations for an expanding homogeneous and isotropic solution of the Einstein Equations, including a cosmological constant, Λ, and having matter with an arbitrary equation of state relating density ρ and pressure p (see Section 5.9). If we consider two cases of a model having zero pressure matter of density ρ_m: the first case having a non-zero cosmological constant and the second having a field of 'X-material' having density ρ_X and pressure $p_X = w\rho_X c^2$, but with a zero cosmological constant:

$$\Lambda \neq 0, \quad p = 0 \quad \Rightarrow \quad \frac{1}{a}\frac{d^2a}{dt^2} = -\frac{4\pi G\rho_m}{3} + \frac{\Lambda}{3}, \tag{2.10}$$

$$\Lambda = 0, \quad p = w\rho c^2 \Rightarrow \quad \frac{1}{a}\frac{d^2a}{dt^2} = -\frac{4\pi G}{3}(\rho_m + \rho_X + 3p_X/c^2). \tag{2.11}$$

We see that these are identical if we set

$$\rho_X = \Lambda/8\pi G \quad \text{and} \quad w = -1. \tag{2.12}$$

In other words, the Λ-term is equivalent to a strange kind of matter with $w = -1$: the Λ-stuff has positive mass density and negative pressure. Since, in our every day experience, pressure is always positive, this makes the Λ-material quite extraordinary.

2.8.2 The Evidence for Λ

First evidence: large scale galaxy clustering

The first statement in modern times that data implied the need for a non-zero cosmological constant was the paper of Efstathiou *et al.* (1990) which discussed the sky-projected two point clustering correlation function for galaxies found in the APM survey by Maddox *et al.* (1990).[37] Maddox *et al.* (1990) had discovered significant extra power on large scales over and above what could be expected on the basis of the until-then successful

---

[37] The sky-projected 2-point correlation function, $w(\theta)$, describes the excess number of pairs of galaxies separated by an angular distance θ, as seen on the sky, relative to what would have been seen if the distribution

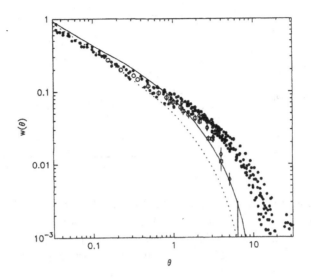

Fig. 2.6 The projected two point correlation function from the APM survey. This measures the amount of clustering on various angular scales θ: if there were no clustering we would have $w(\theta) = 0$. The continuous and dotted curves show the expected clustering according to two $\Lambda = 0$ cold dark matter models. There is a clear excess of clustering on large scales $\theta > 2°$. This excess power on large scales was recognised as being inconsistent with the then-popular CDM model. It was suggested by Efstathiou *et al.* (1990) that this could be explained by having a cosmological constant contributing 70% of the mass density of the Universe.
(From Maddox et al. (1990, Fig. 3. p.45P). © Royal Astronomical Society, with permission.)

numerical experiments simulating the Cold Dark Matter (CDM) theory in which there was no cosmological constant (see Figure 2.6). The final analysis of the APM data came with the impeccable analysis of Maddox *et al.* (1996), a very long paper well worth studying.

The major proponents of the CDM model wrote a paper entitled *The end of cold dark matter?* (Davis *et al.*, 1992) and there were several suggestions as to how CDM might be preserved without resorting to Λ. The papers of Couchman and Carlberg (1992) and Bower *et al.* (1993) are good examples of such alternative solutions to the large scale clustering dilemma posed by the APM data. These exploited the notion of 'biasing', which describes the possibility that the distribution of the luminous material is in some way different from the underlying dark matter distribution: light does not trace mass.

Hubble Diagram for Supernovae

In more recent times, supernova surveys have provided compelling evidence that the Universe is now in a state of accelerated expansion.[38] The turning point came with the

---

was an un-clustered Poisson distribution. So if the mean projected density of galaxies in a catalogue is $\bar{n}$, the probability of finding a galaxy in a small area $\delta\Omega$ located at a distance θ from a randomly chosen galaxy in the catalogue is $\delta P = \bar{n}[1 + w(\theta)]\delta\Omega$ (Peebles, 1980, Section 45, p174).

[38] There is an interesting discussion of Λ from the pre-1989 perspective by Weinberg (1989). A fine review of the history of Λ is given by Calder and Lahav (2008), and a very useful overview of the relevant equations by Carroll *et al.* (1992).

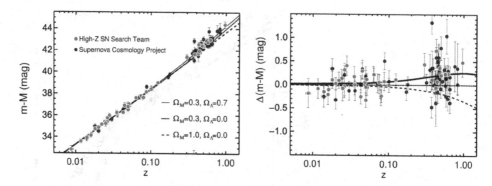

Fig. 2.7 The Hubble diagram constructed from analysis of Type Ia supernova data. The low redshift data is from the Calan-Tololo survey (Hamuy *et al.*, 1996) and the higher redshift data is from the SCP survey (Perlmutter, 1999) and the Hi-z survey (Riess and Filippenko, 1998). The left-hand plot is the classic distance:redshift Hubble diagram, where the distance is expressed in magnitudes. The right hand plot shows the deviation of the data points from a $\Lambda = 0$ model having $\Omega_m = 0.3$. The data strongly favours a model with $\Omega_\Lambda = 0.7$ and $\Omega_m = 0.3$. This diagram, from Riess (2000), is possibly the first time that the SCP and Hi-z data had been published together on the same plot: the agreement between the SCP and Hi-z surveys is excellent.
(From Riess (2000, Fig. 1 p.1286), © 2000 The University of Chicago Press, with permission.)

Nobel Prize winning work of Perlmutter (1999) and Riess and Filippenko (1998) (see Figure 2.7) who realised the idea that the Hubble diagram could decide the nature of the cosmological model that best fitted our Universe. This required considerable resources that would have been unimaginable only 30 years earlier. Subsequent work has served only to strengthen this conclusion.

What the acceleration of the expansion is due to is not known. It is attributed to what is popularly referred to as *dark energy*. On the basis of the supernova observations the dark energy would have to constitute some 70% of the total energy density in the Universe. The obvious candidate for this accelerating dark energy is Einstein's cosmological constant, Λ. As is frequently pointed out, we have no idea what that is a manifestation of. Nevertheless, we have to either prove the supernova result wrong, or find out what this dark energy is. It seems increasingly unlikely that the inference that the cosmic expansion is accelerating will be shown to be incorrect.

Strong support for the presence of this dark energy has come from quite different directions. The dominant discussion comes from studies of the cosmic microwave background anisotropy. To understand what we see requires dark energy, and when models including Λ are fit to the anisotropy data it is found that Λ accounts for $\sim 70\%$ of the energy density of the present day Universe. The simplest dark energy models are parameterised by a parameter w that is the ratio of effective pressure to density of the dark energy: $w = p/\rho c^2$. For Einstein's Λ, $w = -1$ and is independent of time. Observation can now determine a value for w, and detect any trend with time.

In more recent times there has been further evidence in support of the accelerating Universe hypothesis from what are called *baryonic acoustic oscillations*, a slightly enhanced clustering of the galaxies on a scale of $\sim$150 Mpc that is due to the physical processes taking place at the time before the Universe became neutral. These have been discovered and, from their appearance, give strong support for the accelerating expansion and for dark energy. Further support comes from studies of gravitational lensing of distant objects and from studying the evolution of galaxy clusters.

3 The Cosmic Microwave Background

The 'Renaissance in Cosmology'[1] took place with the discovery in 1965 by Penzias and Wilson (1965) of the Cosmic Microwave Background Radiation, now known variously as the 'CMB', 'CMBR', 'CBR', 'MWB' or simply 'Microwave Background' or 'Relict Radiation'. The paper had the rather unprepossessing title of 'A Measurement of Excess Antenna Temperature at 4080 Mc/s' and was less than two pages. In the journal, the Penzias-Wilson paper is immediately preceded by the paper of Dicke, Peebles, Roll and Wilkinson (Dicke *et al.*, 1965) having the title 'Cosmic Black-Body Radiation'.

This explained that they too had been searching for this radiation field and, most importantly, explained what the significance of the discovery would be in terms of our physical understanding of our Universe (see Figure 3.3).

These two papers changed the course of cosmology.

3.1 Discovery of the CMB

During the 1950s and early 1960s the main issue was the great debate between the Steady State and Big Bang theories. This debate had centred around the apparently discrepant radio source counts in deep surveys made at different frequencies by radio astronomers in Cambridge and in Sydney.

The discovery of the cosmic microwave background radiation by Penzias and Wilson in 1965 has turned out to be decisive in establishing a new paradigm in physics: the Hot Big Bang theory (Penzias and Wilson, 1965). However, it should be recognised that the source of the excess radiation they had discovered was still under dispute even in the 1970s and it was not until the first results of the COBE-FIRAS experiment in 1990 that the issue was finally settled.[2]

---

[1] This description is due to Dennis Sciama.

[2] Science demands a very high level of evidence and argument for supporting or rejecting a hypothesis that is put forward to explain any natural phenomenon. This is particularly important for discoveries that are potentially paradigm-shifting, as was the discovery of the CMB. The decade of debate about the nature of the CMB that followed on the discovery was essential, despite the fact that, by 1975, perhaps more than 99% of the scientific community would have supported the interpretation of the CMB as a relict of the Hot Big Bang. Of course, there are still those who question this interpretation: it is now their task to show that whatever explanation they propose can do as well as what has now become the standard interpretation, and, importantly, to suggest an experimental test whereby one can distinguish their theory from the now-standard one.

While the discovery of Penzias and Wilson was itself serendipitous, in that they were not looking for the cosmic background radiation, the idea that the cosmic background radiation was there to be found was well entrenched in both Princeton in the US and Moscow in the (then) Soviet Union. The theoretical framework for this had been set up by George Gamow, who had suggested in the late 1940s that the chemical elements were created in a Hot Big Bang and that the evidence for this would lie in the discovery of the relict radiation. Gamow and his collaborators had written several papers on this starting in the late 1930s (Gamow and Teller, 1939a,b; Alpher, 1948; Gamow, 1949; Gamow and Critchfield, 1949; Gamow, 1952, 1956). Most of these papers centred around the synthesis of the chemical elements in a Hot Big Bang and were important contributions towards later developments.

Short biography: George Gamow (1904–1968) was born in Odessa, then part of the Russian Empire. After completing his undergraduate degree in Odessa he went to Leningrad to work with Friedmann, but Friedmann's early death in 1925 caused Gamow to switch supervisors. He spent the years 1928–1931 at the Institute of Physics of the University of Copenhagen, now known as the Niels Bohr Institute. In 1933 he defected with his wife and by 1934 he was a professor at the George Washington University, Washington, DC, where he went on to hire Edward Teller.

In the late 1930s Gamow and Teller together wrote several papers on nucleosynthesis in stars and on galaxy formation within the context of the expanding Universe model. After the second World War, Gamow wrote a series of papers, several in collaboration with his student, Ralph Alpher, that were to set the scene for the definitive model of the Hot Big Bang theory. For a variety of reasons, this work did not get the attention it deserved until the discovery of the cosmic microwave background radiation by Penzias and Wilson, in 1965.

Gamow (1954b) was the first, following on the discovery in 1953 of the double helix structure of DNA by Crick, Franklin and Watson, to suggest a mechanism for the formation of proteins. As it turned out the idea was not correct, but his friend Francis Crick always said that this proved an inspiration for his own research on the subject.

What had emerged from this work was a variety of predictions for the temperature of the radiation left over from the primordial explosion. The variation in predictions was largely a consequence of a rather poor knowledge of the density and expansion rate of the Universe, and of a rudimentary knowledge of the underlying nuclear physics parameters.[3]

It seems clear that Penzias and Wilson were not influenced by any of this: they were working on advanced telecommunications. Through meticulous experimental technique, they had discovered an isotropic residual source of 'noise' in their work on satellite communications at Bell Labs in New Jersey. The story is well documented that they asked the experts at Princeton whether they had any idea what this might be and, as they say, the rest is history (Peebles et al., 2009).

However, the story of events prior to 1965 is of great historical interest and is documented in the book *Finding the Big Bang* (Peebles et al., 2009). The book contains the recollections of scientists who had been closely involved in the quest to establish the Hot Big Bang theory by discovering the relict radiation left over from the Big Bang. The book by Partridge (2007, re-issued after its first publication in 1997) also provides detailed discussion of the history of this great discovery.

---

[3] See Peebles (2014) and Peebles et al. (2009, Section 3.1) for a detailed discussion of the nucleosynthesis aspects of what was happening at this time.

3.1.1 Pre-1965

The pre-discovery search for evidence of the Big Bang was primarily centred on Princeton in the USA and Moscow in the then Soviet Union. It might be fair to say that the focus in Princeton was to search for the left-over radiation from the Big Bang, while in Moscow the focus was on the nuclear physics of the primordial formation of the elements.

The Princeton group had been searching for this signal from the Universe for a period of time using a small experiment on the roof of the Joseph Henry Laboratories, the home of the Physics Department at that time, and had already found evidence for the background radiation, albeit not at a high level of statistical significance. It is interesting in this context that the Peebles (1965) article entitled *The Black-Body Radiation Content of the Universe and the Formation of Galaxies* was submitted for publication on 8 March 1965, two months before the submitting of the discovery papers, and, inevitably, revised shortly thereafter (1 June). Peebles was already working on the new cosmology even before it was announced.[4]

The Moscow group, headed by Yakov Zel'dovich, also had expectations that the radiation field could be found, though they did not themselves build an experiment to find it. Zel'dovich was already a promoter of the Big Bang theory even before the cosmic background radiation was discovered, though he was strongly inclined towards a cold singular origin (the basis for this being that a hot origin would produce more deuterium and helium than had been observed in stars (Zel'dovich, 1964)).

Zel'dovich had apparently commanded[5] Smirnov to recompute the nucleosynthesis calculation on the assumption of a Hot Big Bang with a view to getting a better handle on the expected temperature (Smirnov, 1965). Smirnov *loc. cit.* concluded that 'The theory for the "hot" state for pre-stellar matter fails, then, to yield a correct composition for the medium from which the first-generation stars formed ...', citing Zel'dovich (1964).[6] There are some remarks on this in Peebles *et al.* (2009, footnote p.36. and remarks on p.95).

It seems that the Zel'dovich group were unaware of the important paper of Osterbrock and Rogerson (1961), which had concluded, on the basis of their study of the spectra of planetary nebulae, and from earlier data on the Sum and HII regions, that the abundance of helium was everywhere in excess of $Y = 0.24$, i.e. 24% by mass, and might in places be as

---

[4] The interesting review of the status of cosmology, *Gravitation and Space Science*, by Dicke and Peebles (1965) was also submitted for publication on 8 March 1965, and appeared in the June issue of the journal. A short endnote added 'in proof' (p.460, referring to page 448) remarked on the publication of the papers of Penzias and Wilson (1965, dated 13 May) and Dicke *et al.* (1965, dated 7 May). The review gives an interesting perspective on pre-1965 cosmology and discusses such issues as the wider consequences of having a varying gravitational constant and the cosmological consequences of having extra fields in the Einstein equations, with particular reference to the theory of Brans and Dicke (1961). Interestingly, on p.451, the paper cites Alpher and Herman (1953), but does not mention George Gamow.

[5] That is how Zel'dovich did things!

[6] At that time Zel'dovich was a strong, almost outspoken, supporter of the notion of the cold Big Bang. In Zel'dovich (1963) he criticises the scenario of Gamow, Alpher and Herman by saying [from the English translation] that they propose a state ... *at such high temperatures that the radiation density is gigantically in excess of the nucleon density. This point of view, on the basis of which the authors attempted even to reconstitute the presently observed abundance of the elements, leads to insurmountable contradictions. In the pre-stellar stage they they obtain a large amount of helium (about 10–20%) and deuterium (about 0.5 %).*

high as 0.32. They cited Gamow (1949) as a potential source for this otherwise difficult to explain amount of helium. The helium production was to be one of the difficulties that the Steady State theory had to confront. Later, Truran *et al.* (1965a,b) firmly established that stars of the kind we observe could not make the observed helium. However, that was shortly before the announcement of the discovery that would cause a revolution in cosmology, at which point the debate about where the helium was made would cease.

The paper by Doroshkevich and Novikov (1964) was perhaps the first published article to argue that this radiation could be detected by the technology then available. They obviously did not know that Bob Dicke and his group in Princeton, had already embarked on the hunt. Andrei Doroshkevich and Igor Novikov, scientists in the Zel'dovich group, were expected to hunt through the entire literature looking for someone who may have unwittingly found it without recognising it. Later on the Zel'dovich group would go on to produce many important papers in cosmology.

Japan was a country with a significant history of research into cosmology through the early work of Nariai and Hayashi, and more recently Tomita. Big Bang nucleosynthesis had been studied in important papers by Hayashi (Hayashi, 1950; Hayashi and Nishida, 1956). Tomita and Hayashi (1963) had already written a paper on the cosmological constant. In the years shortly after the Penzias and Wilson announcement many fine papers on cosmology and the origin of galaxies came out of Japan.

So it is hardly surprising that the study of cosmology developed rapidly after 1965: everyone was ready for it! The discovery of Penzias and Wilson had opened up a new vista in physics.

Prior to 1965, cosmology had been seen by many as a branch of general relativity backed by relatively few, very difficult, observations that were possible only with scarce data gathering resources. Cosmology had been described as a mere 'search for two numbers': the Hubble constant and the value of the deceleration parameter describing the cosmic expansion. After the Penzias and Wilson discovery in 1965 cosmology became a branch of physics.

This radical change is characterised by the difference in approach between the important books of McVittie (1965a) on *General Relativity and Cosmology* and of Peebles (1971) on *Physical Cosmology*. The first is a complete summary of what was known about the Universe prior to the discovery of the CMB, containing all kinds of useful formulae derived from the Einstein equations. By contrast the second hardly mentions general relativity and is characterised by a remarkable span of topics in physics.

Volume 142 of the *Astrophysical Journal* is in itself a remarkable testament to this revolution. Not only does it contain the papers of Dicke, Peebles, Roll and Wilkinson and of Penzias and Wilson, it contains two other papers that show this revolution in action. The first is the classical paper of McVittie (1965b) discussing, from the relativistic point of view, the consequences of the earlier discovery that same year of objects with redshifts in excess of unity (Schmidt, 1965). The second paper is the aforementioned paper of Peebles (1965) which effectively set the tone for the following decades of cosmology: it contains no reference to general relativity nor even to specific cosmological models.

3.1.2 Discovery

In 1978 Arno Penzias and Robert Wilson were awarded the Nobel Prize in Physics for their discovery of the cosmic background radiation in 1965 (Penzias and Wilson, 1965).[7]

> **The Nobel Prize in Physics for 1978 is awarded to Arno A. Penzias and Robert W. Wilson**
>
> for their discovery of cosmic microwave background radiation.
> The discovery of Penzias and Wilson was a fundamental one: it has made it possible to obtain information about cosmic processes that took place a very long time ago, at the time of the creation of the Universe.
> `http://www.nobelprize.org/nobel_prizes/physics/laureates/1978/press.html`

Penzias and Wilson were in the process of calibrating a highly sensitive radiometer (see Figure 3.1) that was intended for use in satellite and balloon telecommunications. What they reported was the discovery of an unaccounted for signal in their experiment, which they described in the title of the paper as an 'excess antenna temperature at 4080 Mc/s'[8] with an excess temperature of 3.5 K ± 1.0 K. They assert that:

> **Quotation from Penzias and Wilson (1965)**
>
> This excess temperature is, within the limits of our observations, unpolarized, and free from seasonal variations (July, 1964–April, 1965). A possible explanation for the excess noise temperature is the one given by Dicke, Peebles, Roll and Wilkinson in a companion letter in this issue.

At the time of publication few people doubted that this was indeed the radiation left over from the Big Bang. However, getting to the point of scientifically acceptable corroboration, and the Nobel Prize, was going to be a long and difficult job involving many experimental groups. The first step was the article of Roll and Wilkinson (1966) presenting the results of the Princeton measurement of $T = 3.0 \pm 0.5$ K at a wavelength of 3.2 cm: this provided support for the black-body nature of the radiation spectrum.

Shortly after the awarding of the prize Robert Wilson wrote a fine article about the search for and discovery of the radiation for *Review of Modern Physics* (Wilson, 1979) in which he describes the Penzias–Wilson experiment. He also noted several occasions on which the radiation had been seen prior to 1965 without being recognised for what it was. The book *Finding the Big Bang* by Peebles *et al.* (2009) is another important reference presenting the story of the quest to discover the relict radiation.

---

[7] They shared the prize with Pyotr Kapitsa who was awarded half the prize 'for his basic inventions and discoveries in the area of low-temperature physics'. Kapitsa had discovered, in the 1930s, how to make liquid helium in large quantities, and went on to discover the phenomenon of superfluidity in 1937. Kapitsa had strong links with Cambridge University, having been a James Clerk Maxwell student (1923–1926) and then Assistant Director of Magnetic Research at Cavendish Laboratory (1924–1932). He returned to the Soviet Union in 1934, but was not permitted to return to Cambridge. He created and directed the Institute for Physical Problems in Moscow, but was fired when he refused to work on the Soviet Hydrogen Bomb project.

[8] 4080 Mc/s corresponds to a wavelength of 7.35 cm.

Fig. 3.1   The microwave radiometer with which Penzias and Wilson (1965) first detected the Cosmic Microwave Background
Radiation in 1965. The telescope, located in Holmdel, New Jersey, was built by Bell Labs as a part of a project to detect
radio waves bounced off an Echo communications balloon. In 1990 the horn was dedicated to the National Park
Service as a National Historic Landmark.
(Credit: NASA: GPN-2003-00013 in the public domain, see page 695)

The corroboration of the discovery would come from an entirely different direction: the
excitation of interstellar spectral lines of the cyanogen molecule. The rotational excitation of the CN molecule in diffuse interstellar clouds is primarily due to the absorption
of millimetre wavelength photons that come mainly from the microwave background
radiation.

3.1.3 Corroboration

In 1940, McKellar (1940) had identified the transitions seen by Dunham and Adams (1939)
and Adams (1941) in the blue part of the spectrum of the star ζ *Ophiuchi* as coming from
the CN and CH molecules in the interstellar space between us and the star.[9] Then, in 1941,
McKellar (1941) analysed in more detail the data on the line intensities provided by Adams

[9] The lines are the $R(1)$ line at $\lambda 3874.0$ Å, and the $R(0)$ line at $\lambda 3874.6$ Å. The CN molecule can be in one of
a number of rotational states, and in each rotational state it can vibrate along the axis joining the two atoms.
The result is a spectrum comprising a series of bands corresponding to each rotational state. The molecular

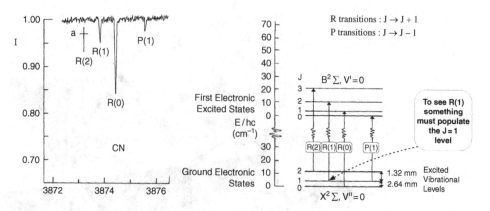

Fig. 3.2 CN lines observed in the direction of the star ζ *Ophiuchi* (Bortolot *et al.*, 1969). The figure on the left shows the spectrum and identifies the lines from the CN molecule. The figure on the right (a 'term diagram') depicts the energy levels associated with the various observed transitions. The line R(2) is the transition from the $J = 1$ to the $J = 2$ rotational state. When the line was first identified there was no known mechanism for populating the $J = 1$ level in order that this transition could take place. We see from the diagram that photons of wavelength 2.64 cm are required. McKellar (1941) remarked that whatever caused this it would have an effective temperature of ~ 2.3 K. It is the ratio of the strengths of the R(0) and R(1) lines that determines the spin temperature of the molecule, which in turn provides an upper limit on the temperature of the CMB.

(Reproduced from Bortolot et al. (1969, Fig. 1 p.308) with permission, © 1969 American Physical Society.)

and came to the conclusion that the lines were due to the molecules that were in a state of rotation, the equilibrium 'spin temperature' being 2.3 K. This was undoubtedly the first 'sighting' of the cosmic background radiation, though at the time, or until the discovery of Penzias and Wilson, there was no reason to identify this temperature with the temperature of an ambient radiation field.[10]

The link between the cosmic background radiation and the spin temperature of the CN molecule was soon made in 1966 by Field and Hitchcock (1966a,b) and by Thaddeus and

---

rotational temperature is found from the ratio of the intensity of the $R(1)$ line to the intensity of the $R(0)$ line, see Figure 3.2.

[10] In a book review, Hoyle (1950) made the connection between the Hot Big Bang and the McKellar temperature of 2.3 K. Hoyle was reviewing a book by Gamow and Critchfield (1949, Appendix VI, p.336) in which the authors consider nucleosynthesis in a Hot Big Bang model. Hoyle objects to the underlying cosmological model on the grounds that it has the wrong age and comments that ... *it would lead to a temperature of the radiation at present maintained throughout the whole of space much greater than McKellar's determination for some regions within the Galaxy.* This was the first suggestion that a Hot Big Bang Universe would be pervaded by an observable radiation field (and in this case via excitation of molecules).

Hoyle, as was frequently the case, was perfectly correct in this assertion.

Most of the review focuses on a critique of the 52 pages that form Chapter 10: *Thermonuclear Reactions*, which has a heavy emphasis on element production in stars. Hoyle had already published two papers on this subject (Hoyle, 1946, 1947) and would later go on to publish two further, iconic, papers on the subject (Hoyle, 1954; Burbidge *et al.*, 1957).

Clauser (1966).[11] To be in the rotational state indicated by the CN spectrum would require large numbers of photons having wavelength 2.64 cm: these would have to come from some hitherto unknown source. The obvious source now presented itself, the microwave background radiation discovered by Penzias and Wilson. If this were the only contributor to the rotational state of the CN molecule, the temperature of the radiation would have to be 3.12 ± 0.15 K.[12] However, although starlight could not itself be responsible for the observed rotational state of the CN, there might be contributions from photons coming from other sources such as collisions, fluorescence, Bremsstrahlung and non-thermal radio sources (Thaddeus and Clauser, 1966). Taking all the uncertainties into account, Field *et al.* (1966); Field and Hitchcock (1966b) put forward a value of $T = 3.2^{+0.2}_{-0.5}$ K for the cosmic radiation temperature at 2.64 cm (strictly speaking, these molecules provide upper limits on the temperature of the exciting radiation). Shortly after this, Bortolot *et al.* (1969) provided similar measurements for CN, CH and CH^+ lines at wavelengths of 2.64, 1.32, 0.559 and 0.359 mm, found in new spectra of ζ *Ophiuchi* suggesting a temperature of $T = 2.83$ K. Almost 25 years later, Roth *et al.* (1993) were able to give the high accuracy result $T = 2.729^{+0.023}_{-0.031}$ K, in good agreement with the direct determination of the FIRAS experiment on board the COBE satellite (Fixsen *et al.*, 1996; Fixsen, 2009).

For many years these short wavelength upper limits from interstellar molecules provided strong constraints on the intensity of the cosmic radiation at wavelengths short of the peak of the spectrum.[13]

3.1.4 The Thermal History of the Universe

Figure 3.3 from Dicke *et al.* (1965) is fundamental to our understanding. It divides the evolution of the Universe into a number of distinct physical regimes. Here is what we see in that figure:

1 **Fireball** when the temperature $T \gg 10^{10}$K. During this period there are only neutrons and protons in equilibrium with thermal electrons and neutrinos.
2 **Nucleosynthesis** the short period during which the neutrons and protons combine to create the elements deuterium and helium, with traces of other light elements. At that time the mass density of the Universe is dominated by photons and the density of baryonic material is very small by comparison.
3 **The radiation era**: the period during which the temperature falls from $T \sim 10^9$K to around $T \sim 10^3$K. This era can be subdivided into two phases: the first phase when

---

[11] The equivalent widths of the corresponding lines in ζ *Ophiuchi* were found to be 3.37 ± 0.1 and 9.20 ± 0.1 (Field *et al.*, 1966). The observed ratio of the populations of the $R(1)$ and $R(0)$ states is $N_1/N_0 \sim 0.3$, which is far too high for almost all other plausible mechanisms of excitation (Field and Hitchcock, 1966b).

[12] Field *et al.* (1966) also presented data for the star ζ *Persei*, getting a temperature of 3.0 ± 0.6 K. The agreement between the temperature values determined from the two stars was support for the notion that the result was not due to the environment in which the two stars are found.

[13] Three decades later, Srianand *et al.* (2000) were able to use molecules to estimate the temperature of the background radiation at a redshift of $z = 2.337$ and show that the temperature at that redshift was between 6 K $< T_{z=2.337} < 14$ K, the actual value being $T_{z=2.337} = 9.1$ K. Subsequent to that there have been many determinations of the radiation temperature at high redshift using observations of the Sunyaev Zel'dovich effect in distant galaxy clusters (de Martino *et al.*, 2012).

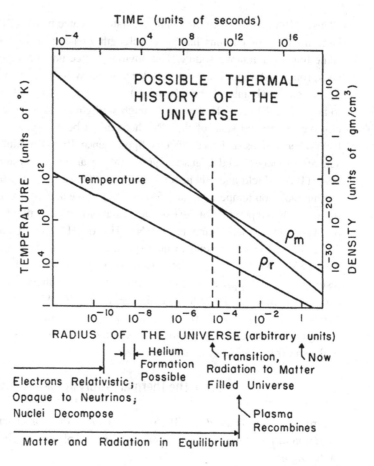

TIME (units of seconds)

POSSIBLE THERMAL HISTORY OF THE UNIVERSE

RADIUS OF THE UNIVERSE (arbitrary units)

Helium Formation Possible

Transition, Radiation to Matter Filled Universe

Now

Electrons Relativistic; Opaque to Neutrinos; Nuclei Decompose

Plasma Recombines

Matter and Radiation in Equilibrium

Fig. 3.3 The history of the Universe from the seminal paper of Dicke *et al.* (1965). This diagram delineates the principal physical regimes during the expansion of the Universe from the first millisecond after the Hot Big Bang to the present day. This was a considerably updated version of Gamow's plot 1.6. Since 1965, a lot of detail has been filled in regarding the earlier epochs (see, for example, Figure 7.4). The filling-in of this diagram describes the story of the development of physical cosmology.
(From Dicke et al. (1965, Fig. 1. p.417) reproduced by permission of the AAS.)

the radiation field density exceeds that of the baryonic matter and a second phase when the matter dominates. The time when the densities of matter and radiation are more or less equal is described on the diagram as the transition from a radiation filled Universe to a matter filled Universe. During this period the matter is ionised by the photons.

4 **The epoch of recombination**: An instant when the matter becomes neutral: the epoch of recombination. This in fact takes place over a relatively short period of time, it is not instantaneous. After that time the photons no longer collide with electrons and they travel freely to the observer. This is where the observed cosmic background radiation comes from, and so when we study the radiation field we are essentially studying the

Universe at this time. It is where the structure of the Universe leaves its imprint on the map of the cosmic background radiation.

5 **The era of structure formation**: The period up until the present where the Universe is neutral and during which cosmic structures can form. Today we would divide that era into a number of regimes.

By today's standards this is a somewhat broad overview: we can now subdivide these regimes and even add some new ones. Moreover, the earlier views of the Universe contained only matter and radiation.

We now know of two other constituents of the Universe that were not part of the scene in the 1960s: the dark matter and the dark energy. We shall come to those later. We also recognise that there are a number of phenomena between us and the time of recombination that distort the radiation pattern that comes from the last scattering surface. These distortions have to be taken into account when computing the properties of the Universe at recombination, but in doing so we learn about the phenomena responsible for the distortions.

3.1.5 The Relict Radiation Hypothesis

To establish firmly that the radiation discovered by Penzias and Wilson and by Dicke, Peebles, Roll and Wilkinson *loc. cit.* was in fact the left-over radiation from a primordial Hot Big Bang would require several tests.

Perhaps foremost among these tests would be to demonstrate that the spectrum of the radiation was accurately Planckian. That would involve measuring the spectrum in the vicinity of the expected peak and at higher frequencies. At the time this was pushing the technology to its limits.

In addition, it would be necessary to establish that the radiation field was isotropic. Proving that would require accurate comparisons of temperatures in different directions. The dipole component of the temperature field due to the motion of the Earth would have to be removed, so ultimately a map of the sky at the relevant frequencies would have to be produced.[14] But even then, the contributions to the radiation flux from the Galaxy would have to be removed.[15] Then there was the related issue that there might be additional contributions from the embryonic inhomogeneities that would lead to the formation of the galaxies and large scale cosmic structures that we see today. This was the essence of the work of Sachs and Wolfe (1967) and Silk (1967), who provided the first estimates for the likely amplitude of such primordial inhomogeneities.

[14] The dipole due to the Earth's motion involves not only the Earth motion about the Sun, or the motion of the Sun about the Galaxy, but also the motion of the Galaxy in the Local Group and relative to the more distant galaxy environment. Without assuming that there was not an intrinsic component in the cosmic dipole, this is an observational problem of considerable difficulty.

[15] Even today, eliminating backgrounds is among the key problems that limit our ability to see the CMB clearly when pushing the limits of our technology.

A Non-cosmological Origin for the CMB?

There was also the important issue of providing evidence that the radiation field was not due to some other phenomenon. There was no lack of alternative candidate theories. Hoyle and collaborators had noted that the total energy density of the CMB is coincidentally of the same order of magnitude as the energy of starlight, of magnetic fields, of turbulent motions and of cosmic rays in the Galaxy, raising the issue as to whether the radiation could be somehow related. In the 1970s the idea arose that the CMB might be due to the re-thermalisation of starlight by 'whiskers' or 'conduction needles' of graphite-like material (e.g. carbyne). In order to explain a thermal spectrum of radiation and to provide the required optical depth, these needles would have to have been extremely long. [16] However, as time passed it became ever clearer that any theory other than a cosmic origin would have to face increasingly difficult challenges of explaining the spectrum of the radiation and accounting for its large scale isotropy or any smaller scale anisotropies that might be discovered.

The Extragalactic Background Light

There was, moreover, one over-riding factor that these alternative theories would have to contend with: they would have to provide the signal strength. The situation is illustrated in Figure 3.4 which shows the flux from all radiation backgrounds known or speculated in 1969 and 1999. It is clear that the dominant background is the cosmic microwave background.[17]

There had been suggestions that the radiation might have originated locally, within the Galaxy, or within the local super-cluster, but the further the sources the more powerful they would have to be, and if they were too close there would be issues with isotropy. Many ideas were proposed and, in the course of time, tested. The cosmological origin of the radiation field was in fact the most straightforward explanation and has since survived many tests. The radiation field has become the main tool for determining the nature of our Universe and for providing the parameters that define it with remarkable precision.

3.1.6 A Black-Body Radiation Spectrum?

Proving that the radiation discovered by Penzias and Wilson and by Dicke, Peebles, Roll and Wilkinson *loc. cit.* was in fact the radiation left over from the Big Bang became the

---

[16] See Li (2003) for a detailed discussion of such ideas.

[17] The plots of Figure 3.4 are an experimental test of what for a long time was referred to as *Olbers' Paradox*. The paradox (attributed to Heinrich Olbers but perhaps first put by Digges in 1576 and Kepler in 1610) asked why the sky at night is dark. If the Universe were infinite and static, every line of sight would end on the surface of a star, and so the sky should be as bright as, say, the Sun, even at night! This has been one of the most important paradoxes of cosmology.

We now know that the observable Universe is neither infinite nor static, so the premise of the paradox is no longer valid and Figure 3.4 tells us the answer to the more modern phrasing of the question: *how bright should the night sky be?* The sky *is* bright at night, and the brightness is dominated by the CMB.

There are excellent discussions of Olbers' Paradox by Harrison (1989, 2000), where he beautifully illustrates the paradox by considering the view looking into a dense forest, and by Overduin and Wesson (2004).

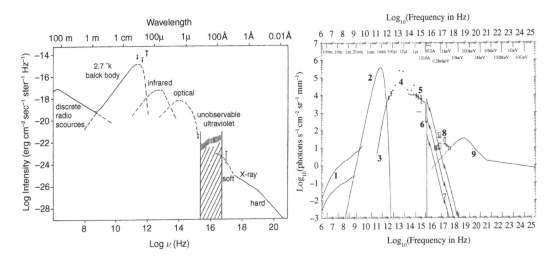

Fig. 3.4   The energy of the extragalactic background light contained in various frequency bands ranging from long-wave radio to γ-ray frequencies. On the left is the diagram showing the radiation flux as a function of frequency as first compiled by Longair and Sunyaev (1969). The solid lines are based on data available at the time, whereas the dashed lines were derived from theoretical estimates. This shows the dominance of the cosmic microwave background radiation over all other backgrounds. Three decades later this was updated by Henry (1999) who provided the right hand panel, but this time the number of *photons* s^{-1} cm^{-2} str^{-1} nm^{-1} is plotted on the vertical axis. Plotted in this way an equal plotted value means an equal amount of energy per logarithmic interval of frequency.
See the original papers for an explanation of the individual data sources and the more recent discussions by Fukugita and Peebles (2004) and Overduin and Wesson (2004).
(Left: From Longair and Sunyaev (1969, Fig. 2 p.66), with permission. See p.695.
Right: From Henry (1999, Fig. 1 p.L50), reproduced by permission of the AAS.)

next priority. Two aspects of the radiation field would provide compelling evidence for the cosmic origin of the radiation: establishing the Planckian shape of the spectrum and establishing its isotropy. The first part of the proof that this radiation field was the aftermath of the Big Bang was to establish whether or not the radiation spectrum was a black body at a temperature of around 3 K. The peak of a black-body spectrum at temperature T is at a wavelength of $\lambda_{\mathrm{peak}} \sim 5.1/T$ mm, corresponding to a frequency of $\nu_{\mathrm{peak}} \sim 59T$ GHz. The measurement of the temperature was an absolute measurement which would require accurate calibration of the equipment being used (Partridge, 2007).

To firmly establish the radiation field as being the left-over radiation from the Big Bang it would be essential to measure the radiation flux at wavelengths short of the peak. This was a formidable task with the technology available at the time: the highest frequencies, at wavelengths short of the peak of the spectrum were largely inaccessible.[18] So the first

[18] Even as late as 1981, the result of a set of measurements in the far infrared by Woody and Richards (1981) caused some consternation. Their balloon-borne experiment had reported a significant excess of radiation, relative to a black body, in the vicinity of the peak of the Planck curve. There were many attempts to reconcile this measurement with the standard Big Bang, most of which suggested that this excess radiation might be the consequence of a period of very early star formation. The measurements were never confirmed, and ultimately

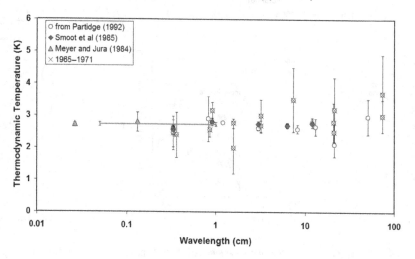

CMB Temperature measurements (up to 1990)

experiments concentrated on mapping out the intensity of the spectrum at the longer wavelengths, i.e. in the Rayleigh–Jean part of the spectrum. By 1971 it was possible to give a value to the temperature,[19]

$$T = 2.72 \,\text{K} \pm 0.08, \tag{3.1}$$

which is very close to our present value of $T = 2.7255 \pm 0.0006$ (Fixsen, 2009).

The consistency of the temperature measurements in the Rayleigh–Jeans part of the spectrum (see Figure 3.5) was supportive of the radiation having a black-body spectrum, but there remained the possibility that it might also have been a dilute black body. This could only be tested by looking beyond the peak of the Planck distribution. It was still not certain in 1970 that the cosmic microwave background was in fact the radiation from the last scattering surface of the Big Bang.

it was the FIRAS experiment on board the COBE satellite that settled the matter once and for all: the spectrum was accurately Planckian (Mather and the COBE collaboration, 1990; Wright *et al.*, 1994b).

[19] Peebles (1971, Table V-1, p134) gives an incisive analysis of the 17 temperature measurements available up until 1971. He shows (*loc. cit.*, p141) that the χ^2 for the fit of the data to a Rayleigh–Jeans law with $T = 2.72 \pm 0.08$ was $\chi^2 = 7.1$ with an expected value of $\chi^2 = 13.5 \pm 5$. So while there was no inconsistency, the low χ^2 suggested that the experiments were not entirely independent. The problem with an absolute (as opposed to relative) measurement of this kind is to remove all systematics from the measurement. After all recognised sources of noise are subtracted the experimentalists look very hard for further sources of noise, and only feel that they may stop looking when they get within range of or get a lower value than previous experiments.

3.2 Detection of Anisotropies

It was important to set limits on the isotropy of the background radiation field. This would reveal whether the source was galactic, or whether it was due to a cosmic distribution of point sources, or whether it was in fact a diffuse background.

It had been clear at the outset that the dipole component of the anisotropy due to the Earth's motion relative to the Universe at large would have the largest amplitude: this is the Doppler effect. However, because of other dipole contributions, notably from the zodiacal light and our galaxy, this was never going to be easy. If this motion were only on the order of hundreds of kilometres per second, it would be undetectable given the sensitivity of the early experiments, but the process of putting limits on the degree of isotropy could begin. So following the discovery of the CMB, early work also focused on putting constraints on the isotropy of the radiation field: that would be an important part of establishing its cosmological origin and would put constraints on the peculiar motion of the galaxy relative to the Local Group and the Virgo Supercluster.

The first attempts to put experimental limits on the level of anisotropy were those of Partridge and Wilkinson (1967), Conklin and Bracewell (1967) and Epstein (1967). Partridge and Wilkinson *loc. cit.* were able to put an upper limit of $\pm 0.1\%$ on the dipole component, which corresponds roughly to a velocity of $\sim 0.001c = 300 \, \mathrm{km \, s^{-1}}$. This was low enough to be important for constraining the motion of the Sun relative to the cosmic background. Indeed, Sciama (1967), using this upper limit on the dipole component, inferred an important upper limit of $\sim 300 \, \mathrm{km \, s^{-1}}$ on the Sun's motion relative to the background by using models for the dynamics of the Local Group and for the Virgo Supercluster.[20]

One of the early scientific priorities for CMB isotropy studies was to determine our motion relative to the frame in which the CMB appeared to be isotropic. There followed a period of some 15 years of experiments at a variety of frequencies and angular resolutions, using a variety of different techniques (Partridge, 1988). According to Partridge (2007, p.198) the first reliable detection of the dipole was that of Corey and Wilkinson (1976). The first determination of the dipole from a full sky map was that of Klypin *et al.* (1987) and Strukov *et al.* (1987) using data from the Relikt-1 satellite. The history of these occasionally contentious measurements of large angular scale anisotropies is reviewed and put in perspective by Lineweaver (1997) and by Partridge (2007). On smaller angular scales (around say 5°–10°), the first detection of anisotropy was reported, or hinted at, in 1991 after the first year analysis of the results of the DMR experiment on board the COBE satellite. The definitive announcement came with the papers of Górski *et al.* (1994) and of Wright *et al.* (1994a).

---

[20] Somewhat later, Sciama (1972) wrote an interesting short commentary on the significance of these observations, which in effect and for the first time established the existence of a cosmic standard of rest. In this sense not all inertial frames are equal. The supposed existence of a relationship between local inertial frames and this fundamental inertial frame is referred to as *Mach's principle*.

3.2.1 Anisotropies on Large Angular Scales

The large angular scale anisotropies would give a contribution to the temperature field that could be described by the sum of a monopole, dipole and quadrupole component:

$$
\begin{aligned}
T(\alpha, \delta) = {} & T_p + T_x \cos \alpha \cos \delta + T_y \cos \alpha \sin \delta \\
& + \tfrac{1}{2} Q_1 (3 \sin^2 \delta - 1) + Q_2 \sin 2\delta \cos \alpha + Q_3 \sin 2\delta \sin \alpha \\
& + Q_4 \cos^2 \delta \cos 2\alpha + Q_5 \cos^2 \delta \sin 2\alpha.
\end{aligned}
\tag{3.2}
$$

The coefficients Q_i are defined relative to the angular coordinates (α, δ), right ascension and declination. This will be recognised as a spherical harmonic expansion of the sky in a general spherical polar coordinate system.[21] If the anisotropy is measured relative to galactic coordinates we have the same expression, but with (l, b) in place of (α, δ). The dipole anisotropy is frequently referred to as the '24-hour' component of the distribution of the temperature on the sky. The quadrupole component was referred to as the '12-hour' component.

The amplitude of the quadrupole component is generally quoted as

$$
Q_{\text{rms}} = \tfrac{4}{15} \left(Q_1^2 + Q_2^2 + Q_3^2 + Q_4^2 + Q_5^2 \right),
\tag{3.3}
$$

which is independent of whichever angular coordinates are used.

3.2.2 Dipole and Quadrupole Components

Anisotropies can be recognised by looking at variations in the temperature of the radiation field in different directions. The variation could be quantified by measuring how much, on average, these variations depend on the angular distance between the measured points: in other words, determining the correlation function of the temperature variations.

The way this is done on a sphere is to make a map and resolve the temperature fluctuations into its dipole, quadrupole, octupole and higher order multipoles.

The primary difficulty here is to generate a map of the cosmic radiation field which is free of contributions from sources that are not the cosmic background radiation, such as the zodiacal light, the radiation from the galaxy and the pollution and distortion of the signal by sources between us and the last scattering surface. So, for example, in the early measurements the observed patches of sky were chosen to be relatively free of radio sources, or scans of the sky were made away from and parallel to the galactic plane. Full sky maps had no alternative but to model the entire sky and remove the unwanted signals by removing known sources and combining maps made at different frequencies. More recently there have been attempts to build detailed sky reference models from the astronomical data describing the unwanted sources.

---

[21] See Appendix F *Functions on a Sphere*. These are the terms involving the spherical harmonics $Y_0^0, Y_1^0, Y_1^1, Y_2^0, Y_2^1, Y_2^2$ (with $b = -\theta$ and $l = \phi$). The higher order terms, i.e. large l, represent contributions from increasingly smaller angular scales.

Interpreting the CMB Dipole

If the Universe were indeed homogeneous and isotropic, an observer moving with velocity v relative to the frame in which the radiation was isotropic would observe a temperature distribution in a direction making an angle θ with the apex of this motion given by

$$T_{\text{obs}} = T_{\text{CBR}}\sqrt{1 - \frac{v^2}{c^2}}\left[1 - \frac{v}{c}\cos\theta\right]^{-1} \simeq T_{\text{CBR}}\left[1 + \frac{v}{c}\cos\theta\right], \qquad (3.4)$$

where in the second of these equations we have ignored terms in $(v/c)^2$. [22] The Earth moves around the Sun and so there has to be a Doppler correction applied to bring the reported $\Delta T/T$ into a Solar System reference frame.

It was recognised early on that the main contributor to the dipole component of the CMB anisotropy would in fact be the motion of the Solar System relative to the Universe at large: due to the Doppler effect, the Universe would appear warmer in the direction towards which we were moving. Eliminating that component would leave a series of multipoles from the quadrupole and upwards whose amplitude should be consistent with, and certainly no bigger than, what is suggested by our precepts about how cosmic structure formed.

The motion of the Local Group of galaxies, or even our local neighbourhood, relative to the CMB, is likely to be due to the gravitational influence of the surrounding distribution of matter on very large scales (say, $50h^{-1} - 100h^{-1}$ Mpc or more). To determine this cosmological component of our motion we need to remove from the observed velocity all those components that are due to local effects. Hence we need to know the Solar motion about our galaxy, the motion of our Galaxy in the Local Group, the motion of the Local Group relative to nearby matter and so on. These numbers become more uncertain the larger the volume we consider.[23]

3.2.3 Early Anisotropy Measurements

The existence of a dipole component to the cosmic background radiation was established by the papers of Conklin (1969)[24] and Henry (1971).[25] Conklin's experiment, which

[22] The relativistic Doppler correction provides an extra term of order $(v/c)^2$ in $\Delta T/T$ proportional to $\frac{v^2}{2c^2}(2\cos^2\theta - 1)$ (Peebles and Wilkinson, 1968; Kamionkowski and Knox, 2003). This additional term has a quadrupole angular distribution and is non-zero even when $\theta = \pi/2$. It is referred to as the *transverse Doppler shift*. We see that the ratio of the dipole to this quadrupole is $O(v/c)$ which for the CMB is $\sim 1.2 \times 10^{-3}$. This term appears as a correction to the quadrupole anisotropy on the order $O(v/c)^2 \sim 1.4 \times 10^{-6}$.

[23] Lilje *et al.* (1986) identified a significant quadrupole component in the inferred non-Hubble components of galaxy velocities. Interpreting this as being due to tidal forces by large scale inhomogeneities, they identified the possible source of the tides as being the Hydra-Centaurus cluster, located at three times the distance of the Virgo Cluster. The question of the convergence of the CMB and galaxy distribution dipoles became an important topic during that period of time.

[24] See Webster (1974, Section 4.3.2), who expressed doubts about this measurement on the basis of possible background contamination.

[25] Henry was one of a long line of PhD students of David Wilkinson in the Joseph Henry Laboratories of the Princeton Physics Department. Wilkinson was perhaps the greatest pioneer of CMB experimentation: over a period of some four decades he developed a wide variety of instruments to search for and analyse the CMB. He had a reputation as a great teacher and during that time he worked with a large number of notable PhD students.

Table 3.1 Measurements of CMB Solar dipole.

Source[a]	ΔT (mK)	Longitude, $l°$	Latitude, $b°$	Experiment
Conklin (1969)	1.6 ± 0.8	96 ± 30	85 ± 30	ground
Henry (1971)	3.3 ± 0.7	270	24 ± 24	balloon
Corey and Wilkinson (1976)[b]	2.7 ± 0.7	308 ± 30	38 ± 20	balloon
Smoot et al. (1977)	3.5 ± 0.6	248 ± 15	56 ± 10	U2 spy-plane
Fixsen et al. (1983)	3.18 ± 0.17	265.7 ± 3.0	47.3 ± 1.5	balloon
Lubin et al. (1985)	3.44 ± 0.17	264.3 ± 1.9	49.2 ± 1.3	balloon
Strukov et al. (1987)	3.16 ± 0.07	266.4 ± 2.3	48.5 ± 1.6	RELIKT-1
Smoot et al. (1992)	3.36 ± 0.1	264.7 ± 0.8	48.2 ± 0.5	COBE-DMR-1yr
Lineweaver et al. (1996)	3.358 ± 0.023	264.31 ± 0.17	48.05 ± 0.10	COBE-DMR 4yr
Hinshaw and Weiland (2009)[c]	3.335 ± 0.008	263.99 ± 0.14	48.26 ± 0.03	WMAP 5yr
Planck Collaboration (2015)[d]	3.3645 ± 0.0020	264.00 ± 0.03	48.24 ± 0.02	Planck 2015
Fixsen et al. (1994)[e]	3.374 ± 0.008	265.6 ± 0.75	48.36 ± 0.5	FIRAS 1yr
Fixsen et al. (1996)	3.372 ± 0.005	264.14 ± 0.17	48.26 ± 0.16	FIRAS 4yr

[a] Extracts from Lineweaver (1997) with WMAP data added.

[b] Never formally published as a paper.

[c] This value is a combination of results using two different methods of background subtraction. The quoted error reflects the level of inconsistency between the two methods. See their Table 4.

[d] Note that we are measuring at the micro-Kelvin level. See also Planck Collaboration I (2015a, Table 1).

[e] Determined only from the redshift of the CMB spectrum determined by the FIRAS experiment.

measured the right ascension of the anisotropy, was ground-based, while Henry's, which determined the declination,[26] was balloon based. These detections were the first step in a long line of experiments of detecting and measuring anisotropies in the cosmic radiation background.

The detection and measurement of an anisotropy by Corey and Wilkinson (1976) was announced at an AAS meeting (Haverford, 1976) but never published as a paper. The cosmic microwave background radiation provides a frame of reference equivalent to that defined by the matter by which it was last scattered. In 1977, Smoot et al. (1977) had used the U2-spy plane to carry their equipment at 20 km altitude in a series of eight flights measuring the temperature of some two dozen discrete patches in the north celestial hemisphere. Each flight yielded about $3\frac{1}{2}$ hours of data. A further three flights made during 1978 were added later (Gorenstein and Smoot, 1981).

An overview of the history of the CMB dipole measurements is summarised in Table 3.1. In the table ΔT represents the amplitude of the dipole, and (l, b) the direction, in galactic coordinates, in which the Sun is moving relative to the CMB These quantities are reduced to the Solar frame, so this describes the motion of the Solar System.[27]

[26] The determination of the declination was considerably more difficult than the determination of the right ascension owing to the experimental constraints imposed by balloon flight.

[27] By 1977, the sensitivity of the more recent measurements was such that we could detect the Doppler effect contribution of the motion of the Earth about the Sun in the data. More recent measurements at the micro-Kelvin level require additional corrections taking account of higher order corrections due to the aberration of light.

3.2.4 The Source of the Cosmic Dipole

The measured cosmic dipole is due to our motion relative to the cosmic background radiation.[28] It represents a combination of the direction in which the overall gravitational attraction from the general matter distribution at the position of the Local Group of galaxies is dragging the Local Group, and of our own motion as a part of the Local Group. In the cosmological context, we are less interested in our peculiar motion within the Local Group than we are in the velocity of the Local Group relative to the Universe at large.

In determining the motion of our Local Group relative to the Universe at large, we have to subtract the local, non-cosmic, contributions like the motion of the Earth around the Sun, the Sun around the galaxy and the galaxy relative to the Local Group. The first two of these are known rather well, the motion relative to the Local Group is less certain since it depends on knowing the precise membership of the Local Group.

We can make a rough estimate of the motion of the Local Group of galaxies relative to the CMB by combining the following approximate velocities:

$$\begin{aligned} \mathbf{V}_{\text{Sun-LG}}: & \qquad 306\,\text{km s}^{-1} \;\; l = 99° \quad b = -4° \\ \mathbf{V}_{\text{Sun-CMB}}: & \qquad 370\,\text{km s}^{-1} \;\; l = 264° \quad b = 48°. \end{aligned} \qquad (3.5)$$

The first is an estimate of the Sun's velocity relative to the Local Group from Courteau and van den Bergh (1999), based on the Solar motion relative to more than 30 Local Group members, and the second is the measured velocity of the Sun relative to the cosmic background (using Equation 3.4). This gives us a velocity for the Local Group relative to the background radiation of $\mathbf{V}_{\text{LG-CMB}} = (\mathbf{V}_{\text{Sun-CMB}} - \mathbf{V}_{\text{Sun-LG}})$:

$$\mathbf{V}_{\text{LG-CMB}}: \qquad 623\,\text{km s}^{-1} \;\; l = 267° \quad b = 28.5°. \qquad (3.6)$$

This corresponds to motion towards right ascension and declination $\alpha = 10^h 35^m$, $\delta = -25°$, with a large error due to the uncertainty in $\mathbf{V}_{\text{Sun-LG}}$.

At first sight we would think that the Local Group should be moving primarily towards our nearest great mass distribution: the Virgo cluster. However, the Virgo cluster is located at $l = 284°$, $b = +74°$ at a distance of 16.05 Mpc: so the situation is somewhat more complicated than simple gravitational motion towards our nearest great mass concentration. Moreover, a peculiar velocity of $620\,\text{km s}^{-1}$, over and above that due to cosmic expansion, at that distance might seem at first sight rather high given that the Virgo Cluster itself is not a particularly rich cluster of galaxies. There is clearly an influence from more distant aggregates of galaxies beyond the Virgo supercluster.[29]

There is a considerably more complex model presented for the Local Group motion in Tully *et al.* (2008), who present a detailed study of the local environment and succeed in

[28] It is conceivable that there is a component of the CBR dipole that is intrinsic to the underlying cosmology. In that case we would not be living in a standard homogeneous and isotropic Robertson–Walker Universe. However, it is unlikely that even a substantial part of the measured dipole component would be due to that: the significant disparity between the magnitudes of the quadrupole and dipole would make this a strange model. In any case, we can account for most of the measured dipole in terms of the observed distribution of matter.

[29] For a discussion of dipole convergence based on the 2MASS Extended Source Catalogue see Bilicki and Chodorowski (2010).

accounting for the peculiar motion of the Local Group in terms of the gravitational effect of three components: the local void, the Virgo Cluster and a more distant contribution whose direction is identified with the Centaurus Cluster. There is a nice review of this in Tully (2007) and a detailed accounting of the local matter distribution by Courtois *et al.* (2013) who produce their 3-dimensional maps of the local Universe in a frame of reference that is at rest relative to the CMB.

Haugbølle *et al.* (2007), Neill *et al.* (2007) and Gordon *et al.* (2008) take a radically different approach and use low redshift Type Ia supernovae as probes of the cosmic flow. Haugbølle *et al.* (2007) find that, out to a distance corresponding to a redshift of $4500\,\mathrm{km\,s^{-1}}$ the flow is dominated by the Great Attractor, but beyond that the direction swings around towards the Shapely concentration of galaxy clusters.

3.2.5 Anisotropy: Cosmic Structure Formation

It had long been realised that, within the context of the Big Bang theory, the formation of galaxies would require the existence of small amplitude density fluctuations that could grow gravitationally into galaxies and the structures we see today.[30]

This gravitational instability hypothesis required that the amplitude of those pre-galactic density fluctuations would have to have been large enough at the time of recombination such that they could form the galaxies and structure we now see. Where these primordial seeds came from was an issue that would only be addressed with the later development of the theory of cosmic inflation (Guth, 1981). Nevertheless, this approach raised the possibility that we might be able to see the primeval structure that led to galaxy formation by observing the temperature fluctuations of the background radiation (Sachs and Wolfe, 1967).

The gravitational evolution of small amplitude inhomogeneities had earlier been calculated for a variety of cosmological models having a variety of equations of state.[31] Now, with the discovery of the CMB the physical processes could be taken into account in greater detail: this is what Peebles (1965) had started in his discussion of isothermal density fluctuations in the presence of radiation, and the role of the sudden recombination of the primeval plasma.

3.3 The Signature of the Universe

In 1970 two papers was published that have radically changed our ability to determine the parameters that describe our Universe. Prior to that time the most precise cosmological datum was the temperature of the Cosmic Background Radiation (CBR) field: $T = 2.72 \pm 0.08K$ (Peebles, 1971, CH. V). The papers were written by Peebles and Yu (1970) and

---

[30] An alternative to galaxy formation via gravitational instability, first suggested in the 1950s, was the idea that the early Universe could be perturbed by cosmic turbulence. There were numerous reasons why this could not work. For a pre-1965 review and history see Layzer (1964), and for post-1965 see Jones (1973, 1976).

[31] The main pre-1965 works on perturbations of homogeneous and isotropic cosmological models were the general relativistic treatment of Lifshitz (1946) and the Newtonian version by Bonnor (1954).

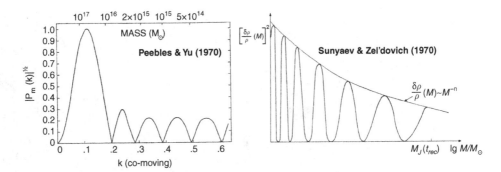

Fig. 3.6 The two predictions of the existence of features in the spectrum of the temperature fluctuations in the cosmic microwave background which were discovered some three decades later and that have been instrumental in making precision cosmology a reality. The figure on the left is from Peebles and Yu (1970), and the one on the right is from Sunyaev and Zel'dovich (1970a). These are perhaps among the most important diagrams in the physics of the Universe.

What is plotted here is the *power spectrum* of the temperature fluctuations, ΔT, relative to the mean temperature, T. This measures the contribution of each length scale, in frequency units, to the total energy in the fluctuations. The first peak arises from the very largest scales and is due to local compression of the cosmic plasma on that scale and the consequent increase in $\Delta T/T$. The second peak is due to local expansion, i.e. rarefaction, and so corresponds to a decrease in $\Delta T/T$, i.e. negative temperature fluctuations, ΔT.

(Left: From Peebles and Yu (1970, Fig. 5 p.831), reproduced by permission of the AAS.
Right: From Sunyaev and Zel'dovich (1970a, Fig. 1b p.9), with permission. See p.695.)

by Sunyaev and Zel'dovich (1970a) see Figure 3.6 and 3.7[32]). What they had recognised was that the physical processes taking place just prior to the recombination of the primeval plasma would leave a subtle imprint on the structure of the random temperature distribution that we see today: the 'signature of the Universe'.[33] To see this imprint would require the analysis of very sensitive temperature maps of the microwave sky.[34]

The papers were quite different in character: the first wrote down the Boltzmann equation for the propagation of radiation and solved it on a computer, while the second described the phenomenon by simple analytic techniques.

What these papers recognised, even before it was generally acknowledged that there would be any detectable fluctuations, was that the statistical nature of the fluctuations in

---

[32] Picture from `http://www.astro.rug.nl/~bernard60/confphoto_breaks.php`.

[33] The modulation of the perturbation amplitude as a function of scale has frequently been attributed to an article of Sakharov (1966), which is about the quantum origin of fluctuations in cold universes. However, the paper, being about cold universes, makes no direct reference to phase-correlated oscillations in the context of cosmic structure formation. See Sunyaev (2009) for an extensive historical discussion of Sakharov's contribution, in particular in Sunyaev's amusing reminscences about the naming of the oscillations, *loc cit* p.131. Here we shall refer to this phenomenon as being due to 'baryon acoustic oscillations', or BAOs for short.

[34] The predicted level of fluctuations was such that Zel'dovich did not believe that this would ever be measurable. See the recollections of Sunyaev on this point (Sunyaev and Chluba, 2009). A few years later, Doroshkevich *et al.* (1978, Figs. 4 & 5) analytically computed the temperature fluctuation power spectrum for two plausible open cosmological models, one having adiabatic perturbations and the other having entropy perturbations. Their Figure 4 was perhaps the first time that the temperature power spectrum had been shown in a form that resembles the currently measured spectrum.

Fig. 3.7   Rashid Sunyaev (left) and Jim Peebles (right) in discussion in Valencia in 2006, on the occasion of the author's 60th birthday celebration. Both Peebles and Sunyaev have, on different occasions, won many honours and awards among which are the Bruce Medal, the Crafoord Prize, the Gold Medal of the Royal Astronomical Society and the Gruber Prize.

the cosmic background radiation could reveal a feature-rich signature that directly reflected the physical conditions at the cosmic time that they were being observed.

3.3.1 Seeing Evidence in the CMB for the Dark Energy

If the Universe were pervaded by dark energy, perhaps in the form of Einstein's cosmological constant, Λ, it would reveal itself in this CMB signature in several ways. This was first realised by (Blanchard, 1984) who presented equations for the angular diameter subtended by a known length scale at a given redshift. In a $\Lambda \neq 0$ Universe, a feature in the CMB signature would appear at a smaller angular scale than in a $\Lambda = 0$ Universe. Since the length scale corresponding to the largest peak in the signature is determined by the well-understood laboratory physics governing the recombination process, it is a standard measuring rod. The angular scale at which the peak is seen determines Λ directly. Stelmach *et al.* (1990) give an extensive set of formulae for calculating the angles subtended by measuring rods in diverse cosmological models.

It turns out that the angular position of the first peak is currently the most accurately determined number in cosmology. The dark energy also reveals itself in several physical processes taking place after the recombination epoch. The integrated Sachs–Wolfe effect is sensitive to the existence of dark energy, as is the growth rate of fluctuations.

3.4 The Pre-COBE Years

The 15 years following the discovery of the CMB saw large numbers of experiments to measure temperatures at various wavelengths and to put limits on anisotropies. The launch of the COBE satellite in 1989 would see the first map of the CMB sky and provide the most accurate measurement of the spectrum and determination of the temperature within the first ten minutes of its data acquisition.[35] The maps did not come until much later, but established the existence of anisotropies on angular scales down to its resolution limit of 7°. COBE was the start of decades of map-making experiments having ever higher sensitivity and resolution.

The First CMB Sky Maps – Balloon Flights

By 1980 it was clear that there was a significant dipole component in the anisotropy of the CMB, and some evidence for small scale anisotropy at a low level. To get a convincing measurement of the dipole and higher components would require a map of the background radiation. The first requirement in map-making would be to remove as many of the unwanted foreground contributions to the radiation as possible, not only foregrounds from our cosmic environment, but also from the Earth-based environment. The latter problem could be resolved either by using balloon-borne detectors and going to very high altitude, or even better, by going into space. But, even then, the removal of contributions from our Galaxy, and all the stuff that lay between us and the surface of last scattering of the CBR would be an unavoidable problem that had to be faced. This would at least require multi-frequency mapping.

Although evidence for a significant dipole component in the angular distribution of the CMB on the sky was obtained from balloon flights at a very early stage (Fixsen *et al.*, 1983; Lubin *et al.*, 1985), there were only upper limits placed on the inhomogeneity of the Universe on various scales.[36]

Relikt-1: the First Space Mission

The first, almost full sky map obtained from a space mission allowing unambiguous determination of the dipole and a firm limit on the quadrupole was from the data of the *Relikt-1* experiment on board the *Prognoz 9* satellite launched in July 1983, see Figure 3.8. Early results were published in the papers of Strukov *et al.* (1987) and Klypin *et al.* (1987), followed by a more detailed upgrade of the latter by Klypin *et al.* (1992) shortly after the first results from the COBE satellite had been reported.[37]

---

[35] COBE was launched on 18 November 1989. The results of the temperature measurements from those first few minutes of data gathering by the FIRAS experiment on board COBE were first presented in January 1990: $T = 2.735 \pm 0.06$ K.

[36] The situation of successive measurements continually lowering the upper limit on the existence of fluctuations caused a certain amount of concern. Partridge (1988, Figure 16.) showed upper limits of $\Delta T/T < 10^{-4}$ on angular scales $3'-90°$, with some experiments imposing limits of $\Delta T/T < 2 \times 10^{-5}$. Was it conceivable that anisotropies would never be found? Of course it was possible, even likely, that reheating of the cosmic plasma between the time of recombination and the present would attenuate the smaller scale anisotropies. But that would be unlikely on the larger scales, say $> 10°$, which correspond to the largest structures we see today.

[37] Although a detailed Relikt-2 concept had been put forward, it never happened because of the dissolution of the Soviet Union in 1991 and lack of funds. Relikt-2 would have had five channels at $22, 34.5, 59, 83, 193$ GHz, but, not surprisingly, working at lower angular resolution than most post-COBE experiments (Klypin *et al.*, 1992, which reviews the experiment and data reduction method in Relikt).

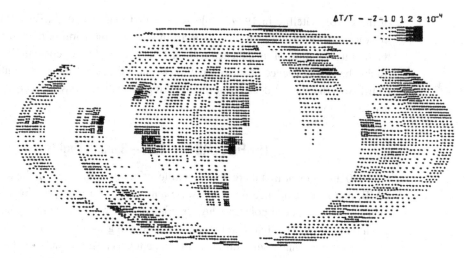

Fig. 3.8 The Relikt-1 sky map from Klypin *et al.* (1987, 1992). The map and results from the Relikt-1 data had already been published in 1987 in Soviet journals (Strukov *et al.*, 1987; Klypin *et al.*, 1987) and subsequently presented and discussed at international conferences (Partridge, 1991, 1992). The Relikt-1 data analysis had confirmed the results on the dipole and quadrupole components of the anisotropy deduced from sky maps made during earlier balloon-borne flights, and also put tight limits on the quadrupole and higher order components. All these results were later confirmed with greater precision by the subsequent analysis of the COBE-DMR data.
(Figure from Klypin et al. (1992, Fig. 4 p.71), © Royal Astronomical Society, with permission.)

The FWHM beamwidth of the measuring antenna was 5.8° and it operated at 37 GHz (8 mm) with a sensitivity of around 0.6 mK (Klypin *et al.*, 1992; Strukov *et al.*, 1992). In 1987, Strukov *et al.* (1987) had reported a detection of the dipole of 3.16 ± 0.12 mK towards $\alpha = 11^h 17^m \pm 10^m$ and $\delta = -7.5° \pm 2.5°$. This corresponded to a velocity of 350 ± 15 km s^{-1} and corresponded rather closely, though with better formal accuracy, to the measurements that had already been reported. Strukov *et al.* (1992) also reported a region of ~ 1 steradian of anomalously low brightness temperature: $\Delta T = -71 \pm 43\,\mu$K centred on a position of galactic coordinates $l = 150°$, $b = -70°$.

3.5 COBE and After

The COBE (Cosmic Background Explorer) satellite was launched on 18 November 1989, into a Polar Earth orbit at an altitude of 900 km. It carried on board two important cosmological experiments: FIRAS (Far InfraRed Absolute Spectrophotometer), for which the Principal Investigator was John Mather, and DMR (Differential Microwave Radiometer), for which the Principal Investigator was George Smoot. The task of FIRAS was to measure the spectrum, and quantify any deviations from a Planckian form, while DMR was to measure and quantify the fluctuations in temperature and produce a full sky map.[38] In 2006 George Smoot and John Mather shared the Physics Nobel Prize in Physics.[39]

> **The Nobel Prize in Physics for 2006 is awarded to John C. Mather and George F. Smoot**
>
> for their discovery of the basic form of the cosmic microwave background radiation as well as its small variations in different directions. The very detailed observations that the Laureates have carried out from the COBE satellite have played a major role in the development of modern cosmology into a precise science.
> http://www.nobelprize.org/nobel_prizes/physics/laureates/ 2006/popular-physicsprize2006.pdf

Although the COBE data has since been superseded by many experiments, it is important to appreciate the amount of work that went into the experiment by some 1000 scientists and engineers over a period of several decades. The COBE team made important discoveries and set the methodology that was to be adopted by later CMB mapping experiments.

In the following sections, we shall look briefly at what the COBE FIRAS and DMR experiments did and the methods that emerged from the COBE data analysis. The details are covered in later chapters.

3.5.1 COBE FIRAS

The FIRAS experiment depended on having a cold reference load at 1.5 K, and so ceased operation when the helium on board COBE ran out in September 1990. The first results from FIRAS were made available in January 1990 when the spectrum of the CMB, based on a mere 9 minutes of data, was shown at a conference (see Figure 3.9, *right plot*), and reported in the journals (Mather and the COBE collaboration, 1990).

It is interesting that the motion of the Earth relative to the CMB, resulting in a dipole component of the temperature anisotropy, was detected by FIRAS while measuring the temperature of the Cosmic Microwave Background (Fixsen *et al.*, 1994). FIRAS has sufficient accuracy in determining the spectrum of the radiation field that directional variations

---

[38] There was a third important experiment on board COBE: the Diffuse Infrared Background Experiment (DIRBE), for which Mike Hauser was the Principlal Investigator. This was an experiment to survey Cosmic Infrared Background (CIB), the diffuse infrared sky in ten spectral channels covering a range of frequencies from 1.25 μm to 240 μm. Although not directly connected to the CMB experiments, it nevertheless played a role in understanding the foregrounds that would obscure the CMB. DIRBE mapped the sky over a period of ten months from November 1989 to September 1990 when the cryogenic system on COBE ran out of liquid helium, and was eventually turned off in December 1993. Long before this, Hauser had been a collaborator of Peebles on a paper that established the principles of spherical harmonic analysis of clustered data on a sphere (Hauser and Peebles, 1973).

[39] During the 1990s both John Mather and George Smoot wrote insightful books about the COBE project and the historical quest that led there (Smoot and Davidson (1993, *Wrinkles in Time*) and Mather and Boslough (1996, *The Very First Light*)). Although they cover almost the same ground, they do so from rather different points of view.

George Smoot donated his share of the Nobel Prize to the charitable Oakland-based East Bay Community Foundation.

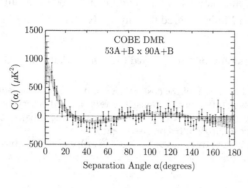 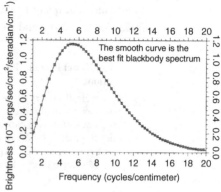

Fig. 3.9 Left: the correlation function of the temperature fluctuations. This was the first result from the COBE-DMR experiment published two years after the launch. It showed for the first time, that there were correlated fluctuations in the temperature of the CMB on all angular scales > 7°.
(From Smoot et al. (1992, Fig. 3 p.L4) reproduced by permission of the AAS.)
Right: the Cosmic Background spectrum determined from 9 minutes of COBE FIRAS data as presented by Mather and the COBE collaboration (1990). The fitted curve is a blackbody having a temperature of $T = 2.735$ K ± 0.06. The errors are smaller than the individual points on the graph. Subsequent analysis using all of the FIRAS data would improve on this measurement and give $T = 2.725$ K ± 0.002 (Fixsen *et al.*, 1994, 1996). Later, recalibration of FIRAS using the WMAP satellite gave 2.7260 ± 0.0013 K (Fixsen, 2009).
(From Mather and the COBE collaboration (1990, Fig. 2 p.L39), reproduced by permission of the AAS.)

in the spectrum could be measured. The result of this analysis gave results that are consistent with the COBE-DMR results and have errors of only 0.5° in both l and b (see Section 3.2.2 and Table 3.1 for the history of dipole measurements).

3.5.2 COBE DMR

The COBE DMR worked at three frequencies: 31.5, 33, and 90 GHz (9.5, 5.7 and 3.3 mm), each frequency channel having a beam width of 7°. At each frequency there were two independent radiometers pointing in directions 60° apart, and reporting the temperature differences in those directions. The entire sky was scanned as the satellite rotated on its axis. It was possible to reconstruct the temperature differences on all scales > 7° from this differential data.

The DMR results reporting the evidence for temperature fluctuations in the cosmic background radiation were not announced until April 1992. The analysis, based on the first year of data, was presented by Smoot *et al.* (1991, 1992). The first of these papers presented the measurement of the dipole component and put a limit on the quadrupole amplitude. The second presented the total fluctuation amplitude on angular scales $15° < \theta < 165°$ and derived a root mean square temperature fluctuation amplitude of $\Delta T/T \simeq 6 \times 10^{-6}$.

The plot of the angular correlation function of the temperature fluctuations[40] on these angular scales is reproduced from the second of these papers in Figure 3.9 (*left plot*), where we see a temperature excess on all scales $\alpha < 30°$. This was the first detection of the temperature irregularities that had been predicted by Sachs and Wolfe (1967), albeit at a substantially lower amplitude. From the shape of $C(\alpha)$ it was possible to determine the *spectral index*, n, of the primordial temperature fluctuations as $n = 1.1 \pm 0.5$, which was consistent with the predictions of Harrison (1970) and Zel'dovich (1972) that $n = 1$.

3.6 Early Post-COBE Experiments

Data collection on the COBE experiment covered the years 1989–1993: it would be almost a decade before WMAP would produce measurements of high sensitivity over a wide range of angular scales. COBE had found direct evidence for structure in the temperature variations on angular scales $> 7°$, the higher resolution scales would reveal the structure that had been predicted by Peebles and Yu (1970). That structure would provide a direct measurement of the physical state of the Universe at and during the time of recombination of the primeval plasma.

Science does not sleep. During the intervening post-COBE and pre-WMAP years there were many ground-based and balloon-borne experiments covering relatively small patches of sky but yielding measurements of the cosmic background temperature fluctuations on much smaller scales than previously. By the time the WMAP experiment flew in 2001, those experiments had quantified much of the structure in the temperature distribution and revealed what Peebles and Yu *loc. cit.* had calculated. There is a summary of all known experiments at NASA-LAMBDA website (2012).

3.6.1 The Peaks of the Power Spectrum

The first task of the post-COBE era was to explore the power spectrum of the temperature fluctuations. These arise from temperature variations associated with primordial sound waves travelling through the Universe at a time when the cosmic medium is still ionised. In essence, the power spectrum shows how irregular are the temperature fluctuations on different angular scales. Figure 3.10 shows the theoretically expected shape of the power spectrum of these fluctuations and divides it into four ranges of l-values that reflect different physical processes taking place at or after the epoch of the recombination of the primeval plasma. The details of the shape would determine the parameters of the cosmological model that best represented the Universe.[41]

---

[40] The correlation function $C(\alpha)$ displays the mean square temperature difference between two points on the sky separated by an angular distance α.

[41] The power spectrum can be thought of in terms of an analogy with the sound of a bell, or a musical instrument. The quality of the sound we hear is the consequence of a mix of frequencies: a cracked bell has a different mix than a perfect bell, and a bell sounds different from a didgeridoo because of its different frequency mix. In music these mixes are known as 'formants'. The inset in Figure 3.10 shows how this works for two notes

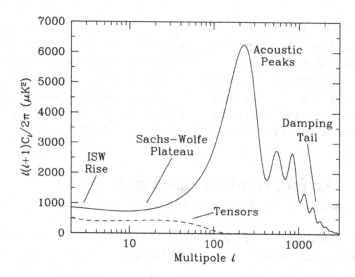
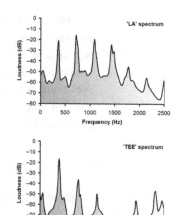

Fig. 3.10 The theoretical CMB power spectrum of temperature fluctuations due to primeval adiabatic density inhomogeneities at the epoch of recombination, for the standard ΛCDM model. The curve shows the spectrum of the fluctuations as a function of the multipole l which translates approximately into an angular scale $\alpha_l \simeq 180/l$ degrees on the sky. The vertical axis is the power per unit logarithmic interval in l, as is appropriate for the logarithmic horizontal axis. The shape of the curve is defined by a variety of physical processes taking place at different times. These are identified on the graph and discussed in the main text.
(From Scott and Smoot (2014, Fig. 27.1), with permission.)
The inset to the right shows the same kind of plot for two different notes being sung, [LA] and [TEE]. The peaks are resonances in the singer's vocal system. The intensity of each of the frequencies that make up each sound is plotted to make a spectrum of the "phoneme" that identifies the sound. This is an important aspect of speech recognition.
(With permission of Dr. Katherine Lynch.)

The spectrum, shown schematically in Figure 3.10, is dominated by a high peak at $l \sim 200$ where the strength of the fluctuations is the greatest. At higher l-values we see a succession of roughly equally spaced peaks of lower amplitude. These peaks are due to acoustic density waves in which the temperature varies as a function of the density. The fact that we see discrete equally spaced peaks is due to the fact that these sound waves have different wavelengths and are 'born' at different times during the expansion of the Universe: the shorter wavelength (higher l) ones were born earlier.

The general shape of the curve can be split into regions where different physical processes are acting. Thus at high frequencies, large l, the 'silk damping regime', the strength falls off towards higher l due to the radiative damping of the perturbations before the recombination epoch. The part labelled 'acoustic peaks' corresponds to the longest wavelength primordial sound waves that are hardly damped. Importantly, the position and relative heights of these peaks carry information about the state of the cosmic plasma at

sung by a singer. The structure of the curve reflects the character of the vocal system. The CMB powers spectrum gives the relative amplitudes of the temperature fluctuations of different frequencies, i.e. on different wavelength scales, and thereby reveals details of the underlying physical state.

and during the period of recombination: the length scale corresponding to the position of the highest peak turns out to be a standard cosmological ruler with which we can measure the distance to the last scattering surface. The structure of the 'Sachs–Wolfe plateau' and 'ISW rise' are largely determined by the physics of the Universe after the recombination epoch.

The tensor mode shown in the figure corresponds to the expected spectrum of temperature fluctuations due to primordial gravitational waves emanating from the epoch of cosmic inflation. These are of low amplitude: the ratio of the amplitudes of the tensor and scalar modes is one of the important cosmological parameters that can be constrained or possibly measured by sensitive detectors. When detected these gravitational wave tensor modes will be probing the Universe right back to the very earliest instants.

3.6.2 Discovering the First Peak of the Power Spectrum

The power spectrum of the distribution of the temperature fluctuations on the sky was expected to peak at an angular scale of around $1°$, or an ℓ-value of $\ell \sim 200$ on the scale shown in Figure 3.10.

COBE, with its $7°$ resolution, fell far short of that, but it did identify the Sachs–Wolfe component of the anisotropy at these larger angular scales. The quest to reveal the peak was on and the following decade saw many balloon-borne and ground-based experiments to map out the power spectrum to ever smaller angular scales.

The power spectrum data from experiments during that inter-COBE-WMAP period are summarised in Figure 3.11, where we see a compilation of results from a variety of experiments by Gawiser and Silk (2000) on the left and the power spectrum as revealed by the Boomerang Balloon-borne experiment (Netterfield, 2002).[42]

The two panels in Figure 3.11 are simply chosen to illustrate the progress in the measurements. Around the year 2000 there were many experiments all displaying their own measurements in conjunction with those of others (see, for example, Xu *et al.* (2002)). The mapping out of the CMB power spectrum from non-space experiments during that decade or so was a very exciting period for cosmology, during which cosmologists watched the expected peaks and troughs appearing one by one with ever increasing statistical significance. Of course, much of the impetus for these small and elegant experiments came from the imminent launch of NASA's *Microwave Anisotropy Probe* (MAP) satellite: the goal was to discover the cosmological parameters before MAP flew, which Boomerang did.

The MAP project was selected by NASA in 1996, following a competition among several alternate proposals. The launch of MAP took place on 30 June 2001, only five years after the project was started. The first results were announced in February 2003, after which MAP continued to produce data until it was switched off in 2010.

---

[42] The names, or acronyms, of some of the experiments contributing to the data plotted in Figure 3.11 are listed in footnote 5 on page 85. Among the first experiments to show multiple peaks in the power spectrum were Boomerang, DASI, MAXIMA, VSA and WMAP, all of which had data extending to $l > 800$.

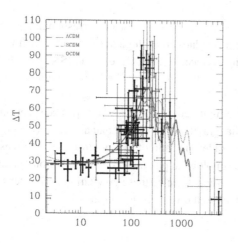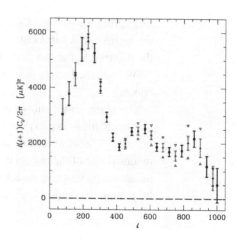

Fig. 3.11   Progress in revealing the peak in the power spectrum of the temperature fluctuations.
Left: the compilation of data up to 2000 by Gawiser and Silk (2000): despite the large error bars in some of the measurements the evidence for a main peak around $\ell \sim 200$ was unequivocal. The multipole wavenumber, ℓ, is plotted logarithmically and the vertical axis is the temperature fluctuation amplitude ΔT for that ℓ-value. (See the original reference for identification of the individual experiments.)
(From Gawiser and Silk (2000, Fig. 4 p.17) with permission. See page 695.)
Right: the power spectrum as first mapped by the Boomerang balloon borne experiment (Netterfield, 2002), accurately defining the location of the first peak at $\ell \sim 200$, and clearly showing the second and third peaks of the power spectrum. The multipole wavenumber, ℓ, is plotted linearly and the vertical axis is the mean square temperature fluctuation for each ℓ-value.
(From Netterfield (2002, Fig. 2 p.610), reproduced by permission of the AAS.)

The satellite was renamed as the *Wilkinson Microwave Anisotropy Probe* (WMAP) in 2003 to honour David Wilkinson who had died in 2002. Not only had Dave Wilkinson been a member and key driver of the WMAP mission science team, he had, as a founder member, also been instrumental in bringing the COBE project into existence.

4 Recent Cosmology

It could be said that the discovery and subsequent exploration of the cosmic microwave background radiation marked a transition from cosmology as a branch of astronomy that was largely a philosophical endeavour to cosmology as an astrophysical discipline that is now a fully fledged branch of physics. The CMB established a secure physical framework within which we could undertake sophisticated experiments that would define the parameters of that framework with ever greater precision. As understanding has grown, more and more of traditional astronomy has been embraced to provide evidence in support of this new paradigm: we have garnered evidence from the study of stars, such as supernovae, of galaxies and their velocity fields, and from observations of galaxy clusters and clustering. These studies have involved ground-based and space-based experiments at all wavelengths from radio to gamma-rays.

'Precision Cosmology' was born and we now have the recognised disciplines of 'astro-particle physics', 'astro-statistics' and 'numerical cosmology', to name but a few.

Fifty years on from the discovery of the CMB a number of issues have been clarified, but many more remain. Among the numerous areas of active research in cosmology today, there are three which have particular bearing on the first half million years: dark matter, dark energy and gravitational waves.

4.1 In the Aftermath of the CMB

The discovery of the CMB not only served to establish our cosmological paradigm as the Hot Big Bang theory, it also stimulated a growth in cosmology as a branch of physics. Fifty years later cosmology has reached a depth and precision of understanding that would have been inconceivable prior to 1965. The 50 years from 1965–2015 saw remarkable advances on both theoretical and data-acquisition and analysis fronts, which are described briefly in this chapter. One important consequence is the blurring of the boundary between theory and observation. Theoretical advances have driven experiments to gather and analyse data through which the theory could be exploited, while ever more sophisticated experiments demand improved methods of data analysis and impose constraints on the values of the parameters of the theories.

It has previously been said that the essence of physics, and science in general, is that it should be possible to perform experiments and to have other groups repeat those experiments, thus providing verification of the results and the consequent conclusions.

Cosmology as a branch of science poses two problems in this respect. Firstly, we only have a single, unique, example of the object of our study, there can be no study of another Universe and so all conclusions and generalisations have to be based on that one sample. Cosmology is more like a forensic science than a usual experimental science: we can measure and come to substantial conclusions, but no more than that. Secondly, we cannot experiment on our one object of study, we can only observe it. We cannot poke and prod it, we cannot interfere with it to see how it reacts. We can merely observe. In that respect cosmology is more akin to forensic archaeology – an understandable analogy since all cosmological observations are essentially a look-back in time.

To some extent these limitations have been circumvented by learning to work with 'surrogates' for our Universe in terms of numerical simulations of diverse cosmological models. 'Numerical cosmology' has made remarkable progress in the 50 years since we had the first computers that were capable of doing the relevant calculations on a large enough scale. It is still in its developmental infancy in the sense that while we can model particular aspects of the Universe based on specific physical premises, we cannot at present model all of cosmic physics even if we knew what the relevant physics was. Nevertheless, having models based on a limited and well-defined number of hypotheses provides templates against which we can ask what is the degree of similarity between what we observe and what we model.

In the following sections of this chapter we present a sample of the advances that have been made both on theory and on data acquisition and analysis. Entire books have been written on each of the developments, so the discussions are inevitably brief.

4.1.1 50 Years of Research – a Short Summary

With the basic paradigm established, research developed on several fronts, which can be broadly expressed as follows:

1 **The physics of microwave background radiation field**

   Within a year of the discovery of the cosmic background radiation, it was recognised by Sachs and Wolfe that we might be able to see protogalaxies and other primordial structure directly by detecting small anisotropies in the temperature of the radiation field. In turn that could constrain other cosmological parameters. This was to become one of the most significant tools in studying our Universe through ground-based and satellite experiments. The Wilkinson Microwave Astronomy Probe (WMAP) set the scene for what has become known as 'precision cosmology'.

2 **The physics of the recombination and re-ionisation.**

   The recombination of the primeval plasma marks the end of the radiation era and the start of the era of structure formation. It is also the place from which the cosmic background radiation emanates. The first task is getting the correct description of how this happens and its influence on what we see. It turns out that there must have been a more recent period when the Universe once more became ionised: this is thought to be due to the turning on of star formation and so studying this can give direct clues as to when and how the process of galaxy and star formation took place.

3 **Observable effects in the CMBR**

There are some effects of considerable importance to be seen in the CMBR, though at the time of their conception the possibility of discovering them seemed somewhat remote. Many of these were recognised by Sunyaev and Zel'dovich soon after the discovery of the radiation field. One of the most important of these is the effect that bears their name in which extremely distant galaxy clusters are seen against the general background by virtue of a temperature dip due to the hot gas in the clusters. Martin Rees had suggested looking for polarisation of the scattered radiation: it would be over 30 years before that became possible to detect and an important tool in probing the nature of our Universe. Peebles and Yu in 1970 had shown the Baryonic Acoustic Oscillations (BAO) that would take almost 40 years to be discovered and exploited as a key probe of the Universe.

4 **Establishing the underlying cosmological model**

In 1965 we did not know whether the Universe was open or closed, nor did we know the value of the Hubble constant determining the scale of the Universe to within a factor two. We had little idea of what the material content was except for loose bounds on the baryon density from nucleosynthesis and observations of luminous material. However, the measurements of the radiation temperature gave a rather precise value for the density of photons. Later it was realised that there had to be a substantial amount of dark matter in the Universe, and later still it became apparent that there was yet another dominant component: the so-called dark energy.

5 **Nucleosynthesis**

It was the Big Bang as a potential source of the elements that drove physicists to computing the nuclear reactions that would take place in a Hot Big Bang. That was one of the driving forces behind searching for the relict radiation. The development of that aspect of the Big Bang enabled us to eliminate many alternative cosmological models and posed important questions for the origin of light elements in stars.

6 **The inflationary model of the Big Bang**

Having established our baseline cosmological model, it seemed to pose almost as many questions as it had answered. Why was the Universe so homogeneous and isotropic in the large? Where did the fluctuations that gave birth to galaxies come from? Why is the geometry of the Universe flat? These were questions that prior to the discovery of the Big Bang would have been relegated to the realms of philosophy. Now, with the pioneering idea of inflation introduced by Alan Guth, known (or at least plausible) physics was able to provide answers.

7 **Early stage galaxy formation**

Historically, through the work of Lifshitz, Bonnor and others, it was thought that the gravitational amplification of small amplitude, statistical, density fluctuations would be too slow to create the structures we see today. Galaxy formation appeared to be a problem for the new theory: it seemed that it would be necessary to paint the required initial conditions onto the Big Bang. The inflationary model came to the rescue: it was possible to find a mechanism within that framework which would generate perturbations of the required amplitude. As realised by Sachs and Wolfe, these primordial pertur-bations would leave their imprint on the map of the cosmic background radiation: we could in principle measure them directly at a time when the Universe was not even a million years old. This would ultimately lead to direct measurements of the amplitude of fluctuations at the time of recombination.

8   Late stage galaxy formation

Even by 1965 there had been a lot of work on the formation of galaxies from collapsing gas clouds. Perhaps the most influential of all papers on this was by Eggen, Lynden-Bell and Sandage. The establishment of the new paradigm provided information about the likely initial conditions for galaxy formation and allowed us to address the question of the origin of galaxy rotation and the galaxy mass function.

9   N-body simulations

Within a decade of the discovery of the cosmic background radiation, small N-body simulations were validating our notions about the Universe, and in particular our under-standing of the origin of large scale structure. Today, such simulations have grown considerably in sophistication and provide a test bench of our understanding of what we observe. They also provide a platform in which to simulate datasets so as to test algorithms for data analysis.

10   The great galaxy catalogues

Creating catalogues of distant galaxies had already begun in the 1950s, first with the *Lick Survey* of Shane and Wirtanen, and then with exploitation of the national *Geo-graphic Palomar Observatory Sky Survey* (POSS). The POSS became a source of data for Zwicky and his collaborators, cataloguing galaxies down to around magnitude[1] 15, for the construction of the *Abell Catalogue of Rich Galaxy Clusters*. Later the *Palomar Sky Survey*, and its Southern Hemisphere counterpart, the UK *Schmidt Survey*, formed the basis of catalogues of objects located by automatic plate scanning machines, the *APM Survey* is an important example. The sample of galaxies drawn from the IRAS satellite all-sky infrared survey played a major role.

11   Redshift surveys

Up until the 1970s the largest catalogue of distant galaxies was the *HMS Catalogue* (Humason *et al.*, 1956): it was aimed at establishing the shape of the velocity–distance relationship and so did not cover the sky uniformly. The sample of radial velocities for relatively nearby galaxies, catalogued in the *Shapley-Ames Catalogue* and the *de Vaucouleurs Catalogue of Bright Galaxies*, provided a fairly complete redshift sample down to around magnitude 12. The first steps towards improving that situation came with the production of 'redshift slices': strips of sky in which all galaxies down to some limiting magnitude had their radial velocities measured. To name but a few there was the *Centre for Astrophysics* (CfA) *Survey* (see Figure 4.1), the *Lick Catalogue*, and the 2dF (*2-degree field*) slice. The IRAS satellite galaxy samples provided the first deep all-sky redshift surveys, and this is now enhanced by the *2MASS Survey*.

Following on from there we have now a number of important wider area redshift surveys, pre-eminent among which is the *Sloan Digital Sky Survey* and a number of

---

[1] For a discussion of measuring brightness and expressing the answer in magnitudes see Appendix B.

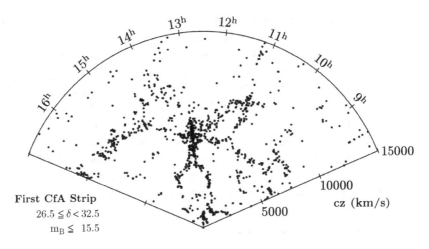

Fig. 4.1 The *de Lapparent Slice*, which in 1986 was the first slice to be published (de Lapparent *et al.*, 1986) from the CfA redshift survey of Geller and Huchra. The slice was 6° by 130° on the sky and was made up from 1100 galaxy redshifts, plotted as distances using the Hubble law. Although voids and filaments had already been detected in 'pie diagrams' directed towards rich galaxy clusters, this was the first time that they could be seen to be a generic feature of what was to become known as the *Cosmic web*. The CfA survey acted as an important stimulus for a succession of ever larger redshift surveys.
(From de Lapparent et al., 1986. Reproduced by permission of the AAS.)

surveys that have been derived from that, notably the *Baryon Oscillation Spectroscopic Survey* (BOSS) which reveals the signature of the early Universe Baryonic Acoustic Oscillations (BAO) at redshifts out to $z \sim 1$. The scale of the BAO feature is comparable with the scale of the entire CfA slice!

12 **Categorising large scale structure**

Peebles quickly recognised that tests of the model's ability to account for the origin of cosmic structure would come in part from the analysis of the observed cosmic structure. To this end he introduced correlation analysis of the galaxy distribution and applied it to the then existing catalogues. This seeded a vast industry in the generation and analysis of catalogues, and the development of alternative mathematical descriptors of the clustering.

13 **The formation of large scale structure**

Since present day structures such as galaxies and galaxy clusters were clearly not small perturbations to the background density, there had to have been significant nonlinear evolution of matter inhomogeneities. Development in this area was stimulated by a remarkable analytic model for large scale structure formation due to Zel'dovich (1970), and later on by the use of numerical simulations. Numerical simulations have become a prime tool in understanding the development of cosmic structure.

14 **Voids and the cosmic web**

Today we recognise that the large scale structure of the Universe is dominated by 'voids': large volumes that are almost devoid of galaxies and surrounded by sheets, filaments and clusters that host most of the luminous matter in the Universe. The first to argue on the basis of data catalogues for such a cellular structure were Jöeveer *et al.* (1977). Their fundamental structural unit was a void of diameter $\sim 50-75\,h^{-1}$Mpc, surrounded by 'walls'. This was strongly motivated by the 'pancake' theory of Zel'dovich (1970), later to be refined by Zel'dovich *et al.* (1982). However, it was not until the *CfA Survey* of de Lapparent *et al.* (1986) that the bubble-filled image of the Universe could be fully appreciated (See Figure 4.1).[2] This view of the Universe developed rapidly with voids being thought of as astronomical objects in their own right. The first realistic simulation of void evolution came with the constrained random field simulations of van de Weygaert and van Kampen (1993), and the phrase 'cosmic web' became part of the language of cosmology with the seminal paper of Bond *et al.* (1996).[3]

15 Galaxy rotation curves

See Figures 4.2 and 4.5. At first sight it would not be thought that galaxy rotation curves would have much impact on cosmology. However, the discovery by Bosma and by Rubin and her collaborators that galaxy rotation curves were flat out to very large distances from the centre of rotation established the existence of substantial amounts of dark matter in the regions surrounding galaxies. Later it was realised that the dark matter was not confined to the environs of galaxies, but existed in still larger amounts in the

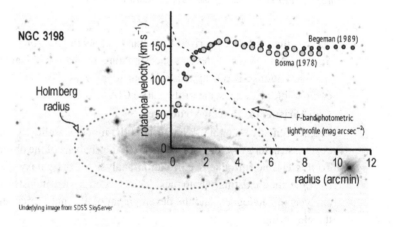

Fig. 4.2 The flat rotation curve of the galaxy NGC 3198 as derived from 21cm line (HI) mapping of the galaxy velocity field by Bosma (1978, 1981, the first of these is his widely circulated PhD thesis.). The curve extends out to almost twice the galaxy's Holmberg radius (26.5 mag. arcsec.$^{-2}$) and was later further extended by Begeman (1989). This is one example of several extensive flat rotation curves in Bosma's thesis study of the velocity field maps of seven galaxies. See also Figure 4.5 ⓒ

---

[2] Following the revelation of the CfA map, there came two important models for this bubble- or foam-like structure. There was the kinematic Voronoi model of Icke and van de Weygaert (1987); van de Weygaert (1991), and the 2-dimensional adhesion model of Kofman *et al.* (1990). Kofman *et al.* (1990) described this structure as the *skeleton of the Universe*.

[3] See the engaging historical account of Einasto (2014) and the lectures of van de Weygaert (2002).

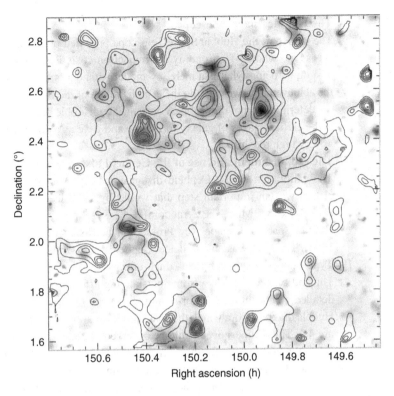

Dark matter in the HST COSMOS field as mapped out by gravitational lensing. The structure shown by the contours is mostly made up of non-luminous matter that is at a redshift of $z \sim 0.7$. The structure reveals its existence because it distorts the light coming from luminous galaxies in more distant parts of the Universe. The images of those galaxies are distorted by the gravitational lensing. Mapping those distortions maps out the dark matter.
(Source: Massey et al. (2007, Nature). With permission from the publisher © 2007. (See page 695))

Universe at large. From that point on cosmology had to incorporate a new component in cosmological models and simulations: dark matter.[4]

16 Gravitational lensing

Following on from the studies of galaxy rotation curves revealing evidence for a substantial amount of dark matter in galaxies, it became crucial to find independent evidence of its existence. How could it be detected on cosmic scales and how might we find out what it was? The answer to the first question came from gravitational lensing: the gravitational distortion of images of distant objects by intervening mass distributions, see Figure 4.3. This was first seen in rich galaxy clusters and then in deep maps of the Universe. The answer to the second question is being sought in a number of searches for dark matter.

---

[4] Zwicky (1937) is usually credited with the realisation, based on data he had acquired, that galaxy clusters had virial masses far in excess of what could be accounted for by the galaxies. This led to a large number of proposals that galaxy clusters were unbound and dispersing, including a paper discussing this possibility in the framework of Λ-cosmology by Lemaître (1961).

17 **The Hubble Space Telescope**

The launch of the Hubble Space Telescope provided opportunities for new discoveries across the whole of astronomy. Its impact on cosmology has been highly significant, despite its being what is a mere, by modern standards, 2.4 meters in aperture. Among the achievements we can cite the establishment of the extragalactic distance scale with the determination of the Hubble constant. The 'Hubble deep fields' provided a first look at the earliest stages of galaxy formation and gave us a direct view of galaxy clustering at very high redshift.

18 **The establishment and use of the Cepheid distance scale**

The calibration of the Cepheid distance scale presented by Madore and Freedman (1991) was an important step that would by 1998 lead to their reporting a value of $H_0 = 72\,\mathrm{kms}^{-1}\mathrm{Mpc}^{-1} \pm 2$ (random) ± 12 (systematic) using data acquired during the Hubble Space Telescope Key Project on the distance scale (Madore and Freedman, 1998). The HST was able to detect Cepheids in external galaxies out to $20-25\,\mathrm{Mpc}$.

19 **Type Ia supernova samples – dark energy**

The end of the 1990s saw some outstanding efforts by two groups towards acquiring detailed and accurate data for a sample of Type Ia supernova. With this they created a Hubble diagram from which the cosmological parameters could be determined, see Figure 4.4. These supernovae could be observed at great distances and were thought to be good standard candles for this purpose. The Hubble diagram derived from these studies revealed the existence of another hitherto unsuspected constituent of the Universe that would account for 75% of the cosmic mass density and that causes the expansion of the Universe to accelerate. This constituent is referred to simply as 'dark energy', and as of this time, we have no idea what it is.

20 **RELIKT-1, COBE and other CMB experiments**

The first satellite-borne experiment was the Russian RELIKT-1 experiment, launched aboard the Prognoz 9 satellite, in July 1983. This was an all-sky mapping experiment.

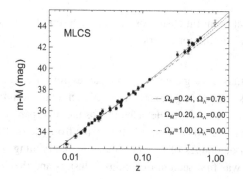
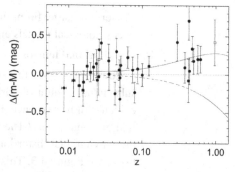

Fig. 4.4   The first SN Ia supernova cosmology Hubble diagram showing the evidence that $\Lambda \neq 0$. The expected Hubble lines for three cosmological models are shown. The detail of the fit is shown in the right-hand panel where the deviation of the data points from a low density, $\Omega_M = 0.2$ model having $\Lambda = 0$ is plotted. The curves for models having $(\Omega_M, \Lambda) = (0.2, 0.8), (1.0, 0.0)$ are also shown. The 'MLCS' method was used for renormalising the supernovae sample to standard candles. This method had the advantage of dealing with possible absorption by intervening dust. *(Source: Riess and Filippenko (1998). Reproduced by permission of the AAS.)*

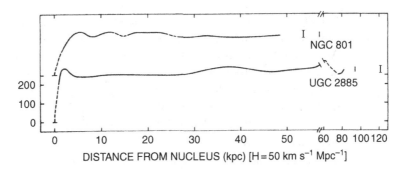

Fig. 4.5 Rubin *et al.* (1980) published rotation curves derived from optical long slit spectra for the brighter regions of 21 type Sc galaxies using long slits extending typically to 83% of the 25 mag. arcsec.$^{-2}$ isophote. The figure shows two examples of flat rotation curves from that sample. They extend out to over 50 kpc, maintaining a constant rotational velocity that is symptomatic of the existence of unseen matter in the outer regions (Rubin, 1983). See also Figure 4.2. *(Reproduced by permission of the AAS.)*

The subsequent analysis of the data by Klypin *et al.* (1987, 1992) put tight upper limits on the amplitude of the fluctuations, but did not find evidence for any fluctuations.

The COBE satellite was launched in November 1989. On board were three experiments: FIRAS, whose goal was to measure the spectrum of the cosmic microwave background radiation, DMR, whose goal was to map the spatial distribution of the temperature of the CMB, and DIRBE, whose role was to search for background radiation in the far infrared. With an angular resolution of $\sim 6°$ the DMR was able to detect, for the first time, the small fluctuations in the background temperature that had been predicted by Sachs and Wolfe (1967), albeit at a much lower level.

COBE was shut down at the end of 1993, having successfully completed the job it was designed to do. It was the trail-blazer for the next generation WMAP and Planck satellites, but while awaiting the launch of WMAP there were many ground-based and balloon-borne experiments that were seeking to improve on the COBE results by reaching higher sensitivity and going to higher resolution over smaller areas of sky. The situation at the year 2000 was reviewed by Gawiser and Silk (2000) (see Figure 3.11).[5]

21 COBE, WMAP and Planck

These three satellites are successive generations of instruments designed largely for a single purpose: to map the cosmic microwave background radiation field. COBE

---

[5] Largely because of the long lead times in space-based experiments it is possible for small scale, non-space based, experiments to compete with a planned space mission.

    Individually, such experiments might have provided no more than a map and a few points on a graph, but taken together they did have a high impact. By 2000 there were many that had taken and published data or were about to take data. They all had identifying tags that were so numerous that the names were hard to remember or to distinguish for anyone not directly involved: ACE, ARGO, ATCA, BAM, BIMA, BOOMERANG, CAT, CBI, DASI, IRS, HACME, IACB, JBIAC, MAT, TOCO, MAX, MAXIMA, MSAM, OVRO/RING, PYTHON, QMAP, SK, SP, SUZIE, TENERIFE, TOPHAT, ULISSE, VIPER, VLA, VSA, WD. These small experiments made serious contributions, and in many ways were a tribute to the ingenuity and enthusiasm of the physicists who worked on them. By 2000 these experiments were able to determine the angular diameter distance to the surface of last scattering with high precision.

carried an additional experiment called FIRAS that was to establish the Planckian nature of the spectrum with unparalleled accuracy. WMAP, the Wilkinson Microwave Anisotropy Probe, provided data that launched cosmology into an era of precision determination of the cosmological parameters. In thinking of these space-based experiments, we should not forget the achievements of ground-based experiments mapping smaller areas of sky, for example VSA, CBI, ACBAR, and balloon borne experiments like BOOMERANG and ARCHEOPS to name but a few.

22 Gravitational waves

The announcement by the LIGO-VIRGO Consortium (Abbott *et al.*, 2016) of the direct detection of a gravitational wave pulse emitted from two merging black holes establishes Einstein's theory of gravitation in previously untested extreme regimes. Einstein's theory lies at the very foundation of cosmology, establishing both the physical, or geometric, nature of space time and the equations that govern gravitational phenomena. It gives credence to the idea that extreme phenomena close to the time of the cosmic singularity could have generated a background of gravitational waves. These might be detected by observations of the B-mode polarisation of the cosmic background radiation provided that the interference from foregrounds can be understood. The detection of cosmic gravitational waves provides the strongest support for our current cosmological paradigm: the Big Bang.

The culmination of all this was a remarkable growth of activity in cosmology: cosmology was no longer the search for two numbers. It became a search for what by analogy with particle physics we might call the standard model of the Universe. The search was fuelled by great data gathering experiments, both from the ground and from space. The exploitation of those datasets has driven us to a new standard in the analysis of very large datasets. We have arrived at 'precision cosmology' in which our Universe is characterised by over a dozen parameters, most of which are now known to within a few percent.

If the discovery of the cosmic background radiation marks a renaissance in cosmology, the advent of precision cosmology surely marks a revolution. Indeed, some have suggested that with the revolution, cosmology has been reduced to filling in the next decimal place of the various parameters. However, that ignores mysteries which go far deeper: the nature of dark matter and of dark energy. We can say we know about our Universe with great precision, but until those issues are resolved we will not be able to say we understand our Universe.

4.2 Dark Matter and Dark Energy

Two surprises that have arisen since 1965 are the evidence for a substantial amount of 'dark matter' and a dominant 'dark energy'. The first evidence for dark matter came primarily from galaxy rotation curves: the high rotation velocities at great distances from the centre of galaxies indicated the presence of some unseen matter. Corroborating evidence was to come from many directions. The evidence for 'dark energy' came from research on the Hubble diagram that revealed the acceleration of the cosmic expansion: to account for this acceleration would require that 70% of the material in the Universe be in some form of dark energy (see Section 2.8.2). The history of this will be covered in detail in Chapter 28.2.

Today, we believe that taken together these two dark constituents of the Universe make up of $\sim$ 95% of the content of the Universe. Our familiar, well understood, baryonic component makes up most of the rest. This is an astonishing conclusion and if the evidence came from a single source we would not be as confident. But now there is abundant corroboration for this from a variety of rather independent directions. The evidence for copious amounts of both unseen dark matter and unseen dark energy is considered to be very convincing.

4.2.1 Galaxy Rotation Curves

Roberts (1976) was perhaps the first to remark on the flatness of spiral galaxy rotation curves. The first direct evidence for a significant quantity of dark matter in galaxies came from studies of the rotation curves of spiral galaxies. The pioneering work in this field was due to Bosma (1978, 1981, the first of these is his widely circulated PhD thesis) and to Rubin et al. (1978, 1980). Both presented sets of rotation curves for a wide ranging sample of disk galaxies, all of which showed extensive flat rotation curves. (see Figures 4.2 and 4.5).

Bosma, in his PhD thesis, had mapped and analysed the 2-dimensional HI velocity field of disk galaxies to almost twice their visible Holmberg radius,[6] finding evidence for warping of the disks and extensive flat rotation curves. Rubin[7] had obtained long slit optical spectra taken along the major axes of a large sample of Sc galaxies, and although not going as far out in radius as Bosma, concluded that this was strong evidence for the existence of substantial amounts of dark matter.

With an ever growing number of rotation curves of disk galaxies tracing galactic rotation to ever greater distances from the galaxy centres, it became apparent that there was far more matter in galaxies than could easily be accounted for by what we see in the form of stars. The brightness of galaxy disks tend to fall off exponentially, and so we would have expected the rotation velocity to fall off more or less like a Kepler Law with distance from the centre.

From simple Newtonian dynamics the rotation speed $V(R)$ about a sphere of radius R containing mass $M(R)$ is given by $V^2(R) \sim GM(R)/R$. If $V(R) = V_{\rm rot}$ is a constant independent of R then the mass $M(R)$ and mean density $\rho(R)$ within a sphere of radius R vary as

$$M(R) \propto R, \qquad \rho(R) \propto R^{-2}. \tag{4.1}$$

These flat rotation curves show that there is a lot of unseen mass driving the rotation of these galaxies. The dark matter cannot all be baryonic: there is an upper limit on the

---

[6] The Holmberg radius corresponds to the radius about a galaxy at which the surface brightness falls to a specific value of 26.5 mag. arcsec.$^{-2}$ in blue light (1-2% of the dark night sky brightness). Figure 4.2 illustrates the Holmberg radius for the galaxy NGC 3198: this is well outside the commonly recognised disk of the galaxy.

[7] Vera Rubin was a PhD student of George Gamow, obtaining her PhD in 1954.

baryonic density of the Universe imposed by nucleosynthesis and by measurements of the anisotropy of the cosmic background radiation.

4.2.2 Galaxy Clusters

Strong evidence for the existence of dark matter on larger scales comes from galaxy clusters that have been imaged in X-rays and for which there is a gravitational lensing map of the gravitational potential. The first example of such a cluster was the so-called *Bullet Cluster*,[8] consisting of two colliding galaxy clusters. The X-rays trace the baryonic matter while the lensing map traces the source of the main gravitational force, in this case the dark matter associated with the system. What is observed is that the X-rays and lensing do not reflect the same distribution: the distribution of dark matter is displaced from the distribution of baryonic matter.

If there were no dark matter and the baryonic matter were entirely responsible for the gravitational potential and the lensing, then the maps would coincide. In statistical terms the displacement is highly significant. There is a second cluster showing the same phenomenon: this cluster is designated MACS J0025.4-1222.

4.2.3 Alternative Ideas

Most of the alternative ideas for avoiding the need to invoke some kind of ill-understood dark matter involve modifying the law of gravitation on large scales. The most prominent, and most straightforward, of these is the Modified Newtonian Dynamics (MOND) theory of Milgrom (1983) and of Sanders (1986, 2005). Bekenstein (2004) provided a relativistic framework for modified gravity and an excellent review of modified gravity theories (Bekenstein, 2010). Some might argue that this is no less ad hoc than introducing some form of mysterious dark matter.

4.3 Gravitational Waves

Gravitational waves can be described as ripples in the geometrical fabric of space-time. This is an astonishing concept that is entirely removed from anything that had gone before,[9] and it was inevitable that it would be the subject of some controversy. The idea was due to Einstein himself. The first definitive paper on gravitational waves was that of Einstein

---

[8] The object is designated as 1E 0657-56. This refers to the position of the object on the sky $\alpha = 0^h 56^m 37.9^s$, $\delta = -55^0\ 57'$

[9] There is an interesting parallel with the search for electromagnetic radiation, which arguably started with Faraday's demonstration of action at a distance in 1831. The theoretical understanding for this was formulated in 1861 by Maxwell in his Theory of Electromagnetism, one consequence of which was a prediction for the existence of electromagnetic waves. Maxwell's theory and the waves were for several decades widely regarded by most of his colleagues as mere theoretical speculation. But, in 1887, at the behest of Helmholtz, Hertz conducted some experiments that finally demonstrated the existence of Maxwell's waves. The discussion then arose as to the nature of the medium in which these waves were manifested: the 'luminiferous æther'. It would take another four decades before we had radio and television.

(1918b).[10] This was the paper in which Einstein gave the quadrupole formula (*loc. cit.* Eq. 30) for the rate at which a system radiates gravitational wave energy.

It was widely considered that such waves might be nothing more than an artefact of particular coordinate systems. Eddington (1923, *The Mathematical Theory of Relativity,* Section 57, p.130) is equivocal when discussing the gravitational wave solution of the weak field Einstein equations: *The statement that in relativity theory gravitational waves are propagated with the speed of light ... is only true in a very conventional sense* and *We can "propagate" coordinate changes with the speed of thought, ...* He goes on to say: *Must we then conclude that the speed of propagation is necessarily a conventional conception without absolute meaning? I think not.* Eddington is prepared to believe these are real waves in the geometry of space-time, but he is rather cautious.

There was a long period of debate among relativists and it was only in the 1950s that it was generally accepted that gravitational waves were an essential part of the theory of general relativity. The turning point was the publication of the papers of Pirani (1957), Bondi *et al.* (1959), and Sachs (1961). With this reassurance, there was considerable motivation to consider attempts to detect gravitational waves. Such a discovery would be a remarkable testimony to Einstein's extraordinary idea that there could be ripples in the geometric fabric of space-time.

4.3.1 Joe Weber's Quest for Gravitational Waves

The quest to discover gravitational waves was started and led, almost single handedly, by Joe Weber. In 1958 and 1959 Weber submitted essays to the annual Gravity Research Foundation essay competition, winning 3rd prize in 1958 and 1st prize in 1959 (Weber, 1958, 1959). These were the basis for his seminal paper on Gravitational Wave detection (Weber, 1960) and a book (Weber, 1961, *General Relativity and Gravitational Waves*).

Not long after this, Weber (1966) built his first detector in his laboratory at the University of Maryland. It consisted of a single $1\frac{1}{2}$ ton aluminium cylinder, 150 cm in length and 61 cm in diameter suspended in a vacuum at its mass centre.[11] The 'Weber bar' had a resonant vibrational frequency in the region of 1657 Hz, at which the bar would ring if excited. These vibrations were detected using strain gauges bonded to the bar. Weber *loc. cit.* reported that he was able to detect the thermal noise at a level of $\sim 2 \times 10^{-14}$ cm, which corresponded to a strain measurement of one part in 10^{16}. Weber estimated that a gravitational wave passing through the bar would shake it by a mere 10^{-13} cm, the radius of the proton! After two years of running this experiment Weber (1967) reported the detection of ten putative gravitational wave events.

---

[10] The earlier paper of Einstein (1916b, see Section 2 *et seq.*) had for the first time introduced the notion of gravitational waves. However, that paper was seriously flawed. As Einstein himself pointed out in the Einstein (1918b) paper *I have to return to the subject matter since my former presentation is not sufficiently transparent, and, furthermore, is marred by a regrettable error in calculation.* The *regrettable error* was his conclusion that a spherically symmetric body could radiate gravitational waves.

[11] Joe Weber's team appears to have consisted principally of Robert Forward, Joel Sinsky and David Zipoy. Robert Forward would later on be involved in the construction of the first laser interferometer gravitational wave detector.

Eliminating the local sources of noise was the key issue, and so Weber operated two or more bars separated by as large a distance as he could manage. During February and March 1968 two detectors separated by a distance of 2 km were running and reporting coincident excitations, Weber (1968). A year later Weber (1969) extended his baseline to 1000 km with a detector at the Argonne National Laboratory and an array of five detectors in Maryland. Seventeen coincident events were recorded over a period of about three months in early 1969. Among these there were five triple detector coincidences and three quadruple coincidences. Weber concluded that *all of the coincidences are neither accidental nor due to seismic or electromagnetic effects. This is good evidence that gravitational radiation has been discovered.* By 1970 Weber (1970, see Fig.1) reported the detection of some 311 coincidences, and suggested that there was an anisotropy in the sky distribution favouring the galactic centre. Indeed, this anisotropy, in which the event count reached a maximum twice in a sidereal day but not in a solar day, may have been the best evidence that what Weber was seeing was not an artefact of some unexplored terrestrial phenomenon.

However, not all were convinced. Field *et al.* (1969), for example, gave astronomical limits on the amount of mass loss from the galaxy through gravitational radiation that would have resulted from Weber's detections. In their 1972 review on 'Gravitational Wave Astronomy', Press and Thorne (1972) remarked that *We (the authors) find Weber's experimental evidence for gravitational waves fairly convincing. But we also recognise that there is as yet no plausible theoretical explanation of the waves' source and observed strength.*

During the following decade many groups around the world built 'Weber bars' in order to provide independent confirmation of Weber's results. None detected any relevant signals despite higher sensitivity, better data processing methodologies and so on. The lack of detections was all the more puzzling given that the number of events at Weber's level of sensitivity was so large. Scepticism was also fuelled by the 'theoretical argument' that such a high rate of events could not be explained by any known astronomical sources. In his end-1976 review of the subject Drever (1977)[12] remarked that *... the consensus view that Weber's results are not due to gravitational radiation seems to me so likely to be correct that it is more profitable to concentrate now on development of detectors of very much greater sensitivity.*

This was depressing, and particularly so for Weber himself. Nevertheless, Weber had started something and he had inspired a new quest to detect gravitational radiation.

4.3.2 The Rise of the Interferometers

Renewed stimulus for trying to detect gravitational waves came when Hulse and Taylor (1975) discovered the pulsar PSR 1913+16. This pulsar was both a fast pulsator (period 59ms) and a component of a tight binary with an orbital period of only 0.323 days. The other component is a neutron star. The orbit could be determined with high precision and within a year it was seen that the periastron of the orbit was precessing at the rate predicted

---

[12]   Ronald Drever was at that time running the Glasgow gravitational wave detector, a modified Weber bar. During his time in Glasgow he had been investigating interferometric detectors as had several others, notably Rainer Weiss (1972) at MIT, with whom Drever would later collaborate on LIGO.

by general relativity (Taylor *et al.*, 1976). By 1979 it was clear that the orbit was decaying at the rate expected if the binary were emitting gravitational radiation (Taylor *et al.*, 1979).

One of Weber's team, Robert Forward, had already looked into the possibility of detecting the radiation using a Michelson interferometer.[13] Michelson and Morley, and subsequent experiments, had shown that the light travel time was independent of the orientation of the arms of the interferometer (see Eisele *et al.*, 2009, for a review). As a gravitational wave travelling perpendicular to the interferometer passes across the interferometer one arm would in effect be stretched while the other would be compressed by the same amount: the light travel time would, for a short while, be different along the arms. This shift in length was just as small as in Weber's experiment, so the goal would be to measure changes in the lengths of the arms of the size of a proton. To achieve this the interferometer arms would have to be measured in kilometres rather than metres: the scale of the experiment would have to change and so would the cost. This would no longer be a case of one pioneer working in his lab, this was an experiment on the scale of high energy physics accelerators.[14]

In the USA, experimentalists Rainer Weiss (MIT) and Ronald Drever (at Caltech since 1979) teamed up with theorist Kip Thorne (Caltech) in the late 1970s to consider building two geographically separated 4 km interferometers. In 1980 the NSF[15] funded the building of prototype interferometric antennas at both Caltech and MIT, and a technical study to look into the feasibility of building a 4 km interferometer. Finally, in 1991, after much, and at times acrimonious, public debate and political wrangling[16] The Laser Interferometer Gravitational-wave Observatory (LIGO) was funded with a congress approved budget of $395M. By 1994 the building of the project sites could begin (see Figure 4.6). One site would be at the DOE Hanford site near Richland in Washington State and the other at Livingston in Louisiana. The distance between the sites is 3002 km. Most of the noise in the signal arises from micro-seismic events and human activities such as traffic, logging and so on. The separation of the sites serves to isolate such local events from events taking place on a global scale. In 1997, the LIGO project had been split into two: LIGO itself and the LIGO Scientific Collaboration (LSC). LSC was formed as an international consortium of science institutions involved in the science and technology of detecting gravitational waves. By the time of the discovery of GW1509014, LSC comprised around 1000 scientists and engineers from almost 100 universities and institutes from all over the world.

---

[13] Forward was at Hughes Research Laboratories at that time. His team succeeded in building a prototype gravitational radiation antenna and measuring a vibration of 10^{-11} cm between the mirrors, the smallest displacement measured with a laser at that time (Moss *et al.*, 1971; Forward, 1978).

[14] To put this in perspective, the 1960s had been a period of growth in building particle accelerators. SLAC (the Stanford Linear Accelerator Center) came on-line in 1966 with an underground tunnel some 4 km in length on a site of 200 hectare, at a 1966 cost of $\sim$ $100M. The 21st century Large Hadron Collider (LHC), which has a 27 km circular tunnel, was estimated to have cost around $\sim$ $5B by 2012. In addition to that there is a similar amount for the experiments and a running cost of $\sim$ $1B per year (*Forbes* July 2012). The total LHC cost over a 14 year period was about 2.2% of the US military budget for 2013.

[15] The US National Science Foundation, widely referred to simply as the NSF.

[16] The evolution of gravitational wave detectors has been a long and often controversial story described in the book of Collins (2004, particularly Chs.31, 32.). The review of that book by Smolin (2013) is worth reading.

 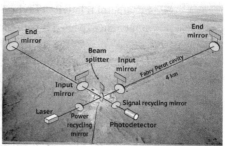

Fig. 4.6   *Left:* The Hanford LIGO Observatory (Washington State). The two arms are 4 km long 1.2 m diameter ultra high vacuum tubes that can carry several high intensity beams. *Right:* Sketch of the interferometer layout. The beams bounce up and down the tubes several hundred times between mirrors that are suspended on wires to isolate them from vibrations in the local environment. For maximum response, the number of times the light bounces up and down the arms is such that the total distance traversed is comparable with the wavelength of the gravitational wave. *(LIGO images from Caltech and MIT websites are in the public domain (see p. 695)).*

The first version of LIGO was in place in 2002, with some data gathering for the following few years, and further enhancements in 2008. These enhancements to the LIGO technology and subsequent upgrades substantially increased the LIGO sensitivity and by September 2015 the total project cost was in excess of $\sim$ \$1000M. The NSF were the sole funding source for the project throughout the years 1980–2016, showing a remarkable tenacity given that they did not know for certain if anything would come of the experiment. This was a high risk high reward project, and on 14 September 2015 it paid off with the detection of a signal known as GR150914.

On 14 September, four days before the official LIGO start-up, Marco Drago, a postdoctoral researcher at the Albert Einstein Institute in Hannover working on LIGO data, noticed the signal that was later to be called GW150914.[17] After five months of exhaustive checking the result of the first detection of gravitational radiation was announced to the world.

4.3.3 GR150914

The detection of gravitational waves on 14 September 2015 by the LIGO collaboration was announced on 12 February 2016 (Abbott *et al.*, 2016). The signal, of duration $\sim$ 0.2 s, was received almost simultaneously by the two on-line LIGO interferometers. The detected signals and the fitted models are shown in Figure 4.7. This detection heralded the birth of

---

[17]   On 14 September LIGO was in 'engineering mode'. It was around 4:50 am local time when the signal arrived at the LIGO antennas, and the engineers had gone home less than an hour earlier. The time of the event was 09:50:45 CET, when members of the LIGO Collaboration in Europe were already at work. At the time of the detection of GR150914, neither of the European VIRGO and GEO600 detectors were on-line.

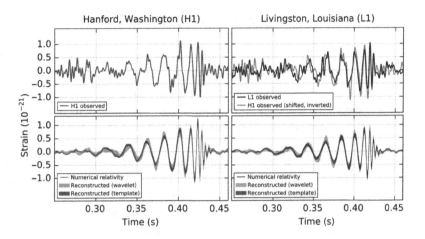

Fig. 4.7 The filtered LIGO signal detected at the Hanford (*left*) and Livingston (*right*) Observatories on 14 September 2015. The top row displays the filtered strain signal measured at each of the observatories, the right hand panel showing both signals superposed (with a time shift for the 6.9 ms difference in arrival time between the observatories). The coincidence between the waveforms is compelling evidence that the received signal was of extra-terrestrial origin. The signal has the rising frequency and amplitude characteristics of a classic 'chirp' signal. The lower two panels show the numerical relativity waveform for the two black hole inspiral and merger orbit model fitting the data.
((*Abbott et al. 2016, fig.1, edited for grey-scale reproduction), Public Domain*)

a new branch of astronomy, one that does not make an observation in the electromagnetic spectrum.[18]

The detected signal was interpreted as the consequence of the merger of two black holes that had been in orbit around one another. The masses assigned to the original black holes were $36^{+5}_{-4}\,M_\odot$ and $29^{+4}_{-4}\,M_\odot$, leaving a remnant black hole of $62^{+4}_{-4}\,M_\odot$. The distance of the event corresponds to a redshift $z \sim 0.09^{+0.03}_{-0.04}$ (410^{+160}_{-180} Mpc) and the event is simply referred to as GW150914.

The interpretation was based on a set of some 250 000 templates for the waveform shown in Figure 4.7, derived from simulations of numerous binary stellar and black hole systems. The simulations covered component mass ranges from $1\,M_\odot$ to $99\,M_\odot$ having a wide range of spin parameters. Using models for the noise signal, this 'chirp' was estimated to be significant at the 5.1σ level, corresponding to a probability of $< 2\times10^{-7}$ that this event was a false alarm. The fitted model provided the amplitude of the signal and hence the estimate of the distance to the source (LIGO Scientific Collaboration and Virgo Collaboration, 2016). The event released fifty times the energy of all of the stars in the Universe put together. The scientific community readily accepted the conclusions that: (a) LIGO had detected a gravitational wave pulse; and (b) the pulse was due to a binary black hole merger.

---

[18] We might say the same of the detection of solar neutrinos and the subsequent detection of neutrinos from the supernova SN1987A. So far these are the only astronomical sources from which neutrinos have been detected. These two discoveries were an important part of the foundation of 'astro-particle physics'.

4.3.4 Numerical Simulations of Black Hole Mergers

The importance of computing the merger between two collapsed objects was first stressed by Dyson (1969, last paragraph). This was an extremely difficult technical problem involving choices of coordinate systems and numerical techniques and limited by available computing resources. It became the holy grail of numerical relativity during the next decades and into the 21st century, during which one of the main problems was in attaining algorithmic stability.

An important breakthrough came in 2005 and 2006 from the direction of numerical relativity. The numerical simulations of binary black hole coalescence had made significant technical progress so that a definitive prediction could be made concerning what the received gravitational wave signal would look like as a consequence of such an event. Up until that time getting stable simulations of the coalescence of orbiting black holes had not been possible, despite decades of effort. Then quite suddenly, three independent groups using different computational schemes came up with successful simulations and predictions of the expected gravitational waveform: the chirp signal. First came the work of Pretorius (2005, 2006) followed by the papers of Baker *et al.* (2006a,b) and Campanelli *et al.* (2006)), with strong agreement between the groups. This is beautifully described in the review articles of Centrella *et al.* (2010a,b).[19]

It is this chirp of the black hole merger that allows us to test Einstein's theory and provide the interpretation of the LIGO detection as being due to a black hole merger. It is also a resounding success of some four decades of research into numerical relativity.

4.3.5 The Aftermath of GW150914

This is the first time we have seen our Universe without using the electromagnetic spectrum, and in that sense this is as big a discovery as when Galileo turned his home-made telescopes on the Moon and planets of the Solar System. A new branch of astrophysics has been born. Understanding precisely what these sources are will require greater positional accuracy to locate gravitational wave events for follow-up with optical telescopes. That will allow the environment of the source to be studied, and it will clarify the nature of the merger event itself.

But the success of LIGO is not only a revolution in astrophysics. The demands of the experiment and data interpretation have benefited from and made substantial contributions to opto-electronics, numerical relativity and the understanding of black holes.

---

[19] Joan Centrella was, at the time, chief of the Gravitational Astrophysics Laboratory at NASA's Goddard Space Flight Center. Commenting to the New York Times in 2005 on the work of her team she said *... we have confidence that these results are the real deal, that we have the true gravitational wave fingerprint predicted by Einstein for the black hole merger.* (Chang, 2006). Her confidence was fully justified.

PART II

NEWTONIAN COSMOLOGY

5 Newtonian Cosmology

We will use Newtonian versions of solutions to the Einstein field equations to describe a series of model universes that provide a framework within which we can better understand how our Universe works. Newton's theory of gravitation has been replaced by Einstein's, but in many respects Newton's theory is a pretty good approximation: good enough that we depend on it in our everyday lives. The Newtonian view is certainly easier for us to relate to and exploit, but we must understand the inherent limitations.

Here, we highlight the fundamental differences between the Newtonian and Einsteinian theories: Newton with his absolute space and universal time, and Einstein with his geometrisation of gravity. Fortunately, there are some relevant solutions of the Einstein equations that have direct Newtonian analogues. Those Newtonian analogues are lacking some important features, notably a lack of a description of how light propagates. Fortunately, we can graft the information from some of the Einstein models onto the Newtonian models to produce what we might call Newtonian surrogates of the Einsteinian models.

In this chapter we introduce the simplest of a series of homogeneous and isotropic cosmological models formulated within the limited framework of Newtonian gravity. These models contains only 'dust': pressure free matter made up of particles that are neither created nor destroyed, and that do not interact with one another. The Universe evolves under the mutual gravitational interaction of those particles.

While this model is not realistic it nevertheless allows us to develop a full cosmological model that is the template for the more complex models that follow. We develop these models in considerable detail since much of what is done will be repeated for the other, more realistic, models.

It is worth remarking that these models are fundamental to numerical N-body simulations of the Universe in which the constituent particles are 'dust'. Such dust models can also be used to study the growth of the structure in the Universe.

5.1 Why Bother with Newton?

5.1.1 Two Views of Gravity

The expansion of the Universe is dominated by the gravitational force. The best theory we have for the gravitational force is Einstein's Theory of General Relativity (Einstein,

1916a), which relates geometry with the material content of the space-time containing that matter. Schematically we can write

$$G_{ab} = \kappa T_{ab}, \tag{5.1}$$

$$\text{Geometry} = \text{Matter}. \tag{5.2}$$

The objects G_{ab} and T_{ab} are neither scalars nor vectors, but *tensors*. G_{ab} describes the geometry and T_{ab} describes the material content of the space-time. κ is a universal coupling constant that mandates the strength of this link. General relativity also provides a prescription for calculating the motion of particles and light in the geometry of the space-time.

The equivalent equation in Newtonian gravity is the Poisson equation for the gravitational potential ϕ:

$$\nabla^2 \phi = 4\pi G \rho, \tag{5.3}$$

$$\text{Potential} = \text{Matter}. \tag{5.4}$$

We compute the motion of the particles by calculating the force field $\mathbf{F} = \nabla \phi$ acting on the particle and then invoking Newton's law of motion: $m\ddot{\mathbf{r}} = \mathbf{F}$.

5.1.2 Two Views of Space-time

General relativity contains Newtonian gravity in the limit of weak gravitational fields and has passed numerous tests in areas where Newton's theory does not even make predictions. So, why do we go back to Newtonian gravity to understand the Universe?

The answer is simple: we do this because the Newtonian world view is intuitively simple and provides a framework for gravitational physics that accords most easily with our intuition.

In the Newtonian world view we have an absolute space of three dimensions and a universal time which is the same for all observers in that space. Although there is no Newtonian theory of light propagation, we think of light in the Newtonian context as having infinite velocity.[1] Communication between observers is instantaneous, whatever the distance separating them. Likewise, within the Newtonian context, gravity is thought to act instantaneously over all distances: we have the concept of 'action at a distance' in which an action in one place influences behaviour at another instantaneously and with no manifest intermediary. Newton himself expressed concern about this point. This makes physics rather easy: we set up coordinate systems in space to describe where things are happening and our theories describe the evolution within that coordinate system relative to this universal time. Time and space are independent.

[1] When we come to considering light propagation in these Newtonian models, we will have to impose by *fiat* a finite and universal velocity of light that is the same in all inertial frames, c.

The Einstein world view is that space and time are not separate: space and time are inextricably intertwined. We can set up any coordinate system we like in that 4-dimensional world. Some coordinate systems are more convenient than others, but the physical quantities we describe must themselves be independent of any particular choice of coordinates we use to describe them. Of course, our measurement of a quantity will certainly depend on the coordinates used.

A solution to the Einstein equations is a geometry, described in the chosen coordinate system. In general, to get from that geometry to the physics of how a particle moves or light ray propagates in that geometry requires algebraic or computational effort. In cosmology we see this in the attempt to discuss how far a galaxy is from an observer. Unlike in Newtonian theory with its absolute space, there is no absolute concept of 'distance' in Einstein's world. The 'answer' depends on how you make the measurement, in other words what you mean by 'distance' or what your definition of distance is.

5.1.3 Two Views of Coordinates

In Newtonian theory the choice of coordinates that are not in uniform motion relative to an inertial frame will give rise to so-called 'fictitious forces'. For example, if a frame of reference is located on a rotating carousel, a person trying to walk on the carousel is subjected to centrifugal and Coriolis forces, which arise only because we choose to describe what is going on from the point of view of the rotating frame. We have to add these forces explicitly into our equations of motion. In Newtonian cosmology we simply refer to an absolute space and universal time for our dynamical description. Those things that are not covered by Newtonian theory, like the propagation of light, we have to impose by fiat.

In Einstein's theory there are no universal reference points and no preferred global coordinate systems. There may be coordinate systems that are more convenient or more suited to a particular problem, such as choosing a spherical coordinate system centred on a particle to describe the gravitational field about the particle. But even then there arises the question as to the choice of radial coordinate. In relativity there is no sense that one radial coordinate is preferable to another.[2]

Fortunately, there is a sensible way around this dilemma. There are many solutions to the Einstein equations that describe cosmologies having direct Newtonian analogues (Ellis, 1971). These tend to describe the simple models that we would like to use as a basis for understanding cosmology. We can graft concepts from Einstein's theory onto these Newtonian models that have analogous Einsteinian solutions.

---

[2] The outstanding example is Schwarzschild's solution to the Einstein field equations for the gravitational field in the vacuum surrounding a non-rotating mass (Schwarzschild, 1916). In the coordinates he used there are singularities at $r = 0$ and at a finite value of r, $r = r_S$, the Schwarzschild radius. While nobody doubted the reality of the $r = 0$ singularity, there was much controversy about what was happening at $r = r_S$. The issue was resolved by Lemaître (1933) who showed it to be no more than a coordinate singularity. In 1939 H.P. Robertson recognised r_S as an event horizon.

5.2 Newtonian versus Relativistic Models

5.2.1 Newtonian Gravity

There are considerable differences between Newton's theory of gravitation and Einstein's. Newton's theory is embodied in his law of universal gravitation, published in his *Principia* of 1687, that the gravitational force between two bodies varied as the product of their masses and inversely as the square of the distance between them. While it is probable that the notion of an inverse square law was not entirely due to Newton, it is certainly the case that Newton used the law to compute various physical phenomena like the trajectories of bodies under the influence of gravity. In particular, he proved the important result that extended spherically symmetric bodies acted as though their masses were concentrated at their centres.

Like many great scientific discoveries there were claims by others that they had done this first, for example by Robert Hooke who went as far as to accuse Newton of plagiarism. We know that the inverse square law had been proposed decades earlier by the distinguished French amateur astronomer Boulliau (latinised name Bullialdus) who, in his *Astronomia Philolaica* (Boulliau, 1645) argued in favour of the inverse square law as a better alternative than Kepler's own inverse distance proposal. Boilliau was one of the first people to accept Kepler's first law of planetary motion, that the orbits of the planets were ellipses with the Sun at one focus. Newton did acknowledge the contribution of Boilliau in the *Principia*.[3] Boulliau, like Newton, Huygens and Hooke, was an early Fellow of the Royal Society of London (elected 1667) and his work was generally known to that community.

In the final analysis, Newton wrote about gravitation, extended our understanding and used it to produce fundamental results such as the proof that the orbits of the planets were conic sections.[4] Newton showed, in his *Principia*, how the orbit of the comet of 1680 could be fitted by a parabola, a technique that was notably used by Halley some years later to compute several cometary orbits and to identify the comet that bears his name.

Newton's theory of gravitation has been an outstanding success and is still used in our everyday lives. Its first great prediction was the existence of the planet Neptune by Adams and by Le Verrier in 1846. By that time some discrepancies had already surfaced with detailed study of the orbit of the planet Mercury.[5]

---

[3] The following is the quote from Newton (1726, 1999): *By treading in the footsteps of procedures devised by the very celebrated Dr. Seth Ward ... and Ismael Boulliau we achieve the same end in this way moreover.* Ward (1653) had in fact strongly criticised Boulliau for making many errors in his work, and held to the notion that the orbits were circular. It was Ward who, in 1671, proposed Newton for election to the Royal Society. See also North (2008).

[4] This proof had been requested by Halley, who had paid for the publication of *Principia*. The proof that Newton published was that the conic section orbit was consistent with his $1/r^2$ law of gravity, but he did not prove the converse, that the $1/r^2$ law would necessarily lead to a conic section orbit.

[5] Le Verrier had been working on the stability of the orbits of the planets since 1839. He was able to predict the time of the 1845 transit of Mercury with an error of only 16 seconds. He was dissatisfied with this and later

That led to a number of proposals for modifications of Newton's inverse square law, most of which were based on *ad hoc* modifications of the $1/r^2$ force law designed specifically to produce the right number. Some of these changed the radial dependence of the force, while others introduced velocity dependent terms. Perhaps the most interesting, and controversial, of these was the theory of Gerber (1898, 1917) who wrote down equations that would limit 'the speed of gravitational influence' (*sic*) to the speed of light.[6]

It was not until Einstein came up with the General Theory of Relativity in 1915 that the issue was resolved, though that did not stop the flow of pseudo-Newtonian alternative theories. Within a month of the publication of the Einstein's great paper Schwarzschild (1916) published a very important solution to the Einstein equations for a point mass. The puzzle of the anomalous precession of the orbit of Mercury had already been solved by Einstein in an earlier draft of the Theory of General Relativity, but the Schwarzschild solution opened up the notion of space-time singularities and black holes. Considering its complexity, Einstein's Theory of General Relativity developed at a remarkable pace over the following decade.

5.2.2 Einstein's gravity

The Special Theory of Relativity (Einstein, 1905) made a difference to the calculated perihelion advance of Mercury, but the difference could only account for a smaller part of the $43''$ anomaly. The issue remained unresolved until Einstein brought in the General Theory of Relativity (see Section 12.1 for more on the development of this theory during the period 1912–1916). Although Einstein's final version of the General Theory did not appear until 1916 (Einstein, 1916a, 1917a, 1918b), Einstein had already calculated the perihelion advance in earlier versions of the theory and came up with the required $43''$ (Einstein, 1915). That, perhaps more than anything else,[7] set the scene for the establishment of the General Theory of Relativity. Newton's gravitational theory was recovered from general relativity in the 'non-relativistic limit' of weak gravitational fields. To quote from Tolman (1934b, p.213):

'It is also remarkable that Einstein's development of relativity was in no sense the result of a mere attempt to account for a small known difference between the observed orbit of Mercury and that predicted by Newtonian Theory, but was the full flowering of

---

studied a number of such transits to conclude in 1859 that the perihelion of Mercury was precessing more than would be expected, by $38''$ per century. He also came to the conclusion that Newtonian Gravity alone could not explain this anomaly. Later, in 1895, Newcombe used revised data on the planets to repeat Le Verrier's work and came up with an anomaly of $43''$ per century. By that time people had stopped looking for some missing matter that might be perturbing the orbit of Mercury and there was a growing acceptance that there must be something not quite right with Newton's theory.

[6] There is a good overview of Gerber's somewhat unclear hypotheses at
`http://mathpages.com/home/kmath527/kmath527.htm`
where the Gerber-based perihelion advance is calculated in detail. The deflection of light under Gerber's hypothesis is calculated at
`http://mathpages.com/home/kmath656/kmath656.htm`

[7] Of the three empirical pillars of general relativity, the advance of the perihelion of Mercury, the deflection of light by gravitational fields and the gravitationally induced redshift, only the advance of the perihelion of the orbit of Mercury was known at the time the theory was developed.

a complicated theoretical structure, growing from fundamental principles whose main justification seemed to lie in their inherent qualities of reasonableness and generality.'

The application of the General Theory of Relativity to cosmology came soon after, and before there were any relevant observations. The first homogeneous and isotropic solutions were static solutions: the Einstein Universe (Einstein, 1917a), the de Sitter Universe (de Sitter, 1917a) and the empty and flat space-time of Minkowski space. The Einstein Universe did not display the redshift phenomenon and was unstable. The de Sitter model was empty but did display a redshift and was adopted as the likely explanation of the redshift phenomenon soon after Slipher's list of galaxy velocities was published (Slipher, 1915, 1917). It was only a matter of 5 years before the Russian mathematician Friedmann (1922) would publish non-static solutions containing matter, though these were not interpreted until the work of Robertson (1928) and of Lemaître (1927, 1931a) who independently derived them (see Sections 2.1.4 and 2.1.2). Robertson (1933) provided a summary of all these models and in doing so clarified a number of important technical issues.

5.2.3 Newtonian Surrogates for GR Models

The key issue was that the solutions derived directly from Einstein's theory had no problems with handling infinite model universes, and the propagation of light in such a model Universe was automatically accounted for from within the theory itself.

Fortunately it is straightforward to derive Newtonian models that are counterparts to homogeneous and isotropic models derived from Einstein's theory. Models having zero pressure material ('dust') have straightforward counterparts. Newtonian models with non-zero pressure have to be 'adjusted' with an eye on the corresponding relativistic solution by arbitrarily adding explicit pressure terms to the Newtonian equation. This parallelism between Newtonian and Einsteinian models is well described by Ellis (1971).[8]

We can use such surrogate models to understand the nature of the expansion and to define various important time scales. Since in Newtonian theory the velocity of light is infinite, there is no direct analogue with the redshift phenomenon and we have to paint this onto the models with the usual eye on the relativistic model. Likewise, quantities that involve the motion of bundles of light rays through the cosmic gravitational field have no Newtonian analogue and so cannot in general be meaningfully discussed. That rules out, for example, discussion of apparent angular diameters of distant objects. The surrogates

---

[8] In Newtonian theory force exerted on particles by the gravitational field is simply described in terms of the second derivative, $E_{\alpha\beta} = \partial^2\phi/\partial x_\alpha \partial x_\beta$ of the gravitational potential, $\phi(\mathbf{x})$. The tidal shear tensor, $E_{\alpha\beta}$, is symmetric in α, β and so has six independent components, while ϕ itself is determined by the Poisson equation $\nabla^2\phi = 4\pi G\rho$. In general relativity the tidal shear tensor appears as the symmetric part of the conformal tensor. The skew part of the conformal tensor has no analogue in Newtonian theory (Ellis, 1971).

This is related to the fact that gravitational waves are possible in Einstein's theory, but not in Newton's theory. The discovery of the binary pulsar B1913+16 by Hulse and Taylor (1975) and the long-term analysis of the orbit (Weisberg and Taylor, 2005; Will, 2014) provided the first indirect yet compelling evidence for the existence of gravitational waves. The first direct detection of gravitational waves was announced by the LIGO Collaboration in 2016 (Abbott et al., 2016).

will also be useful when discussing the origin and evolution of the large scale structure of the Universe, but there are important constraints on just what can be meaningfully done (Callan *et al.*, 1965). This too is a constraint deriving from the finiteness of the speed of light. Fortunately, those constraints are not over-restrictive in discussing most of the cosmic history in relation to cosmic structure.

5.3 Light Propagation and Redshifts

5.3.1 The Dangerous Notion of Recession Velocities

For the sake of completeness, we briefly reiterate some remarks on the interpretation of redshift as a velocity that were previously made in Section 2.2.3. The redshift, z, of a galaxy is defined as the measured relative shift in the wavelength of its spectral lines,

$$z = \frac{\lambda_{obs} - \lambda_{em}}{\lambda_{em}},$$
(5.5)

where λ_{obs} is the measured wavelength of the spectral line whose rest wavelength is λ_{em}. The redshift z is an observable, we can measure it directly using the spectrum of an object. However, in general relativistic cosmology we cannot directly interpret this as a Doppler velocity since there might also be a gravitational contribution.

In general relativistic cosmology, the redshift is due to two factors: the familiar Doppler shift due to the relative movement of source and observer, and a gravitational redshift due to the fact that the photons of light are falling through a varying gravitational field.[9] Thus in Einstein theory we can have non-expanding cosmological models that display a redshift.[10]

Until the 1960s most measured redshifts were small, $z \ll 1$, and it was simple to interpret z as a velocity in the classical, non-relativistic way, via the expression $v = cz$, c being the speed of light. However, this must fail at higher redshifts $z > 1$ since it gives velocities exceeding the speed of light. In order to quote an 'equivalent velocity', it is common practise to use Equation (2.3). However, as emphasised earlier, that is incorrect in the general relativistic context.

In general relativity there is an additional gravitational contribution to the redshift due to the fact that there is a gravitational potential difference between the place of emission of the photon and the place at which it is received (i.e. the Einstein shift). If that potential difference is denoted by $\Delta\phi$, the corrected version of the schematic

---

[9] We see both these effects in our everyday lives when using GPS systems. The pulsed clock ticks from the GPS satellites arrive at the Earth having been subjected to *both* Doppler and gravitational redshifts. Without making appropriate corrections our GPS systems would not report the correct positions.

[10] In the 1930s people were very careful about interpreting redshifts as recession velocities. That may well have been a factor contributing to Hubble's reluctance to state unequivocally that the Universe was expanding, though he did present his results in terms of velocity calculated as a classical Doppler shift.

Equation (2.3) becomes[11]

$$1 + z = \frac{\lambda_{\text{obs}}}{\lambda_{\text{em}}} = \sqrt{\frac{1 + v/c}{1 - v/c}} \left(1 + \frac{\Delta\phi}{c^2}\right). \tag{5.7}$$

We only observe a redshift and we do not know a priori how to make the division between the two causes of the of the observed redshift. This is discussed in detail by Bondi (1947, p.420 eq.50).[12]

5.3.2 Expansion Factor and Redshift

We encounter another problem when relating redshift to epoch. It seems reasonable to ask that, if today we observe a galaxy whose light is redshifted by z, how long is it since that light was emitted? In the Newtonian view of the Universe we have an absolute time and light travels with infinite velocity, unlike in the relativistic world. The difficulty arises because in the Newtonian world we do not have a theory for the propagation of light at a finite speed. On the other hand, the propagation of light is a fundamental part of general relativity and so the issue is more straightforward, albeit technically more complex.

In Newtonian cosmology we could simply make an assumption about how light is redshifted as it travels from the emitter to the observer, again casting an eye on the relativistic solution. That leads to the *ansatz* that if the distance[13] between two observers increases from ℓ_{em} at the time the light was emitted to ℓ_{obs} when is was received, then

$$1 + z = \frac{\ell_{\text{obs}}}{\ell_{\text{em}}}. \tag{5.8}$$

This of course implicitly assumes that the speed of light is finite. The usual 'justification', or motivation, for this is that the photons travelling through the Universe are somehow stretched out by the cosmic expansion.[14]

So knowing the redshift tells us how much the Universe has expanded during the travel time of the light from the emitter. If we were to combine this (correct) relationship with the

---

[11] Formally, Ellis (2009b, Equation 6.14b) writes the redshift of a beam of light passing between two observers, who are comoving with the cosmic expansion, at a distance dl apart, as the sum of two terms

$$cz = (dl)^{\cdot} + (\dot{u}^a n_a)dl. \tag{5.6}$$

The first term is the Doppler term and the second term the gravitational (geometric) term. In this second term $\dot{u}^a n_a$ is the relative acceleration, $\dot{u}^a$, between the observers projected onto the line of sight, n_a, between them.

[12] The Bondi (1947) paper is, despite its somewhat 'old-fashioned' appearance, a model of clarity with regard to explaining the physics that underlie a very general solution for spherically symmetric, but inhomogeneous, solutions to the Einstein equations, both with and without the cosmological constant. The paper is still very much cited with a growing number of citations per year. Hermann Bondi had an outstanding ability to lecture technically difficult aspects of mathematics and physics at all levels. He was certainly the best lecturer that I experienced during my time as an undergraduate at Kings College, London, or during my postgraduate years in Cambridge.

[13] We are in the Newtonian world, so there is no ambiguity with the definition of distance.

[14] The alternative, Doppler-like, explanation would be that an observer moving away from a light source crosses the peaks of the light wave with lower frequency than an observer heading towards the source. However, the fact that the velocity of light in vacuo is a universal constant for all observers, and the consequent failure of the law of addition of velocities for light, makes this 'explanation' no less dubious.

(undesirable) Equation (2.3) we would arrive at a specific relationship between velocity (*sic*) and distance. But remember, we agreed not to use (2.3)!

Taking the observer to be at the present epoch ($\ell_0 = \ell_{obs}$) and the light to have been emitted when the Universe's age was t when the distance between the observers was $\ell(t)$, we can define the expansion factor $a(t)$ at time t to be

$$a(t) = \frac{\ell(t)}{\ell_0} = \frac{1}{1+z}, \qquad (5.9)$$

where $a(t)$ describes the 'expansion history' relative to the current time. The wavelength then grows with ℓ (Equation 5.8) and hence with a, and so the frequency changes as $1/a$. The redshift as defined here is merely a measure of the amount by which the Universe has expanded. Here we have used the normalisation $a(\text{now}) = 1$ so that $a(t) \to 1$ as $z \to 0$. As the model expands into the future the values of z become negative. There is nothing wrong with using redshift values in the range $-1 < z < 0$ in order to discuss future cosmic evolution. However, for dynamical reasons, some cosmological models may not allow values as negative $z = -1$.

All we need to answer the question as to how long is it since the light was emitted is the function $a(t)$. That is done by solving either Newton's dynamical equations or Einstein's equation which determine the rate of expansion of the Universe in terms of its material content. The following sections derive this expansion history for a wide variety of simple models for the material content of the Universe.

5.4 Models with Zero Pressure Dust

5.4.1 Homogeneity and Isotropy

The principal assumption underlying these simple models is that the matter in the Universe, in the large, is distributed homogeneously and isotropically. The simplest justification for this is the impressive isotropy of the Cosmic Microwave Background radiation. The isotropy of this radiation field, taken with the notion that we are not at the centre of the Universe, implies homogeneity.

Of course, the present day energy density of the background radiation is very small in comparison with other constituents of the Universe, and it is conceivable that those other constituents are not themselves homogeneously and isotropically distributed. We do, after all, see enormous structures in the galaxy distribution on scales of megaparsecs. Perhaps the best evidence for the isotropy of the galaxy distribution lies in the 2MASS survey (Jarrett, 2004) which looked at the sky in the near infrared. The 2MASS sky does indeed look isotropic in the large, and invoking the Copernican Principle, it must be homogeneous in the large too unless we are at the centre of this Universe.

5.5 Expansion Dynamics: the Scale Factor

The properties of our simple Newtonian dust model are defined entirely by the *scale factor* $a(t)$: the function that tells us how big the Universe was at some time t in relation to its present size. Since it consists only of dust particles that are neither created nor destroyed the density must simply change as the inverse cube of the size of the Universe: $\rho \propto a(t)^{-3}$. Since the pressure is zero, there are no temperature related effects. The model is simple and it only remains to determine how $a(t)$ varies under the influence of gravity as a function of time.

5.5.1 Isotropic Expansion of a Uniform Sphere

Consider the motion of a galaxy in the Universe that today (t_0) is at distance ℓ_0 from us[15] and that at time t was at a distance $l(t)$. It is convenient to define the *scale factor* $a(t)$ by

$$a(t) = \frac{\ell(t)}{\ell_0}. \tag{5.10}$$

Since the Universe is presumed homogeneous and isotropic, $a(t)$ depends on neither position nor direction. Moreover the equation holds for all values of ℓ_0. Equation (5.10) merely describes how relative distances change as this homogeneous and isotropic model Universe expands. We will normalise lengths relative to their present day value and so put the present value of $a(t)$ to

$$a(t_0) = 1. \tag{5.11}$$

If we consider a uniform sphere of constant mass M and radius $\ell(t)$ at time t, expanding under the influence of the Newtonian gravitational force, then the evolution of the sphere under the influence of gravity is described by the equations

$$\ddot{\ell} = GM/\ell^2, \quad M = \frac{4}{3}\pi\ell^3\rho, \tag{5.12}$$

where the time dependence of ℓ and ρ is implied and has been suppressed for clarity. The first of these is the famous inverse square law.[16] The second of these shows that the density changes inversely as the volume: $\rho(t) \propto \ell^{-3}$ and merely expresses conservation of mass.

If we replace $\ell(t)$ by the scale factor $a(t)$ according to Equation (5.10) and express the mass density $\rho(t)$ in terms of the scale factor, we get:

Acceleration of the expansion

$$\frac{1}{a}\frac{d^2a}{dt^2} = -\frac{4\pi G}{3}\rho. \tag{5.13}$$

---

[15] This is Newtonian cosmology so there is no ambiguity with this notion of distance.
[16] Newton himself showed that a spherically symmetric body affects external objects gravitationally as though all of the mass were concentrated at a point at the centre. He also showed that there is no gravitational force inside a uniform spherical shell.

This is supplemented by an equation expressing the conservation of matter:

Conservation of dust

$$\frac{d\rho}{dt} + 3\frac{\dot{a}}{a}\rho = 0, \tag{5.14}$$

which is equivalent to

$$\rho(t) = \rho_0 a^{-3}. \tag{5.15}$$

This Newtonian derivation of Equations (5.13) and (5.14) governing the expansion of a finite homogeneous sphere of constant mass is of course perfectly valid. A little extra thought is required when talking about the relative motion of two points anywhere within the sphere. However, it is not entirely clear how this extrapolates to an infinite homogeneous model Universe in which the gravitational potential $\phi = \int (\rho/r)dV$ is undefined.[17]

Ex 5.5.1 Prove that a spherically symmetric body affects external objects gravitationally as though all of its mass were concentrated at a point at its centre.

Ex 5.5.2 Prove that for a spherically symmetric shell (a hollow ball), no gravitational force is exerted by the shell on any object inside, regardless of the object's location within the shell. The same is true of electrostatic forces within a conduction spherical shell.

Ex 5.5.3 Discuss the applicability of Equations (5.12) to relative motion $l(t)$ of any two points within a uniform expanding sphere of finite mass.

The Einstein equations, in the simple case of homogeneous and isotropic dust models, give the same equations, though without any issues in the case of an infinite model Universe.

Note that Equations (5.13) to (5.15) need to be modified if there is any substantial pressure due to the matter in the Universe or to some dark energy field. We shall make these modifications at a later time when needed. For the moment we are only discussing the simplest (Newtonian) model for the Universe.

5.5.2 The Hubble Parameter of the Expansion

The *Hubble Parameter* is defined as

$$H = \frac{\dot{a}}{a} = \frac{\dot{\ell}}{\ell}, \tag{5.16}$$

---

[17] Seeliger (1895) proposed modifying Newton's inverse square law with an exponential $e^{-\lambda r}$ term that effectively gave a repulsive force that dominated at great distances. The λ notation appears explicitly in Seeliger's paper (his Equation (2)). (See North (1965, p.16) and Norton (1999).) However, the idea of introducing an exponential cut-off in a potential appears to go back to Neumann (1874), who wrote more on this later (Neumann, 1896). Such an exponentially cut off potential is now referred to as a 'Yukawa potential'. The technical issues surrounding Newtonian cosmology were put on a firmer footing for zero pressure cosmological models by McCrea and Milne (1934) and later by Layzer (1954) and by Heckmann and Schücking (1955).

and is a function of time. H describes the rate of expansion of the Universe and has units of inverse time, although it is generally quoted in observational units of $\text{km s}^{-1} \text{ Mpc}^{-1}$.

The time scale H^{-1} is a characteristic time scale on which the Universe is evolving. It is experimentally measured as a velocity increment per unit distance since it describes the expansion through the relationship between velocity and distance: $\dot{\ell} = H\ell$, or in more familiar notation $v = Hr$. Equations (5.13) to (5.15) have a trivial first integral:

Expansion rate

$$H(t)^2 = \left(\frac{\dot{a}}{a}\right)^2 = \frac{8\pi G\rho}{3} - \frac{kc^2}{a^2}. \tag{5.17}$$

Here k is a constant of integration with dimensions of inverse length, squared: $[L]^{-2}$ (the scale factor $a(t)$ is dimensionless). Multiplying k by c^2 gives us an integration constant that has dimensions of $[T]^{-2}$, as required by the left-hand side of (5.17). We shall evaluate this constant for our Universe later. We write the constant of integration in this form because in the derivation of Equation (5.17) from general relativity, k is the curvature of the space time which has dimensions $[L^{-2}]$. [18]

In Newtonian theory, Equation (5.17) is the analogue of an energy integral with the left hand side $(\dot{a}/a)^2$ term representing the kinetic energy and the $8\pi G\rho/3$ term on the right-hand side representing the potential energy. In this Newtonian scenario, the constant of integration, k, represents the total energy.

Equation (5.17) is generally referred to as the 'Friedmann equation', though, strictly speaking, that terminology is only appropriate to the derivation from the Einstein equations where k has a different meaning. Equation (5.17) is the special case for a dust Universe. This equation has a closed form solution for all k, though before going on to that we shall write k in a different, more useful, form.

Ex 5.5.4 Consider a uniform expanding sphere of dust having fixed mass M and time varying radius $\ell(t)$ and density $\rho(t)$. Verify that the density satisfies an equation like Equation (5.14), with ℓ in place of a.

Ex 5.5.5 Write down the equation of motion of a zero mass particle fixed to the surface of the sphere. (Hint: If the particle's distance from the centre of the sphere is $\ell(t)$ it moves under an attractive force GM/ℓ^2 by Newton's inverse square law.)

Ex 5.5.6 Integrate the equation of motion to give an equation for the radial velocity $\dot{\ell}$ of the particle,

$$\frac{1}{2}\dot{\ell}^2 = \frac{GM}{\ell} + E,$$

where E is a constant of integration.

Ex 5.5.7 Interpret Equation 5.17 in the light of this equation.

---

[18] In the older cosmological literature we see that $a(t)$ is assigned the dimensions of $[L^{-1}]$ such that k is dimensionless, taking on the discrete values $k = +1, 0, -1$. These values of k correspond respectively to what are generally referred to as closed, flat or open models.

From this we infer that, in Equation (5.17), if $k > 0$ there will come a time when $\dot{a} = 0$ and the expansion stops before turning into a collapse. If $k < 0$ the expansion will continue forever. So the case $k = 0$ is an important case dividing these two regimes.

5.5.3 The Density Parameter, Ω

There is an important value of the density, ρ_c, that can be derived from the Hubble parameter (the Hubble parameter has dimensions $[T]^{-1}$). This is the density such that a uniform self-gravitating sphere of density ρ_c isotropically expanding at rate H has equal kinetic and gravitational potential energies:

Critical density $\rho_c(t)$:

$$\rho_c = \frac{3H^2}{8\pi G}. \qquad (5.18)$$

This is also known as the 'Einstein de Sitter critical density'.

Since H is a function of time, then so is ρ_c. This is just Equation (5.17) with $k = 0$, reflecting the balance of the kinetic energy of the expansion and gravitational potential energy.[19]

We can give a dimensionless measure the density of the Universe by expressing the density at a given time in units of ρ_c. This dimensionless measure is referred to as the *density parameter* and assigned the symbol Ω:

Cosmological density parameter, Ω:

$$\Omega = \frac{\rho}{\rho_c} = \frac{8\pi G\rho}{3H^2}. \qquad (5.19)$$

From Equation (5.17) we see that $\Omega = 1$ implies $k = 0$.

In general, Ω depends on time and we shall denote the present day value of Ω by Ω_0.[20]

Ex 5.5.8   Show from Equation (5.17) that a Universe with $k = 0$ has $\Omega = 1$ at all times.

Several different types of matter contribute to the total cosmic density ρ. There are baryons, photons and perhaps some exotic elementary particles and some dark matter. Each of these individually has a density that can be normalised relative to ρ_c, so each contributor has its own Ω. We will, for example, denote the contribution of baryonic material to the critical cosmic density by Ω_B.

---

[19]   This refers to kinetic and potential energies and so is a strictly Newtonian statement. In the relativistic context, the model would be described as a 'zero curvature', or 'flat' model.

[20]   Pre-1970s it was usual to use the parameter $\sigma_0 = \frac{1}{2}\Omega_0$. This was also referred to as the 'density parameter'.

5.5.4 The Present-day Hubble Parameter, H_0

The present day value, H_0, of the Hubble parameter (5.16) is called 'the Hubble constant' and represents the present day cosmic expansion rate. In the strict Newtonian view of the Universe, a galaxy at a distance ℓ_0 would have an assigned recession velocity $\dot{\ell}_0 = H_0 \ell_0$. Although a few decades ago the Hubble constant, i.e. the cosmic expansion rate, was not known to within a factor of two, the situation has vastly improved since the establishment of the extragalactic distance scale as a Key Project for the Hubble Space Telescope, Freedman and Madore (2010). By 2001 the HST Key Project had established a value of $72 \pm 8\,\mathrm{kms}^{-1}\mathrm{Mpc}^{-1}$, most of the possible error being due to systematics (Freedman *et al.*, 2001). Since then the systematics have been reduced leaving the value at $73 \pm 2\,\mathrm{kms}^{-1}\mathrm{Mpc}^{-1}$. The measurements of the CMB anisotropy provide an alternative means for getting at the Hubble constant: the features seen in the power spectrum of the temperature fluctuations contain a scale whose observed size depends on the cosmic distance scale, i.e. the Hubble constant. After collecting 9 years of data the WMAP satellite provided a value for the Hubble constant of $69.32 \pm 0.80\,\mathrm{kms}^{-1}\mathrm{Mpc}^{-1}$ (Bennett, 2013), while the Planck satellite data gave a value of $67.80 \pm 0.77\,\mathrm{kms}^{-1}\mathrm{Mpc}^{-1}$ (Planck Collaboration I, 2013).

The CMB results and the HST results are only marginally consistent and so, unless otherwise stated, we shall use the 'numerically simple' value of:

'Benchmark' Hubble constant

$$H_0 = 70\,\mathrm{kms}^{-1}\mathrm{Mpc}^{-1}, \quad h = \frac{H_0}{100\,\mathrm{kms}^{-1}\mathrm{Mpc}^{-1}} = 0.7, \qquad (5.20)$$

with which the critical density is

$$\rho_c = 1.88 h^2 \times 10^{-29}\,\mathrm{gr.\,cm}^{-3}. \qquad (5.21)$$

In SI units this is $\rho_c = 1.88 h^2 \times 10^{-26}\,\mathrm{kg\,m}^{-3}$.

It should be noted that there are two ways of reporting the density of the Universe in dimensionless units. We can quote the value of the physical density Ωh^2, (see Equation 5.18), or we can quote the density as a fraction Ω of the critical density (see Equation 5.19). The important point is that everyone should agree that h is the Hubble parameter value in units of $100\,\mathrm{kms}^{-1}\mathrm{Mpc}^{-1}$, and not some other value.

5.5.5 The Curvature Term

For our dust model, using definitions (5.16), (5.18) and (5.19) turns Equation (5.12) into the relationship

$$\frac{kc^2}{a^2} = H^2(\Omega - 1), \qquad (5.22)$$

where H, a and Ω are all time dependent. Obviously $\Omega = 1$ means $k = 0$. We can evaluate this equation at the present time, where $a_0 = 1$, to give

$$kc^2 = H_0^2(\Omega_0 - 1), \qquad (5.23)$$

which will allow us to eliminate the constant k from many of the equations that we shall be using in what follows.

It is convenient to define another Ω-parameter, Ω_k, as

$$\Omega_k = \frac{kc^2}{a^2 H^2}.$$

(5.24)

In our dust model, described by Equation (5.17),

$$\Omega_k + \Omega = 1.$$

(5.25)

Later on we shall see generalisations of this important equation to a model containing several different types of matter.

Equation (5.22) allows us to define what we might call the *curvature length scale*

$$R_\dagger = cH^{-1}|1 - \Omega|^{-\frac{1}{2}}.$$

(5.26)

As expected, we have $R_\dagger \to \infty$ as $\Omega \to 1$ and $k \to 0$: the curvature length scale of a flat model Universe is infinite.

5.5.6 The Expansion History

The behaviour of the various model universes as a function of Ω can be seen by looking at the dynamical Equation (5.17) for the expansion rate $H(t)$. Using the definitions (5.19) for Ω and (5.23) for k we have:

The expansion rate in terms of present-day measurable quantities

$$H(a)^2 = \left(\frac{\dot{a}}{a}\right)^2 = \Omega_0 H_0^2 a^{-3} - H_0^2(\Omega_0 - 1)a^{-2}.$$

(5.27)

The right-hand side involves only the present-day values of the Hubble parameter, H_0, and the density parameter, Ω_0, and is only a function of the scale factor a. For that reason we have written the Hubble parameter on the left side as a function $H(a)$ of a rather than as a function of time. In the following sections we will solve this equation for $a(t)$.

We can recall that the first term on the right-hand side comes from the gravitational effect of the cosmic mass distribution, while the second term, the curvature term, appeared as a constant of integration that we have now expressed in terms of the measurable quantities H_0 and Ω_0. The curvature is implicit in the $(1 - \Omega_0)$ term.

As a final remark here we should note that, using $1 + z = 1/a$, we can rewrite (5.27) as

$$H(z) = H_0(1 + z)\sqrt{\Omega_0(1 + z) + 1 - \Omega_0} = H_0(1 + z)\sqrt{1 + \Omega_0 z},$$

(5.28)

which replaces the scale factor a with the observable quantity z.

5.6 Connecting with Observation

We now turn to the important issue of how we would use observations to test a model for the expansion of the Universe described by an equation such as (5.17), or its alternative form (5.28). The former equation provides us with an expansion history, $a(t)$, expressed as a function of the density, ρ, of the Universe and a number, k, that appeared as a constant of integration and was interpreted in the Newtonian context as the total energy of the expanding sphere. Neither ρ nor k are directly observable quantities. The latter replaces ρ and k with the dimensionless parameter Ω_0, the material dust content of this model.

In the early 20th century the key tool for observational cosmology was the Hubble diagram, plotted as a magnitude – redshift (m:z) diagram for galaxies that were considered to be standard candles. Hubble and Humason (1931), in their extension of Hubble's original diagram, had established a relationship between the recession velocity, V, of a galaxy and its apparent brightness, m_{pg}, measured in magnitudes:

$$\log V = 0.2 m_{pg} + 0.5, \quad V = cz, \tag{5.29}$$

(see their Figure 4 and their Equation (1)). The subscript 'pg' indicates that they were estimating the brightness from photographic plates. The recession velocity V was inferred from the measured redshift, z, using the equation $V = cz$.[21]

In the following we shall clarify the connection between Equations (5.28) and (5.29).

5.6.1 The Hubble Relationship: a Naïve Model

We start by considering a naïve model in which the recession velocity, V, of a galaxy is exactly proportional to its distance, D, from us: $V = HD$, for some constant H.[22] We aim to relate the quantity $0.2 m_{pg}$ appearing in (5.29) to the quantity HD.

For historical reasons the intrinsic brightness of an object is the luminosity it would have if placed at a distance of 10 parsecs (10pc). If L is the intrinsic luminosity of the galaxy and l is its apparent luminosity as seen from a distance R (measured in parsecs) then:

$$\frac{l}{L} = \frac{1}{(R/10\text{pc})^2} = \left(\frac{H_{(\text{km s}^{-1}\text{pc}^{-1})}}{c_{(\text{km s}^{-1})}z} 10\text{pc} \right)^2. \tag{5.30}$$

Here we measure $V = cz$ in units of km s^{-1} and H in units of $\text{km s}^{-1}\text{pc}^{-1}$ rather than the usual $\text{km s}^{-1}\text{Mpc}^{-1}$. This makes the expression dimensionally correct, albeit somewhat odd because of the explicit 10pc length-scale which arises in the definition of intrinsic luminosity.

---

[21] Hubble and Humason comment on relativistic corrections (cf. their footnote 1) but do not apply them since such corrections would be smaller than other likely errors.

[22] According to Equation (5.17) a test-particle (galaxy) at distance $D(t)$ in empty Universe with $k < 0$ moves away following the equation $\dot{D}/D = 1/t$. In that case $v = \dot{D} = HD$, with $H = 1/t$. This model could be viewed as a ballistic expansion in which the individual particles move at constant velocity, starting from a singular state at $t = 0$. The fastest moving galaxies travel the furthest.

Converting l and L to the *apparent magnitude*, m, and the absolute magnitude, M, respectively, requires using the definition:[23]

$$m - M = -2.5 \log_{10} \frac{l}{L}, \qquad (5.31)$$

where M is the magnitude the object would have if it were at a distance of $10\,\text{pc}$. With Equation (5.30) this gives

$$\mu = m - M = 5 \log_{10} z + 5 log_{10} \frac{c}{H} + 25, \qquad (5.32)$$

where the constant, 25, absorbs the velocity of light and the units of length, Mpc, are such that H is in the standard units of $\text{km s}^{-1}\,\text{Mpc}^{-1}$. The quantity μ is called the *distance modulus* and is expressed in magnitudes. Using a set of objects that are *standard candles*, i.e. have the same value of M, Equation (5.32) is the *m:z* relationship (5.29) for those objects.

Note that in Equation (5.32) we can only know the absolute magnitude, M, of an object if we know its distance. That requires the establishment of the *cosmological distance scale* using redshift independent distance indicators. Towards this end, a *Cosmic Distance Ladder* is constructed, each step of the distance determinations being based on the previous steps. Estimating the distances to galaxies has been a monumental and often controversial endeavour, starting with the work of Hubble (1926, p.356 *et seq.*) and culminating in the Hubble Space Telescope Key Project (Freedman *et al.*, 2001, 2012). The outcome of this is a determination of the Hubble constant, i.e. the current cosmic expansion rate, H_0, for the local Universe where such redshift independent distance determinations are possible. This is the value of H that goes into Equation (5.32).

Importantly, the relationship (5.29) is only valid for objects that are standard candles and with these standard candles, the local, $z \ll 1$, slope is always 0.2. The dynamical information about the cosmic expansion is contained in the deviations from linearity that arise as we look to high redshifts, $z \sim 1$, where the Universe was substantially younger than it is now. We shall see more about this in Section 5.6.3. This means that learning about the dynamics of the Universe requires observing objects at as high redshifts z as possible, $z \sim 1$ or greater.

The galaxy redshift catalogue of Humason *et al.* (1956) had 474 galaxies of which only 11 had velocities in excess of $10\,000\,\text{km s}^{-1}$. The best one could hope to do was perhaps to detect evidence for deviation from a straight line Hubble law. That would provide evidence for deceleration of the expansion and so determine whether the Universe was open or closed. Even with all that, there remained the issue of calibrating the distance scale itself: determining the Hubble constant. But, for its time, HMS was a monumental work.

The early 1970s marked the advent of a number of detectors that would render the use of photographic plates obsolete. [24] By 1978 Kristian *et al.* (1978) had extended the Hubble diagram beyond $z \simeq 0.5$ and reported the detection of weak evidence for deviations from the simple Hubble Law.

---

[23] There is more on defining the magnitude scale in Appendix B. See Section B.2.

[24] Westphal *et al.* (1975) gave a brief list of photon detecting devices that were operational or coming on-line at the time and published a spectrum and redshift of $z = 0.392$ for the galaxy cluster 0024+1654.

5.6.2 Cosmological Parameters Pre $\sim$ 1970

The early days of observational cosmology saw the cosmic expansion as characterised by three parameters:

$$\text{Hubble parameter:} \quad H = \frac{\dot{a}}{a}, \tag{5.33}$$

$$\text{Density parameter:} \quad \sigma = \frac{4\pi G\rho}{3H^2} = \frac{1}{2}\frac{\rho}{\rho_c}, \tag{5.34}$$

$$\text{Deceleration parameter:} \quad q = -\frac{a\ddot{a}}{\dot{a}^2} = -\frac{1}{H^2}\frac{\ddot{a}}{a}. \tag{5.35}$$

The quantity ρ_c appearing in (5.34) is the critical density (5.18). These relate directly to the expansion factor $a(t)$ through their definitions. From the point of view of observation the quest was to discover the values of these parameters at the present time: H_0, σ_0 and q_0. H_0 would define the scale of the Universe. Determining the mass function of galaxies and clusters would then provide the estimate of the density parameter σ_0. Finally, plotting the Hubble diagram to higher redshifts would determine q_0 and thereby specify the cosmological model.

Obtaining an accurate value for H_0 was fundamental and required construction of an accurate distance ladder. As we see from the above definitions, H_0 plays a fundamental role in determining the scale of the present Universe and the present values of of σ_0 and q_0. By 1970, Sandage presumably felt he had determined H_0 and so he wrote an important paper (Sandage, 1970) in which he famously described the task as 'cosmology: a search for two numbers'.[25]

Now, with precision cosmology, we have come to expect far more than a mere description of the cosmic expansion history. Nevertheless, the search for two numbers was an important major program consuming an enormous amount of observing time on the largest telescopes available. As an example of what might have been achieved we can think of the Steady State theory, the great alternative to the Big Bang theory.

The expansion law and deceleration parameter for the Steady State theory had

$$a(t) \propto e^{H_0 t} \quad \sigma_0 = \tfrac{1}{2}, \quad q_0 = -1. \tag{5.36}$$

This value for q_0 is radically different from most other theories based on the Friedmann–Lemaître models. This was potentially a verifiable prediction of the Steady State theory (Hoyle and Sandage, 1956). The *prediction* that $\sigma_0 = \tfrac{1}{2}$, i.e. that the Universe has the critical density, was interesting since the alternative models had at the time no prior reason

---

[25] Prior to the launch of the Hubble Space Telescope, Allan Sandage had been one of the main drivers of the quest for the cosmological parameters. In 1968 he proposed a value of $H_0 = 75.3^{+19}_{-15}$ km s^{-1} Mpc^{-1}, suggesting that H_0 might be as low as 50 km s^{-1} Mpc^{-1} (Sandage, 1968). Sandage was concerned about the ages of globular clusters. By the early 1970s, he was suggesting a lower value of $H_0 = 55 \pm 7$ km s^{-1} Mpc^{-1} with $q_0 = +1 \pm 0.4$ based on observations of first ranked cluster galaxies. It was not possible to decide whether the Universe was closed or open since $q_0 = \tfrac{1}{2}$ was the dividing line between closed and open models. The uncertainties among the community at this time are best seen in the discussion which follows Sandage's presentations at two conferences: Sandage (1972, 1971). The arguments at that time for a higher value of $H_0 \simeq 100$ km s^{-1} Mpc^{-1} are well summarised by van den Bergh (1975).

to expect any particular value for this parameter. The Steady State Theory was a wonderful alternative theory and it was falsifiable by experiments.

5.6.3 The Search for q_0

The magnitude redshift relationship in a 'dust' cosmology having $\Lambda = 0$ was first given by Mattig (1958) in terms of the parameter q_0:

$$m_{\text{bol}} - M = 5 \log \frac{1}{q_0^2} \left\{ q_0 z + (q_0 - 1)[(2q_0 z + 1)^{1/2} - 1] \right\} + \text{constant}, \quad q_0 > 0 \quad (5.37)$$

(see also Sandage (1961a,c, 1962)). Here, m_{bol} is the bolometric magnitude of the galaxy in the sample. M is the absolute magnitude of the object: this measures its intrinsic brightness and has to be determined from criteria other than the redshift. If the objects in a sample are all of equal brightness ('standard candles' like first ranked cluster galaxies or supernovae) then they have the same M.

Ex 5.6.1 By expanding Equation (5.37) or second order on q_0 show that for small z

$$\mu = m_{\text{bol}} - M = \begin{cases} 5 \log z(1 + z) + \text{const.} & q_0 = -1, \\ 5 \log z(1 + \frac{1}{2}z) + \text{const.} & q_0 = 0, \\ 5 \log z(1 + \frac{1}{4}z) + \text{const.} & q_0 = \frac{1}{2}, \\ 5 \log z + \text{const.} & q_0 = 1. \end{cases} \quad (5.38)$$

Note that the solutions for $q_0 = 0$ and $q_0 = 1$ are not in fact small z approximations, they are exact results. That suggests that we can expect the approximations in the range $0 < q_0 < 1$ to be very good.

Ex 5.6.2 Produce a plot of $m_{\text{bol}} - M : z$ according to Equation (5.37) for $0 < z \leq 1.5$ for $q_0 = -1, 0, \frac{1}{2}, +1$. Use a linear z scale and choose the constant so that $m_{\text{bol}} - M = 35.00$ at $z = 0.02$.

Ex 5.6.3 Add the supernova data from the Union 2.1 data compilation[26] to the Hubble diagram produced in the previous question, adjusting the vertical position of the data by eye.

Ex 5.6.4 What does this suggest to you?

It is, in principle, possible to fit a sample of objects plotted on a magnitude redshift diagram with the relationship (5.37) for some value of q_0. The relationship differs from the simple inverse-square law expression (5.32) because of general relativistic effects on the propagation of light in a curved space-time containing one form or another of matter. This is the key to establishing the nature of our cosmology.

---

[26] The Union 2.1 data can be found on the Supernova Cosmology Project (SCP) website at
`http://supernova.lbl.gov/Union/figures/SCPUnion2.1_mu_vs_z.txt`
or on the supplementary data repository for this book.

From Equations (5.38) we see that the magnitude redshift diagram is well fitted by the relationship

$$\mu = m_{\text{bol}} - M \simeq 5 \log z[1 + \tfrac{1}{2}(1 - q_0)z] + \text{const.}, \quad q_0 > 0 \qquad (5.39)$$

over the entire range $0 \le q_0 \le 1$ and probably beyond.

Terrell (1977) gave a somewhat more useful form of the Mattig Equation (5.37):

$$m_{\text{bol}} - M = 5 \log z \left\{ 1 + z \frac{1 - q_0}{\sqrt{1 + 2q_0 z} + 1 + q_0 z} \right\} + \text{constant.} \quad q_0 \ge 0. \qquad (5.40)$$

Ex 5.6.5   Derive Terrell's Equation (5.40) from the Mattig Equation (5.37).[27]

Equations (5.37) and (5.40) require that $q_0 > 0$ and so cannot be used for the Steady State Theory which has $q_0 = -1$. For that theory, the $m - z$ relationship has to be calculated from scratch (Sandage, 1961a):

$$\mu_{SS} = m_{\text{bol}} - M = 5 \log z(1 + z) + \text{const.}, \qquad (5.41)$$

which fits with the approximation in Equation (5.39) with $q_0 = -1$.

The key Equation (5.37) contains no correction for any evolution in the properties of the galaxies that might produce a systemic bias with redshift. For the details of the effect of evolutionary corrections see Robertson (1955). While, in principle, this is simple to use, the work going into determining the quantity m_{bol} for each sampled galaxy is quite formidable.

5.7 The Simplest Cosmological Models

At the time when there was no compelling evidence for the acceleration of the cosmic expansion, which is most of the 20th century, the recent past of the Universe was considered as being dominated by pressure-free matter, together with some radiation having a comparatively small density.

If the Universe were flat, i.e. the density was the critical density ρ_c, then there would have to be a substantial quantity of dark unseen matter. The luminous cosmic material could account for at most a few percent of the critical density. The nature of this 'missing mass' was unknown. There was no lack of suggestions: 'dark baryons' as in cosmic billiard balls, 'cold dark matter' (CDM) composed of some massive non-baryonic and as yet unknown particle, warm or hot dark matter (WDM or HDM) composed of light particles such as massive neutrinos, which again were undetected.

Some preferred to argue that what we saw was what there was, in which case our Universe would be an open Universe, set to expand forever. The situation in 1995 is beautifully summarised by the divergent opinions of the cosmologists shown in Figure 2.5.

---

[27] Hint: use the identity $[\sqrt{2q_0 z + 1} - (1 + q_0 z)][\sqrt{2q_0 z + 1} + (1 + q_0 z)] = -q_0^2 z^2$.

5.7.1 The Einstein de Sitter Model

The density ρ_c has a special significance. A Universe whose density is $\rho_c(t)$ when its expansion rate is H is referred to as an *Einstein de Sitter* Universe. For such a model, Equation (5.27) tells us that $a^3H^2 = \Omega_0 H_0^2$, a constant independent of time. For our dust model the density varies as $\rho \propto a^{-3}$ and so $a^3H^2 \propto H^2/\rho$ is also independent of time. Hence the quantity $H^2/8\pi G\rho = \Omega(t)^{-1}$ is a constant, independent of time and so this model has $\Omega = 1$ at all times.

Model Universes that are denser than $\rho_c = 3H^2/8\pi G$ when their expansion rate is H will stop expanding and contract down to a future singularity. Models that are less dense will expand forever. The $\Omega = 1$ Einstein de Sitter Universe is a limiting case dividing two classes of behaviour and that is why the parameterisation of the density in terms of ρ_c is so useful. We shall give solutions for these cases later.

5.7.2 Simple Cases: Empty and Flat Models

In Section 5.8 we give the solutions of Equation (5.27) for general values of Ω_0. But first note that the cases $\Omega_0 = 0$ and $\Omega_0 = 1$ simplify the right-hand side of this equation. The solutions are then trivial:

$$a(t) = H_0 t \qquad\qquad \Omega_0 = 0, \qquad (5.42)$$

$$a(t) = \left(\frac{3}{2}H_0 t\right)^{\frac{2}{3}} \qquad\qquad \Omega_0 = 1. \qquad (5.43)$$

The case $\Omega_0 = 0$ can be called the 'empty model', while the case $\Omega_o = 1$ is called the 'Einstein de Sitter' or 'Flat' model. Since $a(t) = (1+z)^{-1}$ this tells us that when we look back to a redshift z in an Einstein de Sitter Universe we are seeing the Universe when its age is a fraction $t/t_0 = (1+z)^{-\frac{3}{2}}$ of its present age, t_0.

5.7.3 Expansion as a Function of Redshift

Equation (5.27) gives the Hubble expansion rate as a function of time. It is frequently useful to express the expansion rate as a function of redshift.

Replacing the scale factors a on the right-hand side of Equation (5.27) with redshifts $(1+z)^{-1}$ as defined in (5.8) we can write

$$H(z)^2 = \left(\frac{\dot{a}}{a}\right)^2 = H_0^2\left[\Omega_0(1+z)^3 - (\Omega_0 - 1)(1+z)^2\right]. \qquad (5.44)$$

The term $\Omega_0(1+z)^3$ reminds us that this is a pressure free dust Universe where $\rho \propto (1+z)^3$.

Ex 5.7.1 Go through the derivation of Equation (5.27) from Equation (5.17). We shall need to be able to do that for more complex models later on.

Ex 5.7.2   Use Equation (5.27) to verify the solutions (5.42) and (5.43).

Ex 5.7.3   Verify Equation (5.44).

Ex 5.7.4   Show that in this pressure free dust Universe the density parameter $\Omega(z)$ at redshift z is related to its present value by the equation

$$\Omega(z) = \Omega_0 \frac{1}{\Omega_0 + (1 - \Omega_0)(1 + z)^{-1}}. \tag{5.45}$$

Hint: Write down $\Omega(z)/\Omega_0$ using the definition (5.19) of Ω.

Equation (5.44) is the simplest of an important family of equations that describe the expansion rate as a function of redshift for model universes containing a variety of forms of material.

5.7.4 Transition to Undecelerated Motion

The two terms on the right-hand side of Equation (5.44) are respectively the gravitational and energy terms (or curvature-term in the relativistic models). Consider the case when Ω_0 is neither 0 or 1. These terms have different redshift dependencies: at very large z, i.e. in the past, the gravitational term $\Omega_0(1 + z)^3$ dominates, while as $z \rightarrow -1$, the future, the curvature term dominates. They are equally important at redshift $z_\dagger$ given by:

$$1 + z_\dagger = \Omega_0^{-1} - 1. \tag{5.46}$$

The gravity term acts to decelerate the expansion, so this redshift marks the transition to undecelerated motion. This can only happen if $\Omega_0 < 1$.

The expansion of the Universe when it is in undecelerated motion is linear with time: $a(t) \propto t$ and its expansion rate $\dot{a}/a$ becomes a constant.

We shall see that in models with a non-zero cosmological constant, the transition away from decelerated motion may occur because of the repulsive effect of the Λ-term.

5.7.5 Ages and the Hubble Parameter h

The relationship between time t since the Big Bang and the value of the scale factor, $a(t)$ or redshift z in the Einstein de Sitter flat dust model ($k = 0$) is given by Equation (5.43).

The age of a flat ($\Omega = 1.0$) dust Universe at a redshift z is calculated from Equation (5.17) to be

$$t(z) \overset{\text{EdS}}{=} \frac{2}{3} H_0^{-1} (1 + z)^{-3/2}. \tag{5.47}$$

The present age of this model Universe is thus $t(0) = \frac{2}{3} H_0^{-1}$. Our estimate of the value of the Hubble constant affects our estimate of its age.

The look-back time (see Figure 5.1), $\tau(z)$, to redshift z is the difference between the present time, $t(z = 0)$ and the age of the Universe $t(z)$ at redshift z:

$$\tau(z) \overset{\text{EdS}}{=} \frac{2}{3} H_0^{-1} \left[1 - (1 + z)^{-3/2} \right]. \tag{5.48}$$

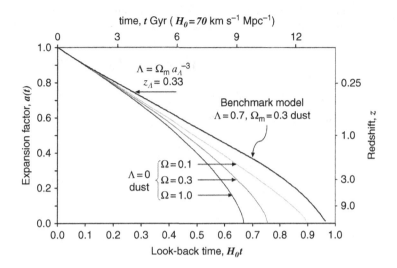

Fig. 5.1 Look-back time for three $\Lambda = 0$ dust cosmological models and the 'benchmark model' having $\Lambda = 0.7$ and $\Omega_m = 0.3$. The lower axis of the figure expresses the look-back time, $H_0 t$, in terms of the present Hubble time H_0^{-1}. The left axis shows the scale factor $a(t)$ and the right vertical axis the redshift z to which we are looking back. The top axis is labelled in Gyr, assuming a Hubble constant of $H_0 = 70$ km s^{-1} Mpc^{-1}. In the benchmark model, the contributions to the energy density of the dust and of Λ were equal at $a_\Lambda = 0.75$ and $z_\Lambda \simeq 0.33$. ⓒ

> **Ex 5.7.5** Derive Equations (5.47) and (5.48) for the age of the Universe at redshift z and the look-back time from the present to a redshift z in the flat dust model.
>
> **Ex 5.7.6** Repeat this for a dust Universe with $\Omega = 0$.
>
> **Ex 5.7.7** Explain why, physically, the Ω_0 model is older than the $\Omega = 1$ model for a given value of H_0.

Obtaining the explicit solution (5.48) for the look-back time was particularly simple since in this flat zero-pressure dust model we have a simple analytic form, (5.43), for the scale factor $a(t)$. In the more general case of a non-flat model (i.e. $\Omega_0 \neq 1$) we must resort to integrating (5.27):

$$t(a) = \int_0^a \frac{da}{aH(a)} = \frac{1}{H_0} \int_0^a \frac{da}{\sqrt{\Omega_0 a^{-1} + (1 - \Omega_0)}}, \qquad (5.49)$$

after some minor manipulations.

5.8 Solutions for Arbitrary 'Curvature'

Equation (5.17), or equivalently Equation (5.27), has closed form solutions expressing the scale factor a as a function of time t. The solution can be expressed in two different forms. The parametric form provides an explicit solution for the scale factor or redshift as a function of time using an intermediate, parametric, variable. The explicit form gives the time directly as a function of redshift or scale factor.

5.8.1 Parametric Form of Solution

The simplest form of the solutions is written down in parametric form using a parameter θ called the 'development angle':

$$\Omega_0 < 1 \quad \left\{ \begin{array}{l} a = \dfrac{\Omega_0}{2(1 - \Omega_0)}(\cosh \theta - 1) \\[2mm] t = \dfrac{\Omega_0}{2H_0(1 - \Omega_0)^{3/2}}(\sinh \theta - \theta) \end{array} \right. \tag{5.50}$$

and

$$\Omega_0 > 1 \quad \left\{ \begin{array}{l} a = \dfrac{\Omega_0}{2(\Omega_0 - 1)}(1 - \cos \theta) \\[2mm] t = \dfrac{\Omega_0}{2H_0(\Omega_0 - 1)^{3/2}}(\theta - \sin \theta). \end{array} \right. \tag{5.51}$$

This form of the solution is particularly useful when looking at spherically symmetric inhomogeneous solutions. Note that we can relate the time and redshift through the substitution $1 + z = 1/a$.

5.8.2 Explicit Form of Solution

We can relate the redshift and time directly by first casting Equation (5.27) into the form

$$H_0 t(z) = \int_0^{a=(1+z)^{-1}} \frac{da}{\sqrt{\Omega_0 a^{-1} + (1 - \Omega_0)}}, \tag{5.52}$$

and then evaluating the integral. There are two cases for $\Omega_0 \neq 1$ depending in the sign of $(1 - \Omega_0)$:

$$H_0 t(z) = \frac{\Omega_0}{2(1 - \Omega_0)^{3/2}} \left[\frac{2\sqrt{(1 - \Omega_0)(1 + \Omega_0 z)}}{\Omega_0(1 + z)} - \cosh^{-1}\left(1 + 2\frac{1 - \Omega_0}{\Omega_0(1 + z)}\right) \right],$$
$$\Omega_0 < 1 \tag{5.53}$$

and

$$H_0 t(z) = \frac{\Omega_0}{2(\Omega_0 - 1)^{3/2}} \left[\cos^{-1}\left(1 + 2\frac{1 - \Omega_0}{\Omega_0(1 + z)}\right) - \frac{2\sqrt{(\Omega_0 - 1)(1 + \Omega_0 z)}}{\Omega_0(1 + z)} \right]$$
$$\Omega_0 > 1. \tag{5.54}$$

5.8.3 The Age of the Dust Universe

The present value of the Hubble constant, H_0, and the density parameter, Ω_0, together determine the present age of the Universe. The age is given simply by putting $z = 0$ in Equations (5.53) and (5.54). In the case of an $\Omega_0 \leq 1$ model Universe:

$$t_0 = \frac{1}{H_0} \left[\frac{1}{(1 - \Omega_0)} - \frac{\Omega_0}{2(1 - \Omega_0)^{3/2}} \cosh^{-1}\left(\frac{2}{\Omega_0} - 1\right) \right] \quad \Omega_0 \leq 1. \quad (5.55)$$

This has the limits

$$t_0 \rightarrow H_0^{-1}, \qquad\qquad \Omega_0 \rightarrow 0$$
$$\rightarrow \frac{2}{3} H_0^{-1} \qquad\qquad \Omega_0 \rightarrow 1.$$

If this were the appropriate model to describe our Universe there should not be any objects older than this in the Universe. Determining ages used to be an important way of constraining the values of H_0 and Ω_0 and testing the consistency of our models.

5.9 Simple Models with Pressure

So far the models we have considered have contained zero-pressure 'dust' as the material component contributing to the overall gravitational forces that control the cosmic expansion. In addition to the dust we have considered models having non-zero curvature and models containing dark energy in the form of a cosmological constant. We have measured their relative contributions to the expansion rate via the parameters Ω_m, Ω_k and Ω_Λ, respectively. We now take the next step and consider the role of pressure forces.[28]

In Newtonian gas dynamics[29] the pressure only exerts its influence in the equation of conservation of momentum through pressure gradients. The (Newtonian) Euler equations of motion contain a ∇p force term which, for homogeneous cosmological models, would be zero. The Newtonian expression for the conservation of energy contains the pressure explicitly in a term describing the work done by the pressure forces. The gas-dynamic equations are supplemented by an equation of state which, in simple cases, relates the thermodynamic state of the gas to the density. At that level it is possible to build Newtonian models in which the pressure is non-zero and directly related to the density by a simple gas law.

5.9.1 The Role of Pressure

The equations for the cosmic expansion should, of course, be derived directly from the Einstein field equations. But for didactic purposes we can use an ad hoc 'fix' to the Newtonian

---

[28] The Euler and Navier–Stokes equations of gas dynamics, and their special relativistic analogues are reviewed in the On-line Supplement on *Gas Dynamics*.

[29] By this we mean 'non-relativistic' gas dynamics. The equations of fluid flow were not known at the time of Newton (Truesdell, 1983). The foundations of hydrodynamics were laid down by Daniel Bernoulli (1738) in his *Hydrodynamica* and by Euler (1757) in his *Principes generaux du mouvement des fluides*. Bernoulli's work, written in Latin, was mainly about hydrostatics where he referred to pressure as 'elasticity'. Euler's work, written in French, discussed how to describe fluid motion. His main result was what we now call *the convective derivative* but he knew nothing about pressure, let alone the action of pressure gradients. Pressure was first correctly described in 1846 by Waterston (1892), but not published for nearly 50 years (see Lord Rayleigh's introduction to the 1892 version of the paper).

equations. We need the equations containing pressure that are analogous to Equations (5.13) and (5.14). At the level of special relativity, the fix is quite straightforward and arises from the simple requirement that the equations of motion be invariant under Lorentz transformations.

Introducing pressure into relativistic cosmological models brings with it several non-Newtonian issues. In the special relativistic version of the Euler momentum equation the density ρ appears combined with the pressure, $(\rho + p/c^2)$.[30] This arises from the requirement that the equations be locally Lorentz invariant. For a gas in which the random particle motions that give rise to the pressure are non-relativistic, this is a small correction. However, if the gas particles are moving at relativistic speeds the p/c^2 contribution cannot be neglected.

In general relativistic cosmological models the gravitational force is computed via the Einstein equations, not from the Poisson equation. Creating a pseudo-Newtonian model that mimics the relativistic solutions thus requires an ad hoc fix-up. The fix-up is to say that the source of the gravitational forces is $\rho + 3p/c^2$ rather than simply ρ. That the pressure contributes to the gravitational field is not surprising, all forms of energy have a gravitational effect. What we cannot convincingly do is explain away the factor 3.

5.9.2 Relativistic Newtonian Surrogates

First there is the equation for the action of gravity, which in the case of zero cosmological constant is

$$\frac{1}{a}\frac{d^2a}{dt^2} = -\frac{4\pi G}{3}\left(\rho + \frac{3p}{c^2}\right). \tag{5.56}$$

Compare with Equation (5.13): we have an additional contribution $3p/c^2$ to the gravitational force. There is no analogue of this in Newtonian theory, but in Einstein's theory all forms of energy, even the random kinetic energy of the particle in a gas, are a source of gravity. Here p represents that kinetic energy, but why it should be a factor of $3p/c^2$ can only be answered properly in terms of relativity. The simplest 'excuse' is that the random motions have 3 degrees of freedom and that is the origin of this factor 3.

This is supplemented by an equation expressing the conservation of energy:

$$\frac{d\rho}{dt} + 3\frac{\dot{a}}{a}\left(\rho + \frac{p}{c^2}\right) = 0. \tag{5.57}$$

Compare with Equation (5.14), which was presented as an equation for conservation of particles. In Einstein's theory this equation in fact comes as a statement of energy conservation and for zero pressure it just looks like an equation for conservation of particles.

---

[30] For a gas in which the random thermal motions of the particles have root mean square velocity v_{th}, the pressure is on the order of $p \simeq \rho v_{th}^2$: this contribution from the pressure is negligible in everyday circumstances.

Viewed as an energy equation we can propose that the appearance of the p/c^2 terms reflects the 'pdV' work done on an elemental volume.[31]

We will also need an equation relating the pressure to the material content: an *equation of state*. The simplest such equation that we can consider is

$$p = w\rho c^2, \tag{5.58}$$

for some factor w that, in the first instance, we shall take to be a constant.

We are more accustomed to an equation of state being something akin to $p \propto \rho^\gamma$, where γ is the adiabatic index. In fact Equation (5.58) is just the relativistic version of $p \propto \rho^\gamma$ with $w = \gamma - 1$. Some examples may put this in perspective:

$$p = 0 \qquad \text{dust} \tag{5.59}$$

$$p = \tfrac{1}{3}\rho c^2 \qquad \text{radiation} \tag{5.60}$$

$$p = \rho c^2 \qquad \text{stiff matter} \tag{5.61}$$

$$p = -\rho c^2. \qquad \text{cosmological constant} \tag{5.62}$$

A radiation gas has $\gamma = 4/3$ so $w = 1/3$. The case $w = -1$ describes negative pressure, which is normally considered to be non-physical, but in fact describes a Universe with a constant cosmological constant. The case $w = +1$ describes what is called 'stiff matter': matter in which the sound speed is equal to the speed of light. Stiff matter is used for simple models for the internal structure of neutron stars.

Values of $w < 0$ would normally be considered to be unphysical, though a Universe pervaded by an Einstein cosmological constant can be regarded as being filled with matter having $w = -1$. We shall discuss that at length in Chapter 6.

Ex 5.9.1 Using Equation (5.58), eliminate p in Equation (5.57) to show that, for constant w

$$\rho \propto a^{-3(1+w)}. \tag{5.63}$$

If $w = -1$ the density of the field remains constant, while for $w < -1$ the density increases as the Universe expands. That makes the value $w = -1$ a critical dividing line between what may be easily acceptable and what might appear to be bizarre.

5.9.3 Simple Radiation Model: $p = \tfrac{1}{3}\rho c^2$

To make it clear that in this section we are dealing with a radiation-only Universe, we shall write the density with a subscript r, as in ρ_r.

---

[31] Consider a volume V of gas bounding a fixed number of immutable fundamental particles. Its energy content is $E = \rho V$. The work done by the pressure forces in changing the volume by an amount dV is pdV, and so conservation of energy leads us to $dE + pdV = 0$, which is Equation (5.57). This is the form of the energy equation within the framework of special relativity: see the On-line Supplement on *Gas Dynamics*.

Eliminating the pressure from Equations (5.56) and (5.57) gives the pair of equations

$$\frac{1}{a}\frac{d^2 a}{dt^2} = -\frac{8\pi G}{3}\rho_r, \tag{5.64}$$

$$\frac{d\rho_r}{dt} + 4\frac{\dot{a}}{a}\rho_r = 0. \tag{5.65}$$

From Equation (5.65) we see that the density of the radiation field falls off as

$$\rho_r \propto a^{-4}. \tag{5.66}$$

This is to be compared with the behaviour of pressure-free matter which falls off as $\rho_m \propto r^{-3}$. If the matter were baryons this would say that the number of baryons in a given co-expanding volume remains constant: baryons are conserved. We shall see later that, in a radiation gas of photons, the *number* of photons per unit volume falls off as r^{-3} and thus photons are conserved just like baryons.

Substituting Equation (5.66) into (5.65) and integrating then gives

$$\left(\frac{\dot{a}}{a}\right)^2 = \frac{8\pi G\rho_r}{3} - \frac{kc^2}{a^2}. \tag{5.67}$$

This looks like Equation (5.17), which should occasion no surprise. The only difference is the a dependence of the density: $\rho_m \propto a^{-3}$ versus $\rho_r \propto a^{-4}$. The curvature constant k can be written in terms of present day parameters by evaluating this equation at the present epoch:

$$kc^2 = H_0^2(1 - \Omega_r), \quad \Omega_r = \frac{8\pi G\rho_r(\text{now})}{3H_0^2}. \tag{5.68}$$

Using this we get the equation for the evolution of the scale factor:

$$\dot{a}^2 = H_0^2\left[\frac{\Omega_r}{a^2} + (1 - \Omega_r)\right]. \tag{5.69}$$

Compare this with Equation (5.27). There is an analytic solution of this:

$$a(t) = (2H_0\Omega_r^{1/2}t)^{1/2}\left[1 + \frac{1 - \Omega_r}{2\Omega_r^{1/2}}H_0 t\right]^{1/2}. \tag{5.70}$$

The early Universe is radiation dominated and so this solution has relevance to early cosmic history.

Ex 5.9.2   Verify the solution (5.70).

Ex 5.9.3   Show that if $\Omega_r > 1$, $a(t)$ reaches a maximum value of

$$a_{\max} = \left(\frac{\Omega_r}{\Omega_r - 1}\right)^{1/2} \quad \text{at time} \quad t_m = \frac{\Omega_r^{1/2}}{\Omega_r - 1}H_0^{-1}.$$

Hint: look at $a^2(t)$.

Ex 5.9.4   Show that if $\Omega_r = 1$, then $a(t) \propto (2\Omega_r^{1/2}H_0 t)^{1/2} \propto t^{1/2}$.

Ex 5.9.5   Show that if $\Omega_r < 1$, then $a(t) \to (1 - \Omega_r)^{1/2} H_0 t$ for $t \gg 2 H_0^{-1} \Omega_r^{1/2} /$
  $(1 - \Omega_r)$.

Ex 5.9.6   Discuss why for very large times $a(t) \propto t^{1/2}$ if $\Omega_r = 1$, while $a(t) \propto t$ if
  $\Omega_r < 1$.

5.10 Light Propagation in Newton's World

Despite our earlier admonishments that we should not consider light propagation in Newtonian cosmological models, we shall do just that using a trick from Bondi (2009, Section
9.5) that exploits relativistic hindsight. This will provide us with a relativistically correct
Newtonian analogy to the notion of 'proper distance' (Equation 5.72) and gives the relativistically correct redshift formula (Equation 5.73). With that we can enter into the subject
of 'horizons' in Newtonian cosmological models.

We shall consider a radial pulse of light received from an emitter located at a (Newtonian) distance $r(t)$ that is receding from us with the Hubble expansion at velocity Hr, where
H is the expansion rate $H(t) = \dot{a}/a$. Here, $a(t)$ is the cosmic scale factor with current value
$a(t_O) = 1$.

Recall that for a co-expanding point the value of $x = r(t)/a(t)$ is a constant, independent of time. The *comoving coordinate* x is a unique fixed label for each of the
particles along a line of sight. In the Newtonian sense the value of x is the distance of the particle when $a(t) = 1$, i.e. the ruler-measured distance to the particle
at the current epoch. In relativistic models this is called the *proper distance* of the
particle.

5.10.1 Light Pulses in a Newtonian Framework

Using the Newtonian law for addition of velocities, the equation of motion of the light
towards us is

$$\frac{dr}{dt} = \frac{\dot{a}}{a} r \pm c \qquad \Rightarrow \qquad a \frac{d}{dt}\left(\frac{r}{a}\right) = \pm c, \tag{5.71}$$

where c is the constant velocity of light.[32] The sign of c depends on whether the direction
of the light is towards $(-)$ or away from $(+)$ the observer.

Suppose the emitter sent out the light pulse at time t_E when at a distance r_E and that we
received it at t_O. Integration of (5.71) then gives:

---

[32] This is, in effect, a ballistic model for the propagation of light. As Bondi (2009, Section 9.5) remarks, the
cosmological interpretation of this equation would require a luminiferous æther for the propagation of light:
a 19th century concept which, following the Michelson–Morley experiment, was rejected in favour of special
relativity.

Light cone equation

$$x_E = \frac{r_E}{a(t_E)} = \pm c \int_{t_E}^{t_O} \frac{d\tau}{a(\tau)}.$$ (5.72)

The $+$ sign defines the backward light cone (i.e. coming towards the observer).
The $-$ sign defines the forward light cone (i.e. emitted by the observer).

Given $a(t)$ this relates the time of reception of the light pulse, t_O, to the distance of the emitter, r_E, at the time the pulse was emitted. Put another way: x_E labels the particle whose light was emitted at t_E and received at t_O. This equation is derived within the framework of general relativity in Section 18.2.3.

Now imagine that the emitter sends out a second pulse at time $t_E + \Delta t_E$ which is received by the observer at time $t_O + \Delta t_O$, where the Δts are very short time intervals. Equation (5.72) then tells us[33] that

$$\frac{\Delta t_O}{\Delta t_E} = \frac{a(t_O)}{a(t_E)}.$$ (5.73)

If Δt_E is the local time interval between the successive crests of an emitted light wave, then Δt_O is the received time interval. This is just the standard redshift equation.

So, with Bondi's trick, we are able to 'derive' the relativistically correct Equations (5.72) and (5.73). This works because in general relativity it is possible to choose to work in a coordinate system where Equation (5.71) is correct (*cf.* Equation 16.21). In that coordinate system the radial coordinate is referred to as the *(Newtonian) proper distance*.[34] The integral in (5.72) is a measure of time called the *conformal time*, and is a measure of the light travel time (see Equation 16.8).

Ex 5.10.1   Discuss why Equation (5.71) should not be adopted as an argument in support of the notion that Newtonian cosmology can replace GR models.

For the relativistic approach see Sections 16.1.4, 16.1.6 and 16.1.7.

5.10.2 Event and Particle Horizons

We distinguish two different horizons: the *particle horizon* and the *event horizon*. The event horizon separates events we are able to see at some time in our history from events that we never see. The particle horizon is the proper distance to the most distant object we can observe, and so divides the Universe into particles an observer has already seen from those that have not yet been seen. The de Sitter Universe (and the Steady State model) have an event horizon, but no particle horizon.

---

[33] Differentiate (5.72) using Leibniz's rule, or simply note that the change in the value of $I = \int_{x_1}^{x_2} f(u)du$ is $\Delta I = f(x_2)\Delta x_2 - f(x_1)\Delta x_1$. Note also that the value of $r_E/a(t_E) = x_E$ is a constant for a co-expanding emitter.

[34] The proper distance is not the light travel time distance from the object (the emitter) to the observer.

Particle Horizons

Consider the simple case of an Einstein de Sitter Universe, where $a(t) = (t/t_{now})^{2/3}$, t_{now} being the present time. If we evaluate Equation (5.72) for $t_E = 0$ and $t_O = t_{now}$, i.e. from the Big Bang at $t = 0$ to the present time we get

$$x_E(0) = \frac{r_E}{a(t_E)} \overset{\text{EdS}}{=} ct_{now} \int_0^{t_{now}} u^{-2/3} du = 3ct_{now} = 2cH_0^{-1}. \tag{5.74}$$

Since $t_E \geq 0$, this is the maximum value, for the Einstein de Sitter model, that the integral (5.72) can take. In accordance with the previous discussion the particle with this value of the label x_E is the furthest particle from which we could have received a light signal at the present time. This represents a *horizon* beyond which we cannot receive a light signal at the present time, see Figure 5.2. We give this distance a special symbol, $D_p(t)$: the proper distance to the *particle horizon* at time t:

Particle horizon

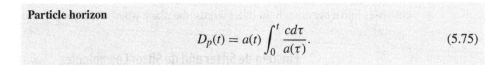

$$D_p(t) = a(t) \int_0^t \frac{c d\tau}{a(\tau)}. \tag{5.75}$$

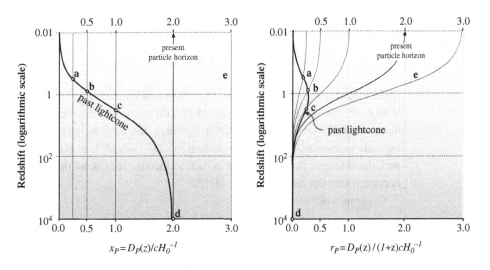

$$x_P = D_P(z)/cH_0^{-1} \qquad\qquad r_P = D_P(z)/(1+z)cH_0^{-1}$$

Fig. 5.2 Pseudo-Newtonian view of particle horizon in an Einstein de Sitter Universe.
The left panel is a sketch of our past light cone (Equation 5.72). The horizontal axis is the ruler (proper) distance to a particle measured in units of the current horizon distance, $2cH_0^{-1}$. The world lines of particles (galaxies) **a, b, c, d, e** are each assigned a fixed value of the coordinate label x as per Equation (5.72). We currently see these galaxies where their world lines intersect the light cone (points marked by dots). World line **d** is the farthest particle we can see, we do not yet see **e**.
The right panel depicts this in expanding coordinates $r = a(t)x = x/(1 + z)$. ©

This is the proper distance to the most distant particle an observer at time t can see: the *particle horizon*.[35] It therefore represents the scale of the volume of the observable material at time t. Note, however, that the most distant material in this volume is observed as it was in the past when the received light was emitted. The light travel time distance from that point is ct, which is smaller than the horizon distance (5.74).[36]

Event Horizons

The *event horizon* describes the boundary of all the parts of the Universe that we will never see and so this is just the opposite of (5.75):

Event horizon at time t:

$$D_e(t) = a(t) \int_t^{t_{\max}} \frac{c d\tau}{a(\tau)}, \tag{5.76}$$

where $t_{\max}$ is either the far infinite future ($t_{\max} = \infty$) or the time at which the Universe collapses into a big crunch. In other words, the place where the light rays stop propagating.

Einstein de Sitter and de Sitter Cosmologies

We can look at two familiar models: the Einstein de Sitter model in which $a(t) \propto t^{2/3}$ and the de Sitter Universe in which $a(t) \propto e^{Ht}$. In the Einstein de Sitter model the proper distance of a particle at time t, or redshift z, is given by the integral (5.74):

$$D_p(z) = 2cH_0^{-1} \left[1 - (1+z)^{-1/2} \right]. \tag{5.77}$$

Evaluating this at $z = \infty$ gives the current particle horizon: $D_p^{\text{EdS}} = 3ct_0 = 2cH_0^{-1}$. There is no event horizon since the integral (5.76) diverges: the observer eventually sees everything in the Universe. In the de Sitter Universe, the particle horizon $D_p^{\text{dS}}(t) \to cH^{-1}e^{Ht}$ expands exponentially, but the event horizon $D_e^{\text{dS}} = cH^{-1} = $ const.: it does not expand with the Universe. So the observer in such a Universe sees, at all times, only the events that happen within the constant radius cH^{-1}.

To summarise:

model	particle horizon	event horizon	
Einstein de Sitter:	$D_p^{\text{EdS}} = 3ct = 2cH(t)^{-1}$	$D_e^{\text{EdS}} = \infty$	(5.78)
de Sitter:	$D_p^{\text{dS}} = cH^{-1}e^{Ht}$	$D_e^{\text{dS}} = cH^{-1}.$	

The horizon structure was an important feature of the steady state theory where the observed Universe had a constant horizon, as required by a theory such as this, while the

---

[35] Harrison (1991) comments that the particle horizon of an observer at some instant of time separates the world lines into two classes: those which will intersect the observer's past light cone and those that do not. He also remarks that to refer to this as a *world line horizon* would be more appropriate.

[36] The proper distance and the light travel distance to the particle horizon are frequently confused.

galaxies expanded away to disappear from sight, over the horizon. It is also an important aspect of inflationary cosmology (see Sections 9.2.3 and 9.2.4).

5.10.3 Particle Horizon When $p = w\rho$

It is possible to handle more complex models, but the integrals become somewhat intractable. We generally know the evolution of the scale factor, $\dot{a}$ in terms of the redshift, $a(z)$ and so it is convenient to rewrite (5.75) in terms of the scale factor $da/\dot{a} = dt$,

$$D_p = a(t) \int_0^{a(t)} \frac{da}{a\dot{a}}. \tag{5.79}$$

As a specific example we can look at a model having current value of the density parameter Ω_0 pervaded only by a gas having equation of state $p = w\rho$. The rate of expansion for this model is derived in Section 6.4 (see Equation 6.83 with $(1 + z) = a^{-1}$):

$$\dot{a}^2 = H_0^2 \left[(1 - \Omega_0) + \Omega_0 a^{-(1+3w)} \right]. \tag{5.80}$$

Here we have put $a_0 = 1$, and H_0 is the current expansion rate. The particle horizon at time t corresponding to a redshift z is then:

Particle horizon for a single component with equation of state $p = w\rho c^2$:

$$D_p(z) = \frac{1}{1+z} cH_0^{-1} \int_0^{a(z)} \frac{da}{\left[(1 - \Omega_0)a^2 + \Omega_0 a^{1-3w} \right]^{1/2}}. \quad w > -1/3. \tag{5.81}$$

The limit on $w > -1/3$ arises because otherwise the integral is divergent: the particle horizon is not defined for $w \le -1/3$.

The integral has some obvious simple cases for a flat model, $\Omega_0 = 1$. For a flat model containing pressure free dust, $w = 0$, we have

$$D_p(z) \stackrel{\text{flat dust}}{=} 2cH_0^{-1}(1 + z)^{-3/2}, \tag{5.82}$$

and as $z \to 0$ we recover the result we had before that the particle horizon in this flat dust case is $2cH_0^{-1}$. For a generic flat model having equation of state $p = w\rho$, we have from (5.81):

$$D_p \stackrel{\text{flat}}{\propto} a^{\frac{3}{2}(1+w)}. \tag{5.83}$$

This also follows simply from $H^2 \propto \rho \propto a^{-(1+3w)}$.

If we put $w = -1/3$ in this last expression we see that $D_p \propto \int_0^{a(z)} da/a$, which has a logarithmic divergence. In fact the integral diverges for all values $w < -1/3$ and so there is no particle horizon for these models.

Dark Energy Cosmological Models

The evidence that the cosmological constant, Λ, is non-zero is overwhelming: the cosmic expansion is accelerating. Whether or not this is due to Einstein's famous constant, or some variant of it, i.e. 'dark energy', it is the major constituent of the Universe. Currently, we know almost nothing about it. The dominance of Λ, or whatever it is, at redshifts $z < 1$ means that we can study it via its effect on the cosmic expansion. The Newtonian models provide a good framework for modelling the expansion at these recent times, though we have to remember that we cannot study the propagation of light in the Newtonian context without importing some relativistic concepts.

A particularly relevant model is the 'benchmark model', which is geometrically flat and contains both pressure-free matter and a cosmological constant. This model, and its generalisations to more exotic forms of dark energy, lie at the basis of the interpretation of modern cosmological data. We study it here via the Friedmann–Lemaître equations that describe the evolution of the Hubble expansion rate, $H(z)$. We then move onto generalising the simple constant Λ to simple models for redshift-dependent dark energy.

As in the previous chapter, the exercises here serve as diversions that fill in details.

6.1 The Accelerating Cosmic Expansion

Important Remark

The evidence for the existence of an agent that causes the acceleration of the cosmic expansion comes from several quite diverse sources of data: supernovae, the power spectrum of the CMB, the present day manifestation of primordial structures ('BAOs'), large scale cosmic structure, to name but a few. Each of these pieces of evidence has received ample confirmation from repeated experiments by independent groups. These sources of evidence agree not only as to the existence of this agency, but also to its magnitude. Even though we have little or no idea what this agency is, its discovery surely stands as one of the great scientific milestones of the past century.

Einstein presented a version of his famous field equations containing an additional constant of nature, the cosmological constant, Λ. Einstein's Λ was independent of both spatial position and time and has had a rather chequered and controversial history. It is now widely accepted that Λ, or whatever else it might be, is non-zero and has the effect of causing an acceleration of the cosmic expansion. If the acceleration of the expansion is due to a simple Einstein Λ-term, then $\Lambda > 0$. We shall have more to say about that and related dark energy models later on.

The consequence of introducing this ad hoc term into the equations can be seen by studying the dynamical equations with the Λ-term. The main practical problem with even the simplest $\Lambda \neq 0$ model containing matter is that the differential equations describing the models can no longer be integrated in a simple closed form[1] except in a few special cases.

The simplest Newtonian model having zero pressure matter of density ρ_m and a cosmological constant Λ, is described by the equation

$$\frac{1}{a}\frac{d^2 a}{dt^2} = -\frac{4\pi G \rho_m}{3} + \frac{\Lambda}{3}. \tag{6.1}$$

This is the generalization of Equation (5.13) and corresponds to the Friedmann–Lemaître equation derived from the Einstein equations. We see immediately that a positive Λ-term increases the acceleration of the expansion, and counters the effect of the first term. There is now strong evidence from a variety of sources that Λ is not zero, and is positive: we can detect this extra acceleration.

The solutions of (6.1) are generally referred to as 'FLRW models'. The botany of FLRW models with $\Lambda \neq 0$ is considerable: quite apart from the diverse nature of the ρ-matter, there are separate cases to consider for $k = 0, k > 0, k < 0$ and for $\Lambda = 0, \Lambda < 0, \Lambda > 0$. Among all the combinations of these cases we can distinguish six classes of solution. There are I: static solutions, II: monotonically expanding solutions that started their expansion from a state of infinite density a finite time in the past, III: solutions that start their expansion from a finite radius an infinite time in the past, IV: solutions that start a finite time in the past and expand to reach a finite radius, V: solutions that oscillate between a singular state and a finite radius, and VI: solutions that collapse to a finite density and 'bounce' to expand to infinity. The classes are briefly summarised in Bondi (2009, Section 9.4).[2]

Most of these solutions are now mainly of didactic interest: the data indicates that we belong to Class II. We can now measure the curvature contribution, k and the value of Λ. k turns out to be close to zero and $\Lambda \sim 0.7$. This means that the dynamical manifestations of Λ are restricted to recent times: matter wins out at high redshift. The good news is that, as far as the directly observable Universe is concerned, this eliminates most of the FLRW solution space. However, what happened at the earliest times is still open to debate.

---

[1] There are closed form analytic solutions involving elliptic integrals, but these are not particularly edifying. See Kharbediya (1976, 1983). The scale factor has an analytic expression in a few important cases, but calculations using them more often than not have to be done numerically.

[2] The wealth of cosmological models that can arise through simply introducing Λ is discussed in the great ancient books by Tolman (1934b) and Bondi (2009, first published in 1952). Ryden (2003) gives an excellent and well-organised overview of the Λ-cosmology zoo.

6.1.1 The Emergence of Λ Models

Because of the importance of the early work of Einstein and of de Sitter, the cosmological constant was retained in formal developments and so appears in the later work of Friedmann and Lemaître. However, it was dropped from consideration in observational papers because there was little or no prospect of determining it. It was to resurface with the data on distant type Ia supernovae: their Hubble diagram seemed to reveal a curvature indicating that the expansion was accelerating. The obvious candidate was the cosmological constant. Introducing a non-zero cosmological constant, despite its somewhat shaky basis in physics, seemed to solve a number of other problems associated with large scale structure, and it was also indicated by CMB maps of the sky.

The cosmological constant is seen by physicists as having two problems for which there does not seem to be any reasonable answer:

- The energy density associated with the cosmological constant, Λ, is constant with time, whereas the energy density of the matter component has been decreasing from an infinitely dense state. It seems like a remarkable coincidence that today these two energy densities are almost the same. Of course, coincidences do happen.
- The fact that, by definition, the cosmological constant is unchanging suggests that it is a property of space-time itself. If so we should be able to calculate it from quantum field theory as a vacuum energy density. Doing this provides estimates for Λ that are some 120 orders of magnitude too big. Of course, we could be doing the wrong calculation.

While these two points could be debated, it seems like a good idea to explore other avenues. A frequently exploited idea is that the Universe is pervaded by material having negative pressure: this material is referred to as *dark energy*. There are numerous suggestions as to what dark energy might be.

6.1.2 Cosmology with a Cosmological Constant

We can write down the first integral of Equation (6.1). To do that we need an equation describing the matter component ρ_m. For pressure free matter this is just Equations (5.14), and (5.15), which for convenience we repeat here:

$$\frac{d\rho_m}{dt} + 3\frac{\dot{a}}{a}\rho_m = 0, \quad \rho_m(a) = \rho_{m0}a^{-3}. \tag{6.2}$$

These express the conservation of matter. Putting this simple ρ_m into Equation (6.1) and integrating we get the analogue of Equation (5.17) for the expansion rate, $\dot{a}/a$, with a time independent Λ-term:

$$\left(\frac{\dot{a}}{a}\right)^2 = \frac{8\pi G\rho_m(t)}{3} - \frac{k}{a^2} + \frac{\Lambda}{3}. \tag{6.3}$$

The constant of integration k is, as before, referred to as the curvature constant. The scale factor $a(t)$ is dimensionless and by convention we take its current value to be $a_0 = 1$.

As before we write the expansion rate as

$$H(t) = \frac{\dot{a}}{a}. \tag{6.4}$$

The time dependence of the expansion rate can be written as a function of z, a or t. The present value of this is the Hubble constant H_0.

The expansion rate is the primary cosmological parameter at any epoch. We use it to define a fiducial density:

$$\rho_{\text{crit}} = \frac{3H(t)^2}{8\pi G}. \tag{6.5}$$

This is sometimes referred to as the *Einstein de Sitter critical density*. In the following the densities of the various components that make up the cosmic medium will be non-dimensionalised relative to this density to define the *cosmological density parameters*.[3]

The models we consider here have three constituents: pressureless matter ('dust'), a cosmological constant and spatial curvature. We introduce three Ω-parameters: Ω_m, Ω_Λ and Ω_k, whose values at the present time and at arbitrary value of the scale factor a (corresponding to redshift $z : a = 1/(1 + z)$) are defined as:

$$\Omega_{m0} = \frac{8\pi G\rho_0}{3H_0^2}, \qquad \Omega_m(a) = \frac{8\pi G\rho(a)}{3H(a)^2}, \tag{6.6}$$

$$\Omega_\Lambda = \frac{\Lambda}{3H_0^2}, \qquad \Omega_\Lambda(a) = \frac{\Lambda}{3H(a)^2}, \tag{6.7}$$

$$\Omega_{k0} = -\frac{k}{H_0^2}, \qquad \Omega_k(a) = -\frac{k}{a^2 H(a)^2}. \tag{6.8}$$

$H(a)$ is the Hubble expansion rate at scale factor $a = (1 + z)^{-1}$ and H_0 is its value at the present time, $H(0)$. $\rho(a)$ is the density of the pressure-free matter (dust) and ρ_0 is the current value. In this model, Λ is taken to be a constant of nature. We will later consider models where Λ depends on redshift.

The terms on the right-hand sides of Equations (6.4)–(6.8) appear in Equation (6.3). Making these substitutions,

$$\Omega_m(a) + \Omega_k(a) + \Omega_\Lambda = 1 \tag{6.9}$$

for all values of a. This is just a rewrite of the Friedmann–Lemaître Equation (6.3). In particular, at the present epoch, $a = 1$, gives

$$\Omega_{m0} + \Omega_{k0} + \Omega_\Lambda = 1. \tag{6.10}$$

The Ωs are the fractional contributions of the cosmic constituents of the model to the total gravitating mass/energy density, and, in this sense, Ω_k, Ω_m and Ω_Λ describe the relative contribution to the expansion rate made by the three components (the curvature, the matter and the cosmological constant).

---

[3] We previously did this in Section 5.5.3 when we introduced the density parameter Ω for a single component cosmology. When describing multi-component models it is convenient to introduce a number of such density parameters.

Equation (6.10) shows us how to determine the present-day value of the curvature constant k from the observationally determined values of Ω_m and Ω_Λ:

$$\Omega_{k0} = 1 - \Omega_{m0} - \Omega_\Lambda. \tag{6.11}$$

The Ω-parameters are not independent of one another.

Ex 6.1.1 The quantity k appeared as a constant of integration. Using (6.8), show that this last Equation, (6.11), is equivalent to

$$k = H_0^2(\Omega_{m0} - 1) + \frac{\Lambda}{3}. \tag{6.12}$$

Determining k from the Hubble diagram was seen as the prime goal of cosmology in the 1950s and 1960s (Sandage, 1970). Since then these parameters have been determined with considerable accuracy using a variety of methods. This will be the subject of much of the discussion of the subsequent chapters.

Using $\rho_m(a) = \rho_{m0}a^{-3}$ expresses (6.9) in Hs and Ωs. We get

$$H(a)^2 = H_0^2\left[\Omega_{m0}a^{-3} + \Omega_{k0}a^{-2} + \Omega_\Lambda\right]. \tag{6.13}$$

This simple yet important equation forms the basis of much of present day observational cosmology.

For didactic purposes we will define a *benchmark model*[4] to have parameter values as follows:

Benchmark model parameters

$$\Omega_k = 0, \quad \Omega_m = 0.3, \quad \Omega_\Lambda = 0.7, \quad H_0 = 70\,\text{km s}^{-1}\,\text{Mpc}^{-1}. \tag{6.14}$$

See Table 6.1. These are the definitions adopted by Ryden (2003, her Table 6.2, p.67).

Some derived quantities using the ratio Ω_Λ/Ω_m are moderately sensitive to the second decimal place of Ω_Λ and Ω_m.

We generally eliminate k and Ω_k from equations, expressing them directly in terms of Ω_m and Ω_Λ. Thus Equation (6.3) takes on the simple-looking forms

$$H(a)^2 = H_0^2\left[\Omega_{m0}a^{-3} + (1 - \Omega_{m0} - \Omega_\Lambda)a^{-2} + \Omega_\Lambda\right], \tag{6.15}$$

$$H(z)^2 = H_0^2\left[\Omega_{m0}(1+z)^3 + (1 - \Omega_{m0} - \Omega_\Lambda)(1+z)^2 + \Omega_\Lambda\right], \tag{6.16}$$

where Ω_{m0} has been used in order to emphasise that we are using the present-day value of Ω_m. Equations (6.15, 6.16) are the $\Lambda \neq 0$ analogue of Equations (5.27, 5.28).

[4] Kiang (2003) refers to this, not inappropriately, as the *30/70 model*.

Table 6.1 Density parameters (from Planck/WMAP).

Parameter	value[a]	Name	Benchmark[b]
Ω_m	0.315	density of non-relativistic matter	0.3
Ω_B	$0.022h^{-2}$	density of baryonic matter	0.045
Ω_c	$0.119h^{-2}$	density of dark matter	0.255
Ω_γ	4.7×10^{-5}	photon radiation density	5.03×10^{-5}
Ω_ν	3.2×10^{-5}	neutrino radiation density	3.47×10^{-5}
Ω_Λ	0.69	cosmological constant parameter	0.7
Ω_k	0	curvature parameter	0

[a] $h = H_0/100\,\mathrm{km s^{-1} Mpc^{-1}}$, with $h = 0.67$.
[b] Benchmark Hubble constant $= 70$ km s^{-1} Mpc^{-1}, i.e. $h = 0.70$.

6.1.3 Ωs in multi-component models

We now make a small diversion into the subject of multi-component models. We can clearly keep adding cosmic constituents to the Ωs defined above (Equations (6.6–6.8)) provided we understand how their densities vary as a function of the scale factor $a(t)$. We can, for example, add in Ω_r for the cosmic radiation and Ω_ν for neutrinos. For massive neutrinos, the scaling of Ω_ν depends on whether the neutrinos are relativistic. In addition, we can split the matter density parameter Ω_m into two constituents: dark matter and baryonic matter, $\Omega_m = \Omega_c + \Omega_b$. Equations (6.11) and (6.15) are modified simply by adding in the Ωs and their scaling factors.

The usual constituents are listed in Table 6.1. In this section we shall distinguish between the material components, which will be labelled by an index i, from the curvature which will be denoted, as usual, by subscript k. The cosmological constant Λ is, for this purpose, a material contributor.

If at a time corresponding to an expansion factor $a = 1/(1 + z)$, a material component of the Universe had density ρ_i, we can define the non-dimensional density parameter for that component as

$$\Omega_i(a) = \frac{\rho_i(a)}{\rho_{crit}(a)} = \frac{8\pi G \rho_i(a)}{3H(a)^2}. \tag{6.17}$$

If in Equation (6.3) we replace ρ_m with $\sum \rho_i$, we can divide through by the critical density to get

$$1 = \sum \Omega_i(a) + \Omega_k, \tag{6.18}$$

(where the sum includes the cosmological constant), which is the same as

$$\sum \Omega_i(a) = 1 - \Omega_k. \tag{6.19}$$

For the special *flat Universe* case, $k = 0$, we get

$$\sum \Omega_i(a) \stackrel{\text{flat}}{=} 1, \qquad k = 0. \tag{6.20}$$

Now we return to the simple case of models with only pressure-free matter, Λ and curvature.

6.1.4 Redshift Evolution of the Ω Parameters

If we combine both equations in (6.6) we get $H_0^2/H(a)^2 = (\rho_0/\rho(a))(\Omega_m(a)/\Omega_{m0})$, and since $\rho(a) = \rho_0 a^{-3}$ we have

$$\frac{H_0^2}{H(a)^2} = a^3 \frac{\Omega_m(a)}{\Omega_{m0}}. \tag{6.21}$$

Substituting this into (6.15) then gives the history of the matter density parameter, $\Omega_m(a)$:

$$\Omega_m(a) = \frac{\Omega_{m0}}{\Omega_{m0} + (1 - \Omega_{m0} - \Omega_\Lambda)a + \Omega_\Lambda a^3}. \tag{6.22}$$

We can do the same thing with the curvature contribution, $\Omega_k(a)$ using Equations (6.8). This gives

$$\frac{H_0^2}{H(a)^2} - a^2 \frac{\Omega_k(a)}{\Omega_{k0}}, \tag{6.23}$$

and then substituting into (6.15) gives

$$\Omega_k(a) = \frac{\Omega_{k0}}{\Omega_{m0}a^{-1} + \Omega_{k0} + \Omega_\Lambda a^2}. \tag{6.24}$$

As $a \to 0$, the contribution of the curvature to the cosmic expansion tends to zero. Thus FLRW models containing pressure-free matter and a constant Λ originate having zero curvature $\Omega_k(a) \sim (\Omega_{k0}/\Omega_{m0})a$: they are, in effect, flat. Since today $\Omega_{k0} \ll \Omega_{m0}$, the early-time values of Ω_k would have had to be finely tuned to a value very close to zero. This is the essence of the *fine tuning problem*.

6.1.5 Time, Redshift and Expansion Rate

The epoch of the cosmic expansion is characterised in numerous ways: the redshift, z, the expansion factor $a(t)$ relative to the present, the time t since the Big Bang, and the look-back time to an object in the past, t_L. We frequently need to make transformations between these. First we recall some definitions,[5] starting with the redshift, z, and the Hubble expansion rate, $H(t)$, at some time t:

$$1 + z = \frac{a(0)}{a(t)}, \tag{6.25}$$

$$H(t) = \frac{\dot{a}(t)}{a(t)}. \tag{6.26}$$

We normalise so that $a(0) = 1$, and then $a(z) = (1 + z)^{-1}$. The time when the expansion factor $a(t)$ takes a particular value can be found by writing (6.26) in the form $da = Hadt$ and integrating:

$$t(a) = \int_a^1 \frac{da}{aH(a)}, \tag{6.27}$$

[5] Here we follow Linder (2003) and particularly Barnes *et al.* (2005).

$$t(z) = \int_0^z \frac{dz}{(1+z)H(z)}. \tag{6.28}$$

We have taken the current value of the expansion factor to be $a_0 = 1$. We now need to get $H(z)$: that requires the Einstein equations and knowledge of what the material constituents of the Universe are.

> **Ex 6.1.2** Using Equations (6.25) and (6.26), show that
>
> $$\frac{dz}{dt} = -(1+z)\,H(z). \tag{6.29}$$
>
> This little equation is very useful when changing from rates of change with respect to time to rates of change with respect to redshift.

Equations (6.25)–(6.29) are generic to any cosmological model: they simply use the definitions of the various quantities involved.

6.1.6 The Deceleration and Density Parameters, σ and q

For completeness we also go back to the classical descriptors of our $\Lambda \neq 0$ models in terms of the density parameter σ and the deceleration parameter, q, which were the subject of the famous 'search for two numbers' that drove early observational cosmology (see Section 5.6.2).

Relating q_0 and σ_0 to Ω_m and Ω_Λ

Early, pre-1980s, cosmology was concerned with describing the present day Universe. The older literature did not use the Ωs, but instead used two parameters: q_0, *the deceleration parameter*, and σ_0, the *density parameter*:

$$\sigma_0 = \frac{4\pi G \rho_0}{3H_0^2}, \quad q_0 = -\frac{1}{H_0^2}\frac{\ddot{a}_0}{a_0}. \tag{6.30}$$

The zero subscript indicates that they are evaluated at the present day. In this notation the cosmological constant and the curvature constant are

$$\Lambda = 3H_0^2(\sigma_0 - q_0), \tag{6.31}$$

$$k = H_0^2(3\sigma_0 - q_0 - 1), \tag{6.32}$$

and so

$$\Omega_\Lambda = \Lambda/3H_0^2 = \sigma_0 - q_0 \quad \Omega_0 = 2\sigma_0. \tag{6.33}$$

The move to using Ωs probably started in the Russian cosmology literature of the 1960s.

Deceleration Parameter

Recalling that the deceleration parameter q at any time t is defined as

$$q(t) = -\frac{a\ddot{a}}{\dot{a}^2}, \tag{6.34}$$

we can show that the rate of change of the Hubble expansion rate, H, is

$$\dot{H}(t) = -(1 + q)H(t)^2. \tag{6.35}$$

This relationship is merely a redefinition of the deceleration parameter and so holds true in all Friedmann–Lemaître models. If we know q as a function of t or z this equation can be integrated.

It should be noted that the relationship (6.35) does not contain any dynamical information: it all comes from the definition of the quantities H and q in terms of the expansion factor $a(t)$.

Ex 6.1.3   Starting with Equation (6.26), derive the relationship (6.35).

Ex 6.1.4   Rewrite Equation (6.35) using redshift z in place of the time t. You will need to use Equation (6.25) to do that.

Ex 6.1.5   Show that if q is given as a function of redshift, $q = q(z)$, the Hubble expansion rate at redshift z is given by

$$H(z) = H_0 \exp \int_0^z \frac{1 + q(u)}{1 + u} \, du. \tag{6.36}$$

Ex 6.1.6   Taking as a model

$$q(z) = \frac{1}{2} - \frac{a}{(1 + z)^b}, \tag{6.37}$$

show that

$$H(z) = H_0(1 + z)^{3/2} \exp\left[(a(1 + z)^{-b} - 1)/b\right]. \tag{6.38}$$

This equation was used by Xu *et al.* (2007) to determine, from the data on type Ia supernovae, that the Universe made the transition to accelerated expansion at around $z_q \simeq 0.35$.

The following exercises extend the previous relationships to all redshifts and are generic to all models:

Ex 6.1.7   Show that Equation (6.1) can be written in terms of the σ and q parameters (Equations (5.34) and (5.35)) as

$$\Lambda = 3H^2(\sigma - q). \tag{6.39}$$

Ex 6.1.8   Show that

$$q = \frac{1}{2}\Omega_m - \Omega_\Lambda, \tag{6.40}$$

$$\Omega_k = 1 + q - 3\sigma. \tag{6.41}$$

So a flat Universe has $3\sigma = 1 + q$.

The Steady State Universe effectively has zero density and so $q_{Steady\ State} = -1$.[6] Sandage (1970) was able to give a value of $q_0 = 1 \pm 1$ when fitting the Hubble diagram with a $\Lambda = 0$ model, while by 1978 Kristian et al. (1978) were quoting a value of $q_0 = 1.7 \pm 0.4$. However, these determinations of q_0 took no account of galaxy evolution as Gunn and Tinsley (1975) had advocated and as had been done by Gunn and Oke (1975) when they gave q_0 in the range $q_0 = +0.33$ to $q_0 = -1.27$.

6.1.7 Age at Redshift z

As before, we can formally write the solution of (6.15) expressing the time as a function of the expansion parameter a as[7]

$$t(a) = H_0 \int_0^a \frac{da}{a\sqrt{\Omega_m a^{-3} + \Omega_k a^{-2} + \Omega_\Lambda}}. \qquad (6.42)$$

Edwards (1972) and Heath (1989) give closed form solutions of this equation in terms of Jacobian elliptic functions.

The *age of the Universe* at the present time is the integral up to $a = 1$:

$$t_0 = \int_0^1 \frac{da}{a\sqrt{\Omega_m a^{-3} + \Omega_k a^{-2} + \Omega_\Lambda}}. \qquad (6.43)$$

Peacock (1999, Section 5.1) gives a useful approximation for the age which works to within a few percent for $0.1 < \Omega_m < 1$ and $|\Omega_\Lambda| < 1$:

$$t_0 \simeq \frac{2}{3} H_0^{-1} [0.7\Omega_m + 0.3 - 0.3\Omega_\Lambda]^{-0.3}. \qquad (6.44)$$

In these last Equations (6.42) and (6.43), the time is expressed as a function of the scale factor. It would be more useful to express this in terms of the time to a particular redshift z corresponding to that value of a. Making the transformation $a = (1 + z)^{-1}$ we can write an equation for the age of the Universe at a redshift z:

$$t(z) = H_0^{-1} \int_z^\infty \frac{dz}{(1 + z)\sqrt{(1 + z)^3 \Omega_{m0} + (1 - \Omega_{m0} - \Omega_\Lambda)(1 + z)^2 + \Omega_\Lambda}}, \qquad (6.45)$$

where we have eliminated Ω_k in favour of Ω_{m0} and Ω_Λ using Equation (6.11).[8]

---

[6] For the Steady State model the Hubble parameter is independent of time and so $\dot{a}/a = H_0$ which gives a scale factor $a(t) \propto e^{Ht}$. A direct calculation of q from the definition (6.34) then gives $q = -1$. We note that for the flat benchmark model having $\Omega_m = 0.3, \Omega_\Lambda = 0.7$, Equation (6.40) gives a value $q_0 = -0.55$.

[7] Remark on notation: it would be usual to give the as under the integral another symbol such as a', or even x. However, to directly indicate that we are integrating over a scale factor by using the variable a gains more than would be lost by obfuscating the notation.

[8] This is sometimes written as:

$$t(z) = H_0^{-1} \int_z^\infty \frac{dz}{(1 + z)\sqrt{(1 + z)^2(1 + \Omega_{m0}z) - z(2 + z)\Omega_\Lambda}}. \qquad (6.46)$$

Solutions for this form of the age equation in terms of computationally advantageous incomplete Legendre elliptic integrals have been discussed by Thomas and Kantowski (2000).

Ex 6.1.9 Derive Equation (6.46) from (6.43).

Ex 6.1.10 Show that for a flat model with $\Omega_m = 0$, the solution of (6.46) is

$$t = \frac{1}{H_0 \Omega_\Lambda^{1/2}} \sinh^{-1} \left[\frac{1}{(1+z)\sqrt{\Omega_\Lambda^{-1} - 1}} \right].$$

6.1.8 Look-back Time to Redshift z

The *look-back time* to an observed object is the difference between the age of the Universe now and the age at the time when the photons we are receiving were emitted. This is simply the difference between the ages at two different redshifts or scale factors:

$$t_L(a) = \int_a^1 \frac{da}{a\sqrt{\Omega_m a^{-3} + \Omega_k a^{-2} + \Omega_\Lambda}}. \tag{6.47}$$

The look-back time is particularly important when fitting models of galaxy evolution.

Usefully, in a flat model $\Omega_k = 0$ where $\Omega_m + \Omega_\Lambda = 1$ the integral can be evaluated in closed form:

$$t_L(a) \stackrel{k=0}{=} \frac{2H_0^{-1}}{3\sqrt{1 - \Omega_{m0}}} \ln \left\{ \left(\frac{a}{a_{m\Lambda}} \right)^{3/2} + \sqrt{1 + \left(\frac{a}{a_{m\Lambda}} \right)^3} \right\}, \quad a_{m\Lambda} = \left(\frac{\Omega_{m0}}{\Omega_\Lambda} \right)^{1/3}.$$
$$\tag{6.48}$$

Here, Ω_{m0} is the present density of the Universe.[9] Figure 6.1 shows the function $a(t_L)$ for a variety of flat $k = 0$ models for various values of $\Omega_\Lambda - 1 - \Omega_m$. Some supernova data is superposed for guidance: that has been renormalised to a Hubble constant of $70\,\mathrm{km s^{-1} Mpc^{-1}}$. We see that the curves for various Λ values start to diverge beyond $a \sim 0.75$, i.e. $z \sim 0.33$ and that the separation is clear by $z \sim 1$.

6.2 The Flat Benchmark Model

6.2.1 Recent Times: Λ and Dark Matter

The flat Λ-models are of considerable interest since they broadly reflect the data on the cosmic expansion. For these models $\Omega_m + \Omega_\Lambda = 1$ (see Equations (6.10)) and the scale factor $a(t)$ can be expressed in terms of familiar functions. For the flat $k = 0$ model Equation (6.15) simplifies to

---

[9] This form is from Ryden (2003, Section 6.2, Eq. 6.28). This equation is frequently given in a form using hyperbolic functions rather than the logarithm shown here, as for example in Waga (1993, Section 3.1.2 Eq. 31, the required transformation is $\mathrm{arcsinh}\, x = \ln\{x + (1 + x^2)^{1/2}\}$).

$$\frac{1}{H_0^2}\left(\frac{\dot{a}}{a}\right)^2 \stackrel{\text{flat}}{=} \frac{\Omega_{m0}}{a^3} + \Omega_\Lambda, \quad \text{for } k = 0 \text{ and } \Omega_\Lambda = \frac{\Lambda}{3H_0^2}. \tag{6.49}$$

This is an important equation since there is evidence that the geometry of the Universe is in fact flat. Moreover, this equation can be solved in closed form. We return to the flat models in Section 6.2.

Like their flat $\Lambda = 0$ counterpart, the Einstein de Sitter Universe, analytic results are available for many observation related relationships.

There are three cases to consider according as to whether $\Lambda < 0$, $\Lambda > 0$ or $\Lambda = 0$:

$$a(t) = \left(\frac{\Omega_m}{\Omega_\Lambda}\right)^{1/3} \sinh^{2/3}\left(\Omega_\Lambda^{1/2}\frac{3}{2}H_0t\right), \qquad \Lambda > 0 \tag{6.50}$$

$$= \left(\frac{3}{2}H_0t\right)^{2/3}, \qquad \Lambda = 0 \tag{6.51}$$

$$= \left(\frac{\Omega_m}{|\Omega_\Lambda|}\right)^{1/3} \sin^{2/3}\left(|\Omega_\Lambda|^{1/2}\frac{3}{2}H_0t\right). \qquad \Lambda < 0 \tag{6.52}$$

The case of a flat $\Lambda = 0$ model is the Einstein de Sitter case since $\Omega_\Lambda = 0$ and $\Omega_k = 0$ entails $\Omega_m = 1$ by Equation (6.11). [10]

The surprise of the last decades has been that observation has suggested that Equation (6.50) might be a good first order description of the recent expansion with $\Omega_\Lambda/\Omega_m \sim 2.3$. We should note that the expansion described by (6.50) grows as $t^{2/3}$ for small values of $\frac{3}{2}\Omega_\Lambda^{1/2}H_0t$ and then goes into exponential growth. The critical redshift at which this happens is around the time the hyperbolic sine term gets to around 1, which corresponds to a redshift

$$z_\Lambda \sim \left(\frac{\Omega_\Lambda}{\Omega_m}\right)^{1/3} - 1 \stackrel{\text{benchmark}}{=} 0.326 \tag{6.54}$$

for the benchmark model. This is the redshift at which the Universe makes *the transition to Λ-dominated expansion*. Here z_Λ is a remarkably low redshift, well within the range of observations. It means that tracking the Hubble diagram out to a redshift of $z \sim 1$ should reveal the transition from power law expansion to exponential expansion.

Note that the usual way of determining z_Λ is to compare the terms on the right-hand side of the acceleration Equation (6.1), rather than to look at the actual solution as we have just done. Doing the comparison that way gives what might be called a dynamical estimate for the redshift of the transition to accelerated expansion:

$$z_{\Lambda,\text{dyn}} \sim \left(\frac{2\Omega_\Lambda}{\Omega_m}\right)^{1/3} - 1 \stackrel{\text{benchmark}}{=} 0.671 \tag{6.55}$$

---

[10] Bondi (2009, Section 9.5 p.82) gives an alternative form to (6.50):

$$a(t)^3 = \frac{1}{2}\left(\frac{\Omega_{m0}}{\Omega_\Lambda}\right)\left[\cosh(\sqrt{\Omega_\Lambda}\,3H_0t) - 1\right], \tag{6.53}$$

(this uses the half-angle formula $\sinh^2 x = \frac{1}{2}[\cosh(2x) - 1]$).

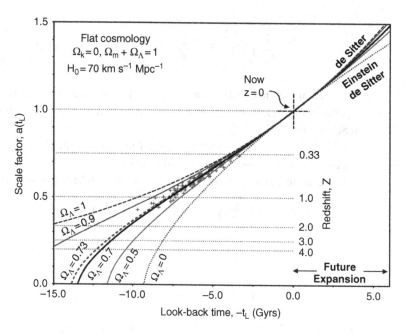

The history of the cosmic expansion for flat Universes with 'dust' and Ω_Λ in the range 0 to 1, i.e. $\Omega_m + \Omega_\Lambda = 1$. The horizontal axis shows the look-back time, t_L: the difference between the age of the Universe now and the age at the time the light we see was emitted (plotted as $-t_L$ so that time increases to the right). The vertical axis is the cosmic expansion factor, $a(t_L)$, at that time: this depends on the cosmological model in the manner shown in the curves. The present time, marked by a cross, is at $(t_L = 0, a(t_L) = 1)$ and the place where the model curves intersect the time axis $(a = 0)$ corresponds to the age of the Universe in that model. The limiting cases are the Einstein de Sitter Universe when $\Omega_\Lambda = 0$ and the de Sitter model when $\Omega_\Lambda = 1$. Lines of constant redshift $z = 1/(1 + a)$ are indicated.

The data points correspond to the supernova sample analysed by Davis *et al.* (2007) using selected data from Riess and Strolger (2007) and Wood-Vasey (2007) (with suitable renormalisation). ©©

for the Benchmark model, The difference in the formulae is the factor '2' in front of the Ω_Λ which comes from comparing $4\pi G\rho$ versus $8\pi G\rho$ with Λ in Equations (6.1) and (6.3).

This is in any case only a rough guide to what is happening. The message is clear: the turn-on of the cosmological constant falls well within the domain of observation. This is easily seen in the divergence of the curves away from the Einstein de Sitter expansion as shown in Figure 6.1.

The complication with using galaxies to trace the evolution of the cosmic expansion is that they evolve over this kind of redshift range. However, there appear to be far fewer problems in using supernovae as the cosmic candles for the Hubble diagram, and that is why it is the distant supernova programs that have revealed the existence of a significant Λ-term.

One might be concerned that it is a surprising coincidence that the Λ-term is just taking over at the current epoch of the expansion: the ratio $(\Omega_\Lambda/\Omega_m)$ just so happens to have a value that renders this transition to undecelerated expansion visible now. We shall discuss this issue later on when we come to other methods of inferring the value of Λ.

6.2.2 Early Times: Matter and Radiation

The generalisation of Equation (6.42) to include the radiation component of a flat Universe is clearly

$$t(a) \stackrel{\text{flat}}{=} H_0 \int_0^a \frac{da}{a\sqrt{\Omega_{r0}a^{-4} + \Omega_{m0}a^{-3} + \Omega_\Lambda}}. \tag{6.56}$$

This reflects the fact that the energy density of the radiation field varies as $\rho_r \propto a^{-4} \propto (1+z)^4$ from its present day value, $a = 1$, expressed in terms of Ω_{r0}. As far as we know the radiation component today consists of photons having a density ρ_γ and neutrinos having a density ρ_ν. Because of events taking place in the earliest seconds of the cosmic expansion we know that $\rho_r = \rho_\gamma + \rho_\nu = 1.691\rho_\gamma$, (see Section 7.4.1, Eq. 7.71). For the benchmark model we have $\Omega_{r0} = 8.5 \times 10^{-5}$.

If we consider that Λ is indeed a time-independent constant, then the Ω_Λ term in the integrand becomes negligible in comparison with the matter and radiation terms at high redshifts ($a \ll 1$). We also recognise that the radiation and matter densities will be equal at a_{eq} such that $\Omega_{r0}a^{-4} = \Omega_{m0}a^{-3}$. That happens when $a_{eq} = \Omega_{r0}/\Omega_{m0} = 2.83 \times 10^{-4}$ using the benchmark values from Table 6.1. This corresponds to a redshift of

$$z_{eq} \simeq 3530, \qquad \text{benchmark model.} \tag{6.57}$$

At higher redshifts the gravitational dynamics of the expansion is dominated by the radiation components, while at lower redshifts the matter components dominate. Until the cosmic plasma becomes neutral at $z \sim 1100$ the radiation pressure is the force driving the expansion.

In order to find the time corresponding to z_{eq} we must integrate (6.56). As it happens this can be done in closed form:[11]

$$H_0 t \stackrel{\text{flat}}{=} \frac{4a_{eq}^2}{3\Omega_{r0}^{1/2}} \left[1 - \left(1 - \frac{1}{2}\frac{a}{a_{eq}} \right) \left(1 + \frac{a}{a_{eq}} \right)^{1/2} \right]. \tag{6.58}$$

Putting $a = a_{eq}$ gives the time when the radiation and matter densities were equal:

$$t_{eq} = H_0^{-1} \frac{4a_{eq}^2}{3\Omega_{r0}^{1/2}} \left(1 - \frac{1}{\sqrt{2}} \right) = 3.40 \times 10^{-6} H_0^{-1} \simeq 47500\,\text{yr}. \tag{6.59}$$

Translating Equation (6.58) from a to z gives the limiting solutions:

$$1 + z \simeq \begin{cases} \left(2\,\Omega_{r0}^{1/2}\, H_0 t \right)^{-1/2} & t \ll t_{eq} \\ \left(\frac{3}{2}\,\Omega_{r0}^{1/2}\, H_0 t \right)^{-2/3} & t_\Lambda \gg t \gg t_{eq}. \end{cases} \tag{6.60}$$

The first $\sim 40000\,\text{yr}$ of the cosmic evolution is radiation dominated, while the next $\sim 340\,000\,\text{yr}$ are matter dominated.

---

[11] Gradshteyn and Ryzhik (2007, 2.222.2, p.73).

6.3 Dark Energy Models

The spatial uniformity and time independence of Einstein's cosmological constant would suggest that it is a property of the space-time that might be derived from a better understanding of physics. An alternative view is to think of the cosmic acceleration as being due to some unknown type of matter or energy that pervades all of space. In this view, the dark energy becomes a dynamic component of the Universe that, like any other material field, evolves with time. In the absence of any concrete evidence as to the nature of this material, all we can do is build simple models for it, compute what the Universe would like and test this observationally.

In the simplest cases such matter might be described as having a mass density ρ_X and having an effective pressure p_X. In the absence of any detailed notion of what this material is, a relationship between ρ_X and p_X has to be assumed: the dark energy is modelled as an exotic gas whose properties are to be explored by experiment.

The simplest case to consider is the linear relationship $p_X = w\rho_X c^2$ for a constant w. This brings us back to the discussion of Section 5.9 and, in particular Section 5.9.2 Equations (5.58–5.62). We will consider cases where $w = $ constant and $w = w(z)$ (see Table 6.2),[12] but first we discuss the trivial yet important case where $w = -1$.

6.3.1 Simple Dark Energy: $w = -1$

Let us turn our attention to an important case: a Universe pervaded only by matter having equation of state $w = -1$:

$$p = -\rho c^2. \tag{6.61}$$

We drop the subscript 'X' since we are supposing that this is the only material content. The appearance of a negative value for the pressure may cause some surprise, since pressure as we normally know it (a force acting per unit area) is always positive. Negative pressure should perhaps be thought of as a quantum phenomenon that masquerades in the classical equations as a pressure-like term, only it is negative. We have abundant evidence that the quantum vacuum behaves in this way.

It will turn out that such a Universe behaves exactly as though it had only an Einstein cosmological constant. Our starting Equations (5.56) and (5.57) for this simple case are

$$\frac{1}{a}\frac{d^2a}{dt^2} = \frac{8\pi G}{3}\rho, \tag{6.62}$$

$$\frac{d\rho}{dt} = 0. \tag{6.63}$$

The second of these tells us that the density is a constant. We shall write this as

$$\rho = \rho_\Lambda, \quad \text{a constant.} \tag{6.64}$$

[12] See Barnes *et al.* (2005) for a nice overview of these models.

With this the first equation can be rewritten as

$$\frac{1}{a}\frac{d^2a}{dt^2} = \frac{8\pi G}{3}\rho_\Lambda = \frac{\Lambda}{3} \quad \text{on defining} \quad \Lambda = 8\pi G\rho_\Lambda. \tag{6.65}$$

Solving this is very straightforward, we have seen the procedure many times before:

$$\frac{\dot{a}^2}{a^2} = H_0^2\left[\Omega_\Lambda + \frac{1}{a^2}(1-\Omega_\Lambda)\right], \quad \Omega_\Lambda = \frac{\Lambda}{3H_0^2}. \tag{6.66}$$

This integrates to

$$a(t) = \begin{cases} \sqrt{\Omega_\Lambda^{-1}-1}\,\sinh(\Omega_\Lambda^{1/2}H_0t), & \Omega_\Lambda < 1 \\ e^{H_0(t-t_0)}, & \Omega_\Lambda = 1 \end{cases} \tag{6.67}$$

with similar solutions for $\Omega_\Lambda > 1$. The solution shows an exponential growth of the expansion factor $a(t)$ as $t \to \infty$: this can be referred to as 'inflationary expansion', though the term is generally reserved for the same phenomenon happening at the birth of the Universe.

Ex 6.3.1 Use Equation (6.66) to write an equation for the evolution $H(z)$ of the Hubble parameter in this model.

Ex 6.3.2 Define $\Omega_\Lambda(z) = \Lambda/3H(z)^2$ and show that the evolution of Ω_Λ is described by

$$\Omega_\Lambda(z) = \Omega_\Lambda(0)\frac{1}{\Omega_\Lambda(0) + (1-\Omega_\Lambda(0)(1+z)^2}. \tag{6.68}$$

Hint: note Equations (6.7) and see Section 6.1.4.

Ex 6.3.3 Deduce that $\Omega_\Lambda(0) = 1$ means that $\Omega_\Lambda(z) = 1$, so Ω_Λ is a constant.

Ex 6.3.4 Rearrange (6.68) to express $\Omega_\Lambda(0)$ in terms of $\Omega_\Lambda(z)$:

$$\frac{1-\Omega_\Lambda(0)}{\Omega_\Lambda(0)} = \frac{1}{(1+z)^2}\frac{1-\Omega_\Lambda(z)}{\Omega_\Lambda(z)}. \tag{6.69}$$

Ex 6.3.5 Show that if $\Omega_\Lambda(0) \neq 1$, $\Omega_\Lambda(z)$ has grown from 0 at $z \to \infty$ to the present value $\Omega_\Lambda(0)$.

Notice that we have introduced a reference time, t_0, into the $\Omega_\Lambda = 1$ case of Equation (6.67). The reason for this is that the scale factor $a(t)$, in that model, is never zero at any time: the empty $\Omega_\Lambda = 1$ Universe is infinitely old. The other curious aspect of this solution is that even though such a Universe is getting exponentially bigger with time, the expansion rate is always H_0^{-1} and so the distance to the horizon is always $d_H = cH_0^{-1}$. (See Section 5.10.2 about horizons.)

6.3.2 The Energy Equation

The starting point for writing down the equations of cosmic evolution is the Einstein equations, under the assumptions that the model is homogeneous and isotropic and that the material content has a specified relationship between pressure, p, and density, ρ. In the classical non-relativistic case this relationship is, in the simplest case, the perfect gas law in which $p \propto \rho^\gamma$, for some adiabatic index γ.

Table 6.2 A guide to the various models in this section.		
Components	w	Section
ρ_X alone	constant	6.4
	$w = w(z)$	6.4.2
ρ_m and ρ_X non-interacting	$w = w(z)$	6.5.1
ρ_m and ρ_X interacting	$w = w(z)$	6.5.3
Simple plausible model	$w = w(z)$	6.5.2

The relativistic situation is, however, somewhat different. There, the simplest case, equivalent in the non-relativistic limit to the perfect gas law, is to use a relationship of the form $p = (\gamma - 1)\rho c^2$, or as it is now usually written, $p = w\rho c^2$, with $w = \gamma - 1$.

The Einstein equations yield three equations for momentum conservation (which in the case of homogeneity and isotropy reduce to a single equation) and an equation for energy conservation. These are the famous *Friedmann equations*.

For a single component medium with no dissipation, a *perfect fluid*, the adiabatic energy equation reads

$$\frac{d}{dt}(\rho c^2 a^3) + p\frac{d}{dt}(a^3) = 0, \qquad (6.70)$$

(see footnote 31, p.123). This is equivalent to

$$\frac{d\rho}{dt} = -3H\left(\rho + \frac{p}{c^2}\right). \qquad (6.71)$$

Rewriting this and using our equation of state $p = w\rho c^2$ gives

$$\frac{\dot{\rho}}{\rho} = -3H\left(1 + \frac{p}{\rho c^2}\right) = -3H(1 + w), \quad p = w\rho c^2. \qquad (6.72)$$

If we suppose $w = w(z)$ we can integrate (6.72) to derive an equation for the density of the Universe as a function of redshift. In this case we can use the definitions (6.25) and (6.26) to eliminate the time t in favour of the redshift z and write Equation (6.72) as

$$\frac{1}{\rho}\frac{d\rho}{dz} = 3\frac{1 + w(z)}{1 + z}, \qquad (6.73)$$

which has solution

$$\rho(z) = \rho(0)\exp\left(3\int_0^z \frac{1 + w(u)}{1 + u}\,du\right). \qquad (6.74)$$

We shall, in the course of exploring the various models see many solutions looking like this. For constant w this last equation integrates simply to

$$\rho \propto a^{-3(1+w)} \propto (1 + z)^{3(1+w)} \quad w = \text{constant}. \qquad (6.75)$$

For the special case $w = 0$, normal pressure free baryonic matter, this reduces to the familiar equation $\rho_m \propto (1 + z)^3$, while for $w = 1/3$ it gives the familiar equation $\rho_r \propto (1 + z)^4$ for the history of the radiation density.

6.3.3 The Friedmann Equation

In the case of a cosmology having a single constituent, the Friedmann equation is

$$\frac{\ddot{a}}{a} = -\frac{4\pi G}{3}\left(\rho + \frac{3p}{c^2}\right). \tag{6.76}$$

This equation make no assumptions about the number of different components that contribute to the pressure or the density, nor about the relationship between the pressure and density of these components. If we assume that this relationship can be written in the form

$$p = w\rho c^2, \tag{6.77}$$

where w could be a function of the epoch z, then we have

$$\frac{\ddot{a}}{a} = -\frac{4\pi G}{3}\rho(1 + 3w). \tag{6.78}$$

The sign of $\ddot{a}$ in this equation is determined by the sign of $\rho + 3p = (1 + 3w)\rho$. For $w > -\frac{1}{3}$ the Universe decelerates, while if $w < -\frac{1}{3}$ the expansion accelerates. The marginal case $w = -\frac{1}{3}$ has $\dot{a} = constant$.

Since the left side of Equation (6.78) is related to the deceleration parameter we can express the deceleration parameter directly in terms of w:

$$q = \frac{1}{2}(1 + 3w). \tag{6.79}$$

This makes no assumptions about what the functional form of w is. Note that this equation holds for all times while the equation of state is $p = w\rho c^2$, but it is restricted to a single component Universe.

6.3.4 The Flat $w = constant$ Model

For a flat model Universe pervaded by material with constant w, the deceleration parameter at a general time t can be written as

$$q(t) = -\frac{a\ddot{a}}{\dot{a}^2} = -\left(\frac{\ddot{a}}{a}\right)\left(\frac{a}{\dot{a}}\right)^2 \stackrel{\text{flat}}{=} \frac{1}{2}\frac{\rho + 3p}{\rho} \tag{6.80}$$

$$= \frac{1}{2}(1 + 3w), \quad k = 0. \tag{6.81}$$

We note that this is a constant, so the deceleration parameter in such a Universe is independent of z. To get a negative deceleration parameter, as indicated by the Supernova Hubble diagram, would require $w < -1/3$ and to get $q = -1$ would require $w = -1$, i.e. a Λ term. We shall come back later to more complex models involving both dark energy and regular matter.

6.4 Single Component, Constant w

For constant w the Friedmann Equation (6.78) for a Universe with equation of state $p = w\rho$ integrates to

$$H^2 = \frac{8\pi G}{3}\rho - \frac{k}{a^2}, \tag{6.82}$$

where k, the *curvature constant*, is a constant of integration. We will return to evaluating it in terms of present observables later.

For a single component Universe with constant w, Equation (6.82) shows that the Hubble parameter varies as

$$H(z) = H_0(1 + z)[\Omega_0(1 + z)^{1+3w} + (1 - \Omega_0)]^{1/2}. \tag{6.83}$$

Here, Ω_0 denotes the present value of the density parameter: $\Omega_0 = 8\pi G\rho_0/3H_0^2$, ρ_0 being the current density.

Ex 6.4.1   Derive Equation (6.75) from (6.72).
Ex 6.4.2   Derive Equation (6.82) from (6.78).
Ex 6.4.3   Should we be surprised that the Hubble parameter $H(z)$ as given by Equation (6.82) does not contain explicit reference to w, while the acceleration as given by Equation (6.78) does?
Ex 6.4.4   Derive Equation (6.83).

There are two well-known cases for w: (a) dust, for which $p_m = 0$, $w_m = 0$, and (b) a radiation gas for which, at temperature T_r, $p_r = \frac{1}{3}aT_r^4$ and the mass density of the radiation field is $\rho_r = aT_r^4$. So, for the radiation field, $p_r = \frac{1}{3}\rho_r$ and $w_r = \frac{1}{3}$. In these cases Equation (6.72) or (6.75) yields the familiar results:

$$w = 0: \qquad\qquad \frac{\dot\rho_m}{\rho_m} = 3H = 3\frac{\dot a}{a} \quad \Rightarrow \rho_m \propto a^{-3}, \tag{6.84}$$

$$w = 1/3: \qquad\qquad \frac{\dot\rho_r}{\rho_r} = 4H = 4\frac{\dot a}{a} \quad \Rightarrow \rho_r \propto a^{-4}. \tag{6.85}$$

We list in Table 6.3 some of the constant w values that have been written about in relation to dark energy. For constant values of w we had $\rho \propto a^{-3(1+w)} \propto z^{3(1+w)}$ (cf. Equation 6.75).

Models with $w < -1$ (phantom matter) behave strangely since, by Equation (6.75), the density increases as the Universe expands and the expansion ends in a 'big rip' when everything gets torn apart. This model is discussed in some detail by Caldwell *et al.* (2003) and by Gibbons (2003). What is curious is that if we live in a model with $w = -1$ we are

Table 6.3 w-values for specific dark energy paradigms.	
w	Energy type
$\frac{1}{3}$	radiation
0	dust-like
$-\frac{1}{3}$	strings
$-\frac{2}{3}$	topological defects
-1	lambda
< -1	phantom matter

living on the brink of a disaster and it is fortunate that the constants of nature are accurately constant![13]

6.4.1 Flatness of the Initial Conditions

From here it is straightforward to show, analogously with the derivation of Equation (6.69), that the history of Ω for such a Universe is described by

$$\Omega(z) = \frac{\Omega_0}{\Omega_0 + (1 - \Omega_0)(1 + z)^{-(1+3w)}}. \tag{6.86}$$

The first thing to notice about this equation is that if $\Omega_0 = 1$, then $\Omega(z) = 1$ at all times. However, if $\Omega_0 < 1$, then at very early times ($z \to \infty$):

$$\Omega(z) \to 1 - \left(\frac{1 - \Omega_0}{\Omega_0}\right)\frac{1}{(1+z)^{1+3w}}, \quad z \to \infty. \tag{6.87}$$

$\Omega(z)$ gets arbitrarily close to 1 for $w > -1/3$. All such model Universes look like the Einstein de Sitter model close to the singularity. Thus, for Universes pervaded by normal matter, ($w \geq 0$), the initial conditions are finally tuned to having $\Omega(z)$ arbitrarily close to 1. This is referred to as *the flatness problem*.

Ex 6.4.5 A Universe containing matter, radiation and dark energy with densities ρ_m, ρ_r and ρ_w has curvature k. The dark energy is characterised by its equation of state $p_w = w\rho_w$ with $w = constant$. Use the integral of the Friedmann equation to show that

$$k = H_0^2(\Omega_{k,0} - 1), \tag{6.88}$$

where $\Omega_{k,0} = 1 - (\Omega_{m,0} + \Omega_{r,0} + \Omega_{w,0})$, the '0' subscripts indicating present day values.

---

[13] The science fiction author Stephen Baxter wrote an excellent short story, *Last Contact*, about the 'big rip' (Baxter, 2007). The story was nominated for a Hugo Award in 2007.

Ex 6.4.6   Use this to show that $\Omega_k(z)$ evolves as

$$1 - \Omega_k(z) = \frac{1 - \Omega_{k,0}}{\Omega_{r,0}(1+z)^2 + \Omega_{m,0}(1+z) + \Omega_{w,0}(1+z)^{1+3w} + 1 - \Omega_{k,0}}. \quad (6.89)$$

Ex 6.4.7   Show that any Universe that is not made up only of dark energy will be very flat near to the origin, $z \to \infty$.

6.4.2 w Varying with Epoch

In the general case, w would be a function of time, or redshift. A simple model for such evolution is due to Linder (2003):

$$w = w_o + w_a(1 - a) = w_0 + w_a \frac{z}{1+z}, \quad (6.90)$$

which is well behaved in both limits $z \to 0$ and $z \to \infty$. There are a number of variations on this theme.

In the general case when the matter has w dependent on epoch, $w = w(z)$, Equation (6.72) for the evolution of the density of the matter can be written as a simple integral:

$$\rho(z) = \rho_0 \exp\left[3 \int_0^z \frac{1 + w(z')}{1 + z'} dz'\right], \quad (6.91)$$

where ρ_0 is the present, $z = 0$, value. The integral can be evaluated in closed form for the case of the model of Equation (6.90).

Ex 6.4.8   Verify that Equation (6.91) is correct for the cases $w = 1/3, 0, -1$.

Ex 6.4.9   Derive Equation (6.91) (hint: Equation (6.29) might be useful).

Ex 6.4.10   Show that, for the one-component dark energy model (6.91) having zero curvature ($k = 0$)

$$\frac{H(z)^2}{H_0(z)^2} \overset{\text{flat}}{=} (1+z)^{3(1+w_0+w_a)} \exp\left(-3w_a \frac{z}{1+z}\right). \quad (6.92)$$

6.5 Two Components

Following Linder (2005) we can consider a two component model by explicitly writing the density ρ as a sum of the densities of the two components:

$$\rho = \rho_m + \rho_X, \quad (6.93)$$

where the 'X' subscript denotes the dark material whose equation of state is

$$p_X = w_X(z)\rho_X. \quad (6.94)$$

When 'X' represents the Einstein cosmological constant, Λ, we shall explicitly use the notation with Λ subscripts: p_Λ and ρ_Λ.

With this, Equation (6.82) becomes

$$H(z)^2 = \frac{8\pi G}{3}(\rho_m + \rho_X) - \frac{k}{a^2}. \tag{6.95}$$

We can rewrite this last equation in a form that does not explicitly involve k by dividing through by $H(z)^2$:

$$1 = \Omega_m(z) + \Omega_X(z) + \Omega_k, \tag{6.96}$$

where the z-dependent density parameters are

$$\Omega_m(z) = \frac{8\pi G\rho_m}{3H(z)^2}, \quad \Omega_X(z) = \frac{8\pi G\rho_X}{3H(z)^2}, \quad \Omega_k = k\frac{(1+z)^2}{3H(z)^2}. \tag{6.97}$$

In the case of a flat Universe, $k = 0$ we have

$$\Omega_m(z) + \Omega_X(z) = 1, \quad k = 0. \tag{6.98}$$

6.5.1 Non-interacting Components

If the components are non-interacting, the energy Equation (6.72) becomes two equations, one for each component:

$$\dot{\rho}_m = -3H\rho_m, \tag{6.99}$$

$$\dot{\rho}_X = -3H\rho_X[1 + w(z)]. \tag{6.100}$$

Using the relationships (6.25) and (6.28) between time and redshift, these can be cast as equations in redshift rather than time:

$$H^2(z) = \frac{8\pi G}{3}(\rho_m + \rho_X) - \frac{k(1+z)^2}{a_0^2}, \tag{6.101}$$

$$\frac{d\rho_m}{dz} = \frac{3\rho_m}{1+z}, \tag{6.102}$$

$$\frac{d\rho_X}{dz} = \frac{3\rho_X[1 + w_X(z)]}{1+z}. \tag{6.103}$$

In the case of constant w:

$$\rho_m \propto (1+z)^3, \quad \rho_X \propto (1+z)^{3(1+w)}, \quad \frac{\rho_X}{\rho_m} \propto (1+z)^{3w}. \tag{6.104}$$

The last of these Equations (6.104) shows that for $w < 0$ the dark energy term becomes less important as you go back in time.

In particular, for a standard cosmological constant, $w = -1$, $\rho_\Lambda = constant$ and there comes a time when the cosmic expansion is Λ-dominated. For the currently determined values of the densities of the various components, this happens around $z \sim 0.5$. This is the source of the comment, previously made in Section 6.2, that it seems like a remarkable coincidence that we happen to be observing the Universe around the time when Λ is just taking over the control of the expansion. The most sensitive tests for dark matter are those using relatively low redshifts in the range $0.2 < z < 2$.

6.5.2 The Simple Plausible Model

The expansion history for a flat $k = 0$ Universe filled with independently evolving matter and dark energy having equation of state $p = w(z)\rho c^2$ is described by

$$\frac{H(z)}{H_0} \overset{\text{flat}}{=} \left[\Omega_m(1+z)^3 + (1 - \Omega_m) \exp\left(3 \int_0^z \frac{1 + w(z')}{1 + z'} dz' \right) \right]^{1/2}. \tag{6.105}$$

The importance of this equation is that it is a plausible representation of the Universe as we see it today, though instead of simply having a cosmological constant it has a more general form of dark energy parameterised by the function $w(z)$.

Using the simple model of Equation (6.90), the integral in (6.91) can be done to give

$$H(z)^2 \overset{\text{flat}}{=} H_0^2[\Omega_m(1+z)^3 + (1 - \Omega_m)f(z)], \tag{6.106}$$

where $f(z) = (1 + z)^{3(1+w_0+w_a)} \exp\left(-3w_a \frac{z}{1 + z} \right)$.

Ex 6.5.1   Using Equation (6.92), or otherwise, derive Equation (6.106).

See Linder and Jenkins (2003) and Amendola *et al.* (2008), who propose testing this model using weak lensing.

6.5.3 Interacting Components

The situation is complicated if the components interact with one another and so transfer energy. A simple model for this interaction would be to introduce an interaction term γ:

$$\dot{\rho}_m = -3H\rho_m + \gamma, \tag{6.107}$$

$$\dot{\rho}_X = -3H\rho_X[1 + w(z)] - \gamma. \tag{6.108}$$

On adding these equations the γs cancel and so total energy conservation is ensured. Using the relationships (6.25) and (6.28) between time and redshift, these can be cast as equations in redshift rather than time:

$$H^2(z) = \frac{8\pi G}{3}(\rho_m + \rho_X) - \frac{k(1+z)^2}{a_0^2}, \tag{6.109}$$

$$\frac{d\rho_m}{dz} = \frac{3\rho_m}{1 + z} - \frac{\gamma}{(1 + z)H(z)}, \tag{6.110}$$

$$\frac{d\rho_X}{dz} = \frac{3\rho_X[1 + w_X(z)]}{1 + z} + \frac{\gamma}{(1 + z)H(z)}. \tag{6.111}$$

These should be compared with Equations (6.101)–(6.103).

To make any progress we need a model for γ and that depends on the physics of X. In the case of simple ionised matter and radiation, the interaction through Thomson scattering is well understood. This is important when discussing the damping of acoustic perturbations prior to and during recombination: a central issue in our understanding of the Universe. In the case of exotic matter, X, getting γ is a problem for particle physics.

It should be remembered that many models of dark energy do not have $w = constant$, and some cannot even be expressed as a simple equation of state with some redshift dependence $w(z)$. However, from the point of view of using observational data to determine the nature of the dark energy, we can at best hope to fit a two or three parameter model such as $w(z) = w_0 + w_a z/(1 + z)$.

7 The Early Universe

The study of the very early Universe has given birth to a discipline that is broadly referred to as *astro-particle* physics, a subject that sits on the border of high energy particle physics, nuclear physics and astrophysics. Astro-particle physics has proven to be a wonderful symbiosis of experimental physics and experimental astrophysics in which the early Universe is, in effect, a high energy physics laboratory. Arguably, it was Gamow and his collaborators, in the immediate post World War II decade, who took the first tentative steps in this direction by arguing that the early Universe was the site of synthesis of the chemical elements.

The physics domain of cosmic nucleosynthesis is the first few minutes of the Big Bang. Hayashi (1950) took the first step further back into the past and towards the Big Bang itself with his exploitation of weak interactions to describe what had happened just prior to the period of nucleosynthesis. Since then, particle experiments have pushed back our understanding of the physics of matter to the point where we can now discuss the period before the first micro-seconds of the cosmic origin within the context of known high energy physics.

7.1 Early Thermal History

Astro-particle physics inevitably brings in an even wider domain of physics than the classical cosmology of general relativistic models. Some fine texts have been written on this subject from a variety of points of view and at different levels.[1] The approach used here is to explain the physics of the early Universe with a view to understanding what precision cosmology has to say about the particle and nuclear physics aspects of the earliest moments after the Big Bang.

The first steps in this direction were taken by Gamow and his co-workers during the first decade after the Second World War. Gamow had realised that the wartime research

---

[1] The discussions of Roos (2003, Ch.5) and Ryden (2003, Ch.10) are superb introductions. The texts of Kolb and Turner (1990, Sections 3.3–3.5) and Weinberg (1972, 2008) are thorough expositions of the thermal history of the early Universe. The older ones may be somewhat dated by now (e.g. the cosmic parameters are far better known now and we now know of three types of neutrino), but they are nonetheless almost unequalled in clarity. The superbly readable review on *The Origin and Abundances of the Chemical Elements* by Trimble (1975) focuses mainly on synthesis of the elements via stellar processes (which is where the elements carbon and heavier come from) and provides a fine historical review of both cosmological and stellar nucleosynthesis.

into nuclear physics could answer some important questions about his concept of the early Universe that he had started working on prior to 1939. We can trace back Gamow's interest in a Big Bang Universe to Gamow and Teller (1939a,b) in which he had described a simple idea for the formation of galaxies, and no doubt his interest extended back to his brief association with Friedmann.

Finding an explanation for the origin of all the chemical elements had been a driving force for Gamow and his collaborators in getting to the Hot Big Bang theory. They suggested that the observed abundances, as they were known at the time, could have been the consequence of a hot, radiation dominated phase of the early expansion of the Universe: a Hot Big Bang. Their aim was to show that all the elements we see could have been made in the Hot Big Bang. Subsequent work showed that this was not in fact possible and that we had to look to the stars as the factories of most of the elements beyond beryllium and boron in the periodic table.

Most of that work was published in the period 1946–1953 following the intense research into nuclear physics that had gone into producing the first nuclear weapon during the Second World War: the 'Manhattan project'. Gamow himself had not been involved in that, though he was later to become involved in the development of the H-bomb. In 1949, the book by Gamow and Critchfield (1949, *Theory Of Atomic Nucleus And Nuclear Energy Sources*) detailed the non-classified understanding of nuclear reactions,[2] and included an important Appendix on cosmology (*loc. cit.* Appendix VI, pp.334 *et seq.*: *The Origin of the Elements*).

7.1.1 Physics Preliminaries

In the discussion that follows we will be at the junction of cosmology and particle physics. This brings with it a confusion of units: particle physics expresses energies in electron volts (eV), or multiples of electron volts (keV, MeV, GeV, ...), while cosmology uses temperatures measured in Kelvin (K). The conversion is simple:

$$1\,\text{eV} = 1.160 \times 10^4\,\text{K} \sim 10^4\,\text{K}, \tag{7.1}$$

where the last is frequently used as a convenient rough-and-ready conversion for assessing graphs with axes labelled in different ways.

At the very early times to be considered here, a considerable simplification of the cosmology arises because we can ignore the curvature of the Universe, and considerations of any cosmological constant or dark energy. Moreover, the expansion is dominated by

---

[2] On 12 August 1945, Gamow had written to Vannevar Bush asking what were the boundaries of what could be talked about and published in the aftermath of the Manhattan Project: '*I wonder where the boundary between what can and cannot be told should be placed*'. Bush replied quickly in a letter dated 14 August, where he refers to a press release that had been issued the day Gamow had written his letter. Bush had persuaded Roosevelt to undertake a crash program to develop a nuclear weapon, and was the person who in 1942 recommended Oppenheimer to be the scientist in charge of the project. Bush later became a member of Truman's Interim Committee which advised on the targets for the first atomic weapons.
http://blog.nuclearsecrecy.com/2013/01/18/george-gamow-and-the-atomic-bomb/
(Alex Wellerstein, historian of science, Stephens Institute of Technology: page may be volatile).

relativistic particles that behave like a radiation gas. The expansion rate, H, of the Universe depends directly on the density, ρ, of the Universe through the Friedmann equation: $3H^2 = 8\pi G\rho(T)$ (i.e. the 00 component of the Einstein equations).

The density $\rho(T)$ depends on the temperature of the cosmic gas in a way that depends on the physical processes taking place at an atomic or sub-atomic level in the gas. It also depends on what kind of particles are present, whether they are fermions or bosons, and their spin states. Photons are bosons with two spin states, whereas there are three species of neutrinos and their corresponding anti-neutrinos, which are fermions.

The Relativistic Cosmic Plasma: g-factors

For relativistic gas, the density $\rho(T) \propto aT^4$, where a is the Stefan–Boltzmann constant, and the pressure is $p = \frac{1}{3}\rho c^2$ (we shall show this in Section 7.1.3). The constant of proportionality depends in detail on the particles that make up the cosmic fluid, and that is a function of temperature. To make this explicit we introduce a temperature dependent proportionality factor $\frac{1}{2}g_*(T)$, and rewrite the Friedmann equations:

Friedmann equation for a gas of relativistic particles

$$3H^2 = \tfrac{1}{2}g_*(T)\, 8\pi G\, aT^4, \tag{7.2}$$

where $g_*(T)$ counts the number of degrees of freedom of the particles that make up the gas at temperature T (see Equation (7.3) below).

The physical nature of the gas at temperature T is absorbed into the function $g_*(T)$, which, in the case of the early Universe, is to be calculated from particle physics. For a single component relativistic fluid, g_* is the number of distinct particles of a given type, and if the constituent particles do not change g_* is simply a fixed number which we denote by g.

For a single component relativistic gas, calculating g is a simple counting exercise. At a given temperature, g counts the number of types of particles of a given species:

Number of degrees of freedom of particles in a gas

$$g = n_{\text{types}}\, n_{\text{antiparticles}}\, n_{\text{spin}}. \tag{7.3}$$

This reflects the number of types of particle, n_{types}, the number of spin states n_{spin} and takes account of whether there are any corresponding antiparticles through $n_{\text{antiparticles}} = 1$ or 2.

For a photon gas, $g_\gamma = 2$ since there are two spin states for the photon (left and right handed). The photon has no charge and no antiparticle.[3] For a gas of neutrinos we have to take account of the fact that there are three different flavours of neutrino (ν_e, ν_μ, ν_τ), and each flavour has a distinct antiparticle ($\bar\nu_e, \bar\nu_\mu, \bar\nu_\tau$). However, all neutrinos have the same

---

[3] This is why the factor $\frac{1}{2}$ appears with the g_* in Equation (7.2). In a standard photon dominated cosmological model, $\frac{1}{2}g_* = 1$ and we recover the familiar Friedmann equation from (7.2). The factor $g_*(T)$ for photons changes at temperatures above $T \sim 1\,\text{MeV}$ when electron–positron pairs are spontaneously created, and changes again at even higher temperatures where more exotic particles are created (as depicted in Figure 7.1).

spin state (left-handed), so Equation (7.3) for neutrinos has $n_{\text{types}} = 3$, $n_{\text{antiparticles}} = 2$, and $n_{\text{spin}} = 1$, so $g = 3 \times 2 \times 1 = 6$.

For mixtures of relativistic particles we sum the g-values. Much of what follows will be concerned with the function $g_*(T)$ evaluated at various epochs during the evolution of the early Universe when mixtures of different particles make contributions to the total energy density. For the cosmic gas of photons and neutrinos we will show later (Section 7.2.4) that

$$g_* = g_\gamma + 3g_\nu = 2 + 6\frac{7}{8}\left(\frac{4}{11}\right)^{4/3} \simeq 3.363. \qquad (7.4)$$

The number 6 appears because we have three flavours of neutrino, each with its own antiparticle, the $7/8$ appears because neutrinos are fermions, and the $4/11$ is the ratio of the neutrino and photon temperatures at the present day. This last parameter is determined from the equilibrium that existed before the neutrinos decoupled from other cosmic material at $T \sim 10^{10}$ K.

The function $g_*(T)$ for $T < 10^{12}$ eV$(\sim 10^{16}$ K$)$ has been computed by Coleman and Roos (2003) and is shown in Figure 7.1, labelled with the various changes in the constituent particles that make up the cosmic plasma and that cause $g_*(T)$ to change rapidly. The same curve is plotted in a somewhat different way in Figure 7.2, where the vertical axis is the value of T^3 as compared to what it would have been in the absence of all that physical complexity when the particle species would always have been the same throughout cosmic history. We shall see that T^3 is a measure of the cosmic entropy.

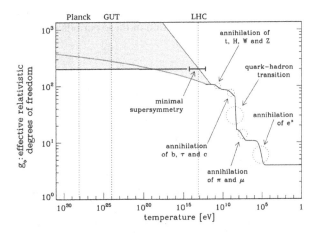

Fig. 7.1   Effective number of relativistic degrees of freedom, $g_{*s}(T)$ for $T < 10^{12}$ eV as derived from the model of Coleman and Roos (2003). At $T = 10^{12}$ eV, $g_{*s}(10^{12}) = 106.75$, above that temperature we can only speculate how $g_{*s}(T)$ grows. The shaded area represents the possible extrapolation for g_{*s}. The vertical dotted line marked LHC denotes the energy probed by the Large Hadron Collider at the time of the announcement of the discovery of the Higgs particle. The value of $g_{*s}(T)$ decreases each time families of particles annihilate, these transitional phases are depicted by dotted circles. The curve is based on the Standard Model of particle physics. See also Figure 7.2 and footnote 17, p.167. *(Adapted from Egan and Lineweaver (2010). Reproduced by permission of the AAS.)*

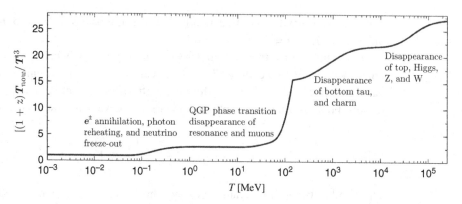

Fig. 7.2 Changes in the number of degrees of freedom since the time when the temperature of the cosmic plasma was ~ 10^{15} K (~ 100 GeV). The shape of the curve is determined by the number and nature of the relativistic particle species present at a given temperature. If there were no changes in the cosmic constituents the curve would be flat and horizontal. The shape of the curve demarcates the different physical regimes during the cooling of the cosmic plasma. Note that the transitions are not always sharp: they take time.
(From Rafelski and Birrell (2014) with permission.)

We see in Figure 7.1 that there are two rapid declines in $g_*(T)$, one occurring at around $T \sim 3 \times 10^{12}$ K, $(3 \times 100\,\text{MeV})$ and the other starting at around $T \sim 10^{10}$ K $(10^6\,\text{eV})$ and being complete at around $T \sim 10^9$ K $(10^5\,\text{eV})$. In Section 7.1.2 we shall use these transitions in $g_*(T)$ to identify and delineate periods of cosmic evolution that are dominated by different particle physics. Except in the cases just mentioned these may not be sharp boundaries, but the identification of these periods is nevertheless useful in describing what is going on.[4]

Temperature-time Relationship

In a radiation Universe the expansion rate varies as $H = 1/2t$ and we can derive the temperature–time relationship in a relativistic gas from the Friedmann equation in the form (7.2):

Expanding cosmic gas of relativistic particles: temperature relates to time

$$\frac{T}{10^{10}\,\text{K}} = (\tfrac{1}{2}g_*)^{-1/4}\, 1.52 \left(\frac{t}{1\text{s}}\right)^{-1/2}, \quad \text{or} \quad \frac{T}{1\,\text{MeV}} = (\tfrac{1}{2}g_*)^{-1/4}\, 1.31 \left(\frac{t}{1\text{s}}\right)^{-1/2}.$$

$$(7.5)$$

The dependence on g_* is rather weak, and for rough calculations taking $\tfrac{1}{2}g_* = 1$ is both simple and adequate. (Above $T \sim 200\,\text{GeV}$, i.e. $T \sim 10^{15}$ K, where all known particles are relativistic, $g_* \sim 100$.)

---

[4] This is like the palaeontological distinction between various geological periods where there are many periods defined, some of which are demarcated by mass extinctions of species. The Permian-Triassic extinction (251 *Ma*) and the Triassic-Jurrasic extinction (200 *Ma*) define the boundaries of the Triassic period.

Period	Time span (seconds)	Characterised by ...	Terminated by ...
Table 7.1 Periods of early cosmic evolution.			
		The Speculative Era	
	$< 10^{-12}$	Speculative physics	Electroweak interaction separates into weak and electromagnetic
		The Fireball Era	
Quark	$10^{-12} - 10^{-6}$	Quark-Gluon plasma	Quarks bind together, form hadrons
Hadron	$10^{-6} - 1$	Hadron anti-hadron pairs	Anti-hadrons annihilated Residual hadrons
Lepton	$1 - 10$	Leptons dominate mass density	Anti-leptons annihilated Residual leptons
Photon	$10 - 10^{13}$	Nucleosynthesis of elements Radiation pressure dominates	Electrons recombine with protons Cosmic plasma almost neutral
		The Stellar Era	
	$> 10^{13}$	Formation of Stars & Galaxies Large scale cosmic structure	

7.1.2 Cosmic Thermal History: Physics

The early Universe goes through periods of evolution which are dominated by different families of particles and which are characterised by the interactions that define the forces between those particles. The end of a period is generally signalled with an extinction of some major family of particles (see Figures 7.3 and 7.4). The major periods are summarised in Table 7.1, where everything that is thought to happen prior to the time when the Universe is $\sim 10^{-12}$s old is collected into a *speculative era* which laboratory experiments have hardly probed.[5] Between successive periods is a transition in the physics. Following our nomenclature these transitions correspond to *epochs*. The sequence of periods listed in the table that cover the time interval from 10^{-12}s to 10^{13}s comprise the *fireball era*, which is ended by the *epoch of recombination*. The period that follows the end of the fireball era is sometimes called the *stellar era* during which there are epochs of star formation, galaxy formation and large scale structure formation.

In the following there is a brief description of each of the fireball periods. When times are given, or corresponding temperatures, it should be appreciated that many of the

---

[5] The geological nomenclature for the division of units of time into subunits is set by the International Commission on Stratigraphy (ICS). Thus an *era* is composed of *periods* which are themselves divided into *epochs*.
While invoking this geological-style terminology we might more appropriately refer to the *speculative era* which is to be subdivided into a number of separate periods such as an inflationary period and a period of baryongenesis. The *fireball era* consists of a quark period followed by a hadron period, etc.
From http://dictionary.reference.com/browse/epoch:
epoch: a noun meaning 'a particular period of time marked by distinctive features, events, etc.'

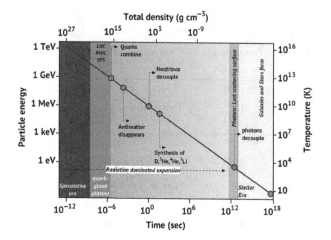

Representation of the thermal history of the Universe from the first nanosecond to the present day. This is drawn from the point of view of the underlying physics that define a succession of eras. The quark-gluon plasma regime has been successfully probed by the LHC, RHIC, and SPS experiments.

(LHC = Large Hadron Collider (CERN), RHIC = Relativistic Heavy Ion Collider (BNL), SPS = Super Proton Synchrotron (CERN))

(Based on Letessier and Rafelski (2005, Fig 1.2.), with permission.) ©

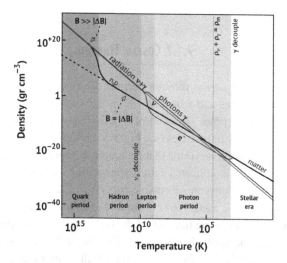

History of the composition of the Universe showing the times when different families of particles are annihilating with their anti-particles. The first great extinction occurs when the baryons annihilate with the anti-baryons, leaving a small fraction of baryons that we see today. Before that time the numbers B of baryons and anti-baryons are almost equal, with a very small excess, ΔB, of baryons over anti-baryons: $|\Delta B| \ll B$, while after the annihilation the number of baryons is simply $B = |\Delta B|$. The second great extinction is the annihilation of the electrons and positrons to leave a smaller number of electrons that will much later recombine with the protons to form neutral atoms. One of the key questions in cosmology is why $\Delta B \neq 0$.

Compare with Figure 7.3 which shows the falloff in temperature with time.

(Based on Harrison (1968a,b, Figures 2 and 3 respectively).) ©

events described are not in fact instantaneous: the values used are merely nominal values. Annihilation of particles of mass m with their anti-particles will begin when the temperature falls to around $kT \sim mc^2$, but may not be complete for a number of Hubble expansion times. The energy cited alongside the name is a nominal energy at which the period starts.

The Quark Period: $T \sim 125\,\mathrm{GeV}$

The end of the speculative era is defined by the time when the weak and electromagnetic forces become distinct: the time of the *electroweak phase transition*. We enter a period that is potentially accessible through experiment. The discovery of the Higgs Boson with a mass $\sim 125\,\mathrm{GeV}$ is a major breakthrough in understanding this period since we can map out its interactions with other particles. Moreover, we also have a successful working theory, the so-called *standard model of particle physics*, which provides a framework for describing the Universe during the time period from $10^{-12}-10^{-6}$ seconds. So while some of the physics of the quark period is still uncertain, we are nevertheless in a domain where we can test ideas through experiments using large accelerators like LHC.

A key part of the standard model is the theory of the strong force, the force that describes the interactions between the quarks and gluons which are the building blocks of hadrons and mesons. This theory is called *quantum chromodynamics*, or 'QCD' for short. The QCD analogue of electric charge is called *color*[6] and the gluons take on the role of the interactions, just like the photons do for the electromagnetic force in quantum electrodynamics. Within the scope of QCD there is a state of matter that is referred to as the *quark-gluon plasma*, or alternatively, *quark soup* which has been intensively studied by a number of experiments. The early Universe during the quark period is thought to be a quark soup, and the period ends when the soup is turned into hadrons.

The Hadron Period: $T \sim 150\,\mathrm{MeV}$

The end of the quark period is signalled by the temperature falling to a low enough value that quarks can bind together to make hadron anti-hadron pairs that are in thermal equilibrium: the *quark–hadron transition* (Tawfik and Harko, 2012). There are two kinds of hadron: *baryons*, made up from three quarks and which are fermions (so they have half-integer spin), and *mesons*, made up of two quarks and which are bosons (and so they have integer spin). Examples of baryons are the proton and the neutron, examples of mesons are pions and kaons.

Opinion is that the number of hadrons and anti-hadrons cannot have been equal, since otherwise we would have today a baryon–anti-baryon symmetric Universe (however, see Section 2.3.1, footnote 17, p.27). There had to be a very slight excess of particles over anti-particles, enough of an excess to account for the baryon density we observe today. The reason for this asymmetry is not well understood, but it almost certainly originates in the *speculative era* of cosmic evolution.

---

[6] The US spelling of 'color' is intended: this is merely the name for a particular attribute of an elementary particle. The attribute represents three quantum states of quarks, anti-quarks have anti-color. Quarks may combine to form baryons or mesons, both of which are 'color-neutral'.

The Lepton Period: $T \sim 3\,\mathrm{MeV}$

The end of the hadron period occurs with the extinction of the anti-hadron population when the hadrons and anti-hadrons annihilate. This leads to the period of leptons during which the dominant particles are leptons and anti-leptons. Leptons are light particles having spin 1/2, and fall into three categories: the *electronic leptons*, e^-, ν_e, the *muonic leptons*, μ^-, ν_μ, and the *tauonic leptons*, τ^-, ν_τ, and the corresponding anti-particles. The μ and τ particles and anti-particles have a very short lifetime and decay to form electrons, and consequently the Universe during the lepton period is dominated by electronic leptons and their anti-particles (e^+, the positron and its associated anti-neutrino). These remain in equilibrium until the temperature drops too low when there is a mass extinction through particle–anti-particle annihilation leaving only a small residue of electrons. The annihilation of the electron–positron population leaves a background of almost collisionless neutrinos and creates a background of photons that we now see as the cosmic background radiation.

This is the period of the neutrino and the associated weak interaction. Those cosmic neutrinos play a key part in the fitting of CMB data and hence in our getting at the parameters of the standard cosmology. This is because they make a substantial contribution to the early cosmic energy density and hence they directly affect cosmic timescales and a variety of important physical processes. Without a full understanding of those neutrinos, it would be difficult to interpret the data.[7] From the time when the weak interactions are in equilibrium, we are in the domain of known laboratory physics.

The Photon Period: $T \sim 1\,\mathrm{MeV}$

We finally enter the photon period that lasts for 380 000 years. During all of that period, the pressure force is dominated by the radiation field, which serves to mould the evolution of primordial density fluctuations that lead to cosmic structure formation. When we study the structure of fluctuations in the CMB temperature map, we are seeing the imprint of the radiation field.

The photons are the main contributors to the total mass density during only the first 77 000 years or so of this long period: this is the *radiation dominated era*. After that the baryonic matter dominates the mass density: this is the *matter dominated era*.

Early on during the photon period, about 3 minutes into the cosmic expansion, the epoch of nucleosynthesis starts and lasts for some 14 more minutes. During this era the surviving neutrons are built into most of the helium, ^4He, we see in our present Universe. Small, though important, trace amounts of deuterium, tritium and lithium are also produced. The observation of these elements in the quantities expected from this

---

[7] There is still a lot to be learned about neutrinos. One of the most important issues arises from the fact that there are two models for the neutrino: the Dirac model and the Majorana model. In the Dirac model there must be yet another neutrino that does not interact with matter: the right-handed *sterile neutrino*. In the early 1930s, Majorana tried to avoid this conclusion. In the Majorana model the neutrino and its anti-particle are identical; this carries with it a violation of lepton number conservation but has the advantage that it might ultimately provide a mechanism for understanding the observed baryon anti-baryon asymmetry. This is obviously an intense area of experimental research (Exo-200 Collaboration, 2014).

era of nucleosynthesis was a vital first step forward in providing evidence for the hot singular origin of our Universe, even before the discovery of the cosmic background radiation.[8]

Cosmic nucleosynthesis of the elements helium, deuterium and lithium has provided an important constraint on the parameters of a wide variety of plausible cosmological models and is an important complement to CMB model fitting.

7.1.3 The Physics of Cosmic Fermions

We now look at the physics of a relativistic fermion gas which is particularly relevant to the lepton period of cosmic evolution. The regime of interest here is the period during which the Universe cooled from 10^{10} K $\simeq 1$ MeV to 10^9 K $\simeq 0.1$ MeV. The dominant particles that concern us here are the leptons: the electron, muon and tau, with their corresponding neutrinos and anti-particles. These particles are all fermions and so, when in equilibrium, their energy distribution is the Fermi–Dirac distribution. At temperature T, those with masses $mc^2 > kT$ are effectively relativistic and, in a first approximation, can be assumed to be moving at the speed of light.

The electron has a mass of $m_e \simeq 0.511$ MeV and so makes the transition from being relativistic to non-relativistic during this period. The μ and τ have masses of 106 MeV and 1777 MeV respectively, and have short lifetimes of 2×10^{-6} s and 3×10^{-13} s. The associated neutrinos all have non-zero mass but their individual masses are still undetermined.[9] The cosmological limits on the sum of the three neutrino masses are the tightest bounds available (Planck Collaboration XIII, 2015, Section 6.4.1, Eq.54b):

$$\sum m_\nu < 0.21 \, \text{eV}, \tag{7.6}$$

as reported by the Planck CMB survey team (this value used both BAO and Planck data).[10] This limit is derived from a combination of experiments.

---

[8] The article of Hoyle and Tayler (1964), *The Mystery of the Cosmic Helium Abundance*, published some 8 months before the announcement of the discovery of the cosmic background radiation, concluded that *The fact that the observed He/H values never differ from* 0.14 *by more than a factor* 2, *combined with the fact that the observed values are of necessity subject to some uncertainty, could be interpreted as evidence that the Universe did have a singular origin* .... This was all the more remarkable a statement given that Fred Hoyle was one of the originators and strongest proponents of the Steady State theory. Hoyle realised that saving the Steady State theory would require invoking special helium-producing environments, such as super-massive objects of a kind that had not at that time been observed.

[9] There are experimental upper limits putting their individual masses at less than 0.2 eV.

[10] Determining individual neutrino masses is not easy. It is possible to obtain the differences of the squared masses from data on oscillations of solar neutrinos. (Solar neutrino oscillations are due to the fact that the neutrinos have different non-zero masses.) The tightest limit on individual neutrino masses is the upper limit from direct measurement. For the electron neutrino mass: $m_{\nu_e} < 2.2$ eV (95% confidence level): so the ν_e particle is relativistic in the cosmic weak interaction regime. The 95% confidence limit on the mass of ν_τ is $m_{\nu_\tau} < 19.2$ MeV and the 95% limit on the ν_μ is $m_{\nu_\mu} < 190$ KeV (Olive and Particle Data Group, 2014).

The Fermi–Dirac Distribution

We can calculate the density of a fermion gas at temperature T knowing that the equilibrium probability density function is the Fermi–Dirac distribution.[11]

Consider the 6-dimensional phase of particle positions and momenta as being divided up into elemental 6-volumes of spatial size d^3V and momentum size d^3p. The volume element of this 6-space is d^3Vd^3p and, importantly, it is invariant under the Lorentz transformation. By the quantum uncertainty principle, given d^3V, we must have $d^3Vd^3p > h^3$, where h is Planck's constant. We can think of the momentum part of the phase space as being divided into small 6-cells the momentum part of which have volume h^3. If the particles are fermions, no two identical particles can occupy the same cell.

The Fermi–Dirac distribution for the number density of particles, $n(p)$ having momentum of magnitude lying in the range $p, p + dp$ is:

Fermi–Dirac distribution function

$$n(p)\, dp = g\, \frac{1}{e^{(E-\mu)/kT} + 1}\, \frac{4\pi p^2 dp}{h^3}. \tag{7.7}$$

This gives the number of fermions having momentum of magnitude p in the interval $p, p + dp$ where μ is the Fermi constant, and h is Planck's constant.

The $4\pi p^2 dp$ is the volume of a spherical shell of radius p and thickness dp and h^3 is the natural (quantum) elemental volume of 6-dimensional phase space. So $4\pi p^2 dp/h^3$ is the number of elemental volumes in the momentum shell per unit volume of coordinate space, and g is the number of distinguishable particles that can be in the same volume element.

Here E is the particle energy, which is related to the particle momentum, p, via the special relativistic relationship:[12]

$$E^2 = p^2c^2 + m^2c^4, \tag{7.8}$$

$$E \simeq \begin{cases} pc & \text{ultra-relativistic particle} \\ p^2/2m & \text{non-relativistic particle.} \end{cases} \tag{7.9}$$

The quantity μ in (7.7) is the chemical potential of the distribution, known in this context as the Fermi constant. It is important at low temperatures and energies, but can be ignored in the case of a relativistic gas.

The chemical potential μ is calculated from the details of the reactions that take place (Dolgov *et al.*, 2002). Particles and their antiparticles in equilibrium have chemical potentials that are equal and opposite, so there is an inherent asymmetry built into the distribution

---

[11] Also commonly known as Fermi–Dirac *statistics*. Here I follow the dictum of Chandrasekhar (2003, Footnote 5, p.384) that this is somewhat of a misnomer.

[12] The kinetic energy, E_k and velocity v of a relativistic particle are given by

$$E_k = mc^2\left[\left(1 + \left(\frac{p}{mc}\right)^2\right)^{1/2} - 1\right], \qquad v = \frac{p}{m}\left[1 + \left(\frac{p}{mc}\right)^2\right]^{-1/2}.$$

Working in the highly relativistic limit provides considerable simplification of integrals over the distribution function!

function unless the chemical potential is zero.[13] Since the particles are relativistic their velocity is c and their energy, E, appearing in this equation is $E \simeq pc \gg mc^2$. In that case the chemical potential μ can be ignored.

Macroscopic Quantities – Single Particle Species

Our cosmic plasma is homogeneous and isotropic and so the macroscopic quantities defining the state of the gas are the gas density, n, its energy density, ρ and its pressure p. These are calculated by taking suitable averages over the distribution function (7.7). Using the approximations just discussed in (7.7), the particle number density per unit volume is

$$\bar{n} = \int_{p=0}^{\infty} n(p)\, dp = g \int_{p=0}^{\infty} \frac{1}{e^{pc/kT} + 1} \frac{4\pi p^2 dp}{h^3} = g\, 4\pi \left(\frac{kT}{hc}\right)^3 \int_0^{\infty} \frac{x^2 dx}{e^x + 1}. \quad (7.10)$$

Particle number density for equilibrium relativistic fermionic gas at temperature T,

$$\text{i.e.} \quad \bar{n} = g\, 4\pi \frac{3}{4}\Gamma(3)\zeta(3) \left(\frac{kT}{hc}\right)^3, \quad (7.11)$$

where we have substituted $x = pc/kT$ to simplify the integral. Here, $\Gamma(x)$ is the standard Gamma-function and $\zeta(x)$ is the Riemann zeta-function. $\Gamma(3) = 2$ and $\zeta(3) = 1.202$.[14]

The statistical weights depend on the number and type of particle species present at temperature T: mapping out $g(T)$ is an important part of understanding the physics of the early Universe.

The energy density of the relativistic fermionic gas is given by summing the energies of the particles over the distribution function:

$$\rho c^2 = \int_{p=0}^{\infty} pc\, n(p)\, dp = g \int_{p=0}^{\infty} \frac{pc}{e^{pc/kT} + 1} \frac{4\pi p^2 dp}{h^3} = g \frac{4\pi c}{h^3} \frac{k^4 T^4}{c^4} \cdot \int_0^{\infty} \frac{x^3 dx}{e^x + 1}. \quad (7.14)$$

---

[13] For an equilibrium distribution of photons, which satisfy Bose–Einstein statistics, the chemical potential is zero because the photon is its own anti-particle.

[14] From Gradshteyn and Ryzhik (2007, GR 3.411.3):

$$I_-^n(\mu) = \int_0^{\infty} \frac{x^{n-1} dx}{e^{x\mu} + 1} = \frac{1}{\mu^n}\left(1 - \frac{1}{2^{n-1}}\right)\Gamma(n)\zeta(\nu) \quad \mu > 0, n > 0. \quad (7.12)$$

For integer n, the Γ-function is $\Gamma(n+1) = n!, n \geq 0$, and the Riemann ζ-function has values $\zeta(2) = \pi^2/6, \zeta(3) = 1.202057, \zeta(4) = \pi^4/90$ and $\zeta(6) = \pi^6/945$ (see also Section 19.1.5):

This is to be compared with the similar integrals that appear in Bose–Einstein statistics (*loc. cit.* GR 3.411.1):

$$I_+^n(\mu) = \int_0^{\infty} \frac{x^{n-1} dx}{e^{x\mu} - 1} = \frac{1}{\mu^n}\Gamma(n)\zeta(\nu), \quad \mu > 0, n > 0. \quad (7.13)$$

The integrals differ only in the factor $\alpha_n = (1 - 1/2^{n-1})$ which, in the important cases $n = 3, 4$, gives $\alpha_3 = 3/4$ and $\alpha_4 = 7/8$. The factor $7/8$ accounts for the difference between the densities of relativistic (or mass zero) fermion and boson gases at the same temperature.

For $n = 4$ and $\mu = 1$, it is straightforward to show, without having done these integrals, that $I_-^4(1) - I_+^4(1) = \frac{1}{8}I_-^4(1)$, from which the result $I_+^4(1) = \frac{7}{8}I_-^4(1)$ follows. (Hint: write out $I_-^4(1) - I_+^4(1)$ and simplify.)

Energy density for equilibrium relativistic fermionic gas at temperature T

$$\rho c^2 = g \frac{1}{2\pi^2} \frac{k^4 T^4}{\hbar^3 c^3} \frac{7}{8} \Gamma(4)\zeta(4) = \frac{7}{8} \frac{1}{2} g \, aT^4, \qquad a = \frac{\pi^2 k^4}{15\hbar^3 c^3}, \qquad (7.15)$$

where $a = \pi^2 k^4/15\hbar^3 c^3$ is the Stefan–Boltzmann constant.

We can do the same calculation for a gas of bosons using the Bose–Einstein distribution in place of the Fermi–Dirac distribution, the integrals are very similar in form.[14] If we choose units in which $\hbar = 1, c = 1, k = 1, a = \pi^2/15$ and

$$\rho_{\text{bosons}} = \frac{\pi^2}{30}gT^4 \quad \text{and} \quad \rho_{\text{fermions}} = \frac{7}{8}\frac{\pi^2}{30}gT^4. \qquad (7.16)$$

These results differ by the famous factor $7/8$.

For both bosonic and fermionic particle species, the number density varies as T^{-3} and the energy density (and pressure) varies as T^{-4}. This means that we can simply add up the contributions to the number density or energy density from the individual particle species when considering a mixture, and define an overall g-factor for the mixture:

$$g_{*n}(T) = \sum_{\text{bosons},i} g_i \left(\frac{T_i}{T}\right)^3 + \frac{7}{8} \sum_{\text{fermions},i} g_i \left(\frac{T_i}{T}\right)^3, \qquad (7.17)$$

$$g_{*\rho}(T) = \sum_{\text{bosons},i} g_i \left(\frac{T_i}{T}\right)^4 + \frac{7}{8} \sum_{\text{fermions},i} g_i \left(\frac{T_i}{T}\right)^4. \qquad (7.18)$$

This assumes that the species are all in equilibrium at the same temperature. T_i is the temperature at which the coefficient g_i is determined.

With this we can, for example, write the evolution of the total density ρ_* as

$$\rho_*(T)c^2 = g_{*\rho}(T) \, aT^4. \qquad (7.19)$$

There is a similar expression for the entropy of a mixture which is discussed later (see Equation (7.35) for the corresponding Equation for g_{*s} and Figure 7.1 for a diagram of the behaviour of the function $g_{*s}(T)$).

7.2 Cosmic Thermal History

7.2.1 Thermodynamics

The 1st Law of Thermodynamics

In classical thermodynamics, the 1st law is often formulated by expressing the change in the thermodynamic parameters of a system as

$$\Delta Q = \Delta U + \Delta W, \qquad (7.20)$$

where ΔW is the work done on the environment of the system in making the change. The work done causes an amount of heat ΔQ to be given to the system and changes its internal energy by ΔU. Looked at the other way around, if you add an amount of heat ΔQ to a system, some of it goes into changing the internal energy ΔU and what is left over does an amount of work ΔW on its surroundings. This simply expresses, in a very general way, the conservation of energy.

However, the expression is so general as to be of little use as it stands. We need more information about the system and the nature of its internal energy, we need to know what this 'work' refers to and, in particular, we need to know what 'heat' is in physical terms.[15] We also need to know the nature of the change expressed by the symbol Δ.

The 2nd Law of Thermodynamics

The 2nd law causes a lot of problems: in its basic form it merely makes a statement about something that cannot happen. One way of stating it is to say that it is impossible to make a process that can have as its sole result the transfer of heat from a cooler body to a hotter body. In motivating this, one frequently invokes concepts such as 'engine' and 'refrigerator' which, although important devices for the application of thermodynamics, are not well tuned to concepts from other areas of physics.

The physical realisation of the 2nd law comes with the introduction of a concept called 'entropy', which is simply a quantitative measure, S, of disorder.[16] The 2nd law is then invoked simply by stating that, *when all parts of a system are included in the calculation*, its entropy either remains the same or increases. If we view entropy as a measure of disorder, the 2nd law says that order cannot grow from disorder without the action of some external agency and it cannot happen by chance. That is what the '... when all parts of a system are included ...' aspect of the previous statement of the 2nd law is about.[17]

The change in the entropy, ΔS associated with a change ΔQ in energy at temperature T is defined as

$$\Delta S = \frac{\Delta Q}{T}, \quad \Delta S \geq 0. \tag{7.21}$$

The first statement is the definition of entropy and the statement $\Delta S \geq 0$ is the mathematical statement of the 2nd law.

The Entropy Equation

In the case of the thermodynamic state of a gas we can combine the 1st Law (7.20) and 2nd Law (7.21) into a simple and useful equation. First we identify the incremental work done ΔW with the action of the pressure p in changing a volume V by an amount ΔV: the

[15] Newton propagated the idea that 'heat' was a particular form of matter. It was not until the advent of Carnot's work clarifying the nature of heat (around 1824) that this idea was dropped and progress could be made. The concept of 'entropy' came later with the work of Clausius in 1855.

[16] See On-line Supplement *Kinetic theory* for a more detailed discussion from the point of view of kinetic theory.

[17] The requirement that '... all parts of a system are included ...' raises an interesting question as regards the Universe, which is not a closed system: do we modify the concept of entropy to apply to an unbounded system like the Universe, or do we simply declare that the concept of entropy is not meaningful in that context? This was discussed in an interesting paper by Egan and Lineweaver (2010, Section 3) who suggested that the entropy of the cosmic event horizon should be taken into account: the event horizon defines a natural volume V for the Universe.

so-called 'pdV work'. The change in internal energy ΔU depends on the nature of the gas, for example, on how many internal degrees of freedom it has. Then we invoke the 2nd law to identify ΔQ with $T\Delta S$ (as suggested by Equation 7.21).

The entropy S is to be thought of as a function of both the volume, V, and the temperature, T, while the pressure and density are taken to depend only on the temperature. With this, the entropy $S(V, T)$ of a volume V of gas in equilibrium at temperature T, is subject to the constraint

$$TdS = dU + pdV. \tag{7.22}$$

This is essentially a consequence of the two laws of thermodynamics (7.20) and (7.21) and is generally referred to as the *thermodynamic identity*.[18]

7.2.2 Cosmic Entropy and Isentropic Expansion

Padmanabhan (2010, p.60) remarks that an equation like (7.22) should need no modification if one can make a suitable definition of the entropy S in terms of the underlying variables U, V.[19] The suitable definition will depend on the physics of the gas. We will identify U with the total energy ρV of the volume, where ρ includes not only the rest mass energy but the internal energy due to molecular or atomic motions.

One of the important relationships between thermodynamic variables is the *Gibbs–Duhem equation*, which, in the absence of chemical potentials, reads:

$$U = TS - PV. \tag{7.24}$$

In the context of an equilibrium relativistic gas, the internal energy of a small volume V is $U = V\rho c^2$, where ρ is the energy density measured in a local inertial frame. In a frame where the fluid is locally at rest, this is $U = V\rho_0 c^2$ where ρ_0 is the rest-mass density.

With this, we can provide an expression for the entropy $S(V, T)$ of a volume V of gas at temperature T, given that the gas has pressure p and density ρ:

$$S(V, T) = \frac{1}{kT}(\rho c^2 + p)V, \tag{7.25}$$

where k is Boltzmann's constant. This quantity is referred to as the *entropy* of the volume V. Note the S is specified as a function of the volume, V and temperature T, so we can consider what happens under variations of T and V.

[18] In Equation (7.22) it has been assumed that the *chemical potential*, μ, of the gas is zero, which is appropriate for the physical regime under consideration here. The full form of the relationship is

$$TdS = dU + pdV - \mu dN, \tag{7.23}$$

where N is the number of particles in the volume. This says that, even at constant volume and constant entropy, changes in the number of particles in the volume may change the energy U. In our relativistic gas, zero chemical potential is equivalent to saying that the number of particles and anti-particles is equal.

[19] We have an immediate perceptual idea of the meaning of quantities like pressure, temperature, and volume. However, this is not the case for quantities like entropy and energy, which are neither observable nor measurable and do not conjure up an innate feeling of what they might represent.

Isentropic and Adiabatic Flows

A flow under which S is a constant is referred to as being *isentropic*. Clearly, from the 2nd Law, (7.21), at finite temperature $dS = 0$ if, and only if, $dQ = 0$. An isentropic flow is one in which no heat is transferred into or out of the volume under consideration. Since an irreversible flow has $dS > 0$, an isentropic flow is necessarily reversible.

An *adiabatic* flow is a flow in which there is neither heat loss nor heat gain in a volume, but which is irreversible due to the action of frictional forces. In an adiabatic process, the entropy is not necessarily a constant. An isentropic flow is a frictionless adiabatic process, i.e. a reversible adiabatic process.[20]

Isentropic Expansion of a Relativistic Gas

Consider a gas consisting of a single species of relativistic particles. For a process in which $dS = 0$, we have $dU = -pdV$, which, with (7.25) is $d(\rho c^2 V) = -pdV$. Hence, for a relativistic gas in which $p = \frac{1}{3}\rho c^2$, we recover the familiar relationship

$$\rho c^2 \propto V^{-4/3} \propto R^{-4} \tag{7.26}$$

from the laws of thermodynamics. It follows that

$$T \propto (\rho c^2 + p)V \propto \tfrac{4}{3}\rho c^2 V \propto V^{-1/3} \propto R^{-1} \tag{7.27}$$

for the relativistic gas at constant entropy. The constant of proportionality in this relationship, $T \propto R^{-1}$, depends on the physics determining the internal state of the gas.

Another useful way of looking at the isentropic expansion, $S = $ constant, of a relativistic gas having $p = \frac{1}{3}\rho c^2$ follows directly from Equation (7.25) and the scaling relations (7.26) and (7.27):

$$S = \frac{(\rho c^2 + p)V}{kT} \propto g(RT)^3 = \text{constant.} \tag{7.28}$$

The total energy density ρ of the cosmic material and its entropy density s at temperature T are then:

Density–temperature and entropy–temperature relationships for single component relativistic fermion gas,

$$\rho(T)c^2 = g_\rho \frac{7}{8}\frac{\pi^2}{30}\hbar c \left(\frac{kT}{\hbar c}\right)^4, \qquad \frac{s(T)}{k} = g_s \frac{7}{8}\frac{2\pi^2}{45}\left(\frac{kT}{\hbar c}\right)^3, \tag{7.29}$$

as calculated from the Fermi–Dirac distribution.

The factors of $7/8$ are absent in the case of a relativistic boson gas.[14]

---

[20] It should be remarked that the term 'adiabatic' is used in a variety of ways. In the thermodynamics literature, the term *isocaloric* is frequently used to emphasise the absence of heat transfer from or to the fluid. Processes taking place in a closed and insulated box are adiabatic. Processes that happen so quickly that heat transfer has no time to take place are also adiabatic.

Gravitational Entropy

Applying an equation like (7.22) to a self-gravitating system is not without its problems. The quantities S, U and V are *extensive quantities* that should simply add up when two systems are brought together. Because of the long-range nature of the gravitational force this does not happen in self-gravitating systems: the gravitational force is a long range force that, unlike the electromagnetic force, is unshielded. Moreover, in general relativity, the concept of 'energy at a point' is not properly defined since gravity can be locally removed by viewing the system from a local freely falling frame (i.e. the Principle of Equivalence).[21]

7.2.3 Effective Number of Relativistic Species, g_*

The quantity $g_*(T)$ appearing in Equation (7.2) plays a role in determining the expansion rate of the Universe, but more importantly it relates the physical properties of the cosmic gas, the material energy density, pressure and entropy, to the temperature, T at any given time. The curve $g_*(T)$ shown in Figure 7.1 reflects the type of elementary particles that are present at any given time, and that is determined from the Standard Model of particle physics and any variants that may be deemed necessary (Coleman and Roos, 2003). We see in the figure how $g_*(T)$ demarcates the different periods that define the physical processes taking place during the cosmic expansion.

We shall work with the *entropy density*, i.e. the entropy per unit volume, which is defined using (7.25) as

$$s = \frac{S}{V} = \frac{\rho c^2 + p}{kT} \tag{7.32}$$

for a single relativistic particle species having density ρ and pressure p at temperature T.

---

[21] See Section 11.3.4. In his treatise *Relativity, Thermodynamics and Cosmology*, Tolman (1934b, Section 71 and Ch. IX) sought to clarify relativistic thermodynamics and addressed the issue of entropy. He argued that the classical entropy density S was a part of an entropy 4-vector $S^a = \phi_0 u^a$ (his Equation 119.3), in which ϕ_0 is the rest frame entropy and u^a is the 4-velocity of the observer. The 2nd law of thermodynamics would then read $S^a_{;a}\sqrt{-g}\,d^4x \geq \delta Q_0/T_0$, where δQ_0 is the (scalar) amount of heat, as measured by a local observer, flowing at temperature T_0 into an element of fluid (*loc. cit.* Section 121). The quantity $\delta Q/T$ is then Lorentz invariant.

The question was first addressed on the basis of kinetic theory by Müller (1967a,b). Stewart (1971, Section 3.3, p.48) argued against Tolman's *ansatz* on the grounds that his definition of entropy did not agree with results derived from kinetic theory for non-equilibrium distributions. Stewart put forward (*loc. cit.* Eq. 3.30, p.47) an alternative definition for the entropy 4-vector:

$$s^a = su^a + q^a/T, \tag{7.30}$$

where s and T are the usual rest frame entropy density (Tolman's ϕ_0) and temperature, and q^a is the heat flux 4-vector (q^a/T is the entropy flux vector). q^a is the heat flux as appears in the Eckart (1940) formulation of the energy momentum tensor for a fluid having heat flux and viscosity (see Section 11.3.4, p.286 *et seq.*). With this local formulation, the 2nd law of thermodynamics, or, more precisely, the H-theorem, simply reads

$$s^a_{;a} \geq 0, \qquad \text{Relativistic expression of the 2nd law of thermodynamics.} \tag{7.31}$$

This is also discussed in some detail in Misner *et al.* (1973, p.567, Exercise 22.7, Eq. 22.16e). See also Israel and Stewart (1979, 1980). The entropy flux 4-vector (7.30) is sometimes called the *Müller–Israel–Stewart entropy*.

Entropy is an additive property, so a number of single-species entropies can simply be added to give the entropy of a mixture of species, each at their own temperature and each with their own g-factors:

$$s = \sum_i \frac{(\rho_i c^2 + p_i)}{kT_i}. \tag{7.33}$$

Since $s \propto T^3$ (Equation 7.28) we would like to write, for some fiducial temperature T an expression of the form:

$$s(T) \propto g_{*s}(T)T^3, \tag{7.34}$$

for some generalised g-factors g_*. Note the extra subscript s on the g_{*s}: it is necessary to emphasise that this renormalisation has been done for the entropy density. Had we done this for the energy density we would have written $\rho(T)c^2 \propto g_{*\rho}T^4$ and, as we shall see, the two g-factors, g_{*s} and $g_{*\rho}$, are not the same.

g_* will inevitably depend on the temperatures T_i and the T, and must be such that it gives the correct answer when only one of the mixture of species is present. For a mixture of bosons and fermions, this leads us to the equation:

Effective number of degrees of freedom for the entropy density of a mixture of relativistic particle species

$$g_{*s}(T) = \sum_{\text{bosons},i} g_i \left(\frac{T_i}{T}\right)^3 + \frac{7}{8} \sum_{\text{fermions},j} g_j \left(\frac{T_j}{T}\right)^3, \tag{7.35}$$

where, at the time when the fiducial temperature is T, the ith particle species is at temperature T_i. The g_i is the statistical weight of that species (number of spin states) as computed from Equation (7.3).[22]

The temperatures T_i of the various species that contribute to the total entropy need not be the same. The main complication is to compute the statistical weights g_i for the particle species that are present at each temperature T: this depends on our detailed understanding of high energy physics and was calculated for $T < 10^{12}$ eV by Coleman and Roos (2003). Their function $g_{*s}(T)$ is shown in Figure 7.1.

If we follow the same procedure for the energy density of a mixture of relativistic particle species we get:

Effective number of degrees of freedom for the energy density of a mixture of relativistic particle species,

$$g_{*\rho}(T) \equiv g_{*\rho}(T) = \sum_{\text{bosons},i} g_i \left(\frac{T_i}{T}\right)^4 + \frac{7}{8} \sum_{\text{fermions},j} g_j \left(\frac{T_j}{T}\right)^4. \tag{7.36}$$

---

[22] Equation (7.3) contains the number of types of particle in the given species, n_{types}, which is obviously an integer. However, in calculating the statistical weight g_* it is usual to recognise that non-equilibrium processes are taking place during the evolution of the system and to compensate for this by using an effective number of particle types, N_{eff} rather than the actual number. Detailed computations show that for neutrinos $n_{\text{types}} = 3$ and $N_{\text{eff}} = 3.046$ (Mangano *et al.*, 2005). See also Section 7.2.4 and Equation (7.66) *et seq.*.

The convention is to simply call this $g_*(T)$ without the extra subscript.

As an example we can consider the phase of the expansion when the Universe has dropped below $\sim 100\,\text{MeV}$, but the electron–positron pairs have not yet annihilated ($T \sim 1\,\text{MeV}$). The μ, ($m_\mu \approx 105\,\text{MeV}$), and τ, ($m_\tau \approx 1777\,\text{MeV}$), particles have become non-relativistic. That leaves the electron and positron, and the ν_τ, ν_μ, ν_e neutrinos with their antiparticles as the only relativistic fermions. While $T > 1\,\text{MeV}$ these particles are all coupled by the weak interaction and are in equilibrium at the same temperature. We then have

$$g_* = \underset{\gamma}{2} + \frac{7}{8}\left(\underset{e^-}{2} + \underset{e^+}{2} + \underset{\nu,\bar\nu}{6} \right) = \frac{43}{4} = 10.75 \quad T > 1\,\text{MeV}. \tag{7.37}$$

These numbers arise as follows: the photon, electron and positron each have two spin states and there are six neutrinos, including the corresponding anti-neutrinos, all having the same spin. A similar counting exercise leads to $g = 106.75$ at temperatures $T > 200\,\text{GeV}$ when all known particles are relativistic.

The neutrinos decouple from the cosmic plasma when the temperature falls to around $\sim 1\,\text{MeV}$. During the period down to $T \sim 0.5\,\text{MeV}$, even though there is no interaction between electrons and neutrinos, the entropy of each species is conserved and the e^+e^- plasma and the neutrinos maintain equal temperatures. Following (7.35) we can write $g_{*S} = 2 + \frac{7}{8}(2+2) = 11/2$ for the e^+e^- plasma.

At around $T \sim 0.5\,\text{MeV}$ the electrons and positrons annihilate each other and inject photons into the Universe (see Section 7.2.4). We then get $g_* = 3.63$ as per Equation (7.4). Using Equation (7.5) we have the following time–temperature relationships:

Time temperature relationships in different physical regimes,

$$t = 0.74 T_{\text{MeV}}^{-2}, \quad g = 10.75, \quad 100\,\text{MeV} > T > 1\,\text{MeV} \tag{7.38}$$

$$t = 1.32 T_{\text{MeV}}^{-2}, \quad g = 3.36, \quad 0.5\,\text{MeV} > T. \tag{7.39}$$

Remember that these assume equilibrium throughout the temperature range and instantaneous changes between neighbouring periods. But they serve to associate an approximate time with the temperature.

7.2.4 Electron–Positron Pair Annihilation

The neutrinos decouple from the cosmic plasma when the temperature of the Universe falls to around $T_\nu \sim 1\,\text{MeV}$ and at that point the Universe is effectively a two-component fluid, one a gas of collisionless neutrinos and the other a gas of electron–positron pairs in equilibrium with photons. Shortly thereafter, when the temperature falls to around $kT_e \sim m_ec^2 = 0.5\,\text{MeV}$, the electron–positron pairs can annihilate to produce photons that are added to the pool of pre-existing photons. As a consequence, the temperature of the photon gas increases relative to the neutrino gas temperature.

In the following we shall compare the entropy of the Universe at two times t_0, just prior to the decoupling of the neutrinos, and t_1, just after the annihilation of the electron–positron pairs. During that period the Universe changes from an equilibrium plasma of electrons, neutrinos and photons, described by a single temperature T_0 to a two component plasma, one component being a gas of collisionless neutrinos at temperature T_ν and the other an electron–photon gas at temperature T_γ. We can relate T_ν and T_γ using the g-factors for the different plasmas and the assumption that the expansion is isentropic during this transformation. We shall consider the quantity RT, where R is the expansion factor, so that during isentropic expansion $RT = $ const. (see Equations 7.27 and 7.28).

Consider first the time when the temperature T_0 is higher than $T_e \sim 0.5\,\mathrm{MeV}$. The plasma of electron–positron pairs and photons has entropy (see Equation 7.35)

$$S_0 = g_0(RT_\gamma)_0^3 = \left[\underbrace{2}_{g_\gamma} + \underbrace{2\frac{7}{8}}_{g_{e^-}} + \underbrace{2\frac{7}{8}}_{g_{e^+}} \right] (RT_\gamma)_0^3 = \frac{11}{2}(RT_\gamma)_0^3, \quad T_0 > T_e. \tag{7.40}$$

After the electron–positron pairs have annihilated to create photons,

$$S_1 = g_1(RT_\gamma)_1^3 = \left[\underbrace{2}_{g_\gamma} \right] (RT_\gamma)_1^3 = 2(RT_\gamma)_1^3, \quad T_1 < T_e. \tag{7.41}$$

If the expansion is isentropic then these are equal and

$$\frac{(RT_\gamma)_1^3}{(RT_\gamma)_0^3} = \frac{11}{4}. \tag{7.42}$$

Consider now the collisionless neutrinos, for them

$$(RT_\nu)_0^3 = (RT_\nu)_1^3. \tag{7.43}$$

But at the time prior to annihilation the photon and neutrinos temperatures are equal, $(RT_\nu)_0^3 = (RT_\gamma)_0^3$ and so

$$(RT_\gamma)_1^3 = \frac{11}{4}(RT_\gamma)_0^3 = \frac{11}{4}(RT_\nu)_0^3 = \frac{11}{4}(RT_\nu)_1^3, \tag{7.44}$$

whence we get the final result concerning the heating of the photon gas by the annihilating electrons:

Ratio of temperatures of neutrinos and photons in present Universe:

$$T_\gamma = \left(\frac{11}{4}\right)^{1/3} T_\nu = 1.401 T_\nu \tag{7.45}$$

was established in the first seconds of the cosmic expansion.

If we could ever measure T_ν this simple result would be a remarkable test of our understanding of the early Universe.

In the low temperature limit when the only relativistic particles are photons and three species of neutrinos,

$$g_* = g_\gamma + g_\nu \frac{7}{8}\left(\frac{4}{11}\right)^{4/3} \simeq 3.363, \tag{7.46}$$

$$g_{*s} = g_\gamma + g_\nu \frac{7}{8}\left(\frac{4}{11}\right) \simeq 3.938, \tag{7.47}$$

where

$$g_\gamma = 2, \qquad g_\nu = 2N_{\text{eff}}, \ \text{with } N_{\text{eff}} = 3.046. \tag{7.48}$$

There are two spin states for the photon, so $g_\gamma = 2$. The three flavours of neutrino are at temperature $T_\nu = (4/11)T_\gamma$, and the factor $7/8$ comes from the fact that neutrinos and their anti-particles are fermions. The number of neutrino species is given as 3.046 and not simply 3: this absorbs a correction factor due to the fact that the neutrino decoupling is not instantaneous as was assumed in deriving these equations (see Section 7.2.4).

7.3 Nuclear Processes

The early models for nucleosynthesis in the Hot Big Bang all assumed that the early Universe consisted of a sea of neutrons dominated by a very hot radiation field, so the main process was the free decay of the neutrons to form the protons and electrons which would become the building blocks of atomic nuclei. The cosmologically relevant nuclear processes for this last step were set out by Gamow and Critchfield (1949) and in the unpublished (government classified) work of Fermi and Turkevich (1949). Gamow and his colleagues, in particular his PhD student Ralph Alpher, produced a series of early papers on this before Hayashi (1950) pointed out that at even higher temperatures neutrons and protons would coexist in an equilibrium that would determine the ratio of neutrons to protons. Hayashi recognised that this ratio depended only on the ambient temperature and further recognised that the key factor was the value of this ratio, n/p, at the temperature when the first deuterium could be formed. He came up with the answer $n/p = 0.25$ using a neutron half-life of around 30 minutes.

Short biography: **Chushiro Hayashi** (1920–2010) earned his BSc degree in physics in 1942 at the University of Tokyo and then, in 1945, went on to work in his home town, Kyoto, with the then future Nobel Prize (1949) winner, Hideki Yukawa. In 1950 he wrote a paper on weak interactions and the neutron:proton equilibrium in the very early Universe. This paper has often been referred to simply as a 'contribution to the Gamow theory of the Big Bang', but it was in fact a far more significant insight than that: he had pushed the application of physics to cosmology back one important step towards the Big Bang itself. In 1957 he became a Professor of the University and went on to study pre-main sequence stellar evolution, writing in 1961 and 1962 papers describing what is now known as 'the Hayashi Phase' for collapse of a protostar to the main sequence. He went on to work on planetary formation (the 'Kyoto Model'). During his long academic career he won many distinguished medals and prizes.

Hayashi had made some assumptions that needed correction,[23] and by 1953 the early history of the Hot Big Bang theory, almost as we know it today, was effectively finalised in the paper of Alpher *et al.* (1953). They had repeated Hayashi's work in greater detail and explored the effects of varying the neutron half-life. They found that they could vary the n/p ratio from $0.17-0.22$ simply by varying the neutron half-life over what they considered to be the probable range.

There still remained some important issues relating to the weak interaction which, in the early 1950s, was poorly understood. A decade after the Alpher, Follin and Herman paper (Alpher *et al.*, 1953) was published, Hoyle and Tayler (1964) used the by then improved understanding of the weak interaction to repeat the Big Bang nucleosynthesis story. Even then, there were also computational difficulties, computers as we know them today hardly existed. With the arrival of the very first commercially available mainframe computers Wagoner *et al.* (1967) were able to consider and handle 144 nuclear reactions.

7.3.1 Weak Interactions

Protons and neutrons are each made up of three quarks. The proton is made up of two up-quarks and a down-quark (uud) while the neutron is made up of two down-quarks and an up-quark (ddu). The up-quark has a charge of $+\frac{2}{3}e$ and the down-quark has a charge of $-\frac{1}{3}e$, and so the proton (uud) has charge $+e$ and the neutron (ddu) has no charge. Just as the electromagnetic force is mediated by the exchange of a photon (it is an 'exchange force'), the weak interaction is mediated by the exchange of the $W^{\pm}$ and Z^0 intermediate vector bosons.[24] Neutrinos feel only the weak interaction, and the weak interaction is the only force that can change the flavour of a quark. Neutrinos are the particles that can change neutrons into protons and vice versa.

At very high temperatures, the weak interaction (see Figure 7.5) controls the equilibrium between neutrons and protons through the reactions

$$n + e^+ \rightleftarrows \bar{\nu}_e + p, \tag{7.49}$$

$$n + \nu_e \rightleftarrows p + e^-. \tag{7.50}$$

---

[23] The outcome of these earlier works by Gamow and colleagues is succinctly summarised in footnote 2 of the important Alpher *et al.* (1953) paper. They refer to the work of Hayashi (1950, p.1348) saying 'A more detailed calculation was made by Hayashi, ...'. However, this somewhat minimises the inspirational insight behind the Hayashi paper that we could better understand the state of the Universe by understanding the phsyics at even earlier epochs. It is true that Hayashi had made the assumption that the neutron half-life was 30 minutes, but that was the only number available to him. In fact he may have seen or been told of the earlier comment by Snell and Miller (1948) giving, without explanation, a value of 30 minutes. See Footnote 27 p.177. Hayashi could not have known about the work of Fermi and Turkevich (1949, remains unpublished since the work is classified).

[24] The W-boson exists in two charge states, W^+ and W^-, while the Z-boson is neutral.

The W and Z particles were discovered in 1983, as a result of which Rubia and van de Meer were awarded the 1984 Nobel Prize in Physics. Their existence was an essential part of the *Electro-Weak* theory of Glashow, Salam and Weinberg for unifying the weak and electromagnetic interactions, for which they were awarded the 1979 Nobel Prize in Physics. W has a mass of $\sim 80.4\,\mathrm{GeV}/c^2$ and Z has a mass of $\sim 91.2\,\mathrm{GeV}/c^2$ and at temperatures above $kT \sim 100\,\mathrm{GeV}/c^2$ the electromagnetic and weak forces are unified into a single force. The Z-boson is the mediator of the interaction between electrons and neutrinos.

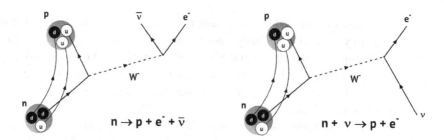

Fig. 7.5 Examples of the weak interaction involving proton, neutron and electron and mediated by the W^- particle. The left figure depicts beta decay (7.52) and the right figure depicts the interaction between a neutron and a neutrino (7.50). In both pictures, the neutron is transformed into a proton by changing one of the neutron's down quarks into an up quark via a W^- vector boson.
Time can be viewed as flowing upwards when interpreting these diagrams. ©

These two reactions maintain a neutron to proton ratio, $n{:}p$, that is determined by the thermal equilibrium between these reactions at a cosmic temperature T, but only for as long as the reaction rate, Γ, is higher than the expansion rate, H: $\Gamma \gg H$. During that period the equilibrium $n{:}p$ ratio is given by[25]

$$n/p = \exp\left(-\Delta mc^2/kT\right), \quad \Delta m = m_n - m_p = 1.2934\,\text{MeV}. \tag{7.51}$$

In the standard model for the expansion we can show that this equilibrium is maintained until the temperature falls to around $kT = kT_f \sim 0.8\,\text{MeV}$, i.e. some $\sim 1.5\,\text{s}$ after the Big Bang. At that point the $n{:}p$ ratio has a value around $\sim 1/5$.

The temperature T_f is called the *freeze-out temperature*. Increasing T_f increases the $n{:}p$ ratio, and makes more neutrons available for turning into helium.

The details of this scenario depend on the weak interaction rate at a given density and temperature, $\Gamma = \langle n\sigma v \rangle$, and the cosmic expansion rate, H : $H^2 = \frac{8}{3}\pi G\rho$, where in the present situation $\rho = (\pi^2/30)g_* T^4$ with which $H \propto T^2$. For our relativistic radiation gas the theory of weak interactions gives $\sigma \propto T^2$ and $v \sim c$. Comparing the reaction rate to the expansion rate: $\Gamma/H \propto T^5/T^2 \propto T^3$. If we put in the numbers we get the simple result that for the weak interactions, $\Gamma \sim T/1\,\text{MeV}$ (see, for example, Mathews *et al.* (2005)). So this decoupling occurs when the temperature falls to $\sim 1\,\text{MeV}$, i.e. around ~ 1s after the Big Bang.

Shortly after the decoupling of the neutrinos, the electrons and positrons annihilate when the temperature falls to $T \sim 0.5\,\text{MeV}$. This happens when the Universe is $\sim 3-5\,\text{s}$ old. The energy from the electron–positron pair annihilation produces the photons of the cosmic background radiation that we see today.

Neutron Decay

Once the neutrinos decouple, the remaining neutrons can decay freely via the reaction:

$$n \rightleftarrows p + e^- + \bar{\nu}_e. \tag{7.52}$$

---

[25] The mass of the neutron is $m_n = 939.6\,\text{MeV}$ and the mass of the proton is $m_p = 938.3\,\text{MeV}$.

This neutron decay is stopped at the time when the timescale for nuclear reactions to make deuterium becomes shorter than the expansion timescale.[26]

The crucial parameter is the neutron half-life. After more than 50 years of experimentation this is still one of the most poorly known physical parameters. The first observation of neutron decay to be reported was that of Snell and Miller (1948), who gave a preliminary estimate of 'about 30 minutes'.[27] The first measurement came with Robson (1950) who then gave a range of 9–18 minutes, and who was soon after to narrow the value to 12.8mins $\pm$ 2.5 mins. (Robson, 1951).

The value reported by Yue *et al.* (2013) for the neutron half-life is 10.255mins $\pm$ 0.014 with a nominal systematic error of ± 0.022. Modern results are generally quoted as the *mean neutron lifetime* τ_n which is related to the half-life $\tau_{1/2}$ simply by $\tau_{1/2} = (\ln 2)\tau_n$. The mean neutron lifetime is $\tau_n = 887.7s \pm 1.2(stat) \pm 1.9(syst)s$.[28]

7.3.2 Building Light Elements – Nucleosynthesis

The paper by Hoyle and Tayler (1964), in a mere three pages, gave a calculation of the helium abundance in a Hot Big Bang Universe that compared with the abundance data available at the time (Osterbrock and Rogerson, 1961). Hoyle saw this as the best evidence yet for a hot singular origin of the Universe. Zel'dovich, on the other hand, had at the same time instructed Smirnov (1965) to compute the nucleosynthesis of light elements in a Hot Big Bang, apparently with a view to disproving the hot singular origin and as a means of supporting his preference for a cold singular origin. After the discovery of the cosmic microwave background radiation in 1965, definitive papers were then written by Peebles (1966a,b) and Wagoner *et al.* (1967).[29]

The approach used in the following is didactic rather than rigorous.[30] In what follows we shall, for simplicity, divide the cosmic evolution into a few distinct phases during the period when the Universe was $\sim 10^{-2}$ s old, at which time the populations of neutrons

---

[26] If there were no further nuclear reactions, then the neutron decay would continue until all the remaining neutrons decayed to protons. The early Universe would consist only of hydrogen, and the only elements present now would be those made in stars. There would be very little helium.

[27] The 'paper' of Snell and Miller (1948) was in fact only a brief abstract of a talk given at a meeting of the American Institute of Physics, no details were given. To quote: .. *the number of neutrons in the sample* $(4x10^4)$ *give for the neutron a half-life of about 30 minutes. It is at present much safer however to say that the neutron half life exceeds 15 minutes.* Hayashi (1950) used 30 minutes because that was the only value available to him at the time.

[28] Olive and Particle Data Group (2014, p.933, and in particular the compilation on p.1380) gives a 'world average' of $880 \pm 0.9s$, but adds (*loc. cit.* p.933) that 'a movement back to a longer neutron lifetime ($\sim 886s$) is likely'. There are two experimental methods for determining the neutron half-life: 'bottle' and 'beam'. The results of Yue *et al.* (2013) are for a beam experiment. There is a systematic difference between the results found by the two techniques of 8.4 ± 2.2 s, which is significant at the 3.8σ level. The values generally quoted are an average over both kinds of experiment.

[29] It is important to remember that in these very early papers, while the theory is the same as today, all calculations were done with a longer neutron half-life (~ 1100 s prior to 1970) and only one family of neutrinos.

[30] The three excellent books of Weinberg (1972, 1993, 2008) discuss the early Universe and, in particular, nucleosynthesis with differing levels of detail, the first of these being the most detailed. While the values of some of the fundamental parameters of physics have changed since that was written, the development of the subject is nonetheless valid and very clear.

and protons were equal, to the end of the nucleosynthesis period, which is around 10^3 s. So we shall, for example, speak of 'the time of neutrino decoupling' when in fact the process extends to and influences the physics at the later 'time when e^+e^- annihilation started' (see p.184) . The only way to do the calculation is to solve the relevant differential equations.

A Simple First Look

The difference in mass between the neutron and proton is $Q = (m_n - m_p)c^2 = 1.2934$ MeV. In equilibrium at temperature T, the neutron:proton ratio is

$$n/p = e^{-Q/kT} \simeq 1/5 \quad kT_f \sim 0.8 \, \text{MeV}, \tag{7.53}$$

where T_f is a rough estimate of the *neutron freeze-out temperature* when the reactions are too slow, compared to the cosmic expansion time, to maintain the equilibrium. After a further ~ 200 s, the temperature falls to ~ 0.08 MeV when the remaining neutrons will all be turned into deuterium. Since the mean neutron life-time is ~ 880 s, neutron decay during that 200 s reduces the n:p ratio by a further factor $e^{-200/885}$ to a value around $n/p \sim 1/6$. This is, approximately, the neutron:proton ratio at the time the neutrons are converted to deuterium. The deuterium is rapidly converted to ^{4}He, leaving behind traces of D, ^{3}He, ^{7}Li and other light elements.[31]

If, for the sake of simplicity, we consider two neutrons and 12 protons, so that the n:p ratio is 1/6, the two neutrons will merge with two protons to make a ^{4}He and leave ten protons. We can sketch this as

$$
\begin{array}{ccccc}
\blacklozenge\blacklozenge & & \blacklozenge\blacklozenge & & \lozenge\lozenge\lozenge\lozenge\lozenge \\
\lozenge\lozenge\lozenge\lozenge\lozenge\lozenge & \longrightarrow & \lozenge\lozenge & + & \lozenge\lozenge\lozenge\lozenge\lozenge \\
\lozenge\lozenge\lozenge\lozenge\lozenge\lozenge & & & & \\
2 \times n + 12 \times p & & {}^4\text{He} & & 10 \times p
\end{array} \tag{7.54}
$$

from which it is obvious that we have 28% of ^{4}He by mass and that $1/11 \simeq 9\%$ of the atoms in the Universe are ^{4}He. This argument leads to the general approximation for the ^{4}He abundance mass fraction, Y_p:

$$Y_p \simeq \frac{2n}{n+p} = 2\frac{n}{p}\frac{1}{(1+n/p)}. \tag{7.55}$$

The second equality simply expresses Y_p in terms of the expected neutron:proton ratio at the time of the start of nucleosynthesis ($t \sim 200$ s).

Nuclear Reactions – Helium Production

Once the weak interactions are turned off, the fast strong and electromagnetic interactions take over the process of element building. A large number of nuclear processes are listed in Wagoner *et al.* (1967, Table 2(b)). The reactions use by Peebles (1966a,b) were

---

[31] I will adhere to the classical (20th century) notation of using D, or d, for deuterium, T, or t, for tritium instead of the more natural ^{2}H and ^{3}H, which makes it explicit that these are isotopes of hydrogen.

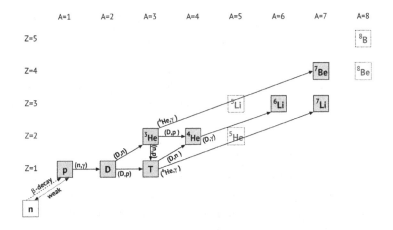

Fig. 7.6 Essential nuclear reactions building up light elements. These were the main reactions as identified by Smirnov (1965); Peebles (1966a,b), and as listed in Equations (7.56–7.64). Subsequently, Wagoner *et al.* (1967, Table 2(b)) used a greatly increased list of 144 reactions. There are no stable nuclei with atomic mass numbers 5 or 8 (shown greyed out) so only small amounts of ^{7}Li and ^{6}Be are produced. The A = 8 mass gap is bridged by the reaction ^{7}Be(^{4}He, γ)^{11}C. ⓒⓒ

$$n + p \rightleftarrows D + \gamma \qquad n(p, \gamma)D, \qquad (7.56)$$

$$D + D \rightleftarrows {}^3\text{He} + n \qquad D(D, n){}^3\text{He} \qquad (7.57)$$

$$D + D \rightleftarrows T + p \qquad D(D, p)T, \qquad (7.58)$$

$${}^3\text{He} + n \rightleftarrows T + p \qquad {}^3\text{He}(n, p)T, \qquad (7.59)$$

$$T + D \rightleftarrows {}^4\text{He} + n \qquad T(D, n){}^4\text{He}. \qquad (7.60)$$

Smirnov (1965, published in the Russian original in 1964) also included

$${}^3\text{He} + D \rightleftarrows {}^4\text{He} + p \qquad {}^3\text{He}(D, p){}^4\text{He}. \qquad (7.61)$$

These reactions[32] are depicted in Figure 7.6, displayed as a function of atomic number, Z (the number of protons in the atom) and the atomic mass number, A (the total number of neutrons and protons) in the atom.

Nuclear Reactions – beyond Helium

Whereas the Universe makes copious amounts of helium, ^{4}He, manufacturing elements beyond helium is not as efficient. We cannot easily get to higher atomic numbers since elements of atomic number 5, like ^{5}He and ^{5}Li are unstable and so cannot be used as stepping-stones to heavier elements. So we have to jump to the synthesis of ^{6}Li, ^{7}Li and ^{7}Be which are produced by reactions such as

$${}^4\text{He} + D \rightleftarrows {}^6\text{Li} + \gamma \qquad {}^4\text{He}(D, \gamma){}^6\text{Li}, \qquad (7.62)$$

---

[32] The two columns in this list of reactions (7.56–7.61) are simply alternative notations for the same reactions.

$$T + {}^4He \rightleftarrows {}^7Li + \gamma \qquad T({}^4He, \gamma){}^7Li, \tag{7.63}$$

$$^3He + {}^4He \rightleftarrows {}^7Be + \gamma \qquad {}^3He({}^4He, \gamma){}^7Be. \tag{7.64}$$

We again meet a barrier because, as there was with atomic number A = 5, there is another bottleneck at A = 8 where the nuclei, like ^8Be, are unstable and so cannot be stepping-stones to heavier species. The main route beyond $A = 8$ is to convert ^7Be to ^{11}C via the reaction

$$^7Be + {}^4He \rightleftarrows {}^{11}C + \gamma \qquad {}^7Be({}^4He, \gamma){}^{11}C. \tag{7.65}$$

Gamow and his collaborators had noticed that the measured relative abundances (see Figure 7.7) of the elements (Goldschmidt, 1938; Brown, 1949)[33] showed an exponential decrease in abundance with increasing atomic weight for atomic weights A < 100. Importantly, there was also a clear correlation with neutron capture cross-sections. This formed the basis of the 'non-equilibrium' neutron capture theory of Gamow (1946) for the origin of the elements (Alpher, 1948).[34] This would provide a normalisation for the cross-sections at an energy of 1 MeV. This is discussed in great detail by Alpher and Herman (1950, Section IV and, in particular, Figures 12 and 13). The work of Gamow and colleagues was motivated and supported by the correlation of relative abundances of elements with their nuclear binding energy: this was the 'non-equilibrium theory for the origin of the elements'. Alpher and Herman (1950, Section III) discuss this theory in considerable detail.

The alternative theory for the formation of the elements was that they were formed by nuclear reactions in stars. This was an older idea set in motion by the important recognition by Bethe and Critchfield (1938); Bethe (1939) that thermonuclear processes could power stars over long periods of time. The idea that the origin of the chemical elements was due to nucleosynthesis in stars and not the Hot Big Bang was put forward by Hoyle (1946, 1947, 1954)[35] whose work was to lead, a decade later, to the famous paper, by E.M. Burbidge, G.R. Burbidge, W.H. Fowler and F. Hoyle (Burbidge *et al.*, 1957, known as BBFH). This was one of the papers that led to Fowler being awarded the Nobel Prize in Physics in 1983 (but, surprisingly, his collaborator Hoyle, who had started this work, did not share in the prize).

---

[33] This data was compiled from solar, stellar and meteoritic data and did not contain data on Li, Be, B.

[34] 1948 was a remarkably productive year for Gamow's group, which has been the subject of an article by Peebles (2014, *Discovery of the Hot Big Bang: What happened in 1948*). The article cited, Alpher (1948), is the paper based on Alpher's PhD thesis, which he completed under the direction of George Gamow.

The discussion about correlations involving the distribution of abundances occurs repeatedly throughout a number of the group's papers over the period 1946–1953. The estimation of cross-sections from the observed abundance is key to their research, where their goal was to argue that all elements were synthesised via non-equilibrium neutron capture during the first minutes of the Big Bang. See Peebles (2014) for an insightful retrospective and critique of this work.

The alternative view was that of Hoyle (1946, 1947) which proposed that nucleosynthesis of the elements would take place in stars (see also Hoyle (1950)).

[35] It was in the year following these first two papers on stellar nucleosynthesis that Hoyle put forward the theory that the Universe was in a Steady State (Hoyle, 1948): a cosmological model without a hot singular origin (see also Bondi and Gold (1948)). In such a theory it was essential that the chemical elements be synthesised in stars.

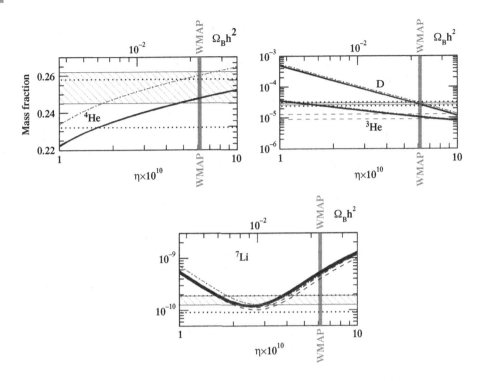

Fig. 7.7 Observed abundances of elements D, ^{3}He, ^{4}He and ^{7}Li compared with computations of Big Bang nucleosynthesis for a variety of values of η_{10} (the normalised number of photons per baryon), or, equivalently, $\Omega_B h^2$. The shaded horizontal bands depict the plausible range of observationally determined estimates for the abundance of each of the elements (note the values on the vertical axes). The vertical line labelled 'WMAP' is the value of η_{10} (or, equivalently, $\Omega_B h^2$) as derived from the WMAP data for the CMB anisotropy. The labelled curves are the determinations of the abundances from latest nucleosynthesis computations, expressed as a function of η_{10}, or $\Omega_b h^2$. We see that the thoeretical values for D, ^{3}He, ^{4}He are satisfactory, while there is a substantial discrepancy in the abundance of ^{7}Li. This is the 'Lithium problem'.

Adapted from Coc (2013, Figure 3.), with permission. ©

The BBFH paper laid out the entire story of the synthesis of the elements in stars, with one notable exception: helium. By the early 1960s it had become evident that over 25% of the mass of the Universe was in the form of the element helium. Such a huge amount could not be made in stars. In 1964, Hoyle collaborated with Roger Tayler to write what in many ways was a remarkable paper: *The Mystery of the Cosmic Helium Abundance* (Hoyle and Tayler, 1964).[36]

[36] Here we see Hoyle, the scientist who had brought up the idea that the Universe might be in a Steady State with no initial singularity, collaborating on a paper reviving the notion of cosmic nucleosynthesis in a Hot Big Bang. This is in many ways a testament to Hoyle's stature as a world-leading scientist who was open to discussing diverse points of view. After the discovery of the CMB in 1965, Hoyle would collaborate in writing another paper on Big Bang nucleosynthesis, Wagoner *et al.* (1967).

In their paper, Hoyle and Tayler gave appropriate acknowledgement to the early work of Alpher, Gamow and Herman, and also to Hayashi who had settled the crucial issue of the initial conditions for such a

7.3.3 Observed Light Element Abundances

The elements D, ^{3}He, ^{4}He and ^{7}Li potentially provide an important validation for the cosmological paradigm: they probe the state of the Universe back to the first minutes of its existence. In the past, the abundances of these light elements were used to constrain the likely values of the cosmic radiation density, $\Omega_\gamma h^2$, and the cosmic density of baryons $\Omega_B h^2$. The cosmic microwave background (CMB) provided an independent and accurate measure of the first of these, and analysis of angular fluctuations of the temperature of the CMB provided accurate values for the latter. The nuclear abundances provide an independent check of the CMB-determined value of $\Omega_B h^2$, and also provide strong clues into the chemistry of the early evolution of stars and galaxies (anything not made in the Big Bang must be made in subsequent stellar evolution).

The current situation as regards determination of the cosmic abundances for these elements is shown in Figure 7.7, which summarises the detailed analysis of Coc (2013). The ^{7}Li abundance is the most discrepant value, the Big Bang nucleosynthesis value is about a factor ~ 3 greater than the spectroscopic observations of galactic halo stars.

7.4 Neutrino Cosmology

The existence of the neutrino was first postulated by Pauli in 1930 though it was not until 1933, after Chadwick's discovery of the neutron, that he felt the confidence to present the hypothesis to a conference of many (or most) of the world's most distinguished physicists (the 1933 7th Solvay Conference in Physics).[37] In 1945, Pauli was awarded the Nobel Prize in Physics 'for the discovery of the Exclusion Principle, also called the Pauli Principle'.

The first detection of a neutrino, the electron neutrino ν_e, was not until 1953 by Reines and Cowan (discovered by Reines and Cowan, 1953, confirmed by Cowan *et al.* (1956)), for which they were awarded the 1995 Nobel Prize in Physics.[38]

---

calculation. They came up with what would be considered a high value for the cosmic helium abundance (32%), but Roger Tayler was always keen to point out, correctly as it turned out, the role of the uncertainty of the neutron half-life in all these calculations.

[37] This was in an open letter (dated 4th Dec. 1930) to Hans Geiger and Lise Meitner, and other delegates, attending a meeting in Tubingen in December 1930: *Dear Radioactive Ladies and Gentlemen, ... I have hit upon a desperate remedy to save the 'exchange theorem' of statistics and the law of conservation of energy. Namely, the possibility that there could exist in the nuclei electrically neutral particles, that I wish to call neutrons, which have spin 1/2 and obey the exclusion principle.*
`http://neutrino.ethz.ch/Group_Rubbia/Neutrino.html`
In the letter he also explains that *But so far I do not dare to publish anything about this idea, ...* Pauli had justifiably been concerned about the apparent non-conservation of energy and of spin and statistics in beta decay. He evidently wrote the letter because he was unsure whether his radical suggestion of an unobservable particle would be generally acceptable. As it turned out, it was not acceptable at that time: he presented the idea at the American Physical Society meeting in Pasadena in June 1931, only to be met with extreme scepticism from his colleagues. We see that, in the letter, Pauli had referred to the particle as a 'neutron'. However, the name was used by Chadwick in 1932 for the particle he had just discovered, and it was Fermi who rescued the situation when in 1934 he coined the name 'neutrino'.

[38] The muon neutrino, ν_μ, was discovered by Danby *et al.* (1962). The tau neutrino had been postulated after the discovery in 1975 of the τ-lepton (Perl, 1975). Following that, there was a lot of indirect evidence for

Neutrinos have no charge and have almost no mass, so hardly interact with matter: they easily pass through the Earth almost without interaction. A neutrino detector must therefore have an enormous mass of material that is suitable for detecting neutrinos, and be shielded from other forms of radiation by another huge mass: this strongly suggests putting a vast detector as deep underground as possible. The next problem is to find a well understood source of neutrinos.

In 1965 Ray Davis decided to look for solar neutrinos[39] by building his detector 1.6 km underground in the Homestake gold mine in South Dakota.[40] His detector was a tank of some 400 000 litres of ordinary dry-cleaning fluid, perchloroethylene. A neutrino hitting a ^{37}Cl chlorine atom would produce an atom of argon ^{37}Ar, which would decay back to ^{37}Cl, emitting a detectable X-ray in the process. In effect, Davis' strategy was to syphon off the few argon atoms from the tank, collect them in a liquid nitrogen cooled graphite trap and count them![41] Davis could detect some 15 argon atoms per month in this way, and with predictions that he should detect about 8 SNU (Solar Neutrino Units: 1 SNU = 10^{-36} neutrino interactions per target atom per second),[42] he could be optimistic that his experiment could detect some solar neutrinos. He detected neutrinos, but only at a flux of only 2.56 ± 0.25 SNU. This huge discrepancy went a long way towards discrediting the experiment, and, in the absence of any other experiment to confirm or disprove Davis' measurement, the subject seemed doomed. The history is well reviewed by Nakahata (2011).

The situation was made no better by the Kamiokande I and II experiments in Japan. Started in 1982 by Masatoshi Koshiba and becoming operational in 1985, the detectors could even determine which general direction a neutrino had come from. This would clearly be an asset for detecting and studying solar neutrinos. By 1988 they had enough solar neutrinos to say that the flux was half of what was expected on the basis of the somewhat uncertain solar models, and in contradiction to Davis' experiment which was still reporting one third of the expected flux. Whether the problem lay in the solar models, the

---

the existence of ν_τ, but it was not 'seen' until 2001 when the specifically designed DONUT experiment went online at Fermilab (DONUT Collaboration *et al.*, 2001).

[39] Solar neutrinos are the product of the proton–proton chain of nuclear reactions taking place in the central regions of the Sun. The basic process is the exothermic reaction

$$2e^- + 4p \rightarrow {}^4He + 2\nu_e + 26.7\,\mathrm{MeV},$$

of which $\sim 3\%$ is carried off by the neutrinos. So one neutrino is created per ~ 13.5 MeV of energy generated.

[40] According to the *New York Times*, Davis had built his first neutrino detector in 1961 and located it in Ohio in a limestone mine at a depth of 2300 feet. This turned out to be not deep enough and his experiment was overwhelmed with cosmic rays. *New York Times* 2 June, 2006:
http://www.nytimes.com/2006/06/02/nyregion/02davis.html?_r=0

[41] Davis writes about the experiment and the years he spent perfecting this remarkable technique in his Nobel Lecture:
http://www.nobelprize.org/nobel_prizes/physics/laureates/2002/
davis-lecture.pdf
He reports in a letter to Fowler (11 August 1967, reproduced in the article) that he fished out 16 ± 4 atoms of argon from a 10^5 (US) gallon tank: a small swimming pool. Perhaps it is understandable that some people did not quite believe that he could do this, and that his count would be low because he'd missed some.

[42] There was some uncertainty in this expectation, in part due to uncertainties about the rate of the important reaction $^3He + {}^4He \rightarrow {}^7Be + \gamma$ which provided the ^7Be for the production of neutrinos via $^7Be + e^- \rightarrow {}^7Li + \nu_e$.

detectors or in the theory of neutrinos could not be determined: the *Solar Neutrino Problem* was born. As other experiments came on line, it became clearer that the problem lay with the neutrino physics.

It was not until the end of the 20th Century that there was any confirmation of the low solar flux. The Japanese Super-Kamiokande detector and the Sudbury Neutrino Observatory (SNO) both became operational around 1999. As it happened, there was a theory-in-waiting: Pontecorvo and Gribov had suggested that neutrinos could behave differently than suggested by the then-standard model of particle physics, an idea that had not up until then been taken seriously.

Finally, in 2002, Ray Davis and Masatoshi Koshiba were awarded the Nobel Prize for Physics.[43]

7.4.1 Effective Neutrino Number, N_{eff}

We showed Section 7.2.4 that the temperature of the neutrino component, once decoupled, is related to the temperature of the photon component by

$$T_\nu = \left(\frac{4}{11} \right)^{1/3}, \; T_\nu = 0.71377 \, T_\gamma, \tag{7.66}$$

where T_γ is the temperature of the cosmic background radiation field. This holds at all epochs subsequent to the neutrino decoupling, and as long as the neutrinos are relativistic. With the present CMB temperature $T_{\gamma 0} = 2.725$ K this gives a neutrino temperature of $T_{\nu 0} = 1.945$ K.

This was derived under the simplifying assumption that the decoupling of neutrinos from the electron–positron plasma is instantaneous, which is an idealisation of the real situation. When this and other effects are considered,[44] the value of g_* relating the energy density ρ_r of the radiation fields to the temperatures of the individual components (Equation 7.46) needs a small modification. This is absorbed into the number of neutrino species, N_ν, replacing it with the 'effective neutrino number' N_{eff} (Mangano *et al.*, 2005):

$$N_{\text{eff}} = 3.046. \tag{7.67}$$

This is a small but significant change in the world of precision cosmology. This value is variously referred to as the 'standard model' or the 'minimal model'. What this means is that inference about the number of neutrino species from cosmological data sets tells us

---

[43] John Bahcall, who had been the driving force behind constructing solar models and the quest for neutrino astrophysics, wrote a fine article for the official website of the Nobel Prize in commemoration of Davis' pioneering work on the search for solar neutrinos,
http://www.nobelprize.org/nobel_prizes/themes/physics/bahcall/
Many scientists were of the opinion that John Bahcall should have shared the prize with Davis and Koshiba. Ray Davis died in 2006 aged 91.

[44] The neutrinos decouple from the electrons at a temperature, $T_{\nu_e}^{\text{dec}} \simeq 2.4$ MeV which is significantly higher than the electron rest-mass temperature. Hence the neutrino decoupling was not complete at the time $e^+ e^-$ annihilation started. That means there is a slight feedback of energy to the neutrinos from the annihilating electrons. There are a number of additional technical complications, one of which is the need to consider the different physical properties of the three neutrino flavours. The details are in Mangano *et al.* (2005, Section 2) and Hannestad (2006, Section 3.1).

N_{eff} rather than N_ν. If there is additional physics to be considered, such as the existence of a fourth, sterile, neutrino which partakes in no interactions, then our experiments would measure the quantity

$$\Delta N_{\text{eff}} = N_{\text{eff}} - 3.046. \tag{7.68}$$

The value for N_{eff} reported by the Planck consortium (Planck Collaboration XIII, 2015, Section 6.4.2, Eq. (60b)) using combined Planck and BAO data analysis was

$$N_{\text{eff}} = 3.15 \pm 0.23. \tag{7.69}$$

Since the energy density of both radiation and massless neutrinos falls off in the same way, we can write for the density of the component made of massless particles, i.e. photons and relativistic neutrinos,

$$\rho_r = \rho_\gamma + \rho_\nu = \left[1 + \frac{7}{8} \left(\frac{4}{11} \right)^{4/3} N_{\text{eff}} \right] \rho_\gamma \tag{7.70}$$

$$= 1.6918 \rho_\gamma \quad \text{for } N_{\text{eff}} = 3.046, \tag{7.71}$$

where N_{eff} is the effective number of neutrino species. If we had used N_ν in place of N_{eff}, the factor 1.6918 would have been 1.681.

With this it is simple to work out the ratio of photons to neutrinos in a unit volume and thence the contribution, $\Omega_\nu h^2$, of neutrinos to the total cosmic density:

$$\Omega_\nu h^2 = \frac{7}{8} \left(\frac{4}{11} \right)^{4/3} N_{\text{eff}}^\nu \Omega_\gamma h^2, \tag{7.72}$$

where N_{eff}^ν is the effective number of neutrino species. This tells us that the neutrinos make of the order of twice the contribution to the total cosmic mass density as the photons in the microwave background.

7.4.2 Neutrino Flavours and Helium

Since the CMB fixes both the baryon density and the radiation density, the predicted abundance of light elements, when compared with observational measures, provides a check on the physics that underlies the nuclear processes. In particular, cosmic nucleosynthesis provides a consistency check on the number of flavours of neutrino since increasing the number of neutrino flavours, N_{eff}, increases the Hubble parameter.[45] By virtue of Equation (7.51) in Section 7.3.1, this increases the $n{:}p$ ratio and hence the helium abundance. The following table shows this:

Neutrino flavours	^{4}He prediction
$N_\nu = 2$	$Y \sim 0.233$
$N_\nu = 3$	$Y \sim 0.247$
$N_\nu = 4$	$Y \sim 0.260$

[45] The Hubble parameter, H, is given by $H^2 = \frac{8}{3}\pi G \rho$, where $\rho = \frac{7}{8}\frac{1}{2}g_* T^4$ (Eq. 7.15).

The values are for a baryon density such that $\eta = 6.1$, (from Cyburt $et\ al.$ (2016, Fig.7)). The best estimate for N_{eff} is $N_{\text{eff}} = 2.994 \pm 0.012$ from the decay of Z^0 into a neutrino and anti-neutrino.

The relationship between cosmological neutrinos and Big Bang nucleosynthesis is reviewed by Steigman (2012).

8

The Inhomogeneous Universe

In this chapter we deal with the equations that govern the evolution of inhomogeneous cosmological models within the framework of Newtonian theory. After a brief overview we discuss how to describe an inhomogeneous Universe in which the density varies randomly from point to point. Then we go on to look at slightly inhomogeneous models. First we cover pressure-free models, and then small amplitude perturbations where pressure might play a part both prior to and after the recombination epoch.

Finally we discuss the brief period during which, after 380 000 yr, the Universe emerges from the fireball. The decoupling of the matter and radiation fields leaves an important imprint on the observed structure, the so-called 'Baryonic Acoustic Oscillations' or BAOs. These baryonic acoustic oscillations are seen both in the cosmic background radiation field and in the distribution of galaxies and provide important cosmological information with high accuracy.

8.1 Evolution of Inhomogeneous Models

The Universe today is manifestly inhomogeneous: it is filled with stars, galaxies and clusters of galaxies. From observations of the cosmic background radiation, we know that it was far more homogeneous in the past than it is now. Indeed, at a redshift of ~ 1100 when the Universe emerged from the fireball phase of the expansion, the degree of inhomogeneity on all scales was far less than 1%. However, the degree of inhomogeneity at that time was measurably not zero: there was some structure at that time. The galaxies and clusters of galaxies must have grown from that initially almost homogeneous state.

The force responsible for that growth was the force of gravity acting on the small fluctuations so as to amplify them until they separate out from the background Universe and collapse to form the large scale systems we see today.

8.1.1 Gravitational Instability

The idea of gravitational instability being responsible for the origin of cosmic structure goes back to the work of Poincaré and Jeans, but the issue of the stability of general relativistic cosmological models can be traced back to the work of McCrea and McVittie (1930); McVittie (1931, 1932, 1933) who built directly on the work of Lemaître (1927) and Robertson (1929).[1] The paper of McVittie (1932) was the first to build a so-called *Swiss cheese* model for an inhomogeneity.[2]

Such models were developed by Lemaître (1931b) and Tolman (1934a), and later by Einstein and Straus (1945, 1946) who modelled the inhomogeneity by cutting out non-overlapping spheres of matter in an otherwise homogeneous cosmology, and replacing the cut out material with an equivalent point mass at the centre of its sphere. By Birkhoff's theorem this would not affect the solution in the remaining homogeneous substratum, and hence the overall dynamical evolution of the model would be unchanged. Bondi (1947) built on the models of Lemaître and Tolman *loc. cit.* and, importantly, showed the correspondence with a Newtonian solution to the same problem. This important solution is variously known as the *Bondi–Tolman* solution or the *Bondi–Lemaître–Tolman* ('BLT' or 'LTB') solution. Most, though not all, of this early work on BLT considered models in which the cosmic medium was pressure free 'dust' (see Section 16.4).

The first work on studying inhomogeneous models via perturbation theory was that of Lifshitz (1946), which was later clarified within a Newtonian context by Bonnor (1954, 1957). The use of perturbation theory opened up the possibility of dealing with more general cosmic constituents for which an equation of state was given, and a wide variety of physical processes that could take place in such a medium. The linearity assumption also allowed the possibility of considering density fields having random fluctuations. We were no longer confined to dust solutions made up of non-overlapping spherical 'vacuoles'.

The discovery of the cosmic background radiation changed everything because it established that the early Universe was hot. At about the time of the discovery, the groups in Princeton and Moscow independently established a many decades long programme to fill in the details of the new paradigm (see Section 3.1.1). One of the prime goals was to understand the origin and evolution of cosmological structure. Interestingly, the groups set off in rather different directions.[3] The Princeton group of Dicke, Peebles and Wilkinson set about refining measurements of the background radiation and understanding the growth of perturbations in relation to the presently observed cosmological structure (Peebles, 1965, 1967). To that end Peebles' group initiated the analysis of catalogues of galaxies using correlation and Fourier techniques (Peebles, 1980). The Moscow group of Zel'dovich focused

---

[1] It is interesting that McCrea and McVittie (1930, line 3 *et seq.*) begin their paper with the words: *It appears (Eddington) that our actual Universe may probably be considered as expanding from an initial state something like 'Einstein's World' towards a limiting state like 'de Sitter's world'.* In the following paper, McVittie (1931) begins by acknowledging that: *It has been shown by Lemaître that the phenomenon known as the recession of the nebulae can be explained, on the theory of relativity, if the Universe is regarded as a three-dimensional sphere of expanding radius.* The reference here is to the paper of Lemaître (1927) and so these papers serve to indicate at what stage the general relativity experts were convinced that the Lemaître solution described the expanding Universe. Of course, Friedmann (1922, 1924) had already discovered the same formal solution, but it was perhaps the combination of a general relativistic model coupled to an incisive astronomical analysis in the 1927 paper that gave added weight to Lemaître's vision.

[2] The Swiss cheese model is widely ascribed to Einstein and Straus (1945). However, in a subsequent paper, 'Correction and Additional Remarks', Einstein and Straus (1946) graciously acknowledge that they had been unaware of the earlier work of McVittie *loc. cit.*

[3] There was nonetheless a considerable overlap in the topics covered by the two groups. Communication between the groups was minimal, this was the cold war period. However, two very important opportunities for meetings took place under the auspices of the IAU: the symposia at Tallinn, Estonia in 1977 (Longair and Einasto, 1978) and Balatonfured, Hungary in 1987 (Audouze *et al.*, 1988).

on understanding the physics of the cosmic radiation field (e.g. Zel'dovich *et al.*, 1969) and the nonlinear growth of the large scale structure (Doroshkevich and Shandarin, 1974).

Much later, direct evidence for the validity of the small amplitude approximation came with the first measurements of the anisotropy of the cosmic background radiation, which revealed a level of inhomogeneity of a few percent at a redshift of $z \sim 1000$: linear theory was applicable to the early Universe.

8.1.2 Growth of Present-day Structure

The present day Universe is far from homogeneous and, except for the very largest scales, cannot be described in terms of small amplitude fluctuations in density. Gravity has driven the linear primordial fluctuations in density to nonlinear levels such that stars, galaxies and clusters of galaxies can be formed.

In 1970 Zel'dovich (1970) provided a fully nonlinear solution within the framework of zero pressure Newtonian cosmological models that made possible a description of the transition from the linear regime to the highly nonlinear regime. One of the predictions of this theory was that structure in the present day Universe would be dominated by large voids surrounded by thin walls (*pancakes*) in which galaxies would form. The physics of this *pancake theory* were worked out by Doroshkevich and Shandarin (1974) and others. The first visualisation of what such a model for large scale structure might look like came with the 2-dimensional Vlasov simulation of Doroshkevich *et al.* (1980). The dominance of voids in the large scale structure was discussed from an observational point of view by Zel'dovich *et al.* (1982) and strikingly revealed in the Centre for Astrophysics 'Slice of the Universe' by de Lapparent *et al.* (1986).

The original Zel'dovich solution applied to non-interacting 'dust' particles, i.e. it described the motions of particles in a collisionless cold gas, so that counter-streaming particles would simply pass through one another. In a major development of the Zel'dovich model, Gurbatov *et al.* (1985b, 1989) were able to provide an analytically tractable model based on the Burgers equation in which counter-flowing streams would dissipate energy in shock waves. This important model was first presented by Gurbatov *et al.* (1985a, 1989) and is referred to as the *adhesion model* (see also the review of Gurbatov *et al.* (2012)). The resulting cosmic structure was first visualised by Kofman *et al.* (1990) in a 2-dimensional simulation of the adhesion process.

8.1.3 Model Predictions

The microwave background temperature fluctuations tell us about the state of the Universe as it emerges from the fireball phase of the expansion. This is about as far as we can look back directly: we look back to a redshift of $z \sim 1100$. Despite the very low level of inhomogeneity revealed in the CMB maps, there is a rich structure there which will lead to the formation of galaxies, clusters of galaxies and the cosmic web. Observing the way in which galaxies are distributed in space, their clustering properties and the general cosmic web-like structure that they trace out, brings additional information about the nature of the initial conditions. Providing a quantitative characterisation of that distribution has been a major goal when analysing the relevant data.

The present day structural organisation of cosmic material is the result of a complex nonlinear gravitational clustering process that is best understood via numerical simulations, N-body models. Given the initial conditions and the cosmological model, such simulations follow the evolution of the Universe from the linear theory phase to the present day. The 'catch' here is that the basic N-body models only have gravitational forces and so only model collisionless dark matter components. Modelling the evolution with a luminous baryonic component requires additional physics that is difficult to incorporate into the basic models. Notwithstanding that, N-body models have been the predominant tool for modelling the evolution of the Universe from the time of recombination to the present day.

8.2 Describing Random Density Fields

The fluctuating density field that gives rise to the stars, galaxies and larger scale cosmic structures we see emerged from small amplitude fluctuations in the cosmic density field at the time when the cosmic plasma recombined and the radiation field decoupled from the matter. Those fluctuations are thought to have originated in the earliest instants of the cosmic expansion, perhaps during the period of inflation. Here we shall confine attention to a single component field of either dust (which might be non-relativistic dark matter), baryons, or baryons tightly coupled to radiation.

There are many ways of characterising structure. This is a difficult task for the present day structure in which we see a highly organised web-like pattern in the distribution of luminous matter (Jones *et al.*, 2004). However, it is known that the amplitudes of the primordial density fluctuations, which lead to the development of the structure we see, are very small and that their underlying statistical distribution is probably Gaussian. By definition the mean of the density fluctuations is zero, so the intensity of the fluctuations is described by the variance of the deviation about the mean, normalised relative to the mean density. In a Gaussian random field, the statistics of a zero mean random distribution is entirely determined by the two-point correlation function of the field values (Equation (8.13)).

8.2.1 Defining the Perturbations

We write the density field, $\rho(\mathbf{x})$, as the sum of the mean field value, ρ_0, and of the fluctuating part of the field, $\delta(\mathbf{x})$, which is defined by

$$\rho(\mathbf{x}) = \rho_0(1 + \delta(\mathbf{x})), \quad \langle \delta(\mathbf{x}) \rangle = 0, \quad \langle \rho(\mathbf{x}) \rangle = \rho_0. \tag{8.1}$$

The angular brackets $\langle . \rangle$ are taken to mean statistical averages over the probability distribution of δ.[4] In the limit where $|\delta(\mathbf{x})| \ll 1$, the mean ρ_0 can be unambiguously defined, and we are said to be in the *linear regime* where effects of order $O(\delta^2)$ are to be ignored.

---

[4] This, and other technical issues, are discussed in greater depth in the On-line Supplement *Random Fields*.

The condition that the fluctuations $\delta(\mathbf{x})$ are small is simply

$$| \delta(\mathbf{x}) | \ll 1. \tag{8.2}$$

It should be noted that the positivity of the density also requires $\delta(\mathbf{x}) \geq -1$. Condition (8.2) is assumed to adequately fulfil the positivity requirement.

8.2.2 Gaussian Random field of Density Fluctuations

The field $\delta(\mathbf{x})$ is considered to be a field whose values are specified statistically. One of the simplest such random fields is the *Gaussian (random) field* whose values at a point are taken from a Gaussian distribution of zero mean and constant variance. In other words there is an underlying probability density, $\mathbb{P}[\delta]$, from which the value of $\delta(\mathbf{x})$ at a randomly selected point $\mathbf{x}$ is drawn:[5]

$$\mathbb{P}[\delta] = \frac{1}{\sqrt{2\pi}\sigma} \exp\left(-\frac{\delta^2}{2\sigma^2}\right). \tag{8.3}$$

In a Gaussian random field, the joint distribution of the field values $\delta(\mathbf{x}_1), \delta(\mathbf{x}_2)$ at pairs of points, $\mathbf{x}_1, \mathbf{x}_2$ separated by a distance $r = | \mathbf{x}_1 - \mathbf{x}_2 |$ is given by :

$$\mathbb{P}(\delta_1, \delta_2, r) = \frac{1}{2\pi\sigma^2\sqrt{1 - C^2(r)}} \exp\left\{-\frac{\delta_1^2 + \delta_2^2 - 2C(r)\delta_1\delta_2}{2\sigma^2[1 - C^2(r)]}\right\},$$

$$\delta_1 = \delta(\mathbf{x}_1), \quad \delta_2 = \delta(\mathbf{x}_2), \quad r = | \mathbf{x}_1 - \mathbf{x}_2 |. \tag{8.4}$$

The function $C(r)$ is called the *2-point covariance function* of the field $\delta(\mathbf{x})$, and, in an *isotropic Gaussian field*, it depends only on the separation, r, between the two points.[6] In addition, in a Gaussian random field the N-point probability density function must be a multi-variate Gaussian. The hierarchy of N-point functions is determined by the mean and covariance that appear in the the 2-point function (8.4) (Adler, 1981, Section 1.6). Gaussian random fields, or 'Gaussian fields' as they are commonly referred to, are very special.

It is useful to note that Equation (8.4) can be written in an alternative form:

$$\mathbb{P}(\delta_1, \delta_2) = \frac{1}{\sqrt{(2\pi)^2|\mathbf{M}|}} \exp\left[-\frac{1}{2}\delta^T \mathbf{M}^{-1}\delta\right], \tag{8.5}$$

where δ is the column vector $(\delta_1, \delta_2)^T$, and $\mathbf{M}$ is the *covariance matrix*. In 2-dimensions:

$$\mathbf{M} = \sigma_\delta^2 \begin{pmatrix} 1 & C(r) \\ C(r) & 1 \end{pmatrix} \quad \text{and} \quad \mathbf{M}^{-1} = \frac{1}{\sigma_\delta^2}\frac{1}{1 - C(r)^2} \begin{pmatrix} 1 & -C(r) \\ -C(r) & 1 \end{pmatrix}. \tag{8.6}$$

Here $| \mathbf{M} | = \sigma_\delta^2[1 - C(r)^2]$ denotes the determinant of the matrix $\mathbf{M}$. In cosmology we refer to the function $\xi(r) = \sigma_\delta^2 C(r)$ as the *2-point correlation function* (see Section 8.2.4).

---

[5] There is an implicit simplification here: for a random field $\delta(\mathbf{x})$ there is no such thing as 'the value at a point'. We can only measure the average of $\delta(\mathbf{x})$ over an infinitesimal volume. Technically we can deal with this by specifying the cumulative distribution for the underlying probability density.

[6] It should be noted that δ_1 and δ_2 are samples from the same random field, and so their statistical distributions have the same variance σ. Equation (8.4) is the special case of the bivariate Gaussian distribution for two zero mean random variates that are identically distributed.

It should be noted that it is perfectly possible to have a Gaussian 1-point function as in Equation (8.3) and yet have a non-Gaussian two-point function, for example:

$$\mathbb{P}[\delta_1, \delta_2, r] = \mathbb{P}[\delta_1]\delta^D(\delta_1 - \delta_2)C(r) + \mathbb{P}[\delta_1]\mathbb{P}[\delta_2][1 - C(r)], \qquad (8.7)$$

where $\delta^D((x)$ is the Dirac delta function.[7] Even if $\mathbb{P}[\delta]$ is a Gaussian, the two-point function $\mathbb{P}[\delta_1, \delta_2, r]$ does not have the form (8.4). Equation (8.7) describes a *step surface, or volume* consisting of abutting cells of different constant density having different sizes. There is a density discontinuity between adjacent cells. The density values are distributed according to $\mathbb{P}[\delta]$. The distribution of the scale of the cells is described by the correlation function $C(r)$ (Berry, 1973, Section 3(ii)).

8.2.3 Fourier Representation of Density Fluctuations

It has been usual since the pioneering work of Peebles (1965) to use the Fourier representation of the field of random density fluctuations $\delta(\mathbf{x})$. This resolves the fluctuating component of the density field into a superposition of sinusoidal waves[8] which in the small amplitude 'linear regime' evolve independently of one another. The initial phases of the sinusoidal waves are considered to be random, but they become correlated through the action of nonlinear gas dynamical processes and gravity. While the fluctuation amplitudes are small, they can be studied using linearised theories.

Fourier Definitions and Conventions

For simplicity of notation we focus on the density fluctuation field at a given time, and so write $\delta(\mathbf{x})$ in place of $\delta(\mathbf{x}, t)$. The field $\delta(\mathbf{x})$ can be written as a Fourier integral:[9]

$$\delta(\mathbf{x}) = \int_\mathcal{V} \delta(\mathbf{k})\, e^{i\mathbf{k}\cdot\mathbf{x}}\, d^3\mathbf{k}, \qquad (8.9)$$

which has the inverse transformation

$$\delta(\mathbf{k}) = \frac{1}{(2\pi)^3}\int_\mathcal{V} \delta(\mathbf{x})\, e^{-i\mathbf{k}\cdot\mathbf{x}}\, d^3\mathbf{x}. \qquad (8.10)$$

Since $\delta(\mathbf{x}, t)$ is real it is easy to show that $\delta(-\mathbf{k}) = \delta^*(\mathbf{k})$. The $\delta(\mathbf{k}, t)$ are to be thought of as being random complex numbers taken from some statistical distribution

---

[7] The superscript D has been used to distinguish the delta function from the conventionally used symbol $\delta(\mathbf{r})$ for the density excursion relative to the mean.

[8] The subject of Fourier analysis is described in some detail in On-line Supplement: *Fourier Analysis*, and the mathematical description of random fields is in On-line Supplement: *Random Fields*.

[9] Here we follow the notation and conventions of Bertschinger (1992, Section 5.8). See also On-line Supplement: *Random Fields*.

　　There is the question of where to put the normalisation factor $(2\pi)^3$. More generally we can write the Fourier transform pair as

$$\delta(\mathbf{x}) = \frac{1}{V}\int_\mathcal{V} \delta(\mathbf{k})\, e^{i\mathbf{k}\cdot\mathbf{x}}\, d^3\mathbf{x}, \qquad \delta(\mathbf{k}) = \frac{V}{(2\pi)^3}\int_\mathcal{V} \delta(\mathbf{x})\, e^{-i\mathbf{k}\cdot\mathbf{x}}\, d^3\mathbf{k}. \qquad (8.8)$$

The common alternatives are $V = 1$, as above, or $V = (2\pi)^3$, though in the engineering literature we often see $V = (2\pi)^{3/2}$, which provides an apparent symmetry between the transform and its inverse.

(e.g. Equation (8.3)). The Fourier transform equations are linear, and so we can differentiate the left-hand sides to give

$$\nabla \delta(\mathbf{x}) = i \int_{\mathcal{V}} \mathbf{k} \, \delta(\mathbf{k}) \, e^{i \mathbf{k} \cdot \mathbf{x}} \, d^3 \mathbf{k}, \quad \nabla^2 \delta(\mathbf{x}) = - \int_{\mathcal{V}} k^2 \, \delta(\mathbf{k}) \, e^{i \mathbf{k} \cdot \mathbf{x}} \, d^3 \mathbf{k}, \qquad (8.11)$$

where $k = |\mathbf{k}|$. This follows simply because $\nabla_{\mathbf{x}} e^{i \mathbf{k} \cdot \mathbf{x}} = i \mathbf{k} e^{i \mathbf{k} \cdot \mathbf{x}}$. Taking a gradient in real $\mathbf{x}$-space is equivalent to simply multiplying the Fourier transform by $i \mathbf{k}$. This is one of the great advantages of working with individual Fourier components.

Finally we give the definition of the Dirac delta function $\delta^D(\mathbf{k})$:

Definition of the 3-dimensional Dirac delta function:

$$\delta^D(\mathbf{k}) = \frac{1}{(2\pi)^3} \int e^{\pm i \mathbf{k} \cdot \mathbf{x}} d^3 \mathbf{x}, \qquad (8.12)$$

from which we see that $\delta^D(\mathbf{k})$ and 1 are a Fourier transform pair.

8.2.4 The 2-point Correlation Function

We can define the 2-*point correlation function*, $\xi(r)$, for two randomly chosen points $\mathbf{x}$ and $(\mathbf{x} + \mathbf{r})$ a distance $r = |\mathbf{r}|$ apart, as the statistical expectation

$$\xi(|\mathbf{r}|) = \langle \delta(\mathbf{x}) \delta(\mathbf{x} + \mathbf{r}) \rangle. \qquad (8.13)$$

This characterises the structure in the field $\delta(\mathbf{x}, t)$ on a scale $|\mathbf{r}|$. It is important to note that this is the correlation function of the *fluctuating part* of the density field relative to the mean density. It is not the correlation function of the density field itself. See also Section 28.5.2, where this is discussed in relation to revealing large scale structure in the present Universe.

8.2.5 The Power Spectrum

The Fourier transform of the correlation function $\xi(\mathbf{r})$ is called the *power spectrum*:

$$P(\mathbf{k}) = \frac{1}{(2\pi)^3} \int \xi(r) e^{-i \mathbf{k} \cdot \mathbf{x}} d^3 \mathbf{x}, \qquad \xi(r) = \int P(\mathbf{k}) e^{i \mathbf{k} \cdot \mathbf{x}} d^3 \mathbf{k}. \qquad (8.14)$$

We note that if we put $r = 0$ in the second of these, we get $\xi(0) = \int P(\mathbf{k}) d^3 \mathbf{k}$. The value of $\xi(0)$ is the total variance of the random field $\delta(\mathbf{x})$ and $P(\mathbf{k})$ is therefore the contribution of the Fourier modes $\mathbf{k}$ in the volume element $d^3 \mathbf{k}$ to the total variance.

Just as $\xi(r)$ is the statistical average of the product of field values at two points, the power spectrum $P(k)$ is the statistical average of a product of the Fourier amplitudes at wave-vector $\mathbf{k}$. We can show this with the following technical manipulation:

$$\langle \delta(\mathbf{k}) \delta^*(\mathbf{k}') \rangle = \left\langle \int_{\mathcal{V}} \delta(\mathbf{x}) \, e^{i \mathbf{k} \cdot \mathbf{x}} \, d^3 \mathbf{x} \times \int_{\mathcal{V}} \delta(\mathbf{x}') \, e^{-i \mathbf{k}' \cdot \mathbf{x}'} \, d^3 \mathbf{x}' \right\rangle$$

$$= \int_{\mathcal{V}} d^3 \mathbf{x} \left(\int_{\mathcal{V}} \langle \delta(\mathbf{x}) \delta(\mathbf{x}') \rangle \, e^{i(\mathbf{k} \cdot \mathbf{x} - \mathbf{k}' \cdot \mathbf{x}')} d^3 \mathbf{x}' \right).$$

Transforming the integration variable $\mathbf{x}' = \mathbf{x} + \mathbf{u}$ in the integral in parentheses separates the integral into two pieces and using (8.12) leaves

$$\langle \delta(\mathbf{k})\delta^*(\mathbf{k}') \rangle = P(k)\delta^D(\mathbf{k} - \mathbf{k}').\tag{8.15}$$

The value of the power spectrum at some wavenumber k is the mean square of the Fourier amplitude $\delta(\mathbf{k})$ for wavenumber $\mathbf{k}$.[10] The power spectrum must be a positive function, and so this last equation tells us that not all functions can be correlation functions.

8.3 Fluid Equations

The (Newtonian) Euler equations for the motion of a simple fluid having density ρ and pressure p moving with velocity $\mathbf{v}$ under the influence of a gravitational potential ϕ, are

$$\underbrace{\frac{\partial \rho}{\partial t} + \nabla \cdot \rho\mathbf{v} = 0,}_{\text{continuity equation}} \qquad \underbrace{\frac{\partial \mathbf{v}}{\partial t} + (\mathbf{v} \cdot \nabla)\mathbf{v} = \frac{\nabla p}{\rho} - \nabla\phi,}_{\text{momentum equation}} \qquad \underbrace{\nabla^2\phi = 4\pi G\rho.}_{\text{Poisson equation}}$$
$$\tag{8.16}$$

These need to be supplemented by an equation for the pressure, i.e. an equation of state. If the pressure is a function only of the density then we can write

$$p = p(\rho) \quad \Rightarrow \quad \frac{\nabla p}{\rho} = c_S^2\frac{\nabla\rho}{\rho}, \quad \text{where} \quad c_S^2 = \left(\frac{\partial p}{\partial \rho}\right).\tag{8.17}$$

As can be easily verified by solving these equations for a small amplitude one-dimensional sound wave, c_S is the sound speed in the fluid.

These equations need further modification when the motions are relativistic, or when the pressure makes a significant contribution to the mass-energy of the fluid (as in the case of the early Universe).

8.3.1 Co-expanding (Comoving) Coordinates

Our zero-order solution is a simple homogeneous and isotropic Newtonian cosmological model. The model is in a continual state of expansion and so it is best to work in co-moving coordinates that expand with the Universe. In that way an observer at rest maintains the same value of the coordinates describing her/his position. The expansion scale factor $a(t)$, normalised to its present value $a(t_0) = 1$, rescales all lengths $\mathbf{r}$ relative to the present value according to $\mathbf{r}(t) = a(t)\mathbf{r}(t_0)$.

In our simple model, the expansion velocity is $\mathbf{v} = H(t)\mathbf{r}$, where $\mathbf{r}$ is the position vector of the point relative to the observer at the origin and $H(t)$ is the (Hubble) expansion rate. Since $\mathbf{v} = \dot{\mathbf{r}}$ we have $\dot{\mathbf{r}} = H\mathbf{r}$ and so $H(t) = \dot{a}/a$.

[10] Equation (8.15) is often written, with lack of mathematical rigour, as $P(\mathbf{k}) = \langle |\delta_\mathbf{k}|^2 \rangle$. This is in fact mathematically incorrect! See Bertschinger (1992, Section 5.8.1) for a good discussion of this. However, it does motivate the interpretation that the power spectrum $P(k)$ is the mean square of the Fourier amplitude of the kth mode. It is also the contribution, per unit wavenumber, of wavenumber k to the total variance of the random field.

The expansion, described by a scale factor $a(t)$, scales all lengths by the factor $a(t)$. Hence for a point located at one time at distance $\mathbf{r}(t)$ relative to the observer at the origin, the value of $\mathbf{x} = \mathbf{r}/a(t)$ remains constant throughout the expansion. The vector $\mathbf{x}$ is the comoving position of the point relative to the observer at $\mathbf{x} = 0$.

In a slightly inhomogeneous Universe the particles will be displaced from their original positions by the gravitational forces arising from the inhomogeneity in the matter distribution, and also by fluid pressure forces if there are any. We shall assign each particle at all times the coordinate $\mathbf{x}$ it had originally. In this way each particle is labelled by the position $\mathbf{x} = \mathbf{r}/a(t)$ it would have had in an otherwise homogeneous Universe. Here $\mathbf{x}$ is the co-expanding (comoving) coordinate assigned to each particle, and we have to re-phrase Equations (8.16) accordingly. Even though in real space the particles may be moving relative to the cosmic expansion, in the $\mathbf{x}$-space they remain in their original positions.

Such coordinates are referred to in fluid mechanics as *Lagrangian coordinates*. The time derivative appropriate to such a coordinate system is the derivative following the motion: $\partial/\partial t + \mathbf{v} \cdot \mathbf{V_x}$ (see Section 13.6.4 and, in particular Equations 13.83, 13.84).

A particle at $\mathbf{x}$ moving relative to the expansion has a velocity $\dot{\mathbf{x}}$ in the co-expanding coordinates. This translates to a velocity $H\mathbf{r} + a\dot{\mathbf{x}}$ in the original coordinates. Because we are using Lagrangian coordinates the particle retains the value of $\mathbf{x}$ despite having a velocity.

Equations (8.16) in the $(\mathbf{x}, t)$ coordinates become (*cf.* Peebles (1980, Eqs. 9.17)):

Equations of fluid motion against a cosmological background (zero pressure):

$$\frac{\partial \delta}{\partial t} + \frac{1}{a}\mathbf{V_x} \cdot \left[(1+\delta)\mathbf{v}\right] = 0, \quad \frac{\partial}{\partial t}(a\mathbf{v}) + (\mathbf{v} \cdot \mathbf{V_x})\mathbf{v} = -\frac{\partial \phi}{\partial \mathbf{x}}, \quad \frac{\partial^2 \phi}{\partial \mathbf{x}^2} = 4\pi G \rho_0 a^2 \delta(\mathbf{x}, t),$$

$$\tag{8.18}$$

$$\delta(\mathbf{x}, t) = \frac{\rho(\mathbf{x}, t) - \rho_0(t)}{\rho_0(t)}. \tag{8.19}$$

The Friedmann equations for the background model determine $a(t)$ and $\rho_0(t)$.

These equations are exact for a flow relative to a homogeneous and isotropic expanding background solution. They hold even if $\Lambda \neq 0$ since the effects of Λ are implicit in the function $a(t)$.

8.3.2 The Linear Regime: Small Amplitude Perturbations

If we neglect second order terms such as $\delta\mathbf{v}$ and $(\mathbf{v} \cdot \nabla)\mathbf{v}$ we arrive at the basis for all the following discussions concerning perturbation growth in the linear regime:[11]

$$\frac{1}{a^2}\frac{\partial}{\partial t}t\left(a^2\frac{\partial \delta}{\partial t}\right) = \frac{\partial^2 \delta}{\partial t^2} + 2H\frac{\partial \delta}{\partial t} = 4\pi G \rho_0 \delta. \tag{8.20}$$

The left-hand side is the growth rate, while the right-hand side is the driving force of gravity. If other forces come into play, like pressure or friction forces, they must be added to the right-hand side in the correct order of perturbation theory.

---

[11] Take the t-derivative of the first of (8.18) and the divergence of the second.

8.4 Perturbations to Zero Pressure Models

The differential equation for the evolution of the relative density fluctuation $\delta(\mathbf{x}, t)$ in a zero pressure Universe is (Equation (8.20)):

$$\ddot{\delta} + 2H\dot{\delta} - \tfrac{3}{2}\Omega_0 H_0^2 \, \delta/a^3 = 0, \tag{8.21}$$

where $H = \dot{a}/a$ is the expansion rate as determined by the Friedmann–Lemaître equation:

$$H^2 = \frac{\dot{a}^2}{a^2} = H_0^2 \left(\Omega_m a^{-3} + (1 - \Omega_m - \Omega_\Lambda)a^{-2} + \Omega_\Lambda \right), \tag{8.22}$$

and where we have used the definition of Ω_0: $4\pi G\rho_0 = \tfrac{3}{2}\Omega_0 H_0^2$. The cosmic scale factor $a(t)$ is normalised so that its value at the present epoch is $a(t_{\text{now}}) = 1$.

In the absence of cosmic expansion, the gravity term $4\pi G\rho_m \delta$ drives up the amplitude δ: gravity is attractive. The presence of the $2H\dot{\delta}$ term slows down the growth. This term is occasionally referred to as the *drag term*: although this nomenclature describes its effect it somewhat belies the way it actually works.

For a simple zero pressure flat model with no Λ-term[12] we have $a(t) \propto t^{2/3}$ and $\rho_0 \propto a^{-3} \propto t^{-2}$. Equation (8.21) is homogeneous in t with solution

$$\delta(t) = At^{2/3} + Bt^{-1}, \quad k = 0, \ \Lambda = 0. \tag{8.23}$$

The constants of integration A and B are fixed by the initial conditions. Asymptotically, at large times, the $t^{2/3}$ term dominates the evolution and, in this model, the density contrast will eventually become nonlinear and perturbation theory breaks down. The t^{-1} term gets bigger in the past. If such a term were relevant today, the Universe would have been more inhomogeneous in the past. We know from observing the angular fluctuations in the cosmic microwave background temperature that this was not the case and so we conclude that for our Universe, $B = 0$.

In theoretical discussion it is usual to use the Fourier representation of the field $\rho(\mathbf{x}, t)$. The Fourier transform is linear operation and so it does not matter whether we transform the equations first or solve them first.

8.4.1 Growth Rate Dependence on Cosmology

It is convenient to separate out the time dependence of the fluctuation δ from the space dependence and define the *growth factor*, $D(t)$ by

$$\delta(\mathbf{x}, t) = \hat{\delta}(\mathbf{x})D(t), \tag{8.24}$$

where the 'hat' on the $\hat{\delta}(\mathbf{x})$ is there to emphasise that this quantity does not depend on time. In fact $\hat{\delta}(\mathbf{x})$ represents the spatial distribution at any fiducial time. $D(t)$ and $\delta(\mathbf{x}, t)$ satisfy the same equation for temporal evolution.

---

[12] The solution of Equation (8.21) is trivial in the case $k = 0$ and $\Lambda = 0$, but analytic solutions for more general cases are hard to come by. Tomita (1969) and Heath (1977, 1981, 1989) presented closed form solutions for some special cases.

In linear theory the irregularities in the density field are simply amplified by the factor $D(t)$: the places where the density field has the mean value, $\hat{\delta}(\mathbf{x}) = 0$, remain at that value.

The growth rate is so fundamental to testing cosmological models on the basis of the recent evolution of structures that it is worth deriving explicitly for general cosmological models. Here we follow Lahav and Suto (2004, Section 2.7) and start from Equation (8.21) which includes the cosmology via the appearance of the Hubble expansion rate, $H(t) = \dot{a}/a$ as the coefficient of *delta*.

The first step is to write the Hubble parameter in terms of the expansion factor $a(t)$ and the present-day constants Ω_m, Ω_Λ:

$$H^2 = H_0^2 \left(\Omega_m a^{-3} + (1 - \Omega_m - \Omega_\Lambda)a^{-2} + \Omega_\Lambda \right). \tag{8.25}$$

Differentiate this with respect to the time t to get

$$2H\dot{H} = H_0^2 \left(-3\Omega_m a^{-3} - 2(1 - \Omega_m - \Omega_\Lambda)a^{-2} + \Omega_\Lambda a^{-2} \right), \tag{8.26}$$

from which

$$\ddot{H} = H_0^2 \left(\tfrac{9}{2}\Omega_m a^{-3} + 2(1 - \Omega_m - \Omega_\Lambda)a^{-2} \right) H. \tag{8.27}$$

On summing (8.26) and (8.27) the curvature terms cancel and, denoting the mean cosmic matter density at expansion factor a by $\bar{\rho}$, we get

$$\ddot{H} + 2H\dot{H} = H_0^2 \frac{3\Omega_m}{2a^3} H = 4\pi G\bar{\rho}H, \tag{8.28}$$

using the definition of $\Omega_m = 8\pi G\bar{\rho}_0/H_0^2$ and $\bar{\rho} = \bar{\rho}_0 a^{-3}$.

Perhaps surprisingly, if we replace δ in Equation (8.21) with H, we have Equation (8.28). In other words $H(t)$ is one of the two solutions of (8.21), and since $H(t)$ decreases with t this is the decaying solution corresponding to the term Bt^{-1} in (8.23). The other solution is then found by constructing the Wronskian[13] $(\dot{D}H - D\dot{H})$ where $D(t)$ is the other solution of (8.21). From (8.21) and (8.28) we have

$$\frac{d}{dt}\left(\frac{D}{H}\right) = \frac{1}{H^2}(\dot{D}H - D\dot{H}) = \frac{\text{const}}{a^2 H^2} \quad \Rightarrow \quad D(t) \propto H(t) \int_0^t \frac{dt'}{a^2(t')H^2(t')}. \tag{8.29}$$

We can rewrite this in terms of the redshift and calculate the constant from the fact that $D(z) \to (1+z)^{-1}$ for $z \to \infty$:

Linear growth in general cosmological models:

$$D(z) = \frac{5}{2}\Omega_m H_0^2 \, H(z) \int_z^\infty \frac{(1+z')}{H^3(z')} dz'. \tag{8.30}$$

---

[13] If both $y_1(x)$ and $y_2(x)$ solve $y'' + a(x)y' + b(x) = 0$, then it is straightforward to show that $W(x) = y_1(x)y_2'(x) - y_2(x)y_1(x)$ satisfies $W' + a(x)W = 0$. The solution of this is $W(x) = ce^{s(x)}$ where c is a constant determined by the initial conditions and $s(x) = \int_0^x a(u)du$. Using this we get $W(x) = y_1 y_2' - y_2 y_1'$ with which we have a first order equation for the second solution $y_2(x)$.

The case $\Omega_m = 1, \Omega_\Lambda = 0$ (i.e. the Einstein de Sitter Universe) gives $D(z) = (1 + z)^{-1}$.
For a spatially flat model, in which $\Omega_m + \Omega_\Lambda = 1$,

$$D(z) \propto \left(\frac{2 + x^3}{x^3}\right)^{1/2} \int_0^x \left(\frac{u}{2 + u^3}\right)^{3/2} du, \quad x = \frac{2^{1/3}(\Omega_m - 1)^{1/3}}{1 + z}. \tag{8.31}$$

There are fitting formulae for this expression (Lahav and Suto, 2004, Section 2.7), but in practice performing this integral numerically is just as easy.

8.4.2 Alternative Formulations

We can replace the time, t in Equation (8.21) with $a(t)$ to get

$$\frac{d^2\delta}{da^2} + (2 - q)\frac{1}{a}\frac{d\delta}{da} - \frac{3\Omega_m}{2a^2}\delta = 0, \quad q = -\frac{a\ddot{a}}{\dot{a}^2}, \tag{8.32}$$

(Linder and Jenkins, 2003). Here, q is the usual deceleration parameter (5.35). This can alternatively be written as

$$\frac{d^2\delta}{da^2} + \frac{1}{a}\left[3 + \frac{d\ln H}{d\ln a}\right]\frac{d\delta}{da} - \frac{3}{2a^2}\Omega_m(a)\,\delta = 0. \tag{8.33}$$

The *growth factor* $D(a)$, defined on the basis of (8.24) as

$$D(a) = \frac{\delta(a)}{\delta(1)}, \tag{8.34}$$

satisfies the same equations as does δ and likewise expresses the value of δ at some past epoch in comparison with the value it would have today if linear growth were to continue. The *growth rate* is then defined by

$$g(a) = \frac{d\ln D}{d\ln a} = \frac{a}{D}\frac{dD}{da}. \tag{8.35}$$

The growth rate depends on the history of the expansion, as expressed here through $a(t)$, and that in turns depends on the equation of state of the matter pervading the Universe.

The equation of state is usually parameterised through the function $w(a) = p/\rho c^2$. Hence measuring the growth rate at different redshifts would be a way of constraining the cosmological parameters and the equation of state.

There is a long history of approximating the solution of this Equation (8.33) in such a way that the solution can reflect any kind of dark energy equation of state. A useful approximation is obtained by parameterising the growth rate $g(z)$ as

$$g(z) \propto [\Omega_m(z)]^\gamma, \quad \text{with} \quad \gamma \simeq 0.6. \tag{8.36}$$

This is a rough compromise fit, the exponent γ depends on the theory. In the standard ΛCDM scenario, the growth rate is well approximated by

$$\frac{1}{a}D(a) = \exp\int_1^a [\Omega_m(b)^\gamma - 1]\frac{db}{b}, \quad \gamma = 0.545, \tag{8.37}$$

over a wide range of parameters. Wang and Steinhardt (1998), Linder (2009) and others have suggested a variety of alternative approximations.

8.5 Perturbations with Pressure: Adiabatic Modes

In this section we follow the relatively simple treatment of Heath (1991) which is based on Bonnor's equations. Density perturbations in a gas with non-zero pressure involve pressure gradients which, on small scales, counteract the effects of gravity. To illustrate this we need to consider a variant of Equation (8.21) that applies to a single Fourier component of the density field. In other words, a sound wave.

There are two cases. Firstly there is the post-recombination Universe where the radiation and matter no longer interact and the pressure is just the gas pressure. Secondly there is the Universe prior to recombination when the matter and radiation are tightly coupled.

8.5.1 Perturbations with Gas Pressure $p = nkT$

In this case the only addition to the pressure-free Equation (8.21) is the additional force contributed by the gas pressure gradient. For the purpose of comparison with the following section, denote the ambient matter density field by ρ_m and the matter pressure by p_m. It will be supposed that the matter has a simple equation of state of the form $p_m = p_m(\rho_m)$.

The pressure gradient force per unit volume is

$$\mathbf{F}_{\text{pressure}} = -\frac{1}{a}\nabla_{\mathbf{x}}p = -\frac{1}{a}\left(\frac{\partial p}{\partial \rho}\right)\nabla_{\mathbf{x}}\rho_m. \tag{8.38}$$

Here, the pressure gradient is the comoving coordinate gradient $\nabla_{\mathbf{x}} = a\nabla_{\mathbf{r}}$.

Finally we note that $\nabla_{\mathbf{x}}p_m = (\partial p_m/\partial \rho_m) = c_s^2$, the sound speed in the gas. Hence, putting $\nabla_{\mathbf{x}}\rho_m = \rho_m\nabla_{\mathbf{x}}\delta_m$ we get

$$\mathbf{F}_{\text{pressure}} = -\frac{1}{a}c_s^2\rho_m\nabla_{\mathbf{x}}\delta_m. \tag{8.39}$$

This contribution from the pressure can now be included in Equation (8.21) to give

$$\ddot{\delta}_m + 2H\dot{\delta}_m - 4\pi G\rho_m\delta_m = -\frac{c_s^2}{a^2}\nabla_{\mathbf{x}}^2\delta_m. \tag{8.40}$$

For a co-expanding sinusoidal matter wave of wavelength λ we have $\delta_m \propto \cos(2\pi a\mathbf{x}/\lambda)$. With this we can replace the sinusoidal space dependence $\nabla_{\mathbf{x}}^2 \rightarrow (2\pi a/\lambda)^2$,

$$\ddot{\delta}_m + 2H\dot{\delta}_m = \left[4\pi G\rho_m - \left(\frac{2\pi c_s}{\lambda}\right)^2\right]\delta_m. \tag{8.41}$$

The behaviour of the solution is determined by the sign of the term on the right-hand side. If it is positive, gravity dominates the behaviour while if it is negative the gas pressure dominates. The two terms are equal when the wavelength has the critical value:

The Jeans length

$$\lambda_J = \left(\frac{\pi c_s^2}{G\rho_m}\right)^{\frac{1}{2}}. \tag{8.42}$$

8.5.2 Perturbations with Radiation Pressure $p = \frac{1}{3}\rho c^2$

Here two factors play a role. The radiation has an energy density comparable with the matter density and so contributes to the gravitational potential. The pressure is mainly due to the radiation scattering off the free electrons in the cosmic plasma.

A word of warning is needed when moving on to discuss the effects of radiation pressure on the evolution of cosmic density perturbations. When the mass density of the Universe is radiation dominated, the sound speed gets close to the speed of light and the Jeans length discussed above is close to the scale of the cosmic horizon. This means that relativistic effects should be taken into account when discussing the evolution of very large scale density fluctuations in a radiation dominated Universe. We shall return to this later, but for the moment focus on the physics that is relevant to sub-Jeans length fluctuations.

To proceed we need a few definitions and a bit more notation. The matter density is denoted by ρ_m and the radiation field energy density will be denoted by $\rho_r c^2 = aT_r^4$, where T_r is the temperature of the radiation field and, while the Universe is ionised, the electrons. We also introduce some notation for the relative density fluctuations in the radiation density and for the fluctuations in the matter density:

$$ s = \frac{\delta\rho_r}{\rho_r}, \qquad \delta = \frac{\delta\rho_m}{\rho_m}. \tag{8.43} $$

We impose a relationship between the fluctuations in the radiation and the matter to ensure that the entropy of the cosmic fluid is not changed by the compression or expansion of the perturbation:

$$ \frac{\delta\rho_r}{\rho_r} = \frac{4}{3}\frac{\delta\rho_m}{\rho_m}, \quad \text{i.e. } s = \frac{4}{3}\delta. \tag{8.44} $$

Such a perturbation is referred to as an *adiabatic density perturbation* in the coupled matter radiation gas.

The modified version of Equation (8.21) that includes radiation density and the radiation pressure gradient forces is

$$ \ddot{\delta} + 2H\dot{\delta} - \underbrace{4\pi G\rho_m\delta}_{\text{gravity matter}} = \underbrace{4\pi G(\rho_r + 3p_r)s}_{\text{gravity radiation}} - \underbrace{\frac{1}{a^2}\frac{\nabla^2 p_r}{\rho_r + p_r + \rho_m}s}_{\text{radiation pressure}}. \tag{8.45} $$

Note that $p_r = \frac{1}{3}\rho_r c^2$, so the gravitational terms contribute $8\pi G\rho_r\, s$. The quantity $\rho_r + p_r + \rho_m + p_m$ is the relativistic mass-energy, where we have $p_m \ll p_r$ and so neglect matter pressure p_m. With $p_r = \frac{1}{3}\rho_r c^2$ this term is $\frac{4}{3}\rho_r + \rho_m$.

If we consider an adiabatic sound wave of comoving wavelength $\lambda = 2\pi a(t)/k$:

$$ \ddot{\delta} + 2H\dot{\delta} - 4\pi G\rho_m\delta = \left[8\pi G\rho_r - \frac{1}{3}\frac{k^2}{a^2}\frac{\rho_r c^2}{\frac{4}{3}\rho_r + \rho_m} \right]s, \tag{8.46} $$

where s is the relative density fluctuation in the radiation field energy density. Eliminating s in favour of δ using Equation (8.43) we get

$$\ddot{\delta} + 2H\dot{\delta} - \left[4\pi G \left(\rho_m + \frac{8}{3}\rho_r \right) - \left(\frac{2\pi}{\lambda} \right)^2 \frac{1}{3} \frac{c^2}{1 + \frac{3\rho_m}{4\rho_r}} \right] \delta = 0. \qquad (8.47)$$

Compare this with the simpler case of Equation (8.41).

The nature of the solution is determined by the sign of the term in square brackets. There is a critical case when the quantity in square brackets vanishes; this happens at a particular value of $\lambda = \lambda_J$, the *Jeans length*. For $\lambda \ll \lambda_J$ the term is positive and the perturbation behaves like a sound wave. For $\lambda \gg \lambda_J$ the term is positive, so the solution is not oscillatory and the amplitude δ grows under the influence of gravity. In the limit that $\lambda \to \infty$, we recover Equation (8.21), but with a different gravitational term since the radiation field contributes to the energy density.

The important quantity to emerge from this is the *adiabatic sound speed* in the gas:

$$c_s^2 = \frac{1}{3} \frac{c^2}{1 + \frac{3\rho_m}{4\rho_r}}. \qquad (8.48)$$

In terms of the sound speed the Jeans length is

$$\lambda_J \simeq \left(\frac{\pi c_s^2}{G\rho} \right) \simeq c_s t_{\exp}, \qquad (8.49)$$

where $t_{\exp}$ is the expansion timescale (roughly $a/\dot{a}$). Here $c_s t_{\exp}$ is the distance a sound wave can travel on the expansion timescale and expresses physically how on scales less than the Jeans length the fluctuating pressure forces can effectively balance the force of gravity. On scales larger than the Jeans length, gravitational amplification is possible because the opposite sides of the fluctuation cannot communicate via pressure forces (sound waves). The Jeans length is also referred to as the *sound horizon*.

The equations of this section are valid provided the matter and radiation are tightly coupled through Thomson scattering, that is, provided transport processes like viscosity and heat transport do not carry energy out of the sound wave. Once the Universe starts to become neutral, the epoch of recombination, this is no longer the case and the problem becomes substantially more complex. Equation (8.43) breaks down in this regime.

8.6 Growing and Oscillatory Solutions

It is usual to consider the evolution of a single spatial Fourier component of the density field in coexpanding coordinates (a plane wave):

$$\delta(\mathbf{x}, t) = \delta(t) e^{i\mathbf{k}.\mathbf{x}/a(t)}. \qquad (8.50)$$

The wave-vector **k** is considered to be fixed and the wavelength is

$$\lambda = \frac{2\pi a(t)}{k}. \tag{8.51}$$

In other words, the perturbation is just a sinusoidal plane wave disturbance whose wavelength grows with the expansion. The wavenumber, k, is fixed throughout the expansion.

Under the given circumstances the behaviour of the Fourier amplitude $\delta(t)$ of the perturbation in the density is described by the following equations:

$$\frac{1}{a^2}\frac{d}{dt}\left(a^2\frac{d\delta}{dt}\right) = -\left(c_s^2\frac{k^2}{a^2} - 4\pi G\rho\right)\delta. \tag{8.52}$$

The time derivative operator reflects the effect of the cosmic expansion, and the sign of the coefficient of δ determines the nature of the solution – whether it is oscillatory or not. The case $k \to 0$ reduces to Equation (8.21).

8.6.1 Limiting Cases

Equation (8.47) has simple limiting solutions in the cases $\lambda \gg \lambda_J$ and $\lambda \ll \lambda_J$. In the former case gravitational forces dominate and the equation is simply the homogeneous equation

$$\ddot{\delta} + \frac{4}{3t}\dot{\delta} - \frac{2}{3t^2}\delta = 0, \tag{8.53}$$

from which we get the famous solution (8.23), $\delta(t) = At^{2/3} + Bt^{-1}$, describing a linear combination of a growing mode and a decaying mode. The decaying mode corresponds to a nonlinear initial condition in the past and since we assume that the Universe in the past was more homogeneous than it is now it is usual to set this decaying term equal to zero.

The equation for the other case, $\lambda \ll \lambda_J$, in which the pressure gradients stabilise the perturbation against gravitational amplification is more difficult unless we consider a didactically useful, albeit contrived, case.

Suppose that a situation could exist in which the Universe was dominated by radiation pressure even while the dominant contribution to the total density was the baryonic component.[14] Then the sound speed would be proportional to $\rho_m/\rho_r \propto a^{-1}$ and Equation (8.52) could be written in the pressure-dominated limit $\lambda \ll \lambda_J$ as

$$\ddot{\delta} + \frac{4}{3t}\dot{\delta} + \frac{1}{t^2}\alpha^2\delta = 0, \quad \alpha = \frac{8\pi}{3}\frac{ct}{\lambda}\bigg|_{eq}, \tag{8.54}$$

---

[14] This would have been an acceptable model in the 1970s at which time it was thought that the major constituents of the Universe were photons and baryons. In such a spatially flat model the baryons would make up the bulk of the present day mass density and the epoch of equality of mass and radiation densities would have occurred long before recombination at a redshift around 10^5. Such a model of course would require a considerable fraction of undiscovered 'missing baryons'. There were also issues with the age of the Universe if the Hubble constant was as high as 100 km s^{-1} Mpc^{-1}. See the blackboard in the 1995 picture reproduced as Figure 2.5.

where α is a constant indicating the scale of the perturbation relative to the horizon at the time when the densities of matter and radiation were equal.

This is again a homogeneous equation and for large α (short wavelengths) it has solution

$$\delta = t^{-1/6} \left[C \cos \left(\alpha \log \frac{t}{t_{eq}} \right) + D \sin \left(\alpha \log \frac{t}{t_{eq}} \right) \right]. \qquad (8.55)$$

This is an oscillatory solution of constant period when plotted in terms of $\log t$. Viewed as a function of t, the oscillations slow down as time goes on. The frequency of the oscillations depends on the wavelength (inversely proportional to α). Thus small wavelengths oscillate with higher frequency.

The factor $t^{-1/6}$ is interesting: this represents a decrease in the amplitude due to the fact that the wave is oscillating in an expanding medium. This is not, strictly speaking, a damping of the sound wave, despite the fact that the amplitude decreases with time. The reason for the decline of the amplitude is that the wave moves against a changing background in which the only dynamical invariants are adiabatic invariants. (An example of this is the oscillation of a pendulum while the string slowly changes length.)

A perturbation starts out its life in the regime governed by Equation (8.53) and later becomes governed by (8.55). The solution for the first has two arbitrary constants we called A and B. The oscillatory solution has different constants C and D. The means for relating these is to solve the equation using the WKB method.

8.6.2 WKB Method

We can study solutions of (8.52) under more general circumstances using the WKB method. Starting with the linear equation for the perturbation amplitude in the form (8.52), we can make a transformation of the time variable,[15]

$$dx = a(t)^{-2}dt. \qquad (8.56)$$

To help follow the argument we can think of the specific case $a(t) \propto t^{2/3}$ which gives

$$x = \left(\frac{t_\dagger}{t} \right)^{1/3}. \qquad (8.57)$$

We have chosen to normalise x to unity at the time $t_\dagger$ when, for a given scale k the pressure gradient and gravitational forces balance. For $t < t_\dagger$ gravitational forces dominate the perturbation, while for $t > t_\dagger$ pressure forces dominate and $t_\dagger$ occurs when the scale of the perturbation is the Jeans length. We shall derive an equation for the number of oscillations the perturbation makes between that time and the epoch of recombination.

---

[15] This is suggested by the term $a^2 \frac{d}{dt}$ appearing in Equation (8.52).

Equation (8.52) for the perturbation evolution then takes the form[16]

$$\frac{d^2\delta}{dx^2} + Q(x)\,\delta = 0,$$ (8.59)

where the frequency $\sqrt{Q(x)}$ is given by

$$Q(x) = a^4 \left(c_s^2 \frac{k^2}{a^2} - 4\pi G\rho \right).$$ (8.60)

$Q(x)$ is expressed as a function of the new variable x and the wavenumber k. In our simple case where $a(t) \propto t^{2/3}$, $Q(x)$ is just a constant depending on the scale k of the perturbation. Our normalisation is such that $Q(x) = 0$ at $x = 1$. In the general case the timescale is scale dependent.

Given the purely growing initial solution, the WKB method tells us which is the corresponding oscillatory solution for $Q(x) > 0$:

WKB solution of variable frequency oscillator:

$$\delta(x) \propto Q(x)^{-1/4} \cos\left[\int_{s=1}^{x} \sqrt{Q(s)}\,ds - \frac{\pi}{4} \right].$$ (8.61)

When $Q(x) > 0$, the solutions are oscillatory, and $Q(x) \sim ac_s k$ for large k.[17]

The first thing to note is that when $a(t) \propto t^{2/3}$, using $Q(x)$ from (8.60) gives $\sqrt{Q} \propto t^{2/3}$ and so the term $Q^{-1/4} \propto t^{-1/6}$. The amplitude of the sound waves decreases as the Universe expands (compare Equation 8.55).

The next thing to note is that, at a given instant of time, the phase of a wave depends only on its wavenumber k. The situation at the epoch of recombination is that waves of specific frequencies have attained their maximum compression. For our simple model these are wavelengths

$$\lambda \sim \frac{\lambda_J}{n}$$ (8.62)

for larger values of n. The mean square amplitude of fluctuations (the power spectrum) is thus a function with well-separated peaks. These peaks are generally referred to as *acoustic peaks* and are seen in the power spectrum of the CMB.[18] They give rise to

---

[16] The reader may notice the similarity between this equation and the Schrödinger equation for a particle in a potential well $V(x)$:

$$\frac{d^2\psi}{dt^2} + \frac{2m}{h} [E - V(x)]\,\psi = 0.$$ (8.58)

Such equations were an essential part of the Bohr–Sommerfeld 'old' (i.e. pre-1925) quantum mechanics which aimed at providing a basis for selecting allowed discrete states of a classical system.

[17] This is simply the oscillatory part of the solution when $Q(x) > 0$. There is another, non-oscillatory and increasing part. These parts are joined via the WKB connection rules which approximate the join between the solutions on either side of $Q(x) = 0$. See Bender and Orzag (1999, Section 10.4).

[18] The peaks are sometimes referred to as 'Sakharov oscillations' (see footnote 33, p67). Similar phenomena occur in low temperature physics, as demonstrated in an interesting laboratory experiment by Hung *et al.* (2013). The experiment was designed to mimic the early Universe using ultra-cold caesium atoms in a vacuum chamber by exploiting the strong similarity of the equations of the early Universe and the equations of an

the baryonic acoustic oscillations that are detected in the distribution of galaxies in deep redshift surveys.

These acoustic oscillations occur as a consequence of the interference between synchronously generated sound waves. The fact that the initial conditions are specified uniquely by the growing mode is of utmost importance in getting the phase-correlated oscillations in the sub-Jeans length regime. If we imagine that perturbations are produced by some other means, for example if there were a forcing term on the right-hand side of Equation (8.52), then there may be no phase correlations – the asymptotic WKB Equation (8.61) need not be valid. In particular, models of the early Universe in which inhomogeneities are created by topological defects may not show these oscillations in the power spectrum (Albrecht *et al.*, 1996).

8.6.3 Evolution of Density Fluctuations

By 1970 there was a plethora of papers on the subject of what at the time was referred to as 'galaxy formation'. The seminal papers of Sachs and Wolfe (1967) and Silk (1968) respectively set down the basis for understanding the pattern of temperature fluctuations and the evolution of primeval adiabatic fluctuations. However, for simplicity, they assumed that the recombination would take place instantaneously.

Peebles (1968) and Zel'dovich *et al.* (1969) recognised that the rate of recombination to the ground state of hydrogen was inhibited by the large flux of Lyman-α wavelength photons present in the CMB radiation field. In that case the rate of the recombination is controlled by the 2-photon decay of the $2s$ state of the hydrogen atom (see Section 20.2 for the details of this process). Because of this complication the recombination did not follow the equilibrium Saha formula for ionisation. The duration of the recombination was around 10% of the expansion time.

Following that, Sunyaev and Zel'dovich (1970) and Peebles and Yu (1970), noted that the finite duration of the recombination process would have significant consequences for the evolution of the perturbation during the recombination period and on the amplitude of the observed temperature fluctuations. A picture of the evolution of the amplitude of primordial density perturbation through the period of recombination from an unpublished calculation by Michie (1969) is shown in Figure 8.1.[19]

---

atomic super-fluid in which long wavelength excitations are described by phonons. The experiment was able to display the onset and evolution of sound waves and the appearance of these oscillations.

[19] Richard Michie died aged 37 in March 1969, having submitted the paper to the *Astrophysical Journal* in September 1967. He had been terminally ill and unable to deal with the referee's remarks, so the paper was never published and only appeared as a preprint of the Kitt Peak National Observatory.

He was the first to attempt the highly technical problem of describing what happened to perturbations during the recombination process, treating the matter and radiation separately and writing an explicit relationship for their mutual interaction. Although he recognised that the re-ionisation itself would not be described by the equilibrium Saha equation, he chose to use a modification proposed by Strömgren (1948) in relation to the ionisation state of the interstellar medium.

The other problem faced by Michie was the description of the transport coefficients as the optical depth of the medium changed rapidly. For this he used an interesting second order moment expansion of the radiation field to solve the equation of radiative transfer, as opposed to the more classical diffusion approximation.

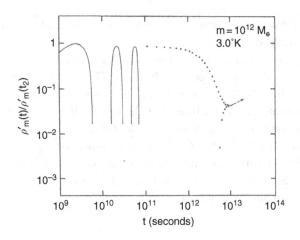

The evolution of a primordial density perturbation through the period of recombination of the primeval plasma as computed by Michie (1969, Figure 2).[19] The computation was done for a perturbation described by mass $M \times 10^{12}\, M_\odot$, assuming a CMB temperature of 3.0 K, and an open cosmological model with matter density corresponding to $\Omega_m h^2 \simeq 0.05$ (the cosmological parameters were not at all well-known at that time). The graph depicts the oscillating amplitude of the acoustic oscillations of the perturbation relative to its pre-recombination amplitude. At the time the paper was written there could be no consideration of the influence of copious amounts of dark matter on perturbation evolution.
(With permission of the Kitt Peak National Observatory.)

8.6.4 Silk Damping of Fluctuations

The effect of damping of perturbations just before and during the period of recombination is observable in the spectrum of the CMB temperature fluctuations. The change in the expansion rate in the presence of a substantial contribution from neutrinos, for example, affects the photon diffusion timescale and thereby affects the damping of perturbations by photons.

The distance, r_d, a photon can random walk in the expansion time $t \sim H^{-1}$ when the photon mean free path to Thomson scattering is λ_T is $\sqrt{ct.\lambda_T}$. If the free electron density is n_e then $\lambda_T \sim (\sigma_T n_e)^{-1}$ and so $r_d^2 \sim c/\sigma_T n_e H$. Doing the calculation in detail by integrating over the history of the Universe and taking the details of Thomson scattering into account, we get that the diffusion distance at recombination is given by

$$r_d^2 = \pi^2 \int_0^{a_*} \frac{da}{a^3 \sigma_T n_e H} \left[\frac{R^2 + \frac{16}{15}(1 + R)}{6(1 + R^2)} \right], \quad R = \frac{p_b + \rho_b c^2}{p_\gamma + \rho_\gamma c^2} \simeq \frac{3\rho_b}{4\rho_\gamma}, \qquad (8.63)$$

My own sense of this important paper is that Michie simply ran out of the time that would be needed to complete the work as he might have wished.

I am grateful to Tod Lauer and Sharon Hunt of the Kitt Peak National Observatory for their substantial help in locating a copy of this paper and finding out about Richard Michie.

where ρ_r and ρ_b are the densities of the radiation and matter. The damping of the fluctuations attenuates the density amplitude of fluctuations on scale λ by a factor

$$\alpha \sim \exp - \left(\frac{2r_d}{\lambda}\right)^2 . \tag{8.64}$$

We have loosely described the sound horizon as the distance a sound wave can travel in an expansion timescale. We can now write down a definition in terms of the parameter R:

Sound horizon at recombination

$$r_s = \int_0^a \frac{da}{\sqrt{3(1+R)}} . \tag{8.65}$$

9

The Inflationary Universe

The concept of the inflationary Universe has become one of the cornerstones of modern cosmology. It describes processes taking place in the very early Universe which may have been responsible for ensuring that the Universe we see today is geometrically flat, and that may provide an explanation for the origin of all the cosmic structures we see. There is no single theory of inflation, but the phenomenon appears to be a feature common to most theories of those very early times that are based on the physics of matter at ultra-high energies as we know it today.

We now have ways of constraining such theories through observations of the cosmic background radiation: the Big Bang has left us its signature. We can look back to ultra-early times and collect archaeological evidence of the Big Bang. This imposes constraints on these fundamental theories and at the same time provides an explanation of how the Universe has come to be the way we observe it.[1]

9.1 Inflationary Models

9.1.1 The Pre-inflation Era

Classical physics breaks down at the time when quantum mechanics and general relativity merge: i.e. when the Schwarzschild radius of a volume containing mass M_P is on the same order as its Compton wavelength. This determines the size and density of the volume, and the density tells us when in the past history of the Universe this state was achieved: $\sim 5.4 \times 10^{-44}$ s after the Big Bang.[2] Clearly, physics at or around this time must be regarded as speculative, or simply unverifiable, though that does not stop theoretical physicists from writing papers about it!

If we look some time after the Planck era we are in the era of classical general relativity. During the 1960s, Moscow was a centre for studies of this very early phase

---

[1] The ADS database counts over 10 000 papers on cosmological inflation published between 1980 and 2015 (search criterion: 'inflation' AND 'cosmology').

[2] The Planck epoch is defined in terms of three fundamental constants: the Planck constant, $\hbar$ from quantum mechanics, the gravitational constant G from general relativity, and the speed of light, c. It is only defined up to a dimensionless multiplicative factor, but that multiplicative factor has been fixed so that we can consistently adopt what is known as the Planck System of Units. See Sections 9.4.1 and A.5.
See the NIST website at http://physics.nist.gov/cuu/Constants/index.html

of the cosmic expansion with the work of the Khalatnikov group. This culminated in important papers on models for the dynamic evolution of the Universe near the initial singularity by Belinskiĭ and Khalatnikov (1969) and Belinskiĭ, Khalatnikov and Lifshitz (1970, the BKL model). It was during that period that Sakharov wrote the first paper on quantum inhomogeneities in the very early Universe, Sakharov (1966a). In that paper Sakharov addressed the origin of density fluctuations in the framework of a cold Big Bang.[3]

At the same time, in a pair of important papers, Misner (1968, 1969) specifically addressed the question of why the Universe was so isotropic. The dogma previous to that time was pretty much that the Universe is now what it is simply because the initial conditions were what they were. In the first of these he proposed that transport processes acting at the time when the primordial neutrinos decoupled from the cosmic plasma at around ~ 1s after the Big Bang, would isotropise an anisotropic Universe. This was based on a homogeneous general relativistic Kasner model for the expansion. In the second, he noted that the homogeneous Bianchi IX models did not in general have a particle horizon, which, it was suggested, would provide the means for generating global isotropy rather than simply within a horizon of scale $\sim ct$. This latter scheme Misner dubbed the *Mixmaster Universe*.

An alternative idea arose in the 1970s that the Universe might have been created out of a quantum fluctuation in a vacuum (see, for example, Starobinsky (1979)), an idea which possibly originated with Dennis Sciama.

Sciama–Tryon inspiration (1973) In 1969 Dennis Sciama gave a seminar at Columbia University on the hypothesis that the entire Universe could have originated from a single particle that itself arose out of a quantum vacuum: the so-called 'Vacuum Genesis Hypothesis'.[a] In the audience was Edward Tryon, who had written his PhD at Berkeley in 1967 on *Classical and Quantum Field-Theoretic Derivations of Gravitational Theory* under the auspices of Steven Weinberg. During the seminar Tryon made a remark that 'Maybe the Universe is a vacuum fluctuation', which apparently engendered some general amusement.[b] In 1973 he published a paper in Nature (Tryon, 1973) explaining how this might happen.

[a] Prior to 1965 Sciama had been a strong supporter of the Steady State theory, a theory which demanded a mechanism for the creation of matter *ex nihilo*.

[b] See Ferris (2003, Ch.18. The Origin of the Universe) and Tryon (1984).

However, quantum vacuum fluctuations was not the only way to explain the present state of the Universe.

9.1.2 Inflation

During the early 1980s, the notion of *Cosmic Inflation*, stimulated by the paper of Guth (1981), took cosmology by storm: it promised to bring some apparent cosmological

---

[3] Zel'dovich, at that time, believed that the Universe was initially cold. Sakharov wrote that paper shortly after returning to academic research in 1965, following 15 years of work on fusion weapons. Within a few years he was being identified as a dissident, winning the Nobel Peace Prize in 1974, and being exiled to the closed city of Gorky in 1980.

coincidences within the framework of a single idea based in physics. There were two issues which Guth sought to address: why the Universe in which we live is within a factor of 10 or so of being flat,[4] and why it is remarkably homogeneous on large scales that were causally disconnected a long time in the past. Guth also remarked that cosmic inflation might solve the mystery of why there were no magnetic monopoles in the Universe.

Guth's simple idea was that there were reasons to think that the early Universe might have gone through a period of very rapid exponential expansion, i.e. *inflation*, during which different parts might causally connect and isotropise. The outcome would have been that at the end of this expansion, the density parameter would be extremely close to unity.

The idea of inflation, in the sense of generating a prolonged period of exponential expansion during the earliest moments after the Big Bang, did not come out of nowhere. Already Brout *et al.* (1978) (see also Englert (1999)) had presented what was effectively a theory of inflation based simply on the idea of creating everything from a primordial vacuum state (see also Vilenkin (1982)).

9.2 Cosmological Conundrums

At the time of this inspired realisation, the Hot Big Bang notion for the origin of the Universe was firmly entrenched. The then-disbanded Steady State theory had no need to address the first question (there was no question of 'cause') and, importantly, it addressed the second by predicting that the Universe should be geometrically flat, *i.e.* have precisely the Einstein de Sitter critical density. Within the framework of the Big Bang theory these are two central issues and are respectively referred to as the *horizon problem* and the *flatness problem*.

There were further problems that arose from particle physics. These arose from early studies of grand unification, the quest to unify the strong, electromagnetic and weak interactions. A beautiful model for unification had been suggested by Georgi and Glashow (1974) in which all forces would be unified before the so-called GUT-epoch when the cosmic temperature was some 10^{15} GeV, but would emerge as separate forces below that temperature. This corresponds to $t_{GUT} \sim 10^{-35}$s, which is well above the Planck time. This is in the arena of classical gravitation and away from any complication such as quantum gravity.

However, the Georgi–Glashow ('GG') model had several important consequences for cosmology that were to constrain its viability. As shown shortly after by Zel'dovich and Khlopov (1978) and by Preskill (1979), the GG model contained a high abundance of stable super-heavy magnetic monopoles: at least one monopole per horizon volume at the GUT epoch. Such a density of monopoles was in conflict with Big Bang cosmology. This would at least require modification of the elegant Georgi–Glashow model, or a

---

[4] In 1980 the total mass density of the Universe was unknown to within a factor of 10–30 (i.e. $\Omega_b = 0.03-1.0$), and there was no notion of dark matter. The evidence for large scale anisotropy was based on the relatively nearby distribution of galaxies. Now we know that the Universe has $\Omega = 1.0$ (Table 27.1), and that the Universe is isotropic in the large to better than a fraction of a percent.

radical change in the Big Bang theory. In addition, in this version of the grand unification theory, there would be a baryon/anti-baryon symmetry which is not observed.

9.2.1 Grand Unification of the Forces: GUT

In 1979, Glashow, Salam and Weinberg won the Nobel Prize for their work in the late 1960s
on unifying the electromagnetic and weak forces of physics: the 'electroweak theory'. The
question was whether one could go a step further and unify this electroweak force with the
strong interaction in some kind of *grand unified theory*, 'GUT' for short, wherein all three
forces merged at some extremely high energy.

Georgi and Glashow (1974) had already suggested a mechanism whereby this might be
able to work based on SU(5) as the underlying symmetry group. In terms of gauge theory
the idea was that the unified group should contain the elements of quantum chromodynamics ('QCD'), and of SU(2) x U(1), the gauge group that unified electromagnetic and weak
interactions. The choice of Georgi and Glashow for a unifying group was to use a group
that would contain SU(3) x SU(2) x U(1), and the logical candidate was SU(5). That is,
SU(5) would become an exact symmetry at ultra-high energies where the three interactions
would be unified.[5]

The prevalent idea was that around the GUT time, $t_{GUT} \sim 10^{-35}$s, there may have
been a first order phase transition out of the unified state, splitting the strong and electroweak interaction. During that phase transition the equation of state of the matter would
mimic a cosmological constant and so the Universe would go into a state of de Sitter-like
exponential expansion. If the de Sitter state lasted long enough it would serve to flat-
ten out any curvature wrinkles in the Universe. This was the key realisation of Kazanas
(1980) and Guth (1981). Since that time there have been many variants of the inflation
theory, all proposing to create circumstances in which the Universe goes through a phase
of exponential expansion.

The SU(5) based GUT had many things going for it, but it ran into a number of problems.
In fact, most GUTs break baryon number symmetry and produce magnetic monopoles,
neither of which is observed: that would be a problem that any such theory would have
to solve. Moreover, in doing so they allow the proton to decay. The proton decay time is
at least $\sim 10^{34}$yr and this imposes a lower limit on the energy scale at which the GUT
unification can take place of at least 10^{15} GeV. This put the problem firmly in cosmological territory at $t_{GUT} \sim 10^{-35}$s. So what happened when the cosmic temperature fell to
10^{15} GeV?

9.2.2 The GUT Phase Transition

The answer was that as the Universe cooled through the GUT era it would break into
bubbles. The separating out of the strong interaction, i.e. the breaking of the SU(5) symmetry, was a first order phase transition that would split the Universe into discrete bubbles,
'domains'. There would be bubbles of baryons and bubbles of anti-baryons producing on

---

[5] The idea that the unification could be achieved via SU(5), or any other gauge group, has never succeeded.

average equal amounts of matter and anti-matter. Magnetic monopoles, none of which have ever been observed, would be the dominant form of matter in the Universe.

The solution to this problem was recognised by Sato (1981a,b,c)[6] who realised that, if the transition to bubbles were indeed a first order phase transition, the bubbles that formed would go into a state of exponential expansion. The expansion could be so great that our Universe might find itself inside one of these domains made up entirely of either baryons or anti-baryons, and there would be very few monopoles.

What Alan Guth did while pursing this research into the monopole problem was to recognise that this has a significant cosmological consequence: 'inflation', with the important consequences that the Universe would isotropise and become geometrically flat during such a period. There was one stumbling-block. It was Alan Guth himself who recognised the flaw in his theory of inflation:[7] there was nothing to stop the inflation going on forever and so the outcome would not be the Universe we live in. The solution for this problem came within a short time: a new inflation theory was independently and simultaneously put forward by Albrecht and Steinhardt (1982) and Linde (1982a,b). This was dubbed *new inflation* and itself quickly gave rise to a plethora of other variants (see Figure 9.1). Other issues were resolved, such as the origin of the perturbations that would lead to the large scale structure we see now, and a variety of testable predictions. There was no doubt that inflation was here to stay.

9.2.3 The Horizon Problem

In intuitive terms, when the age of the Universe is t, light can have travelled a distance ct, and so the points within regions of size $R_H \sim ct$ are in causal contact. The scale R_H is referred to as the 'horizon'. This concept was discussed in more detail in Section 5.10.2 where it was shown that, in an Einstein de Sitter model $R_H \simeq 3ct$. As explained in that section this is the 'proper distance': the distance measure by ruler on a constant time slice of the Universe. When we look out at the CMB the radiation comes directly from a very high redshift $z_{rec} \sim 1100$, the *last scattering surface*. We are seeing a constant time snapshot of

---

[6] The 3 year period 1979–1982 was a remarkably active period for writing papers on GUT inflation, with both the high energy physics community and the much smaller cosmology community publishing papers in their own journals and giving talks at conferences. Fortunately, some people sat on the boundary of the two disciplines ('astro-particle physics') and were able to synthesise. Alan Guth, who was one of the leading high energy physics researchers in this field, commented that until he attended a seminar given by Bob Dicke, he had no idea that the value of the Einstein de Sitter critical density, $\Omega = 1$, was in any way special.

Katsuhiko Sato *loc. cit.* published a series of remarkable papers on the physics of the GUT phase transition in the cosmological context, concluding that GUT theories could solve the baryon and monopole problems. He also appreciated that the phase transition would result in exponential expansion and produce super-horizon perturbations.

But no one, until the advent of Guth's paper, remarked that this would put the Universe into a homogeneous and isotropic state with $\Omega = 1.000\ldots$.

[7] Guth was always very clear about the limitation of his model. In the abstract of his paper he states that *Unfortunately, the scenario seems to lead to some unacceptable consequences, so modifications must be sought.* But cosmic inflation was nevertheless a game-changing concept. It was at a seminar given by Guth at Harvard that Paul Steinhardt was inspired to fix the problem (Albrecht and Steinhardt, 1982), and, meanwhile, the same task was undertaken by Linde (1982a,b) in Moscow.

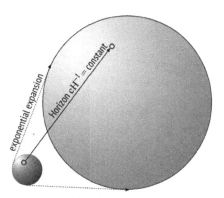

Fig. 9.1 A 'balloon' depiction of inflation in which the scale factor grows exponentially $a(t) \propto e^{Ht}$ while the expansion rate $H = \dot{a}/a$ remains constant. The particle horizon size during a de Sitter phase of expansion is $D_p^{\mathrm{dS}} \sim cH^{-1}$ and remains constant (see Equations 5.78). Within that horizon the curvature is decreasing and the Universe evolves towards a flat state. Lineweaver and Davis (2005) give a clear discussion of the balloon analogy. ⓒ

the Universe, and so the proper distance is the appropriate measure. The angular scale of a causally connected region is $\sim 2°$.

The observed radiation received from that time is highly isotropic: causal regions seen in different directions look almost identical. It would seem that there cannot have been any causal communication between patches on the sky separated by distance of more than a few degrees. So why would physically unconnected patches of the Universe look almost identical? Within the framework of the standard Big Bang theory, this would require a very homogeneous start to the cosmic expansion.

The scale of the causally connected regions depends on the expansion scale $a(t)$, as in Equation (5.75). For $a(t) \propto t^{2/3}$, i.e. the Einstein de Sitter model, that integral evaluates to $R_H^{\mathrm{EdS}} = 3ct$. However, in the de Sitter model where $a(t) \propto e^{Ht}$ for constant expansion rate H, we have $R_H^{\mathrm{EdS}} = cH^{-1}e^{Ht}$. The size of the causally connected regions increases exponentially, despite the expansion rate remaining constant, and, in a Newtonian view of things, does so apparently faster than the speed of light (see Section 5.10.2).

Guth's proposition was that at a very early stage in the cosmic expansion, the equation of state of the cosmic medium might have been such as to drive a phase of de Sitter-like exponential expansion. The exponential expansion of the horizon, $\propto e^{Ht}$ would persist for as long as the de Sitter phase lasted.

The situation is sketched in Figure 9.2 where the forward and past light cones at a number of fiducial times are shown (see Section 5.10 and Equation 5.72). The figure depicts a model in which there are three regimes: a de Sitter phase followed by an Einstein de Sitter phase which is at first radiation dominated and then matter dominated.

9.2.4 The Flatness Problem

Cosmological coincidences in which a parameter takes a value that is close to being special in some way are more often than not just that: coincidences. We shall discuss one of those

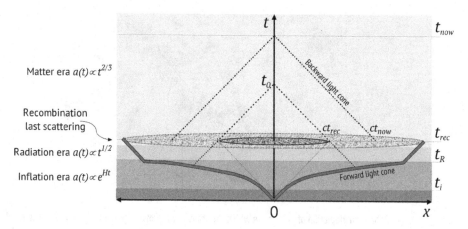

Fig. 9.2   The horizon problem and how it is explained by inflation. The sketch depicts the Universe in physical coordinates so that light rays travel on straight lines. During the inflationary expansion, the particle horizon, depicted by the shaded line, expands exponentially to embrace a volume far larger than the volume determined by the speed of light. All the scales which we observe today and that extend beyond the horizon at the time of recombination were once within the same causal volume at 0. Figure 9.5 shows how different scales that were once within their causal horizon are inflated to a greater size during inflation and re-enter their horizon at a later time. ©

coincidences here: why the Universe now appears to be well approximated by the special flat model cosmological model, the Einstein de Sitter model. Would that not have required extraordinary initial conditions?

In Section 6.4.1 we showed that for a single component Universe with equation of state $p = w\rho c^2$, the density parameter $\Omega(z)$ varied as

$$\Omega(z) \rightarrow 1 - \left(\frac{1 - \Omega_0}{\Omega_0}\right) \frac{1}{(1 + z)^{1+3w}}, \quad z \rightarrow \infty, \tag{9.1}$$

so, provided $w > -\frac{1}{3}$, the value of $\Omega(z)$ at a modestly high redshift like $z \sim 10^{10}$ would be extremely close to unity, no matter what the current value of Ω_0. For a radiation gas with $w = \frac{1}{3}$ the value of $1 - \Omega(z) \sim z^{-2} \sim 10^{-20}$ at $z \sim 10^{10}$. One part in 10^{20} is extraordinarily fine tuning of the high-z state.[8]

9.2.5 The fine tuning problem

There is another problem that could not have been appreciated in the 1980s: we now know that the dark energy density, $\Omega_\Lambda = 0.7$ and the matter density $\Omega_m = 0.3$ are almost equal at the present time. The observed acceleration of the Universe is a recent phenomenon. In those models where $\Lambda = $ constant, i.e. the Einstein cosmological constant, Λ is a constant of nature having value of $\Lambda = 1.3 \times 10^{-123}$ when expressed in fundamental Planck

---

[8]   This, incidentally shows that a closed Universe ($\Omega_0 > 1$) is always closed, and an open one ($\Omega_0 < 1$) is always open. $\Omega = 1$ is special in that Ω retains the same value at all epochs. This led some cosmologists to use this as an argument as to why $\Omega_0 = 1.00\ldots$: we live in an Einstein de Sitter Universe.

units, $\hbar = G = c = 1$.[9] It is difficult to imagine a physical theory in which one of the fundamental constants is so small, so this value of Λ appears to be finely tuned to cause the transition to accelerated expansion to occur at about the present epoch. Had Λ been 10^{-122} or 10^{-124} our Universe would have looked quite different.

One way around this is to propose that Λ is in fact a function of time. Cosmic material which gives rise to a dynamic Λ-term is generally referred to as *Quintessence* (see Sahni (2002) for a review), Quintessence is most simply modelled as a medium with equation of state $p = w(t)\rho c^2$, having time dependent w such as given in Equation (6.90) and Section 6.5. This is particularly convenient when using observational data to constrain w. A more formal approach is to associate quintessence with a scalar field φ, referred to as the *inflaton field*, which evolves under the influence of a potential $V(\varphi)$. There are numerous choices for possible functions $V(\varphi)$, some of which will be discussed below.

9.3 Inflationary Cosmological Models

9.3.1 Exponential Expansion

The quest is to create a period of exponential expansion during some period of time, see Figure 9.3. This can be achieved if the equation of state of the matter is:

$$p_V = -\rho_V c^2, \tag{9.2}$$

when the scale factor and expansion rate, H_V, would evolve as:[10]

$$a(t) \propto e^{H_V t}, \qquad H_V = \frac{8\pi}{3} G \rho_V = \text{constant}. \tag{9.3}$$

This is depicted in Figure 9.1.

The fact that the pressure p_V has to be negative is strange within the context of our familiar world, but such pressures could exist as a consequence of the physics of the vacuum at the high temperatures prevailing in the early Universe.[11]

The exponential expansion phase is referred to as the '*de Sitter phase*' and would continue for as long as the material had this peculiar equation of state. The Universe then has to make a transition to an expansion with a 'normal' equation of state, otherwise it would not resemble our Universe.

During this 'de Sitter' expansion phase the curvature tends to zero (see Equation 6.69), and the longer the duration of this inflation period the closer is Ω_Λ to unity. With this

---

[9] See Section 9.4.1 and Appendix A.5.

[10] See Section 6.3.1 for a description of this one-component simple model and Section 6.4.1 for discussion of the evolution of flatness, $\Omega_k(z)$.

[11] See On-line Supplement *The Hilbert Stress-energy Tensor* for a brief discussion of the properties of the vacuum.

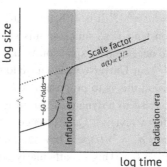

A schematic view of the cosmic expansion around the time of inflation when the expansion is exponential. During the period of exponential expansion there are some 60 e-foldings of the scale factor, corresponding to a growth factor of $\sim 10^{25}$. Once inflation stops we enter a radiation era with power law expansion. ⓒ

'explanation' for the flatness problem, the idea is that all anisotropies and inhomogeneities will tend rapidly to zero during this phase of phenomenal expansion.[12]

The other side of the coin is that it is necessary to generate the primordial inhomogeneities that will eventually give rise to the formation of galaxies and large scale structure. These can be generated during the inflationary era and are generally thought to arise out of quantum fluctuations in the vacuum state (Press, 1980; Hawking, 1982; Guth and Pi, 1982; Starobinsky, 1982). In the simplest models the spectrum of fluctuations is the Harrison–Zel'dovich spectrum with Gaussian distributed fluctuations (Press, 1980).

There is a huge variety of models for inflation, they all have a de Sitter phase. In fact it is not difficult to create models that go into exponential expansion. The problem is to sustain the exponential phase for long enough and then to emerge from that phase into what we see today.

These diverse models come from different starting points about the fundamental nature of the matter in the early Universe. It is fortunate that these models tend to have different consequences that are observable. Two of these consequences are the shape of the spectrum of primordial inhomogeneities, which can be directly observed as the power spectrum of the fluctuations of the temperature in the cosmic background radiation, and the generation of gravitational waves that can be observed in the B-mode polarisation of the cosmic background radiation.

9.3.2 General Inflationary Models

By now there is an entire zoo of inflationary models ranging from variants of the older ones to string theory models. For an overview of many of the models and how they stand with respect to the microwave background data see Linde (2014).

[12] The expansion appears to be happening faster than the speed of light: a sphere smaller than the size of a pea is expanding to the size of the current Universe in something less than 10^{-33} s! In one Hubble expansion time H^{-1} a light ray can only travel a distance cH^{-1}, which is substantially less than a co-expanding length-scale changes in the same time. The particles in the Universe stay in their comoving positions, they partake in the expansion of the underlying space-time (they are in free-fall). No information is transferred during the expansion.

The goal of inflation theories is to achieve a huge expansion of the Universe in a very short time: so we literally have a 'Big Bang' at the earliest moments of the Universe's history. A natural requirement for this is that this should be a period during which the expansion is accelerating, so the scale factor $a(t)$ has the property

$$\frac{\ddot{a}}{a} = -\frac{4\pi G}{3}\rho(1 + 3w) > 0 \tag{9.4}$$

throughout this period (see Section 6.3.3). The condition for accelerating expansion, $\ddot{a}/a > 0$, is seen to be equivalent to having material with an effective equation of state that has

$$\rho + 3p < 0, \quad w < -\tfrac{1}{3}. \tag{9.5}$$

While no 'normal' material has this property, it is possible in a quantum mechanical context to have an effective pressure term that behaves this way.

Conventionally, the dimensionless measure of the acceleration or deceleration of the expansion is the deceleration parameter which for a one-component model having this equation of state $p = w\rho$ is:

$$q = -\frac{1}{H^2}\frac{\ddot{a}}{a} = \frac{1}{2}(1 + 3w). \tag{9.6}$$

This was was defined in (6.34) and (6.81), and was shown in Section 6.3.3 to follow directly from the Friedmann equation.

In the discussion of inflationary models it is more usual to express the deceleration directly in terms of the Hubble expansion rate, $H(t)$, and its rate of change $\dot{H}$, via the dimensionless parameter

$$\epsilon_w = \frac{\dot{H}}{H^2} = -\frac{1}{H^2}\frac{\ddot{a}}{a} + 1 = q + 1 = \tfrac{3}{2}(1 + w). \tag{9.7}$$

When $w = -1$, as in the classical cosmological constant Λ, we have $\epsilon_\Lambda = \epsilon_{(w=-1)} = 0$.

9.4 Driving Inflation

9.4.1 The Planck Time and Planck Units

The equations for inflationary cosmology are frequently written in *Planck units*. These are defined in terms of the universal constants $G, \hbar$ and c. This system of units is discussed in Appendix A.5, but it is worth saying something about them here.

One of the fundamental quantities derived from $G, \hbar$ and c is the Planck time, $t_P = (\hbar c/G^5)^{1/2}$s $= 5.391 \times 10^{-44}$s. This is the time when quantum mechanics and our notions of space-time become blurred.[13] Associated with this are other fundamental quantities, like the Planck mass M_P and the Planck length, l_P. In this way we can define physical quantities in terms of Planck units.

---

[13] We cannot, using known physics, describe what happens at or before this time. That is of course no impediment to field theorists building models of that era! This is the basis of the search for the *Theory of Everything*.

Planck units come in two types: *regular Planck units* in which $\hbar = G = c = 1$, so that $M_P = 1$, and *reduced Planck units* in which $\hbar = c = 1$ and $M_P = (8\pi\hbar c/G)^{1/2} = 1/\sqrt{8\pi G}$. Provided we agree on setting $\hbar = c = 1$, we can write all equations using the Planck mass M_P and, depending on our preference, put $M_P = 1$ or $1/\sqrt{8\pi G}$.

There are three fundamental constants at our disposal, $G, \hbar, c$ from which we can construct quantities having dimensions of mass, length and time, and from there construct the corresponding temperatures, energies, units of resistance and so on. By mixing various combinations of $G, \hbar, c$ we can define these up to an arbitrary constant (like π).[14]

Formally, these multiplicative constants are simply fixed as follows.[15] The values of these quantities start from the definitions of the reduced Compton wavelength $\lambda_C = \hbar/mc$ of a mass m and its Schwarzschild radius, $R_S = 2Gm/r$. Their product, $\lambda_C R_S = 2G\hbar/c^3$, is independent of m and involves only the natural constants, and this is conventionally set to be $2l_p^2$, where l_P is the *Planck length*:

$$l_P = \sqrt{\frac{\hbar G}{c^3}} = 1.616199 \pm 0.000097 \times 10^{35}\,\text{m} \qquad (9.8)$$

This is supported by the American National Institute of Standards and Technology as the basis of *Planck units*.[16]

Fundamental Planck units

$l_P = \left(\frac{\hbar G}{c^3}\right)^{\frac{1}{2}}$	$t_P = \left(\frac{\hbar G}{c^5}\right)^{\frac{1}{2}}$	$M_P = \left(\frac{\hbar c}{G}\right)^{\frac{1}{2}}$
$\simeq 1.616 \times 10^{-33}\text{cm}$	$\simeq 5.391 \times 10^{-44}\text{s}$	$\simeq 2.176 \times 10^{-5}\text{gr}$

Note that $t_P = l_p/c$. From there we can define other physical quantities.[17] It should be noted that cosmologists frequently use the reduced Planck units in which the Planck mass is $(\hbar c/8\pi G)^{1/2}$ since this simplifies the Einstein equations.

9.4.2 The Scalar Inflaton Field

We need a model for the driver of inflation that is based on known or at least plausible quantum physics. In fact there are many possibilities and so in the present simplistic development we will focus on a hypothetical scalar field φ which pervades all of space and

---

[14] In 1874 George Johnstone Stoney (1881, this is the write-up of his earlier 1874 talk) put forward a similar strategy for defining electromagnetic units. Of course he could know nothing of Planck's constant, he set $c = G = e = 1$. This defined the fundamental unit of charge which he later, in 1891, named the *electron*.

[15] There are numerous ways of defining the Planck scale, often giving different multiplying factors and so differing from one another by factors involving a multiple of π.

[16] In that sense the choice of pre-multiplying factors is not arbitrary.

[17] The Planck density and temperature in familiar cgs units are:

$$\rho_P = \frac{M_P}{l_P^3} \simeq 5.4 \times 10^{93}\,\text{gr cm}^{-3}, \quad T_P = \frac{M_P c^2}{k_B} \simeq 1.4 \times 10^{32}\,\text{K} \simeq 1.2 \times 10^{19}\,\text{GeV}, \qquad (9.9)$$

where k_B is the Boltzmann constant. Likewise, in cgs units, we get a mass that corresponds to the present-day cosmological constant $\Lambda \sim 10^{-56}\,\text{cm}^2$: $m_\Lambda \sim \hbar\Lambda^{1/2}/c \sim 3 \times 10^{-66}g$.

which can, under some circumstances, cause accelerated expansion. If we want to have a physical system in mind that can do this we should think of the quantum vacuum in which our φ is the classical expectation value of the ground state of the quantum vacuum. Such a field φ could be referred to as the *inflaton field*. We shall see in Section 11.4 that, if the vacuum is to be thought of as a medium pervading all of space then Lorentz invariance would require it to have equation of state $p_V = -\rho_V c^2$. This is the case for Einstein's cosmological constant, though this is not yet the consequence of any theory involving known physics.

When discussing the field φ that causes the inflation, the *inflaton field*, we model it as though it were a simple, possibly complex-valued, scalar field with a known potential $V(\varphi)$. Moreover we shall treat it as a classical field, perhaps to be thought of as the vacuum expectation of a quantum field: $\langle \varphi \rangle = \langle 0 | \varphi | 0 \rangle$. This mean field drives the expansion and its quantum fluctuations will be responsible for the creation of structure.

The simplest Lagrangian for the inflaton field φ can be taken as[18]

$$L = \left[\tfrac{1}{2}\dot{\varphi}^2 - (\nabla\varphi)^2 - V(\varphi) \right], \tag{9.10}$$

(see Section 11.2.1). The $\tfrac{1}{2}\dot{\varphi}^2 - (\nabla\varphi)^2$ term is the kinetic energy term. The energy-momentum tensor (from Noether's theorem, see Section 11.1.5, Equation 11.49) is

$$T^{ab} = \partial^a\varphi\,\partial^b\varphi - g^{ab}L. \tag{9.11}$$

We can get an effective pressure and density for this field using (11.29) and assuming homogeneity (i.e. $\nabla\varphi = 0$). This gives $T^{ab} = \mathrm{diag}(\tfrac{1}{2}\dot{\varphi}^2 + V, \tfrac{1}{2}\dot{\varphi}^2 - V, \tfrac{1}{2}\dot{\varphi}^2 - V, \tfrac{1}{2}\dot{\varphi}^2 - V)$ and:

Equivalent inflationary fluid parameters

$$\rho = \tfrac{1}{2}\dot{\varphi}^2 + V(\varphi), \quad p = \tfrac{1}{2}\dot{\varphi}^2 - V(\varphi) \quad w = \frac{p}{\rho} = \frac{\tfrac{1}{2}\dot{\varphi}^2 - V(\varphi)}{\tfrac{1}{2}\dot{\varphi}^2 + V(\varphi)}. \tag{9.12}$$

If the field is independent of time, or varies only very slowly, then we have $\rho = V$ and $p = -V$ and the equation of state becomes $p = w\rho$ with $w = -1$, as with Einstein's cosmological constant. It is this that gives rise to exponential expansion required by the theory of inflation.

---

[18] We could even add in terms coupling the field φ directly to the space-time geometry. These might be of the form $f(R)\varphi^2$. where $f(R)$ represents a polynomial in the invariants generated from the Riemann tensor: $f(R) = a_1 R + b_1 R^2 + b_2 R_{ab}R^{ab} + b_3 R_{abcd}R^{abcd} + \dots.$

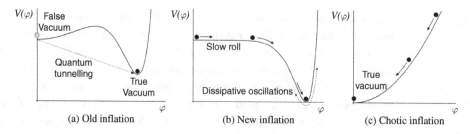

(a) Old inflation   (b) New inflation   (c) Chotic inflation

Fig. 9.4   Three classic model potentials for the inflaton field $V(\varphi)$. The scales of the axes are totally arbitrary. Although the potentials are depicted for pedagogical reasons as being fixed, the shape could in fact be time dependent. So we could dream up a model in which the potential (a) evolved into (b). ⓒⓒ

The classical[19] Euler-Lagrange equation (see Section 11.1) for this Lagrangian is:[20]

Scalar inflaton field evolution

$$\ddot{\varphi} + 3H(t)\,\dot{\varphi} - \frac{1}{a^2}\nabla^2\varphi = -\frac{dV(\varphi)}{d\varphi}. \tag{9.14}$$

$\nabla^2\varphi$ is the Laplacian in comoving coordinates $\mathbf{x} = \mathbf{r}/a(t)$.

This can also be derived from the Friedmann equation for energy conservation (5.57) and Equations (9.12). It looks like the dynamical equation for the motion of a quantity φ that is driven by a force $V'(\varphi)$ against a friction force $3H\dot{\varphi}$. This gives rise to images of balls rolling down hills described by a potential of shape $V(\varphi)$, as in Figure 9.4.

9.4.3 Accelerated Expansion

The expansion of a Universe pervaded by this field is described by the usual Friedmann equations (e.g. (6.3) with pressure and density given by (9.12)):

---

[19] The field φ in these equations is to be interpreted as the vacuum expectation value of a quantum field. Formally we should write this equation as:

$$\langle\varphi\rangle^{\cdot\cdot} + 3H(t)\,\langle\varphi\rangle^{\cdot} = -\langle V'(\varphi)\rangle.$$

Under some circumstances the term on the right-hand side might be treated as $-V'(\langle\varphi\rangle)$, in which case we can simply replace $\langle\varphi\rangle$ by φ and so get to Equation (9.14). It is the quantum fluctuations in the field φ that are responsible for the origin of curvature fluctuations and gravitational waves. The key quantity is $\langle\varphi^2\rangle$ and we have to deal with the quantised field (see Starobinsky (1986) and Sasaki *et al.* (1988)).

[20] For the simple potential $V(\varphi) = \frac{1}{2}m^2\varphi^2$ this is the Klein–Gordon equation in a box expanding at rate H. We should also note that imposing homogeneity and isotropy, Equation (9.14) becomes

$$\ddot{\varphi} + 3H(t)\,\dot{\varphi} = -V'(\varphi). \tag{9.13}$$

We shall make considerable use of this form when discussing inflationary cosmological models.

Expansion under the influence of a scalar field:

$$H^2 = \frac{1}{3M_P^2}\left[\frac{1}{2}\dot{\varphi}^2 + V(\varphi)\right] - \frac{k}{a^2}, \tag{9.15}$$

$$\dot{H} = -\frac{1}{2M_P^2}\dot{\varphi}^2 + \frac{k}{a^2}, \tag{9.16}$$

where, as usual, $H = \dot{a}/a$ and k is a constant of integration.[21]

The expansion is accelerating when $\ddot{a}/a > 0$. We can write $\ddot{a}/a$ in terms of the Hubble expansion rate H (see Equation 9.7):

$$\frac{\ddot{a}}{a} = \dot{H} + H^2 = H^2(1 - \epsilon_H), \quad \epsilon_H = -\frac{\dot{H}}{H^2}, \tag{9.17}$$

from which we see that the condition for accelerated expansion is $0 < \epsilon_H < 1$. Parameters such as ϵ_H serve to characterise the various types of potential $V(\varphi)$.

The case $\epsilon_H = 0$ corresponds to $H = $ const, which is easily verified to give $a(t) \propto e^{Ht}$. The scale factor is growing exponentially while the Hubble expansion rate H remains constant. This is a very important limiting case as it tells us that, during this expansion the field φ remains constant if $k = 0$, and that, if $k \neq 0$, the k-term becomes negligible during the expansion and φ becomes a constant. The constancy of the field φ means that the effective equation of state $p = w\rho$ has $w = -1$ (see Equations 9.12).

We notice that for accelerated expansion the e-folding rate will be determined at any time by the local Hubble constant, H. So we can count the number of e-foldings over a given period of time during inflation from a time t_1 to time t_2 by evaluating [22]

$$N = \int_{t_1}^{t_2} H(t)\,dt \overset{\text{Slow roll}}{\simeq} \frac{1}{M_P^2}\int_{\varphi_{t_1}}^{\varphi_{t_2}} \frac{V(\varphi)}{V'(\varphi)}\,d\varphi. \tag{9.18}$$

Usually, t_2 would mark the end of the inflationary period. In order to explain the observed flatness today, the Universe would have had to have gone through $N \sim 50$–60 e-foldings (a factor 10^{20}–10^{25}). We should note that since temperature varies as $T \propto a^{-1}$, a huge change in scale factor implies a fall to a very low temperature. This implies that there must be a mechanism for getting out of the inflationary era and reheating the Universe so that it has the correct temperature today.

9.4.4 Slow Rolling Approximation

Perhaps the most important aspect of inflation theories in general is that the Universe has to spend a relatively long time inflating in order to get the large number of e-foldings.

---

[21] Equation (9.15) is the Friedmann Equation (6.3), with (9.12) and putting $\Lambda = 0$. (9.16) is derived from the identity $\dot{H} = \ddot{a}/a - H^2$. Recall that in units where $\hbar = c = 1$, M_P is either 1 or $1/\sqrt{8\pi G}$ depending on which Planck units are used (see Section 9.4.1).

[22] This amounts to saying that in an interval in which the log expansion changes by $d\ln a$ the number of e-foldings changes by dN. Hence $dN = -d\ln a = -Hdt = -Hd\varphi/\dot{\varphi}$. The minus sign is there because by convention we count the e-foldings backwards in time from the end of inflation. The last of these equalities tells us how to calculate N for a given potential. The first equality simply tells us that $N = \ln(a_{\text{end}}/a)$.

We have seen from Equations (9.2) and (9.3) that exponential expansion can be achieved while the density and pressure are related by $p = -\rho$. From Equation (9.12) we see this is achieved when $|\dot{\varphi}^2| \ll |V(\varphi)|$: the inflaton field must change slowly.

However, the slow change in the inflaton field has to be maintained for a large number of e-foldings in order to achieve the necessary level of inflation, and so it is essential that $|\ddot{\varphi}|$ also be small and stay small, otherwise $|\dot{\varphi}|$ will increase in magnitude. We see from Equation (9.14) that this can happen if the magnitude of the force term $V'(\varphi)$ driving the evolution of φ remains small. Likewise the magnitude of the curvature of the potential, $V''(\varphi)$, must remain small, otherwise $V'(\varphi)$ changes and so does $\ddot{\varphi}$. The potential $V(\varphi)$ has to be very flat (see Figure 9.4b).

The period during which the field φ changes only slowly over a long period of time is referred to as the *slow roll period*. Not all potential inflationary models have such a period. Equations, (9.14) and (9.15), simplify somewhat during the slow rolling period. While $\ddot{\varphi} \simeq 0$ and $\frac{1}{2}\dot{\varphi}^2 \ll V(\varphi)$ we have the approximations:

Evolution in the slow-roll approximation

$$H^2 = \frac{\dot{a}^2}{a^2} = \frac{1}{3M_P^2}V(\varphi), \qquad 3H\dot{\varphi} = -\frac{dV(\varphi)}{d\varphi}. \tag{9.19}$$

During the inflationary regime we can neglect the curvature term, k/a^2, as this will have inflated away. Given $V(\varphi)$ we may be able solve these for $a(t)$.

From (9.12) we see that when $\ddot{\varphi}$ is very small, we have $\dot{\varphi} \simeq -V'/3H$, so we can achieve a lengthy period of inflation provided both V' and V'' are small when compared with V itself. We infer that the conditions for persistent inflation can be encapsulated in two important dimensionless parameters that characterise the mathematical form of the field $V(\varphi)$:

Slow-roll parameters:

$$\epsilon(\varphi) = \frac{1}{2}M_P^2\left(\frac{V'}{V}\right)^2 \simeq -\frac{\dot{H}}{H^2}, \qquad \eta(\varphi) = M_P^2\left(\frac{V''}{V}\right). \tag{9.20}$$

See, for example, Liddle and Lyth (1992, who use slightly different units). One important advantage of these 'small' theories (i.e. small $\dot{\varphi}$ and $\ddot{\varphi}$), is that the inflation takes place well after the Planck epoch and so mathematical formalisms exist to handle them within the scope of standard quantum field theory and classical general relativity.

There is a simple relationship in the slow-roll approximation between ϵ, η and N:

$$\eta = 2\epsilon - \frac{d\ln\epsilon}{dN}. \tag{9.21}$$

This is derived by writing out (9.19) in terms of the definitions of ϵ, η and N.

9.4.5 Connecting with Observable Quantities

Inflationary theories address not only the horizon and flatness problem; if they did only that there would be little impetus to study them aside from curiosity. The period of inflation is the period during which the quantum fluctuations in the pre-inflationary medium can develop embryonic primordial structure. Because of the inflation, this structure can grow to scales of relevance to the structures we currently observe and thereby open up the possibilities of constraining inflationary models. The CMB anisotropy provides evidence concerning the values of parameters that define inflationary models. The parameters ϵ and η are rather general parameters that can be estimated from CMB observations.

There are two classes of such fluctuation: scalars and tensors. The scalar ones become the density fluctuations that grow into cosmic structures, while the tensor ones generate primordial gravitational waves. The density fluctuations are seen in the angular distribution of the CMB temperature on the sky. We have yet to detect the tensor modes, but detecting them would be *prima facie* evidence for the existence of an early inflationary period.

Both the density fluctuations and the gravitational waves are random fields, each characterised by a power spectrum. It is to be expected that, in a first approximation, these power spectra would be power laws characterised by a model-dependent slope and an amplitude. A valid inflationary theory should predict values for these, and also any deviation from a strict power law. This is a technically complex subject that we shall touch on presently, but for the moment we can present some models that enable us to express what might be measured in terms of the parameters ϵ and η.

We can calculate, for a given inflationary model specified by a potential $V(\varphi)$, values of the parameters ϵ and η and some higher order parameters (see Section 9.5.2). The amplitudes of the curvature (matter) power spectrum, A_S, and the gravitational wave power spectrum, A_T, expressed in terms of the potential V and ϵ, η, are

$$A_S = \left(\frac{H^2}{2\pi \dot{\varphi}} \right)^2 \simeq \frac{1}{24\pi^2} \left(\frac{V}{M_P^4 \epsilon} \right), \quad A_T \simeq \frac{2}{3\pi^2} \left(\frac{V}{M_P^4} \right). \tag{9.22}$$

These are derived later. The power spectrum of the matter distribution should in a first approximation be a power law with amplitude A_S and slope n_S given by $n_s = 1 + 6\epsilon - 2\eta$. The ratio of the power in the tensor and scalar modes is an important quantity:

Tensor to scalar ratio

$$r = \frac{A_T}{A_S} = 16\epsilon. \tag{9.23}$$

Limits on the tensor component impose constraints on the parameters of theories of inflation. We shall return to this in Section 9.5.2 (see Equation 9.41).

9.4.6 Some Choices for $V(\varphi)$

Now we consider some of the simpler functions $V(\varphi)$ that have been considered in the literature. There are innumerable possible choices for the function $V(\varphi)$, Martin *et al.* (2014, Table 1) derive the values of extended parameter for 74 different inflationary models.

Some inflationary potentials

Higgs potential	$\lambda(\varphi^2 - M^2)^2$	Massive scalar field	$\frac{1}{2}m^2\varphi^2$
Power law	$\lambda e^{-\alpha\varphi}$	Chaotic inflation	$\lambda\varphi^\alpha$
Natural inflation	$\Lambda^4\left[1 + \cos\left(\frac{\varphi}{f}\right)\right]$	Hilltop model	$V_0\left[1 - \frac{\gamma}{n}\left(\frac{\varphi}{M_p}\right)^n\right]$

$$\text{(9.24)}$$

We look at three of these models in what follows. The Planck Collaboration papers on inflation[23] provide a concise overview of a number of inflationary models and how they are constrained by the Planck satellite CMB data.

Chaotic Inflation: a Large Field Model

The *chaotic inflation* (Linde, 1983, 2014) models are defined by having only a single term in the potential that is a power law in the field φ:

$$V(\varphi) = (\varphi/\varphi_0)^n. \tag{9.25}$$

Values $n = 2$ and $n = 4$ would be favoured because these correspond to renormalisable field theories. This is referred to as a *large field model* since the initial value of $\phi \sim M_P$.

With this potential, using the first of (9.19) shows that the scale factor grows as

$$a(t) \propto \exp\left[\left(\frac{\varphi}{\varphi_0}\right)^{\frac{n}{2}} \frac{t}{\sqrt{3}M_P}\right], \tag{9.26}$$

and so indeed we do get exponential expansion.

The parameters characterising this model are simple to calculate:

$$\epsilon = \frac{n^2}{2}\left(\frac{M_P}{\varphi}\right)^2, \quad \eta = n(n-1)\left(\frac{M_P}{\varphi}\right)^2, \quad N \simeq \frac{1}{2n}\left(\frac{\varphi_i}{M_P}\right)^2, \tag{9.27}$$

where t_i is the time at which inflation begins and when the field value is φ_i. We note that in order to get a large number of *e*-foldings we need to have $\varphi_i \gg M_P$. In that case $\epsilon \ll 1$ and $\eta \ll 1$: we are in the slow-roll regime. These models are referred to as being 'hot' because the inflationary value of the field is greater than the Planck mass.

So why is this called *chaotic inflation*? Linde's original idea *loc. cit.* was that inflation theories in general did not really require specific initial conditions. One could imagine that different places in the early Universe started their inflation from quite different initial conditions that were due to the stochastic nature of the quantum fluctuations which had

---

[23] Planck Collaboration *et al.* (2014, Section 4.2 and Fig. 1) and Planck Team (2015, Section 6 and Fig. 12) (the figures are difficult to reproduce here in grey-scale). See also Martin (2015).

existed at pre-inflationary times. He imagined a Universe full of different inflating bubbles: we happen to be living in just one of these which is now so big that we cannot be aware of other bubbles outside of our own. Inevitably this caused considerable debate as such an idea, by its very nature, was almost totally untestable.[24]

Power-law Inflation

Power-law inflation is a viable alternative to exponential expansion (Abbott and Wise, 1984; Lucchin and Matarrese, 1985). An exponential potential with an adjustable shape parameter p,

$$V(\varphi) = V_0 \exp\left(-\lambda \frac{\varphi}{M_P}\right), \tag{9.28}$$

gives rise to power law expansion (hence the name)

$$a(t) \propto t^p, \quad p = 2/\lambda^2, \quad \lambda^2 < 6. \tag{9.29}$$

Surprisingly, this is a solution to the full Equations (9.14) and (9.15) (Liddle, 1989, who sets $M_P = 1$). In order to get the required degree of expansion to explain the flatness and horizon problems, we require initial conditions in which the value of the field φ is extremely large and negative.

The slow-roll parameters (9.20) for this model are

$$\epsilon_\varphi = \frac{1}{2}\lambda^2, \quad \eta_\varphi = \lambda^2. \tag{9.30}$$

As we shall see, this is one of the models that, in this its original form, is ruled out by data from the Planck satellite.

The Hilltop Model: Small Field Model

The *hilltop model* for the potential,

$$V(\varphi) = V_0\left[1 - \frac{\gamma}{n}\left(\frac{\varphi}{M_p}\right)^n\right], \quad n \geq 2 \tag{9.31}$$

is a specific example of a class of models that give rise to *small field inflation*, by which is meant that the large number of e-foldings can be achieved with $\varphi \ll M_p$. For this particular model we have

$$\epsilon = \frac{\gamma^2}{2}\left(\frac{\varphi}{M_p}\right)^{2n-2}, \quad \eta = -\gamma(n-1)\left(\frac{\varphi}{M_p}\right)^{n-2}, \quad N \simeq \frac{1}{\gamma(n-2)}\left(\frac{\varphi_i}{M_p}\right)^{2-n}. \tag{9.32}$$

With the constraint that $n > 2$ we can get a large number of e-foldings with small values of φ_i compared with the Planck mass. This situation is known as 'cold' inflation since the temperature associated with the field is less than the Planck temperature.

---

[24] One suggestion for testing this was that we might witness a collision between our bubble and a neighbouring one. This is somewhat reminiscent of the (in my view mediocre) science fiction book *Cosmic Engineers* by Simak (1950) in which our Universe is saved from this disaster by some intrepid scientists.

9.4.7 The End of Inflation

While the Universe goes through this exponential expansion it is cooling to extremely low temperatures. Moreover in our simple model there is nothing to get us out of the inflation phase once it starts. Inflation does not go on forever and the Universe cools: this is not the Universe in which we live.

An important question arises: what stops the inflation and when does that happen? The simple answer is to say that the oscillations of the φ-field are damped by any of a variety of physical processes such as quantum particle creation. We can build on our simple model (9.14) by adding in a dissipative term:

$$\ddot{\varphi} + 3H(t)\dot{\varphi} + \Gamma_\varphi \dot{\varphi} = -\frac{dV(\varphi)}{d\varphi}. \tag{9.33}$$

The term $\Gamma_\varphi \dot{\varphi}$ is to be thought of as a friction term which eventually brings changes in the field φ to a stop. Viewed as a friction term, we can also think that this would be a mechanism for reheating the Universe after the cooling it would experience during this fast expansion: our Big Bang is then a hot one.

9.5 Origin and Evolution of Inhomogeneities

Perturbations arise from quantum fluctuations in the cosmic medium before and during the period of inflation. Two kinds of non-homogeneity appear: curvature perturbations due to irregularities in the matter distribution, and variations in the geometry of the space time due to the creation of gravitational waves. The technicalities of how these fluctuations appear are considerable, so we shall content ourselves with a simplistic discussion.[25]

These perturbations are crucial to our understanding of the Universe: they are responsible not only for the cosmic structure we observe, but also for the fluctuations in temperature of the CMB at the last scattering surface. The fluctuations in density appear on all scales and have a power spectrum which is close to a power law. That observation is not polluted by the complex processes of star and galaxy formation: the radiation arrives at the last scattering surface from the Big Bang in pristine condition. We have not at the time of writing observed the gravitational wave component of the CMB, but when it is seen its power spectrum will also impose string constraints on the earliest moments of the Big Bang.

The announcement of the discovery of gravitational waves from the merging of two black holes by Abbott *et al.* (2016) establishes several fundamental aspects of Einstein's theory that give cause for considerable optimism that evidence for primordial gravitational waves will be discovered in the CMB radiation.

---

[25] The book by Lyth and Liddle (2009) covers many aspects of inflation in some detail, with particular emphasis on the origin of inhomogeneities. Peacock (1999, Section 11.5) presents a clear explanation of the creation of inhomogeneities. See also the excellent lecture notes of Kinney (2002), Baumann (2009) and Langlois (2010). The paper of Souradeep and Sahni (1992) is particularly clear.

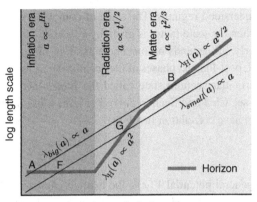

Fig. 9.5 A very schematic view of the evolution of density fluctuations during the inflationary period. The broad line represents the particle horizon, i.e. the range of communication, at a given time. The two thinner lines represent co-expanding density fluctuations on two different scales: a large scale and a smaller scale. Both of these are created in the pre-inflationary era when they are within their causal horizon. See the text for a detailed description. Adapted from Kolb and Turner (1990, Fig. 8.4, p285). ⓒⓒ

9.5.1 The Role of the Horizon

Let us first focus on the key feature of Figure 9.5: the evolution of the scale of the horizon compared with the scale of the Universe. The horizon scale remains constant throughout the inflationary period while particles are getting carried away from one another exponentially.

The density fluctuations we see now on the largest scales were once, prior to the start of the inflationary period, all within a single causal bubble. The inflation stretched them out so that they became larger than the causal horizon, at which time they became 'frozen in' at the amplitudes they had at that point. This is depicted at points **A** and **F** in the figure. After the end of the inflationary period they stay beyond frozen beyond their causal horizon until they re-enter, as depicted by points **B** and **G** in the figure. We notice that the large scales (**AB**) freeze before and re-enter the causal horizon after the smaller scales (**FG**). The very largest scales are fossils of the long distant past when the period of inflation took them outside of their causal horizon.

The pre-inflation fluctuations are supposed to arise from quantum fluctuations in the inflaton field. A small causal patch of Universe at that time will contain our entire Universe, and the scale of the fluctuations that we identify with cosmic structures today will be well within their causal horizon. Calculating these vacuum fluctuations is a matter of particle-in-a-box quantum mechanics.

The relevant fluctuation amplitude on a given comoving scale k is the amplitude it has at the time the scale grows to become the size of the horizon (points **A** and **F** in the figure). After that the amplitude does not change until the scale re-enters the causal horizon (points **B** and **G** in the figure) and we then have the starting conditions for the evolution through well-understood regimes. The scale of comoving wavenumber k reaches the horizon size H^{-1} when the scale factor is given by $k/a = H$.

Clearly, all depends on our understanding of the quantum nature of the vacuum at the earliest, post-Planck, phases of the cosmic expansion. It is generally supposed that this state is a particular vacuum state called the *Bunch–Davies vacuum*[26] (see Birrell and Davies (1982, see Ch.3 and Section 5.4)). For a massless inflaton field, the fluctuations in such a vacuum have correlation function

$$\langle \delta \varphi^2 \rangle = \left(\frac{H}{2\pi} \right)^2. \tag{9.34}$$

For a general potential $V(\varphi)$), the explicit expression for the correlation function will be similar, but multiplied by terms that depend in detail on $V(\varphi)$).

9.5.2 Scalar and Tensor Fluctuations

General perturbations to a homogeneous background can be resolved into three components:

$$\delta g_{ab} = h_{ab}^{\text{scalar}} + h_{ab}^{\text{vector}} + h_{ab}^{\text{tensor}}. \tag{9.35}$$

The scalar component is due to fluctuations in the inflaton field, φ, and is associated with density perturbations, while the tensor component is due to the superposition of two independent polarisations of gravitational wave modes. The vector mode would correspond to vortical perturbations; those cannot be driven by a simple scalar field and so they are taken to be zero in this simple model. The goal is to estimate the amplitudes of the scalar and tensor modes at the time each mode crosses out of its horizon.

We can proceed to do a linear perturbation theory analysis of a slightly inhomogeneous inflating Universe by writing

$$\varphi(\mathbf{x}, t) = \varphi_0(t) + \delta \varphi(\mathbf{x}, t), \tag{9.36}$$

where $\varphi_0(t)$ is the averaged background potential and the magnitude of $\delta\varphi$ is given by (9.34). The scalar component h_{ab}^{scalar} generated by horizon scale fluctuations corresponds to a curvature fluctuation on scale k of

$$R_k = \frac{\delta a}{a} = \frac{\dot{a}}{a} \delta t = H \delta t = H \frac{\delta \varphi}{\dot{\varphi}} = \frac{H^2}{2\pi \dot{\varphi}}. \tag{9.37}$$

Using the slow roll equations (9.19) and the definition (9.20) of ϵ:

$$\langle R_k^2 \rangle = \frac{H^4}{4\pi^2 \dot{\varphi}^2} = \frac{1}{4\pi^2} \frac{H^2}{M_P^4} \frac{V^2}{V'^2} = \frac{1}{24\pi^2 M_P^4} \frac{V}{\epsilon} \Big|_{k/a=H}. \tag{9.38}$$

This is to be evaluated at the time when the scale k exits its horizon.[27]

---

[26] In general there is no natural definition of a quantum vacuum in cosmology. However, in de Sitter space we can define a unique state such that there were no particles in the infinite past. This is the Bunch–Davies vacuum.

[27] There are numerous formal derivations using many different approaches. See for example Lyth and Liddle (2009, Section 25.2), Fujita *et al.* (2013, 2014); Vennin and Starobinsky (2015).

We can repeat a similar argument for the tensor, i.e. gravitational wave fluctuations, however, we must note that in this case the horizon scale fluctuations are

$$T_k = 2 \times 4 \times \frac{H}{2\pi M_P} = \frac{4H}{M_P}. \tag{9.39}$$

The factor 2 is because there are two independent polarisation modes, and the factor 4 is a technical normalisation factor, fixed by convention (Langlois, 2010, Section 4.5).[28] From this we deduce that the variance of the gravitational perturbations is

$$\langle T_k^2 \rangle = \left. \frac{2V}{3\pi^2 M_P^2} \right|_{k/a=H}, \tag{9.40}$$

evaluated at the time the scale k exits from the horizon.

The ratio of $\langle T^2 \rangle$ to $\langle S^2 \rangle$ is the *tensor to scalar ratio*, r, as defined in (9.23):

Tensor to scalar ratio

$$r = \frac{\langle T_k^2 \rangle}{\langle R_k^2 \rangle} = 16\epsilon. \tag{9.41}$$

This is a key quantity in using the CMB to investigate the nature of inflation: once we know the power spectra of the scalar and tensor modes of the CMB fluctuations, we know something about the inflationary potential $V(\phi)$.[29]

9.5.3 Perturbation Theory During Inflation

Here we shall focus on perturbations whose length scale exceeds the causal horizon during their evolution. Formally this would require a discussion within the framework of general relativity where the space-time slices of spatially homogeneous φ do not have uniform curvature. There is, in effect, a time delay $\delta t \simeq -\delta\varphi/\dot\varphi$ between the surfaces of constant φ and the surfaces of constant curvature R. This time-shift difference corresponds to a curvature difference $R \simeq -H\delta t \simeq -H\delta\varphi/\dot\varphi$. See Equation (9.37).

As before, it is convenient to consider individual Fourier components of the fluctuating field $\delta\varphi(\mathbf{x}, t)$, which we will simply denote by $\varphi_{\mathbf{k}}$. Substituting into the slow roll Equation (9.14):

$$\ddot{\varphi}_{\mathbf{k}} + 3H\dot{\varphi}_{\mathbf{k}} + \left(\frac{k^2}{a^2} + V''(\varphi_0) \right) \varphi_{\mathbf{k}} = 0. \tag{9.42}$$

During a period of exponential expansion the expansion rate H is a constant. This is a second order ordinary differential equation in which the coefficients depend on time: the coefficient of $\varphi_{\mathbf{k}}$ is particularly troublesome. Moreover, there needs to be some statement about the initial conditions. As remarked above, this is the Bunch–Davies vacuum.

Consider (9.42) for flat potentials such that the $V''(\varphi_0)$ term can be neglected. The equation looks like a damped harmonic oscillator with frequency k/a. When $k/a \ll H$, the

---

[28] See also the footnote on p.158 of Dodelson (2003, Section 6.4.2).

[29] This is the now-standard first order result that can be derived formally, see, for example Langlois (2010, Section 4.5), among many others. However, the normalisations of the amplitudes may vary between authors.

perturbation length scale is larger than the horizon H^{-1} and the oscillation period becomes longer than the expansion timescale H^{-1} and, in effect, there are no longer any oscillations. Moreover for such super-horizon perturbations the equation is almost $\ddot{\varphi}_{\mathbf{k}} + 3H\dot{\varphi}_{\mathbf{k}} = 0$, a solution of which is $\varphi_{\mathbf{k}} = $ constant. The super-horizon perturbations do not grow in amplitude until they re-enter their horizon (as per Figure 9.5).

The equation can be solved if we neglect the V'' term. This is done with a transformation of the time coordinate to conformal time, τ (see Section 5.10.1 and Section 16.1.4), after which Equation (9.42) becomes

$$(a\varphi)'' + \left(k^2 - \frac{z''}{z}\right)(a\varphi) = 0, \quad z = a\frac{\dot{\varphi}}{H}, \tag{9.43}$$

where the $'$ is the derivative with respect to the conformal time, τ.

Gravitational wave, i.e. tensor, fluctuations are governed by somewhat similar equations. Denoting the Fourier amplitude for wavevector $\mathbf{k}$ by $h_{\mathbf{k}}$, we have

$$\ddot{h}_{\mathbf{k}} + 2H\dot{h}_{\mathbf{k}} + \frac{k^2}{a^2}h_{\mathbf{k}} = 0 \quad \Rightarrow \quad (ah_{\mathbf{k}})'' + \left(k^2 - \frac{a''}{a}\right)(ah_{\mathbf{k}}) = 0. \tag{9.44}$$

There is an important difference between these Equations (9.42) and (9.44): the potential $V(\varphi)$ does not appear in the equation for gravitational waves.[30] This means that while the shape of the curvature power spectrum will be influenced by the nature of the potential $V(\varphi)$, the spectrum of the gravitational waves is not.

In the case of a de Sitter Universe, $a''/a = 2/\tau^2$ and the solution of (9.44) on large super-horizon scales, $k \gg aH$, leads to

$$ah_{\mathbf{k}} = (2k)^{-\frac{1}{2}}e^{-ik\tau)}\left(1 - \frac{i}{k\tau}\right) \propto k^{-3/2}, \quad k\tau \ll 1. \tag{9.45}$$

The key feature of this solution is the $k^{-3/2}$ spatial dependence of the amplitude.

9.5.4 Power Spectra: Amplitudes and Slopes

It is customary to work in terms of the normalised power spectrum $\Delta(k)^2 = k^3 P(k)/2\pi^2$ (e.g. Komatsu *et al.*, 2010, Section 4). This is dimensionless and has the physical interpretation that it is the contribution to the variance per unit logarithmic interval in k. For a power spectrum $P(k) \propto k^{-3}$, $\Delta^2(k)$ is independent of k. A simple power-law $P(k) \propto k^n$ for a constant n is said to be *scale free* since its mathematical form is invariant under a change of scale $k \to \lambda k$. If n depends on k the situation is a little more complicated. A power spectrum of the form $P(k) = A_p k^{n(k)}$, where k_p is some fiducial wavenumber at which the power spectral amplitude is A_p, can be written as

$$\ln P(k) = \ln A(k_p) + n_p \ln \frac{k}{k_p} + \frac{1}{2}\alpha \ln^2 \frac{k}{k_p} + \cdots, \quad \alpha = \left.\frac{dn(k)}{d\ln k}\right|_{k_p}. \tag{9.46}$$

---

[30] Except via any implicit back-reaction influence the high frequency perturbations may have on the scale factor $a(t)$ (Martin and Musso, 2005).

This is just a Taylor expansion about $k/k_p = 1$ and the value of α is such that

$$\frac{d \ln P(k)}{d \ln k} = n(k_p) + \left.\frac{dn(k)}{d \ln k}\right|_{k_p} \ln \frac{k}{k_p} + \cdots. \tag{9.47}$$

If the power spectrum is displayed as a curve on a log:log plot, the curve in this approximation is parabolic around $k/k_p = 1$ with slope $n(k_p)$ at k/k_p. The slope at k_p is referred to as the *tilt* of the spectrum and α is referred to as the *running* of the spectral index.

With this, the normalised curvature spectrum[31] can be written as:

Normalised curvature power spectrum tilt and running

$$\Delta_R^2(k) = \frac{k^3}{2\pi^2}\langle R_k^2 \rangle = A_R \left(\frac{k}{k_p}\right)^{n_s(k_p) - 1 + \frac{1}{2}\left.\frac{dn_s}{d \ln k}\right|_{k_p} \ln \frac{k}{k_p} + O\left(\ln \frac{k}{k_p}\right)^2}. \tag{9.48}$$

The running spectral index is n_s, and the tilt at $k = k_p$ is $n_s(k_p) - 1$.

The term $\langle R_k^2 \rangle$ is the curvature power spectrum (9.38) for inflationary expansion and is $\propto k^{-3}$ for de Sitter expansion. k_p is the *pivot scale* at which these k-dependent quantities are normalised. The Planck data analysis uses $k_p = 0.002\,\mathrm{Mpc}^{-1}$, which is close to the centre of the logarithmic range of wavenumbers spanned by the Planck experiment.

With this, the definition of the spectral index at wavenumber k is

$$n_s(k) - 1 \overset{\text{def}}{=} \frac{d \ln \Delta_R^2(k)}{d \ln k} = n_s(k_p) - 1 + \left.\frac{dn_s}{d \ln k}\right|_{k_p} \ln \left(\frac{k}{k_p}\right) + O\left(\ln \frac{k}{k_p}\right)^2. \tag{9.49}$$

This is the local slope on the log:log plot of $\Delta_R^2(k)$ against $\ln k$.

The quantity is the *running index* and reflects the deviation form a simple power law spectrum:

Running index of the normalised scalar power spectrum, α_s:

$$\alpha_s = \frac{dn_s}{d \ln k} = \frac{d^2 \Delta_R^2}{d \ln k^2}. \tag{9.50}$$

We can do the same for the normalised spectrum of the tensor (*i.e.* gravitational wave) mode:

Normalised gravitational wave power spectrum tilt and running:

$$\Delta_T^2(k) = \frac{k^3}{2\pi^2}\langle T_k^2 \rangle = A_T \left(\frac{k}{k_p}\right)^{n_t(k) = n_t(k_p) + \frac{1}{2}\left.\frac{dn_t}{d \ln k}\right|_{k_p} \ln \left(\frac{k}{k_p}\right) + O\left(\ln \frac{k}{k_p}\right)^2}. \tag{9.51}$$

---

[31] There is, in the literature, an ambiguity of notation here: some papers denote $\Delta(k)$ by $P(k)$.

The spectral indices n_s and n_t are referred to as the *spectral tilt* since they measure the difference between the de Sitter expansion slope and the actual slope. Their values reflect the inflation field parameters ϵ and η. By differentiating (9.38) and (9.40) it is straightforward to show that, to first order:

Spectral index and tensor/scalar ratio as a function of inflation potential parameters:

$$n_s(k) - 1 = 3\eta - 6\epsilon, \quad n_t(k) = -2\epsilon, \quad r = -8\eta_t. \tag{9.52}$$

The expression $r = -8\eta_t$ is a *consistency relation*. If we measured the tensor power spectrum slope the consistency relation would serve to validate or disprove the single field inflation models on which such relationships are based.

By fitting the data with a model based on this equation we can determine both n_s and the running index. The Planck Collaboration (2015) confidence limits on n_S and α are shown in the left panel of Figure 9.6.

9.5.5 The Scalar–Tensor Ratio, r

All acceptable inflationary models generate scalar field fluctuations that become the large scale structure we see today. The most inflationary of these models also generate a fluctuating tensor component which manifests itself as a background of gravitational waves.

Since the spectral index, n_S, of the power spectrum may not be constant (Equation 9.46), it is necessary to pick a fiducial wavenumber at which to make the comparison of the

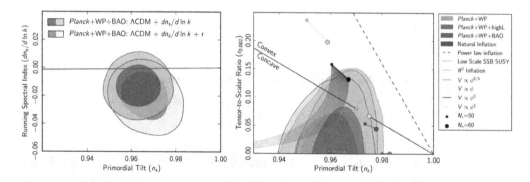

Fig. 9.6   Constraints on the predictions of various inflation models for the parameters describing the observed CMB temperature power spectrum. The confidence contours come from combining data from a number of independent experiments.

Left: Constraints on tilt n_S and running $dn_S/d \ln k$. (From Planck Collaboration *et al.* (2014, Fig.2)).

Right: Constraints on tilt, n_S and scalar-tensor ratio r. The predictions for the various models of inflation listed in the table are shown for a range $N_s = 50-60$ of e-foldings of the expansion. Some of these models are already excluded. (From Planck Collaboration *et al.* (2014, Fig.1).)

(Both figures © *ESA. Reproduced with permission from A&A.*)

amplitudes of the scalar and tensor mode power spectra. The Planck analysis chooses this *pivot scale* to be $k_* = 0.002\,\text{Mpc}^{-1}$ (Planck Collaboration *et al.*, 2014). With this the power spectra are written as

$$P_S = A_S \left(\frac{k}{k_*} \right)^{n_S}, \quad P_T = A_T \left(\frac{k}{k_*} \right)^{n_T}, \quad r = \frac{A_T}{A_S}, \tag{9.53}$$

where n_S, n_T are slowly varying functions of k expressed as an expansion in $\ln(k/k_*)$, as in (9.46). The quantity r expresses the *tensor-to-scalar ratio*, at the pivot point k_*, so with the particular Planck choice for that scale r is written as $r_{0.002}$. The current Planck limit on r at a wavenumber of $k = 0.002\,\text{Mpc}^{-1}$ is

$$r_{0.002} < 0.11. \tag{9.54}$$

The experimental constraints between r and n_S are shown in the right panel of Figure 9.6.

Theoretical estimates for r depend on the details of each particular theory of inflation, and so there is a large number of papers discussing this. There are technical papers from Starobinskii (1983, 1985), which are the great classics of this subject, and more recently from Martin (2015), Meerburg *et al.* (2015) and Spergel *et al.* (2015).

In general terms, the amplitude of the tensor mode depends on the number of e-foldings, N, of the scale of the Universe during the period of inflation. Current thinking is that $N \sim 50-60$. The approximate result relating n_s and N is

$$n_S \simeq 1 + \frac{2}{N}, \quad \frac{dn_S}{d\ln k} \simeq -\frac{2}{N^2}. \tag{9.55}$$

With these, the theoretical prediction for r is

$$r \simeq \frac{12}{N^2}, \tag{9.56}$$

which will be on the order of $\sim 6-8 \times 10^{-4}$. Such low values of r put the tensor mode several orders of magnitude below the scalar mode.

PART III

RELATIVISTIC COSMOLOGY

10

Minkowski Space

Inertial frames are fundamental to special relativity, so much so that Einstein was keen to carry them over into his later general theory of relativity, albeit on a merely local basis. This chapter is about inertial frames from the point of view of Minkowski space and Lorentz transformations. After a brief reminder as to what the Lorentz transformation is, we move on to Minkowski's way of thinking by introducing his 4-space, with its 4-vectors and defining metric. We also introduce the covariant and contravariant representation of vectors in these coordinate systems.[1] Once equipped with this we recast Maxwell's equation in our Minkowski space framework, ending with two tensor quantities: the electromagnetic field tensor (also known as the Faraday tensor) and the energy momentum tensor.

10.1 Why Relativity?

10.1.1 Limitations of Newtonian Theory

Our Universe, on the largest scales, appears to be homogeneous and isotropic. On those vast scales it is the force of gravity that governs its behaviour. Until the beginning of the 20th century, gravitational phenomena such as the motion of bodies in the Solar System were described in terms of Isaac Newton's theory of gravitation, written down by Newton some 250 years earlier. The successes of Newton's theory were truly remarkable. Not only did it provide an explanation of the orbits of bodies about the Sun, but it also made dramatic predictions that were subsequently verified: the return of Halley's comet and the existence of the planet Neptune. What more could one ask of a theory?

Despite those successes, two problems were evident even during the 19th century. There was the famous problem of the precession of the perihelion of Mercury's orbit about the Sun and there was the problem that Newton's theory could not address the cosmology of the time: a homogeneous and static Universe. Seeliger (1895) modified Newtonian gravity at large distances and his resulting cosmological model was perhaps the best one could do at the time.

---

[1] While such a distinction is not essential in special relativity it helps on the route towards general relativity as an acclimatisation to the distinction between covariant and contravariant representations of physical quantities. There is a summary of the matrix notation used here in Appendix C. See also Zangwill (2013, Appendix D).

However, at the end of 1915, Einstein published the first definitive version of his general theory of relativity, which provided a new theory of gravitation and with that came new cosmological models that could be tested against observations and applied to cosmology.

Throughout the rest of the 20th century numerous tests of Einstein's theory have established its overall validity for describing gravitational phenomena. Newton's theory can be retrieved from Einstein's theory as a first approximation that works well in the case of weak gravitational fields: we have every confidence when applying Newton's theory to our every day experience like flying airplanes or playing ball games. In that sense, Newton's theory is a perfectly good model for our everyday experience. Einstein did not invalidate Newton, but provided a more powerful and general framework within which to understand the nature of the force of gravity. It is entirely possible that Einstein's theory will one day be superseded by another yet more powerful theory that unifies all the forces of nature and that will once again open up new vistas in science.

A great drawback of Newton's work was that it provided no theory for the propagation of light: it could hardly do so since there was no real theory of light until Maxwell's theory some 200 years later.[2]

10.1.2 Coordinate Systems

Our perceived notion of space is three dimensional: we require three coordinates to specify the location of any point. If we choose to use Cartesian coordinates those coordinates are conventionally denoted symbolically by (x, y, z). If we use spherical polar coordinates we use (r, θ, ϕ) and everyone knows, through common usage, what these denote. Of course, we can use any symbols we like, but then some careful description of the notation is required. Thus the position of an object on the sky is typically given by its *right ascension*, *R. A.* or α, and *declination*, δ with the relationship that $\alpha \equiv \phi$ and $\delta \equiv 90 - \theta$ (if α and θ are measured in degrees).

If we talk about an event, we need to specify not only where it happened but also when: so there is an additional coordinate required: the time, t. Space-time, as we perceive it, is four dimensional and so requires four coordinates to specify an event. An event in space-time is thus described by a quartet of values (t, x, y, z) or (t, r, θ, ϕ) depending which coordinate system is used for the spatial part.

Einstein's special theory of relativity put space and time on an equal footing. The Lorentz transformation between two inertial frames mixes the space and time coordinates: there is nothing absolute about time. We live in a 4-dimensional space-time.

However, Einstein's view went much further than that: he insisted that the equations of physics should be written in such a way as to be independent of the coordinate system used to evaluate the quantities involved in the equation. The value assigned to a *physical* quantity should be independent of how we measure it or from which frame of reference the measurement is made: the value is a property of the object, not the observer. This is easier said than done!

---

[2] Newton and Huyghens had their respective corpuscular and wave theories of light describing what they thought light was. However, there was nothing quantitative arising from those theories until the work of Maxwell.

The physical length of a ruler does not depend on whether we give the value in meters or inches. Nor does it depend on the value we assign to it in an experiment: observers moving at different speeds relative to the ruler might measure different values, but the physical length is nonetheless an attribute of the ruler and not of the observers.

Any conceptual difficulty here arises from the fact that we have not defined what we mean by a physical quantity. As regards the length of the ruler we could all agree that when talking about the physical length of the ruler we should refer to the value derived by an observer who is at rest relative to the ruler. Likewise, when measuring the density of a volume of fluid, we can agree to measure this in the rest frame of the fluid. However, it is more difficult to agree on how to define the velocity of a particle or physical value for the strength of an electromagnetic field at some arbitrary point of space. It appears that those quantities can only be defined relative to the frame of reference of the observer who has an interest in such values.

The way out of this is simply to assert that physical quantities are those which transform correctly under the Lorentz transformation between inertial frames of reference. At first glance this seems almost like a circular argument: in physics we are only going to consider those quantities that transform according to some specific rule, ignore anything else and then assert that special relativity works provided we stick to describing only those quantities.[3]

There is in fact nothing wrong with this view. On the contrary it is the only view that makes sense when physical quantities can be assigned different measured values by different observers. There is no contradiction provided we can relate those different measurements. There are two important cases of this that come to mind: the measurement of a velocity of a moving object, and the measurement of an electromagnetic field.

10.2 Some Basic Notions

10.2.1 The Change in Our Thinking

Newton's notion of an absolute time, a universal clock or metronome with which all events were timed, was pushed aside by the special theory of relativity. With Einstein, your perceived time depended on which frame of reference you used: there was no absolute time and so different observers using their own watches might come to different conclusions about the time interval separating events. This was an astonishing realisation that led to a number of so-called 'paradoxes' like the famous *twin paradox* and the phenomenon of *time dilation*.

Einstein based his theory of special relativity on two postulates: The first of these postulates required a definition of a particular set of coordinate frames in all of which Newton's first law would apply: bodies under the action of no forces would move at constant speed

---

[3] The notion of *temperature* and the status of the first and second laws of thermodynamics raised a lot of questions over a long period of time in this respect, despite Tolman's in depth discussion in his great book (Tolman, 1934b).

in straight lines. The other assertion of special relativity was that the speed of light as measured by an observer in an inertial frame is the same for all observers: c. This meant that the simple law for compounding velocities by simply adding them failed for light. The simple addition law not only failed for light, it also failed for velocities.[4]

10.2.2 Special Relativity and Inertial Frames

A coordinate system in which Newton's first law,[5] the law of inertia, is valid is referred to as an *inertial frame*. If we have a coordinate system K in which the first law holds, then it is valid also in any coordinate system K' whose origin moves uniformly relative to K:

$$\mathbf{x}' = \mathbf{x} - \mathbf{v}t, \quad t' = t. \tag{10.1}$$

Here $\mathbf{v}$ is the velocity of K' relative to K. Equations (10.1) describe what is referred to as the *Galilean transformation*.

The problem that arose during the 19th century was that while Newton's first law was invariant under a Galilean coordinate transformation, Maxwell's equations were not: inertial frames did not appear to be relevant to electrodynamics.

Ex 10.2.1　Show that the wave equation

$$\frac{\partial^2 \Phi}{\partial x^2} - \frac{1}{c^2}\frac{\partial^2 \Phi}{\partial t^2} = 0 \tag{10.2}$$

is not invariant under the one dimensional version of the transformation (10.1): $x' = x - vt$, $t' = t$. Hint: show that, under this transformation of coordinates, for any function $\Phi(x, t)$:

$$\frac{\partial \Phi}{\partial x} = \frac{\partial \Phi}{\partial x'} + \frac{1}{v}\frac{\partial \Phi}{\partial t'}. \tag{10.3}$$

The resulting equation is not a wave equation. Why would that imply the failure of Galilean invariance for Maxwell's equations?

As we shall see, the Maxwell equations are invariant under the Lorentz transformation, and so Maxwell's theory of electrodynamics is already relativistically correct.

---

[4] Throughout the 20th century there have been people who rejected Einstein's theory of special relativity simply on the grounds of the failure of the law of addition of velocities. This rejection takes place against a remarkable background of evidence supporting special relativity, and so these 'relativity-deniers' are generally dismissed by the scientific community as cranks, or, more politely, as eccentrics

[5] A body continues in a state of rest or moves with uniform speed along a straight line unless subjected to an external force.

10.3 The Lorentz Transformation

Consider first the simpler, and more familiar, case of two coordinate frames $\mathcal{K}$ and $\mathcal{K}'$ such that the origin of $\mathcal{K}'$ is moving with uniform velocity v relative to $\mathcal{K}$ along the common x-axis. The Lorentz transformation between $\mathcal{K}$ and $\mathcal{K}'$ is:

Lorentz transformation in one dimension:
The Lorentz factor $\gamma(v)$ for a relative velocity v is defined as

$$\gamma = (1 - v^2/c^2)^{-\frac{1}{2}}, \tag{10.4}$$

with which the transformation $(x, t) \to ((x', t')$ is

$$x' = \gamma(x - vt), \qquad t' = \gamma(t - xv/c^2). \tag{10.5}$$

From these equations we see that there is no concept of absolute time here, time depends on your frame of reference. It is *space-time* that is important. Here γ is referred to as the *Lorentz factor*. This form of the transformation assumes, without loss of generality, that the origins of the coordinate systems $\mathcal{K}$ and $\mathcal{K}'$ coincide at $t = 0$. If that is not the case some additive constants appear in these equations.

Ex 10.3.1 The Lorentz factor for a particle moving in an arbitrary direction with uniform velocity $\mathbf{u}$ is $\gamma^2 = \left(1 - \mathbf{u} \cdot \mathbf{u}/c^2\right)^{-1}$. Show that

$$1 + \gamma^2 \mathbf{u}^2/c^2 = \gamma^2, \quad \dot{\gamma} = \gamma^3 \mathbf{u} \cdot \dot{\mathbf{u}} \quad \text{and} \quad \dot{\gamma} = \mathbf{u} \cdot \frac{d}{dt}(\gamma \mathbf{u}). \tag{10.6}$$

10.3.1 Arbitrarily Oriented Frames

The more general case of two arbitrarily oriented coordinate systems moving at constant velocity $\mathbf{v}$ relative to one another is less familiar. The Lorentz transformation between two inertial frames described by coordinates $(t, \mathbf{x})$, $(t', \mathbf{x}')$ and moving with relative velocity, $\mathbf{v}$ is

$$\begin{aligned}
\mathbf{x}' &= \gamma(\mathbf{x}^\dagger - \mathbf{v}t), \\
t' &= \gamma(t - \mathbf{v} \cdot \mathbf{x}/c^2), \\
\mathbf{x}^\dagger &= \gamma^{-1}\mathbf{x} + (1 - \gamma^{-1})\mathbf{v}(\mathbf{v} \cdot \mathbf{x})/v^2, \\
\gamma &= \left(1 - \frac{v^2}{c^2}\right)^{-\frac{1}{2}}.
\end{aligned} \tag{10.7}$$

Introducing $\mathbf{x}^\dagger$ here makes the form of the equations look like Equation (10.5). Without this the transformation looks a little more complicated:

$$\mathbf{x}' = \mathbf{x} + \mathbf{v}\left[(\gamma - 1)\frac{\mathbf{v} \cdot \mathbf{x}}{v^2} - \gamma t\right],$$

$$t' = t\left[t - \frac{\mathbf{v} \cdot \mathbf{x}}{c^2}\right]. \tag{10.8}$$

The term $\mathbf{v} \cdot \mathbf{x}$ vanishes for components of $\mathbf{x}$ perpendicular to the velocity vector $\mathbf{v}$.

Ex 10.3.2   Write out Equation (10.8) as a transformation between two coordinate systems $\mathcal{K}(x_1, x_2, x_3)$ and $\mathcal{K}'(x_1', x_2', x_3')$, whose axes are aligned so that at $t = 0$ the x_1 axis coincides with the x_1' axis, and so on. (The relative velocity vector, $\mathbf{v}$ then lies in the x_1-direction.)

Note that these equations are independent of any particular choice of coordinates used in the frames $\mathcal{K}$ and $\mathcal{K}'$. This is a substantial advantage of using vectors rather than coordinates to describe the location of a point: Equations (10.7) are coordinate independent representations of these frames.

Perhaps the most important thing to notice is that the Lorentz transformation (10.7) leaves Newton's first law invariant: it translates straight line motion in one frame into straight line motion in another frame. Moreover, in the limit $v \ll c$ the Lorentz transform becomes the Galilean transform (10.1). Thus we can assert that inertial frames are frames of reference in which Newton's first law holds, and that these frames are related via the set of Lorentz transformations.[6]

10.3.2 The Transformation of a 4-vector

Equation (10.7) is not only important because it tells us how to transform events between inertial frames. The main thrust of special relativity is that all mathematical descriptions of physical phenomena should transform in this same way. As we shall see, the idea is to create mathematical objects that are vector quantities in the 4-dimensional inertial frame describing the space-time.

A 4-dimensional object whose components in one inertial frame are $(\mathbf{A}, \theta)$ transforms into $(\mathbf{A}', \theta')$ in another inertial frame according to

$$\mathbf{A}' = \gamma(\mathbf{A}^{\dagger} - \mathbf{v}\theta), \tag{10.9}$$

$$\theta' = \gamma(\theta - \mathbf{v} \cdot \mathbf{A}/c^2), \tag{10.10}$$

$$\mathbf{A}^{\dagger} = \gamma^{-1}\mathbf{A} + (1 - \gamma^{-1})\mathbf{v}(\mathbf{v} \cdot \mathbf{A})/v^2, \tag{10.11}$$

[6] If you have one inertial frame it follows from this that you can identify all possible inertial frames. However, there is no recipe for finding an inertial frame in the first place! This became a great debate of the 19th century, culminating in the work of Mach who argued that this 'first' inertial frame is the frame of reference in which the system of stars, i.e. the Universe, was not rotating.

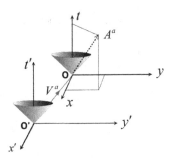

Two frames of reference for observers **O** and **O′** having relative 4-velocity $V^a = \gamma(c, \mathbf{v})$. Observer **O** has a 4-vector $A^a = (\phi, \mathbf{A})$ which can be measured by both observers. ©

with the usual $\gamma = (1 - v^2/c^2)^{-1/2}$ (see Figure 10.1). We shall return to this later when we define the 4-dimensional analogue of velocity. We shall also see a nicer way of writing this down.

10.3.3 Addition of Velocities

Suppose two Lorentz transformations are successively applied with velocities v_a and v_b in the same direction:

$$x_a = \gamma_a(x - v_a t), \qquad t_a = \gamma_a(t - v_a x/c^2), \tag{10.12}$$

$$x_b = \gamma_b(x_a - v_b t_a). \tag{10.13}$$

Then it is easy to eliminate (t_a, x_a) and show that

$$x_b = \gamma_a \gamma_b \left(1 + \frac{v_a v_b}{c^2}\right)(x - Vt), \tag{10.14}$$

with

$$V = \frac{v_a + v_b}{1 + \dfrac{v_a v_b}{c^2}}. \tag{10.15}$$

This is the relativistic law for addition of velocities.

Ex 10.3.3 Show that the pre-multiplying factor $\gamma_a \gamma_b(1 + v_a v_b/c^2)$ appearing in (10.14) is the γ-factor corresponding to the velocity V of Equation (10.15).

Ex 10.3.4 Deduce that Equation (10.14) is the spatial part of a Lorentz transformation from the t, x frame to the t_b, x_b frame.

Ex 10.3.5 Using the Lorentz transformation in the form of Equations (10.5), write down expression for dx' and dt' in terms of dx and dt. If $u = dx/dt$ and $u' = dx'/dt'$, recover the relationship (10.15) for the addition of velocities in the form $u = (u' + v)/(1 + u'v/c^2)$.

For the sake of completeness, the general Lorentz transformation for velocities is

$$\mathbf{u}' = \frac{1}{1 - \dfrac{\mathbf{u} \cdot \mathbf{v}}{c^2}} \left\{ \gamma^{-1}\mathbf{u} + \left[(1 - \gamma^{-1})\frac{\mathbf{u} \cdot \mathbf{v}}{c^2} - 1 \right] \mathbf{v} \right\}, \tag{10.16}$$

where, as usual, the Lorentz factor $\gamma = (1 - v^2/c^2)^{-1/2}$.

10.3.4 The Doppler Effect

Consider two inertial frames $\mathcal{K}$ and $\mathcal{K}'$ moving with relative velocity v along the x-axis of $\mathcal{K}$ and the x'-axis of $\mathcal{K}'$. A light ray propagating along that axis can be viewed either as a series of photons or as a set of wave-crests. Either way, let the time interval between the passage past a point of successive photons or wave-crests be dt in $\mathcal{K}$ or dt' in $\mathcal{K}'$, and let the distances measured between the crests or photons be dr and dr' in $\mathcal{K}$ and $\mathcal{K}'$ respectively. By the principle of the constancy of the velocity of light we have

$$dr = c\,dt, \quad dr' = c\,dt', \tag{10.17}$$

in $\mathcal{K}$ and $\mathcal{K}'$ respectively.

Using the Lorentz transformation in the form (10.5) we have from the equation for t' that

$$dt' = dt\,\gamma(1 - v/c), \tag{10.18}$$

where dx has been eliminated in favour of dt using $dr = c\,dt$. If we identify the measured frequency of the light as the rate at which wave-crests or photons pass the observer, we have $\nu = 1/dt$ in K and $\nu' = 1/dt'$ in K', and then

$$\nu = \nu' \sqrt{\frac{1 - v/c}{1 + v/c}}. \tag{10.19}$$

This is the relativistic version of the famous *Doppler effect*. The classical formula is $\nu = \nu'(1 + v/c)^{-1}$, which differs from the relativistic formula in terms of $O(v^2/c^2)$.

10.4 Lorentz Invariance

The Lorentz transformation mixes up space and time: how you describe a phenomenon depends on the coordinate system within which you choose to work. Nevertheless, the laws of physics must be independent of coordinates, despite that fact that some coordinate systems will be more useful than others.[7]

The question is how to write down the equations of the laws of physics in a way that is manifestly independent of coordinates. The answer sounds simple: use a notation that

---

[7] A circle is best described in polar coordinates, though there is a perfectly adequate description in familiar Cartesian xy-coordinates. Whichever coordinate system we choose, it is still the same circle.

guarantees invariance under Lorentz transformations. We do this by describing the objects which are the subject of the physical laws in terms of Lorentz invariant objects – *tensors*. In doing this we achieve Einstein's first postulate of special relativity, that the laws of physics should be the same in all inertial coordinate systems (Section 10.2.1).

10.4.1 Proper Time and Distance

We need to address Einstein's second postulate of special relativity, the once-controversial notion that the speed of light should be the same in all inertial coordinate systems.

Consider two inertial observers A and B moving with uniform velocity V relative to one another. We can conveniently set up rectangular coordinate axes for each observer so that the relative motion is directed along their x-axes. An *event*, $\mathcal{E}$, is described by each observer saying where and when, in his/her frame of reference, the event took place. That is, for an event taking place on their common x-axis, observer A reports $\mathcal{E}(x_A, y_A, z_A, t_A)$ and observer B reports $\mathcal{E}(x_B, y_B, z_B, t_B)$. Since the coordinate systems of A and B are related by a Lorentz transformation we have:

$$x_A = \gamma_V(x_B + Vt_B), \quad y_A = y_B, \quad z_A = z_B, \qquad (10.20)$$

$$t_A = \gamma_V(t_B + x_B V/c^2), \qquad (10.21)$$

$$\text{where} \quad \gamma_V = (1 - V^2/c^2)^{-1/2}.$$

These are just Equations (10.5).

Suppose now that there is another event close to $\mathcal{E}$ in both space and time. A will observe it at $(x_A + dx_A, y_A + dy_A, z_A + dz_A, t_A + dt_A)$ and B observes it at $(x_B + dx_B, y_B + dy_B, z_B + dz_B, t_B + dt_B)$. We can relate dx_A, dy_A, dz_A and dt_A to dx_B, dy_B, dz_B and dt_B by simply differentiating Equation (10.21) to give

$$dx_A = \gamma_V(dx_B + Vdt_B), \quad dy_A = dy_B, \quad dz_A = dz_B, \qquad (10.22)$$

$$dt_A = \gamma_V(dt_B + dx_B V/c^2). \qquad (10.23)$$

From these it is easy to show that

$$dx_A^2 + dy_A^2 + dz_A^2 - c^2 dt_A^2 = dx_B^2 + dy_B^2 + dz_B^2 - c^2 dt_B^2, \qquad (10.24)$$

which is entirely independent of the relative velocity V of the observers A and B.

Ex 10.4.1   A rod is at rest, lying along the x-axis in the frame of observer A sitting in an inertial frame. A sees the length of the rod is $L_0 = \delta x_A$ by measuring the difference δx_A in the x_A-coordinates of the ends of the rod.

   An inertial observer B moves with velocity V relative to A along A's x-axis. Suppose that B's x-axis is in the same direction as A's, in other words B sees A and the rod moving along the x_B-axis with velocity $-V$. B measures the length of the rod by measuring the time δt_B the rod takes to pass a given point on his/her x-axis and asserts that the rod has length $V\delta t_B$.

Use Equation (10.21) to show that, using this procedure, B measures a length

$$L(B) = \sqrt{1 - \frac{V^2}{c^2}} L_0. \tag{10.25}$$

So for all $V < c$, B sees the rod as being less than its *rest length* L_0.

This is the famous *Fitzgerald–Lorentz contraction*, which can be summed up by saying that bodies appear to contract in the direction of their motion.

10.5 Minkowski Space Concepts

10.5.1 Space-Time Interval and Proper Time

Equation (10.24) can be rephrased by writing the squared distance ds^2 between neighbouring events, as seen from an inertial frame, as:

Minkowski space metric

$$ds^2 = -c^2 dt^2 + d\mathbf{x}^2, \tag{10.26}$$

where ds is referred to as the *space-time interval* between neighbouring events viewed from an inertial frame. This equation is referred to as the equation describing the *line element* of Minkowski's space-time. The separation of two neighbouring events is described as space-like, time-like or null according as to whether

$$
\begin{aligned}
ds^2 &> 0 \quad \text{space-like} \\
ds^2 &= 0 \quad \text{null} \\
ds^2 &< 0 \quad \text{time-like}.
\end{aligned}
\tag{10.27}
$$

We notice that if two events are connected by a light ray their separation is null, since the path of light rays is defined to be $ds^2 = 0$. The convention we have adopted for the Minkowski metric is the *space-like* $(1, 3)$ convention for the interval ds^2. The coordinates of an event are then ordered as (t, x, y, z).

It is convenient to define a quantity that is a measure of the separation of neighbouring events via the equation[8]

$$c^2 d\tau^2 = -ds^2. \tag{10.28}$$

Clearly ds and $d\tau$ are equivalent, this is just a relabelling. The quantity τ is called the *proper time*. Like ds, the value of $d\tau$ is something all inertial observers can agree on. That follows because, as we have seen in (10.24), the metric is Lorentz invariant.

---

[8] The $-$ sign in this definition arises because we have chosen to work with a signature that is $(- + ++)$. If we had chosen to work with a $(+ - --)$ signature we would have defined $c^2 d\tau^2 = ds^2$.

Two important concepts arise: proper time, $d\tau$, between events having *time-like separation* and *space-like separation*, $d\ell$, between events that have space-like separation:

$$d\tau = \sqrt{dt^2 - dr^2/c^2}, \quad ds^2 < 0 \tag{10.29}$$

$$d\ell = \sqrt{dr^2 - c^2dt^2}, \quad ds^2 > 0. \tag{10.30}$$

Expressing the space-time interval between neighbouring events in this way is an important way of describing space times.

The proper time is invariant under Lorentz transformations. It is a quantity whose value all inertial observers can agree upon. The Euclidean distance $d\mathbf{x}^2$ is not invariant: it depends on the relative state of motion of the observers. Following the precepts of special relativity, we regard ds^2 as a proper measure of the physical distance between two events: it is invariant under Lorentz transformation. The space-time interval ds is an aspect of space-time that remains invariant under the Lorentz transformations that describe the different world-views of observers who are in uniform motion relative to one another.

The coordinate democracy under Lorentz transformation of the local geometry as described by the line element ds suggests that, mathematically, we should not make a distinction between t and the spatial coordinates x, y, z: the Lorentz transformation in any case just mixes them up while leaving the underlying geometry unchanged. To express this we replace the time coordinate $ct \rightarrow x^0$ and then our events are labelled by giving their coordinates as (x_0, x_1, x_2, x_3).

10.5.2 Matrix Notation

Another way of writing the proper distance (10.26) between events is as a matrix operation on the space-time interval (dx_1, dx_2, dx_3, dt):

$$ds^2 = \begin{bmatrix} cdt & dx_1 & dx_2 & dx_3 \end{bmatrix} \begin{bmatrix} -1 & 0 & 0 & 0 \\ 0 & +1 & 0 & 0 \\ 0 & 0 & +1 & 0 \\ 0 & 0 & 0 & +1 \end{bmatrix} \begin{bmatrix} cdt \\ dx_1 \\ dx_2 \\ dx_3 \end{bmatrix}. \tag{10.31}$$

The diagonal matrix $\text{diag}\,[-1, +1, +1, +1]$ is known as the Minkowski or Lorentzian metric of the space-time: it is the means by which we calculate distances between neighbouring space time points.

We can make this notation more compact by replacing cdt with dx_0. As previously stated, this has the advantage of emphasising the democracy between space and time coordinates expressed through the Lorentz transformation. With this we can rewrite Equation (10.31) as

$$ds^2 = \sum_{b=0}^{3} \sum_{a=0}^{3} \eta_{ab} dx_a dx_b, \tag{10.32}$$

$$\text{with} \quad dx_a = \{dx_0, dx_1, dx_2, dx_3\}, \tag{10.33}$$

$$\text{and} \quad \eta_{ab} = \text{diag}\,(-1, +1, +1, +1). \tag{10.34}$$

We can make Equation (10.32) look less intimidating by invoking the *summation convention* in which it is assumed that repeated indices are summed over the range 1 to 4 unless specifically stated otherwise:

$$ds^2 = \eta_{ab}dx_a dx_b. \tag{10.35}$$

This form of equation for the line element is very important as it will provide the basis for going beyond describing Minkowski space-time (see Figure 10.2) and describing the more exotic space-times of general relativity.

10.5.3 Coordinates and Conventions: a Diversion

In Equation (10.26) we are implicitly adopting the convention that the space time interval ds^2 should be positive when $c^2 dt^2 < d\mathbf{x}^2$. We implicitly adopted the convention when asserting that the diagonal elements of the matrix η_{ab} were written as $\text{diag}(-1, +1, +1, +1)$, reading from top left to bottom right. This has the merit that, in a given coordinate frame, two events occurring simultaneously, i.e. with $dt = 0$ in that frame, have a separation $d\mathbf{r}^2$ that corresponds with our Newtonian view of the world. On the other hand the time interval dt between events occurring at the same place is then an imaginary number. The commonly seen alternative is the use of $\eta_{ab} = \text{diag}(+1, -1, -1, -1)$.

In the early days of special relativity it was usual to complexify the time coordinate by using $t \to \sqrt{-1}t$ and writing the space time coordinates as (x_0, x_1, x_2, x_3) with $x_0 = \sqrt{-1}\,ct$ (Minkowski, 1908). This way 4-dimensional vectors are treated identically to 3-dimensional vectors, except that one of the components is imaginary. The space-time

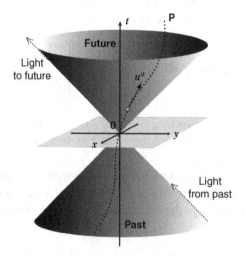

Fig. 10.2 Minkowski space view of future and past light cones of an observer **0**. The z-coordinate has been suppressed. The path of a particle **P** that passes **0** is shown with its 4-velocity vector u^a. The path of the particle is confined to the interior of the cones since it cannot travel faster than light. Light rays coming to **0** and emitted from **0** trace out the surface of the cones in this 4-dimensional depiction. ⓒ

metric then becomes diag$(1, 1, 1, 1)$. Many textbooks on electromagnetism still use this convention, or the similar convention with coordinates (x_1, x_2, x_3, x_4) and $x_4 = ict$.[9]

The number of pluses minus the number of minuses is commonly referred to as the *signature* of the matrix, so sig$(-1, +1, +1, +1)) = +2$.[10]

There is no commonly accepted usage, except perhaps that the relativity community tends to use the $(- + ++)$ signature while the field theory community tends to use $(+ - --)$. The convention that x_0 represents the time-like coordinate, and that the signature of the Minkowski metric should be $+2$ can be referred to as the 'MTW' convention after Misner *et al.* (1973).

10.5.4 Non-orthogonal Coordinates: Another Diversion

So far we have assumed that we are using an *orthogonal coordinate frame* (x_0, x_1, x_2, x_3), by which we mean that the coordinate axes are mutually perpendicular. Since the choice of coordinates should not matter when describing physical phenomena we should be able to work with non-orthogonal coordinate systems. It is useful to discuss this briefly at this point in order to introduce a notational device that will play a major role when we come to the general theory of relativity.

The situation is illustrated in two dimensions in Figure 10.3, which shows two oblique axes defined by unit vectors $\mathbf{e}_1$ and $\mathbf{e}_2$. Two ways of representing the coordinates of the point P with respect to these axes are shown. When we drop perpendiculars from P to the axes we get the points that we have labelled (x_1, x_2). In this case

$$x_1 = \mathbf{r} \cdot \mathbf{e_1}, \quad x_2 = \mathbf{r} \cdot \mathbf{e_2}, \tag{10.36}$$

where $\mathbf{r}$ is the vector $\overrightarrow{OP}$. When we draw vectors from P parallel to $\mathbf{e}_1$ and $\mathbf{e}_2$ we find points that we shall denote by (x^1, x^2). Then using these coordinates

$$\mathbf{r} = x^1 \mathbf{e}_1 + x^2 \mathbf{e}_2. \tag{10.37}$$

There should be no confusion between the use of the symbol x^2 to denote on the one hand a coordinate value, and on the other hand the square of the quantity x.

---

[9] See. for example, Zangwill (2013, see Appendix D, 'Managing minus signs in Special Relativity'). I have chosen to follow the precedent set by Stephani (2008); Ohanian and Ruffini (2013) and avoid using an imaginary time-like coordinate from the outset by using a metric diag$(-1, +1, +1, +1)$. At the same time it is worth introducing the concept of contravariant vectors and their covariant counterparts, i.e. *covectors*, so that the length of the 4-vector v^a is $\sqrt{(v_a v^a)}$. A good alternative would be to work in terms of *forms* as did Misner *et al.* (1973, Sections 2.1- 2.7).

[10] Formally, the concept of *signature* applies to symmetric matrices having real elements (as opposed to complex): the signature is the number of positive eigenvalues minus the number of negative eigenvalues. Such a matrix $\mathbf{A}$ represents quadratic forms $f(\mathbf{x}) = \mathbf{x}^T \mathbf{A} \mathbf{x}$, where $\mathbf{x}$ is a vector of possibly complex numbers. The number of distinct real roots of the equation $f(\mathbf{x}) = 0$ is equal to the signature of the matrix representing the quadratic form.

More usefully, the signature can be given as a pair of numbers telling explicitly how many minus and plus signs there are in the metric. Thus sig$(-1, +1, +1, +1))$ has signature $(1, 3)$. This is more informative: both Minkowski space and plane polar coordinates might otherwise be said to be metrics of 'signature 2'.

It has become more usual to put the time-like coordinate first. This is the practical approach when considering extra spatial dimensions in extended theories of space-time.

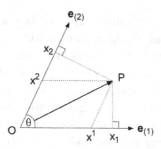

Coordinates of a point P in a non-orthogonal coordinate system with axes defined by unit vectors $(\mathbf{e}_1, \mathbf{e}_2)$. P can be represented either as $P(x_1, x_2)$ or $P(x^1, x^2)$. ⓒⓒ

Note that when adding the two vectors $x^1 \mathbf{e}_1$ and $x^2 \mathbf{e}_2$ to make $\mathbf{r}$ we are transporting the vector $x^2 \mathbf{e}_2$ parallel to itself along the $\mathbf{e}_1$ axis to make the classical vector triangle displaying the sum of two vectors to make a third. The notion of parallel transport plays a fundamental role that becomes less trivial if the space is not Euclidean (Euclidean geometry is often characterised by the axioms of parallelism). In order to make such a *parallel transport* of a vector from one place to another in a non-Euclidean space we shall have to invoke a *connection* between neighbouring points that tells us how to move a vector along a line parallel to itself.

We should further note that the element of area dS bounded by pairs of neighbouring lines that are parallel to each of the axes, is $dS = dx^1 dx^2 \sin\theta$; we have gained a factor of $\sin\theta$ because of the non-orthogonal nature of the coordinates. If we think of the elemental area as being defined by a pair of infinitesimal vectors $d\mathbf{x}^1 = dx^1 \mathbf{e}_1$ and $d\mathbf{x}^2 = dx^2 \mathbf{e}_2$ the element of area can be viewed as a vector $d\mathbf{S} = d\mathbf{x}^1 \times d\mathbf{x}^2 = dx^1 dx^2 \mathbf{e}_1 \times \mathbf{e}_2$.

Moving to Index Notation: Covariant and Contravariant

We shall now investigate the relationship between these two representations of the vector $\mathbf{r} = \overrightarrow{OP}$. It is easy to see from the geometry that

$$x_1 = x^1 + x^2 \cos\theta, \quad x_2 = x^1 \cos\theta + x^2, \tag{10.38}$$

$$\cos\theta = \mathbf{e}_1 \cdot \mathbf{e}_2, \tag{10.39}$$

θ being the angle between the axes. The inverse of this is

$$x^1 = (x_1 - x_2 \cos\theta)/\sin^2\theta, \quad x^2 = (-x_1 \cos\theta + x_2)/\sin^2\theta. \tag{10.40}$$

Ex 10.5.1 Show that $x_1 = x^1$ and $x_2 = x^2$ only if $\mathbf{e}_1 \cdot \mathbf{e}_2$, that is, if the coordinate system is orthogonal.

It is perhaps easier to write these transformations between the (x^1, x^2) and (x_1, x_2) representations in matrix form:

$$\begin{pmatrix} x_1 \\ x_2 \end{pmatrix} = \begin{pmatrix} 1 & \cos\theta \\ \cos\theta & 1 \end{pmatrix} \begin{pmatrix} x^1 \\ x^2 \end{pmatrix}, \tag{10.41}$$

$$\begin{pmatrix} x^1 \\ x^2 \end{pmatrix} = \frac{1}{\sin^2\theta} \begin{pmatrix} 1 & -\cos\theta \\ -\cos\theta & 1 \end{pmatrix} \begin{pmatrix} x_1 \\ x_2 \end{pmatrix}. \tag{10.42}$$

This can be more simply written as

$$x_a = g_{ab}x^b, \quad x^a = g^{ab}x_b, \tag{10.43}$$

where the summation convention over repeated indices is implied and where g_{ab} and g^{ab} denote the matrices

$$g_{ab} = \begin{pmatrix} 1 & \cos\theta \\ \cos\theta & 1 \end{pmatrix}, \quad g^{ab} = \frac{1}{\sin^2\theta} \begin{pmatrix} 1 & -\cos\theta \\ -\cos\theta & 1 \end{pmatrix}, \tag{10.44}$$

as per Equations (10.41) and (10.42). These two matrices must be mutual inverses, so

$$g^{ab}g_{bc} = \delta^a_c = \begin{cases} 0, & a \neq c \\ 1, & a = c \end{cases} \tag{10.45}$$

as can be simply verified by multiplying the matrices in Equation (10.44).

The determinant of the metric $g = \det g_{ab}$ plays an important role in doing calculus on non-Cartesian coordinate grids:

$$g = \det g_{ab}. \tag{10.46}$$

The quantity $\sqrt{|g|}$ enters into the volume element when doing integrals. In the case of the preceding example,

$$g = \det g_{ab} = \sin^2\theta. \tag{10.47}$$

We note that the element of area is $dS = dx^1 dx^2 \sin\theta = \sqrt{g}\, dx^1 dx^2$. This result generalises to higher dimensions and in particular to Minkowski space where the volume element is $dV = \sqrt{|g|}\, dx^1 dx^2 dx^3\, c\, dt$.

The Line Element

Finally, we note that we can write the vector $\mathbf{r}$ as

$$\mathbf{r} = x^1 \mathbf{e}_1 + x^2 \mathbf{e}_2, \tag{10.48}$$

from which

$$d\mathbf{r} = dx^1 \mathbf{e}_1 + dx^2 \mathbf{e}_2, \tag{10.49}$$

and so

$$d\mathbf{r}^2 = (dx^1)^2 + 2\cos\theta\, dx^1 dx^2 + (dx^2)^2. \tag{10.50}$$

We recognise this as being simply the result of the matrix multiplication

$$d\mathbf{r}^2 = g_{ab}dx^a dx^b, \tag{10.51}$$

and with this we can interpret the matrix g_{ab} as being analogous to η_{ab} in Equation (10.32). We refer to g_{ab} as the *metric tensor* and hence we refer to η_{ab} as the *metric tensor of Minkowski space*. The alternative form

$$ds^2 = dx_a\, dx^a \qquad (10.52)$$

is equally valid.

We note one important caveat regarding Equation (10.32): technically it should be written

$$ds^2 = \eta_{ab} dx^a dx^b, \qquad (10.53)$$

with raised indices on the coordinates. However, that did not matter since the coordinates in which we wrote (10.32) were orthogonal (see Figure 10.3). We should also note that with η_{ab} defined as in (10.31), $\det \eta_{ab} = -1$.

Information 10.5.4

The x^1, x^2 are referred to as the *contravariant components* of the vector **r**.
The x_1, x_2 are referred to as the *covariant components* of the vector **r**.
Indices are raised and lowered using the appropriate form of the metric tensor.
We speak of *covariant* and *contravariant* vectors and tensors (matrices).

The summation convention always sums over pairs of repeated indices, one raised and one lowered. Hence the summation convention is never invoked over three repeated indices.

We can think of the contravariant vectors as representing positions and the covariant vectors as representing gradients relative to those coordinates.
Some objects will have mixed contravariant and covariant components.

From now on we shall write the indices in their correct place despite the fact that it makes no difference when the coordinates are orthogonal Cartesian coordinates.

10.5.5 The Lorentz Transformation in Matrix Notation

Now let us review what all this means for our understanding of the Lorentz transformation, as this will be important when we come to discussing general relativity. To simplify a little let us focus on two frames of reference $\mathcal{K}$ and $\mathcal{K}'$ sharing a common x-axis, where the origin of $\mathcal{K}'$ moves at constant velocity v relative to $\mathcal{K}$ along the common axis.

The transformation (10.5) can conveniently be written in a matrix form:[11]

$$\begin{pmatrix} ct' \\ x' \end{pmatrix} = \begin{pmatrix} \gamma & -\gamma v/c \\ -\gamma v/c & \gamma \end{pmatrix} \begin{pmatrix} ct \\ x \end{pmatrix} = \gamma \begin{pmatrix} 1 & -v/c \\ -v/c & 1 \end{pmatrix} \begin{pmatrix} ct \\ x \end{pmatrix}, \qquad (10.54)$$

---

[11] There is a difference in the braces used to denote the matrices in Equations (10.31) and (10.54) because these are quite different objects. The former, (10.31) describes the geometry of space-time, in this case Minkowski space. The latter, (10.54) describes a coordinate transformation and allows us to obtain the values of quantities as measured by different observers.

or, in a more compact notation,

$$x'^a = \sum_{b=0}^{1} L^a{}_b(v)x^b, \quad x^a = \begin{pmatrix} ct \\ x \end{pmatrix}, \tag{10.55}$$

$$L^a{}_b(v) = \begin{pmatrix} \gamma & -\gamma v/c \\ -\gamma v/c & \gamma \end{pmatrix}. \tag{10.56}$$

The Lorentz transformation matrix in (10.54) has been written as $L^a{}_b$: this respects the convention that we sum over repeated pairs of indices, one of which must be 'up' and the other 'down'.

Equation (10.55) can be written even more compactly using the summation convention

$$x'^a = L^a{}_b(v)x^b, \tag{10.57}$$

where the repeated index b is taken to be summed over the entire range of b-values (in this case 0 and 1). This makes the meaning of the equation more transparent as it is no longer dominated by summation signs.

Comment:

It may be helpful to think of the matrix multiplication implied in Equation (10.57) having the upper index, a, counting along the rows and the lower index, b, counting down the columns.

Forward and Backward Transformation

If we do a Lorentz transformation from $\mathcal{K}$ to $\mathcal{K}'$ and then back from $\mathcal{K}'$ to $\mathcal{K}$ we should get back where we started. Denoting the coordinates in $\mathcal{K}$ by x^a and the coordinates in $\mathcal{K}'$ by x'^a we have

$$x'^d = L^d{}_c(v)x^c, \quad x^b = L^b{}_d x'^d, \tag{10.58}$$

for the forward and backward transformations. Combining these two we get

$$x^b = L^b{}_d(-v)L^d{}_c(v)x^c, \tag{10.59}$$

and hence we infer that

$$L^b{}_d(-v)L^d{}_c(v) = \delta^b_c. \tag{10.60}$$

We can verify this by writing Equation (10.59) in matrix form:

$$\begin{pmatrix} x^0 \\ x^1 \end{pmatrix} = \gamma^2 \begin{pmatrix} 1 & -v/c \\ -v/c & 1 \end{pmatrix} \begin{pmatrix} 1 & v/c \\ v/c & 1 \end{pmatrix} \begin{pmatrix} x^0 \\ x^1 \end{pmatrix}, \tag{10.61}$$

which is readily seen to be true by multiplying out the matrices.

Invariance of the Space-time Interval

We showed in Section 10.5.1 that the space-time interval

$$ds^2 = -c^2 dt^2 + d\mathbf{x}^2 \tag{10.62}$$

is invariant under Lorentz transformation. Let us look at this in matrix notation again by considering a Lorentz transformation between two inertial frames $\mathcal{K}$ to $\mathcal{K}'$. The space-time interval can be written in index notation as

$$ds^2 = \eta_{ab} dx^a dx^b. \tag{10.63}$$

This is just a repeat of equation Equation (10.32). If we write the transformation between the coordinate systems x^a and x'^b in $\mathcal{K}$ and $\mathcal{K}'$ as

$$x^a = L^a{}_c(v) x'^c, \tag{10.64}$$

we have

$$ds^2 = \eta_{ab} L^a{}_c(v) L^b{}_d(v) \, dx'^c dx'^d. \tag{10.65}$$

This involves the multiplication of three matrices and it is easily verified that:

$$\gamma^2 \begin{pmatrix} 1 & v/c \\ v/c & 1 \end{pmatrix} \begin{pmatrix} -1 & 0 \\ 0 & 1 \end{pmatrix} \begin{pmatrix} 1 & v/c \\ v/c & 1 \end{pmatrix} = \begin{pmatrix} -1 & 0 \\ 0 & 1 \end{pmatrix}. \tag{10.66}$$

(Note that the transformation (10.64) is from $x'^c \to x^a$ and so is the reverse transformation, which is why the signs of the velocities are reversed in the previous equation.) Hence

$$ds^2 = \eta_{cd} \, dx'^c dx'^d, \tag{10.67}$$

which shows the invariance of the space-time interval, as required.

A point worth noting is the order in which the matrices appear in Equation (10.66). This arises because the expression $\eta_{ab} L^a{}_c(v) L^b{}_d(v)$ says that you combine the η with an L first: you cannot combine two Ls since they share no common indices. Moreover, when you perform the multiplication $\eta_{ab} L^a{}_c(v)$, the index a is the summed index and it ranges over the rows of L and the columns of η. So in the product (10.66) we see an L preceding the η in the order of multiplication, before multiplying the result of that by the second L (see Appendix C for more details).

Addition of Velocities

Consider three inertial frames, $\mathcal{K}$, $\mathcal{K}'$ and $\mathcal{K}''$ all moving along their common x-axis, in which the velocity of $\mathcal{K}'$ relative to $\mathcal{K}$ is v_a, and the velocity of $\mathcal{K}''$ relative to $\mathcal{K}'$ is v_b. Denoting the coordinate systems in $\mathcal{K}$, $\mathcal{K}'$ and $\mathcal{K}''$ by x^a, x'^a and x''^a respectively the coordinate systems of the frames are related via the Lorentz transformations

$$x'^b = L^b{}_c(v_a) x^c, \quad x''^a = L^a{}_b(v_b) x'^b, \tag{10.68}$$

from which we see that

$$x''^a = L^a{}_b(v_a) \, L^b{}_c(v_b) \, x^c, \tag{10.69}$$

linking frames $\mathcal{K}$ and $\mathcal{K}''$. We have to show that $L^a{}_b(v_a)\, L^b{}_c(v_b) = L^a{}_c(v_a + v_b)$. This is done in the following exercise.

Ex 10.5.2   Multiply two matrices, $L(v_a)$ and $L(v_b)$, each having the form given in Equation (10.54), to show that the product $L(v_a)L(v_b)$ has the same form.

Ex 10.5.3   Writing $L(v_a)L(v_s) = L(v_c)$, show that

$$\gamma(v_a)\gamma(v_b)(1 + v_a v_b/c^2) = \gamma(v_c),$$

$$\gamma(v_a)\gamma(v_b)(v_a + v_b) = \gamma(v_c)v_c.$$

Ex 10.5.4   Hence deduce that

$$v_c = \frac{v_a + v_b}{1 + v_a v_b/c^2},$$

which is the same result as given in Equation (10.15). Thus the product of two Lorentz transformation matrices is another Lorentz transform matrix with velocity given by this last equation.

Mathematically speaking, the fact that the product of two Lorentz transformation matrices is also a Lorentz transformation matrix says that the Lorentz transformations form a group: the *Lorentz Group*.

10.5.6 4-vectors

Let us start with a specific example of what we are looking for. In everyday 3-dimensional Newtonian space we talk of the velocity **v** of a particle and we are happy to represent it in our frame of reference, $\mathcal{N}$, as a 3-vector with components (v_1, v_2, v_3). If we move to another Newtonian frame, $\mathcal{N}'$, moving at velocity **V** relative to the first, we adjust our coordinate representation of the particle to have components $(v_1 - V_1, v_2 - V_2, v_3 - V_3)$, where (V_1, V_2, V_3) are the components of **V** in the $\mathcal{N}$ frame of reference. This is the law of addition of velocities in the *Galilean transformation* between two Newtonian frames of reference.

We have already seen that the law of addition of velocities in special relativity is substantially more complicated, we cannot just add two velocities.

We need to deal with a 4-component vector, a *4-vector*. What we mean by this is that if we have a 4-velocity that is represented by the 4-vector $u^a = (u^0, u^1, u^2, u^3)$ in one frame, $\mathcal{K}$, and as $u'^a = (u'^0, u'^1, u'^2, u'^3)$ in another frame, $\mathcal{K}'$, then

$$u'^a = L^a{}_b\, u^b. \tag{10.70}$$

The summation over the repeated index b on the right-hand side is implied because we have agreed to use the summation convention. Written out in full for the simple case of the Lorentz transformation in one dimension this is

$$\begin{pmatrix} u'^0 \\ u'^1 \\ u'^2 \\ u'^3 \end{pmatrix} = \gamma(v) \begin{pmatrix} 1 & -v/c & 0 & 0 \\ -v/c & 1 & 0 & 0 \\ 0 & 0 & 1 & 0 \\ 0 & 0 & 0 & 1 \end{pmatrix} \begin{pmatrix} u^0 \\ u^1 \\ u^2 \\ u^3 \end{pmatrix}. \tag{10.71}$$

Compare this with Equation (10.54) relating the coordinates $(\mathbf{r}, t)$ and $(\mathbf{r}', t')$ of an event in two inertial frames: (10.71) is the same same transformation.

How is the 4-vector u^a related to the three vector $\mathbf{u}$? First we notice that the 3-velocity $\mathbf{u}$ is simply

$$\mathbf{u} = \frac{d\mathbf{r}}{dt}. \tag{10.72}$$

We are going to use, instead of this, the rate of change of position in the 4-dimensional space with proper time τ. The use of proper time is appropriate here since the velocity of the particle cannot be greater than the speed of light. Recall that when $c^2 dt^2 > d\mathbf{r}^2$, $ds^2 < 0$ and the proper time is then defined by $c^2 d\tau^2 = -ds^2$. To make the transformation from dt to $d\tau$ we use Equation (10.26), writing it as

$$d\tau = dt \sqrt{1 - \frac{1}{c^2}\left(\frac{d\mathbf{r}}{dt}\right)^2}$$

$$= dt \sqrt{1 - \frac{\mathbf{u}^2}{c^2}}, \tag{10.73}$$

which can be written as

$$d\tau = \gamma(\mathbf{u})^{-1}\, dt, \quad \gamma(\mathbf{u}) = \left(1 - \frac{\mathbf{u}^2}{c^2}\right)^{-1/2}. \tag{10.74}$$

The idea would be to work entirely in terms of proper time, which means that the spatial component of the velocity 4-vector would be $d\mathbf{r}/d\tau$ instead of $d\mathbf{r}/dt$. So the 4-vector we are looking for is

$$u^a = \left(\frac{dt}{d\tau}, \frac{d\mathbf{r}}{d\tau}\right) = \left(\gamma(\mathbf{u}), \gamma(\mathbf{u})\frac{\mathbf{u}}{c}\right), \tag{10.75}$$

which can be written as:

4-velocity:

$$u^a = \gamma(\mathbf{u})\,(1, \mathbf{u}/c). \tag{10.76}$$

It is straightforward to verify that this is a Lorentz invariant 4-vector. Using the $\mathbf{u}/c$ normalisation of the spatial component of the 4-velocity has the advantage that

$$\eta_{ab}\, u^a u^b = -1, \tag{10.77}$$

which is consistent with the definitions we shall be using in the upcoming discussion of particle motion in general relativity. This normalisation is, of course, equivalent to simply setting $c = 1$ when working in the relativistic context.

Below we shall interpret this 4-vector as a matrix having these components, in which case it will be written as a 1×4 column matrix (see Equation 10.71).

Putting this into Equation (10.71), and keeping only the two relevant dimensions, we get

$$\gamma(u') \begin{pmatrix} 1 \\ u'/c \end{pmatrix} = \gamma(v)\gamma(u) \begin{pmatrix} 1 & -v/c \\ -v/c & 1 \end{pmatrix} \begin{pmatrix} 1 \\ u/c \end{pmatrix}, \tag{10.78}$$

which leads to the two equations

$$\gamma(u') = \gamma(v)\gamma(u)(1 - uv/c^2), \tag{10.79}$$

$$\gamma(u')u' = \gamma(v)\gamma(u)(u - v). \tag{10.80}$$

Dividing one by the other we have

$$u' = \frac{u - v}{1 - uv/c^2}, \tag{10.81}$$

which is the same result as given in Equation (10.15).

Since we now know that $\gamma(\mathbf{u})(1, \mathbf{u})$ is a 4-vector, we can use the generic Lorentz transformation for a 4-dimensional vector given in (10.11) to write down the transformation of velocities directly as

$$\gamma(\mathbf{u}')\mathbf{u}' = \gamma(\mathbf{v})\left[\gamma(\mathbf{v})\mathbf{u}^\dagger - \mathbf{v}\gamma(\mathbf{u})\right], \tag{10.82}$$

$$\gamma(\mathbf{u}') = \gamma(\mathbf{v})\left[\gamma(\mathbf{u}) - \gamma(\mathbf{u})\mathbf{v} \cdot \mathbf{u}/c^2\right]. \tag{10.83}$$

Dividing the first by the second we are left with

$$\mathbf{u}' = \frac{\mathbf{u}^\dagger - \mathbf{v}}{1 - \mathbf{v} \cdot \mathbf{u}/c^2}, \tag{10.84}$$

which is just Equation (10.16) for the addition of velocities.

10.5.7 Minkowski Space Particle Dynamics

It is necessary at this point to touch base with our familiar Newtonian world of the motion of particles. We start with the notion of the particle of rest mass m_0 moving with velocity $\mathbf{v}$ in an inertial frame of reference. We have seen (Section 10.5.6) that the 4-velocity

$$u^a = \gamma(\mathbf{v})(1, \mathbf{v}/c) \tag{10.85}$$

is a 4-vector that transforms properly under Lorentz transformation. From there we can deduce the rules of relativistic addition of velocities. The natural step is to define the 4-momentum of the particle as

$$p^a = m_0 u^a = \gamma(v)m_0(1, \mathbf{v}/c) = (\gamma m_0, \gamma m_0 \mathbf{v}/c), \tag{10.86}$$

where m_0 is the mass it has as measured in its own rest-frame.

If we look at the $v/c \to 0$ limit of $\gamma m_0 = E/c^2$, where E is the *relativistic particle energy*, we see that

$$\gamma m_0 = E/c^2 \underset{v/c \to 0}{=} m_0 \left(1 + \tfrac{1}{2}v^2/c^2 + \cdots\right). \tag{10.87}$$

We recognise this low velocity limit of E as being the sum of the *particle rest mass energy*, $m_0 c^2$, and the familiar kinetic energy $\frac{1}{2} m_0 v^2$. This encourages us to write the 4-momentum (10.86) as

$$p^a = \left(\frac{E}{c^2}, \frac{\mathbf{p}}{c} \right), \quad \mathbf{p} = \gamma m_0 \mathbf{v} = m\mathbf{v}, \tag{10.88}$$

where $m = \gamma m_0$ might as well be called the *relativistic mass*.[12] Again, in using $\mathbf{v}/c$ as the spatial component of 4-velocity we have non-dimensionalised $\mathbf{p}$.

The magnitude of this 4-vector is

$$\eta_{ab} p^a p^b = -m^2 c^2 = -E^2/c^2 + \mathbf{v} \cdot \mathbf{v}, \tag{10.89}$$

which then gives us the important relativistic expression:

$$E^2 - \mathbf{p}^2 c^2 = (m_0 c^2)^2. \tag{10.90}$$

In the low velocity limit we have what is perhaps the most famous equation ever:[13]

Perhaps the most famous equation in history:

$$E_0 = m_0 c^2, \tag{10.91}$$

where the subscript on the energy indicates that this is the rest-frame value of E.

10.6 Minkowski Space Scalars and 4-vectors

We have learned that in Minkowski's space-time scalars and 4-vectors are quantities that, when measured in inertial frames, remain invariant under the action of a Lorentz transformation between different inertial frames. This is generally referred to as *Lorentz invariance*. Since general relativity includes the special theory as a particular case, the concept of local Lorentz invariance is fundamental to modern physics, be it general relativity or quantum mechanics field theory.

The following are examples of four-vectors:

$$x^a = (ct, x, y, z) = (ct, \mathbf{r}), \tag{10.92}$$

$$u^a = \gamma \left(1, \frac{dx}{cdt}, \frac{dy}{cdt}, \frac{dz}{cdt} \right) = \gamma(1, \mathbf{u}/c), \tag{10.93}$$

---

[12] Using $m = \gamma m_0$ is somewhat perverse and often leads to confusion. Think of this as $\mathbf{p} = m_0 (\gamma \mathbf{v})$.

[13] Some would say 'infamous' on the grounds of its association with the atomic bomb. However, it should be remarked that this equation had nothing to do with the bomb: the required physics had been well known for a considerable time. This can be seen in Robert Serber's 1943 lecture notes *The Los Alamos primer: The First Lectures on How to Build an Atomic Bomb*. These notes formed the basis of a course given to physicists at Los Alamos during the start of the Manhattan Project. See, for example, Serber (1992), and the article of Bernstein (2011) which comments on the original Serber notes and puts them in the broader, historical, context of the Frisch–Peierls Memorandum.

$$\nabla_a = \left(\frac{1}{c}\frac{\partial}{\partial t}, \frac{\partial}{\partial x}, \frac{\partial}{\partial y}, \frac{\partial}{\partial z}\right) = \left(\frac{1}{c}\frac{\partial}{\partial t}, \boldsymbol{\nabla}\right). \tag{10.94}$$

The gradient operator ∇_a has a lower index because it is constructed by dividing by a position four-vector, which has upper indices (see Section 10.5.4 *et seq.*).

10.6.1 The Minkowski Space Metric

The Minkowski space metric in Cartesian coordinates is

$$\eta_{ab} = \begin{pmatrix} -1 & 0 & 0 & 0 \\ 0 & 1 & 0 & 0 \\ 0 & 0 & 1 & 0 \\ 0 & 0 & 0 & 1 \end{pmatrix} = \begin{pmatrix} -1 & \mathbf{0}^T \\ \mathbf{0} & {}^{(3)}\mathbf{I} \end{pmatrix}, \qquad \eta^{ab} = \begin{pmatrix} -1 & \mathbf{0}^T \\ \mathbf{0} & {}^{(3)}\mathbf{I} \end{pmatrix}, \tag{10.95}$$

where ${}^{(3)}\mathbf{I}$ is the 3×3 unit matrix having 1s down its main diagonal and zeros elsewhere. The matrix notation is developed and illustrated in Appendix C.

Four-vectors in this representation are 4-component column vectors like the 4-velocity

$$u^a = \gamma \begin{pmatrix} 1 \\ v/c \end{pmatrix}, \quad \gamma = (1 - v^2/c^2)^{-1/2}, \tag{10.96}$$

where $\mathbf{v}$ is the velocity of the particle in the local instantaneous inertial frame . It is convenient to write this in-line as $u^a = \gamma(1, \mathbf{v}/c)$ with the understanding that this is in fact a column matrix (compare Equation 10.76). The metric tensor η_{ab} is used to raise and lower indices and so

$$u_a = \eta_{ab}u^b = \gamma \begin{pmatrix} -1 & \mathbf{0}^T \\ \mathbf{0} & {}^{(3)}\mathbf{I} \end{pmatrix} \begin{pmatrix} 1 \\ \mathbf{v}/c \end{pmatrix} = \gamma \begin{pmatrix} -1 \\ \mathbf{v}/c \end{pmatrix}. \tag{10.97}$$

So lowering the index on u^a changes the sign of the time-like component of the 4-vector.

Using this notation we easily see that

$$\eta_{ab}u^a u^b = \gamma^2 \begin{pmatrix} 1 & \mathbf{v}/c \end{pmatrix} \begin{pmatrix} -1 & \mathbf{0}^T \\ \mathbf{0} & {}^{(3)}\mathbf{I} \end{pmatrix} \begin{pmatrix} 1 \\ \mathbf{v}/c \end{pmatrix} = -1. \tag{10.98}$$

Since η_{ab} is the index-lowering operator, this can be written as

$$u^a u_a = -1. \tag{10.99}$$

The 4-velocity is normalised by the Lorentz γ-factor.

In the rest frame of the particle where $\mathbf{v} = \mathbf{0}$ we have $u^a = (1, 0)$. With this we can make an important splitting of the Minkowski metric into two parts:

$$\eta^{ab} = \begin{pmatrix} -1 & 0 & 0 & 0 \\ 0 & 0 & 0 & 0 \\ 0 & 0 & 0 & 0 \\ 0 & 0 & 0 & 0 \end{pmatrix} + \begin{pmatrix} 0 & 0 & 0 & 0 \\ 0 & 1 & 0 & 0 \\ 0 & 0 & 1 & 0 \\ 0 & 0 & 0 & 1 \end{pmatrix} = -u^a u^b + h^{ab}. \tag{10.100}$$

The matrix h^{ab} is described as the *projection tensor* that projects any vector or tensor onto the space perpendicular to the world line of the observer,

$$h^{ab} = \eta^{ab} + u^a u^b. \tag{10.101}$$

The tensor h_{ab} is the projection tensor into the rest space of an observer moving with velocity u^a.[14] We can derive the relationship

$$h^{ab}u_b = 0 \tag{10.102}$$

from Equations (17.12) and (10.101). This is a formal way of showing that the 4-velocity has no component in the 3-space defined by h^{ab}.

10.6.2 Light Cone (Null) Coordinates

In the 'natural' (ct, x, y, z) coordinate system used up to this point for inertial frames, light rays are described by $ds = 0$ and move on paths in space time having $x = ct$. Using the alternative (x^0, x^1, x^2, x^3) coordinates, this is just $x^0 = x^1$. However, we are entitled to use any coordinate system that may be convenient to the problem at hand.

An important choice of coordinates arises from the transformation

$$x^+ = \frac{1}{\sqrt{2}}(x^0 + x^1), \quad x^- = \frac{1}{\sqrt{2}}(x^0 - x^1). \tag{10.103}$$

We can think of x^+ as the 'time' coordinate and x^- as a spatial coordinate. Using the identity

$$(dx^0)^2 - (dx^1)^2 = 2dx^+dx^-, \tag{10.104}$$

we can write the space time interval ds as

$$ds^2 = -2dx^+dx^- + (dx^2)^2 + (dx^3)^2. \tag{10.105}$$

So if we write this in the form $ds^2 = \hat{\eta}_{ab}dx^a dx^b$ we see the metric representation of the space time in this coordinate system is

$$\hat{\eta}_{ab} = \begin{pmatrix} 0 & -1 & 0 & 0 \\ -1 & 0 & 0 & 0 \\ 0 & 0 & 1 & 0 \\ 0 & 0 & 0 & 1 \end{pmatrix}. \tag{10.106}$$

This transformation of coordinates (10.103) was introduced by Dirac (1949) and today plays a role in areas of field theory such as high energy hadron collisions and string theory.

To see what happens to these coordinates as one moves between inertial frames, we perform a Lorentz transformation from (x^0, x^1) to (x'^0, x'^1):

$$x^0 = \gamma(v)(x'^0 + x'^1 v/c), \quad x^1 = \gamma(v)(x'^1 + x'^0 v/c). \tag{10.107}$$

This gives the remarkably simple equations

$$x^+ = \left(\frac{1 + v/c}{1 - v/c}\right)^{1/2} x'^+, \quad x^- = \left(\frac{1 - v/c}{1 + v/c}\right)^{1/2} x'^-, \tag{10.108}$$

---

[14] The expression for the projection tensor depends on the signature of the metric. When the signature is $(+ - - -)$ we have $u^a u_a = +1$ and $h_{ab} = g^{ab} - u^a u^b$ (compare Equation 17.12).

which simplify further by introducing the hyperbolic angle ψ defined as

$$\psi = \frac{1}{2}\ln\frac{1+v/c}{1-v/c}, \quad v = \tanh\psi, \tag{10.109}$$

with which

$$x^+ = x'^+ e^\psi, \quad x^- = x'^- e^{-\psi}. \tag{10.110}$$

This simple transformation applies to all 4-vectors V^a.

Ex 10.6.1   Derive Equations (10.108) and (10.110).

Ex 10.6.2   Show that the result of two successive Lorentz transformations with velocities v_a and v_b is a Lorentz transformation for velocity v_c with $\psi_c = \psi_a + \psi_b$, where the ψ factors are defined as in Equation (10.109).

10.7 Classical Electrodynamics

We shall use Gaussian electromagnetic units with c being the speed of light in vacuum. These units differ from the Lorentz–Heaviside units that are commonly used in field theory in that the factors of 4π here get absorbed into the charge density ρ and the current $\mathbf{J}$. See Appendices A and D and for more details.[15]

10.7.1 Maxwell's Equations

The early versions of special relativity were largely concerned with Maxwell's equations for the electromagnetic field. Einstein's first papers established the link between the electric and magnetic field, but it was the work of Minkowski that formulated this within the framework of the 4-dimensional space-time that rightfully bears his name: *Minkowski space*. Much of the groundwork leading to Einstein's unification of the electric and magnetic fields had been done following on from Maxwell's original work by Poynting, Heaviside, Lorentz and others whose own work shed light on the physics of the mathematical work of Maxwell.

In the units used here, the Maxwell equations for the electromagnetic field in a vacuum, supplemented with the (non-relativistic) Lorentz Force law, are

$$\nabla \cdot \mathbf{E} = 4\pi\rho, \qquad \text{Coulomb's law} \tag{10.111}$$

$$\nabla \cdot \mathbf{B} = 0, \qquad \text{No magnetic monopoles} \tag{10.112}$$

$$\nabla \times \mathbf{E} = -\frac{1}{c}\frac{\partial \mathbf{B}}{\partial t}, \qquad \text{Faraday's law} \tag{10.113}$$

---

[15] The Maxwell equations in various systems of units are nicely summarised in Table 2 of Jackson (1998). The substitution $4\pi\rho \to \rho$ and $4\pi\mathbf{J} \to \mathbf{J}$ change our Gaussian units to Heaviside units. Note that many texts on electromagnetism use the complex time (x_1, x_2, x_3, ict) convention for 4-vectors (e.g. Zangwill, 2013).

$$\nabla \times \mathbf{B} = \frac{1}{c}\frac{\partial \mathbf{E}}{\partial t} + \frac{4\pi}{c}\mathbf{J}, \qquad \text{Ampère's law} \tag{10.114}$$

$$\mathbf{F} = q\left(\mathbf{E} + \frac{1}{c}\mathbf{v} \times \mathbf{B}\right). \qquad \text{Lorentz force law} \tag{10.115}$$

The electric part $q\mathbf{E}$ comes from the Coulomb law, while the magnetic part, $(q/c)\mathbf{v} \times \mathbf{B}$ was first derived by Heaviside in 1899. Lorentz arrived at this form from his work on the theory of the electron.[16] The Lorentz force law does not come from the Maxwell equations unless further assumptions are made. However, as we shall see below, when the Maxwell equations are written in a covariant form the Lorentz force law comes for free (see Equation 10.139).

Ex 10.7.1   Define a new field vector $\mathbf{G}$ having complex values

$$\mathbf{G} = \mathbf{E} + i\mathbf{B}, \quad i = \sqrt{-1}. \tag{10.116}$$

Show that Maxwell's Equations (10.111–10.114) then reduce to a pair of equations:

$$\nabla \cdot \mathbf{G} = 4\pi\rho, \quad \nabla \times \mathbf{G} - \frac{i}{c}\frac{\partial \mathbf{G}}{\partial t} = \frac{i}{c}4\pi\mathbf{J}. \tag{10.117}$$

This shows that the electromagnetic field components can be combined into a single complex object. However, the Lorentz force law (10.115) does not fall within this framework.

It is convenient to consider a charge density field, $\rho(t, \mathbf{x})$, that is in motion with local velocity $\mathbf{v}(t, \mathbf{x})$. We can then rewrite the Lorentz force Equation (10.115) as a force density. Consider a small volume δV of charge having charge density ρ moving with velocity $\mathbf{v}$. This corresponds to a charge current density $\mathbf{J} = \rho\mathbf{v}/c$ in our units. With this the Lorentz force can be expressed in terms of the fields ρ and $\mathbf{J}$ as a force density:

$$\mathbf{f} = \rho\mathbf{E} + \mathbf{J} \times \mathbf{B}/c. \tag{10.118}$$

We can also eliminate the fields $\mathbf{E}$ and $\mathbf{B}$ from the Maxwell equations to give an equation for the charge and current densities ρ and $\mathbf{J}$:

$$\frac{\partial \rho}{\partial t} + \nabla \cdot \mathbf{J} = 0. \tag{10.119}$$

10.7.2 Electromagnetic Scalar and Vector Potentials

We can introduce potentials ϕ and $\mathbf{A}$ such that[17]

$$\mathbf{E} = -\frac{1}{c}\frac{\partial \mathbf{A}}{\partial t} - \nabla\phi, \tag{10.120}$$

$$\mathbf{B} = \nabla \times \mathbf{A}. \tag{10.121}$$

---

[16] Ampère's law is the $\nabla \times \mathbf{B} = 4\pi\mathbf{J}/c$ part of this, known long before Maxwell did his work. Maxwell's great contribution was in fact the $\dot{\mathbf{E}}/c$ term.

[17] The notation, units and normalisations used here follow those of Jackson (1998).

The vector $\mathbf{A}$ introduced in this way is automatically consistent with $\nabla \cdot \mathbf{B} = 0$ (Equation 10.112). It is important to note that the potentials $\mathbf{A}$ and ϕ are neither observable nor measurable.[18]

However, any vector field of the form $\mathbf{A} + \nabla \psi$ will also be consistent with $\nabla \cdot \mathbf{B} = 0$, and so there is not a unique vector potential that determines a given magnetic field. If we change $\mathbf{A}$ in this way the value of $\mathbf{E}$ in (10.120) will change by $\nabla \dot{\psi}/c$ unless we also adjust ϕ to compensate this. Thus the transformation of the potentials

$$\mathbf{A}' = \mathbf{A} + \nabla \psi, \tag{10.122}$$

$$\phi' = \phi - \frac{1}{c} \frac{\partial \psi}{\partial t}, \tag{10.123}$$

leaves the values of the fields $\mathbf{E}$ and $\mathbf{B}$ as given in Equations (10.120) and (10.120) unchanged.

We have a freedom of choice as regards the field ψ: this freedom is referred to as a *gauge freedom*. The existence of such a freedom tells us something about the physical properties of the field we are describing. The choice we make here is to choose ψ so that

$$\nabla \cdot \mathbf{A} + \frac{1}{c} \frac{\partial \phi}{\partial t} = 0. \tag{10.124}$$

This condition $\nabla \cdot \mathbf{A}$ imposed on the vector $\mathbf{A}$ is referred to as the *Lorenz gauge*.[19] Since the gauge condition (10.124) applies equally to the original and transformed potentials, we see that

$$\nabla^2 \psi - \frac{1}{c^2} \frac{\partial^2 \psi}{\partial t^2} = 0, \tag{10.125}$$

and so if the condition (10.124) holds initially, it holds at all times.

With this gauge choice Equations (10.120) and (10.121) become

$$\nabla^2 \mathbf{A} - \frac{1}{c^2} \frac{\partial^2 \mathbf{A}}{\partial t^2} = -\frac{4\pi}{c} \mathbf{J}, \tag{10.126}$$

$$\nabla^2 \phi - \frac{1}{c^2} \frac{\partial^2 \phi}{\partial t^2} = -4\pi\rho. \tag{10.127}$$

Without imposing the Lorenz gauge condition (10.124) the first of these equations would contain additional terms involving $\nabla \cdot \mathbf{A}$ and $\partial \phi / \partial t$.

10.7.3 The Faraday Tensor

Equations (10.124, 10.126 and 10.127) suggest that we can conveniently combine $\mathbf{A}$ and ϕ into a 4-dimensional vector having components A_a, and the pair $\mathbf{J}$ and ρ into a 4-dimensional vector current having components J_a, with $a = 0, 1, 2, 3$:

---

[18] The Aharonov–Bohm effect asks the question as to whether the potentials are indeed just mathematical devices, or whether one might in fact measure the potentials. The effect is a quantum mechanical effect in which a charged particle can be accelerated even when it is confined to a volume where there is no electromagnetic field.

[19] Named for the Danish physicist Ludwig Lorenz, not the Dutch physicist Henrik Lorentz. The older use of the name 'Lorentz gauge' rather than 'Lorenz gauge' is probably a consequence of familiarity with the name of Henrik Lorentz.

$$A^a = (\phi, \mathbf{A}), \quad A_a = (-\phi, \mathbf{A}), \tag{10.128}$$

$$J_a = (c\rho, \mathbf{J}). \tag{10.129}$$

With this the equations involving the potentials $\mathbf{A}$ and ϕ become

$$\Box A_a = -\frac{4\pi}{c} J_a, \quad \partial^a A_a = 0, \quad \Box \equiv \nabla^2 - \frac{1}{c^2} \frac{\partial^2}{\partial t^2} \equiv \partial_a \partial^a. \tag{10.130}$$

The *d'Alembertian* $\Box$ is the wave operator in $3+1$-dimensions and the notation ∂_a denotes the partial derivative with respect to the coordinate x^a. While this prettifies the equations it would have no particular significance unless the objects A_a and J_a were Lorentz invariant 4-vectors that transform correctly between different Lorentz frames of reference. This is covered in most electromagnetism texts.

Using the derivatives of the potential A_a, we can create a skew-symmetric 4-dimensional matrix F_{ab} of derivatives of the potential field A_a, and the contravariant counterpart F^{ab} from derivatives of A^a:

$$F_{ab} = \frac{\partial A_b}{\partial x^a} - \frac{\partial A_a}{\partial x^b}, \quad F^{ab} = \frac{\partial A^b}{\partial x^a} - \frac{\partial A^a}{\partial x^b}. \tag{10.131}$$

To evaluate the components of the F_{ab} in terms of the fields we note that Equations (10.120) and (10.121) can be written out in a coordinate system (x^0, x^1, x^2, x^3) with $x^0 = ct$.[20]

The electric field involves $\phi = -A_0$ and so only contribute to the $F_{0\mu}$ and $F_{\mu 0}$ components of F_{ab}:

$$F_{0\mu} = \frac{\partial A_\mu}{\partial x^0} - \frac{\partial A_0}{\partial x^\mu} = \frac{1}{c} \frac{\partial A_1}{\partial t} + \frac{\partial \phi}{\partial x^\mu} = -E_\mu, \tag{10.132}$$

where we have used the fact that $A_a = (-\phi, \mathbf{A}) = (A_0, \mathbf{A})$. The magnetic field components are

$$B_1 = \partial_2 A_3 - \partial_3 A_2, \quad B_2 = \partial_3 A_1 - \partial_1 A_3, \quad B_3 = \partial_1 A_2 - \partial_2 A_1. \tag{10.133}$$

Assembling all the components of the matrix in this way, we end up with the skew-symmetric tensor[21]

$$F_{ab} = \begin{pmatrix} 0 & -E_x & -E_y & -E_z \\ E_x & 0 & B_x & -B_y \\ E_y & -B_x & 0 & B_z \\ E_z & B_y & -B_z & 0 \end{pmatrix}, \quad F^{ab} = \begin{pmatrix} 0 & E_x & E_y & E_z \\ -E_x & 0 & B_x & -B_y \\ -E_y & -B_x & 0 & B_z \\ -E_z & B_y & -B_z & 0 \end{pmatrix}. \tag{10.134}$$

---

[20] The components E_μ of the electric field $\mathbf{E}$ and the components B_μ of the magnetic field $\mathbf{B}$ are not the components of a 4-vector, but they are related to the derivatives of the 4-vector potential $A^a = (\phi, \mathbf{A})$. We can write the electric field components as either E_μ or E^μ and likewise for the components of $\mathbf{B}$. So, for notational consistency we denote the components of the electric field as though they were the components of the spatial part of a covariant vector $\mathbf{E} = (E_1, E_2, E_3)$ (i.e. lower indices). This is motivated by the notation $\partial_\mu \phi = \partial \phi / \partial x^\mu$.

[21] F^{ab} is an antisymmetric tensor. Antisymmetric tensors are in many ways rather special and they are often referred to as *bivectors* because they can be thought of as being defined by two vectors that form a 6-dimensional vector space (see Ex.10.7.1). So F^{ab} is occasionally referred to as the *Faraday bivector*. Jordan *et al.* (2009) has a detailed description of the bivector approach to general relativity and fields. For a brief development of 6-dimensional electromagnetism see Stephenson and Kilmister (1962, Ch. 2).

The relationship between these two forms is described in Appendix D. The tensor F_{ab} is now referred to as the *Faraday tensor*, although the expression *electromagnetic field tensor* is still in common use.[22]

10.7.4 The Faraday Tensor and the Maxwell Equations

The Faraday tensor remains invariant under the Lorentz transformations. Under the influence of a Lorentz transformation the components get mixed up, but it is still the same electromagnetic field. The Maxwell equations are

$$F^{ab}{}_{,b} = \frac{4\pi}{c} J^a,$$ (10.135)

$$F_{[ab,c]} = 0.$$ (10.136)

Here, J^a is the 4-current $(c\rho, J_1, J_2, J_3)$ which combines the charge density ρ and current density $\mathbf{J} = (J_1, J_2, J_3)$ of the standard Maxwell equations into a single 4-vector.[23] Since F_{ab} is anti-symmetric, the derivative $F^{ab}{}_{;ab} = 0$ and so the 4-divergence of the current J^a is zero:

$$J^a{}_{,a} = 0.$$ (10.138)

This expresses the conservation of the current J^a.

The Lorentz force (*cf.* Equation (10.115)) exerted on an electron moving with 4-velocity u^a in an electromagnetic field is given by

$$f^a = -eF^{ab}u_b, \qquad f^a u_a = -eF^{ab}u_a u_b = 0.$$ (10.139)

The first of these is easily verified using $u_b = \gamma(v)(-1, \mathbf{v}/c)$. The second of these follows simply from the antisymmetry of F^{ab} and asserts that the force acts perpendicular to the motion of the electron. What is important about the identification of the electromagnetic field with the tensor F_{ab} is that we get the Lorentz force 'for free', whereas in the classical, non-relativistic, theory the force law is an addition to the Maxwell equations.

A tensor, $^*F_{ab}$ that is dual to F^{ab} can be constructed using the 4-index permutation symbol ϵ_{ijkl}:

$$^*F_{ab} = \epsilon_{abcd}F^{cd}.$$ (10.140)

This merely rearranges the elements of the tensor F^{ab}, but with this we have an alternative set of equations

$$F_{[ab,c]} = 0 \quad \Leftrightarrow \quad ^*F^{ab}{}_{,b} = 0,$$ (10.141)

making a pleasant symmetry in the equations.

---

[22] Beware: the signs of the entries in the Faraday tensor depend on the signature of the metric.

[23] The second of these when expanded out reads

$$F_{[ab,c]} = F_{ab,c} + F_{bc,a} + F_{ca,b} = \frac{\partial F_{ab}}{\partial x^c} + \frac{\partial F_{bc}}{\partial x^a} + \frac{\partial F_{ca}}{\partial x^b} = 0.$$ (10.137)

F_{ab} is an anti-symmetric tensor, and so three terms, that would otherwise be present, vanish. This is a good exercise in tensor manipulation.

The scalar quantity

$$F_{ab}F^{ab} = 2(\mathbf{B}^2 - \mathbf{E}^2) \tag{10.142}$$

is a well known invariant of the electromagnetic field and is independent of the choice of coordinates. We can construct another invariant from $^*F_{ab}$ and F^{ab}:

$$^*F_{ab}F^{ab} = -4\,\mathbf{E}\cdot\mathbf{B}. \tag{10.143}$$

These are the two invariants that characterise electromagnetic fields. The second tells us, for example, that if the electric and magnetic fields are mutually perpendicular in one inertial frame, then they are mutually perpendicular in all inertial frames.

We might note that in terms of the complex valued field $\mathbf{F} = \mathbf{E} + i\mathbf{B}$ introduced in Exercise 10.7.1 (Equation 10.116 *et.seq.*) we recover these two scalar invariants from

$$F^2 = (\mathbf{E} + i\mathbf{B})\cdot(\mathbf{E} + i\mathbf{B}) = (E^2 - B^2) + 2i\mathbf{E}\cdot\mathbf{B}, \tag{10.144}$$

(up to a constant of proportionality).

10.7.5 Electromagnetic Energy Momentum Tensor

The energy momentum tensor for the electromagnetic field was first written down by Larmor in 1898 in his Adams Prize winning essay and published in his book *Aether and Matter* in 1900 (Larmor, 2012). Larmor's achievement was to unite the Maxwell stresses and Poynting's energy fluxes into a single object: the energy momentum tensor. But the story of the energy momentum tensor in 4-dimensional Minkowski space starts with the remarkable paper by Minkowski (1908, 1909) in which he laid down the mathematics of the 4-dimensional space-time, which is the arena of the theory of electromagnetism and special relativity.

> **Short biography: Herman Minkowski** (1864–1909) Minkowski was born in what, at the time of his birth, was Lithuania and obtained his doctorate in 1885 from the University of Königsberg. In 1902 he went to Göttingen where he became a close friend of Hilbert and later went to Zurich where, at the Eidgenössische Polytechnikum, he is said to have been one of Einstein's teachers. In 1908 Minkowski introduced the 4-dimensional view of space-time, following on from the earlier work of Lorentz, Poincaré and Einstein on electromagnetism and special relativity, and promoted that with a considerably less technical paper (Minkowski, 1909) simply called *Raum und Zeit* (Space and Time), which was quickly translated into several other languages. This, of course, gave immediate wide access to his view of space-time, which is now called 'Minkowski Space'. Although Einstein was said to be critical of Minkowski's work, dismissing it as mathematical trickery, it can hardly be doubted that Minkowski's geometrisation of relativity must have influenced Einstein's thinking when he turned to his theory of gravitation. Minkowski died from appendicitis shortly after the publication of *Raum und Zeit*.

Already, in 1873, Maxwell had written that the $e\mathbf{E} + \mathbf{J}\times\mathbf{B}$ forces acting on charges and currents arose from what he called stresses in the electromagnetic field. He stated this in the form

$$F_i = \frac{1}{4\pi}\frac{\partial}{\partial x_j}\left(\epsilon\delta_{ij} - E_iE_j - B_iB_j\right), \quad \epsilon = \tfrac{1}{2}(E^2 + B^2), \tag{10.145}$$

where ϵ was identified with the energy density of the electromagnetic field. The quantity $\epsilon/4\pi$, i.e.

$$\mathcal{E} = \frac{1}{8\pi}(E^2 + B^2),\tag{10.146}$$

is the electromagnetic field energy density. It only required Poynting to identify in 1884 the vector that now bears his name,

$$\mathbf{S} = \frac{c}{4\pi}\mathbf{E} \times \mathbf{B},\tag{10.147}$$

as the flux of energy and momentum and all the elements were in place.

The flurry of activity in the first years of the 20th century is well known through the papers of Lorentz (1904), Poincaré (1905) and finally Einstein (1905). But the 4-dimensional setting remained for Minkowski in 1908. Shortly thereafter von Laue (1911) generalised Minkowski's treatment of electromagnetism to derive the energy momentum tensor of a continuous medium.

The energy momentum tensor for the electromagnetic field is given by

$$T_{EM}^{ab} = \frac{1}{4\pi}\left(F^a{}_m F^{mb} - \frac{1}{4}\eta^{ab}F^{mn}F_{mn}\right).\tag{10.148}$$

The normalising factor is chosen so that in the system of units being used the component T_{00} is the energy density of the electromagnetic field (Equation 10.146). It is simple to show that this tensor is trace-free: $T_a{}^a = 0$.

The gradient of the energy momentum tensor is the Lorentz force density. This can be seen as follows:

$$\frac{\partial T^{ab}}{\partial x^b} = \frac{1}{4\pi}F_{am}\frac{\partial F^{mb}}{\partial x^b} = F_{am}J^m,\tag{10.149}$$

using the Maxwell Equations (10.135). Then putting $J^a = eu^a = e(c, \mathbf{u})$ we get the required result on multiplying out the matrix product (see equation (D.24) in Appendix D.3).

The energy momentum tensor T_{EM}^{ab} is derived from a Lagrangian in Section 11.2.3. In Appendices D.4 and D.5 its components are written out explicitly in terms of the the fields $\mathbf{E}$ and $\mathbf{B}$.

11

The Energy Momentum Tensor

The energy momentum tensor, which describes the physical properties of a gravitating medium, is intimately linked with the geometry of space-time. The route to determining the form of the energy momentum tensor for a given medium is via the Lagrangian describing the material that makes up the medium. Here we describe that process starting with the essentials of Lagrangian dynamics for particles and fields and going on to show some basic examples. One of the benefits of going via Lagrangian field theory is that we can also analyse the source of symmetries associated with the medium itself and their associated invariants.

This chapter is based around elementary Lagrangian field theory and can be safely skimmed or skipped. If you decide to skim, sections 11.4 and 11.3.3 are particularly pertinent. There is an On-line Supplement, *'Lagrangians and Hamiltonians'*, to support this.

11.1 Dynamics of Fields

If we move on from classical dynamical systems to physical systems described in terms of fields $\phi(x^a)$ and their gradients $\partial_a \phi(x^a)$ we have to generalise the concept of the Lagrangian and the associated action to deal with fields that, in effect, have an infinite number of degrees of freedom: the field is defined at every point of some volume $\mathcal{V}$. The transition from discrete to continuous systems is well treated in many texts, and in particular Goldstein *et al.* (2001, Ch.12), so we can simply point out the parallels with what has been said above.

The main issue is to deal with a Lagrangian volume density $\mathcal{L}$ so that the action becomes

$$\mathcal{S} = \int_{t_1}^{t_2} \int_{\mathcal{D}} \mathcal{L} \, dV \, dt, \tag{11.1}$$

where dV is an element of spatial volume (i.e. $dxdydz$). In other words, the contribution to the total action from an element of volume dV is $\mathcal{L}dV$. The $dVdt$ is strongly suggestive that this will become an integral over the four-dimensional volume traced out by $\mathcal{D}(t_1)$ as it evolves with time to $\mathcal{D}(t_2)$.

For the sake of simplicity, in the following we shall focus attention on Lagrangians for real scalar fields ϕ. The results generalise straightforwardly to vector fields, like the electromagnetic field 4-potential, A^a, and tensor fields such as the electromagnetic field tensor F^{ab}, and we shall give examples using these. The discussion is framed within the arena of Minkowski space. The corresponding results for the pseudo-Riemannian spaces of general relativity will be handled later, when necessary.

11.1.1 Lagrange Equations in Classical Dynamics

At the heart of the Lagrangian formulation of dynamics lies the *Lagrangian* for the physical system, $\mathcal{L}$. This is an explicit function of what are known as *generalised coordinates*, $\mathbf{q}$, and their time derivatives, $\dot{\mathbf{q}}$. In classical mechanics we usually think of the $\mathbf{q}$ as describing the configuration of the system and the $\dot{\mathbf{q}}$ as being the time derivatives of those descriptors. The space of $\mathbf{q}$-values is referred to as the *configuration space* of the system. The values of $\mathbf{q}$ and $\dot{\mathbf{q}}$ at an instant of time should completely specify the state of the system. The number of elements $\mathbf{q} : \{q_1, \ldots, q_n\}$ needed to describe the configuration of the system is referred to as the dimension, n, of the configuration space.[1]

We can display the explicit dependence of L on the variables $\mathbf{q}$ by writing

$$L = L(\mathbf{q}, \dot{\mathbf{q}}, t). \tag{11.2}$$

In classical mechanics we generally write the Lagrangian as the difference between the kinetic energy T and potential energy V expressed as functions of the $(\mathbf{q}, \dot{\mathbf{q}})$:

$$L = T - V. \tag{11.3}$$

This is a good working model for most dynamical systems, but the Lagrangian does not always have to be of this form. When there is no explicit time dependence in L, the energy of the system, $H = T + V$, is conserved. The *Lagrange equations* for the evolution of the system are derived from variation of the action (Goldstein *et al.*, 2001, Ch. 2):

$$S = \int_{t_0}^{t_1} L(q^a, \dot{q}^a, t) dt. \tag{11.4}$$

The minimum, $\delta S = 0$, taken over all paths of integration with these specified end-points, gives

$$\frac{d}{dt} \frac{\partial L}{\partial \dot{q}^a} - \frac{\partial L}{\partial q^a} = 0, \quad a = 1, \ldots, n, \tag{11.5}$$

where n is the number of variables in the vector $\mathbf{q}$ (the dimension of the configuration space of the system).

In classical mechanics, if the Lagrangian does not depend explicitly on one of the coordinates, say q^m, then we see from the Lagrange Equation (11.5) that

$$\frac{d}{dt} \frac{\partial L}{\partial \dot{q}^m} = 0, \quad \text{for any } m \text{ such that } \frac{\partial L}{\partial q^m} = 0. \tag{11.6}$$

Hence

$$p_m = \frac{\partial L}{\partial \dot{q}^m} = \text{constant}, \quad \text{for any } m \text{ such that } \frac{\partial L}{\partial q^m} = 0. \tag{11.7}$$

---

[1] The Hamiltonian view uses a $2n$-dimensional *phase space* built up from the generalised coordinates q^a and their *conjugate momenta*, $p_a = \partial L / \partial \dot{q}^a$.

So whenever a coordinate such as q_m does not appear specifically in the Lagrangian, there is an associated conservation law. What is important is that we can transform q^m in any way we want and the associated conservation law still holds. Such a coordinate is said to be an *ignorable coordinate*, or *cyclic coordinate*, and reflects a symmetry of the system.[2]

11.1.2 Lagrange Equations for a Particle in Special Relativity

The kinematics of a relativistic particle in an inertial frame were considered in Section 10.5.7. Now we wish to achieve a covariant description in terms of Lagrangian dynamics. This requires a little extra thought since the space-time split of the previous section is no longer valid: the Newtonian notion of absolute time has to be abandoned and we have to rethink d/dt in Equation (11.5). Consider the motion, in an inertial frame, of a particle of rest mass m_0 moving under the influence of a conservative force $F = -\nabla V$. The relativistic equation of motion is

$$m_0 \frac{d(\gamma \mathbf{v})}{dt} = -\nabla V, \quad \gamma = \left(1 - \frac{\mathbf{v}^2}{c^2}\right)^{-1/2}. \tag{11.8}$$

This can be easily verified to come from the Lagrangian

$$L = -m_0 c^2 \sqrt{1 - \mathbf{v}^2/c^2} - V, \tag{11.9}$$

in which we see that the kinetic energy term is no longer the Newtonian $\frac{1}{2}mv^2$: the Lagrangian is no longer of the form $L = T - V$. The relativistic 3-momentum of the particle is $p_\mu = \partial L/\partial v_\mu = \gamma m_0 v_\mu$, as per (10.88). It can be shown that the conserved quantity $H = \mathbf{v} \cdot \mathbf{p} - L = \gamma m_0 c^2 = E$ (see Equation 10.87). If we identify H with the usual $T + V$ we have that the kinetic term, T, in the Lagrangian is $E = \gamma m_0 c^2$.

We are still using the time in a particular inertial frame. How do we make this independent of frame, i.e. covariant? The 4-velocity of the particle is $u^a = \gamma(1, \mathbf{v}/c)$. We might guess that we should replace $1 - \mathbf{v}^2/c^2$ in (11.9) with $-u_a u^a$.[3] We should also replace t with the proper time τ, or indeed any parameter that is proportional to τ.[4]

Our covariant Lagrangian, which we can denote by Λ, and the Lagrangian equations are then[5]

$$\frac{d}{d\tau}\left(\frac{\partial \Lambda}{\partial u_a}\right) - \frac{\partial \Lambda}{\partial x^a} = 0, \qquad \Lambda = \sqrt{-u^a u_a} \text{ or } \Lambda = -\frac{1}{2} u^a u_a. \tag{11.10}$$

This leads to the equations of motion

$$\frac{d\pi_a}{d\tau} = -\frac{\partial V}{\partial x^a}, \quad \text{with} \quad \pi_a = m_0 u_a, \ p^a = \pi^a c^2 = (E, \gamma \mathbf{p}c). \tag{11.11}$$

Here, $p_a = (p^0, \mathbf{p})$ is the 4-momentum of the particle and $E = \gamma m_0 c^2$ is its energy. Since $p_a p^a = -m_0^2 c^4$, we have $E^2 - \mathbf{p}^2 c^2 = (m_0 c^2)^2$ (see Equation 10.90).

---

[2] There is another scaling symmetry that occurs when the potential term $V(\mathbf{q})$ is a homogeneous function of the q^i, i.e. $V(\alpha \mathbf{q}) = \alpha^k V(\mathbf{q})$ for some non-zero k, not necessarily real.

[3] Here we follow Goldstein *et al.* (2001, Section 7-9, p.326 *et seq.*), who gives all the details.

[4] This is closely related to defining an affine parameter on a geodesic in curved space-time, see Section 11.1.3.

[5] Both versions of Λ lead to the same equations. It is in fact easier to use $\Lambda = -\eta^{ab} u_a u_b$. Lagrangians are not unique and all Lagrangians that lead to the same equations of motion are equally acceptable.

11.1.3 Lagrange Equations for a Particle in General Relativity

We can look at free particle motion in general relativity knowing that the elemental distance, ds, between two neighbouring points is given by the generalisation of the Minkowski metric to $ds^2 = g_{ab}(x^a)dx^a dx^b$. Here the elements of the metric $g_{ab} = g_{ab}(x^a)$ depend explicitly on the space-time coordinates x^a. In other words, the Minkowski metric η_{ab} is distorted by becoming coordinate dependent. Einstein's insight was that a particle moving freely in this space-time would take the shortest path between two points.

With the line element $ds^2 = g_{ab}dx^a dx^b$, the elemental distance between two points on a general 4-space path can be written as

$$ds^2 = g_{ab}\frac{dx^a}{d\lambda}\frac{dx^b}{d\lambda}(d\lambda)^2. \tag{11.12}$$

Here $s(\lambda)$ is a parameter defining the position of a 4-space point on the path. The length of a path on a space-time curve $\mathcal{C}$ between two given points labelled λ_1 and λ_2 of the space-time is

$$l(\mathcal{C}) = \int_{\mathcal{C}:\lambda_1,\lambda_2} ds = \int_{\lambda_1}^{\lambda_2} \left| \sqrt{g_{ab}\frac{dx^a}{d\lambda}\frac{dx^b}{d\lambda}} \right| d\lambda, \tag{11.13}$$

where now the parameter λ is an arbitrary parameter describing position along the path. λ takes on values that run between the ends of the path $\mathcal{C}$ at $\lambda = \lambda_1$ and $\lambda = \lambda_2$.[6] At this point the path $\mathcal{C}$ is completely arbitrary, but there is the implicit requirement on the parameterisation λ that two intersecting curves connecting λ_1 and λ_2 should have the same value of λ at the point of intersection.[7]

The idea is to pick out the shortest path running between these two end points, the *geodesic*. This involves minimising (11.13) over all paths joining the points labelled by λ_1 and λ_2. This is a standard problem in the calculus of variations:

$$\delta_{\mathcal{C}} \int_{\mathcal{C}} ds = 0, \tag{11.14}$$

where the $\delta_{\mathcal{C}}$ indicates minimisation relative to variations in the path $\mathcal{C}$ while keeping the end-points fixed. Following Einstein's precept that free particles in curved space times move on geodesics, this can be regarded as the equation of motion of a free particle in curved space-time.

It makes things easier to notice that the shortest path is as well found by minimising the sum of the squares of the elemental line lengths:

$$I(\mathcal{C}) = \int_{\lambda_1}^{\lambda_2} g_{ab}\frac{dx^a}{d\lambda}\frac{dx^b}{d\lambda}d\lambda. \tag{11.15}$$

The integrand is a function of the coordinates x^a and the derivatives $\dot{x}^a = dx^a/d\lambda$, so we can write this as:

---

[6] If we parameterise using proper time, $\lambda = \tau$, the integrand is the function $\Lambda = \sqrt{-u^a u_a}$ of Equation (11.10).

[7] Note that if we make an arbitrary transformation of the parameter λ, the integrand does not change: this is referred to as *re-parameterisation invariance*. Consequently, we get the same path independently of our chosen parameterisation (subject to the usual mathematical restrictions on continuity, single-valuedness and so on).

Lagrangian for geodesics

$$\delta \text{e} \int_{\lambda_1}^{\lambda_2} \Lambda(x^a, \dot{x}^a)d\lambda = 0, \quad \Lambda = g_{ab}(x^c)\dot{x}^a\dot{x}^b, \quad \dot{x}^a = \frac{dx^a}{d\lambda}. \tag{11.16}$$

This greatly simplifies the problem of finding equations for geodesics.[8]

Lagrange showed how to solve this problem and the function $L(x^a, \dot{x}^a)$ is referred to as the Lagrangian of the problem. The shortest path is the solution of the Lagrange equations

$$\frac{d}{d\lambda}\left(\frac{\partial \Lambda}{\partial \dot{x}^c}\right) - \frac{\partial \Lambda}{\partial x^c} = 0, \tag{11.17}$$

(compare equation 11.10). There are four equations here in the space-time of general relativity, but in fact this derivation applies to Riemannian spaces of any number of dimensions. Doing the differentiation gives

$$2\frac{d}{d\lambda}(g_{ac}\dot{x}^a) - \frac{\partial g_{ab}}{\partial x^c}\dot{x}^a\dot{x}^b = 0,$$

which is

$$2g_{ac}\ddot{x}^a + (g_{ac,b} + g_{cb,a} - g_{ab,c})\dot{x}^a\dot{x}^b = 0, \tag{11.18}$$

where, in the second term, we have exploited the index-symmetry of g_{ab} to write $2g_{ac,d}\dot{x}^d\dot{x}^a = (g_{ac,b} + g_{cd,a})\dot{x}^d\dot{x}^a$. Multiplying by g^{ec} and rearranging gives the final result:

$$\ddot{x}^e + \tfrac{1}{2}g^{ec}(g_{ac,b} + g_{cb,a} - g_{ab,c})\dot{x}^a\dot{x}^b = 0. \tag{11.19}$$

This is usually written as:

Geodesic in space time described by =metric g_{ab},

$$\ddot{x}^e + \Gamma^e_{ab}\dot{x}^a\dot{x}^b = 0, \quad \Gamma^e_{ab} = \tfrac{1}{2}g^{ec}(g_{ac,b} + g_{cb,a} - g_{ab,c}). \tag{11.20}$$

This is also the equation of motion of a free particle in that space-time.

The collection of coefficients Γ^e_{ab} is referred to as a *Christoffel symbol*. This is the equation for a geodesic in this space-time described by metric g_{ab} and it is also the equation of motion of the free particle.[9] Note that when g_{ab} is independent of position, i.e. $g_{ab} = \eta_{ab}$, this is just the special relativistic version of force-free Newtonian motion, $\ddot{x}^e = 0$.

---

[8] It also leads to a convenient method for computing the fundamental Christoffel symbols of general relativity. These are presented in Section 13.3.3 and an example of how this facilitates their computation is given in Section 15.5.1.

[9] This derivation does not work for null geodesics. The problem arises in the Lagrangian of Equation (11.13) before we use the trick of using ds^2 in place of ds. When the path is a null geodesic, ds of Equation (11.13) is always zero and the derivative $\partial L'/\partial x^c$ is then infinite. Null geodesics have zero length. To handle null geodesics via the variational principle we have to consider non-null paths that are arbitrarily close to a null geodesic and then take a limit.

11.1.4 The Lagrange Equations for Classical Fields

In the previous discussion the state of the system could be fully defined by a finite number of parameters, $q^a, \dot{q}^a, a = 1, \ldots, n$. For a continuous classical field, such as a fluid flow or an electromagnetic field, the field values at an instant are defined at all points in a volume of space. The Lagrangian then becomes a function of an infinite number of variables: the fields and their gradients. In this section we will, for reasons of simplicity, consider systems in which there is only one *scalar* field $\phi(\mathbf{x}, t)$. This readily generalises to systems comprising a number of scalar fields $\phi_A(\mathbf{x}, t), A = 1, \ldots, M$, but, importantly, the ϕ_M should not be understood as components of a vector field.

The density $\mathcal{L}$ should be expressed as a function of the field values $\phi(\mathbf{x}, t)$ and its derivatives:

$$\mathcal{L} = \mathcal{L}(\phi, \nabla\phi, \dot{\phi}, \mathbf{x}, t). \tag{11.21}$$

Note the explicit dependence on $\mathbf{x}$. The Lagrangian density can conveniently be expressed as a function of the coordinates x^a in 4-dimensional Minkowski space:

$$\mathcal{L} = \mathcal{L}(\phi, \phi_{,a}, x^a), \quad \phi_{,a} = \partial_a \phi. \tag{11.22}$$

Lagrangians can sometimes be separated into the sum of terms, one of which depends only on the ϕ and the other of which depends only on the field gradients $\partial_a \phi$. Furthermore, there may be no explicit dependence on the coordinate x^a. In that case we can, by analogy with the situation in mechanics, write the Lagrangian as

$$\mathcal{L} = \mathcal{T}(\partial_a \phi) - \mathcal{V}(\phi). \tag{11.23}$$

As the notation suggests, we call $\mathcal{V}(\phi)$ the *potential of the scalar field*. In the cases when this split is possible the term $\mathcal{T}(\partial_a \phi)$ is often a quadratic form of the field gradients $\partial_a \phi$, and then it is appropriate to refer to $\mathcal{T}$ as the *kinetic energy* or *kinetic term*.[10]

The stationary value of the action is then found precisely as before and the resulting Lagrange equations are:

Lagrangian field equations:

$$\frac{d}{dx^b}\left(\frac{\partial \mathcal{L}}{\partial \phi_{,b}}\right) - \frac{\partial \mathcal{L}}{\partial \phi} = 0, \quad \phi_{,b} \equiv \frac{\partial \phi}{\partial x^b}. \tag{11.24}$$

If the Lagrangian depends on more than one field, then there is one such equation for each field. The quantity

$$\pi^a = \frac{\partial \mathcal{L}}{\partial \phi_{,a}} \tag{11.25}$$

is referred to as the *canonical momentum*.

---

[10] Examples where this split is not manifest are the Lagrangian for the electromagnetic field when expressed in terms of the field tensor F^{ab} and the Lagrangian for the Einstein vacuum field equations when expressed in terms of the curvature scalar R.

Some Practical Remarks

It is, because of familiarity, perhaps easier to relate to the Lagrangian written in the form (11.21) than in the form (11.22). The following may add some perspective on the relationship between the 3-space view of (11.21) and the Minkowski 4-space view of (11.22).

Consider the simple function $E_{ab} = \phi_{,a}\phi_{,b}$, which in the 4-space view is written variously as

$$E_{ab} = \phi_{,a}\,\phi_{,b} = \partial_a\phi\,\partial_b\phi, \quad \text{with } \partial_a \equiv \left(\frac{\partial\phi}{\partial t}, \nabla\phi\right). \tag{11.26}$$

Then, as an example, the trace $E^a{}_a$ is calculated as

$$E^a{}_a = \eta^{ab}\partial_a\phi\,\partial_b\phi = -\dot\phi^2 + (\nabla\phi)^2. \tag{11.27}$$

Such 'kinetic' terms frequently occur in Lagrangians.

Perhaps the easiest way to see where Equation (11.27) comes from is to write out the Minkowski space representation in matrix form:

$$E^a{}_a = \eta^{ab}\partial_a\phi\,\partial_b\phi = \begin{pmatrix}\dot\phi & \nabla\phi\end{pmatrix}\begin{pmatrix}-1 & \mathbf{0} \\ \mathbf{0} & \mathbf{I}\end{pmatrix}\begin{pmatrix}\dot\phi \\ \nabla\phi\end{pmatrix} = \begin{pmatrix}\dot\phi & \nabla\phi\end{pmatrix}\begin{pmatrix}-\dot\phi \\ \nabla\phi\end{pmatrix} \tag{11.28}$$

$$= -\dot\phi^2 + (\nabla\phi)^2, \tag{11.29}$$

where $\mathbf{I} = \mathrm{diag}(1,1,1)$ is the 3-space identity matrix. Note that the sign of the trace $E^a{}_a$ depends on the relative values of $\dot\phi^2$ and $(\nabla\phi)^2$. For a homogeneous realisation of the field we will have $E^a{}_a < 0$.[11]

A slightly more complicated example is to evaluate a term like the first term in the Lagrange Equation (11.24). Continuing with our example of using $E_{ab} = \phi_{,a}\,\phi_{,b}$ we need to know how to evaluate

$$\frac{\partial}{\partial x^b}\frac{\partial E^a{}_a}{\partial\phi_{,b}} = \frac{\partial}{\partial x^b}\frac{\partial}{\partial\phi_{,b}}\left(\eta^{mn}\phi_{,m}\phi_{,n}\right) = 2\partial_b(\eta^{bm}\partial_m\phi). \tag{11.30}$$

Looking at this in the Minkowski coordinates we can write

$$\partial_b(\eta^{bm}\partial_m\phi) = (\partial_b\eta^{bm}\partial_m)\phi = \begin{pmatrix}\partial_t & \nabla\end{pmatrix}\begin{pmatrix}-1 & \mathbf{0} \\ \mathbf{0} & \mathbf{I}\end{pmatrix}\begin{pmatrix}\partial_t \\ \nabla\end{pmatrix}\phi, \tag{11.31}$$

$$= (-\partial_{tt} + \nabla^2)\phi, \tag{11.32}$$

$$= \Box\,\phi. \tag{11.33}$$

[11] Note, again, the dependence of the sign of $E^a{}_a$ on the signature adopted for the metric.

The 'box' differential operator, $\Box = \partial_a \partial^a = \partial^2$, is called the *d'Alembertian*:[12]

$$\partial^2 \equiv \Box \equiv \partial_a \partial^a = -\frac{\partial^2}{\partial t^2} + \nabla^2. \tag{11.34}$$

11.1.5 Canonical Energy Momentum Tensor

Since we have chosen to express the Lagrangian density in terms of a 4-dimensional space in which all coordinates are on an equal footing, we do not have an immediate analogy with the cyclic coordinates of Equation (11.7). To make progress we need to understand the term $\partial \mathcal{L}/\partial x^a$, which, when zero, would be the analogue of the statement in (11.7) that x^a is a cyclic coordinate.

In classical mechanics it has been known since the work of Routh[13] in the mid-1870s that cyclic coordinates indicate the presence of a symmetry or conserved quantity. In 1915, Emmy Noether (1915) was responsible for establishing this relationship between continuous symmetries and conservation laws at a fundamental level that allowed the principle to be used in quantum field theory and in higher dimensional space-times.

> **Short biography:** **Emmy Noether** (1882–1935). The link between symmetries and conservation laws was established at a fundamental level by Noether (1918): a contribution to the deepest understanding of physics which has had a long-lasting impact, in particular in modern field theory.
>
> Owing to her exceptional ability she was allowed to audit courses at her home town University of Erlangen during 1900 to 1902, despite the fact that women were at that time generally barred from attending universities. After that she visited Göttingen and attended courses by Hilbert, Klein and Minkowski, returning thereafter to Erlangen to do her PhD. A period of highly distinguished work resulted in her being invited back to Göttingen, where despite strong anti-feminist objections, she was 'allowed' to lecture, but only as Hilbert's unpaid assistant. It was during that period that she wrote her seminal paper on invariants. By 1919 she had moved on to make fundamental contributions in other areas of mathematics, but had to leave Germany in 1933 when she went to Bryn Mawr College near Philadelphia. When she died in 1935 Einstein described her as 'a giant among mathematicians'.

We can now indicate how, in a flat space-time context, the energy momentum tensor for a field ϕ, and the associated conservation laws for ϕ, arise.[14] We have

$$\frac{d\mathcal{L}}{dx^a} = \frac{\partial \mathcal{L}}{\partial \phi}\phi_{,a} + \frac{\partial \mathcal{L}}{\partial \phi_{,b}}\phi_{,ab} + \frac{\partial \mathcal{L}}{\partial x^a}, \tag{11.35}$$

---

[12] The coordinate independent definition of the d'Alembertian is $\Box \equiv \partial_a \partial^a$. Its expression in a particular coordinate system depends on the signature of the adopted metric. It is easy to verify that

$$\Box \equiv \partial_a \partial^a = \eta^{ab}\partial_a\partial_b = \begin{cases} -\frac{\partial^2}{\partial t^2} + \nabla^2, & \eta^{ab} = \text{diag}(-1, +1, +1, +1) \\ \frac{\partial^2}{\partial t^2} - \nabla^2, & \eta^{ab} = \text{diag}(+1, -1, -1, -1). \end{cases}$$

[13] Routh exploited cyclic coordinates to come up with a powerful hybrid of Lagrangian and Hamiltonian dynamics which is known as 'Routhian dynamics'. In 1854 Edward Routh had succeeded as Senior Wrangler in the Cambridge Mathematics Tripos examination, just ahead to James Clerk Maxwell. They shared the prestigious Smith Prize which at that time was awarded for outstanding examination performance.

[14] The generalisation to curved space time and the relationship with Killing fields is beautifully described in the pedagogical review by Fleming (1987).

by the chain rule for calculating the total derivative of $\mathcal{L}(\phi, \phi_{,a}, x^a)$. Using the Lagrange field Equation (11.24) this becomes

$$\frac{d\mathcal{L}}{dx^a} = \frac{d}{dx^b}\left(\frac{\partial\mathcal{L}}{\partial\phi_{,b}}\phi_{,a}\right) + \frac{\partial\mathcal{L}}{\partial x^a}, \tag{11.36}$$

which on rearrangement is the result we need:

$$\frac{d}{dx^a}\left\{-\frac{\partial\mathcal{L}}{\partial\phi_{,b}}\phi_{,a} + \mathcal{L}\delta_a^b\right\} = \frac{\partial\mathcal{L}}{\partial x^a}. \tag{11.37}$$

The quantity in curly braces, we can denote by $T_a{}^b$:

Energy momentum tensor defined in terms of the Lagrangian:

$$T_a{}^b = -\phi_{,a}\frac{\partial\mathcal{L}}{\partial\phi_{,b}} + \mathcal{L}\delta_a{}^b. \tag{11.38}$$

The indices on $T_a{}^b$ can be raised and lowered using η_{ab} or η^{ab}. T^{ab} is the *energy momentum tensor* or *stress-energy tensor* of the field ϕ.

Equation (11.37) tells us that if the Lagrangian $\mathcal{L}$ does not explicitly depend on the coordinates x^a, in which case $\partial\mathcal{L}/\partial x^a = 0$, then

$$\frac{dT^{ab}}{dx^b} = 0. \tag{11.39}$$

The energy momentum tensor, under these circumstances, has zero divergence. Physically, this expresses the existence of conservation laws. We shall discuss this in relation to the electromagnetic field in Section 11.2.4.

11.2 Lagrangians and the Energy Momentum Tensor

We will, in the following examples, calculate the energy momentum for a variety of fields given their Lagrangians.

11.2.1 Simple Real Scalar Field

A simple, yet important, example of the use of the Lagrangian for matter is provided by the massive scalar field, $\phi(x^a)$. We shall first discuss the particular case of the real Klein–Gordon scalar field in the context of special relativity and its description in Minkowski space with metric η^{ab}. This will prepare the way to put the scalar field in a Riemannian space-time with an arbitrary metric g_{ab}.

In Minkowski space the Lagrangian for the real Klein–Gordon field is

$$\mathcal{L} = -\tfrac{1}{2}\eta^{ab}\partial_a\phi\partial_b\phi - \tfrac{1}{2}m^2\phi^2, \tag{11.40}$$

and the corresponding action is

$$S = \int \left(-\tfrac{1}{2}\eta^{ab}\partial_a\phi\,\partial_b\phi - \tfrac{1}{2}m^2\phi^2 \right) d^4x. \tag{11.41}$$

Here $\partial_a\phi = (\partial_t\phi, \nabla\phi)$ is the 4-derivative of the field $\phi(t, \mathbf{x})$.

If we generalise the action in Equation (11.41) to curved space-times we invoke the principle of equivalence and simply replace the Minkowski metric with the Riemannian metric g_{ab}, treat all derivatives ∂_a as covariant derivatives and change the element of volume to the invariant form $d^4x \to \sqrt{-g}d^4x$, where $g = \det g_{ab}$. We then have

$$S_{GR} = \int \left(-\tfrac{1}{2}g^{ab}\partial_a\phi\,\partial_b\phi - \tfrac{1}{2}m^2\phi^2 \right) \sqrt{-g}d^4x. \tag{11.42}$$

We shall return to this equation in Section 14.5 when demonstrating a remarkable alternative route to obtaining the energy momentum tensor for general relativistic Lagrangians like this one.

The Lagrangian for the simple real scalar field (11.40) is[15]

$$\mathcal{L} = -\tfrac{1}{2}\eta^{am}\partial_m\phi\,\partial_a\phi - \tfrac{1}{2}m^2\phi^2 = -\tfrac{1}{2}\partial^a\phi\,\partial_a\phi - \tfrac{1}{2}m^2\phi^2. \tag{11.44}$$

The Klein–Gordon equation for the field ϕ comes from the Euler–Lagrange equations using $\partial_a\phi$ as the 'kinetic' terms and ϕ as the 'potential':[16]

$$\partial_a\frac{\partial\mathcal{L}}{\partial(\partial_a\phi)} - \frac{\partial\mathcal{L}}{\partial\phi} = 0 \quad\Rightarrow\quad -\eta^{ab}\partial_a\partial_b\phi + m^2\phi = 0, \tag{11.45}$$

which is, with (11.34), simply:

Klein–Gordon equation for a massive scalar field:

$$(-\partial^2 + m^2)\phi = 0 \quad\text{or}\quad \ddot{\phi} - \nabla^2\phi + m^2\phi = 0. \tag{11.46}$$

The energy–momentum tensor, as defined in (11.38), is calculated as

$$T^{ab} = \frac{\partial\mathcal{L}}{\partial(\partial_a\phi)}\partial^b\phi - \eta^{ab}\mathcal{L}, \tag{11.47}$$

$$= \partial^a\phi\,\partial^b\phi + \tfrac{1}{2}\eta^{ab}\left[\partial^m\phi\,\partial_m\phi + m^2\phi^2\right]. \tag{11.48}$$

Note that this energy–momentum tensor can be written

$$T^{ab} = \partial^a\phi\,\partial^b\phi - \eta^{ab}\mathcal{L}. \tag{11.49}$$

---

[15] This is the form of the Lagrangian when the adopted signature of the metric η_{ab} is $(-+++)$.
In the case that the signature is $(+---)$ the Lagrangian would be

$$\mathcal{L} = \tfrac{1}{2}\eta^{am}\partial_m\phi\,\partial_a\phi - \tfrac{1}{2}m^2\phi^2, \quad \eta^{ab} = \mathrm{diag}(+1,-1,-1,-1). \tag{11.43}$$

Whichever the signature, it is simple to verify that the term $\pm\tfrac{1}{2}\partial^a\phi\,\partial_a\phi$, (with sign appropriate to the signature) produces the same term $\tfrac{1}{2}(\ddot{\phi} - \nabla^2\phi)$, and hence leads to the same Klein–Gordon equation. Physics is independent of the way the world is described.

[16] When differentiating a function $f(\phi, \partial_a\phi, \partial^b\phi, \dots)$ with respect to $\partial_b\phi$ it is a good idea to write every occurrence of $\partial^a\phi$ in the function as $\eta^{am}\partial_m\phi$ before differentiating.

Ex 11.2.1 Derive Equations (11.48) and (11.46).

Ex 11.2.2 Show that

$$\partial_b T^{ab} = (\partial^m \partial_m \phi + m^2 \phi)\, \partial^a \phi = 0. \tag{11.50}$$

So, if the field ϕ satisfies the Klein–Gordon equation, the energy–momentum tensor has zero divergence, and the energy–momentum is conserved.

11.2.2 The ϕ^4 Scalar Field Potentials

An important variant on the previous simple scalar field example is provided by a field $\phi(x)$ having the Lagrangian

$$\mathcal{L} = -\tfrac{1}{2}\partial_a \phi\, \partial^a \phi - \tfrac{1}{4}\lambda \phi^4. \tag{11.51}$$

Suppose that under a rescaling of the variable x the field $\phi(x)$ rescales as

$$\bar{x} = \alpha x, \qquad \bar{\phi}(\bar{x}) = \alpha^{-1}\phi(x), \tag{11.52}$$

then the Lagrangian (11.51) simply rescales as

$$\bar{\mathcal{L}} = -\alpha^{-4}\tfrac{1}{2}\partial_a \phi\, \partial^a \phi - \tfrac{1}{4}\alpha^{-4}\lambda \phi^4 = \alpha^{-4}\mathcal{L}. \tag{11.53}$$

It is clear from the Lagrange equations that multiplying the Lagrangian by a constant factor does not change the field equations. So the ϕ^4 scalar field has a special scale invariant symmetry that is not possessed by the Klein–Gordon field.

This scaling symmetry is broken if we add an explicit mass term, $\tfrac{1}{2}m^2\phi^2$, to the scalar field potential of the Lagrangian (11.51):

$$V(\phi) = \tfrac{1}{2}m^2\phi^2 + \tfrac{1}{4}\lambda\phi^4. \tag{11.54}$$

With this the Lagrangian is

$$\mathcal{L} = -\tfrac{1}{2}\partial_a \phi\, \partial^a \phi - \tfrac{1}{2}m^2\phi^2 - \tfrac{1}{4}\lambda\phi^4. \tag{11.55}$$

11.2.3 The Electromagnetic Field

The Lagrangian for the electromagnetic field is expressed as

$$\mathcal{L} = -\frac{1}{16\pi} F^{ab}\, F_{ab}, \tag{11.56}$$

where F_{ab} is the skew-symmetric electromagnetic field tensor.[17] We can introduce a vector potential field, A_a, such that

$$F_{ab} = A_{b,a} - A_{a,b}, \tag{11.57}$$

and the Lagrangian becomes

$$\mathcal{L} = -\frac{1}{8\pi}\left(\partial^a A^b \partial_a A_a - \partial^a A^b \partial_b A_a\right). \tag{11.58}$$

---

[17] By 1897 Larmor (2012) had already derived the Maxwell equations from a variational principle.

We can compute the energy momentum tensor for the electromagnetic field from Equation (11.38), modified in the obvious way for a vector field instead of a scalar field (see, for example, Landau and Lifshitz (1980, Section 33)).

Exercise: Candidate energy–momentum tensor for the electromagnetic field
(Comment: this exercise involves manipulating indices with η_{ab}).

Ex 11.2.3 Use the Equations (11.56) and (11.57) to derive the Lagrangian (11.58) for the electromagnetic field in terms of the vector potential, A_a.

Ex 11.2.4 Use the Lagrangian (11.58) to show that

$$\frac{\partial \mathcal{L}}{\partial A_{a,b}} = -F^{ab}.$$

Ex 11.2.5 Deduce from the definition of the energy–momentum tensor (11.38) that

$$\bar{T}^{ab} = -F^{am}\partial^b A_m - \tfrac{1}{4}\eta^{ab}F^{mn}F_{mn}. \tag{11.59}$$

Ex 11.2.6 Show that this tensor $\bar{T}^{ab}$ is not symmetric under interchange of a and b.

This equation for the energy–momentum tensor lacks the symmetry on interchange of indices. While not in principle a problem, it is a requirement of general relativity that the source of the gravitational field be a symmetric tensor. However, the technique we have used for generating the energy–momentum tensor from the Lagrangian often leads to non-symmetric forms of the energy–momentum tensor and so the practice is to symmetrise it by adding a term that symmetrises the tensor while at the same time preserving the fact that its divergence is zero. The technique for doing this was discovered by Belinfante (1939, 1940) and Rosenfeld (1940).

This can be achieved by adding the derivative of the rank 3 tensor $K^{abc} = A^a F^{bc}$ and noting that, in a vacuum,

$$\partial_c K^{abc} = \partial_c A^a F^{bc} - \partial_c(A^a F^{bc}) - A_a \partial_c F^{bc} = \partial_c(A^a F^{bc}), \tag{11.60}$$

(the vacuum Maxwell equation gives $\partial_c F^{bc} = 0$). Adding this term to the antisymmetric energy momentum tensor of Equation (11.59) gives:

Maxwell stress tensor:

$$T^{ab} = \frac{1}{4\pi}\left(F^{am}F_{bm} - \tfrac{1}{4}F^{mn}F_{mn}\eta^{ab}\right). \tag{11.61}$$

This is a symmetric, divergence free tensor which has the property that the sum of its diagonal elements $T^a{}_a = 0$. The added term arises from the fact that the electromagnetic field does not have spin zero.

11.2.4 Electromagnetism with a Source Term

In classical electrodynamics the source of the electric field is the charge, e, or charge density, ρ, and the source of the magnetic field is the electric current **j**. In relativistic electrodynamics the charge and current are seen as part of a Lorentz 4-vector,

$$J^a = (c\rho, \mathbf{j}). \tag{11.62}$$

The conservation of charge is expressed as

$$\nabla \cdot \mathbf{j} + \frac{d\rho}{dt} = 0, \quad \Leftrightarrow \quad J^a{}_{,a} = 0. \tag{11.63}$$

In order to include a source for the electromagnetic field, we simply add the Lagrangian for the free field, $\mathcal{L}_{em}$, and the Lagrangian, $\mathcal{L}_{int}$, for the interaction between the field and the electromagnetic 4-current J^a:

$$\mathcal{L} = \mathcal{L}_{em} + \mathcal{L}_{int}, \tag{11.64}$$

where

$$\mathcal{L}_{em} = -\frac{1}{16\pi} F_{ab} F^{ab}, \quad F_{ab} = A_{b,a} - A_{a,b}, \tag{11.65}$$

$$\mathcal{L}_{int} = A_a J^a. \tag{11.66}$$

It should be recognised that $\mathcal{L}_{em}$ depends only on the derivatives of A_a, while $\mathcal{L}_{int}$ depends only on A_a. So the only extra term this Lagrangian provides is the contribution from $\partial \mathcal{L}_{int}/\partial A^a = J^a$. The field equation including the source contribution is thus:

Maxwell equations with source term:

$$F^{ab}{}_{,b} = 4\pi J^a. \tag{11.67}$$

Notice that since F^{ab} is skew-symmetric, this is consistent with $J^a{}_{,a} = 0$, the conservation of current.

11.2.5 Massive Vector Field

Electromagnetism is a massless *vector field theory*: the field can be totally specified in terms of the vector potential A^a. A massive vector field would have to include a mass term in the field Lagrangian. If the mass of the field is m the additional term is

$$\mathcal{L}_m = \frac{1}{8\pi} m^2 A_a A^a, \tag{11.68}$$

and the total Lagrangian density is then

$$\mathcal{L} = -\frac{1}{16\pi}(A^{a,b} - A^{b,a})(A_{a,b} - A_{b,a}) + m^2 A_a A^a. \tag{11.69}$$

This, with a slightly different normalisation for the mass term, is referred to as the *Proca Lagrangian*. So the derivation of the field equations is very similar to the case of electromagnetism with a source term in that the added term does not involve derivatives of the vector potential. For a massive vector field and current J^a the field equations are easily shown to be

$$F^{ab}{}_{,b} + m^2 A^a = 4\pi J^a. \tag{11.70}$$

Of course, we can express $F^{ab}{}_{,b}$ in terms of the gradients of the vector potentials to get

$$A^{b,a}{}_b - A^{a,b}{}_b + m^2 A^a = 4\pi J^a.$$

Writing $A^{a,b}{}_b = \Box A^a$ and restricting attention to the case $A^a{}_{,a} = 0$ gives the simple equation

$$\Box A^a - m^2 A^a = -4\pi J^a. \tag{11.71}$$

The choice restriction $A^a{}_{,a} = 0$ is a gauge choice.

An interesting aspect of this is that while the first (electromagnetic-like) term of the Lagrangian is gauge invariant,[18] the total Lagrangian including the mass term is not gauge invariant. The implication of this is that a spin 1 particle that is described by gauge invariant fields is necessarily massless: this is the photon which is the quantum of the field A^a. There do exist spin 1 particles that are not massless: their masses are thought of as arising from the breaking of this gauge symmetry.

11.3 Energy–Momentum Tensor: Fluids

The preceding derivation of the energy–momentum tensor from the Lagrangian needs some physical insight as to what its components mean. We now look at a few simple examples: the free particle, a distribution of particles, and a perfect fluid with and without dissipation. The electromagnetic field is discussed in Appendix D.5. When deriving expressions for T^{ab} we work in the local rest frame of an observer moving with the particle, group of particles, or fluid. This simplifies things considerably because in a local rest frame, the 4-velocity is $u^a = \gamma(c, \mathbf{v}) = (c, \mathbf{0})$. The corresponding 4-momentum is then just $p^a = m_0 u^a = m_0(c, \mathbf{0})$, and, if we are talking about a particle at rest, its energy $E = p^0 c = m_0 c^2$, the rest mass energy. Our starting point is the definition of the energy–momentum tensor (11.38).[19]

11.3.1 The Equations of Classical Gas dynamics

It will be useful to recall the equations of classical gas dynamics. The state of the gas is described by its density and pressure, ρ and p, its temperature T, and its velocity $\mathbf{v}$:

---

[18] By this we mean that the Maxwell equations are unchanged under the transformations

$$A'_a = A_a + \psi_{,a}. \tag{11.72}$$

[19] The subject of deriving the energy–momentum tensor within a relativistic framework from fundamental physics is somewhat technical and some issues are not as yet settled. Most of these issues revolve around systems in which viscous and thermal dissipation take place, and, as we shall see, are associated with the concepts of entropy and temperature.

$$\frac{d\rho}{dt} = -\rho \nabla \cdot \mathbf{v}, \tag{11.73}$$

$$\rho \frac{dv_i}{dt} = -\frac{\partial p}{\partial x_i} + \frac{\partial}{\partial x_j}\left[\mu \left(\frac{\partial v_i}{\partial x_j} + \frac{\partial v_j}{\partial x_i} - \frac{2}{3}\delta_{ij}\nabla \cdot \mathbf{v} \right)\right] + \rho f_i, \tag{11.74}$$

$$\rho \frac{dE}{dt} = -p\nabla \cdot \mathbf{v} + \mu \left[\left(\frac{\partial v_i}{\partial x_j} + \frac{\partial v_j}{\partial x_i} \right)^2 - \frac{2}{3}(\nabla \cdot \mathbf{v})^2 \right] + \nabla \cdot (\kappa \nabla kT). \tag{11.75}$$

Here μ is the coefficient of shear viscosity, κ is the thermal conductivity, and k is the Boltzmann constant. We assume here that the coefficient of *bulk viscosity* is zero: this is the case for a normal monatomic gas, but may not be the case for molecular gases, nor for relativistic gases. In simple gases both the shear viscosity, μ, and the thermal conductivity, κ, depend on the product of the particle mean free path, l_c, and their thermal velocity $\bar{v}$. The rough approximation for κ is $\kappa \simeq n l_c \bar{v}$, where n is the particle number density. Similarly, the approximation for μ is $\mu \simeq \rho l_c \bar{v}$.[20] In addition to these equations we need constitutive equations describing the relationships between ρ, p, T.

The task in this section is to derive the energy–momentum tensor that leads to the relativistic equivalent of these equations.

11.3.2 Energy–Momentum Tensor: Free Particles

In the case of a free particle we use the Lagrangian $L = -\frac{1}{2}p^a p^b$ (11.10) and from (11.38) we have $T^{ab} = u^a p^b - L\eta^{ab}$. Since $p^a p_a = -m_0^2$, the term $L\eta^{ab}$ is a constant on the particle trajectory and can be ignored. We can rewrite $p^a = m_0 u^a$ to get[21]

$$T^{ab}(\mathbf{x}) = m_0 u^a u^b\, \delta^{(3)}\big(\mathbf{x} - \mathbf{x}^{(P)}(t) \big). \tag{11.76}$$

The $\delta^{(3)}$-function puts the particle at the 3-space point $\mathbf{x}^{(P)}$ and says that the energy–momentum is zero everywhere except at the point where the particle is located. The energy–momentum tensor of a collection of stationary particles is simply the unedifying sum of such terms taken over the positions $\mathbf{x}^{(K)}$ of the particles.

It is instructive to consider such a collection of stationary particles in the *fluid approximation*. This approximation involves taking an average over all particles in the local rest frame of a small fluid element having 4-volume ΔV. What we mean by 'small' here is 'bigger than the distance between particles and smaller than the scale of the features of the gas or fluid'. Working in the local rest frame we can write $\Delta V = \Delta^{(3)}x^a \Delta x^0$, where $\Delta^{(3)}x^a = \Delta^{(3)}$ is the spatial averaging volume and Δx^0 is a short period of time during which we count the particle trajectories that pass through this volume.

---

[20]   There is a more detailed discussion of this in the On-line Supplement on *"Gas Dynamics"*.

[21]   When discussing the energy–momentum tensor it is usual to work in terms of the momenta rather than the velocities. This is because, when dealing with fields, the 'conjugate momentum' arises naturally from the Lagrangian and is an essential part of the Hamiltonian view of the field dynamics.

Averaging (11.76) over a collection of N particles in the rest frame of the volume $\Delta^{(3)}V$ and doing the integral:

$$\langle T^{ab}\rangle_{\Delta V} = \frac{1}{\Delta V}\int_{\Delta V} T^{ab}d^4x = \frac{1}{\Delta^{(3)}V}\sum_{K=1}^{N} m_0 u_K^a u_K^b. \tag{11.77}$$

The integral is taken over those particles whose trajectory takes them through the elemental 4-volume $\Delta V = \Delta^{(3)}x^a \Delta x^0$ in the time Δx^0. The assumption that all the particles in $\Delta^{(3)}$ are stationary makes that trivial and allows further simplification. This is what we mean when we describe a medium as *dust*.

In the case of dust we have $u^a = (c, \mathbf{0})$. The density in $\Delta^{(3)}$ is its mass divided by its volume: $\rho = \Sigma m_0/\Delta^{(3)}$. So

$$\langle T^{00}\rangle_{\Delta^{(3)}V} = \frac{1}{\Delta^{(3)}V}\sum_{K} m_0 c^2 = \rho c^2, \quad \langle T^{0\mu}\rangle_{\Delta^{(3)}V} = 0, \quad \langle T^{\mu\nu}\rangle_{\Delta^{(3)}V} = 0. \tag{11.78}$$

Hence the fluid-scale energy–momentum tensor for the gas of these particles, i.e. dust, has only one non-zero element, $T^{00} = \rho c^2$. We can write this as

$$T^{ab}_{\text{dust}} = \rho U^a U^b, \tag{11.79}$$

where U^a denotes the macroscopic 4-velocity of an element of the dust gas.[22]

11.3.3 Energy–Momentum Tensor: Perfect Fluid

We now address the question of particles in the volume element ΔV that are not at rest in the instantaneous rest frame of ΔV. The simplest such state is the *perfect fluid*. A perfect fluid is a fluid that flows with no viscous, i.e. dissipative, forces, nor heat transfer.[23] Relative to a frame of reference in which the fluid is instantaneously at rest, the only forces acting on an elemental volume of fluid are the pressure forces that act perpendicular to the surface of the volume element.

Firstly we note that in Equation (11.77) the spatial elements $T^{0\mu}$ are given by

$$\langle T^{0\mu}\rangle_{\Delta V} = \frac{1}{\Delta V}\int_{\Delta V} T^{0\mu}d^4x = \frac{1}{\Delta^{(3)}V}\sum_{K=1}^{N} m_0 u_K^0 u_K^\mu = \frac{1}{\Delta V^{(3)}}\sum_{K=1}^{N} cp_{(K)}^\mu = 0, \quad \mu = 1,2,3. \tag{11.81}$$

The sum of the momenta in the elemental volume on the far right of this equation must be zero since, by construction, the volume is at rest in this coordinate system.

Let us now turn to the 3-space part $T^{\mu\nu}$ of T^{ab}. By construction the local 3-space components can have no preferred direction. This means that $T^{\mu\nu} = P\eta^{\mu\nu}$, where P is the

---

[22] If the particles have random motions for which the phase space density of the particle momenta is $f(p^a)$, we can define the energy–momentum tensor as (Bernstein, 1988, Ch. 3):

$$T^{ab} = \int \frac{p^a p^b}{p_0} f(p^a)\frac{d^3p}{(2\pi)^3}. \tag{11.80}$$

The 4-momentum in the integral is constrained by the relativistic condition $c^2 p^a p_a = -E^2 + \mathbf{p}^2 c^2 = -m^2 c^4$ (see Eq. 10.90). This defines a surface in momentum space referred to as the *mass shell*.

[23] Unlike in classical systems, in relativity heat transfer carries momentum: there is a transfer of energy.

isotropic pressure acting on the surface of the volume element $\Delta^{(3)} V$. We can see this from the 3-space components of (11.77):

$$\langle T^{\mu\nu} \rangle_{\Delta V} = \frac{1}{\Delta V} \int_{\Delta V} T^{\mu\nu} d^4 x = \frac{1}{\Delta^{(3)} V} \sum_{K=1}^{N} m_0 u_K^{\mu} u_K^{\nu}, \quad \mu, \nu = 1, 2, 3. \tag{11.82}$$

We can now gather up the various components we have calculated

$$T^{ab} = \begin{pmatrix} \rho & 0 & 0 & 0 \\ 0 & P/c^2 & 0 & 0 \\ 0 & 0 & P/c^2 & 0 \\ 0 & 0 & 0 & P/c^2 \end{pmatrix} \tag{11.83}$$

$$= \begin{pmatrix} \rho + P/c^2 & 0 & 0 & 0 \\ 0 & 0 & 0 & 0 \\ 0 & 0 & 0 & 0 \\ 0 & 0 & 0 & 0 \end{pmatrix} + \begin{pmatrix} -P/c^2 & 0 & 0 & 0 \\ 0 & P/c^2 & 0 & 0 \\ 0 & 0 & P/c^2 & 0 \\ 0 & 0 & 0 & P/c^2 \end{pmatrix}. \tag{11.84}$$

Since in this comoving frame the 4-velocity is simply $u^a = (1, \mathbf{0})$, we can write this as:

Energy–momentum tensor for perfect gas

$$T^{ab} = \left(\rho_0 + P_0/c^2 \right) U^a U^b + \eta^{ab} P_0/c^2, \tag{11.85}$$

where we have put 0-subscripts on the pressure and density to denote that these are the values in the rest frame.

The gradient of the energy–momentum tensor (11.85) is zero: $\partial_b T^{ab} = 0$. The spatial part of this gradient $\partial_\mu T^{a\mu}$ is

$$\partial_\mu \left[\left(\rho_0 + p_0/c^2 \right) u^a u^\mu + \eta^{a\mu} p_0/c^2 \right] = u^a \partial_\mu \left[\left(\rho_0 + p_0/c^2 \right) u^\mu \right]$$
$$+ \left(\rho_0 + p_0/c^2 \right) u^\mu \partial_\mu u^a + \eta^{a\mu} \partial_\mu p_0 = 0, \quad \mu = 1, 2, 3. \tag{11.86}$$

If we evaluate this in the local fluid rest frame where $u^a = (c, \mathbf{0})$ the spatial part of the term $u^a \partial_\mu [\dots]$ is zero and so we are left with

$$\left(\rho_0 + p_0/c^2 \right) u^\mu \partial_\mu u^a + \eta^{a\mu} \partial_\mu p_0 = 0, \quad \text{where} \quad u^\mu \partial_\mu u^a = \frac{dx^\mu}{d\tau} \frac{\partial u^a}{\partial x^\mu} = \frac{du^a}{d\tau}. \tag{11.87}$$

The spatial components of this equation, i.e. $a = 1, 2, 3$, give

$$\left(\rho_0 + \frac{p_0}{c^2} \right) \frac{du^\nu}{d\tau} + \eta^{\nu\mu} \frac{\partial p_0}{\partial x^\mu} = 0, \quad \nu = 1, 2, 3. \tag{11.88}$$

We recognise this as the relativistic version of the standard force equation for a fluid element moving only under the influence of pressure forces. Since this equation is valid in the local rest frame it is valid in all frames. After some work, we obtain the more familiar special relativistic form of these equations:

Fluid flow equation – special relativity:

$$\gamma^2 \left(\rho + \frac{p}{c^2}\right) \frac{d\mathbf{v}}{dt} = -\nabla p - \frac{\mathbf{v}}{c^2} \frac{\partial p}{\partial t}, \tag{11.89}$$

$$\gamma \frac{d\rho}{dt} + \left(\rho + \frac{p}{c^2}\right) \left[\frac{\partial \gamma}{\partial t} + \nabla \cdot (\gamma \mathbf{v})\right] = 0. \tag{11.90}$$

The second of these comes from the $\partial_\mu[\left(\rho_0 + p_0/c^2\right) u^\mu = 0$ term in Equation (11.86).

11.3.4 Dissipative Processes

The perfect fluid energy–momentum tensor (11.85) was 'derived' on the basis of the isotropy of the pressure as expressed in Equation (11.81). By virtue of those assumptions there were no off-diagonal terms in T^{ab}. These off-diagonal terms would appear when we have atomic, molecular or radiative transport of energy and momentum in and out of our comoving volume element $\Delta^{(3)}$. This local transport of energy and momentum results in dissipation of energy from the elemental volumes that make up the medium.

Viscous forces, for example, arise when momentum is transferred between fluid elements having different velocities: they 'share' their momentum as a result of which the velocity gradients are decreased. The balance of energy flow in a system is usually the province of thermodynamics. However, calculating these off-diagonal terms in a dissipative flow must be a task in kinetic theory which takes account of the nature of the collisions going on in the fluid or gas on atomic scales. The outcome of the kinetic theory calculation would be a set of *transport coefficients* that may depend on local macroscopic conditions such as density and temperature.

Dissipation through Shear Viscosity

The general structure of T^{ab} is as depicted in Figure 11.1. The simplest model for the off-diagonal terms, as was argued in the 19th century, is to say that they are dependent on first order derivatives of the fluid velocity and temperature fields. Technically, the fluid stress tensor arises from a simple model in which the stresses on an element of the fluid comprise an isotropic component $P\delta_\nu^\mu$ and a component that is due to the local shear in the fluid velocity. In the 3-space of our classical mechanics experience we write this as[24]

$$T_{\alpha\beta} = P\delta_{\alpha\beta} + \mu\pi_{\alpha\beta}, \qquad \pi_{\alpha\beta} = \left(\frac{\partial v_\beta}{\partial x_\alpha} + \frac{\partial v_\beta}{\partial x_\alpha} - \frac{2}{3}\delta_{\alpha\beta}\frac{\partial v_\gamma}{\partial x_\gamma}\right), \quad \alpha, \beta = 1, 2, 3. \tag{11.91}$$

The term involving $\frac{2}{3}\delta_{\alpha\beta}$ makes $\pi_{\alpha\beta}$ trace-free.

Dissipation through Heat Transfer

The notions of temperature and of heat diffusion in relativistic theories have a long and controversial history. Planck (1907) thought that the temperature would transform as $T = \gamma T_0$,

---

[24] This is written in the style of classical fluid mechanics, where the indices on vectors are all lowered. The summation convention still applies.

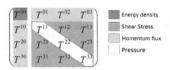

The structure of the energy–momentum tensor. The energy–momentum tensor is symmetric: $T^{ab} = T^{ba}$. Its components can be divided into three groups: the energy density, T^{00}, the momentum flux $T^{0\mu}$, $\mu = 1, 2, 3$ and the stresses $T^{\mu\nu}$, $\mu, \nu = 1, 2, 3$. The diagonal components of the stress component reflect pressure forces, while, in a fluid, the off-diagonal components arise from transport processes in the fluid (e.g. viscosity and heat conduction). ©

while others (Hasenöhrl, 1908, who did not accept the principle of relativity) suggested $T = \gamma^{-1}T_0$ (Møller, 1967, recounts the story).

There were, and still are, issues surrounding the heat diffusion equation in relativity: the diffusion equation is parabolic and so is associated with an infinite propagation velocity.[25] There are still papers being published offering fixes for that. The lesson from Israel and Stewart (1979) was that the only way to address the issue of relativistic thermodynamics was the framework of kinetic theory. The lectures of Maartens (1996) cover most aspects of this tricky subject.

Classically, the heat flux, $\mathbf{q}$, is related to the temperature gradient through Fourier's Law: $\mathbf{q} = -\kappa \nabla T$, where κ is the *thermal conductivity*.[26] In non-relativistic flows, heat flow makes no contribution to the momentum equation, but appears in the energy equation through $\nabla \cdot \mathbf{q}$. We see this term appearing on the right-hand side of (11.75). Using the classical 2nd law of thermodynamics in the form (7.22), we have, with $V = 1/\rho$ for an elemental volume having a fixed number of particles:

$$\rho T \frac{ds}{dt} = \rho \frac{dE + p \, d(1/\rho)}{dt} = \rho \frac{dE}{dt} + p \nabla \cdot \mathbf{v} = \nabla \cdot (\kappa \nabla T) = \nabla \cdot \mathbf{q}, \tag{11.92}$$

where the change from $d(1/\rho) \rightarrow \nabla \cdot \mathbf{v}$ comes using (11.73). The right-hand side is Fourier's expression for the rate of heat loss by thermal conduction, i.e. *Fourier's Law*. The term on the far left can be written as[27]

$$\rho \frac{ds}{dt} + \nabla \cdot \left(-\frac{\kappa \nabla T}{T} \right) = \kappa \frac{(\nabla T)^2}{T^2}. \tag{11.94}$$

[25] A non-relativistic example: if we heat up one end of an infinitely long bar, the temperature starts to increase at all distances along the bar. There is no notion of propagation of the heat signal, and so evidently something is missing from the heat equation, even in the non-relativistic world. See Joseph and Preziosi (1989) for more on this problem in other branches of physics.

[26] In this section we shall absorb Boltzmann's constant, k in the coefficient κ. That saves writing kT where there is a T. The abbreviation $\theta = kT$ is often used in the non-relativistic literature.

[27] This needs the identity

$$\nabla \cdot \left(\frac{\kappa \nabla T}{T} \right) = \frac{1}{T} \nabla \cdot (\kappa \nabla T) - \kappa \frac{(\nabla T)^2}{T^2}. \tag{11.93}$$

Note that the right-hand side of (11.94), which is always positive, shows that *entropy is generated* by the dissipative process of heat conduction. A similar argument can be given for the viscous dissipation. Dissipative processes increase the entropy.[28]

Equation (11.94) suggests that we think of this as expressing the entropy balance within an elemental 3-volume that is subject to heat conduction. In this way $-\kappa \nabla T / T$ is to be interpreted as the *entropy flux*, $\mathbf{J}$:

$$\mathbf{J} = -\kappa \frac{\nabla T}{T} = \frac{\mathbf{q}}{T}. \qquad (11.96)$$

A simple generalisation of the concept of entropy to the relativistic situation now suggests itself:

$$S^a = su^a + J^a, \qquad (11.97)$$

where s here is the *relativistic entropy density* and $J^a = (0, \mathbf{J})$ is the space-like *relativistic energy flux*. The right-hand side is the space-time split of the entropy 4-vector S^a.

Recall, however, that the definition of $\mathbf{q}$ comes from the Fourier law, which is not properly expressed in a relativistic way. Eckart (1940, Eq.37) suggested a relativistic generalisation of the heat flux $\mathbf{q}$:[29]

$$q^a = -\kappa h^{ab}(T_{;b} + T\dot{u}_b). \qquad (11.98)$$

The relativistic heat flux is not parallel to the temperature gradient. Eckart *loc. cit.* says that this is the simplest expression for the heat flux that is consistent with the 2nd law of thermodynamics. He also emphasises that there is no other motivation for the strange acceleration term $T\dot{u}_b$ in this definition of q^a: according to this heat flow occurs in an accelerated volume even in the absence of a temperature gradient. There appears to be no physical understanding of this.[30]

This is important because the Eckart (1940) energy–momentum tensor for a fluid with viscosity and heat transfer is:

Eckart (1940) energy–momentum tensor for a dissipative fluid:

$$T^{ab} = (\rho + p)u^a u^b + p\eta^{ab} + (q^a u^b + q^b u^a) + \pi^{ab}. \qquad (11.99)$$

[28] The usual route at this point (see e.g. Landau and Lifshitz (1987), their Equations (2.6) and (2.7)), is to write

$$\rho \frac{ds}{dT} = \rho \left[\frac{\partial s}{\partial t} + \mathbf{v} \cdot \nabla s \right] = \frac{\partial \rho s}{\partial t} + \nabla \cdot (\rho s \mathbf{v}), \qquad (11.95)$$

which suggests that $\rho s u^a = \gamma(\rho s, \rho s \mathbf{v}) = \gamma \rho s(1, \mathbf{v})$ would be a suitable relativistic entropy 4-vector. This accords with the original suggestion of Tolman (1934b, Section 71 and Ch. IX) that we should use $S^a = s_0 u^a$, with s_0 being the rest-frame entropy density. However, that does not lead to our Equation (11.94).

[29] Here $h^{ab} = g^{ab} + u^a u^b$ is the projection tensor that picks out the 3-space component of a 4-vector, e.g.: $h^{ab}u_b = (0, \gamma \mathbf{v})$. In this way q^a is a spatial vector: $u_a q^a = q^0 = 0$. We define $\dot{u}_b = u^a \partial_a u_b$, the projection of the 4-gradient of u^a on to the 0-axis. $\dot{u}_b$ is the 4-acceleration vector.

[30] Equally problematic is the fact that the diffusion equation is associated with an infinite propagation speed, clearly not a relativistic attribute. There have been numerous ad hoc attempts to fix this by making the heat equation hyperbolic. This is simply achieved by replacing the Laplacian by a d'Alembertian with a subluminal arbitrary propagation speed.

The heat flux, q^a, appears here and would in non-relativistic theories be directly related to the temperature through the Fourier Law.

11.4 The vacuum – Lorentz Invariance

Consider the Friedmann equation for a homogeneous and isotropic Universe with expansion factor $R(t)$,

$$R^3 \frac{d\rho}{dt} + \left(\frac{p}{c^2} + \rho\right) \frac{dR^3}{dt} = 0. \tag{11.100}$$

If ρ represents the energy density of the vacuum, we can write $\rho(t) = \rho_V = constant$: a vacuum is just that, a vacuum, and its energy density must be independent of time. Hence the only way this equation can be satisfied for a vacuum cosmology is to have

$$\rho_V = -\frac{p_V}{c^2}, \tag{11.101}$$

where p_V is the pressure exerted by the vacuum.

There is another, perhaps more profound way of looking at this, which is to say that the vacuum must be Lorentz invariant: a vacuum must look the same when viewed from two inertial frames of reference, one of which is moving at uniform velocity v along the x-axis relative to the other. The expression of that can be put in terms of the energy–momentum density for a homogeneous and isotropic gas having density ρ and pressure p:

$$T^{ab} = \begin{pmatrix} \rho & 0 & 0 & 0 \\ 0 & p/c^2 & 0 & 0 \\ 0 & 0 & p/c^2 & 0 \\ 0 & 0 & 0 & p/c^2 \end{pmatrix}. \tag{11.102}$$

If we write our Lorentz transformation between the two frames as

$$L^a{}_b = \begin{pmatrix} \gamma & -\gamma\beta & 0 & 0 \\ -\gamma\beta & \gamma & 0 & 0 \\ 0 & 0 & 1 & 0 \\ 0 & 0 & 0 & 1 \end{pmatrix}, \quad \beta = v/c, \quad \gamma = (1-\beta^2)^{-1/2}, \tag{11.103}$$

then

$$T'^{pq} = L^p{}_a L^q{}_b T^{ab}, \text{ summation over } a \text{ and } b \text{ implied} \tag{11.104}$$

$$= \begin{pmatrix} \gamma^2\left(\rho + \beta^2\frac{p}{c^2}\right) & -\gamma^2\beta\left(\rho + \frac{p}{c^2}\right) & 0 & 0 \\ -\gamma^2\beta\left(\rho + \frac{p}{c^2}\right) & \gamma^2\left(\rho\beta^2 + \frac{p}{c^2}\right) & 0 & 0 \\ 0 & 0 & p & 0 \\ 0 & 0 & 0 & p \end{pmatrix}. \tag{11.105}$$

In order that this energy momentum tensor should look the same as (11.102) in all inertial frames we require that the off-diagonal terms in (11.105) should vanish. This can only happen if

$$p = -\rho c^2. \tag{11.106}$$

If we substitute this into the expressions for T'^{00} and T'^{11} we find that $T'^{00} = \rho c^2$ and $T'^{11} = p$. In other words

$$T'^{ab} = T^{ab} \quad \text{if } p = -\rho\, c^2. \tag{11.107}$$

This is the same as the condition (11.101) that the vacuum have a constant density during the expansion. This also shows that, other than having an empty space $\rho = 0, p = 0$, the only other way of having a Lorentz invariant vacuum is to have $\rho\, c^2 = -p$.

Lemaître was aware that the cosmological constant of Einstein could be viewed as a medium having this somewhat unusual equation of state. He favoured a positive cosmological constant, $\Lambda > 0$, because he thought it should be associated with a positive energy density.

12 General Relativity

The notion that gravity is a manifestation of the curvature of space-time is a remarkable idea. Going beyond the idea and putting it into a useful mathematical form was perhaps even more extraordinary. The technical difficulties are a huge barrier to understanding and using the theory.

We discuss the discovery and the principles that Einstein had to lay down in order to be able to realise this idea without disrupting all of known physics. It would be essential to keep the idea of local inertial frames. These principles must be put in a geometric framework that builds on local inertial frames and tries to connect local inertial frames that are accelerating relative to one another. Grossmann's suggestion had been that Riemannian space-time would provide such a structure when equipped with a metric and a way of connecting these different local inertial frames.

12.1 The Early Years

Today we see the general theory of relativity as a complete package with rules for calculating gravitational fields and their effects. That is, however, not how it came about. Like the development of electrodynamics during the previous century, the theory of gravitation came in steps. However, the theory of gravitation emerged over a relatively short period of time, reaching a crescendo in 1915 when both Einstein and Hilbert presented, in the same month, different derivations of the gravitational theory that survive to this day.

Einstein's own development of his theory of gravitation was almost a one-person undertaking lasting more than a decade. It followed on from his publication of the *Special Theory of Relativity* in 1905, starting with work on the influence of gravitation on the propagation of light. However, Einstein was not the only one in search of a theory of gravitation. Planck had already asserted that all forms of energy would be influenced by gravitational fields, and by 1912 Mie had produced an impressive theory of gravitation (Smeenk and Martin, 2005), based purely on electrodynamics, but remaining firmly within the framework of special relativity.[1] Mie's work had inspired Hilbert to join in the quest to unify gravitation and electrodynamics (Corry, 1999; Sauer, 2007; Renn and Schemmel, 2012).

---

[1] Mie had been critical of Einstein's move away from Minkowski space and towards the complications of the metric tensor. Mie was not alone in this view.

12.1.1 The *Entwurf* Theory of Gravitation

When Einstein's effort is described as 'almost a one person undertaking' the 'almost' refers to his life-long friend Marcel Grossmann, who played an instrumental mathematical role in the development of general relativity, and who was Einstein's co-author on the important *Entwurf* paper (Einstein and Grossmann, 1913).

> **Short biography: Einstein and Grossmann** Einstein and Grossmann met at the Zurich Polytechnikum where they were studying to acquire qualifications to become high school (gymnasium) teachers. They graduated in 1900 and went their separate ways until coming together once again at the Eidgenössische Technische Hochschule (ETH) in Zurich in 1912 when Einstein was appointed to the chair of theoretical physics. Grossmann had been professor of descriptive geometry there since 1907. In 1913 they wrote their first paper together: known simply as the *Entwurf*, or *Outline*. This was an important first attempt at a covariant theory of gravitation, though it was later to be discarded in favour of Einstein's version of 1915.

What Grossmann brought to the collaboration was a knowledge of the mathematical work of Riemann, Christoffel and Levi-Civita.[2]

The *Entwurf* paper of 1913 (Einstein and Grossmann (1913): 'Entwurf einer verallgemeinerten Relativitätstheorie und einer Theorie der Gravitation') was in fact two rather distinct papers put together as a 'Physical Part', written by Einstein, and a 'Mathematical Part', written by Grossmann (see Figure 12.1). In the mathematical part, Grossmann introduces the Riemann and Ricci tensors that are fundamental to the general theory of relativity and shows how to calculate these using the Christoffel symbols. There followed over the period 1913–1914 a series of six joint papers.

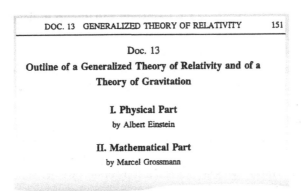

Fig. 12.1 Top of first page of the *Entwurf* paper (Einstein and Grossmann, 1913, English translation). The paper was in fact two papers combined into one. This was perhaps the first try at producing a covariant theory of gravitation based on the structure of a Riemannian space time. *(From the cited paper.)*

---

[2] Riemann's work was published posthumously by Dedekind in 1867, and was taken up by Christoffel who in 1869 published his famous work on the geometric connection that bears his name. Somewhat later, Levi-Civita used the Christoffel connection to define parallel transport on a Riemannian manifold and in 1900 published a book on tensor calculus that Einstein used in his development of the theory of gravitation.

Shortly after Einstein died in 1955, a small notebook was found covering the period 1912–1914. It contained calculations on a variety of subjects, but on the 14th page there appears for the first time the equation for the line element in tensor form and later on the derivation of the equations that were to appear in the *Entwurf*. The book is now referred to as *The Zurich Notebook*.[3]

With his student Michele Besso, Einstein showed that the prediction for the advance of the perihelion motion of the planet Mercury from *Entwurf* theory was only 18″ per century. The Einstein–Besso work on the advance of the perihelion of Mercury was never published, but exists in manuscript form, the original of which was found among the possessions of the Besso family. See Janssen (2007) for a detailed discussion of this. It was Johannes Droste (1914), a student of Lorentz, who independently derived and published that result: one might say that Einstein had been scooped.

As it turned out, the *Entwurf* version of gravitation theory was doomed to failure. But it was not the failure to correctly account for the advance of the perihelion of Mercury that caused Einstein to look anew into revising the theory of the gravitational force: it was more significant that he had failed to fully embrace the notion of covariance.

In a letter to Sommerfeld in July 1915, Einstein wrote his view of Grossmann's contribution to this paper, saying: *Grossman will never claim to be considered a co-discoverer. He only helped in guiding me through the mathematical literature, but contributed nothing of substance to the results.*[4] Although this may sound a little harsh, it should be noted that a few months after this letter was written, Einstein was to dismiss the *Entwurf* paper as incorrect, replacing it with his new version of the theory of gravitation, which is today the currently accepted version. Einstein was giving Grossmann credit for the mathematical basis of the theory, and taking the responsibility for the failure of *Entwurf* upon himself.

12.1.2 Einstein and Hilbert in November 1915

Einstein's move away from the *Entwurf* version of gravity culminated in its retraction in a paper presented to the Berlin Academy on 4 November 1915 in which he presented the first version of his new theory. This was based more closely on the mathematical, geometric, precepts set down by Grossmann. A small correction was added on 11 November, and then on 18 November he submitted the paper getting the correct value for the advance of the perihelion of Mercury. But a week later still, he changed his field equations yet again, and those were the ones we work with today. This final change is significant since it occurred after Hilbert had sent Einstein a copy of his own derivation of the field equations.

[3] Commentary on the book is available at
http://www.pitt.edu/~jdnorton/Goodies/Zurich_Notebook/
The facsimile of the book can be viewed at
http://alberteinstein.info/vufind1/Record/EAR000034421
http://alberteinstein.info/vufind1/Digital/EAR000034421#page/1/mode/2up
A complete transcription and facsimile is presented in Janssen *et al.* (2007b) with detailed commentary by Janssen *et al.* (2007a).

[4] As translated and told by Weinstein (2012a, Section 5.2). Galina Weinstein has written extensively and authoritatively on Einstein's life and works in a series of article that are to be found on the open *arXiv* repository (www.arXiv.com). Her discussions of the *Entwurf* paper are found at Weinstein (2012b,c).

Hilbert (1915) came up with general relativity at almost exactly the same time as Einstein. Hilbert had given a seminar on his gravitational theory in Göttingen on 20 November 1915, having previously sent Einstein a copy of his paper which Einstein had received on the 18th, just a few days before submitting his own paper. There were allegations that Einstein, on seeing Hilbert's draft had corrected some errors, one serious (Corry, 1999), in his own manuscript. Whatever the truth of that, there is no doubt that Einstein had complained to others that his work had been plagiarised by his friend David Hilbert.[5] Notwithstanding all the rancour over this incident, the equations describing the gravitational field are still called 'the Einstein field equations', and we still derive or modify them on the basis of the 'Hilbert action principle'.

12.1.3 Subsequent Development

During the period 1915–1955 the now-classical solutions of the Einstein field equations were derived, disputed and discussed. The important Schwarzschild and de Sitter solutions had come within a couple of years of that first publication, and within 10 years the bending of light by the Sun had been established (though not without some contentious aspects). Further cosmological solutions came from Friedmann and Lemaître with the important later rationalisation of Robertson and Walker. The possibility of testing the geometry of our Universe arose with the advent of large telescopes and the ability to probe the Universe to greater depths. But generating and understanding solutions to the Einstein equations was not easy and progress was relatively slow, especially when compared with progress in that other great pillar of 20th century science: quantum mechanics. It would need something new to put general relativity into a higher gear. That 'something new' came, in large part, with the work of a small but powerful group in Hamburg in the 1950s: they produced a collection of papers that is now referred to as the *Hamburg Bible*. The main authors were Ehlers, Jordan, Kundt, Sachs and Trümper.

12.2 Einstein's Theory of Gravitation

What might have motivated Einstein to think that this geometrisation of gravity might solve the problem of the perihelion of Mercury? The complexity that arises from building a theory of gravity around Riemannian geometry is both formidable and forbidding, involving a serious learning curve.[6] Einstein's original presentation of the special theory of relativity was quite unlike the 4-dimensional view presented later by Minkowski (1908,

---

[5] This gave rise to the controversy of 'who got the general theory of relativity first?'. In fact Hilbert conceded precedence to Einstein by adding a reference to Einstein's work in his paper and adding the comment: *The differential equations of gravitation that result are, as it seems to me, in agreement with the magnificent theory of general relativity established by Einstein in his later papers.* See the discussions of Corry *et al.* (1997), Winterberg (2003) and Todorov (2005) for more on this.

[6] We see Einstein's own effort and frustration in his notes on p.33 of the *Zurich Notebook*, where after several pages of algebra manipulating tensors, with much crossing out and correction, Einstein comments that the work had become 'too involved' ("zu umstaendlich"). There is a clear sense of frustration.

1909), but he was quick to adopt the 4-dimensional description. He realised that the transformation between inertial frames that were not in uniform motion relative to one another would not be the Lorentz transformation, but had to be a transformation that preserved the Minkowski line element. This was essential to his adopting the principle of equivalence as a fundamental basis for his theory of gravitation.[7] The answer to his problem came from Marcel Grossmann in 1912 while they were in Zurich together: the theory should be based on Riemannian geometry with Christoffel's connection providing the link between coordinate frames at different points. This was an extraordinary suggestion, which we see Einstein starting to work with on page 27 of his *Zurich Notebook*, where he explicitly acknowledges, in writing, that he is working with 'Grossmann's tensor'.

This is the same as asking why special relativity, while giving a better answer than Newtonian dynamics, failed to provide the correct answer. Einstein probably saw this as a consequence of the fact that special relativity dealt only with inertial frames in relative uniform (i.e. not accelerating) motion. We associate gravitational forces with accelerations, and so it would be necessary to handle inertial frames in relative accelerated motion. Nevertheless, we should not abandon the fact that, despite their relative acceleration, these frames are nonetheless still local inertial frames.

This reflects our experience that when in free fall in a gravitational field we are in a local inertial frame. This assertion that a freely falling frame is an inertial frame is one manifestation of the so-called principle of equivalence. Inertial frames in relative accelerated motion can no longer be related via a Lorentz transformation, so if we are to consider acceleration due to gravity we must allow more general transformations.

12.2.1 The Underlying Principles

Einstein based his theory of general relativity on two fundamental premises, which in their simplest possible form might be stated as:

---

[7] In 1908, two young mathematicians at St. John's College, Cambridge, Ebenezer Cunningham and Herbert Bateman, independently of Minkowski, recognised that the relativistic transformations between inertial frames could be seen as a rotation in a 4-dimensional space-time. This was possibly the first time that had been noticed. They knew that the Lorentz transformation left the Maxwell equations invariant. They then asked, and answered, the question as to whether there was a larger group of transformations that would leave the Maxwell equations invariant and showed that this was the group of conformal transformations (Bateman (1908, 1910) and (Cunningham, 1908, 1914)). As a part of their proof they showed that the line element

$$ds^2 = -c^2dt^2 + dx^2 + dy^2 + dx^2 \tag{12.1}$$

was invariant under these quite general transformation. In achieving this they had, in effect, generalised the Lorentz transformation to non-inertial frames, answering a question which Einstein was to ask later in 1913 (*Zurich Notebook*).

Cunningham and Bateman were Senior Wranglers in consecutive years in mathematics at Cambridge and both winners of the prestigious Smith Prize. These two mathematicians would not at that time have seen Minkowski's paper, and it is unlikely that Einstein or Grossmann had seen their work. Such was the speed of scientific communication at the start of the 20th century. The book by Cunningham (1914) was possibly the first book to be published on relatvitiy in the English language. Their story, and with it the intriguing story of the reaction of the UK astrophysics community to relativity, is well told in the book by Warwick (2003).

- **The principle of equivalence**
  All bodies in a gravitational field are influenced by the field in the same way.
- **Mach's principle**
  A space empty of matter should not have any gravitational fields.

Added to this were two other requirements:

- **The geometry of space-time is Riemannian**
  To which Einstein added the notion that free particles should move along geodesics in this space and, in particular, light should move along null geodesics.
- **The principle of covariance**
  The laws of physics should be expressed by equations which hold good for all systems of coordinates.

The equivalence principle and Mach's principle were certainly important to Einstein, Today the former is almost taken for granted in one form or another. However, Mach's principle has never been a manifest part of general relativity. Even obtaining a precise formulation as to what Mach's principle really is is problematic.[8] We see from Einstein's *Zurich Notebook* that the issue of bringing in the mathematical complexity of Riemannian spaces at first brought him technical frustration. But it was in the end the correct decision. The principle of covariance posed a major problem for Einstein and at various times he either threw it out or accepted it. We shall say a few things about these issues below.

12.2.2 The Principle of Equivalence

The principle of equivalence has many different formulations, and several versions such as a 'strong' version and a 'weak' version. Some of the standard texts on general relativity cover the principle of equivalence in considerable detail: it is an important principle that defines and constrains the scope of theories of gravity. Having said that, it should be remembered that the realisation that there is such a principle goes back to Galileo when, as the story goes, he dropped rocks off the Leaning Tower of Pisa and showed that gravity affects all bodies in the same way, independently of their mass. This was an extremely important realisation: even if Galileo had not in fact performed the experiment, the discussion paved the way for Newton.

Until Planck argued in 1891 that all forms of energy have inertia, light included, and so should be influenced by gravitational forces, the equivalence principle had been almost entirely concerned with the relationship between gravitational and inertial mass (which is what the Galileo experiment was about). Einstein took up the banner and started discussing the propagation of light in gravitational fields as early as 1907, which was the time he recognised the importance of the principle of equivalence in his quest to produce an all-encompassing theory of gravitation.

---

[8] Good introductions to Mach's principle are Sciama (1953), Dicke (1963) and Narlikar (2011).

Consider an idealised observer freely falling in a uniform gravitational field and compare him/her with an observer who is in uniform unaccelerated motion in a region of space where there is no gravitational field. Then the principle of equivalence states that all experiments that they might perform would lead to identical results.[9] The gravitational field need not be uniform provided the experiments are kept suitably local. The principle of equivalence is therefore asserting that, at any space-time point, the gravitational field can be locally transformed away: the principle asserts a relationship between the gravitational field and coordinates.

This assertion need not of course be true, and indeed, some of Einstein's early attempts to obtain a theory of gravitation while hanging on to the principles of special relativity did not work because the attempted compromise did not satisfy the principle of equivalence. It is a genuine scientific principle in the sense that it asserts something that need not be true. A theory that embraces the principle of equivalence can be falsified.

The *strong equivalence principle* is somewhat different in that it asserts that the gravitational force is due entirely to geometry, and that no additional fields contribute. So not only is there merely a relationship between the gravitational field and coordinates, as asserted earlier, gravitation is geometry. This has an immediate consequence that the Newtonian gravitational constant must be the same everywhere throughout space-time: it is the only coupling constant between gravitation and acceleration. Theories of gravitation that add extra fields to the Einstein equations do not follow the strong principle of equivalence.

A somewhat more modern approach to the principle of equivalence, and one certainly easier to interpret, is to rank the strength of the principle in terms of the phenomena it governs. In the version of the equivalence principle where we state that you can always choose a local inertial frame that locally cancels out the effect of gravity we are in effect stating that inertial and gravitational mass are equal. There is no demand about electromagnetism in that statement, and there is no implied constraint on what the correct version of Maxwell's equations would be. But we could go further and propose a stronger equivalence principle by insisting that, in such an inertial frame, light should travel in straight lines. The extrapolation of this is a very strong principle that says that all laws of physics hold in such freely falling frames. This conforms to what would normally be thought of as the *strong equivalence principle*.

Whichever version of the principle of equivalence one adopts, it tells us little or nothing about the underlying laws that govern gravitation. On the other hand the principle does tell us how to generalise laws that we know to hold in our local inertial frames to the situation of curved space-times. We shall see this later: we simply replace the partial derivative with covariant derivatives that are appropriate to the geometry of curved space-times.

---

[9] This is often formulated in terms of the comparison between an observer in a freely falling elevator and an observer in a similar elevator placed away from all gravitational fields and away from any influences that may cause acceleration. The principle of equivalence asserts that both of these are inertial observers.

12.2.3 Mach's Principle

Mach held that it was no coincidence that the inertial frames of physics were precisely the frames relative to which the system of distant stars (galaxies) was not rotating.[10] He clearly felt that there should be a causative effect between the most distant matter in the Universe and the local physics as determined by inertial frames. What that causative effect might be was never explicitly stated. In fact, the underlying vagueness of Mach's suggestion has created a substantial philosophical discussion about what might be meant by this. Indeed, most of the published discussion concerning Mach's principle over the past century has been philosophical rather than within the domain of physics.

Einstein (1917a) famously stated that *In a consistent theory of relativity there can be no inertia relative to 'space', but only an inertia of masses relative to one another.* This perhaps states more clearly what might be expected of physics when addressing Mach's notion. Einstein felt that the properties of the physical world should be determined from fundamental principles: a particle would have the attribute of mass by virtue of its place in the Universe, not by fiat. Heller (2010, p.19) expresses the opinion that *The intention to create a theory of physics incorporating Mach's principle was one of the main motives behind Einstein's efforts which eventually led to the emergence of the general theory of relativity.*[11] See also Heller (1989) for an insightful commentary on Mach's *The Science of Mechanics* (Mach and McCormack, 2013) in which Mach's principle was first discussed.

Ironically, within months of Einstein's adopting the cosmological constant and coming up with what is now known as the 'Einstein static Universe', in which the self-gravity of matter was precisely balanced by a cosmological constant, de Sitter came up with an empty model Universe that had a gravitational field and showed redshifts of test particles put into the space. This was a manifestly anti-Machian solution in that, according to Einstein's Machian view, there should be no gravitational field without matter to produce it.[12]

The discussions of Raine (1975, 1981) describe in clear terms the sense in which a well-specified Machian notion could be used as a boundary condition, or constraint on the solutions of the Einstein equation. The issue of the influence of the rotation of the system of distant galaxies (i.e. the rotation of the Universe) on local systems has received attention from Lynden-Bell *et al.* (1995) and Bičák *et al.* (2007).

Brans and Dicke (1961) constructed a metric based theory of gravity with the main purpose of explicitly incorporating Mach's principle into gravitational theory. This is now referred to as the *Brans–Dicke Theory*. This was done by adding a scalar field to the

---

[10] Many of the issues surrounding Mach's notion concerned systems in rotation: rotation was perceived as being relative to the 'fixed stars', and so the distant frame of fixed stars must somehow govern our local physics, at least in determining what an inertial frame was. The nice example under discussion was to consider two spheres in an otherwise empty Universe. If one were set into a state of rotation its shape would become oblate. From the point of view of an external observer, what was the reference which determined that one of the spheres should be oblate?

[11] High energy physics of course gives a different perspective on that. We now know that a vacuum is not and cannot be devoid of energy, and we have discovered the Higgs particle that is thought to endow particles with mass by virtue of its interaction with this quantum vacuum.

[12] Schulmann *et al.* (1998, *The Berlin Years: correspondence*, doc. 317).

Einstein equations, in much the same way that Einstein had brought in his cosmological constant. This Brans–Dicke scalar field would provide a background which served as a cosmic preferred frame of reference against which the Machian ideas could be formulated.

Einstein's theory, in the form we know it now, does not obviously have anything to say about Mach's principle, unless we re-interpret it somewhat generously as 'matter determines geometry'. Ultimately it was the explicit dependence of the matter component on the metric which led Einstein to abandon Mach's principle. He realised that, because of this, the matter and geometry were inseparable. This view was expressed by Einstein himself in February 1954 in a letter to Felix Pirani.[13]

Einstein (1918a) had been the first person who brought serious scientific discussion to Mach's principle and was clearly strongly motivated by the notion of a Machian Universe. This was the paper in which he suggested that the cosmological constant might close the space-time and so avoid a need for boundary conditions at infinity, and indeed, Einstein believed for some time that the cosmological constant was demanded by the Machian requirement of his theory. Thereafter the name Mach's principle was assigned to what otherwise would merely have been described as Mach's conjecture.

12.2.4 The Principle of Covariance

The statement of this principle, that the laws of physics should be the same in all frames of reference, belies its power as a way of generalising equations of physics from their Lorentz-invariant form to their being valid within the framework of general relativity. By the time the special theory of relativity was in place in 1905 it was clear that Newton's equations of motion would have to be revised in order to be invariant under Lorentz transformations. The revision would have to be such that in the Newtonian limit of small velocities we recovered Newtonian theory. This was achieved, but while the revised equations made a new, better, prediction for the advance of the perihelion of the orbit of Mercury, it still did not get the correct answer. The Maxwell equations did not need revising as they were already Lorentz invariant.

So modifying the laws of physics was not a new idea when general relativity emerged, though the procedure for modification would not have been evident without the principle of covariance. Calculus on a Riemannian space-time required some changes to familiar notions like differentiation. The ordinary derivative of a function would have to be replaced by a covariant derivative that would compensate for the curvature in the underlying space. Without that the equations of physics would gain extra coordinate specific terms in some

---

[13] Lehmkuhl (2011) translates the key lines of Einstein's letter, written in German, as:

> [Proponents of Mach's principle] think that the field should be completely determined by matter. But this is tricky, for the T_{ik}, which are supposed to represent 'matter', always presuppose the g_{ik} field. [...] [O]ne (sic) should not speak of Mach's principle anymore.

The article of Lehmkuhl (2011) presents an interesting discussion of the issue that ... the T_{ik}, which are supposed to represent 'matter', always presuppose the g_{ik} field. It was this intimate connection between the matter and geometry that stimulated John Wheeler's vision of geometrodynamics, in which 'matter is geometry and geometry is matter', and a hope of finding a path towards quantum gravity.

coordinate systems. The simplest guideline was therefore to take the equations of physics as they appeared in special relativity, and replace all derivatives by their covariant counterpart. While this should not be done blindly it has proved to work rather well. With this kind of generalisation it is always possible to add terms that would be zero in an inertial frame or in the absence of gravitation and generate ad hoc modified equations for physical phenomena which accord with local experience.[14]

12.3 The Geometry of Space-Time

12.3.1 Euclidean Geometry – Perceived Space

Geometry is described by a *metric* which describes how far apart two neighbouring points in a space are. In order to quantify that distance we need to set up coordinates and express, within those coordinates, what we mean by the phrase 'how far apart?'. In other words, we need to express the metric in terms of a coordinate system if we wish to make quantitative statements. If we take familiar Oxy rectangular Cartesian coordinates in the 2-dimensional space of a flat sheet of paper, the distance between two neighbouring points having coordinates (x, y) and $(x + \Delta x, y + \Delta y)$ is $ds^2 = \Delta x^2 + \Delta y^2$. This is what we might call the 'Pythagorean distance' betwen the points.[15]

What is of central importance is that, given the Pythagorean measure, the distance we calculate ds is independent of the way in which we choose our coordinates: the origin can be shifted to another place, or the axes can be rotated without changing the value we calculate for ds. Our maps may have different scales and orientations, but the distance between two cities is nonetheless unchanged. Moreover, if we do not use rectangular Cartesian coordinates, but polar coordinates (r, θ) where $ds^2 = dr^2 + r^2 d\theta^2$, the answer for the value of ds is still the same. Our notion of distance is that it is an attribute describing the relationship (proximity) between two points.

Cartesian and polar coordinates are both orthogonal systems of coordinates. This manifests itself in the fact that the distance is expressed as the sum of two squares of the individual increments: there is no $dxdy$ term in the expression for ds^2 in Cartesian coordinates, and there is no $drd\theta$ term in the expression for ds^2 in polar coordinates. This arises because the Ox and Oy axes are perpendicular. If the axes were not perpendicular, we would have a different equation for ds, but the calculated value would still be the same.

It is fundamental to the Euclidean view of the world that quantities whose value does not depend on the coordinates used to describe them are describing physical attributes. In the 3-dimensional world of our spatial perception the height of a building does not depend on how we measure it.

---

[14] The underlying physics cannot be ignored and later (Section 14.4.2) we shall give a specific example where adding such a term is essential.

[15] There are many alternative definitions of distance. One is the 'Manhattan distance' $ds_M^2 = |\Delta x| + |\Delta y|$, which is used in image analysis and in database technology. A generalisation of the Pythagorean distance is $ds = ((\Delta x)^n + (\Delta y)^n))^{1/n}$ which is used in computational geometry.

12.3.2 Einstein's 4-dimensional World

The relativistic view as enunciated by Einstein changed that because we in fact live in a 4-dimensional world, our local 'space-time' in which the speed of light is a constant, no matter how it is measured, even if the person measuring it is moving. It is there that the worlds of Newton and Einstein part company: the law of addition of velocities that is so familiar in the Euclidean world of Newton where the velocity of light is infinite and space and time are separate is not the same as the Einstein world view as a space-time. In Newtonian theory two points are neighbours if they are close in the 3-dimensional Euclidean space of our perception.

For Einstein, the 'points' are locations in space and time, in other words, a point is an 'event'. The separation has to be computed in space and time, not just in the space of our perception. Two events occurring on the same street a minute apart are closer than two events taking place on the same street a week apart. Events in space-time are labelled by four coordinates specifying location and time: (t, x, y, z).

The key to Einstein's view is that events that can be connected by a light ray have zero separation in space-time. So, in the local Cartesian coordinates of our perception, Einstein's space-time separation between events that are separated by $(\Delta x, \Delta y, \Delta z)$ and take place an interval of time Δt apart, is

$$\Delta s^2 = \Delta x^2 + \Delta y^2 + \Delta z^2 - c^2 \Delta t^2, \qquad (12.2)$$

where c is the velocity of light. In our intuitive (and, as it turns out, misleading) perception of simultaneous events for which $\Delta t = 0$, this reduces to our usual Euclidean view of the spatial distance.

Here we see a deviation from the Pythagorean distance in Euclidean space: the separation is no longer the sum of the squares of the separations: there is a minus sign in front of the contribution from the separation in time, Δt. Such a measure of separation is referred to as a Lorentzian metric (as opposed to our Pythagorean metric).

Following on from the previous argument the task is to make sure that this space-time separation Δs^2 is independent of the coordinate system used to measure it. The set of allowable coordinate transformations that achieve this is the set of Lorentz transformations, which includes frames of reference that are moving in a straight line with uniform velocity relative to one another.

If (12.2) is true in one space-time coordinate system (t, x, y, z) then it is also true in any other set of coordinates (t', x', y', z') related to (t, x, y, z) via a Lorentz transformation. The price we pay for this is that while Δs given by (12.2) remains the same in all Lorentz related frames, we have to give up the familiar notion that the 3-space distance that we perceive and measure, $\Delta x^2 + \Delta y^2 + \Delta z^2$, is the same for all observers making the same measurements from frames that are in uniform straight line motion relative to one another. This was perhaps the hardest lesson to accept from the work of Lorentz and Einstein: this is the theory of special relativity.

12.3.3 Non-Euclidean Geometry

So far we have stressed the word 'Euclidean' when describing our geometry: a Euclidean space is one in which Pythagoras' theorem holds and where the values on the axes are also real numbers. There is only one Euclidean space in one dimension: the real line. In two dimensions it is the plane, and so on. In two dimensions the angles in a triangle add up to π radians, or 180°.

An example of a non-Euclidean geometry is the *surface* of a sphere. On the sphere, the analogue of straight lines are the great circles: they are the shortest distance between two points – the 'geodesics'. The surface of a sphere is clearly non-Euclidean since triangles made up of parts of great circles have angles that do not add up to 180°. In fact the sum of the internal angles of a geodesic triangle on a sphere is always greater than 180°. But we must tread carefully when using our intuition since we 'see' a spherical surface embedded in 3-dimensions: intuitively we embed the spherical surface in the 3-dimensional Euclidean space of our perception.[16] In doing this we intuitively "Euclideanise" the surface of the sphere.

We have to view the surface of the sphere from the point of view of an ancient mariner who saw the Earth from a more global point of view than the resident of a town or village. If the mariner sails along great circles (taking the shortest route between points) 300 km westwards, and then, in succession, the same distances northwards, eastwards and southwards, he does not come back to where he started. If he sails in a triangle and measures the sum of the angles through which he turns, the sum will be greater than 180°. His geometry is certainly not Euclidean. However, his geometry is locally Euclidean in that he can travel small distances in his rowing boat and Euclidean geometry is accurate enough to be entirely useful.[17]

The essence of a non-Euclidean space is that it can be thought of as a patchwork of superposed locally Euclidean spaces. At every point of the non-Euclidean space, we can attach a local Euclidean space that serves the purpose of making local measurements. There is of course the question of what precisely we mean by 'local' here: how local is 'local'? To answer that we need to quantify the large-scale geometry.

12.3.4 Why a Riemannian Space?

A Riemannian space has several good properties:

- Every point in the space can be identified by a set of n coordinates labelled $\{x^a\}$, $a = 0, \ldots, n - 1$. n is the dimension of the space.

---

[16] Embedding spaces in higher dimensional Euclidean geometries is an important trick that early workers in general relativity, notably, de Sitter, Weyl and Robertson, used to understand the nature of solutions to the Einstein equations.

[17] The analogy in general relativity is that it is useful to think of Einstein's space-time as a patchwork of localised Lorentz reference frames. The connection between neighbouring Lorentz frames is determined by the geometry of the space-time, not the Lorentz transformation.

- The labelling of the points is such that we can make transformations between different sets of coordinates: $f : \{x^a\} \leftrightarrow \{\bar{x}^a\}$. The mapping f is one-to-one and continuously differentiable.
- There exists a metric g_{ab} such that the distance between two neighbouring points $\{x^a\}$ and $\{x^a + dx^a\}$ is

$$ds^2 = \sum_{a,b} g_{ab} dx^a dx^b. \tag{12.3}$$

The metric g_{ab} is a continuous and differentiable function of position.

The entire metric structure lies in the function g_{ab}. This says nothing about the global geometry, only the local geometry as expressed via the g_{ab}. Moreover it says nothing about the signature of g_{ab}, the number of positive, zero or negative eigenvalues in the metric G_{ab} (12.3). In addition, if g_{ab} were skew symmetric, this would make no difference to the line element ds^2. Because the geometry is strongly localised, it is possible to cut holes in the space and cut and paste different pieces of space-times provided that any boundary conditions and conditions of continuity are respected.

The concept of a Riemannian geometry provides the ideal framework for specifying local geometries and is not particularly restrictive. Because we want the space to look locally like the metric of special relativity we choose the metric of general relativity to be locally Lorentzian.

Einstein constructed a set of equations that allow us to derive the geometry of the Riemannian space-time associated with a given distribution of matter. These equations are ten nonlinear partial differential equations which require properly specified boundary conditions for their solution. What 'properly specified' might mean has been the subject of a considerable literature over the past decades (see Hawking and Ellis (1975, particularly Ch.7)).

> **Remark:** The metric in general relativity
>
> In general relativity the metric of a space-time is deduced from the Einstein field equations. The physical interpretation of the coordinates used to derive that solution is contained within the metric itself. The coordinates have no a priori meaning except insofar as we might have sought to use coordinates that manifested some particular symmetry.

What this means is that *solving the Einstein equations* is only one part of the process of understanding the gravitational field associated with a given distribution of matter.

This process consists firstly of understanding the resulting geometry in terms of the propagation of light and test particles. To get a complete understanding of the geometry is not so easy since the g_{ab} that are derived from the field equations only describe the local geometry about each point. To get an overview of the entire geometry of space time involves patching these localities together subject to the constraints imposed by boundary conditions and symmetries. We may find that, in the coordinate system we used, the geometry has singularities. It is then necessary to see whether these singularities can be removed by some clever coordinate transformation or whether they are an essential part of the solution.

The second part of the process of understanding a solution is to put observers into the space-time and get them to perform experiments. Again, it must be remembered that the description of the space-time depends on the coordinates that were used in deriving it and that those coordinates may not be optimal for addressing the question of observations.[18] Observers live in locally Minkowski spaces where special relativity holds, and so it may be a good idea for understanding observations in the space-time to remap the coordinate system into some manifestly useful form. In any case, the important point is to remember that such observers only measure physical quantities which, by definition almost, are quantities whose value does not depend on the specific coordinate used.

12.4 Describing Geometries – the Line Element

We can rewrite the space-time separation (12.2) as

$$ds^2 = -c^2 dt^2 + dx^2 + dy^2 + dz^2 = \sum_{i=0}^{3} \sum_{j=0}^{3} \eta_{ij} dx^i dx^j$$

$$= \eta_{ij} dx^i dx^j, \quad \eta_{ij} = \mathrm{diag}(-1, 1, 1, 1) \tag{12.4}$$

where we have written $(t, x, y, z) \rightarrow (x^0, x^1, x^2, x^3)$ using superscripts to denote the different components. The elements of the matrix η_{ij} are zero off the diagonal $(i \neq j)$ and $(-1, +1, +1, +1)$ down the diagonal. In the last expression we have adopted the *summation convention* that repeated indices implies summation. An expression for the elemental distance, ds, between neighbouring points of a space or space-time is referred to as the *line element*. The distance does not depend on the coordinates, but the line element does. Note that we have written the coordinates $x^i = (x^0, x^1, x^2, x^3)$ with upper indices. This is because we are going to treat the coordinates as components of a *contravariant vector*. As we saw when discussing oblique axes in 2-dimensions, (see Section 10.5.4 and Figure 10.3), there are two ways of writing components of a vector **v** in a non-orthogonal frame: we can choose the components that are orthogonal projections of the vector onto the axes, or we can choose the components that are parallel-projected onto the same axes. These two descriptions give the same numbers for the components when the axes are orthogonal. The line element for this choice of coordinates is therefore

$$ds^2 = g_{ab} dx^a dx^b = (dx^1)^2 + 2 \cos \theta \, (dx^1)(dx^2) + (dx^2)^2, \tag{12.5}$$

---

[18] As we shall see, this was a serious problem during the first decades after the publication of Einstein's theory. It is probably fair to say that there was considerable confusion over the interpretation of the important cosmological solution of de Sitter, particularly with regard to the redshift of a distant galaxy that would be measured by an observer. Another outstanding example was the gravitational wave solutions of Brinkmann (1925) and Baldwin and Jeffery (1926), which were almost disregarded or regarded with scepticism until the 1950s. We shall discuss such issues later.

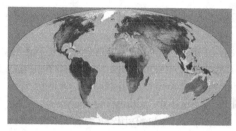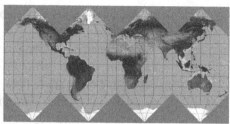

(see Equation 10.50). It should be remembered that this line element is simply a description of the geometry in a particular coordinate system. If we change the coordinates we change the expression of the line element, but the geometry remains unchanged. Same geometry, different description: it is the geometry that is important and we are free to choose the coordinates in any way that is convenient.[19]

So, back to the main issue: if the curved space-time is to be considered a patchwork of localised Euclidean or Lorentz frames of reference, we can represent the geometry simply by making the metric g_{ab} a function of position, $g_{ab}(x)$ describing how the local geometry varies from place to place. The local geometry is described by expanding $g_{ab}(x)$ about a point.

It is one of the underlying principles of general relativity that locally the world should look like the world of special relativity. Indeed, the so-called *principle of equivalence* asserts that we cannot distinguish between freely falling frames of reference and the inertial frames of special relativity.

If, as in Einstein's theory, we believe that the geometry of space-time is distorted by the presence of gravitating material, then the form of $g_{ab}(x)$ will be determined by a set of equations together with appropriate boundary conditions. Einstein provided us with such equations: the *Einstein field equations*.

12.4.1 The Metric Tensor: Coordinates and Reality

A solution to the Einstein equations is a description of the geometry of the space-time that satisfies the boundary conditions specifying the problem. So the Schwarzschild solution is 'the asymptotically flat, spherically symmetric, solution for the gravitation field due to a point mass m at the origin of the coordinates system'. The solution is a mathematical

---

[19] It is possible to describe the geometry of space-time without recourse to coordinates. This greatly simplifies the notation and allows us to focus on the physics rather than the intricacies of dealing with the coordinate system. Schutz (1980) is an excellent source for this approach.

statement of the geometry satisfying the conditions and is generally given in the form of the metric of the space-time in some system of coordinates.

It is the 'some system of coordinates' and the interpretation of the solution in terms of that coordinate choice that causes most of the misunderstandings surrounding the general theory of relativity. Indeed, shortly after Einstein's paper on general relativity, and Schwarzschild's spherically symmetric solution were published, this became a point of contention as people tried to come to terms with this radical new theory and, in particular, the apparent singularities in the solution. The arguments raged over a period of more than 15 years and it was perhaps the work of Lemaître (1933) that brought clarity to the coordinate related issues surrounding Schwarzschild's important solution.

> **Remark:** One of the main difficulties experienced in coming to terms with general relativity arises from the enormous flexibility in choosing a system of coordinates: the same solution, expressed in terms of the metric tensor of space-time, can look different and can even appear to have different properties. This was one of the problems that arose when rationalising Slipher's discovery that 37 of the 41 galaxies in his sample displayed redshifts of their spectral lines, and models of the Universe such as de Sitter's model.
>
> One of the ways of dealing with this is to describe the space-time and its physical properties in a way that does not depend explicitly on coordinate systems.

Figure 12.2 shows two quite different coordinate representations of the geometry of a sphere. Recognising that these are in fact different coordinate representations of the same sphere is a difficult mathematical problem.

12.4.2 Visualising Space-Time Curvature

We can easily visualise the curvature of a 2-dimensional surface: we perceive such a surface embedded in 3-dimensional Euclidean space, as depicted in Figure 12.3. We cannot

Fig. 12.3 A 3-dimensional view of 2-dimensional manifolds having curvature of opposite sign. The curvature can be measured using data on the 2-surface by summing the angles α, β, γ of a 'geodesic triangle' in the surface. If $\alpha + \beta + \gamma < \pi$, as in the figure on the left, the curvature is said to be negative, while if $\alpha + \beta + \gamma > \pi$, as in the right-hand sketch, the curvature is said to be positive. The difference $\alpha + \beta + \gamma - \pi = kA/R^2$ for a geodesic triangle of area A drawn on a surface of curvature R, and $k = \pm 1$ is the sign of the curvature. ©

do the same embedding in order to visualise a 3-dimensional surface in a 4-dimensional space. Going to higher dimension than 2 makes such visualisation difficult and it becomes necessary to exploit spatial symmetries of the space-time to reduce the number of dimensions, or to deal with solutions that are snapshots of the solution at a given value of the time coordinate. So, exploiting spherical spatial symmetry reduces the visualisation of a solution to one spatial coordinate (the radial coordinate) plus the time coordinate. Unlike in Figure 12.3, the time coordinate is generally preserved when depicting space-times, since the key to understanding solutions of the Einstein equations often lies in looking at the way light propagates.[20]

The sketches in Figure 12.3 reveal an important concept. Our 3-dimensional view of the embedded surface reveals its shape, whereas the 2-dimensional being living in the surface can only talk about information based on a restricted number of measurements that must be made within the surface. Mathematically this is expressed by saying that the 2-dimensional inhabitant of the surface measures the *intrinsic curvature* of the surface at a point, whereas we, the 3-D observers, can measure the *extrinsic curvature*.[21]

---

[20] Moschella (2005) and Jonsson (2005) present space-time visualisations of some of the basic solutions to the Einstein equations.

[21] This is beautifully narrated in the book *Flatland: A Romance of Many Dimensions* by Abbott (2015). This short story had an impact on the study of differential geometry after the publication of Einstein's general theory appeared on the scene. Stewart (2008) has an annotated version of the story.

13   Space-Time Geometry and Calculus

Unlike Newton's theory of gravitation, Einstein's theory of general relativity views the gravitational force as a manifestation of the geometry of the underlying space-time. The idea sounds good, it evokes images of billiard balls rolling around on tables with hills and valleys where the balls' otherwise rectilinear motion is disturbed by the geometry of their environment. The difficulty is how to achieve the parallel goal of expressing the force of gravitation geometrically, and to do so without destroying all that we have learned about physics in our local environment.

As we saw in the previous chapter, Einstein saw the principles of covariance and equivalence as a way of formalising that. The theory of gravitation should always admit local inertial frames in which our known laws of physics would hold. Moreover, the mathematical expression of the laws of physics would be the same in all inertial frames. A key step at this point was to argue that physical entities are described by mathematical objects that transform correctly under local Lorentz transformations. This brings us to Minkowski's use of 4-vectors and tensors as the mathematical embodiment of physical quantities.

However, these are local statements, not global ones. They do not tell us how the geometry would affect two widely separated inertial observers in the presence of a gravitational field. The clue was given to Einstein by Marcel Grossmann who suggested that this link would be provided by insisting that the underlying geometry was the geometry of a Riemannian space. This provides the structure to address global issues and to connect different parts of the space.

In this chapter we describe this process and provide the mathematical structure that arises when we follow up on Grossmann's plan. We learn about connecting parts of the space-time and we establish notions of derivatives, geodesics and measures of the curvature.

13.1 A Geometric Perspective

The space time of Einstein's theory is specified by its geometry. A gravitational field is thought of as distorting the space-time away from the no-gravity space-time of Minkowski. So, just as the Minkowski space of special relativity is entirely specified by a *metric tensor* or *line element* telling us what the space time separation of neighbouring points is, the space-time of the general theory is specified by a more general metric. If we work in a coordinate system $x^a = (x^0, x^1, x^2, x^3)$ the general line element has the form:

$$ds^2 = g_{ab}dx^a dx^b \quad \text{(summation over } a \text{ and } b = 0, 1, 2, 3 \text{ implied)}, \qquad (13.1)$$

where the *summation convention* is used to represent the double sum over all four indices. The detail of the geometry of the space-time described in this way is determined by a variety of tensors that are expressible entirely in terms of the g_{ab}.

General relativity insists that whatever legitimate coordinates we choose to use in writing down the metric, it is just an alternate description of the same metric. Some coordinates are simply more useful than others, such as when expressing a symmetry that is intrinsic to the space-time, or because they cover more of the space-time, or because they do not introduce artefacts that are simply coordinate singularities that have nothing do with the underlying space-time.

It is this freedom of choice of coordinate systems in the theory of general relativity that presented conceptual difficulties from the time of inception of the theory.

A 'solution' of the Einstein equations is a metric g_{ab} that satisfies certain boundary conditions. Those boundary conditions define the problem, as in 'we have a single mass in an otherwise empty space-time' or 'the solution must be spatially homogeneous and isotropic'. Any solution can be made to look very different simply by the choice of a legitimate transformations of coordinates. The question that was frequently asked was (and still is) 'which are the correct [*sic*] coordinates to use?'. The answer is, of course, 'all are correct' – the question should perhaps have been 'which is the most useful for my purpose?'.

13.1.1 Scalars, Vectors and Tensors

A general Riemannian space can be described in terms of local coordinate systems about each point of the space via a local metric tensor g_{ab}. To cover the space requires a patchwork of such neighbourhoods. In the case when the space has simple symmetries it may be possible to cover the entire Riemannian space with a single coordinate system and a single metric which would manifest that symmetry, but even then we could make coordinate transforms that would render that symmetry unrecognisable. As an example, a homogeneous and isotropic space-time might be made to look inhomogeneous through the use of distorted coordinates. There are several approaches to the dilemma posed by the generality of coordinate systems in general relativity.

The first approach is to work only in terms of physical quantities, i.e. quantities that we can measure and that transform as scalars, vectors, tensors, spinors and so on under changes of coordinate system. By 'scalars' we mean, in this context, a rank 0 tensor, not just a quantity that has no indices.

The second approach is to avoid explicit use of coordinate systems when writing and manipulating the basic equations of a problem. We are not unfamiliar with this: we use vectors, **r**, in our everyday Euclidean space to describe an arrow connecting two points.[1]

---

[1] It was Willard Gibbs who introduced the use of vectors at the end of the 19th century. Prior to that all equations were written out in specific coordinate frames. There was widespread reluctance to adopt vectors during the

We can talk about the length of a vector, $|\mathbf{r}|$ and write down the angle $\cos^{-1}(\hat{\mathbf{r}} \cdot \hat{\mathbf{s}})$ between two unit-length vectors $\hat{\mathbf{r}}$ and $\hat{\mathbf{s}}$ without reference to any particular coordinate choice. Of course, we ultimately need to make a coordinate choice when facing practical problems such as calculating the values of physical quantities described in this way. In making such a calculation we recognise that there are some coordinate systems that have greater practical value than others: describing a sphere in Cartesian coordinates is not as convenient as describing it in spherical polar coordinates. In doing that we are exploiting any manifest symmetries of the problem.

For objects that cannot be described by vectors we have other, usually more complex and perhaps less intuitive, mathematical descriptors that can be used. The difficulty is often one of recognising the appropriate mathematical framework, which often is to be found in the depths of seemingly esoteric and abstruse mathematics that takes us out of our comfort zones. The use of Riemannian geometry to describe space-times is a fine example of this. Even Einstein in his Zurich days seems to have been reluctant to adopt the Riemannian geometry advocated by his friend and colleague Grossmann in their joint 1913 paper: we see some of his frustration with the complexity this introduced in his Zurich notebook.

13.1.2 Why Riemannian Geometry?

Grossmann had realised that Riemannian geometry would be the ideal tool to describe and work with the geometry of Einstein's space-time. This immediately raises the question as to 'what is Riemannian geometry' and why is it deemed particularly appropriate for describing space-time? Looking up 'Riemannian geometry' we learn that it is a branch of 'differential geometry' that studies 'Riemannian manifolds'. In particular, Riemannian geometry studies smooth manifolds having a 'Riemannian metric', which is expressed as a 'bilinear form' defined on the 'tangent space' at any point. Despite its fundamental importance, this at first sight seems to be somewhat abstract and out of touch with the physical problem at hand.

So we should reverse the process and ask ourselves what, perhaps, did Einstein tell Grossmann that caused Grossmann to come up with the notion of 'Riemannian geometry'? Einstein had already caused a scientific revolution with his special theory of relativity which reconciled Newton's laws with the constancy of the speed of light. Maxwell's theory was already consistent with that and so much of physics known at the start of the 20th century was unified under the banner of 'Lorentz invariance': the laws of physics should be invariant under Lorentz transformations. In thinking about his theory of gravitation, Einstein also believed in the principle of equivalence: that you could locally transform gravity away by being in a special, freely falling, local coordinate system. In that way Lorentz invariance would hold in that locality. The issue was to describe this on a more than local scale.

The notion of patches of space time that have local Lorentz invariance would have rung bells for Grossmann. He knew about Riemannian geometry, which could provide a localised geometric description of a space in terms of a distance function that

first decades of the 20th century. There is a similar reluctance even today to adopt the use of differential forms, spinors or Clifford algebras, even in circumstances when they might be particularly useful.

generalised the concept of Euclidean distance. The distance function was the Riemannian metric, in terms of which it was possible to describe the local properties of this 'Riemannian manifold'. A manifold is a topological space[2] such that the vicinity of each point resembles Euclidean space. Of course, Einstein was interested in locally Lorentzian spaces rather than locally Euclidean spaces, so Grossmann was looking at *pseudo-Riemannian manifolds* rather than simple Riemannian manifolds. The (pseudo) Riemannian manifold could provide the mathematical framework on which to base a theory of gravitation.

One of the benefits of the Riemannian model of space-time is that it provides a definition for quantities called *tensors* which are properties of the space-time and not of the coordinate system used to map the space-time. Such quantities must obey particular transformation laws under change of coordinate system.

13.1.3 Riemannian Manifolds

Firstly, we should state, in rough terms, what a *manifold* is. We need to start with Riemann's famous assertion that almost every mathematical object can be described using n parameters, and hence regarded as a point in an n-dimensional space of such objects. Basic requirements for such a space might be that it has some structure which we can impose by demanding that every point has the concept of an *open neighbourhood*. Without this requirement we enter the world of mathematical structures that might be far removed from the world of our physical experience or perception (unless you are a pure mathematician!). The notion of an open neighbourhood of a point in a set simply says that every point of the set has identifiable neighbours that are also members of the set. At this level our set of points is a topological space: it has no structure other than neighbourliness.

The next level is to impose some structure on the topological space. There are many possibilities at this juncture, but the one we need is what is referred to as a *manifold*. A manifold, $\mathcal{M}$, is a set of points that looks locally like a Euclidean space.[3] A local n-dimensional Euclidean space looks like $\mathbb{R}^n$, and with this we say that the underlying manifold is n-dimensional. The Euclidean space comes with lots of structure: it has vectors, distances, and angles: it is the world in which we live.

Enter the Metric

The next step is to generalise this concept of a manifold and to endow it with a *metric*, i.e. a local distance function. The logical choice of metric is the *bilinear form* which expresses

---

[2] A topological space is a collection of points that have the notion of 'neighbourhood', so it is more than just an unstructured collection of points that bear no relationship to each other. The definition of 'neighbourhood' is somewhat technical, but nontheless there is an underlying intuition of what it means. Knowing who the neighbours are is a good thing, it imposes structure. But we need to be somewhat more quantitative than that, we need to know about things such as the proximity of the neighbours. This is the additional structure gained by being a Riemannian manifold.

[3] We want locally Lorentzian spaces rather than Euclidean spaces. Strictly speaking, we should refer to a *pseudo-Riemannian manifold*, but since we are doing GR the 'pseudo' prefix is generally understood.

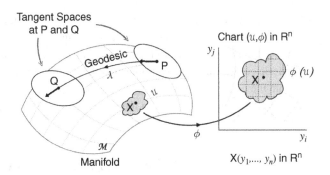

Points **P**, **Q** and **X** of an n dimensional manifold. The geometry local to each point is Euclidean (or Minkowski if the manifold is pseudo-Riemannian). Point **X** has a local neighbourhood $\mathcal{U}$ in which the manifold's coordinate system $\{x_1, \ldots, x_n\}$ is mapped onto a region $\phi(\mathcal{U})$ of a Euclidean space $\mathbf{R}^n$ with coordinates $\{y_1, \ldots, y_n\}$. The manifold has tangent spaces at **P** and **Q** in which we can draw vectors. In order to say that a vector at **P** is parallel to a vector at **Q** we need a connection that links different regions of the manifold. The possibility of linking **P** and **Q** is not essential to being a Riemannian manifold, the connection between two points is provided by an add-on due to Christoffel. ©

the distance between two neighbouring points as a general quadratic form in the distance between the points.

How should we imagine a Riemannian space-time? The answer is to draw pictures as in Figure 13.1, but it is important to remember that such pictures are depicted as having the manifold embedded in a higher dimensional space. Invoking the analogy of an ant living on the surface of a sphere, the ant-view would look locally Euclidean, and on the basis of its local experience it would have no idea that it in fact lived on a sphere. It would have to generate the knowledge of its world in purely two-dimensional terms. Our description of the geometry would rely on seeing the ant-world as a sphere embedded in our 3-dimensional world.

In the world of the Riemannian manifold we are guaranteed the existence of a locally Euclidean space at each point. This gives us a means of using our intuition on the local level, but it does not provide much help at the global level of relating the neighbourhoods of two points that are distant from one another (i.e. points whose Euclidean neighbourhoods do not overlap).

Enter Christoffel's Connection

The definition of a Riemannian manifold provides for there being locally Euclidean coordinate patches at every point, but there is nothing in the definition that connects two points. Additional structure is needed if we are to be able to say that a vector drawn at one point is parallel to a vector drawn at another point. We must make a connection between the points. The mechanism of how to achieve that was provided by Christoffel, and is now called the Christoffel connection.

In Euclidean space we have an almost primitive notion of what we mean by two directions at different points, P and Q, being parallel to one another. To verify parallelism in

Euclidean space we merely connect the two points by a straight line and transport one vector along the line to the other point while maintaining the angle the vector makes with the line. The same happens in a Riemannian manifold except the straight line is replaced by the geodesic connecting the two points and the transport is done by maintaining a constant angle with the geodesic. However, to do this on a curved space time requires the concept of a *connection*: a formula for relating the points in the vicinity of P with those in the vicinity of Q.

13.1.4 Atlases, Maps and Charts

Formalising these geometric ideas is far less simple and requires a number of conceptual terms specifying what is what. Hawking and Ellis (1975, Ch. 2) and Wald (1984, Ch. 3) provide a technical introduction to differential geometry in the specific context of spacetime geometry, while the treatment of Schutz (1980) is oriented to general physics and is rather more pedagogical.

The basic entity is an n-dimensional Riemannian manifold $\mathcal{M}$: this provides us with a continuous set of points that can be labelled in a continuous fashion with arbitrary coordinates $\{x_1, \ldots, x_n\}$. Thus every point P of the manifold has the concept of a local neighbourhood which we can denote by the symbol $\mathcal{U}_P$ (see Figure 13.1).

The Riemannian manifold $\mathcal{M}$ is a space that can be covered by local coordinate patches in each of which the geometry is locally Euclidean: $\mathcal{M} = \bigcup_P \mathcal{U}_P$. Every local subset $\mathcal{U}_P$ can be assigned a Cartesian coordinate system. The word 'local' is important here.

The intuitive picture we have of this is that we have a manifold (say, the Earth) which is described by some general coordinate system, and we also have a collection of images, which we would call 'maps', in which the relative location of points is described by local Cartesian coordinates (i.e. 'map' coordinates). The collection of these 'maps' is of course a World Atlas in which the entire collection of maps depict the entire manifold. We can start to formalise this by drawing two pictures: one of a localised subset $\mathcal{U}_P$ associated with the point P of $\mathcal{M}$, and the other of the local map in which the points are labelled with P at the centre of a Cartesian coordinate system $\{y_1, \ldots, y_n\}$.

In the language of mathematics, it is not done quite that way. Mathematically, our 'map' becomes the function, ϕ_P, that takes the $\{x_1, \ldots, x_n\}$ coordinates inherited from the manifold into the local patch $\{y_1, \ldots, y_n\}$ coordinates. The object $(\mathcal{U}_P, \phi_P)$ is referred to as a *chart*, which in our intuitive discussion was the map of the area around P (Figure 13.1).

13.1.5 Geodesics and Connections

Einstein's other great insight was that the geodesic structure of space-time was vital in understanding how particles and light moved under the influence of the geometry: all matter that was acted upon only by the geometry, i.e. gravitational forces, took the shortest path between any two points. From special relativity he knew that light travelled along a special class of geodesics, null geodesics, the paths of zero length in this locally Lorentzian space-time.

Riemannian geometry provided the means of calculating derivatives and finding the geodesic curves in the space-time. To do that it was necessary to formulate the relationship between different neighbourhoods. It would then be possible to say what one meant in a curved space time that a vector at one point was parallel to some vector at another point.[4] The mathematical device that makes this possible is, not surprisingly, referred to as the *connection*, and the key connection in the case of the geometry of space time was the *Christoffel affine connection*. The Christoffel connection is, in effect, just a collection of numbers that makes it possible to calculate things about the geometry in any given coordinate system used to formulate the Riemannian metric. Christoffel's numbers depend on the coordinate system, they do not respect Lorentz transformations when going from one coordinate system to another.

In particular the Christoffel connection makes possible calculation of covariant derivatives, derivatives that depend only on the geometry and not on the coordinate system describing the geometry. The *values* of the calculated covariant derivatives will depend on the coordinate system of course.

13.2 Tensors, Vectors and Scalars

The fundamental objects that describe physical quantities within the framework of Riemannian geometry are tensors and their generalisation, *tensor densities*. The classical mathematical notation is a symbol with indices, as in the curvature tensor $R^a_{\ bc}$, where the indices represent the components of the object expressed within a generic coordinate system. The horizontal ordering of the indices is important, it reflects values in different coordinate directions. We do not in fact need the indices to say what the object is, but we do need them when it comes down to addressing the issue of what the measured value of the object is in a given coordinate system. If we wish to give a value for or measure the curvature tensor in a given coordinate system we need an explicit expression for it within that coordinate system. The generality of coordinate systems offered within Riemannian geometry inevitably makes the resulting expression for its value somewhat forbidding.

13.2.1 Up and Down Indices

One of the major stumbling blocks in coming to grips with the geometric aspect of general relativity is illustrated by the following questions. If a vector or tensor represents a physical quantity, why does our notation in component form use 'up' and 'down' indices? If the metric is **g** what is the difference between the coordinate representations g^{ab} and g_{ab}? Why do we even need this? The origin of this issue probably arises from the fact that in Cartesian coordinates there really is no such distinction.

---

[4] If you move in a given direction from one point to a neighbouring point, and then move to the next neighbouring point in such a way that you are moving parallel to the previous move, you are moving in the same direction. The curve you trace in so doing is a geodesic.

The components A^a and A_a of vector **A** expressed in a Cartesian coordinate system will have the same numerical values. If we use a non-Cartesian coordinate system this is no longer true, and that is because, in a non-Cartesian system of axes, there is no unique way of specifying the coordinate values. This is illustrated in Figure 10.3 of Section 10.5.4. The metric tensor g_{ab} provides the mechanism for transforming between these 'up' and 'down' indices:

$$A_a = g_{ab}A^b, \quad \text{NB: summation convention on } b. \tag{13.2}$$

Locally, where the geometry is Euclidean, this corresponds precisely to what Figure 10.3 is showing. The metric g_{ab} is used to transform a contravariant representation A^a of a vector **A** into its covariant representation A_a. There is obviously an inverse transformation that maps the numbers A_a into the numbers A^a.

Viewing the set of numbers g_{ab} as elements of an $N \times N$ matrix, we can write the inverse of g_{ab} as g^{ab} which is such that the matrix product is the unit matrix:

$$g^{am}g_{mc} = \delta^a_c = \begin{cases} 0 & a \neq c \\ 1 & a = c \end{cases}, \tag{13.3}$$

and then we have

$$A^a = g^{ab}A_b. \tag{13.4}$$

This procedure generalises to tensors. For a tensor **T** we get the representations

$$T^m_b = g_{bn}T^{mn}, \quad T^n_a = g_{bm}T^{mn} \quad T_{ab} = g_{am}g_{bn}T^{mn}. \tag{13.5}$$

These relationships, which are often given as the definition of what a tensor is, generalise to tensors requiring more than two indices for their description. The total number of indices is called the *rank of the tensor*. The type of tensor representation of rank 4 such as $R^a_{.bcd}$ is described as $\left(\begin{smallmatrix} 1 \\ 3 \end{smallmatrix} \right)$.[5]

13.2.2 Scalars, Vectors and Tensors

Scalar Quantities

The line element ds describes how close two points are in 4-space, and is expressed in terms of the metric g_{ab}. The metric is a continuous and differentiable function of the assigned coordinates such that $ds^2 = \Sigma_{ab}g_{ab}(x^m)\,dx^a dx^b$. The mathematical form of g_{ab} may differ from one coordinate system to another, but the value of the separation ds remains the same. ds is a *scalar*, it is a property of the space, not the parameterisation of the coordinates. However, the determinant of g_{ab} is an example of a number that depends on the coordinates. It is not a scalar, it is a property of the coordinate system rather than of the space-time.

---

[5] The 'dot' below the index a in $R^a{}_{bcd}$ emphasises the position of the index that has been raised. A space is often used in place of a dot, as in $R^a{}_{bcd}$.

Vector Quantities

Now let us turn to the familiar concept of a vector. If we think of a vector as being a direction at a point, an arrow, then that is a property of the space and should be invariant under coordinate transformations. The simplest such vector is the vector pointing to an infinitesimally close-by neighbouring point. This is what we mean by the numbers dx^a. The values of the numbers that make up the object dx^a depend manifestly on the choice of coordinate system used to describe g_{ab}, they are not scalars. But the object that is the combination of all the dx^a that define the arrow pointing to the neighbour is coordinate invariant: it is a *vector*. Since this arrow is not coordinate dependent we can assign to it a notation that does not look manifestly coordinate dependent. In two and three spatial dimensions we have a variety of familiar ways for denoting that an object is a vector: $\mathbf{v}$, $\vec{v}$, $\underline{v}$, and so on. Throughout this book we have used the first of these. The arrow to the nearby point which is represented by the dx^a can unambiguously be denoted by $\mathbf{dx}$.[6]

We know, from elementary calculus, the rule for transforming the coordinate representation of the vector $\mathbf{dx}$ between different coordinate assignments. If we have two coordinate systems denoted by $\{x^a\}$ and $\{\bar{x}^b\}$ such that the relationship between $\{x^a\}$ and $\{\bar{x}^b\}$ has the nice properties of being continuous, differentiable and one-to-one,

$$d\bar{x}^b = \sum_a \frac{\partial \bar{x}^b}{\partial x^a} dx^a = \frac{\partial \bar{x}^b}{\partial x^a} dx^a. \tag{13.6}$$

This is just the chain rule for differentiation in calculus. The last expression here is using the *summation convention* that a repeat of indices denotes summation over the indices unless otherwise stated. There is nothing special about our vector $\mathbf{dx}$, so we can use this as the basis for the general definition that any quantity represented by the set of numbers A^a which transforms according to

$$\bar{A}^b = \frac{\partial \bar{x}^b}{\partial x^a} A^a \tag{13.7}$$

is a vector. Note that the individual numbers A^a are themselves not scalars.

Tensor Quantities

So what, if anything, is next beyond scalars and vectors? Are there objects which are physical objects and therefore have intrinsic properties that do not depend on the coordinate system used to describe them, and that are neither scalars nor vectors? We have two examples of such quantities from elementary mechanics and electromagnetism: the moment of inertia of a solid and the electromagnetic field expressed in terms of the Faraday tensor, F_{ab}.

---

[6] As an aside, we may think of the three components, E_a that make up the electric field, $\mathbf{E}$, of Maxwell's equations. There is no fourth quantity with which they can be combined to make a 4-vector. The only thing they can be joined with to make a Lorentz invariant object are the magnetic field components, B_a, and in that case they cannot form a 4-vector. They can be arranged so as to form a 4-tensor: the Faraday tensor F_{ab} (see Equation 10.134), or even a bivector (see Ex. 10.7.1, p.262).

Consider a physical object **T** that is represented by the quantities T_{ab} in the coordinate system $\{x^a\}$. For example, **T** might be the moment of inertia of a solid or the components of the Faraday tensor. If the components of **T** transform according to

$$\bar{T}^{mn} = \frac{\partial \bar{x}^m}{\partial x^a} \frac{\partial \bar{x}^n}{\partial x^a} T^{ab}, \qquad (13.8)$$

where $\bar{T}^{mn}$ is the representation of **T** in the $\bar{x}^a$ coordinate system, then the numbers T_{ab} represent a physical object. **T** is a property of the space-time and not of the chosen coordinate system.

The transformations between the contravariant and covariant representations of a general tensor **T**, analogous to (13.8) can be summarised as

$$\bar{T}^{mn} = \frac{\partial \bar{x}^m}{\partial x^a} \frac{\partial \bar{x}^n}{\partial x^a} T^{ab}, \quad \bar{T}^m{}_n = \frac{\partial \bar{x}^m}{\partial x^a} \frac{\partial x^b}{\partial \bar{x}^n} T^a{}_b, \quad \bar{T}_{mn} = \frac{\partial x^a}{\partial \bar{x}^m} \frac{\partial x^a}{\partial \bar{x}^n} T_{ab}, \qquad (13.9)$$

and this generalises to higher order tensors that need more indices to describe.

A fundamental example of a tensor is the metric g_{ab} which describes the local geometry of a Riemannian space. The geometry is a property of the space, and not of the coordinates that are used to describe the metric. So, g_{ab} could be described as the *metric tensor* and denoted in a coordinate independent way by the symbol **g**.

However, there are many objects of physical interest that do not fall under this definition. *Tensor densities* are a simple yet important generalisation of this definition of a tensor that broaden the scope of physical entities that can be handled.

13.3 Working on a Riemannian Manifold

The Riemannian geometry only promises us a metric describing the local geometry at every point.[7] Vectors can be defined at each point by giving a direction and a magnitude. The vectors lie in the tangent space of the manifold (see Figure 13.1). However, the problem remains of how to relate and connect the tangent spaces at different points in such a way that we can do dynamics on more than a local scale. Breaking away from the confines of locally Minkowski spaces is an essential part of general relativity. It is also the aspect that makes it technically more complex.

Having the knowledge of the metric and the connection allows us to define physically meaningful quantities that locally describe the geometric properties of the manifold: the Riemann and Ricci tensors and the curvature scalar, and, in particular, a tensor that is a combination of these: the Einstein tensor. It is the Einstein tensor that was ultimately identified with the gravitational field in empty space away from sources of gravitation. Setting the tensor equal to zero gave the vacuum solutions of the full theory of general relativity. It is surprising just how much the simple hypothesis that space-time is to be modelled as a Riemannian manifold with a connection gives us.

---

[7] The main resource for Sections 13.3 and 13.4 is Weinberg (1972, Chapter 4) where everything is worked out in meticulous detail, and on unpublished lecture notes by Ed Bertschinger (http://web.mit.edu/edbert/GR/gr2.pdf, may be volatile).

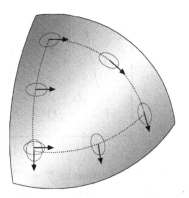

Parallel transport of a vector around a closed curve on a non-flat surface. The vector always lies in the tangent plane to the manifold, which here is the surface. The tangent planes at the various points on the path are depicted as small disks. The vector maintains the same orientation relative to the curve along which it is transported. As it moves from one segment of the path to another, that orientation is maintained relative to the new segment. Since the surface is not flat, when the loop is completed the orientation of the vector has changed relative to its original direction. ⓒⒸ

13.3.1 Parallel Transport

Parallelism of lines plays an important role in the local Euclidean world of our perception. In non-Euclidean spaces the concept of parallelism poses special problems such as the one depicted in Figure 13.2 for the case where the space is the surface of a sphere. The figure shows a closed path consisting of three line segments making up a (curved) triangle on the surface of the sphere. A vector, which lies in the plane tangent to the surface, is moved along each line segment so as to maintain a constant angle between the line and the vector. This is how we would move parallelly on a flat, Euclidean, surface. When our vector gets to the junction with another segment it is moved in the same way, maintaining the angle it makes with that line segment until it meets the next junction between two segments. By the time it gets back to the starting point the vector is no longer making the same angle with the first line segment as it was at the start of the process.

The notion of parallelism on a curved manifold has to be defined carefully because it requires that points at different locations be somehow connected or related (see Section 10.5.4 and Figure 13.1). The metric that is imposed is not by itself sufficient for this task: we have to define a *connection* which provides us with a rule for asserting that two vectors are parallel. The connection mechanism is an important add-on to the concept of a Riemannian manifold that provides us with a sensible notion of differentiation on non-flat spaces: the *covariant derivative*. An alternative strategy would be to cover the manifold with non-intersecting curves (a *congruence* of curves) and define a notion of parallel transport relative to that congruence. This leads to the important notion of the *Lie derivative*.

13.3.2 Parallel Transport and Differentiation

The task is to select a method for parallel transport of a vector that can differentiate between two vectors. To overcome the issue raised by Figure 13.2 we need to be able to say when two vectors on a curve are merely parallelly transported versions of the same vector.

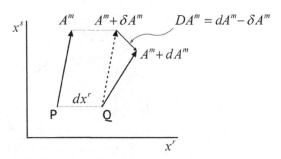

Comparing a vector A^m at **P** with a nearby vector $A^m + dA^m$ at **Q** in order to calculate the covariant derivative. The vector A^m at **P** is parallel transported to **Q** where its new value, $A^m + \delta A^m$ is compared with the vector $A^m + dA^m$. The resulting equation for the difference DA^m must be valid at all points **Q** in the neighbourhood of **P**. This requirement defines what we mean by parallel transport from **P** to **Q**. ©

Consider a vector A^m located at a point **P** (or, more precisely, in the tangent space at **P**). We wish to compare it with a vector at a neighbouring point **Q** located dx^m from **P** (see Figure 13.3). In being moved from $\mathbf{P}(x^r)$ to $\mathbf{Q}(x^r + dx^r)$, the vector of A^m changes to $A^m + \delta A^m$. This change δA^m is to be thought of as being in part due to the non-Cartesian coordinate system $\{x^r\}$, i.e. the spatial curvature. We want to compare this with the vector $A^m + dA^m$ at the point **Q**:

$$DA^m = (A^m + dA^m) - (A^m + \delta A^m) = dA^m - \delta A^m. \tag{13.10}$$

Since the displacement dx^r is infinitesimally small the 'correction' term δA^m must be linear in the displacements dx^r and proportional to A^m. This can be achieved if δA^m is of the form

$$\delta A^m = -\Gamma^m_{rs} A^r dx^s, \tag{13.11}$$

where the coefficients Γ^m_{rs} depend on the local coordinates in a way that is yet to be determined. With this we have

$$DA^m = dA^m + \Gamma^m_{rs} A^r dx^s, \tag{13.12}$$

which can be written

$$DA^m = \left(\frac{\partial A^m}{\partial x^s} + \Gamma^m_{rs} A^r \right) dx^s. \tag{13.13}$$

Since, by construction, DA^m is a vector and dx^s is also a vector, the object in brackets must be a tensor despite the fact that the object Γ^m_{rs} is coordinate dependent and therefore not a tensor. We will find a mathematical expression describing how to calculate the numbers Γ^m_{rs} in a given coordinate system in the next section. At this stage of the argument we only need to know that they enable us to relate the small change in δA^m linearly to the small shift in position dx^a, as per Equation (13.11).

The object in brackets is called the *covariant derivative* of the contravariant vector A^m: it tells us by how much A^m changes for a small change in x^s. It is denoted variously by:

Covariant derivative of vector A^m and its diverse notation:

$$\partial_r A^m \equiv \nabla_r A^m \equiv A^m_{;r} = \frac{\partial A^m}{\partial x^s} + \Gamma^m_{rs} A^r, \qquad (13.14)$$

where Γ^m_{rs} is a Christoffel symbol of the second kind.

If we define the contravariant version of A^m to be $A_n = g_{nm} A^m$ we then have

$$\partial_r A_m \equiv \nabla_r A^m \equiv A_{m;r} = \frac{\partial A_m}{\partial x^r} - \Gamma^s_{mr} A_s. \qquad (13.15)$$

We can go through the same kind of argument to derive the covariant derivative of a second rank tensor T^{ab}:

Covariant derivative of tensor T^{ab}:

$$T^{ab}_{;c} = \frac{\partial T^{ab}}{\partial x^c} + \Gamma^a_{mc} A^{mb} + \Gamma^b_{mc} A_m, \qquad (13.16)$$

with similar results for $T^a_{\ b}$ and T_{ab}. We shall come back to this later in Section 13.6.

It is straightforward to verify that the covariant derivative satisfies the basic rules of ordinary differentiation. The important quantities Γ^a_{bc} are the *Christoffel symbols*.

13.3.3 Christoffel Symbols and the Metric

The issue now is to derive an expression for the coefficients Γ^a_{bc} in terms of the metric. Going back to Equation (13.10) for the definition of the change DA^m (see also Figure 13.3) and the relationship $A_m = g_{mn} A^n$, we have

$$DA_m = D(g_{mn} A^n) = A^n Dg_{mn} + g_{mn} DA^n = A^n Dg_{mn} + DA_m, \qquad (13.17)$$

then, for any A^n, $A^n Dg_{mn} = 0$ and so $Dg_{mn} = 0$. In other words:

The covariant derivative of the metric tensor is zero:

$$\partial_p g_{mn} \equiv g_{mn;p} = 0. \qquad (13.18)$$

We can show this directly by writing out the Christoffel symbols:

$$g_{mn;p} = \frac{\partial g_{mn}}{\partial x^p} - g_{mk} \Gamma^k_{np} - g_{kn} \Gamma^k_{mp} = \frac{\partial g_{mn}}{\partial x^p} - \Gamma_{mnp} - \Gamma_{nmp}, \qquad (13.19)$$

where $\Gamma_{abc} = g_{ap} \Gamma^p_{bc}$ is the index-lowered version of Γ^a_{bc}. If we impose the condition that $g_{mn;p} = 0$, then:

Christoffel symbol of the first kind:

$$\Gamma_{m,np} = \frac{1}{2}\left(\frac{\partial g_{mn}}{\partial x^p} + \frac{\partial g_{mp}}{\partial x^n} - \frac{\partial g_{np}}{\partial x^m} \right), \qquad (13.20)$$

as can be verified by direct substitution. We can raise the index m with $\Gamma^m_{np} = g^{mk}\Gamma_{k,np}$:

Christoffel symbol of the second kind:

$$\Gamma^k_{np} = \frac{1}{2}g^{km}\left(\frac{\partial g_{mn}}{\partial x^p} + \frac{\partial g_{mp}}{\partial x^n} - \frac{\partial g_{np}}{\partial x^m}\right). \tag{13.21}$$

There are 64 possible coefficients Γ^a_{bc}, of which 24 are zero when the metric is diagonal. There are relatively efficient ways to calculate the Γ^a_{bc}, the easiest of which is to use the algebraic computing program *Mathematica*, see Sections 15.2.1 and 15.5.1.

We shall see a different way of approaching the definition of the Christoffel symbols in Section 13.3.4 when we create a local coordinate system at a point that corresponds most closely to a local Minkowski frame.

It is possible to write the derivatives of the metric in terms of the Christoffel symbols. To do this it is convenient to work with the symbol $\Gamma_{abc} = \frac{1}{2}\left(g_{ab,c} + g_{ac,b} - g_{bc,a}\right)$ (Equation 13.20). It is trivial to verify (by substitution) that

$$g_{ab,c} = \Gamma_{abc} + \Gamma_{bac}. \tag{13.22}$$

A corollary of this is that the derivative of the metric tensor vanishes when the Christoffel symbols are zero.

Trace of a Christoffel Symbol

There is a very useful result from the theory of matrices that makes it possible to express the trace Γ^a_{ab} of a Christoffel symbol in terms of the determinant of the metric tensor, thereby obviating the need to evaluate the Christoffel symbols in those cases. The theorem is that for a non-singular matrix $\mathbf{A}$, [8]

$$\mathrm{tr}\left\{\mathbf{A}^{-1}\frac{\partial \mathbf{A}}{\partial x^a}\right\} = \frac{\partial}{\partial x^a}\ln|\det \mathbf{A}|. \tag{13.23}$$

From the definition of the Christoffel symbol Γ^a_{bc} we have

$$\Gamma^a_{ac} = \frac{1}{2}g^{ab}\frac{\partial g_{ba}}{\partial x^c} = \frac{1}{2}\frac{\partial}{\partial x^c}\ln|\det g_{ab}|, \tag{13.24}$$

($g^{am}g_{mb} = \delta^a_c$ and so g^{ab} and g_{ab} are inverse matrices). So we have our result:

---

[8] Weinberg (1972, Ch.4 Section 7) shows this in the following way:

$$\delta \ln \det \mathbf{A} = \ln \det (\mathbf{A} + \delta\mathbf{A}) - \ln \det \mathbf{A} = \ln \det \mathbf{A}^{-1}(\mathbf{A} + \delta\mathbf{A}) = \ln \det (\mathbf{1} + \mathbf{A}^{-1}\delta\mathbf{A})$$

$$\underset{\delta\mathbf{A}\to 0}{=} \ln(1 + \mathrm{tr}\,\mathbf{A}^{-1}\delta\mathbf{A}) \to \mathrm{tr}\,(\mathbf{A}^{-1}\delta\mathbf{A}).$$

Taking the variation with respect to each of the x^a gives the result. If $\det \mathbf{A} < 0$ then we apply this to $|\det \mathbf{A}|$.

Trace of Christoffel symbol:

$$\Gamma^a_{ac} = \frac{\partial}{\partial x^c} \ln \sqrt{g}, \qquad g = |\det g_{ab}|. \tag{13.25}$$

With this, we can eliminate explicit mention of the Christoffel symbols from some equations that would otherwise require them.

13.3.4 Normal Coordinates and Christoffel Symbols

The metric g_{ab} is a local descriptor of the structure of the underlying Riemannian space. There is no specific prescription of how we should choose the coordinates with which to do this, and so we are free to choose whatever is the most convenient.

Suppose the metric is expressed in a coordinate system $\{x^a\}$ and that we wish to make a coordinate transformation $\{x^a\} \rightarrow \{\xi^a\}$, such that, in the neighbourhood of a point P, the metric is a quadratic function of the chosen coordinates. In other words, the value of the metric components g_{ab} at points around and close to P is a quadratic function of the coordinates of the nearby points in the new coordinate system:

$$g_{ab}(\xi^m) = g_{ab}(P) + \text{terms of order } (\xi^a - \xi^a_P)^2, \tag{13.26}$$

where $\{\xi^m_P\}$ are the coordinates of P in the ξ^m-coordinate system. This is a second order approximation to the local geometry and there are no linear terms.[9] We now go on to derive the transform that achieves this.

Once we have made the transformation (13.26) the metric coefficients $g_{ab}(\xi^m)$ expressed in the $\{\xi^m\}$ coordinate system are constant to second order in the distance from P. A simple linear rescaling $\xi^m \rightarrow \bar{\xi}^m$ then locally brings the metric to the Minkowski form η_{ab} in which local distances are given by $ds^2 = \eta_{ab}d\bar{\xi}^a d\bar{\xi}^b$.

The transformation that achieves this in the vicinity of P relates the small displacement from P in both the coordinate systems and will be of the form

$$(\xi^m - \xi^m_P) = (x^m - x^m_P) + \text{a quadratic function of } (x^a - x^a_P) + \text{higher order terms}$$
$$= (x^m - x^m_P) + \tfrac{1}{2}\Gamma^m_{ab}(P)(x^a - x^a_P)(x^b - x^b_P). \tag{13.27}$$

The condition that there be no linear term in the transformation reduces to

$$\frac{\partial g_{ab}}{\partial x^c} = \Gamma^m_{ac}g_{mb} + \Gamma^m_{cb}g_{am}, \tag{13.28}$$

where the summation over the index m is implied. This can be satisfied if

$$\Gamma^m_{ac}g_{mb} = \frac{1}{2}\left(\frac{\partial g_{ab}}{\partial x^c} + \frac{\partial g_{bc}}{\partial x^a} - \frac{\partial g_{ac}}{\partial x^b}\right). \tag{13.29}$$

---

[9] Think of standing on a hillside. You will erect a vertical axis perpendicular to the hillside at the spot where you stand, and then set up a plane perpendicular to that and hence tangential to the surface of the hillside at the spot where you are standing.

(Multiply both sides of the preceding equation by g^{bd} and use the identity $g_{mb}g^{bd} = \delta_m^b$, which is zero unless $b = m$.)

This gives the definition of Γ_{ac}^m in terms of the metric components

$$\Gamma_{ac}^d = \tfrac{1}{2}g^{bd}\left(\frac{\partial g_{ab}}{\partial x^c} + \frac{\partial g_{bc}}{\partial x^a} - \frac{\partial g_{ac}}{\partial x^b}\right), \tag{13.30}$$

$$= \tfrac{1}{2}g^{bd}\left(g_{ab,c} + g_{bc,a} - g_{ac,b}\right), \tag{13.31}$$

or, on relabelling indices,

$$\Gamma_{bc}^a = \tfrac{1}{2}g^{ae}\left(g_{eb,c} + g_{ec,b} - g_{cb,e}\right). \tag{13.32}$$

The quantity Γ_{bc}^a is the *Christoffel symbol* that we met earlier in (13.21) in the context of defining parallelism and the covariant derivative. Recall that Γ_{bc}^a is not a tensor since it depends on a specific location and a specific coordinate system.[10]

The Christoffel symbols Γ_{bc}^a define a particular transformation that expresses the metric locally in a way that has no linear terms in the coordinate displacements from the origin. In the $\{\xi^m\}$ normal coordinates at a point P, the Christoffel symbols vanish since to first order the metric expressed in those coordinates depends quadratically on the deviation from P, and not linearly. They provide a locally quadratic approximation to the metric in terms of the displacements in the coordinate directions.

The Christoffel symbols have the symmetry

$$\Gamma_{bc}^a = \Gamma_{cb}^a, \tag{13.34}$$

and so there are 40 of them in the space-time of Einstein: four for the index a and, in view of this symmetry, only ten from the indices b and c.[11]

Alternative Notations for Christoffel Symbols

There is an alternative notation, $\left\{{}^{\;a}_{b\,c}\right\}$ and, in the older literature, $\{bc, a\}$, both of which look less tensor-like. The definition is

$$\left\{bc, a\right\} \equiv \left\{{}^{\;a}_{b\;c}\right\} \equiv \quad \Gamma_{bc}^a = \tfrac{1}{2}g^{ae}[g_{eb,c} + g_{ec,b} - g_{cb,e}]. \tag{13.35}$$

The Γ_{bc}^a-notation is perhaps more commonly seen, and is in any case less cumbersome. Despite not being a tensor quantity, we use the summation convention over repeated indices, as in $\Gamma_{df}^e\Gamma_{cb}^f$ where summation over the index f is implied.

---

[10] The Γ_{bc}^a do not transform as a tensor should. Under a general coordinate transformation $x^a \to x'^a$ the symbol $\Gamma_{bc}^a \to \Gamma_{qr}'^p$ is given by

$$\Gamma_{qr}'^p = \frac{\partial x'^p}{\partial x^a}\frac{\partial x^b}{\partial x'^q}\frac{\partial x^c}{\partial x'^r}\Gamma_{bc}^a + \frac{\partial x'^p}{\partial x^a}\frac{\partial^2 x^a}{\partial x'^q\,\partial x'^r}. \tag{13.33}$$

The first term on the right is the tensor-like part of the transformation.

[11] It is possible to consider geometries in which $\Gamma_{bc}^a \neq \Gamma_{cb}^a$. The difference $\Gamma_{bc}^a - \Gamma_{cb}^a$ is referred to as the *torsion* of the connection. Non-Einstein theories of gravitation have been built around geometries with non-zero torsion.

13.4 Geodesics – Free Particle Motion

Geodesics are referred to as being time-like, space-like or null depending on whether their defining tangent vectors are time-like, space-like or null. Particles move along time-like geodesics which can be parameterised by a distance measure along the path. That parameter can be taken to be the local proper time along the path, but it need not be. Null geodesics are the paths taken by light rays and have zero length because they are defined by $ds^2 = 0$. So parameterising the position of a light ray along its path is not as straightforward.

13.4.1 Defining Geodesics: Local Straight Lines

The normal coordinate system defined by (13.26) is just a local flat space and so the local geodesics through the point P in this coordinate system are just straight lines. If we parameterise these geodesic straight lines with a parameter λ they can be described simply by the equation

$$\frac{d^2 \xi^m}{d\lambda^2} = 0, \tag{13.36}$$

which, on transforming back to the x^a coordinate system, becomes:

Equation for a geodesic in a curved space-time having metric g_{ab}:

$$\frac{d^2 x^a}{dt^2} + \Gamma^a_{bc} \frac{dx^b}{dt} \frac{dx^c}{dt} = 0, \tag{13.37}$$

where Γ^a_{bc} is the connection given in terms of the metric by (13.32).

This is the 'straight line' motion of a particle moving under no forces in the $\{\xi\}$-coordinates. Hence Equation (13.37) is the equation of a *geodesic* in the $\{x^a\}$ coordinate system.[12]

So we see a physical interpretation of the Christoffel symbols in terms of the forces that appear to act when the coordinate system is not locally Minkowski.[13] Hence the 'fictitious'

---

[12] This brings to the fore the notion that the shortest path between two points, the geodesic path, is a property of the space-time and not of the coordinates used to describe the space-time. So being a geodesic in one coordinate system is being a geodesic in all coordinate systems. Changing the coordinates is merely a labelling process and does not do anything to the space-time itself, it merely affects our view of the space-time.

[13] In his first try at getting a theory of gravitation in Einstein and Grossmann (1913, the *Entwurf* paper), Einstein saw the Christoffel symbols as the embodiment of the gravitational force. Perhaps because Einstein, contrary to his co-author Grossmann, was not yet sold on the idea of general covariance, he had several options for field equations based on the Christoffel connection which he discussed in his 1914 paper, Einstein (1914, footnote p.1060). One possibility was

$$T^a_{b,a} - \begin{Bmatrix} a \\ b \; c \end{Bmatrix} T^a_b = 0.$$

Einstein at that time believed that, in the weak field approximation, the space-time curvature would have to be zero, which was presumably why he had at that time rejected the Ricci tensor from representing the gravitational field. This was part of what is now called 'Einstein's Hole argument', with which Einstein had convinced himself that generally covariant field equations would not be physically interesting, if not permissible.

Coriolis and centrifugal forces of Newtonian mechanics are seen to arise from the choice of non-inertial coordinate systems. The action of those forces is described by the velocity dependent terms coupled by the Christoffel symbols for the coordinate system that is in use. Note, however, that we cannot leap to the conclusion that this is an expression of Mach's principle: this interpretation is purely local and we have no way of linking all the local metrics so as to give a global picture of the space-time. It is, however, a direct expression of the (weak) equivalence principle in the form that we can *locally* transform away a gravitational force field by doing our experiments in a freely falling elevator.

If the curvature of the space-time is due to gravity, as expressed in the Einstein world view, then we see that the Γ^a_{bc} terms encompass that too. There is no longer any distinction between 'real' and 'fictitious' forces.

13.4.2 Affine Parameters

Equation (13.37) determines a trajectory $\Gamma : x^a(t)$ in space-time that is parameterised by the parameter t, which, classically speaking, we think of as 'time'. Unlike its Newtonian counterpart, (13.36), the equation is nonlinear. Since we require that our equations be independent of the coordinate system in which they are written, we are not completely free to choose what we mean by t.

If, for example, we make a transformation to a new parameter, $z = z(t)$, we have

$$\frac{d^2}{dt^2} = z'^2 \frac{d^2}{dz^2} + z''(t)\frac{d}{dz}. \tag{13.38}$$

A first order derivative appears on the right which can only be zero if $z''(t) = 0$, i.e. $z = \alpha t + \beta$, for some constants α, β: a linear transformation. This at the same time reduces the z'^2 terms to a constant scaling factor, α^2. The geodesic equation is invariant under linear transformations of the parameter measuring position along the particle path.[14]

> **Note: Affine parameter**
>
> A parameter that expresses the position of a point on a curve in terms of the fraction of the total curve length that has to be travelled to reach the point from one end is referred to as an *affine parameter*. Accordingly we refer to the independent variable t in (13.37) as an affine parameter.

For a non-null timelike geodesic there is a 'natural' choice for t: the local proper time at each point of the curve. An alternative would be the use of the global time coordinate implied by the metric. However, a null geodesic has zero length and so the existence of a 'natural' parameterisation is not so obvious, notwithstanding the fact that (13.37) is equally valid for null geodesics. We put the issue of dealing with null geodesics aside for the moment (Section 13.4.4).

---

[14] We are nonetheless at liberty to make any (1 to 1 and invertible) transformation $z(t)$ we like. The result is simply a more complicated equation representing the geodesic that can be reversed at any time by transforming back to the variable t.

Tangent Vectors to Time-like Geodesics

The definition of a tangent vector, t^a, to a curve $\Gamma(\lambda) : x^a(t)$ parameterised by an affine parameter λ is

$$t_a = \frac{dx^a(\lambda)}{d\lambda}. \tag{13.39}$$

With this we can write the metric $ds^2 = g_{ab}dx^a dx^b$ as

$$g_{ab}t^a t^b = \text{constant}, \tag{13.40}$$

on choosing our affine parameter to be proportional to the elemental 4-distance, ds.[15] This, taken with the Equation (13.37) for the geodesic, tells us that the tangent vector always has a constant length. Time-like vectors, with our choice of metric signature, have length $ds^2 < 0$, and so we are at liberty to set

$$t^a t_a = -1. \tag{13.41}$$

Our tangent vectors are of constant length along a geodesic. Since 4-velocity, u^a, of a particle is tangential to its world line and is normalised so that $u^a u_a = -1$: the 4-velocity of a free particle is constant.

> **Note: Motion of a free particle in general relativity**
> The space-time trajectory of a free particle moving only under the influence of gravity is a time-like geodesic. The particle's 4-velocity, u^a, is tangential to the geodesic and has constant length $u^a u_a = -1$.

One of the fundamental premises of general relativity is that the path of a free particle, i.e. one moving under the action of no forces except the force of gravity, is a geodesic. We have shown that the free particle's 4-velocity has constant length.

The 4-momentum of a particle of rest mass m is $p^a = mcu^a$. Hence the magnitude of the momentum 4-vector is $p_a p^a = (mc)^2 u_a u^a$, from which we have

$$g_{ab}p^a p^b = -m^2 c^2. \tag{13.42}$$

13.4.3 Connecting with Newtonian Gravity

We can give a simple motivation for what Christoffel symbols are by looking into the familiar Newtonian gravity. There are two approaches here: one starts from Newton's equation of motion of a particle, $\mathbf{x}(t)$, in a gravitational field with potential $\phi(\mathbf{x})$:

$$\ddot{\mathbf{x}} + \nabla\phi = 0. \tag{13.43}$$

This might be designated as the *Newtonian gravitational field equations*. The other, opposite, approach is to write down the weak field metric and derive this equation. Within the framework of Newtonian theory, we want to identify the gradient of the potential with

---

[15] It is at this point that we see why we must exclude null geodesics from this argument: null geodesics have $ds = 0$.

the Christoffel symbols and thereby give an intuitive basis for thinking of the Christoffel symbols in terms of gravitational forces.

Writing (13.43) in component form, the equations for the trajectory of a particle, $x^\alpha(t)$, moving in a gravitational field with potential $\phi(\mathbf{x})$ are

$$\frac{d^2 x^\alpha}{dt^2} = -\eta^{\alpha\beta} \frac{\partial \phi}{\partial x^\beta} = \eta^{\alpha\beta} \partial_\beta \phi, \quad \alpha, \beta = 1, 2, 3. \tag{13.44}$$

Now introduce an affine parameter $t = k\lambda + c$ for some constants k, c. Then we have $dt = kd\lambda$ and

$$\frac{d^2 x^\alpha}{d\lambda^2} + k^2 \eta^{\alpha\beta} \partial_\beta \phi = 0. \tag{13.45}$$

Rewrite this[16] using $k = dt/d\lambda$ and putting $ct = x^0$:

$$\frac{d^2 x^\alpha}{d\lambda^2} + \frac{1}{c^2} \eta^{\alpha\beta} \partial_\beta \phi \left(\frac{dx^0}{d\lambda} \right)^2 = 0. \tag{13.46}$$

Comparing this with the geodesic equation of motion $\ddot{x}^a + \Gamma^a_{bc} \dot{x}^b \dot{x}^c = 0$ (Equation 13.37), we readily make the identification

$$\Gamma^a_{00} = \frac{1}{c^2} \eta^{\alpha\beta} \partial_\beta \phi = \frac{\partial \phi}{\partial x^a}. \tag{13.47}$$

This lends credence to our interpretation of the geodesic equation as an expression for the acceleration of a particle in a gravitational field being proportional to the values of the Christoffel coefficients. In this case Γ^a_{00} would be the only non-zero Christoffel symbol in this way of describing Newtonian gravity. We note that $\Gamma \sim O(1/c^2)$ and thus describes weak gravitational fields.

13.4.4 A Brief Word on Null Geodesics

We approached the definition of geodesics from the starting point of straight line motion in a coordinate system chosen so that the metric was locally expressed as a quadratic function in the coordinates. This was the choice of normal coordinates (Sections 13.3.4 and 13.4.1) in which, locally, free particles would move in straight lines. Likewise, photons would move in straight lines, and so the equation for the path of a photon is given by the same Equation (13.37) as for a particle having non-zero mass:

$$\frac{d^2 x^a}{d\lambda^2} + \Gamma^a_{bc} \frac{dx^b}{d\lambda} \frac{dx^c}{d\lambda} = 0, \quad \text{null} \Rightarrow g_{ab} \frac{dx^a}{d\lambda} \frac{dx^b}{d\lambda} = 0, \tag{13.48}$$

which follows from $ds^2 = g_{ab} dx^a dx^b = 0$ on a null curve. This second equation is simply a constraint on the path $x^a(\lambda)$: its tangent vector, k^a, (the direction of the propagation in 4-space) should be null:

$$k^a = \frac{dx^a}{d\lambda}, \quad g_{ab} k^a k^b = k^a k_a = 0. \tag{13.49}$$

---

[16] Hindsight is useful!

Since the length of the null geodesic is zero ($ds^2 = 0$) we cannot choose λ to be the proper time, but that does not prevent us from finding the paths of light rays from (13.48).

13.5 Covariant Derivative

13.5.1 Covariant Derivative Along a Curve

Figure 13.4 depicts a curve $\Gamma : x^a(\lambda)$ having a tangent vector $u^a(\lambda) = dx^a/d\lambda$. If the curve is a geodesic having tangents $u^a(\lambda)$, then the equation for $u^a(\lambda)$ is

$$\frac{D\,u^a}{D\lambda} \equiv \frac{du^a}{d\lambda} + \Gamma^a_{bc} u^b u^c = 0, \tag{13.50}$$

and, conversely, if this equation is satisfied, then u^a is tangent to a geodesic.

The notion that a vector v^a appears to have a fixed direction in the local space of an observer moving along a geodesic with velocity u^a requires that the angle between u^a and v^a remains fixed. This is as close to the notion of parallelism that we can get on a curved space-time.[17]

It makes sense to ask what is the equation governing a vector v^a that is parallelly transported along the curve by u^a. The covariant derivative, D_λ, acting along the curve $\Gamma : x^a(\lambda)$ is defined as:

Covariant derivative acting along a curve $\Gamma : x^a = x^a(\lambda)$:

$$D_\lambda v^a \equiv \frac{D\,v^a}{D\lambda} = \frac{dv^a}{d\lambda} + \Gamma^a_{bc} u^b v^c, \quad u^b = \frac{dx^b}{d\lambda}. \tag{13.51}$$

It can be shown that this is indeed a tensor by calculating how it transforms under the mapping $x^a \rightarrow x'^a(x^b)$. The full derivation is given in Weinberg (1972, Ch.4 Section 9).[18]

We can also write the covariant equivalent of the directional derivative, $u^b \partial_b$, of a vector X^a as:

Directional derivative of vector field X^a along direction field u^a:

$$u^b X^a_{;b} \equiv u^b \partial_b X^a = u^b \frac{\partial X^a}{\partial x^b} + \Gamma^a_{bc} u^b X^c, \tag{13.52}$$

with the logical extension to directional derivatives of tensors, as in $u^b \partial_b T^{cd}$. This simplifies in the case that u^b is the tangent field of a curve $\Gamma(\lambda) : x^a(\lambda)$ parameterised by an affine parameter λ.

---

[17] See also Section 13.3.2 and Figure 13.3.

[18] This looks somewhat similar to the convective derivative of fluid mechanics. However, this derivative along a curve requires a metric and a connection for its definition. A closer analogy with the fluid flow convective derivative is the Lie derivative, which requires neither a metric nor a connection. This is discussed in Section 13.6.4.

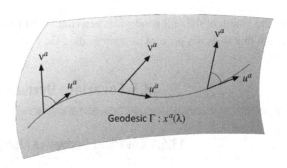

Fig. 13.4   Parallel transport of a vector v^a along a geodesic. The vector u^a is tangent to the geodesic at all points and so it is parallelly transported into itself. Parallel transport of v^a is defined by its making a constant angle with the tangent vector v^a. The orientation of v^a about u^a is controlled by the choice of connection. ©

When the vector v^a is parallelly transported along a curve defined by its tangent vectors u^a, as depicted in Figure 13.4, we have:

Vector v^a parallelly transported along timelike geodesic with tangent u^a:

$$u^b v^a{}_{;b} \equiv u^b \partial_b v^a = u^b \frac{\partial v^a}{\partial x^b} + \Gamma^a_{bc} u^b v^c = 0. \tag{13.53}$$

This defines *parallel transport*.

In particular, when the vector v^a parallelly transports itself:

$$v^b \partial_b v^a = v^b \frac{\partial v^a}{\partial x^b} + \Gamma^a_{bc} v^b v^c = 0, \quad \Rightarrow \quad v^a \text{ is tangent to a geodesic,} \tag{13.54}$$

the curve defined by $v^a = dx^a/d\lambda$ is a geodesic of the manifold with the given metric and connection.

13.5.2 Time-like Geodesic Motion

We have defined a geodesic as a curve that minimises the distance between two points, and we calculated the equation for geodesics using the calculus of variations (Section 11.1.3). We have also seen an alternative definition of the geodesic as a curve that parallelly transports its own tangent vector (Equation 13.54). We now consider motion along time-like geodesics.

Acceleration Along a Time-like Curve

A curve $\Gamma : x^a(\lambda)$, parameterised by an affine parameter λ and tangent vectors $u^a(\lambda)$, might be the path of a particle moving with velocity $u^a(\lambda)$. We can then regard the rate of change of u^a with λ as a measure of the acceleration. We have (see Equation 13.13)

$$\frac{D u^a}{D\lambda} = \frac{du^a}{d\lambda} + \Gamma^a_{bc} u^b u^c = u^b \left(\frac{\partial u^a}{\partial x^b} + \Gamma^a_{bc} u^c \right). \tag{13.55}$$

We recognise the term in parentheses on the right (see Equation (13.52)) as $u^a_{,b}$, and so we have:

Rate of change of the tangent vector u^a as a particle moves along its geodesic:

$$\frac{D u^a}{D\lambda} \equiv \dot{u}^a = u^b \partial_b u^a = u^b u^a_{;b}. \tag{13.56}$$

When u^a is identified with the particle velocity, this is its acceleration.

The notation $\dot{u}^a$ is shorthand for $D u^a/D\lambda$. This equation defines the *acceleration* of the tangent vector along the curve.

In the special case that the curve is a geodesic, then by Equation (13.54)

$$\dot{u}^a = 0, \qquad \text{geodesic motion.} \tag{13.57}$$

So free particles moving under no forces other than gravity move on geodesics and experience no acceleration. This is the expression of the fact that an observer in a freely falling elevator experiences no acceleration.

Geodesic Deviation for a Time-like Congruence

A pair of geodesics through neighbouring points can be described in terms of the path length s along one of them, and the perpendicular distance v to its neighbour (see Figure 13.5). The points on the geodesic are then of the form $\Gamma(v) :\ x^a(s, v)$. Let η^a denote the vector connecting the two neighbouring geodesics *at the same value of s*.

We have the following equations for the tangent u^a, and for the deviation η^a:

$$u^a(s, v) = \frac{\partial x^a}{\partial s}, \qquad \eta^a(s, v) = \frac{\partial x^a}{\partial v}. \tag{13.58}$$

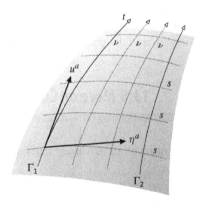

A neighbourhood of time-like geodesics parameterised by affine parameter s and labelled with a parameter v. The geodesic Γ_1 has tangent vector u^a. The vector η^a connects the two neighbouring geodesics Γ_1 and Γ_1 and is used to define the deviation between them. ⓒ

A simple result of partial differentiation tells us that

$$\frac{\partial u^a}{\partial v} = \frac{\partial^2 x^a}{\partial s \, \partial v} = \frac{\partial \eta^a}{\partial s}, \tag{13.59}$$

since the order of differentiation is interchangeable. Using (13.56), the change in η^a with s along a given curve $\Gamma(v)$ can be written

$$\frac{D\eta^a}{Ds} = \frac{d\eta^a}{ds} + \Gamma^a_{bc}\frac{dx^b}{ds}\eta^c = \frac{\partial^2 x^a}{\partial s \, \partial v} + \Gamma^a_{bc}\frac{\partial x^b}{\partial s}\frac{\partial x^c}{\partial v} = \frac{Du^a}{Dv}. \tag{13.60}$$

Thus we have the important relationship

$$\frac{D\eta^a}{Ds} = \frac{Du^a}{Dv}. \tag{13.61}$$

Writing out the derivatives on either side (as in Equations (13.60)) and changing to the semi-colon notation for the directional covariant derivatives yields the more practical variant of (13.61):

$$u^b \eta^a_{;b} = \eta^b u^a_{;b}. \tag{13.62}$$

If, in addition, the curve Γ_1 along which u^a is a tangent is a geodesic, the following argument shows that the projection of η^a onto u^a is constant:

$$\frac{D(\eta^a u_a)}{Ds} = u_a \frac{D\eta^a}{Ds} + \eta_a \frac{Du^a}{Ds} = u_a \frac{Du^a}{Dv} = \frac{1}{2}\frac{D(u^a u_a)}{Dv} = 0, \tag{13.63}$$

where the term $\eta Du^a/Ds = 0$ because the curve defined by u^a is a geodesic and where the derivative of $u^a u_a$ is zero since $u^a u_a = -1$. The connecting vector η^a is parallelly transported along the geodesic.

Hence the scalar $\eta^a u_a$ is transported without change along the geodesic by the tangent vector u^a. In particular, if the connecting vector is orthogonal to the tangent at some point on the geodesic, i.e. $\eta_a u^a = 0$, it is orthogonal to the geodesic at all points.

We would also like to know how the connecting vector η^a evolves in a given gravitational field. That has to wait until after we have introduced the Riemann tensor describing the gravitational distortion of the geodesic congruence (Section 13.7.4).

13.6 More on Covariant Derivatives

The most important result on covariant derivatives is that all covariant derivatives of the metric are zero:

$$\partial_m g_{ab} \equiv g_{ab;m} = 0, \tag{13.64}$$

see Equation (13.18). We can understand why this should be so on the basis of the principle of equivalence, which tells us that we can always choose a local coordinate system in which the metric is the Minkowski metric, η_{ab}. In that locally Minkowski frame the Christoffel symbols vanish and so the covariant derivative of the metric η_{ab} in that frame is zero:

$\eta_{ab;c} = 0$, all components are zero.[19] If all components of a tensor are zero in one frame, they will be zero in every other frame, and so $g_{ab;c} = 0$.

In one way this can be regarded as a consequence of our definition of the connection, i.e. our decision to use the Christoffel connection. But in fact it is a requirement that defines the connection that we propose to use. Without insisting on this requirement we would lose a great deal of the simple structure of our description of the space-time, among which is the key notion of parallel transport of vectors from one place on the manifold to another.

13.6.1 Derivatives of Covariant and Contravariant Tensors

Firstly, we summarise with a set of useful equations for covariant derivatives of various types. In the following, the function ϕ will represent a scalar, u^a, a vector and t^{ab} a tensor. The quantities u_a and t_{ab} represent the corresponding vector and tensor with the indices lowered by use of the metric g_{ab}:

$$\phi_{;a} = \phi_{,a}, \tag{13.65}$$

$$u^a{}_{;b} = u^a{}_{,b} + u^m \Gamma^a{}_{mb}, \tag{13.66}$$

$$t^{ab}{}_{;c} = t^{ab}{}_{,c} + \Gamma^a{}_{mc} t^{mb} + \Gamma^b{}_{mc} t^{am}, \tag{13.67}$$

and, for completeness, the equations for vectors and tensors with lowered indices:

$$u_{a;b} = u_{a,b} - u_m \Gamma^m{}_{ab}, \tag{13.68}$$

$$t_{ab;c} = t_{ab,c} - \Gamma^m{}_{ac} t_{mb} - \Gamma^m{}_{bc} t_{am}. \tag{13.69}$$

An upper-index derivative is a little subtle since we understand the coordinates with respect to which we are differentiating as being contravariant objects, x^a. So to evaluate a derivative of $\phi(x^p)$ such as $\phi^{;a}$, we need to do the following:

$$\phi^{;a} = g^{am} \phi_{;m} = g^{am} \phi_{,m} = g^{am} \frac{\partial \phi}{\partial x^m}, \tag{13.70}$$

and then $\phi(x^p)$ is being differentiated with respect to the coordinates in terms of which it is defined.

13.6.2 Divergence of Tensors: a Simplifying Equation

The covariant divergence of a 4-vector u^a is $u^a{}_{;a}$, which by (13.66) is

$$\text{div} u \equiv u^a{}_{;a} = u^a{}_{,a} + \Gamma^m{}_{mb} u^b. \tag{13.71}$$

We see here the term $\Gamma^m{}_{mb}$, the trace of the Christoffel symbol. This allows an important simplification via the use of Equation (13.25):

$$u^a_{;a} = u^a_{,a} + u^m \frac{\partial \log \sqrt{|g|}}{\partial x^m} = \frac{1}{\sqrt{|g|}} \frac{\partial (u^m \sqrt{|g|})}{\partial x^m}, \tag{13.72}$$

where $|g|$ denotes the absolute value of the determinant of the metric tensor g_{ab}.

---

[19] It is on the basis of the vanishing of the connection coefficients in this local frame that Misner *et al.* (1973, p.466) argued that there could be no such thing as a localised gravitational energy momentum tensor.

There is a similar result for the divergence of a skew-symmetric tensor $s^{ab} = -s^{ba}$. By (13.67),

$$s^{ab}_{;a} = s^{ab}_{,a} + \Gamma^a_{ma}s^{mb} + \Gamma^b_{ma}s^{am}. \tag{13.73}$$

If the tensor is skew symmetric, the last term vanishes since Γ^b_{ma} is symmetric under interchange of the indices m, a. We are left with

$$s^{ab}_{;a} = \frac{1}{\sqrt{|g|}} \frac{\partial}{\partial x^a} \left(\sqrt{|g|}\, s^{ab} \right), \qquad s^{ab} \text{ skew symmetric.} \tag{13.74}$$

If we consider the related case of the covariant derivative of the covariant version s_{ab} of the same skew symmetric tensor, we have, from Equation (13.69),

$$s_{ab;c} = s_{ab,c} - \Gamma^m_{ac}s_{mb} - \Gamma^m_{bc}s_{am}, \tag{13.75}$$

from which we can derive the result

$$s_{ab;c} + s_{bc;a} + s_{ca;b} = \frac{\partial s_{ab}}{\partial x^c} + \frac{\partial s_{bc}}{\partial x^a} + \frac{\partial s_{aca}}{\partial x^b}, \qquad s^{ab} \text{ skew symmetric.} \tag{13.76}$$

13.6.3 Laplacian of a Scalar Field

Consider taking the divergence of the gradient of a scalar field $F(x^a)$. The gradient is just grad $F = F_{;a} = F_{,a}$ and we can apply Equation (13.72) for the divergence of a contravariant vector field by writing

$$\text{div grad } F \equiv \nabla^2 F = \frac{1}{\sqrt{|g|}} \frac{\partial}{\partial x^a} \left(\sqrt{|g|}\, g^{ab} F_{,b} \right). \tag{13.77}$$

The preceding discussion works in any number of dimensions, and for any metric signature and so we can look at this in the more familiar framework of taking the Laplacian of a scalar function in 3-dimensional spherical polar coordinates (r, θ, ϕ). The metric in that situation is simply

$$g_{ab} = \text{diag}\,(1, r^2, r^2 \sin^2 \theta), \tag{13.78}$$

and so $g^{ab} = \text{diag}\,(1, 1/r^2, 1/(r^2 \sin^2 \theta))$ and $\sqrt{|g|} = r^2 \sin \theta$. We then have

$$\nabla^2 F = \frac{1}{r^2 \sin \theta} \frac{\partial}{\partial x^a} \left[r^2 \sin \theta \left(g^{a1} \frac{\partial F}{\partial r} + g^{a2} \frac{\partial F}{\partial \theta} + g^{a3} \frac{\partial F}{\partial \phi} \right) \right]. \tag{13.79}$$

After calculating all the derivatives and much rearrangement we arrive at the familiar expression

$$\nabla^2 F = \frac{1}{r^2} \frac{\partial}{\partial r} \left(r^2 \frac{\partial F}{\partial r} \right) + \frac{1}{r^2 \sin \theta} \frac{\partial}{\partial \theta} \left(\sin \theta \frac{\partial F}{\partial \theta} \right) + \frac{1}{r^2 \sin^2 \theta} \frac{\partial^2 F}{\partial \phi^2}. \tag{13.80}$$

13.6.4 Differentiation: the Lie Derivative

There is another derivative that plays a central role in general relativity, the *Lie derivative*. Indeed, the Lie derivative plays an important role in many fields such as Hamiltonian mechanics, fluid dynamics, and general relativity. The famous mathematician Vladimir

Arnold (1980, p. 198) coined the name *fisherman's derivative* to describe a fisherman in his boat, observing objects in the river in his vicinity. Before going into the general relativistic notion of the Lie derivative, it is instructive to appreciate how and where this derivative appears in classical, i.e. non-relativistic, fluid dynamics. We are then able to appreciate precisely what is the difference between choosing Cartesian or non-Cartesian coordinates for describing a flow. We begin with the discussion of how a scalar field is carried from one location to another by a flow, and then move on to the question of what happens with a vector field. For the latter we use the fluid vorticity as a specific example.

In fluid mechanics we are familiar with the notion of a derivative following the motion of the fluid. A scalar quantity f that is associated with a comoving fluid element moving with the flow at velocity $\mathbf{v}(\mathbf{x})$ evolves as

$$\frac{Df}{Dt} \equiv \frac{\partial f}{\partial t} + (\mathbf{v} \cdot \nabla)f = 0. \tag{13.81}$$

The quantity:

Lie derivative of scalar field $f(\mathbf{r})$ relative to a vector v:

$$\mathcal{L}_{\mathbf{v}}f = (\mathbf{v} \cdot \nabla)f(\mathbf{x}), \tag{13.82}$$

is called the *Lie derivative* of f along $\mathbf{v}$. The more familiar name in fluid dynamics is the *directional derivative*, while Df/Dt is referred to as the time derivative following the motion $\mathbf{v}$, or the *Lagrangian derivative* of f.

We can see how this arises as follows. A flow with velocity field $\mathbf{v}(\mathbf{x}, t)$ carries an element of fluid from a point $P : \mathbf{x}$ to $\bar{P} : \mathbf{x} + \mathbf{v}(\mathbf{x}, t)\delta t$ in a time interval δt. So, considering the values of $f(\mathbf{x}, t)$ at two spatial points in the flow that are $\mathbf{v}(\mathbf{x}, t)\delta t$ apart in space after a time interval δt, we have for the change in f:

$$\delta f = f(\mathbf{x} + \mathbf{v}(\mathbf{x}, t)\delta t, t + \delta t) - f(\mathbf{x}, t) = \frac{\partial f}{\partial t}\delta t + [\mathbf{v}(\mathbf{x}, \mathbf{t}) \cdot \nabla]f\delta t, \tag{13.83}$$

$$\frac{Df}{Dt} = \frac{\partial f}{\partial t} + [\mathbf{v}(\mathbf{x}, \mathbf{t}) \cdot \nabla]f. \tag{13.84}$$

This is simply an application of Taylor's theorem.

Staying with the example of fluid flow, it can be shown that the equation for the transport of vorticity $\boldsymbol{\omega} = \nabla \times \mathbf{v}$ by an inviscid incompressible flow is

$$\frac{\partial \boldsymbol{\omega}}{\partial t} + \underbrace{(\mathbf{v} \cdot \nabla)\boldsymbol{\omega} - (\boldsymbol{\omega} \cdot \nabla)\mathbf{v}}_{\text{Lie derivative}} = 0. \tag{13.85}$$

This gives rise to the following definition of the *Lie derivative* of the vector field $\boldsymbol{\omega}$:

Lie derivative of vector field ω relative to a vector v:

$$\mathcal{L}_{\mathbf{v}}\boldsymbol{\omega} = \mathbf{v} \cdot \nabla\boldsymbol{\omega} - \boldsymbol{\omega} \cdot \nabla\mathbf{v}. \tag{13.86}$$

This is the vector field generalisation of the scalar transport Equation (13.82).

We can now write down the analogous 4-space generalisations of this derivative of scalar, ϕ, vector, v_a, v^a, and tensor, t_{ab}, t^{ab} fields with respect to a vector field ξ^a:

$$\mathcal{L}_\xi \phi = \xi^a \phi_{,a}, \tag{13.87}$$

$$\mathcal{L}_\xi v^a = \xi^b v^a_{,b} - v^b \xi^a_{,b}, \tag{13.88}$$

$$\mathcal{L}_\xi v_a = \xi^b v_{a,b} + v_b \xi^b_{,a}, \tag{13.89}$$

$$\mathcal{L}_\xi t^{ab} = \xi^c t^{ab}_{,c} - t^{cb} \xi^c_{,a} - t^{ac} \xi^c_{,b}, \tag{13.90}$$

$$\mathcal{L}_\xi t_{ab} = \xi^c t_{ab,c} + t_{cb} \xi^c_{,a} + t_{ac} \xi^c_{,b}. \tag{13.91}$$

The formal derivation of these equations is a lengthy process. We recognise the form of our Equations (13.82) and (13.85) in the first two of these. Note that the derivatives that appear here are 'comma' derivatives, they do not involve the connection that appears in the definition of the covariant derivative.[20] So, if the the space-time manifold is endowed with a metric and a covariant derivative:

Lie derivative of a vector ω^a when the covariant derivative is defined:

$$\mathcal{L}_\xi \omega^a = \xi^b \omega^a_{;b} - \omega^b \xi^a_{;b}. \tag{13.92}$$

The Lie derivative of a vector field ξ^a in the direction u^a that is tangent to a geodesic is denoted by $\mathcal{L}_u \xi^a$. We see that Equations (13.62) follow from the relationships $\mathcal{L}_u \xi^a = 0$, or $\mathcal{L}_\xi u^a = 0$.

A particularly important case arises when calculating the Lie derivative of the metric $\mathcal{L}_\xi g_{ab}$, which using (13.91) is

$$\mathcal{L}_\xi g_{ab} = \xi^c g_{ab;c} + g_{cb} \xi^c_{;a} + g_{ac} \xi^c_{;b} = \xi_{b;c} + \xi_{b;c}. \tag{13.93}$$

The first term, $\xi^c g_{ab;c}$ vanishes because the symmetric connection associated with the metric is chosen so as to make all covariant derivatives of the metric tensor vanish. If there exists a vector ξ^a such that $\mathcal{L}_\xi g_{ab} = \xi_{a;b} + \xi_{b;a} = 0$, the vector field ξ^a is referred to as a Killing vector field, or simply a *Killing vector*. ξ^a reflects a symmetry of the metric g_{ab}.

13.7 The Riemann Tensor

We now come to the central issue of space-time: how to characterise the curvature of the space-time in terms of the metric and its Christoffel connection. This was the clue that Grossmann passed on to Einstein.[21]

---

[20] If we simply replace the comma derivatives by semi-colon derivatives in these formulae, and assume that the connection is symmetric, we recover the same equations with covariant derivatives in place of ordinary derivatives.

[21] In his *Zurich Notebook* Einstein is seen working with what he notes as 'Grossmann's tensor', i.e. the Riemann tensor. One gets the impression on reading those pages in the *Notebook* that Einstein was at first frustrated by the evident complexity of the scheme. It cannot have been obvious at that time, 1913, that developing the theory of gravitation around Riemann's tensor would lead to anything useful.

13.7.1 Definition and Conventions

The definition of the Riemann tensor is expressed in terms of the connection Γ^{a}_{bc} as:[22]

Riemann tensor expressed in terms of the connection Γ^{a}_{bc}:

$$R^{a}_{bcd} = \Gamma^{a}_{db,c} - \Gamma^{a}_{bc,d} + \Gamma^{a}_{cf}\Gamma^{f}_{db} - \Gamma^{a}_{df}\Gamma^{f}_{cb}. \tag{13.94}$$

A flat space-time has $R^{a}_{bcd} = 0$. This is also called the *Riemann curvature tensor* or the *Riemann–Christoffel tensor*.[23]

An alternative way of writing the lowered version $R_{abcd} = g_{am}R^{m}_{cbd}$ is

$$R_{abcd} = \frac{1}{2}\left(\frac{\partial^2 g_{ad}}{\partial x^b\,\partial x^c} + \frac{\partial^2 g_{bc}}{\partial x^a\,\partial x^d} - \frac{\partial^2 g_{ac}}{\partial x^b\,\partial x^d} - \frac{\partial^2 g_{bd}}{\partial x^a\,\partial x^c}\right) + g_{am}\left(\Gamma^{m}_{cf}\Gamma^{f}_{db} - \Gamma^{m}_{df}\Gamma^{f}_{cb}\right). \tag{13.95}$$

The main interest in this version is that in linearised gravitation where we can ignore terms in the square of the connection, Γ^2, we have an expression for the Riemann tensor involving only the metric components. It also exposes the symmetries of the Riemann tensor under swapping of indices.

The Riemann tensor satisfies a number of symmetries and identities. The symmetries can be expressed by the equations

$$R_{abcd} = R_{cdab}, \quad\text{and}\quad R_{abcd} = -R_{bacd} = -R_{abdc}, \tag{13.96}$$

which are readily appreciated from Equation (13.94). From the second of these it follows that all components of the Riemann tensor with $a = b$ or $c = d$ will be zero.

From (13.95) we can write the Riemann tensor in a local inertial frame as

$$R_{abcd} = \frac{1}{2}\left(\frac{\partial^2 g_{ad}}{\partial x^b\,\partial x^c} + \frac{\partial^2 g_{bc}}{\partial x^a\,\partial x^d} - \frac{\partial^2 g_{ac}}{\partial x^b\,\partial x^d} - \frac{\partial^2 g_{bd}}{\partial x^a\,\partial x^c}\right), \tag{13.97}$$

$$= \tfrac{1}{2}\left(g_{ad,bc} + g_{bc,ad} - g_{ac,bd} - g_{bd,ac}\right). \tag{13.98}$$

This follows since, as we have seen, the Christoffel symbols are zero in such a local inertial frame.

13.7.2 Bianchi Identities

The identity:

1st Bianchi identity:

$$R_{abcd} + R_{adbc} + R_{acdb} = 0, \tag{13.99}$$

---

[22] Beware, there are several different definitions in use having different ordering of the indices. The ordering makes a difference only in that the inherent symmetries of the Riemann tensor manifest themselves in slightly different ways.

[23] Landau and Lifshitz (1980, Section 92) has a good discussion of the Riemann tensor.

is easily verified to hold in the local inertial frame, and so holds in all frames. This is called the *1*st *Bianchi identity*.

The 2nd *Bianchi identity* is:

2nd Bianchi identity:

$$R^m{}_{abc;d} + R^m{}_{adb;c} + R^m{}_{acd;b} = 0. \qquad (13.100)$$

We can likewise derive this from the local inertial frame expression (13.98) with which we can easily calculate derivatives such as

$$R_{abcd,m} = \tfrac{1}{2}\left(g_{ad,bcm} + g_{bc,adm} - g_{ac,bdm} - g_{bd,acm}\right). \qquad (13.101)$$

From this we deduce that

$$R_{abcd,m} + R_{abmc,d} + R_{abdm,c} = 0, \qquad (13.102)$$

in which the last three indices are cyclically permuted. Since this is true in a local inertial frame it is correct in all frames and so we simply replace comma-derivatives with semi-colon (covariant) derivatives to get the general result that holds in all frames:

$$R_{abcd;m} + R_{abmc;d} + R_{abdm;c} = 0. \qquad (13.103)$$

Taking into account all symmetries and constraints, the Riemann curvature tensor has 20 independent components.

13.7.3 Second Order Covariant Differentiation

The value of the second covariant derivative of a vector depends on the order in which the derivatives are taken: $u_{a;d;c} \neq u_{a;c;d}$. The difference can be written in terms of a tensor R_{abcd}:

Second order covariant derivatives and the Riemann tensor:

$$u_{a;d;c} - u_{a;c;d} = R_{abcd}u^b, \qquad (13.104)$$

which expresses the fact that second order differentiation depends on the order of differentiation.[24]

One consequence of this is that there is an ambiguity when generalising equations of classical physics that involve second order partial derivatives to the general relativistic version (see, for example, Section 14.4.2).

---

[24] The usual way of writing Equation (13.104) is $u_{a;dc} - u_{a;cd} = R_{abcd}u^b$, recognising that everything to the right of the ';' represents a derivative. The order of differentiation is read from left to right following the semi-colon.

13.7.4 Geodesic Deviation and the Riemann Tensor

Now we aim at relating the deviation between neighbouring geodesics to the curvature of the space-time as described by the Riemann tensor. To this end we take a congruence of geodesics whose tangent vectors are u^a, and select two neighbouring geodesics. We will connect two particles freely moving on each of these geodesics by a vector η^a and follow the evolution of η^a along the geodesic pair. The situation is depicted Figure 13.5. Geodesics Γ having tangent vector u^a are parameterised by s. The connecting vector η^a connects points on neighbouring geodesics having the same value of the para meter s.

Our starting point is Equation (13.62) in the form

$$\frac{D\eta^a}{Ds} = u^b \eta^a_{;b} = \eta^b u^a_{;b}. \tag{13.105}$$

We now calculate the derivative

$$\frac{D^2 \eta^a}{Ds^2} = u^c (u^b \eta^a_{;b})_{;c} = u^c (\eta^b u^a_{;b})_{;c} = u^c \eta^b_{;c} u^a_{;b} + u^c \eta^b u^a_{;bc}, \tag{13.106}$$

and then use the definition of the Riemann tensor (13.104) to replace $u^a_{;bc}$ with $u^a_{;cb} + R^a_{pcb} u^p$:

$$\frac{D^2 \eta^a}{Ds^2} = \eta^b \left[u^a_{;c} u^c_{;b} + u^c u^a_{cb} \right] + \eta^b u^c u^p R^a_{pcb}. \tag{13.107}$$

The quantity in square brackets is just $[u^c u^a_{;c}]_{;b} = 0$ since the acceleration $u^c u^a_{;c} = 0$ for free particles which are moving on geodesics. We are left with:

Geodesic deviation equation:

$$\frac{D^2 \eta^a}{Ds^2} = R^a_{pcb} u^c u^p \eta^b. \tag{13.108}$$

This derivation of the equation of geodesic deviation is interesting in that it does not explicitly use the Christoffel symbols to describe the geometry, their role is merely implicit in the definition of the covariant derivatives and in the definition of the Riemann tensor. The alternative derivation involves writing out the geodesic equation at two neighbouring locations, using a first order Taylor expansion to join them (see, for example, Weinberg (1972, Ch.6 Section 10) and Hobson *et al.* (2006, Section 7.13)).

Geodesic Deviation: the Newtonian Analogy

In Newtonian theory the acceleration of a particle, A, at 3-space location x^α in a gravitational field described by a Newtonian potential ϕ is the ever-familiar

$$\ddot{x}^\alpha = -\partial^\alpha \phi, \qquad \alpha = 1, 2, 3. \tag{13.109}$$

A nearby particle, B, a distance ξ^α away in the same potential field experiences acceleration

$$\ddot{x}^\alpha + \ddot{\xi}^\alpha = -[\partial^\alpha \phi]_{x^\alpha + \xi^\alpha} = -\partial^\alpha \phi - \xi^\beta \partial_\beta \partial^\alpha \phi, \tag{13.110}$$

(by Taylor's theorem).[25] The difference between these two equations is

$$\ddot{\xi}^\alpha = -K^\alpha_{\ \beta}\xi^\beta, \quad K^{\alpha\beta} = \partial^\alpha\partial^\beta\phi. \tag{13.111}$$

Thus the *tidal field tensor* K^{ab} determines how particles in the neighbourhood of A move relative to A. The trace of the tidal field tensor, $K^\alpha_{\ \alpha}$, i.e. $\nabla^2\phi$, is determined directly by the density of gravitating material via Poisson's equation.

13.8 The Ricci Tensor and Curvature Scalars

The Ricci tensor is the trace of the Riemann tensor:

Ricci tensor:

$$R_{ab} = R^m_{\ amb} = g^{lm}R_{lamb}. \tag{13.112}$$

Expressed in terms of Christoffel symbols this is

$$R_{ab} = R^m_{\ amb} = \partial_m\Gamma^m_{ba} - \partial_b\Gamma^m_{ma} + \Gamma^m_{mn}\Gamma^n_{ba} - \Gamma^m_{bn}\Gamma^n_{ma}. \tag{13.113}$$

The Ricci scalar, also known as the *curvature scalar*, is the trace of the Ricci tensor:

$$R = g^{lm}R_{lm} = g^{ij}g^{lm}R_{iljm}. \tag{13.114}$$

Another scalar that can be constructed from the Riemann tensor is the *Kretschmann scalar*:

$$K = R_{abcd}R^{abcd}. \tag{13.115}$$

This plays a role in studying singularities of solutions of the Einstein field equations. The Einstein equations have non-trivial vacuum solutions (i.e. solutions with $T_{ab} = 0$ which are not simply Minkowski space) in which $R_{ab} = 0$ and $R = 0$. Those solutions have non-zero curvature and a non-zero Kretschmann scalar. For example, the Schwarzschild solution has $K = 48G^2M^2/c^4r^6$, which is finite at the Schwarzschild radius $r = 2GM/c^2$.

13.8.1 Contracted Bianchi Identity

From the Bianchi identity (13.100) we can derive the very important *contracted Bianchi identity*:

Divergence of the Einstein tensor G^{ab}:
contracted Bianchi identity:

$$G^{ab}_{\ \ ;a} = 0, \quad \text{where } G^{ab} = R^{ab} - \tfrac{1}{2}R\,g^{ab}. \tag{13.116}$$

---

[25] We maintain the up and down indices even though in the Cartesian description of Newtonian gravitation there is no difference.

This is the geometrical basis of Einstein's field equations. By setting G_{ab} proportional to the energy momentum tensor T_{ab}, i.e. $G_{ab} = \kappa T_{ab}$, Einstein was saying that the energy momentum tensor was the source of the curvature of space-time. With this he automatically satisfied the required conservation of energy momentum, expressed as $T^{ab}{}_{;b} = 0$.

In four dimensions, the Einstein tensor is the only divergence-free tensor that can be constructed from the metric tensor and its first and second derivatives (Lovelock, 1972).

13.8.2 The Conformal Curvature Tensor

The Ricci tensor has ten independent components. Since the Riemann tensor has 20 components, what is the object that describes the ten components of the Riemann tensor that are not included in the Ricci tensor? The answer is given by Ellis (1971, reprinted as Ellis (2009b)):

Conformal tensor:

$$C^{ab}{}_{cd} = R^{ab}{}_{cd} - 2g^{[a}_{[c} R^{b]}_{d]} + \tfrac{1}{6} R\, g^{[a}_{[c} g^{b]}_{d]}. \tag{13.117}$$

Ellis (1971) attributes this to the seminal 1960 paper of Jordan, Ehlers and Kundt (translated and reprinted in Jordan *et al.* (2009)).

The tensor $C^{ab}{}_{cd}$ is referred to as the *conformal tensor* or the Weyl tensor. It has ten independent components. As Ellis (1971, (Section 4.1)) remarks, we can think of R_{ab} as the trace part of R_{abcd}, and C_{abcd} as the trace-free part. The symmetric part of C_{abcd} is $E_{ab} = C_{abcd} u^b u^d$ and is referred to as the *electric* part of the Weyl tensor. The dynamical effect of E_{ab} is to shear the particle flow lines: it is the relativistic analogue of the tidal shear tensor.[26]

---

[26] Ellis *loc.cit* points out the analogy with Newtonian theory where the second derivative of the Newtonian gravitational potential, $\phi_{,\mu,\nu}$, can be split into a trace part $\nabla^2 \phi = \phi^{,\sigma}_{,\sigma}$, whose source is the density of matter ρ and that is determined by the Poisson equation, and a trace-free part,

$$E_{\mu\nu} = \phi_{,\mu,\nu} - \tfrac{1}{3} \delta_{\mu\nu} \phi^{,\sigma}_{,\sigma}. \tag{13.118}$$

Here $\delta_{\mu\nu}$ is 1 or 0 according as to whether $\mu = \nu$ or not. $E_{\mu\nu}$ is the Newtonian *tidal tensor*.
In our familiar Newtonian theory, the Poisson equation $\nabla^2 \phi = 4\pi G \rho$ appears to determine only the trace of $\phi_{,\mu,\nu}$. As a small exercise for the reader, we might ask what is it that determines the remaining $\mu \neq \nu$ components of $\phi_{,\mu,\nu}$?

14

The Einstein Field Equations

The Einstein field equations relate the geometry of space-time to the source of the curvature: the material and energy that pervade the Universe. This is the essence of the theory of general relativity, 'GR'. In their full generality the equations cannot be solved analytically, they are nonlinear partial differential equations which can only be solved in situations of high symmetry. In the case of weak gravitational fields we recover Newton's theory, as we should expect. Importantly, the theory also provides us with the local inertial frames of the special theory of relativity. In this way, the well-established classical physics is 'safe', we do not have to rewrite our textbooks.

Locally, we live in the arena of weak quasi-Newtonian gravitational fields. The geometry is slightly curved, otherwise there would be no gravitation at all, and so the metric of the space-time is almost the Minkowski metric. Knowing this we can easily go the one or more steps beyond Newton and look for tests of these post-Newtonian deviations. The classical tests such as gravitational lensing and planetary orbits provide strong evidence for GR at this level. In recent decades we have seen the rise of GPS devices that require the use of GR in order to return correct positions.

Non-Newtonian predictions, such as strong gravitational fields, i.e. black holes, and gravitational waves are now being tested. We have abundant evidence for black holes from the study of the nuclei of galaxies. The existence of gravitational waves was inferred from the decay of binary stellar orbits and now they have been detected directly using gravitational wave antennas.

General relativity, or some modification of the theory, is the framework for gravitational physics. In this chapter we look at the experimental support for Einstein's theory and consider possible alternative theories. We also address the question as to how we should carry over our understanding of the laws of physics to the curved space-times of general relativity.

14.1 The Einstein Equations

Einstein's leap of imagination, seeing the gravitational field as being an aspect of the curvature of space-time, is surely one of the most profound scientific insights ever. The geometric insight was not the only one: Einstein recognised the energy–momentum tensor as being the source of the space-time curvature. But the energy–momentum tensor involves the metric, and in this way geometry and matter are intertwined and inseparable.

Einstein's theory is that the geometry of space-time is determined by the distribution of energy,

$$\text{Geometry} \quad \Leftrightarrow \quad \text{Mass-energy.}$$

His equations, the *Einstein field equations*, stipulate what that relationship should be:

Einstein's field equations:

$$\underset{\text{Geometry}}{G_{ab}} \quad = \quad \underset{\text{matter}}{\kappa T_{ab}}, \qquad \kappa = \frac{8\pi G}{c^4}. \tag{14.1}$$

Here, G_{ab} is an object derived from the space-time metric, g_{ab}: it is called *the Einstein tensor*, we will come back to this later. T_{ab} describes the distribution of mass-energy and κ is the *coupling constant* telling us how effective that mass-energy is in distorting the space-time geometry. The constant κ contains the *Newtonian* gravity coupling constant G: here is the link between Einstein and Newton.

In general relativity the energy–momentum tensor for a perfect fluid having density ρ, pressure p and moving with 4-velocity u_a is

$$T_{ab} = \left(\rho + p/c^2\right)u_a u_b + p g_{ab}. \tag{14.2}$$

This is the simple generalisation of the energy momentum tensor in special relativistic fluid flow. We note that there is a term involving the metric g_{ab} on the right-hand side: the energy–momentum tensor must be expressed within the framework of a curved space-time. With this the geometry of space-time impinges on *both* sides of the Einstein field equations. Matter and geometry are not independent, they are inseparably intertwined. John Wheeler phrased this as:

John Archibald Wheeler's expression of Einstein's gravity:
Space tells matter how to move.
Matter tells space how to curve.

This contrasts starkly with the Newtonian version: 'matter tells matter how to move'.[1]

The energy–momentum tensor, T_{ab}, is symmetric with the indices running from 0 to 3, so Equations (14.1) are ten nonlinear partial differential equations describing the geometry of space-time determined by a given energy–momentum tensor. It must not be forgotten that the solution of these equations depends not only on the distribution and nature of the energy density, but also on a set of suitably defined boundary conditions. Typical boundary conditions might be formulated in terms of some initial conditions (specification of

---

[1] Wheeler often repeated this quotation in slightly different forms, both in print and in conversation and lectures. There are many discussions about what he did in fact say and whether he should have said 'space-time' rather than 'space'. This version is from the margin of Misner *et al.* (1973, p.5), while the adjacent text says *Space acts on matter, telling it how to move. In turn, matter reacts back on space, telling it how to curve*. Wheeler was particularly fond of this quote and probably gave every variant that could be thought of!

the field on some space-like hyper-surface) or as a boundary condition at infinity. In that respect, the theory of the gravitational field is similar to the theory of the electromagnetic field.[2]

Einstein added two further conditions to the theory. One was a specific variant of what is called the *principle of equivalence*: that gravitation is not anywhere locally distinguishable from acceleration by physical experiments. The second, which brought the general theory of relativity into contact with the already established special theory, was that the coordinate frames attached to freely falling particles were inertial frames. He had already presented in his special theory the notion that light propagated along the null geodesics $ds^2 = 0$ between points.

14.2 Testing Einstein's General Relativity

One of the driving forces behind the development of the general theory of relativity, if not the strongest, was the problem of the perihelion of the orbit of the planet Mercury. The special theory of relativity had produced an extra contribution, but it was insufficient to account for the entire effect. Einstein had in June 1913 begun collaborating on the problem with a close friend, Michele Besso,[3] using the 1913 version of his gravitational theory that had been developed with Marcel Grossmann.

The collaboration with Besso involved some historically important written correspondence between the two on the subject of the perihelion advance problem. This correspondence is known as the *Einstein–Besso correspondence*. Some 52 pages of their correspondence on the perihelion problem appears in Volume 4 (Doc. 14) of the Einstein archive. The original pages are in private hands and it is certain that there are more, as yet

---

[2] John Wheeler coined the name *Geometrodynamics* for the theory of general relativity to reflect the notion that the gravitational field was entirely encoded in the geometry of space-time. Wheeler's idea was perhaps that this was the route into quantum gravity. The geometrisation of gravity is an idea that can be traced back to the second half of the 19th century with the work of Riemann and Grassmann and to Clifford (1870) with his talk to the Cambridge Philosophical Society on *On the Space-Theory of Matter*. To quote Clifford's salient points from the talk ...

(1) *That small portions of space are in fact of a nature analogous to little hills on a surface which is on the average flat; namely, that the ordinary laws of geometry are not valid in them.*
(2) *That this property of being curved or distorted is continually being passed on from one portion of space to another after the manner of a wave.*
(3) *That this variation of the curvature of space is what really happens in that phenomenon which we call the motion of matter, whether ponderable or etherial.*
(4) *That in the physical world nothing else takes place but this variation, subject (possibly) to the law of continuity.*

[3] Besso and Einstein had known each other from Einstein's early days in Zurich (1896) and continued to be lifelong friends. Besso got a strong acknowledgement in Einstein's great 1905 paper on the special theory of relativity (Einstein, 1905): *In conclusion, I note that when I worked on the problem discussed here, my friend and colleague M. Besso faithfully stood by me, and I am indebted to him for several valuable suggestions.*

unseen, pages. The correspondence showed that the Einstein–Grossmann theory gave an answer of around $18''$.[4]

The way in which these tests were done and analysed in the early days of general relativity (Eddington, 1923, p.105)) was to write the metric as a simple generalisation of the Schwarzschild metric:

$$ds^2 = -\left(1 + \frac{b_1}{r} + \frac{b_2}{r^2} + \cdots\right)(cdt)^2 + \left(1 + \frac{a_1}{r} + \frac{a_2}{r^2} + \cdots\right)^{-1} dr^2 + r^2(d\theta^2 + \sin^2\theta d\phi^2),$$

(14.3)

and to use Solar System experiments to determine the as and bs. This was the first attempt at parameterising a post-Newtonian approximation to Einstein's theory of gravitation. In order that this metric should include the non-relativistic Newtonian limit, we need $b_1 = -2GM/c^2$.

The experimental data available at the time were consistent with the values:

Test	GR value
Newton's law of gravitation	$b_1 = -2Gm/c^2$
Deflection of light by the Sun	$a_1 = -2Gm/c^2$
Advance of perihelion of Mercury's orbit	$b_2 = 0$

At the time Eddington wrote: *these three tests establish the coefficient of dt^2 to order r^{-2} and the coefficient of dr^2 to order r^{-1}*. The table shows the importance of observing and measuring the gravitational deflection of light rays by the Sun: that is the only handle on verifying the spatial part of the metric. Einstein's own calculation of the perihelion advance had been published even before the final paper on general relativity was published, so the deflection of light was the outstanding piece of the jigsaw.

Such approximations, referred to as 'PPN', *parameterised post-Newtonian approximations*, now play a central role in constraining geometric theories of gravitation that can be expressed as simple power series (Will, 2006, 2009). Whereas Einstein named only three effects that should be tested, the current version of PPN uses ten parameters to describe a wide variety of phenomena that serve to test Einstein's version of gravitation. We shall touch on PPN in due course.

14.2.1 Solar System Orbits

The motion of a planet about the Sun in general relativity is a key test to validating the theory. When the 'planet' is a test particle and the 'Sun' is a point mass m the metric of the space time in which the particle is moving is the Schwarzschild metric (15.132). For non-relativistic orbits, such as orbits of Solar System bodies, the Schwarzschild solution

---

[4] The value given in the Einstein–Besso draft manuscript is $1821''$ due to using an erroneous mass for the Sun. Correcting that error gives the $18''$ value. Droste (1914) had obtained and published the same value, also on the basis of the *Entwurf* theory.

could be expanded in a perturbation series in the small parameter Gm/c^2r, which gave the metric in the form (14.3).

The coordinates in which the approximate metric (14.3) are expressed are the standard 'Schwarzschild' coordinates. In these coordinates the spatial part of the metric is not locally isotropic: in other words it does not reduce to the form $f(r)(dx^2 + dy^2 + dz^2)$ or $dr^2 + r^2(d\theta^2 + \sin^2\theta d\phi^2)$.[5] We can work in locally isotropic spherical polar coordinates by using a different radial coordinate ρ in which the form of the approximate metric, analogous to (14.3), takes on the form

$$
\begin{aligned}
ds^2 = &- \left(1 - 2\frac{Gm}{c^2\rho} + 2\beta\frac{G^2m^2}{c^4\rho^2} + \cdots \right)(cdt)^2 \\
&+ \left(1 + 2\gamma\frac{Gm}{c^2\rho} + \cdots \right)(d\rho^2 + \rho^2 d\theta^2 + \rho^2\sin^2\theta d\phi^2),
\end{aligned}
\tag{14.4}
$$

(see Equation (15.139)). In this order of approximation the radial coordinate ρ is related to the radial coordinate r of (14.3) by the approximate equations

$$
\rho = r\left(1 - \gamma\frac{Gm}{c^2r} + \cdots \right), \qquad r = \rho\left(1 + \gamma\frac{Gm}{c^2\rho} + \cdots \right).
\tag{14.5}
$$

These are just the small Gm/c^2r limit of Equation (15.138). The coefficients β, γ are related to a_1, a_2, b_1 via $b_2 = \beta - \gamma, a_1 = \gamma$. Standard general relativity corresponds to $\beta = \gamma = 1$.

The orbit of the planet which in Newtonian theory would be describing an orbit of semi-major axis a and eccentricity e in this weak field general relativistic approximation is

$$
r = \frac{a(1 - e^2)}{1 + e\cos\left[(1 - \delta\phi/2\pi)\phi\right]},
\tag{14.6}
$$

where the angle ϕ is the true anomaly of the planet's position in its orbit. $\delta\phi$ is the perihelion shift per orbit and is given by

$$
\delta\phi = \frac{6\pi Gm}{c^2}\frac{1}{a(1 - e^2)}\frac{2 - \beta + 2\gamma}{3}, \quad \text{per orbit.}
\tag{14.7}
$$

Note that in this equation the parameters β and γ are the coefficients derived from the power series of the metric expressed in isotropic coordinates (14.4).

Brans–Dicke Theory

In the Brans and Dicke (1961) variant of general relativity, there is a parameter ω that reflects the strength of the Brans–Dicke scalar field. The field was introduced specifically to create an explicitly Machian theory of gravity without completely throwing away Einstein's

---

[5] In the early days of relativity the physical meanings of the different coordinate systems were poorly understood: the coordinates were generally chosen as a matter of mathematical convenience. This gave rise to a certain amount of confusion concerning which coordinate choices were 'better' or more 'meaningful', and in particular, about how to understand the apparent hiatus at $r = 2Gm/c^2$ in the Schwarzschild solution.
See Weinberg (1972) Chapter 8 and in particular Section 3 of that chapter, where the orbit of a planet is worked out in detail. This is certainly one of the clearest discussions of the classical tests.

theory. The metric describing the weak gravitational field due to a point mass in Brans–Dicke theory has

$$\beta = 1, \quad \gamma = \frac{\omega + 1}{\omega + 2}, \qquad \text{Brans–Dicke} \tag{14.8}$$

and so the limit $\omega \to \infty$ gives the general relativistic result (Weinberg, 1972, Ch.8 Section 6). In this way the measured value of the parameter γ imposes constraints on the Brans–Dicke coupling, ω. The only way to reconcile the Brans–Dicke theory with a finite value of ω would be to postulate an additional source that would contribute to the advance of the planet's perihelion.

Dicke's own suggestion was that the Sun's gravitational field might possess a significant mass quadrupole moment, J_2, that would cause an additional contribution to the advance of the perihelion:

$$\delta_{\text{Solar}}\phi = \frac{3\pi R_{\text{Sun}}^2}{a^2} \frac{1}{(1 - e^2)^2}(1 - \frac{3}{2}\sin^2 i)J_2. \tag{14.9}$$

Here, R_{Sun} is the Sun's radius, and i is the inclination of the orbit with respect to the Solar equator.[6] An independent check on this hypothesis was to measure the oblateness of the Sun's figure to see whether there was any shape distortion over and above that due to the Solar rotation. Dicke and Goldenberg (1967) set up an experiment to measure the Solar oblateness and reported an oblateness that could account for 3.26″ per century, requiring the Brans–Dicke theory to account for only 39.7″ per century instead of 43″ per century. If correct, this result would put the Brans–Dicke theory in the frame for an acceptable Machian theory of gravitation. They had derived a value of $J_2 = (2.47 \pm 0.23) \times 10^{-5}$. The effect they had measured amounted to an extra distortion of the Sun's shape by a mere 31.3 ± 2.4 km.

Inevitably, this caused a considerable controversy which continues today. Direct determination of the Sun's oblateness is very complex, even using modern satellite data (Rozelot and Damiani, 2011; Kuhn *et al.*, 2012; Gough, 2012).

Asteroids and Near-Earth Objects

The planet Mercury was an ideal target for this test since it has the shortest period of all Solar system objects and has a highly eccentric orbit. The anomaly in the orbital precession had been noted by Le Verrier at the end of the 19th century. Asteroids also provide potential test objects: the asteroid Icarus, a 'Near Earth Asteroid', or 'NEA', has an orbital period of ~ 409 days and an orbital eccentricity of $e = 0.83$. The measurement of the orbital precession of the asteroid Icarus was first suggested by Shapiro *et al.* (1968, 1971).

The current procedure is to determine the orbital parameters via both optical astrometry and radar observations of these objects. Since the effect of anisotropic re-radiation of sunlight affects the orbit (the Yarkovsky drift rate) the asteroids have to be modelled taking account of such data as their size, rotation rate and albedo. The determination of the orbital precession of several Near Earth Asteroids is reviewed by Margot and Giorgini (2009) who

---

[6] The orbital inclination i is frequently set to $i = 0°$, whereas the inclination of the Earth's orbit to the Solar equatorial plane is $i_{\text{Earth}} = 7.15°$ and Mercury's is $i_{\text{Mercury}} = 3.38°$.

list data for eight objects having orbital precession in excess of $10''$ per century, including Icarus. The status of measurements through space-based geodesy is reviewed in detail by Combrinck (2013).

14.2.2 The Total Solar Eclipse of 1919

Since the biggest deflections of light by the Sun were expected from stars observed closest to the Solar limb (giving a larger value of $2m/r$), the only prospect for doing this experiment in the period shortly after the theory was published was to observe a star field during an eclipse of the Sun and to compare the same star field when the Sun was not there. This task was undertaken by Eddington in 1919 (Dyson *et al.*, 1920), a story that is beautifully recounted by Coles (1996) (see also Coles (2001)). The main result of that expedition is shown in Figure 14.1, which, taken at face value, is an outstanding affirmation of Einstein's prediction.

At present, the Solar gravity light deflection $\Delta\theta$ is measured in terms of a parameter γ that is exactly 1.0 for Einstein's theory. In terms of γ the deflection of a light ray passing the Sun at distance d is given by (Will, 2009)

$$\Delta\theta = \left(\frac{1+\gamma}{2}\right) \times 1.7505 \left(\frac{R}{d}\right) \text{ arcsec,} \qquad (14.10)$$

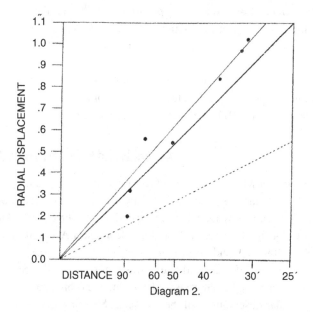

Diagram 2.

Fig. 14.1 First measurements of deflection of light by a gravitational field: the results of the Solar eclipse expedition of 1919 (Dyson *et al.*, 1920). The horizontal axis is the reciprocal of the deflection. The dashed line shows the expectation based on applying Newton's theory to the deflection of light, while the heavy solid line shows the expectation based on Einstein's general relativity. The thin solid line is the fit to the observations. *(From Dyson et al. (1920, Diagram 2, p332)).*

where R is the solar radius. The value was determined from VLBI (very long baseline inter-ferometry) by Shapiro *et al.* (2004), who in 2004 analysed deflection data of 541 compact radio sources over the period 1979–1999, and obtained $\gamma = 0.9998(3) \pm 0.0004(5)$. More recently Lambert and Le Poncin-Lafitte (2011), using data spanning the period 1979–2010, together with improved astrometric and geophysical models in the VLBI data pipeline, reported $\gamma = 0.99992 \pm 0.00012$.

14.2.3 The Gravitational Redshift

The other test mentioned by Einstein was the gravitational redshift, but at that time it was not considered feasible. In 1925, Adams reported the detection of a redshift in the light from the white dwarf companion of Sirius. The value agreed with the prediction of general relativity. However, it was not until 1959 that Pound and Rebka (1960) detected the gravitational redshift in a laboratory experiment, reporting a value for the gravitational redshift that was within 10% of Einstein's predicted value. What made that possible was the discovery by Mössbauer in 1959 of the effect that bears his name and was to win him the Nobel Prize in 1961. Within a few years Pound and Rebka increased their accuracy by a factor of 10.

The value of the gravitational redshift, $z = \delta \nu / \nu$ is expressed in terms of a parameter α and the Newtonian potential difference ΔU between the location of the emitter and receiver:

$$z = (1 + \alpha)\frac{\Delta U}{c^2}, \tag{14.11}$$

where for Einstein's theory $\alpha = 0.0$ (Will, 2006). Today the best value is from Vessot *et al.* (1980) using a clock launched to $10\,000$ km on a rocket, which puts a limit $|\alpha| < 2 \times 10^{-4}$, confirming the Einstein value to within 0.007%.[7]

14.2.4 Going beyond the Three Classical Tests

Einstein's general relativity makes several important predictions that have no analogue within the framework of Newtonian theory: they are effects uniquely associated with general relativity and its variants. Among these are the redshift of light by a gravitational field, the existence of gravitational waves, and the precession of gyroscopes.[8] A complete list and the associated tests are discussed in detail in Will (2014).

A possible generalisation of (14.3) to account for further tests was to calculate the consequences of the static metric

$$ds^2 = -(1 + A)dt^2 + (1 + B)dr^2 + 2Cdrdt + (1 + D)r^2(d\theta^2 + \sin^2 \theta d\phi^2), \tag{14.12}$$

---

[7] This experiment was named 'Gravity Probe-A' (GPA), launched June 1976.

[8] In 1963, NASA funded a project to look into the feasibility of measuring the general relativistic precession of gyroscopes in Earth orbit. It was not until 1994 that the project was given the go-ahead with a launch in April 2004. This mission was called 'Gravity Probe B' (GPB). However, owing to unforeseen sources of noise the mission could not achieve its hoped-for goal of $\sim 1\%$ precision in measuring the rate of precession, and barely achieved $\sim 20\%$ (Everitt, 2011). The mission had cost \$760 M over a period of 47 years. See Will (2015) and articles therein.

and to expand the functions A, B, C, D in a power series in $2m/r$ as is done in (14.3). To second order this would leave eight coefficients to determine. This, however, is not a very practical way of going about testing general relativity, particularly in the case of strong gravitational fields where $2m/r$ is not small. There is, moreover, the issue of the choice of coordinate system used in the metric and the expression of the expected phenomenon in terms of that particular coordinate system. We have seen the Schwarzschild metric expressed in very different looking ways! Bodenner and Will (2003) have an interesting discussion of the coordinate system issue in which they calculate light deflection in the Schwarzschild metric to order $(2m/r)^2$, using three different coordinate representations of the Schwarzschild metric. To first order the calculations agree, but they differ at second order.

The resolution of such problems is to compute observable quantities, and to express the results only in terms of directly measurable quantities. The metric components of general relativity are not observables!

14.2.5 Parameterised Post-Newtonian Gravitation

The *parametrised post-Newtonian* formalism provides more than a test of general relativity, it is potentially a mechanism for constraining theories of gravitation that can be formulated in the terms of PPN. That includes theories like scalar tensor theories. The formalism in its current form is due to Will and Nordtvedt (1972). In the current form of PPN, a gravitational theory is expressed in terms of ten parameters. The values of these parameters are constrained by experimental data, and they are updated as the experiments improve, thereby tightening the strength of the constraints.

Will (2006, 2009, 2014) provides excellent reviews of the general PPN approach, including dealing with specific theories of gravitation and with gravitation in non-vacuum environments.

14.2.6 Relativistic Effects in GPS Systems

A few decades ago it would have surprised people to discover that general relativity has any everyday-life applications. The theory was certainly respected as an important part of 20th century science, and Einstein was certainly the iconic scientist of the 20th century. But, unlike quantum mechanics, general relativity was not regarded as being 'down to Earth'. However, with the advent of consumer *Global Positioning Systems (GPS)*, that has all changed. GPS relies on accurate timing. and if the general theory of relativity had not been developed the resulting positioning errors would render the devices of limited use. Without the general relativistic corrections, positioning errors of around 10 km would build up in a day (Ashby, 2002, 2003).

There are two relativistic contributions to the timing. There is the special relativistic effect of time dilatation which comes into play because the satellites are moving at substantial speeds. This causes a slowing down of on-board clocks by around -7 microseconds per day relative to Earth-based clocks. Then there is the general relativistic effect due to the gravitational redshift which amounts to a gain of some 46 microseconds per day relative to the Earth-based clocks. The size of the general relativity contribution is bigger than the special relativity contribution and acts in the opposite sense, resulting in a timing error of around 39 microseconds per day.

The corrections are applied on-board the satellite: each satellite has a different orbit and so each will require different clock corrections. In effect, the on-board clocks are adjusted to coincide precisely with the International Atomic Time of Earth-bound clocks. In that way the Earth-bound GPS system does not need to know anything about the satellite with which it is communicating. Accuracies of a few nanoseconds are required to make the GPS an effective tool.

14.3 Beyond Einstein

14.3.1 The Cosmological Constant

In 1916 the widely held view was that the Universe, as it was known at the time, was static. Einstein had already shown that general relativity could improve on Newtonian theory as the theory of gravitation, and realised that general relativity would provide a mathematical description of the Universe. However, constructing a cosmological model would require a repulsive force to balance the attraction of gravitation. The solution was to add a term to the equations: the *cosmological constant*:

$$G_{ab} + \Lambda g_{ab} = \kappa T_{ab}, \qquad \kappa = \frac{8\pi G}{c^4}. \tag{14.13}$$

From here, Einstein was able to solve the equations with $T_{ab} = 0$, finding a stationary solution in which this balance was realised. We call that solution the *Einstein static Universe*. The balance between the force of gravity and the pressure from the Λ-term is unstable. In that sense the Einstein Universe could not be a candidate for the static Universe solution he was looking for. Later, de Sitter came up with a different model, also involving the cosmological constant.

14.3.2 What if not Einstein's Theory?

Modifications of Einstein theories are nothing new. Einstein (1917a) himself introduced the cosmological constant, Λ, and in a later paper (Einstein, 1919) tried to create a unified version of general relativity and electromagnetism from which Λ could be derived.[9] The Kaluza–Klein theory (Nordström, 1914; Kaluza, 1921; Klein, 1926)[10] sought to unify

---

[9] Einstein starts with field equations $R_{ik} - \frac{1}{4}Rg_{ik} = \kappa T_{ik}^{EM}$, where T_{ik}^{EM} is the electromagnetic field energy momentum tensor, and then recovers the standard field equations in the form $R_{ik} - \frac{1}{2}Rg_{ik} + \Lambda g_{ik} = \kappa T_{ik}$. He expresses satisfaction with that since *the quantity Λ appears in the fundamental equations as a constant of integration, and no longer as a universal constant peculiar to the fundamental law.* At the time this had the perceived advantage that the negative pressure due to the Λ-term would serve to bind the electron.

[10] Nordström's highly original idea of unifying Minkowski space and electromagnetism seems to have been largely forgotten. Pais (1982, p.204) reports Einstein as saying, in 1950: .. *in Zürich I believed that I had an argument which showed that the scalar theory, you know, the Nordström theory, was in conflict with the*

gravitation and electromagnetism in a 5-dimensional theory. Kaluza was successful in doing this, but had no explanation as to why we had no evidence for this extra spatial 5th dimension. That came with Klein, who suggested that the extra dimension was extremely small: the extra dimension is 'compactified' on such small scales as to be experimentally unobservable. This idea played a significant role in the subsequent development of higher dimensional theories of gravitation.[11] Somewhat later the Steady State theories of Bondi and Gold (1948) and Hoyle (1948) and Hoyle and Narlikar (1964) introduced modifications that would avoid the need for a cosmic singularity. The Brans and Dicke (1961) theory sought to embody a notion of Mach's principle by introducing a new scalar field into the Einstein equations.

There is renewed impetus to modify the Einstein equations following the discovery of acceleration of the cosmic expansion.. That has stimulated a number of different approaches to modifying the field equations. On the one hand, there are the modifications of the Einstein–Hilbert action by adding extra terms to the geometric part of the action. In this sense the Einstein equations are merely a first approximation to the bending of space-time by matter and energy. On the other hand there are theories which aim at doing better than merely adding a Λ-term. These propose a field-theoretic model for the 'dark energy' which we should think of as a medium pervading all of space. There is also an important third branch to this thinking which suggests that our space-time is not 4-dimensional, but is in fact of higher dimension and that the dark energy is perhaps a manifestation of that.

Einstein himself was insistent that any theory of gravitation should not throw out Newton's theory, but should embrace it as a first approximation. The same might now be said of Einstein's theory. Aside from the accelerated cosmic expansion, there is compelling evidence that Einstein's theory is, at the experimental level, the best model for a theory of gravitation that we have at this point. If there were a radical re-working of the theory of the gravitational field, it might be argued that we should not throw out the good things we have gained from Einstein's view.[12] The subject of the theories of modified gravity and their consequences for post-Newtonian tests of relativity and cosmology is exhaustively and clearly reviewed by Clifton *et al.* (2012).

14.4 Physics in Curved Space-Time

The energy momentum tensor[13] is among the most important concepts of physics. Indeed, in the general theory of relativity it plays a key role in determining the geometry of the space-time, and the geometry in turn determines the dynamics of the material described by

---

equivalence principle. But I soon realised I was wrong. In 1914 I came to believe in fact that the Nordström theory was a good possibility.

[11] The Kaluza–Klein theory is nicely reviewed and put into a modern context by Overduin and Wesson (1997).

[12] At a fundamental level, we might think of throwing out Riemannian geometry. However, Riemannian geometry with a connection is a good way of tying together the Lorentz frames of special relativity at all the points of space-time, thereby preserving special relativity locally and hence most of classical physics. Not everyone agrees with that, but those who disagree tend to be in a minority.

[13] Also called the *stress energy–momentum tensor*, or simply the *stress tensor* since it can also encapsulate the internal stresses in elastic media and the dissipative forces in fluid dynamics.

the energy–momentum tensor. General relativity is a nonlinear theory. Why is it that the energy–momentum tensor is the source, i.e. the right-hand side, of the field equations?

The approaches taken by Einstein and Hilbert as to the role of the energy–momentum tensor as a source of space-time curvature were quite different. By 1914 Einstein and Grossmann had derived the vacuum field equations from a variational principle, but they and others were effectively guessing how to include matter as a source of the gravitational field (see Section 13.4.1 and its footnote 12, p.323). Hilbert on the other hand, largely because of his strong support for Mie's view that everything should be explained in terms of electromagnetic fields, wrote down the variational principle so as to include the effect of the electromagnetic field. In that way Hilbert was able to derive the full Einstein field equations, and also derive the energy–momentum tensor for the electromagnetic field through variation of the action with respect to the metric. We shall see more about this below. Hilbert's equations became generally known as the Einstein–Maxwell equations.

14.4.1 The Gravitational Field in Curved Space-Time

Matter in space-time influences the geometry of the space-time. In Newtonian theory the material asserts its gravitational influence through the mass density; the effect of the gravitational force is calculated by solving the inhomogeneous Poisson equation for the potential, $\phi(\mathbf{x})$, of the force field:

$$\nabla^2 \phi = 4\pi G \rho, \tag{14.14}$$

$$m\ddot{\mathbf{x}} = -\nabla \phi. \tag{14.15}$$

The gravitational force on a test particle is then given by $\mathbf{F} = -\nabla \phi$. In Einstein's theory of gravitation, the matter in the Universe exerts its influence on the geometry via the *energy–momentum tensor*, T_{ab}, and the particle trajectories are the geodesics in the resulting geometry (see Section 13.4):

$$R_{ab} - \tfrac{1}{2}R g_{ab} = \kappa T_{ab}, \tag{14.16}$$

$$\ddot{x}^a + \Gamma^a_{bc} \dot{x}^b \dot{x}^c = 0. \tag{14.17}$$

The left-hand side of the first equation is the geometry described in terms of the Ricci tensor R_{an} and the scalar curvature R. The geometry is determined directly through T_{ab}. The second equation is the equation for a geodesic, and the object Γ^a_{bc}, the Christoffel connection, describes the local geometry. The Christoffel symbol Γ^a_{bc} is not a tensor: it depends explicitly on the choice of coordinates. It is necessary to calculate the components of Γ^a_{bc} for any particular coordinate system that might be used to describe the geometry g_{ab}.

How can we motivate Equation (14.17) on the basis of the principle of equivalence in the form 'change comma derivatives into semi-colons'? The Newtonian starting point (14.15) is not even Lorentz invariant.[14] Under no forces the Newtonian path is a straight line, the equation of which is $\ddot{x}^a = 0$:

---

[14] The argument is given in Carroll (2003, Section 4.1). The use of the chain rule is analogous to the familiar trick in mechanics of writing $\ddot{s} = v\,dv/ds$ with $v = \dot{s}$.

$$\frac{d^2 x^a}{d\lambda^2} = \underbrace{\frac{d}{d\lambda}\frac{dx^a}{d\lambda}}_{\substack{\text{chain} \\ \text{rule}}} = \frac{dx^b}{d\lambda}\frac{\partial}{\partial x^b}\frac{dx^a}{d\lambda} \xrightarrow[\substack{\text{equivalence} \\ \text{principle}}]{\partial_{,b} \to \partial_{;b}}$$

$$\frac{dx^b}{d\lambda}\left[\frac{\partial}{\partial x^b}\frac{dx^a}{d\lambda} + \Gamma^a_{bc}\frac{dx^c}{d\lambda}\right] \underset{\substack{\text{chain} \\ \text{rule}}}{=} \frac{d^2 x^a}{d\lambda^2} + \Gamma^a_{bc}\frac{dx^b}{d\lambda}\frac{dx^c}{d\lambda}. \qquad (14.18)$$

This last expression is just the left-hand side of (14.17).

Einstein was impressed by the obvious correspondence between the Poisson equation (14.15) and the geodesic equation (14.17) and so readily identified the expression of the action of gravity through Newtonian $\nabla^2 \phi$ with the action of the geometry $\Gamma^a_{bc}\dot{x}^b\dot{x}^c$. The look and feel of Einstein's work is dominated by the Γ^a_{bc} symbols, as we see in Einstein's *Zurich notebook* of 1913, (he uses the notation $\left\{\begin{smallmatrix} a\ b \\ c \end{smallmatrix}\right\}$ for Γ^a_{bc}). There are comments in the notebook to the effect that Einstein found the equations he was trying to deal with frustratingly difficult.

14.4.2 Electromagnetic Potentials in Curved Space

We have two descriptors of the electromagnetic field: the Faraday tensor F_{ab} and the 4-potential A^a from which F_{ab} is derived. With either of these, the procedure for writing down the curved-space Maxwell equations is fairly well defined. The Minkowski metric is replaced everywhere it appears in the Lorentz invariant equations with the space-time metric g_{ab}, and all derivatives are replaced with covariant derivatives. This latter step simply involves replacing the comma derivatives with semi-colon derivatives, an application of the principle of equivalence. However, we will show that this does not always lead to the correct answer.

Curved Space Maxwell Equations for the Faraday Tensor

The curved-space Maxwell Equations (10.135) are derived from the local Lorentz frame equations of special relativity by replacing comma derivatives with semi-colons. So Equations (10.135) and (10.136) generalise to

$$F^{ab}_{\ \ ;b} = \frac{4\pi}{c}J^a, \qquad (14.19)$$

$$F_{[ab;c]} = F_{ab;c} + F_{bc;a} + F_{ca;b} = 0, \qquad (14.20)$$

$$ma^a = q\,F^{ab}u_b, \qquad (14.21)$$

$$J^a_{\ ;a} = 0. \qquad (14.22)$$

The third of these is the Lorentz force acting on a charge q moving with instantaneous velocity u_b. The fourth, which follows from the first two, is the conservation of charge.

Curved Space Maxwell Equations for Vector Potential

The Maxwell equations in curved space can also be expressed directly in terms of the 4-potential, A^a, but, as we shall show, this is not a simple comma to semi-colon replacement for Equation (10.130).

The Faraday tensor F_{ab} is defined in terms of the 4-potential A_a by the equation

$$F^{ab} = -A^{a,b} + A^{b,a}. \tag{14.23}$$

The Maxwell equations that involve the source term J^a are obtained by taking the divergence of this:

$$F^{ab}{}_{,b} = -A^{a,b}{}_{,b} + A^{b\cdot a}{}_{,b\cdot} = \frac{4\pi}{c}J^a. \tag{14.24}$$

If we were to apply the Lorenz gauge condition $A^b{}_{,b} = 0$ we would immediately arrive at $\Box A^a = -4\pi J^a/c$, but we do not do that for the moment since that term plays a key role in what follows. The notation in (14.24) marks which of the pairs of indices have been lowered: dots have been inserted to denote the 'lowered' index.

A naïve application of the principle of equivalence (substituting $, \rightarrow ;$) would then lead us to write this as

$$-A^{a;b\cdot}{}_{\cdot b} + A^{b\cdot a}{}_{;b\cdot} = \frac{4\pi}{c}J^a, \quad \text{NB: incorrect!} \tag{14.25}$$

in a curved space-time. Misner *et al.* (1973, Section 16.3, Section 22.4) give a thorough discussion of this point.

The problem lies in the order of taking the derivatives in the second term $(A^{b\cdot a}{}_{,b\cdot})$ on the left of Equation (14.24). Had we switched the order of differentiation in that term, $A^{b\cdot a}{}_{,b\cdot} \rightarrow A^{b\cdot a\cdot}{}_{\cdot b}$, we would have written

$$-A^{a;b\cdot}{}_{\cdot b} + A^{b;a\cdot}{}_{\cdot b} = \frac{4\pi}{c}J^a. \tag{14.26}$$

In a curved space-time the order of covariant differentiation matters. This last equation can be manipulated to give the correct version of (14.25):

$$-A^{a;b\cdot}{}_{\cdot b} + A^{b\cdot a}{}_{;b\cdot} + R^{ab}A_b = \frac{4\pi}{c}J^a. \tag{14.27}$$

This is Equation (14.25) with an additional term $R^{ab}A_b$ coupling the Ricci tensor R^{ab} to the potential A_b. If we now bring in the Lorenz gauge condition in the form $A^b{}_{;b} = 0$ we have the curved space Maxwell equation

$$-A^{a;b}{}_b + R^{ab}A_b = \frac{4\pi}{c}J^a. \tag{14.28}$$

This is consistent with the curved space Maxwell equations written in terms of F^{ab}.

This resolution of the ambiguity is known as *factor ordering*. The differential operator

$$\Box_{dR}A^a \equiv -A^{a;b}{}_b + R^a{}_bA^a \tag{14.29}$$

is the curved space-time equivalent of the d'Alembertian $\Box$ operator. $\Box_{dR}$ is called the *de Rham vector wave operator*.

Part of the reason this is necessary is that the quantity A^a is not an observable quantity, the observables are the components of the Faraday tensor F^{ab}. The rule when using the principle of equivalence to generalise locally flat equations of physics is to only apply the generalisation to equations involving observable (measurable) quantities where intuition is a reasonable guide (see Misner *et al.* (1973, Box 16.1, p.390)).

14.5 Action Principles

In the view that the theory of space-time should be derivable from an action principle,[15] the action should consist of a term $\mathcal{L}_{GR}$ describing the geometry, and if the space-time is to have some matter, a term $\mathcal{L}_M$ for the material content. The latter may describe more than one component of the matter content and their mutual interactions:

$$\mathcal{S} = \int \mathcal{L}_{GR}\, \sqrt{-g}\, d^4x + \int \mathcal{L}_M\, \sqrt{-g}\, d^4x. \tag{14.30}$$

The geometric Lagrangian $\mathcal{L}_{GR}$ is a function of the geometry, which in the Riemannian model of space-time is a function of the metric tensor describing the local geometry at each point of space time.[16] The material Lagrangian $\mathcal{L}_M$ is a function of the fields describing the matter and their derivatives and has an implicit dependence on the geometry through the appearance of the metric in the covariant derivative.

The Einstein–Hilbert action describing the dynamics of the geometry is

$$\mathcal{L}_{GR} = R, \tag{14.31}$$

where R is the scalar curvature of the space-time. This is the simplest geometric term that one could have and it leads directly to the Einstein field equations in a vacuum:

$$\delta_g \int \mathcal{L}_{GR}\, \sqrt{-g}\, d^4x = 0 \quad \Rightarrow \quad R_{ab} - \tfrac{1}{2} g_{ab} R = 0. \tag{14.32}$$

The subscript g on δ_g indicates that the variation is with respect to the metric. R_{ab} is the Ricci tensor and $R = g^{ab} R_{ab}$.

The material Lagrangian $\mathcal{L}_M$ depends on the nature of the fields occupying the space-time. If space-time is permeated by a simple scalar field, ϕ, the Lagrangian will depend on the, ϕ, and its covariant derivatives, and hence on the metric: $\mathcal{L}_M = \mathcal{L}_M(\phi, \partial_a\phi, \dots)$. In classical field theory one varies the matter action $\mathcal{L}_M$ with respect to the fields ϕ. This results in the dynamical equations for the fields and ultimately leads to a specification of an energy momentum tensor, T^{ab}. The energy momentum tensor is divergence free, $T^{ab}{}_{;b} = 0$ which leads to a specification of the conserved properties of the field.

Varying the action with respect to the metric also yields the dynamical equations for the field and an expression for the energy momentum tensor:[17]

$$\delta_g \mathcal{S}_M = \delta_g \int \mathcal{L}_M\, \sqrt{-g}\, d^4x = \int T_{ab}\, \delta g^{ab}\, \sqrt{-g}\, d^4x. \tag{14.33}$$

---

[15] See Chapter 11 and the On-line Supplement *Lagrangians and Hamiltonians* for a more detailed discussion of the role of action principles in physics.

[16] $\mathcal{L}_{GR}$ and $\mathcal{L}_M$ are in fact Lagrangian *densities*, the term $\sqrt{-g}\, d^4x$ being the invariant element of volume in a Riemannian space having metric g_{ab} with $g = \det g_{ab}$.

[17] The derivation of these equations is rather technical. See Hawking and Ellis (1975, Section 3.3), Gotay and Marsden (1992, Theorem 4.1) and Forger and Römer (2004, Theorem 4.2), all of whom give detailed examples for a variety of fields.

This is equivalent to

$$T^{ab} = 2\frac{\partial \mathcal{L}_M}{\partial g_{ab}} + g^{ab}\mathcal{L}_M, \qquad T_{ab} = -2\frac{\partial \mathcal{L}_M}{\partial g^{ab}} + g_{ab}\mathcal{L}_M. \qquad (14.34)$$

Clearly, this definition of the energy–momentum tensor will give a symmetric T^{ab}. The appearance of the energy–momentum tensor via this route is quite remarkable and is a consequence of the expression of the action within the framework of Riemannian geometry where the metric is itself a dynamical variable. This provides a route for avoiding having to symmetrise T^{ab} defined via the usual route, as is the case with the electromagnetic field.[18]

As a simple example we can take the Minkowski space Lagrangian (11.40) for massive scalar field, and replace η^{ab} with g^{ab} to give the curved space-time Lagrangian

$$\mathcal{L} = -\tfrac{1}{2}g^{ab}\partial_a\phi\partial_b\phi - \tfrac{1}{2}m^2\phi^2. \qquad (14.35)$$

Using (14.34) easily gives

$$T^{ab} = \phi^{;a}\phi^{;b} - \tfrac{1}{2}g^{ab}\left[g^{pq}\phi_{;p}\phi_{;q} + m^2\phi^2\right]. \qquad (14.36)$$

We can go back to the non-GR version by replacing g^{ab} with η^{ab}.

14.5.1 Scalar Fields Coupled to Curvature

In Section 11.1.5, Equation (11.38), we showed how, within the framework of special relativity, to derive the energy–momentum tensor from the Lagrangian for the fields. The principle of equivalence then tells us that we can take the locally Lorentz invariant energy–momentum tensor we have derived, and simply adapt it to the general theory of relativity by changing all derivatives to covariant derivatives and replacing the Minkowski metric with the space-time metric g_{ab}. However, as shown in Section 14.4.2, this procedure is not without potential difficulties.

Generalising from Minkowski space physics to a general curved space-time relies on the assumption that the field equation in Minkowski space does not gain any terms involving the curvature of the space time. The Klein–Gordon equation affords a simple example of this. The simple generalisation to curved space-time is

$$\partial^a\partial_a\phi - m^2\phi = 0, \qquad (14.37)$$

while a possible alternative, in the absence of any other guiding principle, is

$$\partial^a\partial_a\phi - m^2\phi - \alpha R\phi = 0, \qquad (14.38)$$

for some constant α describing the coupling between the scalar field and the curvature. These equations are the same in a local Minkowski space.

---

[18] In calculating T^{ab} it is often convenient to make use of the equation

$$\frac{\partial g^{pq}}{\partial g_{ab}} = -\tfrac{1}{2}\left(g^{pa}g^{qb} + g^{pb}g^{qa}\right).$$

This follows from the definition of g^{ab} being the inverse of g_{ab}.

Equation (14.38) with $m = 0$ arises from the Lagrangian

$$\mathcal{L} = \tfrac{1}{2}\left(\phi_{;a}\phi^{;a} - \alpha R\phi^2\right), \tag{14.39}$$

where R is the scalar curvature of the space-time. This gives the equation for ϕ:

$$\Box\phi + \alpha R\phi = 0, \tag{14.40}$$

and the energy momentum tensor

$$\begin{aligned}
T_{ab} = {} & \phi_{;a}\phi_{;b} - \tfrac{1}{2}g_{ab}\phi_{;m}\phi^{;m} \\
& - \alpha\left[(R_{ab} - \tfrac{1}{2}g_{ab}R)\phi^2 + g_{ab}\Box(\phi^2) - (\phi^2)_{;ab}\right].
\end{aligned} \tag{14.41}$$

The terms on the second line arise from the $\alpha R\phi^2$ coupling of the scalar field ϕ to the curvature scalar.[19]

14.5.2 Modified Lagrangians

The question arises as to where these Lagrangians come from, or equivalently, how they are discovered. This is particularly pertinent in those cases where we cannot write down $L = T - V$ because we do not know what T and V might be, as in the case of electromagnetism. The basic rule is perhaps that if the proposed Lagrangian yields the correct field equations, then it must be a good candidate. But there are other requirements, perhaps the most important being that, for consistency with the geometric part of the Einstein equations, the energy momentum tensor T_{ab} should be a symmetric tensor in the sense that $T_{ab} = T_{ba}$ and that it should have zero divergence.

While the Equation (11.38) for the energy–momentum tensor does have zero divergence, it does not always result in energy–momentum tensors that are symmetric. We shall see an example of that when we discuss the energy–momentum tensor for the Maxwell electromagnetic field: it will be necessary to add a term that symmetrises it without affecting the zero-divergence condition. We shall also see that in general relativity there is a better way than (11.38) of generating an energy–momentum tensor from the Lagrangian which guarantees both symmetry and zero divergence.

The Lagrangian in Equation (11.38) can be multiplied by an arbitrary non-zero constant and still yield the same field equations and a rescaled energy–momentum tensor that still has zero divergence. We can take advantage of this to ensure that the terms energy–momentum tensor have the identical physical interpretation as they have in our local Newtonian world view. So, for example, the component T_{00} should coincide with the energy density of the field.

Lagrangians are not unique. As we shall see with the electromagnetic field we can add terms to a Lagrangian that do not change the field equations, but that render the Lagrangian more useful. Lagrangians do not even need to be quadratic in the field variables.

[19] Ford (1987) considered this as a model whereby the cosmological 'constant' would be damped to its present value by the action of such a scalar field. The problem with such an approach, as remarked by Ford himself, is that coupling fields to the curvature in this way also causes a decline in the effective gravitational constant G.

The elements available for constructing Lagrangians are generally the fields and their derivatives, or alternatively the potentials that yield the fields and the derivatives of these potentials. A good example is the Lagrangian for the general theory of relativity where the Lagrangian that yields the geometric part of the Einstein field equations is simply the Ricci scalar, R:

$$\mathcal{S} = \int R\sqrt{-g}\, d^4x. \tag{14.42}$$

Here, $\sqrt{-g}d^4x$, where $g = \det g_{ab}$, is the invariant volume element in a curved space-time.

For a variety of reasons people have looked at modifications for the Einstein field equations. The logical approach to generating alternative field equations is via modifications of the action. A simple modification is to add a terms in R^2 to the Lagrangian to give

$$\mathcal{S} = \int (R + \alpha R^2)\sqrt{-g}\, d^4x \tag{14.43}$$

for some constant α, which is to be constrained from experimental data and cosmology.

15

Solutions of the Einstein Equations

The Einstein equations are fundamental for the calculation and interpretation of cosmological data, and underlie our quest for Precision Cosmology. In this chapter we present
some of the most important solutions in general relativity that are relevant to cosmology,
but mostly without the detailed derivations. These can be found in many excellent texts
on the subject. There is a focus on comparing the same solution in different coordinate
systems.

After presenting short-cut methods to calculate Christoffel symbols, the cases of the
Bianchi I and FLRW solutions are worked out in detail. Then the classical solutions are
presented and put in their historical context. We do this in the classic way, using the
Christoffel symbols for each representation of the metric.

15.1 What is a Solution?

The Einstein equations relate the geometry of space-time to its material (i.e. energy)
content:

$$\underset{\text{geometry}}{G_{ab}} = \underset{\text{energy}}{\kappa \ T_{ab}} , \tag{15.1}$$

where G_{ab} is the *Einstein tensor* which depends on the metric of the space-time, and T_{ab} is
the energy–momentum tensor describing the physical nature of the material content. κ is
the gravitational coupling constant. Given the expression for an energy–momentum tensor
in some system of coordinates, determine what the metric of the space-time associated
with that distribution of energy must be. Conversely, given an explicit expression for a
metric tensor in terms of some coordinates, we can compute the Einstein tensor and then
determine what the energy–momentum tensor must be to be the source of that geometry.[1]

In classical electromagnetic theory this is a relatively straightforward process because
the theory is linear: given charges and currents we can determine the field generated. General relativity is nonlinear, so we cannot, except in some weak field limits, add solutions of
the Einstein equations to generate new solutions.

Electromagnetism and general relativity have a feature in common in that not all fields
are consistent with physically realistic sources. In electromagnetism we could specify a

---

[1] In the latter case, there is no prior guarantee that the resulting tensor will be a physically meaningful energy–
momentum tensor.

magnetic field that is purely radial but, since magnetic monopoles do not exist, there would be no physical source associated with that field. Likewise in general relativity there is no guarantee that a given arbitrary metric tensor will be compatible with any physically meaningful energy–momentum tensor.

There is one important advantage in relativity: because of the principle of covariance we are allowed to choose any coordinate system that may be convenient for describing our space-time.

Since the choice of coordinate system in general relativity is entirely arbitrary, we need to set down some guidelines as to what kind of coordinate systems might be 'useful' in addressing physical problems, and what attributes of a solution might be regarded as undesirable. If the sought-after solution has special symmetries, it is as well to choose a coordinate system that makes these symmetries explicit. We will next show two examples of this.

However, there is a downside to this freedom in that having a solution in some arbitrary coordinate system does not mean that we can easily interpret it. Until the 1980s this was the source of some serious misunderstandings about the physical nature of the largest scale perturbations to the cosmic density: were they 'real', or mere artefacts of a strange choice of coordinates?

15.1.1 Minkowski Space Transformed

We start by looking at the Minkowski space solution in three coordinate representations. The Minkowski metric is the simplest exact solution of the Einstein field equations and, in a coordinate system we shall denote by (x^0, x^1, x^2, x^3), it is given by

$$ds^2 = -(dx^0)^2 + (dx^1)^2 + (dx^2)^2 + (dx^3)^2, \tag{15.2}$$

(see Equation 10.31). Minkowski space is one of three static solutions of the Einstein equations. The other two are Einstein's static solution and the de Sitter solution.

If we transform to standard spherical polar coordinates with

$$x^0 = t, \qquad x^1 = r \sin\theta \sin\phi, \tag{15.3}$$

$$x^2 = r \sin\theta \cos\phi, \quad x^3 = r\cos\theta, \tag{15.4}$$

$$r \geq 0, \quad 0 \leq \theta \leq \pi, \quad 0 \leq \phi < 2\pi, \tag{15.5}$$

we get:

Minkowski space in spherical coordinates:

$$ds^2 = -dt^2 + dr^2 + r^2(d\theta^2 + \sin^2\theta\, d\phi^2), \tag{15.6}$$

$$= -dt^2 + dr^2 + r^2\, d\Omega^2, \tag{15.7}$$

where $d\Omega^2 = d\theta^2 + \sin^2\theta\, d\phi^2$ is the element of solid angle. This form of the metric tells us that the separation, $dl = ds_{dt=0}$, of two points of a 3-surface $t = constant$ is simply $dl^2 = dr^2 + r^2(d\theta^2 + \sin^2\theta\, d\phi^2)$, which we recognise as the distance between two points separated by $dr, d\phi, d\theta$, in a Euclidean 3-space that is described in spherical polar coordinates. The standard form of the metrics of most spherically symmetric coordinates are rather similar to this.

If we transform to light-cone (null) coordinates via the transformation

$$v = t + r, \qquad w = t - r, \quad v, w \in \mathbb{R}, \tag{15.8}$$

to get the metric into the form:

Minkowski space in light-cone (null) coordinates:

$$ds^2 = -dv\,dw + \tfrac{1}{4}(v - w)^2 d\Omega^2. \tag{15.9}$$

At first sight this looks somewhat strange since there is neither a dv^2 nor a dw^2 term. But this is why the coordinates are interesting, the surfaces $v = constant$ and $w = constant$ are null surfaces: the surfaces of a past and of a forward light-cone. This example illustrates just how intuitively difficult it can be to interpret the simplest of coordinate systems.

An important transformation of Minkowski space is the transform to *Rindler coordinates* which are defined by the transformation

$$x^0 = x \sinh(gt), \quad x^1 = x \cosh(gt), \quad x^2 = y, \quad x^3 = z, \tag{15.10}$$

which sends the Minkowski space line element into the form:

Minkowski space in Rindler coordinates:

$$ds^2 = -g^2 x^2 dt^2 + dx^2 + dy^2 + dz^2. \tag{15.11}$$

The Riemann tensor for this metric vanishes (as it must). Note that the inverse transformation

$$t = \frac{1}{g} \operatorname{arctanh} \frac{x^0}{x^1}, \quad x = \sqrt{(x^1)^2 - (x^0)^2}, \quad y = x^2, \quad z = x^3, \tag{15.12}$$

requires that $(x^1)^2 - (x^0)2 \geq 0$, or $|x^1| \geq |x^0|$, for x to be real. So this does not map the whole of Minkowski space into the t, x, y, z coordinates, only the interior of the light-cone of the observer at the origin of the Minkowski space. The observer who is at rest at the origin of the Rindler coordinate system must accelerate to maintain his/her position.[2]

---

[2] One-dimensional constant acceleration relative to an inertial observer means that $d(\gamma v)/dt = g$, where $\gamma = (1 - v^c/c^2)^{-1/2}$, which integrates to $\gamma v = gt$. This gives the velocity and position as a function of time:

$$v = \left(1 + (gt/c)^2\right)^{-1/2} gt \quad \Rightarrow \quad x = \sqrt{(c^2/g)^2 + (ct)^2} - c^2/g, \tag{15.13}$$

where the second equality is just the integral of the first. The path can alternatively be expressed as

$$\left(1 + (gx/c^2)^2\right)^2 - \left((gt/c^2)^2\right)^2 = 1. \tag{15.14}$$

This is a hyperbola in the (t, x) space and is the path our accelerating observer must take to maintain constant acceleration.

The instantaneous interval of proper time of the accelerating observer is $d\tau = \gamma^{-1} dt$, which integrates to

The solution was first obtained by Born (1909) and brought to wider prominence by Bade (1953).[3]

15.2 Solutions with a Diagonal Metric Tensor

In this section we are going to go through the steps needed to generate a solution of the Einstein equation in a form that is useful for doing calculations of physical phenomena. We will start with a proposed metric, calculate the Christoffel symbols, and then deduce the equations of motion. The specific examples we will use are the *Bianchi I metric* and the *FLRW metric*.

Most of the solutions dealt with here are described in coordinate systems in which the metric has only diagonal terms, so $g_{ab} = 0$, $a \neq b$. This cannot represent more than a small fraction of the known solutions, but they are the ones of immediate relevance in our context. Using the Lagrangian to calculate the Christoffel symbols is a general procedure that is simpler than evaluating all the derivatives of the metric and doing the sums.[4] The method of determining the connection from the Euler–Lagrange equations was given in Section 11.1.3 Equation (11.16) and is demonstrated in Section 15.5.1. Here we consider an alternative method that works only in the case that the metric is diagonal, $g_{ij} = 0$, $i \neq j$. This considerably simplifies direct calculation of the Christoffel symbols, in part because each can be written out explicitly and there is no need to sum over indices.

15.2.1 Diagonal Metrics: Calculating the $\Gamma^a{}_{bc}$s

> **Note: Turning off the summation convention**
>
> In what follows we shall occasionally need to drop the summation convention, as when discussing Christoffel symbols that have two of the indices the same. We shall indicate non-summed indices with capital letters, thus Γ^A_{Ac} will, for $A = 0, 1, 2, 3$ denote each of the symbols $\Gamma^0_{0c}, \Gamma^1_{1c}, \Gamma^2_{2c}, \Gamma^3_{3c}$ for any value of $c = 0, 1, 2, 3$. (This is not a standard form of notation, capital letter indices are generally reserved for spinor quantities, and we shall only use it for labelling individual Christoffel symbols.)

There are 64 Γ^a_{bc}s, but when the metric is diagonal 24 of them are zero, the ones for $a \neq b \neq c$ are all different. There are four Γ^A_{AA} ($A = 0, 1, 2, 3$) and 12 Γ^A_{Ac} (four values of A and

$gt/c = \sinh(g\tau/c)$ and thence the parameterisation of the path

$$t = \frac{c}{g} \sinh \frac{g\tau}{c}, \quad x = \frac{c^2}{g} \left[\cosh \frac{g\tau}{c} - 1 \right]. \tag{15.15}$$

The metric (15.11) follows by calculating $ds^2 = -c^2 dt^2 + dx^2$.

[3] For a modern discussion see Misner *et al.* (1973, Section 6.2 p166).

[4] The easiest way to compute algebraic expressions for Christoffel symbols of a given metric is undoubtedly to use the *Mathematica* program from http://www.wolfram.com/mathematica/. *Mathematica* does not itself provide the facility, but there are several reliable freely downloadable *Mathematica Notebooks* that do this.

three values of $c \neq A$ for each A). Likewise there are 12 Γ^A_{Ac}, but by symmetry ($\Gamma^a_{bc} = \Gamma^a_{cb}$) these do not have to be calculated. That leaves 12 Γ^c_{AA} because there are four values of A and for each of those three values of $c \neq A$. That accounts for all 64 possibilities.

A considerable simplification occurs because for a diagonal metric:

$$g^{AA} = 1/g_{AA}. \tag{15.16}$$

In the expression for the connection

$$\Gamma^a_{bc} = \tfrac{1}{2} g^{ap} \left(g_{bp,c} + g_{pc,b} - g_{bc,p} \right), \tag{15.17}$$

we note that, for a diagonal metric, the sum over the index p reduces to the one term for $p = a$. So the index p appearing in the brackets must be a. For a specific value of $a = A$, we then have

$$\Gamma^A_{bc} = \tfrac{1}{2} g^{AA} \left(g_{bA,c} + g_{Ac,b} - g_{bc,A} \right), \quad \text{no summation on } A. \tag{15.18}$$

Consider now the terms in which $b \neq c$, when the term $g_{bc,A} = 0$ (i.e. our diagonal assumption). Then only one of the terms $g_{bA,c}, g_{Ac,b}$ can be non-zero, namely the one with either $b = A$ or the one with $c = A$. We are left with

$$\Gamma^A_{Ac} = \tfrac{1}{2} g^{AA} g_{AA,c}, \quad \Gamma^A_{bA} = \tfrac{1}{2} g^{AA} g_{AA,b}, \tag{15.19}$$

again with no summation on the A. These are the same expression: $\Gamma^A_{Ab} = \Gamma^A_{bA}$. That leaves only the case $b = c = B \neq A$ for which we have

$$\Gamma^A_{BB} = -\tfrac{1}{2} g^{AA} g_{BB,A}. \tag{15.20}$$

Summing up, we have the following rules for relatively quick calculation of the values of the Christoffel symbols of a diagonal metric:

Christoffel symbols for general diagonal metric:

$$\Gamma^a_{bc} = 0 \quad \text{all of } a, b, c \text{ different}, \tag{15.21}$$

$$\Gamma^A_{AA} = \tfrac{1}{2} g^{AA} \frac{\partial g_{AA}}{\partial x^A}, \quad \Gamma^A_{BB} = -\tfrac{1}{2} g^{AA} \frac{\partial g_{BB}}{\partial x^A}, \tag{15.22}$$

$$\Gamma^A_{AC} = \tfrac{1}{2} g^{AA} \frac{\partial g_{AA}}{\partial x^C} = \Gamma^A_{CA}. \tag{15.23}$$

The capital Latin indices A, B, C take on values $0, 1, 2, 3$, and there is no summation implied when they are repeated.

The Christoffel symbols all involve products of two of the four diagonal metric components, so changing the signature of the metric has no effect on their numerical values. For the general diagonal metric the simplification is considerable, but the calculation can nonetheless be tedious.

Equations (15.22 and 15.23), for the non-zero components, can be written in a slightly different way:

$$\Gamma^A_{AA} = \frac{1}{2} \frac{\partial \ln |g_{AA}|}{\partial x^A}, \quad \Gamma^A_{BB} = -\frac{1}{2g_{AA}} \frac{\partial g_{BB}}{\partial x^A}, \quad \Gamma^A_{AC} = \Gamma^A_{CA} = \frac{1}{2} \frac{\partial \ln |g_{AA}|}{\partial x^C}, \tag{15.24}$$

since $g^{AA} = 1/g_{AA}$ for a diagonal metric and remembering that there is no summation on the repeated indices.

15.3 Bianchi I Solution

For our first example we look at the Bianchi I solution. This metric was first fully investigated in the cosmology context by Jacobs (1968). The Bianchi I model is one of a large family of homogeneous but anisotropic cosmological models. The Bianchi I solution and its family are beautifully discussed in Ellis *et al.* (2012, Ch. 18, Section 18.2). The application of this model and others to the polarisation of the CMB is covered using a similar metric-based approach to that used here in Sung and Coles (2011).

The Bianchi I model has metric

$$ds^2 = -c^2dt^2 + a(t)^2dx^2 + b(t)^2dy^2 + c(t)^2dz^2, \qquad (15.25)$$

which, when $a(t) = b(t) = c(t)$ reduces to the flat FLRW model. The evolution of these models depends on the nature of the material content.

This metric is particularly simple since the coefficients of the coordinate differentials dx, dy, dz are only functions of time and not the spatial coordinates. We note that the time coordinate here has no pre-multiplying function and so an observer at fixed coordinates (x, y, z), is a comoving observer and has $ds^2 = -c^2dt^2$. The time coordinate ticks off his local comoving proper time.[5] We can guess that when the functions $a(t), b(t), c(t)$ are all different, the coordinate axes will be stretched differently in the (x, y, z) directions. It will remain to be proved that this is not simply the transformation of an otherwise isotropic space into a strange coordinate system (just as (15.11) and (15.9) are in fact the Minkowski metric).

First we need the Christoffel symbols and Ricci tensor for this model.

15.3.1 Christoffel Symbols and the Ricci Tensor Components

Using the procedure described in Section 15.2.1, the Christoffel symbols for (15.25) are easily verified to be

$$\Gamma^0_{11} = \dot{a}a, \quad \Gamma^0_{22} = \dot{b}b, \quad \Gamma^0_{33} = \dot{c}cr,$$

$$\Gamma^1_{10} = \Gamma^1_{01} = \frac{\dot{a}}{a}, \quad \Gamma^2_{20} = \Gamma^2_{02} = \frac{\dot{b}}{b}, \quad \Gamma^3_{30} = \Gamma^3_{03} = \frac{\dot{c}}{c}. \qquad (15.26)$$

As an aside, the motion of a particle moving only under the influence of gravity is described by the geodesic equation

$$\frac{d^2x^m}{ds^2} + \Gamma^a_{pq}\frac{dx^p}{ds}\frac{dx^q}{ds} = 0 \quad \Rightarrow \quad \frac{d^2x^m}{ds^2} = 0, \qquad (15.27)$$

---

[5] Such solutions were first investigated by Kasner (1925b), but did not attract much attention until the 1970s. See Section 15.3.4.

where s is an affine parameter along the geodesic. i.e. proper time. The right-hand equation follows simply because $\Gamma_{00}^m = 0$ for all m. A particle that starts at rest stays at rest. This is indeed a *comoving coordinate system*.

With this it can be verified that the four non-zero Ricci tensor components are the diagonal ones:

$$R_0^0 = -\left(\frac{\ddot{a}}{a} + \frac{\ddot{b}}{b} + \frac{\ddot{c}}{c}\right), \tag{15.28}$$

$$R_1^1 = -\left[\frac{\ddot{a}}{a} + \frac{\dot{a}}{a}\left(\frac{\dot{b}}{b} + \frac{\dot{c}}{c}\right)\right], \quad R_2^2 = -\left[\frac{\ddot{b}}{b} + \frac{\dot{b}}{b}\left(\frac{\dot{c}}{c} + \frac{\dot{a}}{a}\right)\right],$$

$$R_3^3 = -\left[\frac{\ddot{c}}{c} + \frac{\dot{c}}{c}\left(\frac{\dot{a}}{a} + \frac{\dot{b}}{b}\right)\right], \tag{15.29}$$

$$R = -2\left(\frac{\ddot{a}}{a} + \frac{\ddot{b}}{b} + \frac{\ddot{c}}{c} + \frac{\dot{a}\,\dot{b}}{a\,b} + \frac{\dot{b}\,\dot{c}}{b\,c} + \frac{\dot{c}\,\dot{a}}{c\,a}\right). \tag{15.30}$$

In the case $a(t) = b(t) = c(t)$ these reduce to the Friedmann equations.

The corresponding Einstein field equations, with a cosmological constant are

$$R_{ik} - \tfrac{1}{2}R\,g_{ik} = -\kappa\,T_{ik} + \Lambda g_{ok}, \tag{15.31}$$

where κ is the standard gravitational coupling, $\kappa = 8\pi G/c^2$. The energy–momentum tensor is

$$T_{ik} = (\rho + p)u_i u_k + p\,g_{ik}, \tag{15.32}$$

where u_i is the 4-velocity which has $u^k u_k = -1$. We can bring the Λ-term into this by using $p_\Lambda = -\rho_\Lambda = -\Lambda/8\pi G$.

15.3.2 Bianchi I Solution: Kinematics

We can make some progress in studying the kinematics of the solution, i.e. the aspects which do not depend on the energy–momentum tensor, if we introduce new variables

$$H_x = \dot{a}/a, \quad H_y = \dot{b}/b, \quad H_z = \dot{c}/c. \tag{15.33}$$

The H_x, H_y, H_z are the expansion rates in the three directions x, y, z. This is analogous to how the isotropic equations are handled. Define some familiar-looking cosmological variables:

$$V = abc = R(t)^3, \quad H = \tfrac{1}{3}\left(H_x + H_y + H_z\right), \tag{15.34}$$

with which we can define the volume expansion rate Θ:

$$\Theta \equiv \frac{\dot{V}}{V} = H_x + H_y + H_z = 3H. \tag{15.35}$$

Since $V = R^3$:

$$H = \frac{\dot{R}}{R}, \quad \Theta = 3\frac{\dot{R}}{R}. \tag{15.36}$$

Since $\dot{\Theta} = 3\dot{H}$, this gives

$$\dot{\Theta} + \tfrac{1}{3}\Theta^2 = 3\ddot{R}/R, \tag{15.37}$$

which gives the acceleration of the overall expansion.

From H_x, H_y, H_z we can construct the 3×3 expansion rate matrix **H** with components

$$H_{\mu\nu} = \text{diag}\left(H_x, H_y, H_z\right), \qquad \mu, \nu = 1, 2, 3. \tag{15.38}$$

This can be split into a trace-free matrix $\Sigma_{\mu\nu}$ and an isotropic part:

$$\Sigma_{\mu\nu} = H_{\mu\nu} - \tfrac{1}{3}(H_x + H_y + H_z)\delta_{\mu\nu}, \tag{15.39}$$

$$= H_{\mu\nu} - \tfrac{1}{3}\Theta\delta_{\mu\nu}, \tag{15.40}$$

which is

$$\Sigma_{\mu\nu} = \tfrac{1}{3}\,\text{diag}\left(2H_x - H_y - H_z,\ 2H_y - H_z - H_x,\ 2H_z - H_x - H_y\right), \quad \mu, \nu = 1, 2, 3. \tag{15.41}$$

Furthermore, $\dot{V}/V = 3\dot{R}/R$, and so an alternative way of writing this is

$$\Sigma_{\mu\nu} = \text{diag}\left(H_x - H,\ H_y - H,\ H_z - H\right), \quad \mu, \nu = 1, 2, 3, \tag{15.42}$$

with which we have:

Bianchi I spatial shear tensor $\Sigma_{\mu\nu}$:

$$\Sigma_{\mu\nu} = \text{diag}\left(\frac{\dot{a}}{a} - \frac{\dot{R}}{R},\ \frac{\dot{b}}{b} - \frac{\dot{R}}{R},\ \frac{\dot{c}}{c} - \frac{\dot{R}}{R}\right) \quad \mu, \nu = 1, 2, 3. \tag{15.43}$$

This shows $\Sigma_{\mu\nu}$ as being a direct measure of the local anisotropy, measuring the differences in each of the x, y, z directions of the expansion rate relative to the overall expansion rate.

From either of (15.40) or (15.41)[6] we can calculate that

$$2\Sigma^2 + \tfrac{1}{3}\Theta^2 = H_x^2 + H_y^2 + H_z^2, \qquad \Sigma^2 = \tfrac{1}{2}\Sigma_{\mu\nu}\Sigma_{\mu\nu}, \tag{15.46}$$

where we use the summation convention over the 3-indices $\mu, \nu = 1, 2, 3$ of the matrix $\Sigma_{\mu\nu}$. Σ is the magnitude of the shear. Likewise we can show the important result:

Magnitude of the spatial shear, Σ, in Bianchi I Universe:

$$2\Sigma^2 = (H_x^2 - H^2) + (H_y^2 - H^2) + (H_z^2 - H^2). \tag{15.47}$$

Equation (15.47) shows how the shear Σ is directly related to a natural measure of the anisotropy. The ratio of Σ to H, i.e. the relative anisotropy, is directly related to the

---

[6] Doing it from (15.41) requires the identities

$$\tfrac{3}{2}\Sigma^2 = (H_x^2 + H_y^2 + H_z^2) - (H_x H_y + H_y H_z + H_z H_x), \tag{15.44}$$

$$\Theta^2 = (H_x^2 + H_y^2 + H_z^2) + 2(H_x H_y + H_y H_z + H_z H_x). \tag{15.45}$$

The second of these shows that $\Sigma = 0$ if the expansion is isotropic: $H_x = H_y = H_z$.

observed anisotropy of the cosmic background radiation (Barrow *et al.*, 1985; Sung and Coles, 2011).

15.3.3 Bianchi I Solution: Dynamics

To make progress we must invoke the Einstein equations with an energy momentum tensor which represents the source of the gravitational field.[7] We adopt the perfect fluid form for the energy momentum tensor with the usual isotropic pressure term $p = p(\rho)/c^2$,

$$T_{ab} = (\rho + p)u_a u_b + p g_{ab}, \tag{15.48}$$

and include the cosmological term Λ. We get the field equations for the metric (15.25) in the form

$$\frac{\ddot{b}}{b} + \frac{\ddot{c}}{c} + \frac{\dot{b}\dot{c}}{bc} = -8\pi G p + \Lambda, \quad \frac{\ddot{c}}{c} + \frac{\ddot{a}}{a} + \frac{\dot{c}\dot{a}}{ca} = -8\pi G p + \Lambda,$$

$$\frac{\ddot{a}}{a} + \frac{\ddot{b}}{b} + \frac{\dot{a}\dot{b}}{ab} = -8\pi G p + \Lambda, \tag{15.49}$$

$$\frac{\dot{a}\dot{b}}{ab} + \frac{\dot{c}\dot{a}}{ca} + \frac{\dot{b}\dot{c}}{bc} = -8\pi G \rho + \Lambda. \tag{15.50}$$

Jacobs (1968) was the first to give a complete set of solutions, with $\Lambda = 0$, for a variety of equations of state. For a vacuum cosmology with $\Lambda = 0$, the sum of these four equations is $R = 0$: the vacuum model has zero curvature.

We can eliminate both p and Λ from Equations (15.49) to leave two equations for the functions $a(t), b(t), c(t)$:

$$\frac{\ddot{a}}{a} - \frac{\ddot{b}}{b} + \frac{\dot{c}}{c}\left(\frac{\dot{a}}{a} - \frac{\dot{b}}{b}\right) = 0, \quad \frac{\ddot{b}}{b} - \frac{\ddot{c}}{c} + \frac{\dot{a}}{a}\left(\frac{\dot{b}}{b} - \frac{\dot{c}}{c}\right) = 0, \tag{15.51}$$

which can be integrated to give

$$\frac{\dot{a}}{a} - \frac{\dot{b}}{b} = \frac{k_c}{abc}, \quad \frac{\dot{b}}{b} - \frac{\dot{c}}{c} = \frac{k_a}{abc}. \tag{15.52}$$

This contains only terms in the expansion rates H_x, H_y, H_z and the volume $V = abc = R^3$.

The expansion rate is

$$\theta = \tfrac{1}{3}\text{tr}(H_{\mu\nu}) = \tfrac{1}{3}(H_x + H_y + H_z). \tag{15.53}$$

The shear is the trace-free part of $H_{\mu\nu}$:

$$\sigma_{\mu\nu} = H_{\mu\nu} - \theta\delta_{\mu\nu}, \tag{15.54}$$

$$= \tfrac{1}{3}\text{diag}\left(2H_x - H_y - H_z,\ 2H_y - H_z - H_x,\ 2H_z - H_x - H_y\right). \tag{15.55}$$

We then find that the expansion rate $\theta = u^a{}_{;a}$ and the shear σ are given by:

---

[7] Here we are following Singh and Tiwari (2008), who treat this problem for an equation of state $p = w\rho$.

Expansion, θ, and trace-free shear of Bianchi I:

$$\theta = 3H, \qquad 2\sigma^2 = H_x^2 + H_y^2 + H_z^2 - \tfrac{1}{3}\theta^2, \tag{15.56}$$

$$8\pi G\rho + \Lambda = H_x H_y + H_y H_z + H_z H_x. \tag{15.57}$$

15.3.4 Kasner's Solution

The Kasner solution is a subset of the Bianchi Type I models, also referred to as the *Vacuum Bianchi I* solution. The solution finds an important role in the study of cosmological singularities in general relativity since its behaviour is characteristic of many homogeneous but anisotropic solutions, particularly in the early work of Khalatnikov and his co-workers. All the Bianchi I models behave like Kasner's solution near to their singularity.

The Kasner metric is (Landau and Lifshitz, 1980, Section 102, p.354 *et seq.*)

$$ds^2 = -dt^2 + t^{2p_1}\,dx_1^2 + t^{2p_2}\,dx_2^2 + t^{2p_3}\,dx_3^2, \tag{15.58}$$

$$p_1 + p_2 + p_3 = 1, \qquad p_1^2 + p_2^2 + p_3^2 = 1, \tag{15.59}$$

where (x_1, x_2, x_3) are Cartesian coordinates on $\mathbb{R}^3$, $t > 0$ and the constants p_i are real numbers. If we arrange the p_i so that $p_1 < p_2 < p_3$ we have

$$-\tfrac{1}{3} \le p_1 \le 0, \quad 0 \le p_2 \le \tfrac{2}{3}, \quad \tfrac{2}{3} \le p_3 \le 1. \tag{15.60}$$

The solution is a vacuum solution so $R_{ab} = 0$ and $R = 0$. The Kretschmann scalar is

$$R_{abcd}R^{abcd} = \left(p_1^2 p_2^2 + p_2^2 p_3^2 + p_3^2 p_1^2\right) t^{-4}, \tag{15.61}$$

and so is only zero in the special case that two of the p_i are zero. The volume of the solution grows with t, and all solutions have a singularity at $t = 0$.

Generalisations of the Kasner solution are discussed in Misner *et al.* (1973, Box 30.1 p.806).

15.4 The Friedmann–Lemaître Solution ($\Lambda = 0$)

The Robertson–Walker metric expressed in comoving spherical coordinates has the form

$$ds^2 = -dt^2 + a(t)^2 \left[\frac{dr^2}{1 - \kappa r^2} + r^2 d\theta^2 + r^2 \sin^2\theta d\phi^2 \right], \tag{15.62}$$

where $a(t)$ is the dimensionless scale factor, r is a (comoving) radial coordinate having dimension $[L]$, and κ is the curvature constant with dimensions $[L^{-2}]$.

The spatial part of the metric

$$d\ell^2 = a(t)^2 \left[\frac{dr^2}{1 - \kappa r^2} + r^2 d\theta^2 + r^2 \sin^2\theta d\phi^2 \right] \tag{15.63}$$

measures the *comoving distance* between two neighbouring points at the same value of t. The coefficient of dr^2 tells us that the geometry is not flat unless $\kappa = 0$, while the pre-multiplying scaling factor $a(t)$ tells us that the spatial volume is evolving in time.

15.4.1 Christoffel Symbols for the Metric

The non-zero Christoffel symbols can most directly be calculated following the recipe given in Section 15.2 and are[8]

$$\Gamma^0_{11} = \frac{a\dot{a}}{1 - \kappa r^2}, \quad \Gamma^0_{22} = a\dot{a}\,r^2, \quad \Gamma^0_{33} = a\dot{a}\,r^2 \sin^2\theta, \tag{15.65}$$

$$\Gamma^1_{11} = \frac{\kappa r}{1 - \kappa r^2}, \quad \Gamma^1_{22} = -r(1 - \kappa r^2), \quad \Gamma^1_{33} = -r(1 - \kappa r^2)\sin^2\theta, \tag{15.66}$$

$$\Gamma^2_{33} = -\sin\theta\cos\theta, \quad \Gamma^2_{12} = \Gamma^2_{21} = \frac{1}{r}, \quad \Gamma^3_{32} = \Gamma^3_{23} = \cot\theta, \quad \Gamma^3_{31} = \Gamma^3_{13} = \frac{1}{r}, \tag{15.67}$$

$$\Gamma^1_{10} = \Gamma^1_{01} = \Gamma^2_{20} = \Gamma^2_{02} = \Gamma^3_{30} = \Gamma^3_{03} = \frac{\dot{a}}{a}. \tag{15.68}$$

The corresponding non-zero elements of the Ricci tensor R_{ab} are

$$R_{00} = -3\frac{\ddot{a}}{a}, \quad R_{11} = \frac{1}{1 - \kappa r^2}\left(a\ddot{a} + 2\dot{a}^2 + 2\kappa\right), \tag{15.69}$$

$$R_{22} = r^2\left(a\ddot{a} + 2\dot{a}^2 + 2\kappa\right), \quad R_{33} = r^2\sin^2\theta\left(R\ddot{a} + 2\dot{a}^2 + 2\kappa\right), \tag{15.70}$$

and the Ricci scalar $R = R^a{}_a$ is

$$R = 6\left(\frac{\ddot{a}}{a} + \frac{\dot{a}^2}{a^2} + \frac{\kappa}{a^2}\right). \tag{15.71}$$

(To get this it is necessary to first get the mixed-form of the Ricci tensor $R^a{}_b = g^{am}R_{mb}$.)

We shall write the Einstein equations for a perfect fluid in the form

$$R_{ab} = T_{ab} - \tfrac{1}{2}g_{ab}T, \quad T_{ab} = (\rho + p)u_a u_b + p g_{ab}. \tag{15.72}$$

The coordinate system is comoving so a comoving observer has 4-velocity $u^a = (1, 0, 0, 0)$. Hence using mixed indices we simply have $T^a{}_b = \mathrm{diag}(-\rho, p, p, p)$.

Now use the values for the components of the Ricci tensor R_{ab} given above. This results in only two equations, one arising from the 00-component R_{00} and the other arising from the α, β-components ($\alpha, \beta = 1, 2, 3$). The α, β-components provide only one equation since the spatial diagonal entries of the energy–momentum tensor are all the same, and all the off-diagonal components are zero. We get

$$\text{00-component}: \qquad -3\frac{\ddot{a}}{a} = 4\pi G(\rho + 3p), \tag{15.73}$$

---

[8] If we write the metric (15.62) in the slightly more general form $ds^2 = -dt^2 + a(t)^2\gamma_{\mu\nu}dx^\mu dx^\nu$, $\mu, \nu = 1, 2, 3$, the Christoffel symbols are

$$\Gamma^0_{\mu\nu} = a\dot{a}\gamma_{\mu\nu}, \quad \Gamma^\mu_{0\nu} = \frac{\dot{a}}{a}\delta^\mu_\nu, \quad \Gamma^\sigma_{\mu\nu} = \tfrac{1}{2}\gamma^{\sigma\rho}\left(\partial_\mu\gamma_{\nu\rho} + \partial_\nu\gamma_{\mu\rho} - \partial_\rho\gamma_{\mu\nu}\right), \tag{15.64}$$

and all others are zero.

$$\alpha\beta\text{-component}: \quad \frac{\ddot{a}}{a} + 2\frac{\dot{a}^2}{a^2} + 2\frac{\kappa}{a^2} = 4\pi G(\rho - p). \tag{15.74}$$

These are then combined to give the Friedmann equations:

Friedmann equations:

$$\left(\frac{\dot{a}}{a}\right)^2 = \frac{8\pi G}{3}\rho - \frac{\kappa}{a^2}, \qquad \frac{\ddot{a}}{a} = -\frac{4\pi G}{3}(\rho + 3p), \tag{15.75}$$

which can be solved when given an equation of state $p = p(\rho)$.

We can rearrange Equations (15.75) and re-introduce the cosmological constant to give a form of the Friedmann equations that separates the contributions of density and pressure:

Friedmann equations – alternative versions:

$$\left(\frac{\dot{a}}{a}\right)^2 + \frac{\kappa}{a^2} = \frac{8\pi G}{3}(\rho + \Lambda^*), \quad 2\frac{\ddot{a}}{a} + \left(\frac{\dot{a}}{a}\right)^2 + \frac{\kappa}{a^2} = -\frac{8\pi G}{3}(p - \Lambda^*). \tag{15.76}$$

Here $\Lambda^* = \Lambda/8\pi G$ is the effective mass density, ρ_Λ, and negative pressure, $-p_\Lambda$, of the Λ contribution to the Einstein field equations.

In addition to these dynamical equations we have the conservation of mass/energy and momentum as expressed through the contracted Bianchi identity $G^{ab}{}_{;b} = 0$ (see Equation 13.116). This gives us $T^{ab}{}_{;b} = 0$, the $a = 0$ component of which is:

GR energy conservation in a perfect fluid:

$$\dot{\rho} + 3\frac{\dot{a}}{a}(\rho + p) = 0. \tag{15.77}$$

This is the GR equivalent of Equation (5.57), see Section 5.9.2 and footnote 31. The derivation of (15.77) follows easily from $T^a{}_b = \mathrm{diag}(-\rho, p, p, p)$ and the definition of its covariant derivative (13.67).

15.5 Solutions Having Spherical Symmetry

Having spherical symmetry simplifies most problems in general relativity, but nonetheless even simple results are not achieved without considerable effort. Here we shall study a standard metric describing a general spherically symmetric space-time. We show how to derive the Christoffel symbols using the Euler–Lagrange equations and discuss particular solutions and their generalisations.

The most general spherically symmetric metric can be written as[9]

$$ds^2 = -e^{2F(r,t)}dt^2 + e^{2H(r,t)}dr^2 + r^2(d\theta^2 + \sin^2\theta\, d\phi^2). \qquad (15.78)$$

This metric is very important as it underlies the theory of both homogeneous cosmological models and of spherically symmetric fluid models for massive stars and black holes. The functions $F(r,t)$ and $H(r,t)$ are determined by the material content of the space-time as described by the energy–momentum tensor and by the boundary conditions defining the problem. We shall use this form of the metric to demonstrate how the Einstein equations can be written out in a given coordinate system.

15.5.1 Christoffel Symbols – the Connection

To write out the Einstein equations in a given coordinate system requires the calculation of the Christoffel symbols for the given metric. This is generally a tedious procedure resulting in equations that cannot be solved except in very special cases. This accounts for the relatively small number of exact solutions that were known until the last couple of decades of the 20th century, when other techniques for generating solutions were found.

The most efficient way of determining the Christoffel symbols for a metric is to write down the equations for the geodesics from the Euler–Lagrange equations.[10] Geodesics were discussed in Section 13.4. The method of determining the connection from the Euler–Lagrange equations was given in Equation (11.16). We follow that path here.

The Lagrangian for the geodesics in the metric (15.78) is

$$L = -e^{2F(r,t)}\left(\frac{dx^0}{d\lambda}\right)^2 + e^{2H(r,t)}\left(\frac{dr}{d\lambda}\right)^2 + r^2\left(\frac{d\theta}{d\lambda}\right)^2 + r^2\sin^2\theta\left(\frac{d\phi}{d\lambda}\right)^2, \qquad (15.79)$$

where λ is a parameter along the geodesic. From this we can derive four Euler–Lagrange equations, one for each of the variables x^a. The x^0 equation that arises from this is

$$\frac{d^2x^0}{d\lambda^2} + 2F'\frac{dr}{d\lambda}\frac{dx^0}{d\lambda} - \dot{F}\left(\frac{dx^0}{d\lambda}\right)^2 + \dot{H}e^{2H-2F}\left(\frac{dr}{d\lambda}\right)^2 = 0. \qquad (15.80)$$

In terms of the Christoffel symbols, the x^0 geodesic equation is

$$\frac{d^2x^0}{d\lambda^2} + 2\Gamma^0_{r0}\frac{dr}{d\lambda}\frac{dx^0}{d\lambda} + \Gamma^0_{00}\left(\frac{dx^0}{d\lambda}\right)^2 + \Gamma^0_{rr}\left(\frac{dr}{d\lambda}\right)^2 = 0. \qquad (15.81)$$

Now we simply compare (15.80) with (15.81) to get the result

$$\Gamma^0_{r0} = F', \quad \Gamma^0_{00} = -\dot{F}, \quad \Gamma^0_{rr} = \dot{H}e^{2(H-F)}, \qquad (15.82)$$

---

[9] Tolman (1934b, Sections 94,98) shows that any spherically symmetric metric can be reduced to this form. (NB: Tolman uses the $(---+)$ metric signature and a somewhat different notation for the symbols.) He also discusses an alternative form of the spherically symmetric metric:

$$ds^2 = -e^{2A(r,t)}dt^2 + e^{2B(r,t)}[dr^2 + r^2(d\theta^2 + \sin^2\theta\, d\phi^2)].$$

Note that the coordinates (r,t) here are not to be thought of as being the same as in the metric (15.78).

[10] Misner *et al.* (1973, Section 14.6 and Ex. 14.13) present this calculation using differential forms.

where the dot, $\dot{\ }$, and dash, $'$, denote derivatives with respect to the variables t and r respectively. The remaining Christoffel symbols are:

$$\Gamma^r_{00} = F' e^{2(F-H)}, \quad \Gamma^r_{rr} = H', \quad \Gamma^r_{0r} = \dot{H}, \tag{15.83}$$

$$\Gamma^r_{\theta\theta} = -r e^{-2H}, \quad \Gamma^r_{\phi\phi} = -r \sin^2\theta\, e^{-2H}, \tag{15.84}$$

$$\Gamma^\theta_{\phi\phi} = -\sin\theta\cos\theta, \quad \Gamma^\theta_{r\theta} = r^{-1}, \tag{15.85}$$

$$\Gamma^\phi_{\phi\theta} = \cot\theta, \quad \Gamma^\phi_{r\phi} = r^{-1}. \tag{15.86}$$

There are 12 independent Christoffel symbols expressed in terms of two functions, $F(r,t)$ and $H(r,t)$. If we consider *static solutions* in which $F = F(r)$ and $H = H(r)$ the number of Christoffel symbols is nine.

15.5.2 The Einstein Tensor

With these Christoffel symbols and the metric we can now write down the components of the Einstein tensor:

Einstein tensor for the metric (15.78):

$$G_{00} = e^{-2H}\left(\frac{2}{r}H' - \frac{1}{r^2}\right) + \frac{1}{r^2}, \tag{15.87}$$

$$G_{11} = e^{-2H}\left(\frac{2}{r}F' + \frac{1}{r^2}\right) - \frac{1}{r^2}, \tag{15.88}$$

$$G_{22} = G_{33} = e^{-2H}\left(F'' + F'^2 - H'F' + \frac{1}{r}(F' - H')\right)$$
$$- e^{-2F}\left(\ddot{H} + \dot{H}^2 - \dot{H}\dot{F}\right), \tag{15.89}$$

$$G_{01} = G_{10} = \frac{2}{r}\dot{H}\, e^{-(F+H)}. \tag{15.90}$$

(The differences from Tolman (1934b, Eq. 98.3) are due to the different signature of the metric and to his use of the mixed covariant-contravariant version of the tensor.)

15.5.3 Static Spherically Symmetric Solutions

Whereas today there are a large number of exact solutions of the Einstein equations, the first ones published were not easily discovered or understood. The spherically symmetric solutions based on the form of the metric (15.78) played an important part in that early development: the Einstein equations could be explicitly written out and the challenge was to find solutions.

The Einstein Equations (15.87–15.90) for the spherically symmetric metric (15.78) are a set of nonlinear partial differential equations for the functions $F(r,t)$ and $H(r,t)$, together with an energy–momentum tensor describing the matter content and a set of boundary conditions describing the global system. Once a solution has been found there is the non-trivial problem of interpreting it.

To make progress it is essential to impose practical restrictions on the functions $F(r, t)$ and $H(r, t)$. Tolman (1939) gave a systematic discussion of static solutions in which F and H were independent of the t-coordinate, some were vacuum solutions and others were solutions with matter.[11]

The simplest case is the vacuum solution $G_{ab} = 0$ (Misner *et al.*, 1973, Section 32.2). Then Equation (15.90) tells us that H is independent of t: $H = H(r)$. With this Equation (15.87) becomes an ordinary differential equation for $H(r)$ with solution

$$H = -\frac{1}{2} \ln |1 - 2M/r| \tag{15.91}$$

for some constant M. With this we can now solve Equation (15.88), which is a differential equation for $F(r, t)$. So the 'constant of integration' is an arbitrary function of t and we have

$$F(r, t) = \frac{1}{2} |1 - 2M/r| + f(t), \tag{15.92}$$

where the arbitrary function $f(t)$ is our 'constant' of integration. The metric (15.78) is then

$$ds^2 = -(1 - 2M/r) \, e^{-2f(t)} dt^2 + (1 - 2M/r)^{-1} dr^2 + r^2 (d\theta^2 + \sin^2 \theta d\phi^2). \tag{15.93}$$

We see that it is trivial to eliminate the function $f(t)$ by using a new time-like coordinate $dt' = e^{f(t)} dt$ to give:

Schwarzschild's solution:

$$ds^2 = -\left(1 - \frac{2M}{r}\right) dt^2 + \frac{1}{\left(1 - \frac{2M}{r}\right)} dr^2 + r^2 (d\theta^2 + \sin^2 \theta \, d\phi^2), \tag{15.94}$$

where we have dropped the prime on the time coordinate.

15.5.4 Birkhoff's Theorem

Birkhoff's theorem asserts that the geometry of a region of space-time that is a spherically symmetric vacuum is described by the Schwarzschild solution. We see this simply on the basis of the generality of the spherically symmetric metric (15.78) that we started with: the only additional assumption we used was that the material content had $T_{ab} = 0$, i.e. a vacuum.

---

[11] The metric defines a geometry which in turn leads to the Einstein tensor G_{ab} for the space-time described by that metric. The Einstein equations tell us that the energy–momentum tensor that is associated with that solution is given by $G_{ab} = \kappa T_{ab}$, but there is no prior guarantee that this T_{ab} corresponds to anything that is physically meaningful. For example, 'solution IV' of Tolman (1939) describing a spherical relativistic star required the equation of state to be such that the density was quadratic in the pressure.

The importance of this is that if, for example, we consider the spherical collapse of a massive star of finite radius,[12] the geometry of the space-time outside of the star is the Schwarzschild geometry. This means that we can patch together a vacuum solution and a solution for a collapsing spherical ball of fluid to get a solution that extends throughout space-time.

15.6 Classic Spherically Symmetric Solutions

The Einstein Field equations are ten nonlinear partial differential equations, so finding exact solutions is never going to be trivial. However, within a decade of Einstein's great 1915 paper, several exact solutions were known. There were also a number of approximate, quasi-Newtonian, weak field, solutions. Those solutions had been found by exploiting symmetries to reduce the number of equations, or using perturbations of an underlying Minkowski space in order to linearise the equations. Today we have many more mathematical techniques that we can bring to bear on finding solutions to the Einstein equations (MacCallum, 2006). The known solutions up until its date of publication are catalogued in the 732 page book of Stephani *et al.* (2009).

For didactic reasons, we shall in what follows be looking at the classic solutions expressed in different coordinate systems and explain some of the controversies that arose from misinterpretation.

15.6.1 Alternative Coordinate Systems

In general relativity the coordinate system chosen to represent the space-time is arbitrary, the physics of the space-time does not depend on the chosen coordinate system. This freedom is a powerful aspect of general relativity, as we shall see. However, it must be appreciated that when making coordinate transformations the domain of validity of the new metric may not be the same as the original.

Remark: Coordinate transformation
The new coordinates may describe only a part of the space-time described by the original metric, or may even extend to new regions of the space-time. Apparent singularities in the metric may either appear or disappear. The null geodesics play an important part in understanding the relationships between the old and new descriptions.

Space-time diagrams of the coordinate system can play an important part in interpreting the solutions.

In this chapter we shall look at each of the solutions we present in several different coordinate systems. This serves, in part, to illustrate the sources of confusion that arose

---

[12] The star should have no electric charge. If the star has a spherically symmetric electric field, then the geometry will be that of the Reissner–Nordstrom solution: this is a generalisation of the Birkhoff theorem (see Misner *et al.* (1973, Ex. 32.1).

in the early days when people tried to understand the few solutions that existed. But in more recent times, coordinate transformations have played a vital role in understanding and extending solutions.

15.6.2 Solutions: the Role of Λ

The Einstein equations can be written in either of the forms:

$$R_{ab} - \tfrac{1}{2}g_{ab}R = \kappa T_{ab}, \tag{15.95}$$

$$R_{ab} = \kappa\left(T_{ab} - \tfrac{1}{2}Tg_{ab}\right). \tag{15.96}$$

The second form follows from the first since, from (15.95), we have $R = -\kappa T$. The energy–momentum tensor specifies the nature of the matter, and given an energy–momentum tensor we can derive a geometry. The *vacuum field equations* set $T_{ab} = 0$.

The question arises as to what we do about the cosmological constant Λ. Because of the history of the subject, the Einstein equations with cosmological constant are normally written

$$R_{ab} - \tfrac{1}{2}g_{ab}R + \Lambda g_{ab} = \kappa T_{ab}. \tag{15.97}$$

Here, Λ is regarded as a geometric entity that is a constant of nature.

On the other hand, by moving the Λ term to the other side of the equation,

$$R_{ab} - \tfrac{1}{2}g_{ab}R = \kappa T_{ab} - \Lambda g_{ab}, \tag{15.98}$$

we are regarding Λ as a material field that pervades space and contributes to the distortion of the geometry of space-time. Here Λ can be regarded as being a material field with density and pressure given by $\rho_\Lambda c^2 = \Lambda/8\pi G$ and $p_\Lambda = -\rho_\Lambda c^2$.

In the paper introducing the Λ-term, entitled 'Cosmological Considerations on the Theory of General Relativity', Einstein (1917a) recognised this,[13] commenting in the very last sentence of the paper that *That term is necessary only for the purpose of making possible a quasi-static distribution of matter, as required by the fact of the small velocities of the stars.*

In 1917, and within a year of the publication of the theory of general relativity, two important cosmological solutions to the field equations were presented, one by Einstein and the other by de Sitter. We shall discuss them below because they illustrate a number of the key factors that arise when defining, finding and interpreting cosmological solutions. The interpretation was not without considerable controversy and was only resolved by the later work of Friedmann, Lemaître, Robertson and Walker, along with the data of Hubble and his collaborators.

---

[13] The paper is reproduced in translation in Lorentz *et al.* (2000, p.174 *et seq.*). In Section 4 of that paper 'On an Additional Term for the Field Equations of Gravitation' Einstein states that ... *if it were certain that the field Equations (13) which I have hitherto employed were the only ones compatible with the postulate of general relativity, we should probably have to conclude that the theory of relativity does not admit the hypothesis of a spatially finite Universe.* Equation (13) referred to here is the Einstein equations for the metric he proposes (i.e. 15.99) and the following Equation (13a) is the first time the cosmological constant appears in the literature.

15.6.3 Einstein's Static Model

This was Einstein's first cosmological model (Einstein, 1917a). It contained matter of density ρ maintained at a constant value by a positive cosmological constant $\Lambda = 4\pi G\rho/c^2$.

The Einstein static model line element is usually written in the form:

Einstein static model:

$$ds^2 = -dt^2 + \frac{dr^2}{\left(1 - \frac{r^2}{R^2}\right)} + r^2 d\Omega^2, \quad R^2 = \frac{c^2}{\Lambda} \text{ is a constant.} \tag{15.99}$$

Einstein identified R as the 'radius' of the model, and generally the equations are written in terms of R rather than Λ. We shall follow that here.[14]

As a model for the Universe it had the downside that it was unstable: the balance between gravitational attraction and repulsion due to Λ was not a stable situation. Moreover, it could not account for the redshifts in the spectral lines of the nebulae.[15]

The Coordinate System

First a comment about the coordinate system: these are comoving coordinates. We can show this by calculating the orbit of a particle moving from rest in this space-time. The equation of motion is

$$\frac{d^2 x^a}{ds^2} + \Gamma^a_{bc} \frac{dx^b}{ds} \frac{dx^c}{ds} = 0. \tag{15.100}$$

If the particle is initially at rest, then we only have to show that it has zero acceleration, and so it stays at rest. The initial condition is that the components of velocity, $dr/ds, d\theta/ds, d\phi/ds$ are all zero and hence the only derivatives remaining on the second term of (15.100) are those for $b = c = 0$. Hence

$$\frac{d^2 x^a}{ds^2} + \Gamma^a_{00} \left(\frac{dt}{ds}\right)^2 = 0. \tag{15.101}$$

However, for this metric[16] $\Gamma^a_{00} = 0$ and so the acceleration is zero. The particle remains at rest. In this coordinate system the time t is the local proper time of a comoving observer.

Light Propagation: the Redshift

We can calculate the redshift of a photon emitted from the origin quite simply as follows. A radial light ray is defined by $ds = 0$ and $d\theta = d\phi = 0$, and so the equation of motion of a photon of light is

---

[14] We are assigning Λ units of $[\sec]^{-2}$, consistent with the Newtonian treatment (Equation 6.13).

[15] Here we follow the somewhat old-fashioned but nevertheless excellent treatment of Tolman (1934b, Section 140).

[16] See Equation (15.22) which tell us that $\Gamma^A_{00} = -\frac{1}{2} g^{AA} \partial g_{00}/\partial x^A = 0$ for the metric (15.99) where $g_{00} = -1$.

$$\frac{dr}{dt} = \pm\sqrt{1 - \frac{r^2}{R^2}}. \tag{15.102}$$

This is the coordinate velocity of the photon.

If a signal is sent out at $t = t_0$ from $r = 0$ and is received at the origin r at t_1 then

$$t_1 = t_0 + \int_0^r \frac{dr}{\sqrt{1 - r^2/R^2}} = t_0 + \arcsin\frac{r}{R}. \tag{15.103}$$

Since the observer at coordinate r remains at r (these are comoving coordinates), then two pulses emitted δt_0 apart are received δt_1 apart and

$$\delta t_1 = \delta t_0. \tag{15.104}$$

This is easily interpreted as there being no difference in the emitted and received frequencies of the light: there is no redshift. This is hardly surprising since the gravitational redshift is cancelled by the cosmological force due to Λ. If that were not so the Einstein Universe would not be static.

Alternative Coordinate Systems

By virtue of the substitution

$$r = \frac{\varpi}{1 + \varpi^2/4r^2} \tag{15.105}$$

we get to the isotropic form of the Einstein static model line element,

$$ds^2 = -c^2 dt^2 + \frac{1}{[1 + \varpi^2/4R^2]^2}\left(d\varpi^2 + \varpi^2 d\theta^2 + \varpi^2 \sin^2\theta d\phi^2\right). \tag{15.106}$$

This is referred to as 'isotropic' because the spatial part of the metric is isotropic.

A further coordinate transformation can bring us into a $4 + 1$-dimensional view of the solution in which we have four space coordinates and one time coordinate:

$$z_1 = R\sqrt{1 - \frac{r^2}{R^2}}, \quad z_2 = r\sin\theta\cos\phi, \quad z_3 = r\sin\theta\sin\phi, \quad z_4 = r\cos\theta, \tag{15.107}$$

$$z_1^2 + z_2^2 + z_3^2 + z_4^2 = R^2. \tag{15.108}$$

With these substitutions the metric becomes

$$ds^2 = -dt^2 + dz_1^2 + dz_2^2 + dz_3^2 + dz_4^2. \tag{15.109}$$

This embeds the Einstein's static solution in a 5-dimensional Minkowski space with Euclidean spatial sections. Equation (15.108) reveals the model as the surface of a sphere in this extended space. The value of R is the curvature of the space-time.

15.6.4 de Sitter's Static Model

The solution of de Sitter (1917a) appeared just a few months after the Einstein static model. The model is a homogeneous and isotropic vacuum solution of the Einstein equations with a non-zero cosmological constant, Λ:

$$G_{ab} = \Lambda g_{ab}. \tag{15.110}$$

The solution as presented by de Sitter was given in the form:[17]

de Sitter's solution: static coordinates

$$ds^2 = -\left(1 - \kappa^2 r^2\right) dt^2 + \frac{dr^2}{\left(1 - \kappa^2 r^2\right)} dr^2 + r^2 d\Omega^2, \quad \kappa^2 = \Lambda/3. \qquad (15.111)$$

Einstein had criticised this solution on the grounds that it was devoid of any physical possibility. Einstein wanted to insist that the g_{ab} be fully determined by the distribution of matter, and, more strongly, that the metric should be the Minkowski metric in the absence of matter. This was, in effect, a restatement of *Mach's principle*.[18]

Remark: The de Sitter solution in $N + 1$-dimensional space-time has become one of the most important solutions of the Einstein equations, particularly with respect to quantum gravity and the study of higher dimensional theories of gravity.

It was probably the awareness of the discussions surrounding this solution that caused Hubble to interpret his redshift distance relation as what he called 'the de Sitter effect': even though the metric is static the observer at the origin sees light from distant sources as being redshifted.

Alternative Coordinate Systems

The metric (15.111) is static in the sense that the coefficients of the metric do not depend explicitly on the time coordinate. The difficulties with the de Sitter form of the metric were attacked soon after the solution was published. Lanczos, Lemaître, Robertson, and Weyl all provided important insights into this important space-time, often through the use of ingenious coordinate transformations.[19] However, it must be remembered that the new coordinates may describe only a part of the space-time described by the original metric, or may even extend to new regions of the space-time. We are looking at the same space-time, but in a way that might have a quite different interpretation than the original. (See Hawking and Ellis (1975, Section 5.2) for a detailed geometric view.)

The transformation of coordinates $r \rightarrow \bar{r}, t \rightarrow \bar{t}$ (Robertson, 1928):

$$r = \bar{r}e^{\kappa t}, \quad t = \bar{t} - \frac{1}{2\kappa} \log\left(1 - \frac{\bar{r}^2}{R^2}e^{2\kappa\bar{t}}\right), \qquad (15.112)$$

leads to the dynamic form for the metric:

---

[17] The coordinates are *static* because the metric coefficients do not depend explicitly on the t-coordinate.

[18] See Schulmann *et al.* (1998, docs.317, 366) (The collected papers of Albert Einstein. Vol.8: The Berlin years: Correspondence 1914–1918).

[19] Superficially, the de Sitter solution resembles the Schwarzschild metric that had been discovered the previous year by Schwarzschild (1916) and by Droste (1917). The Schwarzschild solution describes the vacuum space time around a mass point (see Equation 15.132): we simply replace the curvature κ^2 with $2m/r^3$ and we have the Schwarzschild solution. This was formalised by Robertson (1927) where, among other things, he presented a coordinate transformation that took one solution into the other. This was further discussed in his great paper of 1928 (Robertson, 1927, 1928, Section 5).

de Sitter solution: dynamical form

$$ds^2 = -dt^2 + e^{2Ht}(dr^2 + r^2d\theta^2 + r^2\sin^2\theta d\phi^2), \quad H = \kappa. \tag{15.113}$$

The alternative form of this displaying the Euclidean nature of the space sections is

$$ds^2 = -dt^2 + e^{2Ht}(dx^2 + dy^2 + dz^2), \tag{15.114}$$

where we have relabelled $\kappa = H$, which is a constant. In (15.113) we have relabelled the coordinates using r, t and in (15.114) we have written the spatial part of the metric in Cartesian coordinates. Robertson regarded this as a better model for describing the kinematics of the observed Universe since the space sections are locally Euclidean and isotropic and all observers share the same proper time as the observer at the origin.

Metrics (15.113) and (15.114) also describe the Steady State cosmology, though in that case the field equations they satisfy are modifications of the Einstein field equation. The Steady State modification is a scalar field, the 'C-field', out of which matter would be created to fill in the gaps as the Universe expanded.

Redshifts in de Sitter Space: I

Here we study the redshift of light rays in the context of the de Sitter metric in the coordinates of the metric (15.111).

Explaining Slipher's observed redshift in the spectral lines of the spectra of galaxies (published in Eddington (1923, p.162)) soon became a key goal for relativistic cosmological models. Showing that the Einstein static model would show no such redshift was straightforward.

However, the situation with regard to de Sitter's solution (15.111) was not so simple. Unlike in the case of the Einstein static model, in the de Sitter model a test particle placed at rest at a coordinate distance r from an observer at the origin would be accelerated away.[20] Furthermore, in the metric (15.111) the proper time in the local frame of the emitter would be different from that measured by the observer at the origin. So there are two contributions to the redshift in the de Sitter model: one from the dispersal of the galaxies and the other from the curvature of the de Sitter space-time.

We can proceed as in the case of the Einstein static model by writing the coordinate velocity of a radial ray of light (i.e. $ds^2 = 0$) at position r as

$$\frac{dr}{dt} = \pm(1 - \kappa^2 r^2). \tag{15.116}$$

---

[20] This is a little tedious: it involves calculating the Christoffel symbols that enter into the geodesic equation for this metric. The acceleration for radial motion is

$$\frac{d^2r}{ds^2} = \frac{1}{3}\Lambda r. \tag{15.115}$$

(See Eddington (1923, Section 70).)

The ray emitted at coordinate time t_0 will therefore arrive at the origin at coordinate t_1 given by

$$t_1 = t_0 + \int_0^r \frac{dr}{1 - \kappa^2 r^2}. \tag{15.117}$$

Doing the integral shows that a pulse of coordinate time duration δt_0 emitted from r is received as a pulse of duration δt_1:

$$\delta t_1 = \delta t_0 + \frac{1}{1 - \kappa^2 r^2} \frac{dr}{dt} \delta t_0. \tag{15.118}$$

Up until this point everything is straightforward. Here we see the two contributions to the redshift: the first term depends on the geometry of space-time, while the second depends on the velocities the test particles have acquired due to their acceleration by the curvature of the space-time.

The complications arise because we now need to convert the coordinate time intervals to proper times. This would not be difficult were it not for the fact that test particles placed at rest in this Universe accelerate away, and hence there is a dependence of the solution on the initial conditions and on the orbits of the particles. In large part the cause of this situation is the fact that the time coordinate in this form of the metric depends on location: the origin of the coordinates is a special place.[21] Tolman (1934b, Section 144(d)) goes through the orbital calculation and derives a Hubble redshift relationship involving arbitrary parameters describing the orbits. His contemporaries, however, transformed to a variety of coordinate systems in which the time was the proper time of comoving observers. We will discuss that below.

Hubble and the de Sitter Effect

There is a remark of some historical importance made by Sandage (1975) about the redshift distance relationship in this form of the de Sitter solution. If for simplicity we suppose that in Equation (15.118) the second term involving dr/dt were always zero, then the test particle would have $dr = d\theta = d\phi = 0$. Of course, keeping the velocity equal to zero at all times is rather artificial and would require that the particles be driven by forces, but they are test particles and we can endow them with any property we like, however unnatural. With this assumption the relationship between the proper times at the time and place of emission and the time and place of reception is simply

$$\frac{\delta \tau_1}{\delta \tau_0} = (1 - \kappa^2 r^2)^{1/2}. \tag{15.119}$$

---

[21] Sandage (1975) in his review of the history of extragalactic redshift measurements, refers to the coefficient of dt^2 in the de Sitter metric (15.111) as *.. the scandalous term* $(1 - r^2/R^2)$. Sandage goes on to remark that Hubble had been aware of the de Sitter prediction that the redshift $\delta\lambda/\lambda \propto r^2$: the so-called *de Sitter effect*. Sandage reminds us that Hubble explicitly stated in the last paragraph of his paper that the relationship he had discovered might be only a tangent to some more general relationship: *The outstanding feature, however, is the possibility that the velocity–distance relation may represent the de Sitter effect* .. and finally ... *it may be emphasized that the linear relation found in the present discussion is a first approximation representing a restricted range in distance*.

The wavelength change from λ at emission to $\lambda + \delta\lambda$ at reception at the origin is

$$\frac{\delta\lambda}{\lambda} = \frac{\tau_1}{\tau_0} - 1 \simeq \kappa^2 r^2. \tag{15.120}$$

So, under these circumstances, we get a redshift distance relationship which to first order in $\kappa^2 r^2$ is $\delta\lambda/\lambda \propto r^2$. According to Sandage, this is the relationship that Hubble was aware of and that led him to believe that he had discovered the 'de Sitter effect' that had been a subject of research by many obervers during the previous decade.

Redshifts in de Sitter Space: II

De Sitter space is empty, so we consider test particles that are stationary relative to the Universe. As with the Einstein static solution, we follow the development of Tolman (1934b, Section 144, p357 *et seq.*).

The geodesics in the metric (15.113) are given by

$$\frac{d^2 x^a}{ds^2} + \Gamma^a_{bc} \frac{dx^b}{ds} \frac{dx^c}{ds} = 0, \tag{15.121}$$

and, as with the discussion of the Einstein static Universe, we suppose the particles have no motion relative to the background Universe: $dr/ds = d\theta/ds = d\phi/ds = 0$, so that the only surviving terms in (15.121) are those with $b = c = 0$. The accelerations $\ddot{x}^a$ are thus determined by the Christoffel symbols Γ^a_{00}. These are all zero for this metric and so the test particles remain comoving. Also, the time t at each particle measures the local proper time. All test particles are equivalent, there is no distinguished 'central' particle.

The photon path $ds = 0$ for a radial light ray is thus

$$\frac{dr}{dt} = \pm e^{-Ht}. \tag{15.122}$$

Hence if the particle at r emits a photon towards the origin at t_0 and this photon is received at t_1, we have

$$\int_{t_0}^{t_1} e^{-Ht} dt = \int_0^r dr = r = \text{constant}, \tag{15.123}$$

since the r-distance to the particle is fixed (both emitter and receiver are comoving). Doing the integral relates the clocks of the emitter and receiver: $e^{-Ht_0} - e^{-Ht_1} = constant$ and so the relationship between an interval δt_0 in t_0 is related to an interval δt_1 at t_1 by

$$\delta t_1 = e^{H(t_0 - t_1)} \delta t_0. \tag{15.124}$$

So pulses separated by δt_0 at emission are received at separation δt_1, which gives rise to a wavelength change in the light of

$$\frac{\lambda_1}{\lambda_0} = \frac{\delta t_1}{\delta t_0} = e^{H(t_1 - t_0)}. \tag{15.125}$$

For small distances (i.e. $H(t_1 - t_0) \ll 1$), putting $\delta\lambda = \lambda - \lambda_0$, we have

$$\frac{\delta\lambda}{\lambda_0} \simeq H(t_1 - t_0) \simeq Hr, \tag{15.126}$$

which we recognise as the Hubble expansion law.

This is, in essence, the derivation by Robertson (1928) within the framework of a general relativistic cosmology, of what we now know as the linear part of the Hubble expansion law. Specifically, this applies to the vacuum solution with a cosmological constant and metric (15.113).

Embedding in Higher Dimensional Spaces

Following Robertson (1928),[22] if we make a further transformation of (15.114) to the five variables

$$\alpha = xe^{\kappa t}, \quad \beta = ye^{\kappa t}, \quad \gamma = ze^{\kappa t}, \tag{15.127}$$

$$\zeta = \kappa^{-1}\sinh\kappa t + \tfrac{1}{2}\kappa r^2 e^{\kappa t}, \quad \eta = \kappa^{-1}\cosh\kappa t - \tfrac{1}{2}\kappa r^2 e^{\kappa t}, \tag{15.128}$$

the de Sitter space is then simply the hyperboloid:

$$\alpha^2 + \beta^2 + \gamma^2 + \eta^2 - \zeta^2 = \frac{1}{\kappa^2}. \tag{15.129}$$

The hyperboloid is embedded in a flat 5-dimensional Minkowski space spanned by the coordinates $(\alpha, \beta, \gamma, \eta, \zeta)$.[23] The sphere centred on the point $(0,0,0,0,1/\kappa)$ corresponds, in the limit $\kappa \to 0$, to the light-cones $x^2 + y^2 + z^2 - t^2 = 0$ in Minkowksi space. The group of coordinate transformations in 5-dimensions that leave the Equation (15.129) invariant is called the *de Sitter group*.

For a wider description of de Sitter space, with discussion of horizons, thermodynamics and quantum fields, see Carroll (2003, Section 8.1), Bousso (2003) and Moschella (2005).

The analogy that comes to mind is that of the famous paradigm for the expanding Universe describing the world of ants living on the surface of expanding balloon. The ants can determine the geometry of their world by studying (spherical) triangles in the surface to which they are confined. They can deduce that they live in an expanding spherical world. However, by embedding the balloon in a higher dimensional space, our 3-dimensional Euclidean space, they would understand their world as a spherical expanding surface in an otherwise Euclidean space: the metric they determine in their living space would be inherited from the embedding Euclidean space. They would instantly see the invariance group of their world. This is precisely what Robertson did in his 1928 paper for what has become known as the Robertson–Walker model of our Universe.

15.6.5 The Einstein – de Sitter Solution

In 1932 Einstein and de Sitter (1932) together published, in a two page article, a cosmological solution which was to become the standard of the post-1960s study of cosmology:

---

[22] Robertson's seminal 1928 paper followed on from the work of de Sitter (1917a) and of Weyl (2009) (*Golden Oldie* reprint of article published in 1923) both of whom had earlier embedded the de Sitter model in a higher dimensional space.

[23] Technically the *de Sitter space* is the hyperboloid of one sheet (15.129) defined by $\kappa^2 > 0$, and the metric on the de Sitter space is thus the metric induced by the embedding Minkowski space. Robertson (1928) showed that the isometry group of the de Sitter space is the Lorentz group $O(1,4)$ consisting of ten generators: a time translation, three Lorentz boosts, and six rotations of which three are simple spatial rotations. The mapping $\zeta \to i\zeta$ converts the de Sitter space into a 5-dimensional sphere, the *Euclidean de Sitter manifold* which is invariant under the group $O(5)$, which plays a role in quantum field theory.

the *Einstein de Sitter solution*. Until the renaissance of a non-zero cosmological constant in the 1990s, this was perhaps the most influential cosmological solution to the Einstein equations: it was very simple and was not in contradiction to any of the data at the time.

The paper was written shortly after Hubble's publication (Hubble, 1929) of the velocity–distance relationship and de Sitters' somewhat remarkable improvement on Hubble's plot (de Sitter, 1930). The opening paragraphs of the paper are most interesting, two statements made there show the evolution in thinking about cosmology. The first is *There is no direct evidence for the curvature, ...* and the second is *Historically, the term containing the 'cosmological constant' λ was introduced into the field equations in order to enable us to account theoretically for the existence of a finite mean density in a static Universe. It now appears that in the dynamical case this end can be reached without the introduction of λ.* There was no reference to the solutions of either Friedmann or Lemaître.

Consequently, they used the metric in the form

$$ds^2 = -c^2 dt^2 + R(t)^2(dx^2 + dy^2 + dz^2), \tag{15.130}$$

(their metric signature has been changed to our adopted 'MTW' convention) and, in the absence of λ and pressure, deduced the equation relating the expansion rate and the density:

$$H^2 = \frac{1}{R^2}\left(\frac{dR}{dct}\right)^2 = \frac{1}{3}\kappa\rho, \tag{15.131}$$

where $\kappa = 8\pi G$ (they used h for what we now call H). From this, using an expansion rate $h = 500\,\mathrm{kms^{-1}Mpc^{-1}}$, they deduced a density of $\rho = 4 \times 10^{-28}\,\mathrm{gr\,cm^{-3}}$. This value of the density accorded with de Sitter's own estimates at the time based on the Hubble distance scale.

The derived density for this model is called the *Einstein de Sitter critical density* for the given value of the Hubble constant (see Section 5.5.3). It would be difficult to overstate the importance of this short and simple article in cosmology.[24]

15.6.6 Schwarzschild Solution

The Schwarzschild (1916) solution, an exact solution of the Einstein equations describing the space-time around a point mass in an otherwise empty space, was found shortly after the publication of Einstein's final paper on general relativity (see also Droste, 1917). The solution is described by the metric:[25]

---

[24] By comparison with the number of times, probably thousands, the phrase 'Einstein de Sitter model' has been used in print, the original paper is hardly cited outside of the historical literature (perhaps a few dozen citations).

[25] Droste (1914) had calculated the motion of the perihelion of Mercury, based on the weak-field approximation to the first (*Entwurf*) version of general relativity (Einstein and Grossmann, 1913). This was a part of Droste's PhD thesis (1916), working under the guidance of Lorentz. Although that was not the final theory of general relativity, it was nonetheless adequate for vacuum solutions. Einstein (1915) did the same calculation within the framework of the newer theory. Shortly after Einstein's publication of the final theory (Einstein, 1915) Droste independently found and published the full solution (Droste, 1917). In this paper Droste (1917, footnote p.203) acknowledges Schwarzschild. Historically, Droste's papers have hardly been noticed, though Eddington did make a passing reference to him in the bibliography of Eddington (1923, p.264), (though this was not referred to in the main text (Section 44, p.95 *et seq.*)).

Schwarzschild solution

$$ds^2 = -\left(1 - \frac{2m}{r}\right)dt^2 + \left(1 - \frac{2m}{r}\right)^{-1}dr^2 + r^2 d\Omega^2, \quad \Lambda = 0. \qquad (15.132)$$

First published by Droste in his PhD thesis (1916) when it was communicated to the Dutch Academy and in a journal the following year (Droste, 1917).

Asymptotically, as $r \to \infty$, the solution looks like Minkowski space as described in Equation (15.7).

The Schwarzschild solution is manifestly singular at the origin $r = 0$ and also at $r = 2m$. However, the latter singularity is merely a coordinate singularity that can be eliminated by suitably different choices of coordinates, as first shown by Lemaître (1925, 1933). The singularity at $r = 0$ is an essential singularity.

Alternative Coordinate Systems

Lemaître's solution was to transform the Schwarzschild's coordinates r, t to new coordinates ρ, τ which converted the metric to:

Schwarzschild solution in Lemaître coordinates:

$$ds^2 = -d\tau^2 + \frac{r}{2m}d\rho^2 + r^2 d\Omega^2, \quad r = \left[6m^2(\rho - \tau)\right]^{2/3}. \qquad (15.133)$$

This form of the metric has no issue at $r = 2m$, which led Lemaître to declare that $r = 2m$ was 'une singularité fictive'. This form of the metric had in fact been proposed earlier, but in the context of the perihelion advance of the orbit of the planet Mercury, by Painlevé (1921a,b) and by Gullstrand (1922).[26]

A light ray is defined by $ds^2 = 0$ and a radial light ray has $d\theta = 0$ and $d\phi = 0$. The result is a relationship between r and t which can be easily solved to give the radial 'orbit' of the light ray in these coordinates:

$$t = \pm[r + 2m \ln|r - 2m| + \text{const.}], \qquad (15.134)$$

and again we see an apparent problem with $r = 2m$. On seeing this equation for the orbit, the problem is easily dealt with if we replace the time coordinate t with

$$t \to \bar{t} = t + 2m \ln(r - 2m), \qquad (15.135)$$

and the metric now becomes:

---

[26] There is a clear description of this at:
`http://en.wikipedia.org/wiki/Gullstrand%E2%80%93Painlev%C3%A9_`
`coordinates`
where there is a reference to Andrew Hamilton's fine video *Inside a Black Hole*:
`http://online.itp.ucsb.edu/online/colloq/hamilton1/`.
Hamilton's web page on this subject describes and explains the video:
`http://casa.colorado.edu/~ajsh/singularity.html`

Schwarzschild solution in Eddington–Finkelstein coordinates:

$$ds^2 = -\left(1 - \frac{2m}{r}\right)d\tilde{t}^2 + \frac{4m}{r}d\tilde{t}dr + \left(1 + \frac{2m}{r}\right) + r^2 d\Omega^2, \qquad (15.136)$$

and this is perfectly regular at $r = 2m$, but the singularity at the origin is still there.[27] This solution was first presented by Eddington and later by Finkelstein, so the coordinates are called *Eddington–Finkelstein coordinates*.

Eddington (1923, p.92) seems to have been the first to discuss the transformation $r \rightarrow r'$ defined by the equation

$$r = \left(1 + \frac{m}{2r'}\right)^2 r', \qquad (15.138)$$

which transforms the metric into a spatially isotropic system of coordinates:[28]

Schwarzschild solution in Eddington–Robertson coordinates:

$$ds^2 = -\left(\frac{1 - \dfrac{m}{2r'}}{1 + \dfrac{m}{2r'}}\right)^2 dt^2 + \left(1 + \frac{m}{2r'}\right)^4 (dr'^2 + r'^2 d\Omega^2). \qquad (15.139)$$

This is generally referred to as the Schwarzschild solution in *isotropic coordinates* (see Adler, *et al.* (1975, Section 6.2 and his Eq. (6.61))).

A more complex variant is due to Kruskal using the transformation

$$u = \left(\frac{r}{2m} - 1\right)^{1/2} \exp\left(\frac{r}{4m}\right) \cosh\left(\frac{t}{4m}\right), \qquad (15.140)$$

$$v = \left(\frac{r}{2m} - 1\right)^{1/2} \exp\left(\frac{r}{4m}\right) \sinh\left(\frac{t}{4m}\right), \qquad (15.141)$$

from which we can reconstruct r by solving the equation

$$\left(\frac{r}{2m} - 1\right) \exp\left(\frac{r}{2m}\right) = u^2 - v^2. \qquad (15.142)$$

The resulting metric in *Kruskal coordinates* is:

---

[27] The way to see whether a singularity is real or not is to calculate quantities that are invariant under any choice of coordinates: such invariants represent physical quantities whose value cannot depend on the coordinates. An example of such an invariant that is relevant to the Schwarzschild solution is the square of the Riemann tensor, $R_{abcd}R^{abcd}$ which has the value

$$R_{abcd}R^{abcd} = 48m^2/r^6. \qquad (15.137)$$

This value is independent of all allowable coordinate transformations. We see that at $r = 2m$ this has a finite value, $3/4m^4$, while at the origin it is singular. This expression was given by Eddington (1923, p.92).

[28] North (1965, p.112) attributes this to Robertson (1928, see p.846 Eq.20), but Robertson (1927) cites Eddington (1923, p.93).

Schwarzschild solution in Kruskal coordinates:

$$ds^2 = \left(\frac{32m}{r}\right) \exp\left(-\frac{r}{2m}\right)(du^2 - dv^2) - r^2 d\Omega^2, \qquad (15.143)$$

where again there is nothing strange happening at $r = 2m$, but $r = 0$ is still a singular point. The interesting point about this solution is that light rays, null geodesics, are simply

$$u + v = \text{const}, \quad u - v = \text{const}. \qquad (15.144)$$

The light rays are diagonal lines when the space time is viewed in the u, v plane.

Schwarzschild (anti) de Sitter Metrics

We earlier remarked on the superficial similarity between the de Sitter metric and the Schwarzschild metric. Both metrics are of the form

$$ds^2 = -f(r)dt^2 + \frac{dr^2}{f(r)} + r^2(d\theta^2 + \sin^2\theta d\phi^2). \qquad (15.145)$$

A specific and important example of this is the Schwarzschild solution including a non-zero cosmological constant Λ:

Schwarzschild de Sitter metric, $\Lambda > 0$
Schwarzschild anti-de Sitter metric, $\Lambda < 0$

$$ds^2 = -\left(1 - \frac{2m}{r} - \frac{\Lambda r^2}{3}\right)dt^2 + \frac{dr^2}{\left(1 - \frac{2m}{r} - \frac{\Lambda r^2}{3}\right)} + r^2(d\theta^2 + \sin^2\theta d\phi^2). \qquad (15.146)$$

(Also known as the Kottler spacetime.)

This was first derived by Kottler (1918). We see that for $\Lambda > 0$ there is another coordinate singularity at $r_\Lambda = \sqrt{3/\Lambda}$, which is the de Sitter horizon. The solution was analysed more recently by Podolký (1999) and has become a subject of interest (and controversy) about the effect of Λ on the bending of light (Perlick, 2010, Section 5.2). The orbits of light rays in the Kottler solution are identical to those in the Schwarzschild solution, which had led to the incorrect conclusion that the cosmological constant has no effect on light bending. Rindler and Ishak (2007) showed that in fact there is a difference in the degree of bending, as might have been expected. The issue arose out of the misconception that having the same equation for the light path implied the same bending angles: however, since the Kottler metric specifically includes Λ, the *observed* bending angles are not the same.

16

The Robertson–Walker Solution

Our standard cosmological model describes a Universe that, in the large, is homogeneous and isotropic and filled with a variety of kinds of matter. Among the various matter fields that we can readily identify are baryons, photons and neutrinos. There are in addition components for which we have indirect yet compelling evidence, namely dark matter and dark energy. The dynamics of the Universe is governed by the law of gravitation: Einstein's general relativity (GR) or perhaps some variant thereof.

The GR Robertson–Walker metric finds its origins in the famous works of Friedmann and Lemaître, but it was Robertson and Walker who independently set the model on a firm mathematical footing. The Robertson–Walker metric is the fundamental general relativistic basis of modern cosmology. It provides the framework within which all observations are interpreted and all observational tests of the underlying cosmology carried out so far can be fit to this model.

Here we will refer to this solution of the Einstein equations as the 'FLRW metric'. We have already encountered the Robertson–Walker metric in Section 15.4.1.

16.1 The Metric in Diverse Coordinate Systems

We are, in general relativity, at liberty to choose whichever coordinate system is best adapted to solving a particular problem. The *physics* of the FLRW solution cannot depend on any specific coordinate choice, but the way in which the values are assigned to measurements will depend on the specific coordinate system.[1] Choosing a particular set of coordinates is a matter of convenience.

One of the issues is to come to terms with the variety of different coordinates systems which have been used to describe the FLRW metric. In this section we present the most important of these representations of the FLRW metric. We go on to discuss the concepts of distance within this framework and touch on the subject of how perturbations to the metric will evolve during the cosmic expansion.

There is no standard notation that differentiates these solutions: the same symbol '*r*' is frequently used for the radial coordinate in a variety of solutions. In order to avoid

[1] We are accustomed to using spherical polar coordinates to describe positions and components of velocity on the surface of a sphere. We could equally use Cartesian coordinates, but the values of the positions and velocity components would then be less intuitively appealing.

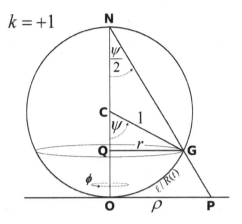

Fig. 16.1 The coordinate systems used in describing the Robertson–Walker metric for a closed, $k = +1$ Universe. The space is shown as a sphere of unit radius cut by a polar slice, $\phi = const$. The null geodesics arriving at the observer **O** are longitudinal lines on the surface of the sphere. The observer **O** is receiving light from a galaxy at **G**. The arc **OG** subtends an angle ψ at the centre of the sphere, **C**, and sweeps out a polar cap as it is rotated through $0 \le \phi < 2\pi$. This cap is part of **O**'s light past light-cone. The base of the cap is a circle of latitude with radius r centred at **Q**. Finally **P** is the projection of **G** onto the plane tangent to the sphere at **O**. The sphere is thought of as being rescaled by a function of time, $R(t)$, which is given by the Friedmann equations. Thus the arc **OG** represents the proper distance $\ell = R(t)\psi$ to the galaxy.
Adapted from Rindler (2006, fig. 16.3). ⓒⓒ

confusion in this text, the radial variables that define a particular version of the coordinates will be denoted by different symbols: $r, \psi, \eta, \ldots$ The relationship between these is depicted in Figure 16.1 and described in the following sections, and summarised in Section 16.1.5.

16.1.1 Classical Co-moving Coordinates

The solution of the Einstein equations that describes this is the *Robertson–Walker metric*:

Robertson–Walker metric: comoving coordinates

$$ds^2 = -c^2 dt^2 + R(t)^2 \left[\frac{1}{1-kr^2} dr^2 + r^2 d\Omega^2 \right]. \tag{16.1}$$

The coordinates used here are what are referred to as *co-moving coordinates*. The *curvature constant* $k = -1, 0, +1$ is a dimensionless number, as is the radial coordinate r which labels the co-expanding particles. $R(t)$ is the *scale factor* and has dimensions of $[L]$. This should be contrasted with the coordinates used in the expression (15.62) for the same metric.[2]

The coefficient of dt^2 is a constant, so test particles have fixed (r, θ, ϕ)-coordinates and the scale of the Universe is determined by the *scale factor* $R(t)$, which is in turn determined

---

[2] This form of the metric is invariant under the scaling transformations $R \to \lambda R, r \to \lambda^{-1} r, k \to \lambda^2 k$.

by solving the Einstein equations appropriate to the specific matter content. To understand why this coordinate system is referred to as 'co-moving' we only need to look at the equation of motion for a free particle located at some point (r, θ, ϕ). Its 4-velocity, u^a in this metric is given by

$$\frac{du^a}{ds} + \Gamma^a_{bc}\frac{du^b}{ds}\frac{du^c}{ds} = 0, \quad u^a = \frac{dx^a}{ds}, \quad u^a u_a = -1. \tag{16.2}$$

It is easy to verify that in this coordinate system $\Gamma^a_{00} = 0$ and so the observer at (r, θ, ϕ) experiences no acceleration: he/she is in a local (freely falling) inertial frame.[3]

This form of the metric is very similar to the metric (15.99) for the Einstein static Universe and was the form of the metric used by Friedmann (1924).

16.1.2 Isotropic coordinates

An alternative form of the line element in which the spatial component of the metric is manifestly isotropic in the 3-space spherical coordinates (ρ, θ, ϕ) is:

Robertson–Walker in spatially isotropic coordinates:

$$ds^2 = -c^2 dt^2 + \frac{R(t)^2}{(1 + \frac{1}{4}k\rho^2)^2}\left[d\rho^2 + \rho^2 d\Omega^2\right]. \tag{16.3}$$

The radial coordinate has been written as ρ to avoid confusion with the radial coordinate r used in (16.1). In the case $k = 0$ these two forms of the metric coincide. The term $(d\rho^2 + \rho^2 d\Omega^2)$ can also be written $(dx^2 + dy^2 + dz^2)$ where (x, y, z) are the usual Cartesian coordinates derived from the familiar (ρ, θ, ϕ) spherical polars. The transformation $r = \rho/(1 + k\rho^2/4)$ brings us back to the standard form (16.1).

In the case of a flat space, $k = 0$, both metrics (16.1, 16.3) have the very simple form

$$ds^2 = -c^2 dt^2 + R(t)^2\left[dx^2 + dy^2 + dz^2\right], \tag{16.4}$$

where, as usual, $R(t)$ is determined by the energy–momentum tensor and the Einstein equation.

16.1.3 Hyper-spherical Coordinates

There is an important variant on metrics (16.1) and (16.3) in which we change the coordinate r to to a variable ψ which is proportional to the radial distance.[4]

---

[3] If the observer was not in a freely falling frame but had a component of velocity relative to the Universe at large, the Universe would not look isotropic and so the metric would manifest an anisotropic angular component. So the isotropy of the spatial part of Equation (16.1) already indicates that the observer is co-moving, without doing any calculations. The time t for an observer at fixed (r, θ, ϕ) is the proper time.

[4] This form of the metric is extensively discussed by Misner *et al.* (1973, Section 27.8 and Section 29.2 eq. 24a,b *et seq.*).

We define a new radius ψ such that[5]

$$
r = \begin{cases} \sin \psi & k = +1 \\ \psi & k = 0 \\ \sinh \psi & k = -1. \end{cases} \tag{16.5}
$$

With this change we get the Robertson–Walker metric into the form:

Robertson–Walker metric: hyper-spherical coordinates:

$$
ds^2 = -c^2 dt^2 + R(t)^2 \left[d\psi^2 + S_k^2(\psi) d\Omega^2 \right], \quad S_k(\psi) = \begin{cases} \sin \psi & k = +1 \\ \psi & k = 0 \\ \sinh \psi & k = -1. \end{cases} \tag{16.6}
$$

The multiplier $S_k^2(\psi)$ in front of the $d\Omega^2$ means that, for $k \neq 0$, the elemental area–distance relationship differs from the Euclidean $dA_{\text{Euclid}} = R(t)^2 \psi^2 d\Omega$ in a way that depends on whether $k = \pm 1$. This form of the metric is particularly useful for calculating observed quantities that relate to the distant Universe. Note that ψ and $S_k^2(\psi)$ are dimensionless, while $R(t)$ has dimensions of length. It is customary to introduce a dimensionless expansion factor $a(t)$ such that $R(t) = a(t)R(t_0)$ with $a(t_0) = 1$.

Proper Distance

The *proper distance*, ℓ, between an observer and a galaxy is the 'ruler distance' (see below) as measured on a hypersurface of constant proper time. Bondi's pseudo-Newtonian version was discussed in Section 5.10.1. In terms of the metric (16.6) the element of proper distance, $d\ell$, between two events is obtained by setting $dt = 0$: i.e. $d\ell = R(t)d\psi$.

The proper distance ℓ is the distance we would get using rigid rulers laid end-to-end and is directly related to the radial coordinate, ψ and the scale factor $R(t)$ via

$$
\ell(t) = R(t)\psi. \tag{16.7}
$$

This is easily appreciated from Figure 16.1 where ℓ is the length of the arc **OG** scaled up by expansion scale factor $R(t)$. The null geodesics defining light rays have $ds = 0$ and so from the metric (16.6) $cdt = R(t)d\psi$ so that $d\ell = cdt = R(t)d\psi$. The velocity of the light ray relative to these co-expanding observers is always c.[6]

16.1.4 FLRW Metrics Using Conformal Time

Finally, there is a particularly important transformation $t \to \eta$ of the time coordinate,

$$
\eta = \int_0^t \frac{dt}{R(t)}. \tag{16.8}
$$

---

[5] There is a way of combining all three cases of k into a single expression: $S_k(\psi) = (\sin \sqrt{k}\psi)/\sqrt{k}$. This works because $\sin(ix) = i \sinh x$. Another possibility is to simply give the function a name like si(x).

[6] See Section 16.1.1 for a different approach.

Here η is called the *conformal time* and it has an important interpretation. If we look at the path of a radial light ray, $ds = 0$ in the FLRW metric (16.6), we see that in an interval of time dt light travels a distance $d\ell = cdt = Rd\psi$, where ψ is to be interpreted as a radial coordinate distance. The total coordinate distance traversed by the light in a time t is $c \int_0^t dt/R(t) = c\eta$. In other words:

> The conformal time η is a measure of the distance a light ray could have travelled since the Big Bang.

This transformation brings the standard metric (16.1) to the form

$$ds^2 = R(\eta)^2 \left[-c^2 d\eta^2 + \frac{r^2}{1 - kr^2} dr^2 + r^2 d\Omega^2 \right], \tag{16.9}$$

and the metric (16.6) to the form

$$ds^2 = R(\eta)^2 \left[-c^2 d\eta^2 + (d\psi^2 + S_k^2(\psi)d\Omega^2) \right]. \tag{16.10}$$

We see from this last equation that along a radial light ray, $ds = 0$, the change in conformal time equals the change in the radial coordinate ψ: $\delta\eta = \delta\psi$. This form of the metric serves to simplify some calculations involving light rays.

16.1.5 Relationships between Coordinate Systems

The Robertson–Walker metric has been presented using a number of choices for the radial coordinate. Three of these are frequently used, often to solve the same problem: r, ρ and ψ in Equations (16.1), (16.3), and (16.6) respectively. This was, in the early days of relativity a source of considerable confusion – which was the 'real radius'? The answer of course is 'none of them', the metric itself is not an observable quantity. If we only deal with measurable quantitites, their values are guaranteed to come out the same, no matter which system of coordinates is used.

Figure 16.1 is a sketch of a closed space-time illustrating the three different radial coordinates, r, ρ and ψ, in a simple way that shows how they are related. The transformations between r, ρ and ψ are easily seen to be:[7]

$$r = \sin\psi = \frac{3\tan(\psi/2)}{1 + \tan^2((\psi/2)}, \quad \rho = 2\tan(\psi/2). \tag{16.11}$$

16.1.6 FLRW Particle Horizon

The notion of a *cosmological horizon* plays an important role in understanding the structure of the Universe, and in particular in interpreting what we see in the cosmic background radiation fluctuations at the surface of last scattering (Davis and Lineweaver, 2004).[8] The

---

[7] See the excellent discussion in Rindler (2006, Ch. 16 and 17). Our variables r and ρ are Rindler's η and r respectively: I preferred to keep the r for the 'classical' form of the metric (16.1).

[8] See Section 5.10.2 for a pseudo-Newtonian discussion of horizons.

cosmological horizon is defined by the distance that photons can travel since the Big Bang. It delineates the volume of space-time that a point, P, can potentially influence. Regions outside of that volume cannot receive information from P. As time elapses, the volume of influence of P increases: more material comes within what might be called P's horizon of influence.

The flat, $k = 0$, case of the Robertson–Walker metric provides a simple illustration of the notion of the particle horizon that is particularly relevant to observations of the cosmic background radiation from the early Universe. The metric (16.1) is then

$$ds^2 = -c^2 d\tau^2 + R(\tau)^2 (dx^2 + dy^2 + dz^2), \tag{16.12}$$

where, for convenience, we have expressed the spatial part of the metric in Cartesian coordinates and have changed the notation of the time coordinate $t \to \tau$ to make the time coordinate look like proper time of an observer at fixed (x, y, z). Now we use the time coordinate transformation

$$\eta = \int_0^\tau \frac{d\tau}{R(\tau)}, \tag{16.13}$$

as in (16.8). The meaning of this transformation was touched on in the discussion following Equation (16.8): η was shown to be a measure of the radial comoving coordinate distance a light ray could travel since the Big Bang. In comoving coordinates, the particles moving with the expansion have fixed coordinates, so we can envisage a sphere of light expanding at the speed of light engulfing ever more particles as time progresses.

This brings the metric (16.12) to the form

$$ds^2 = R(\eta)^2 \left[-c^2 d\eta^2 + dx^2 + dy^2 + dz^2 \right]. \tag{16.14}$$

The arguments η and τ of the functions $R(\eta)$ and $R(\tau)$ in Equations (16.12) and (16.14) are not the same: they are related via the transformation (16.13).

Equation (16.14) is the Minkowski metric multiplied by the time-dependent factor $R(\eta)$. Light rays are defined by $ds^2 = 0$ and in a Cartesian (r, η)-plane plot of the coordinate system (16.14) the light rays are at $45°$ to the axes, just as in Minkowski space. This is illustrated in Figure 16.2 where the light-cones for three observers $\mathbf{O}$, $\mathbf{P}$ and $\mathbf{Q}$ are shown. The Big Bang where $R(\eta) = 0$ is spread out flat in this coordinate system.

From Equation (16.12) the path of a radial light ray, $ds^2 = 0$ is $c d\tau = R(\tau) dr$ and so we can define a distance to this horizon via the equation

$$\int_0^{r_H} dr = \int_0^{t_H} \frac{c d\tau}{R(\tau)} = \tau_H. \tag{16.15}$$

In other words, $r_H = c\tau_H$, which is entirely expected. If $R(\tau)$ had been such that the integral had diverged, r_H would be infinite and there would be no particle horizon. (See also Section 5.10.)

Causal Structure of the Last Scattering Surface

Figure (16.2) depicts an observer $\mathbf{O}$ at the present time looking back along the past light-cone, towards the Big Bang. While the Universe is transparent he/she will see structure

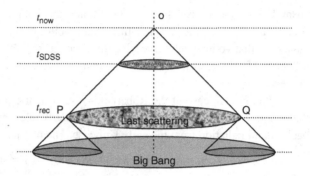

Fig. 16.2 Space-time diagram illustrating particle horizons in Robertson–Walker Universe. This is drawn in 'conformal coordinates' where time, t, runs vertically upwards and space is horizontal. In this model the volumes at a constant t are depicted as horizontal slices. The light rays travel in straight lines inclined at 45° (see Equation (16.14)). In these coordinates the world line of an observer **O** is a vertical line extending back to the singularity at time $t = 0$. Everything observed by **O** lies on the surface of **O**'s past light-cone.

The past light-cone of **O** identifies a volume of the Universe at the time of the Big Bang: we can have no observational knowledge of what lies outside that cone since the light from those regions has not yet reached us.

The CMB is observed coming from the time t_{rec} when the Universe became neutral. **PQ** represents the entire sky as seen from **O**. However, it is clear that **Q** lies outside the past light-cone of **P**, and likewise **P** lies outside the past light cone of **Q**. A causal patch on the sky, defined by the past light-cone of **P** or **Q** subtends an angle of $\sim 1°$ on the sky. So it is surprising that the high degree of isotropy of the CMB is showing us that **P**'s local Universe is virtually identical to **Q**'s, when they have not at that time been causally connected. ©

in the Universe (depicted by the time slice t_{SDSS}) and be able to look as far back as t_{rec} when the Universe became opaque. On the past light–cone of **O** are two observers **P** and **Q**, whose own past light-cones do not intersect. So **P** and **Q** cannot see one another and they are beyond one another's particle horizon.

What the cosmic background radiation emanating from t_{rec} tells an observer at **O** is that the causal neighbourhoods of **P** and **Q** look almost identical. This manifest large scale homogeneity is perhaps surprising in light of the fact the **P** and **Q** have never at that time been in causal contact. Without a physical explanation of this remarkable coincidence we can only say that the 'Universe is as it is because it was as it was'. [9]

16.1.7 Proper Distance Coordinates

Consider the radial motion, $d\phi = d\theta = 0$ of particles and light rays in the Robertson–Walker cosmology. The relevant part of the metric (16.12) can then be written

$$ds^2 = -c^2 dt^2 + R(t)^2 dr^2, \quad d\phi = d\theta = 0. \tag{16.16}$$

Here r is a radial comoving coordinate. We have seen that the transformation $d\eta = dt/R(t)$ transforms this into the conformal form of the metric, (16.14) in which the light rays travel

[9] Hermann Bondi in his undergraduate lectures on relativity and cosmology during my time at King's College London.

in straight lines as in Minkowski geometry. The paths of co-expanding particles in the conformal coordinates are hyperbolæ, just as in Minkowski space.

We can make a transformation to a different radial coordinate ϖ, the *proper distance coordinate* that makes the particle paths into straight lines, by writing

$$\varpi = R(t)r. \tag{16.17}$$

A particle that shares the cosmic expansion has fixed r, and so for such a particle

$$\dot{\varpi} = \dot{R}r = (\dot{R}/R)\varpi. \tag{16.18}$$

The velocity of recession of a comoving particle in ϖ-coordinates is proportional to its ϖ-distance. This is the Hubble Law.

More generally for particles having a variable position $r(t)$, we have $r\,d\varpi = r\dot{R}dt + Rdr = \varpi(\dot{R}/R)\,dt + Rdr$, and so (compare with the pseudo-Newtonian Equations (5.71)):

$$Rdr = d\varpi - \varpi\frac{\dot{R}}{R}dt. \tag{16.19}$$

Substituting this into (16.16), gives the metric in the non-diagonal form:[10]

Robertson–Walker metric: Painlevé-Gullstrand (radial part)

$$ds^2 = -\left[c^2 - \left(\frac{\dot{R}}{R}\right)^2\varpi^2\right]dt^2 - 2\frac{\dot{R}}{R}\varpi\,d\varpi\,dt + d\varpi^2. \tag{16.20}$$

A particle fixed in the cosmic expansion, i.e. a galaxy, is labelled by some value r_0 of r and by (16.17) has ϖ-coordinate $\varpi_0 = R(t)r_0$. The $d\varpi\,dt$ term vanishes at the origin $\varpi = 0$ and the metric about that point is just the Minkowski metric.

We notice that the coefficient of dt^2 can change sign at some value of $\varpi = \varpi_H$. This is the ϖ-coordinate radius of the particle horizon (see also Equation 16.15).

The path of a light ray in these coordinates is given by $ds = 0$ This gives a quadratic equation for $d\varpi/dt$ which can be solved to give the remarkably simple equation for the path of the light ray in (t, ϖ)-coordinates:

$$\frac{d\varpi}{dt} - \frac{\dot{R}}{R}\varpi = \pm c, \tag{16.21}$$

which depends explicitly on the cosmological model through the time dependence of the scale factor $\dot{R}/R$ (again, compare with Equations 5.71). There are two solutions, corresponding to the $\pm c$ choice. The quantity $(\dot{R}/R)\varpi$ is the Hubble expansion velocity at the point with radial coordinate ϖ, while $\dot{\varpi}$ is the speed of the light ray in these coordinates. So this last equation tells us that the relative speed between the light ray and the comoving observer receiving it is c, as we would hope and expect.

---

[10] This is sometimes referred to as the Painlevé–Gullstrand form of the metric, though this term is seen more frequently in relation to the Schwarzschild solution (Painlevé, 1921a,b; Gullstrand, 1922). For a thorough discussion of this metric in relation to light-cones and horizons see Ellis and Rothman (1993, 1995, Eq. (24) of the former).

16.2 Weak Gravitational Fields

Most gravitational fields we meet in physics are very weak, in which case the Newtonian limit applies. Here we shall consider a number of departures from the Newtonian limit.

16.2.1 The Static Newtonian limit

First we consider weak gravitational fields that are weak in the sense that the metric describing the distortion of the underlying Minkowski geometry is close to the Minkowski metric. We can write the metric as

$$g_{ab} = \eta_{ab} + \epsilon h_{ab}, \quad |\epsilon| \ll 1. \tag{16.22}$$

We put the ϵ here simply as an indication that the deviation from the Minkowski metric η_{ab} is small. The ϵ is also of help when looking for approximate solutions to the Einstein equations when we can do a power series expansion in ϵ. The term h_{ab} can be thought of as describing 'wrinkles' in the background space-time, which in this case is the Minkowski space-time.

The idea that h_{ab} is a small change to the geometry of the space-time carries with it the assumption that we can do many of the manipulations of tensors using the 'background' Minkowski space metric. In particular we can raise and lower indices using η_{ab} in place of g_{ab}:

$$h^{ab} = \eta^{am} \eta^{bn} \, h_{mn}. \tag{16.23}$$

It is easy to check that, to order ϵ^2,

$$(\eta_{ab} + \epsilon h_{ab})(\eta^{bc} - \epsilon h^{bc}) = \delta_a^c, \tag{16.24}$$

(just multiply out the left-hand side and drop terms in ϵ^2), where the elements of δ_a^c are zero unless the indices are the same. Then

$$g^{ab} = \eta^{ab} - \epsilon h^{ab}. \tag{16.25}$$

The Static Weak Field Metric

We have already encountered a metric of this kind in Section 14.2.1 (Equation 14.3) when discussing the early tests of Einstein's theory. With the metric in the special form

$$g_{ab} = -(1 - 2\phi) \, dt^2 + (1 - 2\phi)^{-1} \left(dx^2 + dy^2 + dz^2 \right), \quad \phi = \frac{2GM}{c^2 r}, \tag{16.26}$$

the term $g_{00} = -(1 - 2\phi)$ alone explains the advance of the perihelion of the orbit of Mercury. The deflection of light by the Sun's gravitational field requires the $(1 - 2\phi)^{-1}$ modifier of the spatial part of the metric.[11]

---

[11] We shall see that the general weak field metric for the case of a perturbed Universe containing only a perfect fluid and no gravitational waves can be put in the form

$$ds^2 = -\underbrace{(1 + 2\Phi)dt^2}_{\text{Newtonian}} + \underbrace{(1 - 2\Psi)}_{\text{curvature}} a(t)^2 \, (dx^2 + dy^2 + dz^2),$$

The fact that we can get the correct orbits of the planet by fixing only the g_{00} term of the Minkowski metric, while getting the deflection of light correct requires modification of the spatial part of the Minkowski metric, suggests that[12] the deflection of light is not due to Newtonian gravity, but to some physical phenomenon that is not a part of Newtonian theory: i.e. the geometric curvature of space-time.

We can easily calculate the Christoffel symbols for this metric:

$$\Gamma^0_{00} = \frac{1}{2} g^{00} \frac{\partial g_{00}}{\partial x^0}, \quad \Gamma^A_{00} = -\frac{1}{2} g^{AA} \frac{\partial g_{00}}{\partial x^A}, \quad \text{no summation on } A = 1, 2, 3, \quad (16.27)$$

which depend only on g_{00}. Evaluating Γ^α_{00} we get

$$\Gamma^\alpha_{00} = -\frac{1}{2}(1 - 2\phi)(-2\boldsymbol{\nabla}\phi) = \boldsymbol{\nabla}\phi, \quad \text{to first order} \quad (16.28)$$

and so we recover the equation of motion for the particle, i.e. the geodesic equation in this metric, as

$$\ddot{\mathbf{x}} + \boldsymbol{\nabla}\phi = 0. \quad (16.29)$$

This, as expected, is the Newtonian equation of motion. In this way we identify the Christoffel symbol Γ^α_{00} with the force $\boldsymbol{\nabla}\phi$. We note that the spatial part of the metric plays no role at this level of approximation.

Since $|\phi| \ll 1$, we can rewrite Equation (16.26) in the form

$$g_{ab} = -(1 + 2\Phi) \, dt^2 + (1 - 2\Phi)(dx^2 + dy^2 + dz^2), \quad (16.30)$$

where $\Phi = -\phi$ (see footnote 11). In this form the deviation from the Minkowski metric (16.22) is $h_{ab} = \text{diag}(-2\Phi, -2\Phi, -2\Phi, -2\Phi)$ and $\nabla^2 \Phi = 4\pi G\rho$.

Weak Field Space-Time Geometry

If we allow ourselves to look forwards to the definitions of the Riemann and Ricci tensors we can do the following. The Riemann tensor $R^a{}_{bcd}$ is, to order $1/c^2$ (cf. Equation 13.95),

$$R^a{}_{bcd} = \Gamma^a_{db,c} - \Gamma^a_{cb,d} + \text{higher order terms.} \quad (16.31)$$

If Γ^i_{00} is the only non-zero Christoffel symbol and is given by (13.47), then the only non-zero components of $R^a{}_{bcd}$ are

$$R^i{}_j = \frac{1}{c^2} \eta^{ik} \frac{\partial^2 \phi}{\partial x^j \, \partial x^k} + \text{higher order terms,} \quad (16.32)$$

which is the gravitational potential shear tensor, i.e. the familiar tidal field.

where Φ and Ψ are small quantities representing respectively a small correction to the Newtonian potential and a small contribution to the curvature of the spatial geometry. If there are no off-diagonal spatial terms in the energy–momentum tensor, it can be shown that, in this weak field approximation, $\Phi = \Psi$, which is Equation (16.30) in the case $a(t) = 1$.

[12] There is obviously some hindsight in this remark!

The trace of the Riemann tensor, the Ricci tensor $R_{bd} = R^m{}_{bmd}$, has only one non-zero component, which is

$$R_{00} = \frac{1}{c^2}\nabla^2\phi + \text{ higher order terms}, \tag{16.33}$$

which, apart from the normalisation factor $1/c^2$, is the left-hand side of the Poisson equation $\nabla^2\phi = 4\pi G\rho$.

16.2.2 The Weak Field Einstein Equations

We have not yet dealt with the metric perturbation h_{ab} except in the special case when it represents the 'Newtonian' approximation, leading to the metric in the form (16.26). We now need to get an equation for a general h_{ab}. To this end we must write down the vacuum Einstein equations in the form $R_{ab} = 0$ and so we need to compute the Ricci tensor R_{ab} in terms of the h_{ab}.

This involves the Christoffel symbols which are defined as

$$\Gamma^a_{bc} = \tfrac{1}{2}g^{ad}(\partial_b g_{cd} + \partial_c g_{db} - \partial_d g_{bc}), \tag{16.34}$$

and which in our approximation reduce to

$$\Gamma^a_{bc} = \tfrac{1}{2}\eta^{ad}(\partial_b h_{cd} + \partial_c h_{db} - \partial_d h_{bc}). \tag{16.35}$$

The Ricci tensor can be expressed, to this order, in terms of the Christoffel symbols:

$$R_{ab} = \partial_b \Gamma^c_{ac} - \partial_c \Gamma^c_{ab}, \tag{16.36}$$

and so, combining the last two equations, we get the Ricci tensor and the curvature scalar $R = \eta^{ab}R_{ab}$:

Weak field geometry:

$$R_{ab} = \tfrac{1}{2}\epsilon\left[\Box h_{ab} - \partial_k(\partial_a h^k_b + \partial_b h^k_a) + \partial_a\partial_b h\right], \tag{16.37}$$

$$R = \tfrac{1}{2}\epsilon\left[\Box h - \partial_a\partial_b h^{ab}\right], \tag{16.38}$$

where $h^{ab} = \eta^{ac}\eta^{bd}h_{cd}$, $h = \eta^{ab}h_{ab}$ and $\Box = \partial^a\partial_a$.

The derivatives here are simply derivatives with respect to the familiar (ct, x, y, z) coordinates of Minkowski space and so the object variously described as $\partial^2 \equiv \partial_k\partial^k \equiv \Box$ is simply the wave operator

$$\Box \equiv \partial^2 = \left[\nabla^2 - \frac{1}{c^2}\frac{\partial^2}{\partial t^2}\right]. \tag{16.39}$$

Setting $R^{ab} = 0$, yields the linearised equations for h_{ab}:

$$\partial_k\partial^k h_{ab} - \partial_k(\partial_a h^k_b + \partial_b h^k_a) + \partial_a\partial_b h = 0. \tag{16.40}$$

The derivation of these equations appears in most books on general relativity that cover weak fields.

Equations (16.40) look pretty daunting. But we can go several steps further by putting:

Trace-reversed metric perturbation:

$$\bar{h}_b^a = h_b^a - \tfrac{1}{2}\delta_b^a h. \tag{16.41}$$

(This is called trace-reversed because $h = -\bar{h}$.)

and noticing that

$$\partial_k \partial_b \bar{h}_a^k = \partial_k \partial_b (h_a^k - \tfrac{1}{2}\delta_a^k h) = \partial_k \partial_b h_a^k - \tfrac{1}{2}\partial_a \partial_b h, \tag{16.42}$$

$$\partial_k \partial_a \bar{h}_b^k = \partial_k \partial_a (h_b^k - \tfrac{1}{2}\delta_b^k h) = \partial_k \partial_a h_b^k - \tfrac{1}{2}\partial_b \partial_a h. \tag{16.43}$$

The sum of the right-hand sides of these two equation is the last three terms of Equation (16.40) which thereby reduces to

$$\partial_k \partial^k h_{ab} - \partial_k \partial_b \bar{h}_a^k - \partial_k \partial_a \bar{h}_b^k = 0. \tag{16.44}$$

What has been gained by introducing the quantity $\bar{h}_{ab}$?

The answer lies in the fact that if we impose the condition:

de Donder gauge condition:

$$\partial_k \bar{h}_a^k = 0, \tag{16.45}$$

also known as the *harmonic coordinate condition*,

the term $\partial_k \partial_b \bar{h}_a^k + \partial_k \partial_a \bar{h}_b^k$ appearing on the left-hand side of (16.44) becomes zero, leaving us with the simple equation

$$\Box h_{ab} = \partial_k \partial^k h_{ab} = 0 \tag{16.46}$$

for the linearised vacuum Einstein equations (see Equations 16.37).[13]

What makes us think that we can arbitrarily impose a condition like (16.45)? As we shall now show, setting $\partial_k \bar{h}_a^k = 0$ amounts to making a gauge transformation of the 'wrinkle' field, h_b^a. This parallels the situation in electrodynamics where the transformation of the vector potential $\bar{A}_m = A_m - \phi_{,m}$ leaves the fields unchanged, so we can add or remove gradient terms $\phi_{,m}$. How we define the field ϕ is open to choice: we can, for example choose either the Coulomb gauge or the Lorenz gauge.

16.3 Gauge Transformations

We can simplify (16.44) by exploiting *gauge transformations*. These are transformations of the coordinate system, or of the quantity of interest, which make no difference to the

[13] In linear theory, Equation (16.45) is equivalent to $\partial_k h_b^k = \tfrac{1}{2}\partial_a h$ (by Equation 16.41). We can substitute this latter condition directly into Equations (16.37) and get to the result $\partial_k \partial^k h_{ab} = 0$ without going through the quantity $\bar{h}_{ab}$. This 'trick' would, however, obscure the underlying reason for making such a transformation.

underlying physical quantities. The key point to remember is that the field g_{ab} is not an observable quantity: we only observe its derivatives though quantities like the curvature tensors, and in each coordinate system, via the connection Γ^a_{bc}, which, for example, appears as the Newtonian force in weak fields.

There are several familiar examples from electromagnetic theory. If, for example, we represent the magnetic field $\mathbf{B}$ in terms of a vector potential $\mathbf{B} = \nabla \times \mathbf{A}$, we do not change anything if we add a gradient to the vector $\mathbf{A}$: $\mathbf{A} \to \mathbf{A}' = \mathbf{A} + \nabla\phi$. In that case we still have $\mathbf{B} = \nabla \times \mathbf{A}'$ since $\nabla \times \nabla\phi \equiv 0$ for any function ϕ. The arbitrariness of the function ϕ makes this a powerful tool.[14] We shall be generalising this gauge transformation that uses a scalar function ϕ to a gauge transformation based on a 4-vector field ξ_a.

Consider a small coordinate transformation[15]

$$\bar{x}^a = x^a + \epsilon \xi^a(x). \tag{16.47}$$

The metric of the $\bar{x}$ coordinates is related to the original x-coordinate metric by

$$g_{ab} = \frac{\partial \bar{x}^c}{\partial x^a} \frac{\partial \bar{x}^d}{\partial x^b} \bar{g}_{cd}. \tag{16.48}$$

Putting (16.47) into (16.48) gives, to $O(\epsilon)$, the general result

$$\bar{g}_{ab} = g_{ab} - \epsilon \left(\xi_{a,b} + \xi_{b,a} \right), \tag{16.49}$$

with which the weak field metric (16.22) becomes

$$\bar{h}_{ab} = h_{ab} - 2\,\xi_{(a,b)}, \quad \text{where } \xi_{(a,b)} = \tfrac{1}{2}(\xi_{a,b} + \xi_{b,a}). \tag{16.50}$$

The perturbations to the metric h_{ab} and $\bar{h}_{ab}$ are merely descriptions of the same field in different coordinate systems. This says that we can add an arbitrary symmetrised derivative to the h_{ab} and, in effect, still be describing the same field in a slightly altered coordinate system.

This is analogous to the 4-vector potential A_a of electrodynamics where we can add an arbitrary gradient to A, $\bar{A}_a = A_a - \phi_{,a}$, and still be describing the same electromagnetic field. The transformation (16.47) has played a very important role in general relativity.

16.3.1 Exploiting Gauge Freedom

We noted (Equation 16.49) that under the small coordinate transformation

$$\bar{x}^a = x^a + \epsilon \xi^a(x), \tag{16.51}$$

the h_{ab} transforms according to

$$\bar{h}_{ab} = h_{ab} - \xi_{a,b} - \xi_{b,a}, \tag{16.52}$$

[14] The discussion here follows the ancient and now virtually unobtainable lectures of Schild (1967) on general relativity and by Sachs (1967) on gravitational waves, given to the 4th Summer Seminar on *Applied Mathematics*, held at Cornell University, 26 July–20 August 1965.

[15] The indices a on the quantities x^a and $\bar{x}^a$ refer to the components relative to corresponding sets of axes in the two different coordinate systems, and so we occasionally see the indices being written as x^a and $x^{\bar{a}'}$ to emphasise that.

and since h_{ab} is a physical quantity described by a tensor, then $\bar{h}_{ab}$ describes the same quantity. Making this gauge transformation does not affect the physics of our wrinkle on Minkowski space, it merely affects our coordinate dependent description of it.

We can for the moment simplify the notation a little by setting $\epsilon = 1$ and, in effect, absorb it into the vector ξ. Suppose we can chose ξ to be such that

$$\partial_p \partial^p \xi_a = \bar{h}^m_{a,m}. \tag{16.53}$$

Then

$$\partial_k (\bar{h}^k_{a,b} + \bar{h}^k_{b,a}) = \partial_p \partial^p (\xi_{a,b} + \xi_{b,a}), \tag{16.54}$$

and we see that the second and third terms on the left-hand side of (16.44) are merely a manifestation of a gauge transformation and so can be set to zero. However, in doing this we are imposing constraints on the coordinate system.

This leaves our weak vacuum field approximation as

$$\partial_k \partial^k h_{ab} = 0, \qquad \partial_k \partial^k \bar{h}_{ab} = 0, \tag{16.55}$$

which is just the wave equation on a Minkowski background. Our wrinkle is a gravitational wave.[16]

Note that we got to this point by imposing a particular gauge condition on the h_{ab}:

$$h^k_{a,k} = 0. \tag{16.56}$$

We see from Equation (16.53), which defined the transformations, that this gauge condition defines a set of coordinate transformations such that $\Box \xi^a = 0$. This set of transformations ensures that metric perturbations described in the de Donder gauge remain in that gauge under such transformations. This gauge condition imposes four constraints on the ten components of the symmetric tensor h_{ab}.

Physical Interpretation of the Gauge Condition

The gauge conditions impose a restriction on the physical quantities that are being described in a gauge invariant way. However, despite this restriction, we know that the quantity being described is not an artefact of the choice of coordinates. We can interpret the physical nature of the gauge invariant solutions provided by the de Donder gauge choice as follows.

Equation (16.55) has a simple solution:

$$h^a_b = A^a_b \exp(ik_m x^m), \tag{16.57}$$

where A^a_b, the wave amplitude, is a constant determined by the boundary conditions. Differentiating this as in Equation (16.56) we get the important result that

$$h^a_{b,a} = ih^a_b k_a = 0, \quad A^a_b k_a = 0. \tag{16.58}$$

So the choice of this gauge restricts the solution to being a plane wave: there is no component of h^a_b in the direction of propagation. As we shall see, this is directly reflected in

---

[16] This could also be regarded as a mass-zero Klein–Gordon field.

the metric corresponding to h_b^a. This condition says that the perturbation corresponds to a transverse wave.

Weak Field Einstein Equations

The vacuum field equations are just Equations (16.37) in either of the forms

$$\Box \bar{h}_{ab} = 0, \qquad \Box h_{ab} = 0. \tag{16.59}$$

These are equivalent since $h = -\bar{h}$. As we have just seen, this is the plane wave equation. We shall go further into this below.

The field equations for a general non-zero energy momentum tensor T_{ab} are

$$\Box \bar{h}_{ab} = \frac{16\pi G}{c^4} T_{ab}, \tag{16.60}$$

where the components of the energy–momentum tensor here refer to the background Minkowski metric. Here we see the energy–momentum tensor as the source of the gravitational radiation field $\bar{h}_{ab}$ in a way that is precisely analogous with the situation in electromagnetism. In Minkowski space the solutions can be expressed in the familiar retarded potentials for a given distribution of matter described by T_{ab}.

16.3.2 Gravitational Waves: Wrinkles in Space-Time

So far we have a weak field approximation for a small wrinkle on Minkowski space. Apart from the wave Equation (16.55) we have not in fact specified any boundary conditions or coordinates. Firstly we note that h_{ab} is symmetric in a, b and so has ten independent components. We are free to impose additional conditions on the h_{ab}. The simplest choice is to set four conditions:

$$h = h_a^a = 0, \qquad h_{0\alpha} = 0, \quad \alpha = 1, 2, 3. \tag{16.61}$$

The first of these insists that the perturbation h_{ab} is traceless. With the transverse (16.56) we refer to this gauge choice as the *traceless transverse gauge*.[17]

We are left with two degrees of freedom. If we choose the direction of propagation to be along the x^1-axis then we are left with the components

$$h_{ab} = \begin{pmatrix} 0 & 0, & 0 & 0 \\ 0 & 0, & 0 & 0 \\ 0 & 0, & h_+ & h_\times \\ 0 & 0, & h_\times & -h_+ \end{pmatrix}, \tag{16.62}$$

where the equations for the components h_+ and $h_\times$, expressed in terms of the coordinates are

$$\left(\frac{\partial^2}{\partial x_1^2} - \frac{\partial^2}{\partial t^2} \right) h_+ = 0, \quad \left(\frac{\partial^2}{\partial x_1^2} - \frac{\partial^2}{\partial t^2} \right) h_\times = 0, \qquad (c = 1). \tag{16.63}$$

---

[17] The traceless transverse gauge is the direct analogue of the Coulomb gauge in electromagnetism.

Writing out the resulting metric using re-labelled coordinates $x^1 \to x, x^2 \to y, x^1 \to z$ we have:

Weak field gravitational wave in background Minkowski space:

$$ds^2 = -dt^2 + dx^2 + (1 + h_+) \, dy^2 + (1 - h_+) \, dz^2 + 2h_\times \, dydz, \qquad (16.64)$$

expressed in the traceless transverse ('TT') gauge. We see that we have acquired an off-diagonal $dydz$ term. We note that the structure of the wave lies in the y-z plane which is perpendicular to the x-direction of propagation. This is the transverse wave as expected from Equation (16.58). The wave distorts the geometry of space-time differently in the y- and z-directions: this is a shear wave.

16.3.3 Physical Interpretation: LIGO

These two components are independent polarisations of the gravitational wave, and in that respect the situation is identical to electromagnetic waves which also have two polarisations. However, as we shall see, there is a fundamental difference in that these two polarisations are not linear polarisation modes as in the sense of an electromagnetic wave.

First it is important to appreciate that the 'small wrinkles' approximation linearises the Einstein equations, and so the solutions can be simply superposed, as in electrodynamics. The two modes shown in Equation (16.63) are independent, and so one mode can exist without the other. So we have in fact two solutions for the gravitational wave on a Minkowski background:

$$ds^2 = -dt^2 + dx^2 + (1 + h_+) \, dy^2 + (1 - h_+) \, dz^2, \qquad (16.65)$$

$$ds^2 = -dt^2 + dx^2 + dy^2 + dz^2 + 2h_\times \, dydz. \qquad (16.66)$$

Notice that if we apply the transformation

$$y = (y' - z')/\sqrt{2}, \quad z = (y' + z')/\sqrt{2}, \qquad (16.67)$$

to Equation (16.66), and then drop the primes, we get the other mode (16.65). The transformation (16.67) is a rotation about the x-axis of the y-z, plane through an angle of $45°$. This contrasts with the polarisation of an electromagnetic wave in which the modes are $90°$ apart.

The action of the wave on a set of test particles lying in a plane perpendicular to the direction of propagation can be easily deduced from the metric of Equations (16.65) and (16.66). It suffices to look at the h_+ mode, since the behaviour of the $h_\times$ mode is simply a $45°$ rotation of the h_+ mode. Setting $dt = 0$ and $dx = 0$ in Equation (16.65) allows us to solve this: we then have the distance between neighbouring points lying in the plane of the wave and having coordinates (dy, dz).

Consider four neighbouring points, A,B,C,D, with coordinates $A(d, 0)$, $B(0, d)$, $C(-d, 0)$ and $D(0, -d)$, where initially the points lie on a circle of small radius d. Then the proper lengths s_{AC} and s_{BD} of the segments AC and BD vary as

$$s_{AC}^2 = 2d^2[1 + h_+(t)], \tag{16.68}$$

$$s_{BD}^2 = 2d^2[1 - h_+(t)], \qquad h_+ \propto \exp(i\omega t), \tag{16.69}$$

where ω is the frequency of the wave (see Equation (16.57)). So the circle gets distorted into an ellipse with major axis along the y-axis, oscillates back to a circle and then changes to an ellipse with major axis along the z-axis. This is a typical shear flow.

In the LIGO experiment the light travels up and down the interferometer arms, length L, on a time that is much shorter than the time for a crest of the gravitational wave to pass. The effect of the wave on the light in one traverse is therefore small. For the light travelling in the y-direction we have $ds = 0$ and so $cdt = \sqrt{1 + h_+}dy \simeq (1 + \frac{1}{2}h_+)dy$. This means that during the time the light takes to travel down the y-arm, the length of the y-arm has effectively been stretched to $L(1 + \frac{1}{2}h)$. This stretching continues for as long as the arm is acted upon by the pulse, so with each trip up and down the arm the phase lag of the light is being constantly increased. If the light travels up and down the arms a few hundred times it spends only $\sim 10^{-3}$s having its phase boosted, while it takes a 100 Hz wave $\sim 10^{-2}$s to pass by, and so the effect of the wave on the light is a small factor (Faraoni, 2007).

16.4 The Lemaître, Tolman, Bondi Solution

An important exact and analytically solvable solution of the Einstein field equations was derived independently by Lemaître (1933)[18] and Tolman (1934a).[19] Bondi (1947) provided a complete analysis of the solution.[20] The solution describes an inhomogeneous spherically symmetric distribution of dust and, expressed in terms of cosmic time, t, and the comoving radius, r, has metric

$$ds^2 = -c^2dt^2 + \frac{R'(t,r)^2}{1 + f(r)}dr^2 + R(t,r)^2(d\theta^2 + \sin^2\theta d\phi^2), \tag{16.70}$$

where we will use the notation that for any function $X(r, t)$,

$$X(r,t)' \equiv \frac{\partial X}{\partial r} \quad \text{and} \quad \dot{X}(r,t) \equiv \frac{\partial X}{\partial t}. \tag{16.71}$$

Here $f(r)$ is an arbitrary function of r such that $f(r) > -1$. If we specifically put $R(t, r) = a(t)r$ and $f(r) = -kr^2$ we recover the Friedmann–Robertson–Walker metric.

---

[18] This is the famous paper in which Lemaître showed that the radius $r = 2GM/c^2$ in the Schwarzschild metric is not in fact singular. See Equation 15.133 et seq.. There is an extensive discussion of this by Smarr and York (1978, Section 5D).

[19] This solution was an extension of the earlier work of McCrea and McVittie (1930) who considered small deviations from an underlying Einstein Universe. The solution is discussed extensively within a modern relativistic framework in the book by Ellis et al. (2012, Section 15.1 et seq.). Krasinski (1997, p.66 et seq.) discusses this model, which he refers to as the L-T model, within the context of a wider set of inhomogeneous solutions to the Einstein equations.

[20] The paper of Bondi loc. cit. is beautifully pedantic in its derivation and discussion of the general relativistic aspects of the solution. See, in particular, the Appendix of the paper.

For dust, the energy momentum tensor has only one component, $T^{00} = \rho c^2$. The corresponding Einstein equations[21] yield equations for $R(t, r)$ that can be expressed in terms of an arbitrary function $F(r)$ as follows:

$$\frac{F(r)'}{R^2 R'} = 8\pi G\rho, \quad F(r) = -2R^2\ddot{R}, \quad \dot{R}^2 = f(r) + \frac{F(r)}{R} + \frac{1}{3}\Lambda R^2. \tag{16.72}$$

This is sometimes referred to as the *generalised Friedmann equation* (Ellis *et al.*, 2012, Equation 15.3),[22] describing a distribution of nested spherical dust shells of different density. From the last of these, the arbitrary function $F(r)$ is readily understood to represent the initial matter distribution as a function of the comoving radius, r, of the shell. The function $f(r)$ takes on the role of the initial velocity (i.e. the kinetic energy) of the dust shell at radius r. From the first of these equations we see that $dF(r) = 8\pi G\rho R^2 dR$, which suggests that $F(r)'$ is proportional to the mass within the shell $[R, R + dR]$.

16.4.1 LTB Solutions

The solutions fall into three classes depending as to whether $f > 0, f < 0$ or $f = 0$. Given the similarity between the BTL metric (16.70) and the Robertson–Walker metric in the form (16.3), this should occasion no surprise. It is as though each spherical shell of the BTL solution were a Robertson–Walker metric which had emerged from its own singularity at some time $T(r)$ depending on the comoving radius r of the shell.

Indeed, if we write

$$f(r) = 2E(r), \quad \text{and} \quad F(r) = 2GM(r), \tag{16.73}$$

we have from the BLT Equations (16.72) that $F = \dot{R}^2 R - fR$, and so

$$\frac{2GM}{R} = \dot{R}^2 - 2E - \frac{1}{3}\Lambda R^2. \tag{16.74}$$

This shows that we can interpret $M(r)$ as the gravitational mass within a comoving radius r, and $E(r)$ as the 'curvature energy'.

The parallel with the Robertson–Walker model suggests that these equations will have solutions that are similar in form to the solutions for the model, and, indeed, the solutions to (16.72) for $\Lambda = 0$ are given parametrically in terms of a development angle η by:

$$R = \frac{F}{2f}(\cosh\eta - 1), \quad t - T(r) = \frac{F}{2f^{3/2}}(\sinh\eta - \eta), \quad f > 0, \tag{16.75}$$

$$R = \frac{F}{-2f}(1 - \cos\eta), \quad t - T(r) = \frac{F}{2(-f)^{3/2}}(\eta - \sin\eta), \quad f < 0, \tag{16.76}$$

and

$$R = \left(\frac{9F}{4}\right)^{1/3}(t - T(r))^{2/3} \quad \text{for} f = 0. \tag{16.77}$$

---

[21] These are written out explicitly in the case $\Lambda \neq 0$ by Bondi (1947, Appendix, p.422: Eqs. 5 & 7).

[22] Ellis *et al. loc. cit.* put $F(r) = 2m(r)$ and $f(r) = 2E(r)$. See also Misner and Sharp (1965) for a more general discussion of these equations including heat flux in the diffusion approximation.

In these solutions, $T(r)$ is an arbitrary function. These solutions have not only their local cosmic singularity when $t = T(r)$, but also shell-crossing singularities. Shell-crossing singularities are avoided if $R(r,t)' > 0$ for all shells r at all times t, but this condition is difficult to ensure simply from the initial conditions. If there is a shell such that $2GM(r) > rc^2$, then all the material within that shell is trapped in a central black hole.[23]

The LTB equations are important because they are a simple and exact inhomogeneous solution to the zero-pressure (dust) Einstein equations.

16.4.2 Local Shear, Expansion Rate and 3-Curvature

Because the LTB solution is inhomogeneous, the local shear, σ, and expansion rate, θ, of the flow and the 3-space curvature, $^{(3)}R$, are simple functions of the radial coordinate scaling function $R(r,t)$:

$$H = \tfrac{1}{3}\theta = 2\frac{\dot{R}}{R} + \frac{\dot{R}'}{R'}, \quad \sigma^2 = \frac{1}{3}\left(\frac{\dot{R}'}{R'} - \frac{\dot{R}}{R}\right)^2, \quad {}^{(3)}R = -\frac{(2ER)'}{R^2 R'}, \tag{16.78}$$

where the function $E(r)$ is defined as in (16.73) (see, for example, Sussman (2008, 2009) for a discussion of these and other local properties of the LTB solution).

16.4.3 The Curvature Function – Accelerated Expansion

The curvature function $f(r)$ in the LTB metric is an arbitrary function, and the choice $f(r) = -kr^2$ reduces to the FLRW model for curvature constant $k = -1, 0, +1$, when we put $R(t,r) = ra(t)$. The value of kr^2 changes the character of the expansion. Paranjape (2009) noted that if we choose k to be a function of r, so that $f(r) = -k(r)r^2 < 0$, then the asymptotic $t \to \infty$ solution of the open LTB solution (16.75) would lead to an expansion law,

$$R(t,r) \simeq r\sqrt{-k(r)}(t - T(r)). \tag{16.79}$$

So we can design an expansion law simply with a suitable choice of $k(r)$. Paranjape *loc. cit.* computes the expansion law for a simple choice

$$k(r) = -1/(1 + r^\alpha), \quad 0, \alpha < 2 \tag{16.80}$$

for which he shows that this drives an accelerated expansion, as seen by the centrally placed observer.[24]

This is interesting in the context that there is accelerated expansion in this model without the need for a cosmological constant. The accelerated expansion in this model at late

---

[23] This was first discussed by Novikov (1964), whose solution was discussed at greater length in Misner *et al.* (1973, Section 31.4, p.826).

[24] Although the cosmological data on, for example, the supernova Hubble diagram, has been calculated on the basis of such models, the models do of course violate the Copernican principle that we are not sitting at some special place in the Universe. The observational status of more general models, based around the suggestion that the structure might drive the acceleration of the expansion via a back-reaction mechanism has also been analysed (Wiltshire, 2009).

times and at large distances apparently occurs because this model Universe is not homogeneous. However, what is clear is that the effect of non-decelerated expansion occurs simply because of the choice of $k(r)$, with $2E(r) = f(r) = -k(r)r^2 < 0$: the acceleration is simply a consequence of the initial conditions.[25]

16.5 Averages in Inhomogeneous Cosmological Models

The possible influence of *back reaction* on the overall cosmic expansion has been the subject of an important debate since the suggestion of Buchert and Ehlers (1997) in Newtonian cosmology that the nonlinear evolution of density perturbations might have a back-reaction on the overall expansion, possibly driving an exponential expansion. Buchert (2000) went on to repeat the analysis for inhomogeneous general relativistic solutions. This is referred to as *dark energy from structure* (Buchert, 2008). Whether or not it can explain the observed acceleration of the cosmic expansion is still hotly debated, but if we are to do precision cosmology, it is essential that we understand what is the contribution from this effect.[26]

16.5.1 Averaging Quantities and Equations

Averaging a quantity over a finite volume is often taken for granted.[27] When we make a measurement of the 'density at a point' we in fact measure the density over a sampling volume that is limited by the capability of our equipment. If we call the domain of averaging $\mathcal{D}$ the spatial average of a scalar quantity $S(\mathbf{x})$ is usually taken to be

$$S_{\mathcal{D}} \equiv \langle S(\mathbf{x}) \rangle = \frac{\int_{\mathcal{D}} dV S(\mathbf{x})}{\int_{\mathcal{D}} dV}, \tag{16.81}$$

where dV represents some, possibly weighted, element of volume. The angular brackets denote the mean, $S_{\mathcal{D}}$, of the scalar S, where the definition of the mean is given by the ratio of the integrals in the second equality. In the Euclidean space $\mathbb{R}^3 : \{x_1, x_2, x_3\}$ of our experience, dV could simply be $d^3x = dx_1 dx_2 dx_3$, or it could be probability weighted with a probability density function $p(x)$, in which case $dV = p(x)d^3x$.

---

[25] Unless $k(r)$ arises as a consequence of some averaging procedure, this is *not* an example of the 'back-reaction' of the inhomogeneity on the overall cosmic expansion of the kind we discuss in Section 16.5.2.

[26] The work of Isaacson (1968a,b) in which he calculated the back-reaction of an ensemble of high frequency gravitational waves in an FLRW Universe foreshadowed much of this topic. The possibility of a significant back-reaction arose because the contribution of the waves to the energy density increased as the 4th power of the wave frequency. Isaccson's calculation, done within the framework of the WKB approximation, was mathematically rigorous.

After the re-emergence of the concept of back-reaction with the work of Buchert and Ehlers *loc. cit.* the topic became, and still is, widely discussed, and raised considerable controversy.

This was further analysed by Paranjape and Singh (2008); Paranjape (2009, 2008, PhD thesis) from the point of view of perturbation theory, using the exact model as a template against which to compare the successes and failures of perturbation theory.

[27] I am grateful to Roberto Sussman for good discussions about this. I learned a lot.

In highly compressible fluid flow the average is taken using the density $\rho(x)$ as the weighting factor: this is referred to as *Favre averaging* (Favre, 1965). In slightly compressible flow the averaging is generally taken without a weighting factor since the density excursions are then very small: this is referred to as *Reynolds averaging* (Boussinesq, 1877).[28]

Averaging the Einstein Field Equations

In general relativity the volume element should be the invariant 3-volume on the hypersurfaces of constant proper time. This is determined from the spatial part of the metric $^{(3)}g_{ij}$: i.e. $J = \sqrt{^{(3)}g}d^3x$, where $^{(3)}g = |\,^{(3)}g_{ij}\,|$. With this,

$$S_\mathcal{D} = \frac{\int_\mathcal{D} S(\mathbf{x})\sqrt{^{(3)}g}\,d^3x}{\int_\mathcal{D} \sqrt{^{(3)}g}\,d^3x}. \tag{16.82}$$

The first to use this averaging method were Noonan (1984, Section 4) and Futamase (1989, 1988), who studied averages of inhomogeneous cosmologies that were small perturbations to an FLRW model, and Kasai (1992) who put no limits on the degree of inhomogeneity, but sought to construct Universes that were, on average, FLRW models. The question of averaging inhomogeneous cosmologies was taken up once again by Buchert and Ehlers (1997) in the context of Newtonian cosmology, and by Buchert (2000) in the context of general relativistic models.

Means of Functions versus functions of Means

Using Equation (16.82) we can compute the fluctuation in the time derivative of an arbitrary quantity, $S(\mathbf{x})$. Since this is a purely 3-space problem, we can simplify the notation by writing $^{(3)}g \rightarrow g$. We need to differentiate the integral (16.82), and so we need to know the time derivative of $J = \sqrt{g}$. Since the change in the volume element dV with time is simply proportional to $\sqrt{g}$, and since the rate of change of the volume element is simply the local divergence of the velocity, $\nabla \cdot \mathbf{v}$, we have

$$\dot{J} = \theta J \quad \text{where } J = \sqrt{g},\ ^{(3)}g = \det{}^{(3)}g_{ij} \text{ and } \theta = \nabla \cdot \mathbf{v} = \dot{V}/V. \tag{16.83}$$

With this, the volume $V_\mathcal{D} = \int_\mathcal{D} dV$ of the domain $\mathcal{D}$ evolves as:

$$\frac{\dot{V}_\mathcal{D}}{V_\mathcal{D}} = \frac{\int_\mathcal{D} \dot{J}\,d^3x}{\int_\mathcal{D} J\,d^3x} = \frac{\int_\mathcal{D} \theta dV}{\int_\mathcal{D} dV} = \langle\theta\rangle_\mathcal{D}. \tag{16.84}$$

In this case, the result is not surprising, the mean expansion rate of the volume is just the mean of the velocity field divergence.

A situation where the operation order is important is taking the time derivative of the averaged quantity $\langle S\rangle_\mathcal{D}$ versus taking the average of the derivative of S, i.e. $\langle\dot{S}_\mathcal{D}\rangle$:

$$\frac{d\langle S_\mathcal{D}\rangle}{dt} - \left\langle\frac{dS}{dt}\right\rangle_\mathcal{D} = \langle\theta S\rangle_\mathcal{D} - \langle\theta\rangle_\mathcal{D}\langle S\rangle_\mathcal{D}. \tag{16.85}$$

---

[28] In classical fluid dynamics, the idea of splitting flows into a mean flow plus a fluctuating component is important in sub-grid simulations. For a discussion of various weighting schemes see Cloutman (1999).

This follows from taking the time derivative of (16.82) and using the results (16.83), and noting that because this is a time derivative of a comoving volume, we can take the time derivative of the integrand without worrying about the boundary $\mathcal{D}$.

16.5.2 Back-reaction from Stochastic Inhomogeneities

Applying this last result to the expansion rate, θ, gives us the difference between the evolution of the mean expansion rate, $\langle\theta\rangle_\mathcal{D}$ in the volume $\mathcal{D}$, and the expansion of the volume $\mathcal{D}$:

$$\frac{d}{dt}\langle\theta\rangle_\mathcal{D} - \langle\dot{\theta}\rangle_\mathcal{D} = \langle\theta^2\rangle_\mathcal{D} - \langle\theta\rangle^2_\mathcal{D}. \tag{16.86}$$

Note that the right-hand side is positive.[29]

This says that the time derivative of the spatially averaged expansion rate is greater than the average of the time derivatives of the local expansion rates. In other words, the acceleration of the whole volume is greater than the averaged acceleration within the volume. It is as though $\langle\theta^2\rangle_\mathcal{D} - \langle\theta\rangle^2_\mathcal{D}$ were driving a different rate of expansion than the expected $\langle\dot{\theta}\rangle_\mathcal{D}$.

This simple argument is the essence of the back-reaction argument: the mean square expansion rate exceeds the square of the mean expansion rate. The general view on this appears to be that there may well be such an effect, but that it is so small that it 'could not act as the driver for the dominant exponential expansion we observe'. Another possibility is that this is merely an artefact of an inappropriate use of perturbation theory.

16.6 Primordial Perturbations in General Relativity

Perturbations to the FLRW metric were first considered by Lifshitz (1946) and clarified by Bonnor (1954, 1957). One of the problems with that work, and for much of what followed, was the lack of a gauge-independent description of what was meant by the density fluctuation, $\delta\rho$. In other words, the definition of $\delta\rho$ depends on the choice of coordinates. It would always be possible to eliminate $\delta\rho$ simply by transforming the time coordinate to the local proper time.[30] A consequence of this was that there was no proper gauge invariant description of the evolution of non-oscillatory perturbations on super-horizon scales.[31]

[29] Because the variance of any random variable X is always positive: $\mathrm{Var}(X) = \mathbb{E}[X^2] - (\mathbb{E}[X])^2 \geq 0$.

[30] The issue was raised by Olson (1976, in particular see Appendix B.) where he followed the coordinate-free analysis of Hawking (1966).

[31] In fact in 1969 Layzer, doing a very similar analysis to today's analysis, found the correct gauge invariant solution. See Layzer (1975, Equation 4.5.8) and Layzer (1971); Layzer and Burke (1972). Owing to severe publication delays of *Stars and Stellar Systems vol. IX*, this key 1969 paper did not appear until 1975.

The modern approach to the problem was initiated by the seminal work of Bardeen (1980) and Kodama and Sasaki (1984), who resolved the gauge issues by defining gauge-invariant variables to describe the fluctuations about the mean. That work was substantially upgraded and clarified by Ma and Bertschinger (1995), who gave a very complete list of references for the development of this topic. Moreover, they discussed perturbations in flat models that included both dark matter and Λ.

Our background model is flat and permeated by a fluid having equation of state $p = w\rho$: the scale factor grows as $a(t) \propto t^{2/(3+w)}$. With this the expansion rate $H(t)$ at time t is given by $Ht = 2/3(1+w)$.

16.6.1 Primordial Perturbations: the Metric

We consider here only the case of a flat background FLRW cosmological model in a coordinate system where the perturbed metric is written in the *longitudinal gauge* form:[32]

$$ds^2 = -\left(1 + \Phi(t, \mathbf{x})\right)dt^2 + \left(1 - 2\Psi(t, \mathbf{x})\right)a(t)^2\delta_{mn}dx^m dx^n. \qquad (16.87)$$

When $\Phi = \Psi = 0$, this is the flat FLRW model.[33]

What characterises this form is that it is diagonal and the coefficients of dt^2 and $dx^a dx^b$ are different.[34] The advantage of the Newtonian gauge is that there is no gauge freedom left. This is unlike the situation with comoving coordinates, i.e. the comoving gauge, where we have to add an extra condition to fix the coordinates (as we saw in Section 16.3).[35]

The Einstein Tensor Components

We need to set the Einstein tensor equal to the energy–momentum tensor.[36] Given the metric (16.87), we can work out the Christoffel symbols using the technique shown in Section 15.2.1 and from there calculate the Riemann tensor. After a tedious amount of work we end up with the Einstein tensor in the unprepossessing form:

$$G^0{}_0 = 3H^2(1 - \Phi) + 2a^{-2}\nabla^2\Psi - 6H\dot{\Psi}, \qquad (16.89)$$

$$G^\mu{}_\mu = (2\dot{H} + H^2)(1 - 2\Phi) - 2\ddot{\Psi} + 6H\dot{\Psi} + 2H\dot{\Phi}$$
$$- a^{-2}\nabla^2(\Psi - \Phi) + a^{-2}\partial_\mu^2(\Psi - \Phi), \qquad (16.90)$$

[32] Be aware that different authors use slightly different definitions for Φ and Ψ. Occasionally they are swapped, so that the coefficient of dt^2 is $(1 - \Psi)$, or they have sign changes.

[33] We saw a version of this (Equation 16.26), with $\Psi = \Phi = \phi$, for the case that the perturbations were not dependent on time: the static weak field metric. See footnote 11 of Section 16.2.1. Because of this analogy the metric is also said to be in the *Newtonian* gauge.

[34] A frequently used alternative form of the metric is

$$ds^2 = a(\tau)^2\left[-(1 + 2\mu(\tau, \mathbf{x}))dt^2 + (1 - 2\nu(\tau, \mathbf{x}))\delta_{mn}dx^m dx^n\right], \qquad (16.88)$$

which is referred to either as the *conformal* longitudinal, or conformal Newtonian, gauge.

[35] The detailed derivation of the following equations and their ramifications are discussed in depth in the On-line Supplement *Relativistic Perturbation Theory*

[36] Here I follow the excellent discussions of Räsänen (2010) and of Gottlöber (1994). Both use the longitudinal gauge, the latter uses the $(+ - - -)$ metric signature. It is convenient to set $c = 1$ here.

$$G^{\mu}{}_{\nu} = a^{-2}\partial_{\mu}\partial_{\nu}(\Psi - \Phi), \tag{16.91}$$

$$G_{0\mu} = 2\partial_{\mu}(\dot{\Psi} + H\Phi). \tag{16.92}$$

The main technical difficulty in arriving at this point is in deciding which terms involving the Γs in the computation of the Ricci tensor can be neglected because they are of the order c^{-2}. The sure way of getting the right first order result is to follow the perturbations to second order, as Malik and Wands (2009) have done.

The Einstein Equations and Solutions

In the simplest case, we take the energy momentum tensor for the cosmic material to be that of a perfect fluid (again with $c = 1$):

$$T_{ab} = (\rho + p)u_a u_b + p g_{ab}, \qquad p = w\rho, \tag{16.93}$$

and we consider a general equation of state connecting p and ρ of the form $p = w\rho$.

Because $G_{ab} = \kappa T_{ab}$, Equations (16.93) and (16.91) tell us that the difference $(\Psi - \Phi)$ has no spatial dependence, hence it can only be a function of time. However, when we take the statistical view of these perturbations the mean values will have to be zero, by construction. From this we conclude that

$$\Phi = \Psi. \tag{16.94}$$

This results in a considerable simplification of the Einstein tensor. During the period of neutrino decoupling there are anisotropic stresses transferring energy over scales comparable with the horizon scale (Misner, 1968, 1969). Since the contribution of the neutrino energy density is significant at that time, these have a significant effect on the evolution of perturbations. This is a case where, for a short period of time, $\Psi \neq \Phi$. The energy momentum tensor determines the relationship between Φ and Ψ.

Assuming an equation of state, $p = w\rho$, the upshot of this is a pair of relatively simple equations:

$$3H^2 + 2a^{-2}\nabla^2\Phi - 6H\dot{\Phi} = 4\pi G\delta\rho, \tag{16.95}$$

$$\ddot{\Phi} + 4H\dot{\Phi} - 2\dot{H}\Phi + 3H^2\Phi = 4\pi G\delta p = 4\pi G\, b^2(\delta p + \delta\rho), \tag{16.96}$$

where, following Gottlöber (1994), $b = c_s/c$ is the velocity of sound, normalised to the speed of light and is related to w by $b^2 = 4w/3(1 + w)$. These last two equations combine to give

$$\ddot{\Phi} + (4 + 3b^2)H\dot{\Phi} - \underbrace{\left[2\dot{H} + 3H^2(1 + b^2)\right]\Phi - b^2 a^{-2}\nabla^2\Phi}_{\text{Jeans term}} = 0. \tag{16.97}$$

The 'Jeans term' determines the behaviour of the solution depending as to whether it is negative (growing solution) or positive (oscillatory solutions). On the super-horizon scales the pressure forces driving oscillations can be neglected and the evolution is dominated by the term in square brackets. When that term is zero, one of the solutions will be $\Phi =$ constant, and the other will be a decaying mode (as we saw in the Newtonian context).

The background solution evolves as

$$2\dot{H} + 3H^2(1 + b^2) = H^2 \, \frac{w(3w - 1)}{3(1 + w)}. \tag{16.98}$$

We recognise the left-hand side as the term in square brackets in (16.97). Moreover, the right-hand side term $w(3w - 1)/3(1 + w)$ is zero for the special cases $b = 0$ (matter only) and $b = \frac{1}{3}$ (radiation only). So super-horizon perturbations have $\Phi = $ constant and do not evolve until they enter their horizon.

The notion of a curved space-time is sometimes difficult to grasp. One of the ways to understand this is to consider the collection of trajectories, or 'world lines', of non-interacting test particles ('dust') in an otherwise empty curved space time: a vacuum solution of the Einstein equations. The relationship between neighbouring world lines is distorted by the space-time curvature, i.e. the gravitational field. The trajectories are sensors of the geometry that can be easily visualised.

One of the most important equations of general relativity, the *Raychaudhuri equation*, was derived by Raychaudhuri (1955, 1957) where he used it to study singularities of solutions to the Einstein equations. The Raychaudhuri equation has played a fundamental part in the study of singularities in general relativity, particularly in 1970s when it was used as a means of describing the geometry of the space-time. Here we derive it and apply it to bring out the nature of the geometry of space-time which is not manifest in the coordinate driven approach that we have followed up to this point.

17.1 Geometrical Issues: Space-Time Curvature

In Sections 13.4 and 13.5.2 we discussed the notion of geodesics in a 4-dimensional space-time. The discussion at that point concerned the geometry of a Riemannian manifold at a stage where it was merely endowed with a way of relating different points through a connection: the Christoffel connection Γ_{bc}^a. The geodesic in the space time to which we had attached a coordinate system and a metric was given by the important Equation (13.37). The importance of that equation was that it provided a direct link between geodesics in general relativity and particle trajectories in Newtonian theory (Equation 13.47).

The flow of a fluid in 4-dimensional space-time is visualised as the collection of trajectories of the fluid elements in the space-time, the fluid world lines. If the fluid has zero pressure, we call it *dust* and the fluid elements, or particles in this case, move along trajectories that are geodesics in the underlying space-time manifold. The instantaneous 4-velocity of the particle is a tangent to the trajectory. Locally, at least, there is a hyper-plane perpendicular to the trajectory at each point. In our case of a 4-dimensional space-time, the hyper-plane is a 3-dimensional space, which we refer to as a hyper-plane by analogy with our experience in 3-dimensional Euclidean space.

More generally, we think in terms of a *congruence*: a family of non-intersecting curves in an open region of space-time such that through each point there passes a single curve of this family. The fluid trajectories in a region are an example of a congruence, as are timelike geodesics.

17.1.1 Hyper-surfaces and Congruences of Integral Curves

Congruences are intimately related to vector fields. Given a field that defines a vector $v^i(x^j)$ at every point $P(x^j)$ of a space that is endowed with a coordinate system $\{x^j\}$, we can define the *integral curves* associated with the vector field as the curves that are at each point tangent to the vector field, see Figure 17.1. These are the solutions of the ordinary differential equations

$$\frac{dx^i}{d\lambda} = v^i(x^j), \tag{17.1}$$

where λ parameterises the position along each curve that is a solution of this differential equation. The solutions of these equations are the *integral curves* of the vector field v^i.[1] Of course the existence of the solutions requires a sufficient degree of continuity in the vector field, but, given such conditions, the vector field defines a unique integral curve through each point where the vector field is non-zero. The collection of such curves is the congruence associated with the vector field. Conversely, all congruences define a vector field provided certain mathematical conditions are satisfied.

At each point on a given curve of a congruence, there is a hyper-surface wherein all vectors in that hyper-surface are orthogonal to the curve at that point. While there

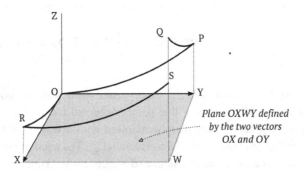

Fig. 17.1 The vectors **OX** and **OY** are tangent to the integral curves *OR* and *OP*. To define the integral surface associated with these two vectors we next construct the tangent planes to the integral curves at *P* and *R* and draw the integral curves through the points *P* and *R*. If there is an integral surface these last two integral curves would have to have met at a point. In this diagram they are depicted as not intersecting and so there is no integral surface that can be defined. *Based on Figure 236 of Arnold (1980).* ©

---

[1] This is where Frobenius comes in. Consider the differential, or 'Pfaffian', $\psi = U(x,y,z)dx + V(x,y,z)dy + W(x,y,z)$ where U, V, W define the components of a vector $\mathbf{u} = (U, V, W)$. The question is under what conditions does there exist a function $f(x,y,z)$, an integrating factor, and a function $\phi(x,y,z)$ such that

$$f\,d\phi = U(x,y,z)dx + V(x,y,z)dy + W(x,y,z). \tag{17.2}$$

In the case $f = 1$ we know that $\nabla \times \mathbf{u} = 0$ is both necessary and sufficient for the existence of a solution $\phi(x,y,z)$, which we call the velocity potential.

is such a hyper-surface defined at each point, there is no prior guarantee that it is possible to find one hyper-surface that is orthogonal in this way to all curves of the congruence.[2]

Congruences of integral curves are studied independently of the Einstein equations and play a role in many branches of physics where there are flows. At the base of the study of congruences of curves is a theorem of Frobenius (1877), which appears in two quite different guises: as a theorem relevant for first order homogeneous partial differential equations and as a theorem on differential forms (Schutz, 1980, Section 3.7 and Section 4.26).[3] Arnold (1980, Appendix 4) gives a nice introduction to this in the language of differential forms. See also Wald (1984, Appendix B.3).

There are several types of physically important congruence on a space-time. Perhaps the most fundamental are the geodesic congruences. In relativity the time-like and null geodesics are of central importance: the former tells us about the motion of particles under the influence of no forces other than gravitation, while the latter tells us about the propagation of light rays. But there are also congruences defined by the flow of a fluid moving in a gravitational field: the defining tangent vector field is the flow velocity.

17.1.2 Vector Fields Normal to 3-surfaces

As usual, we work in a 4-space which is a Riemannian manifold with metric g_{ab}. Consider a time-like vector field N^a that is normal to some space-like 3-surface $\Sigma : f(x^a) = \text{const}$. N^a is given by the gradient $N^a = -\alpha f_{,a}$, where α is chosen so that $N^a N_a = -1$ and will be, in general, a function of position, $\alpha = \alpha(x^a)$. If we know the functional form of $f(x^a)$ then N^a is given by

$$N^a = -\frac{f_{,a}}{|\,g^{ab} f_{,a} f_{,b}\,|^{1/2}}. \tag{17.3}$$

The sign ensures that the values of f on successive surfaces are taken to increase towards the future (i.e. $N^a f_{,a} > 0$).

Using N^a we can then define a tensor which is the metric in the 3-surface Σ:

$$h_{ab} = g_{ab} + N_a N_b, \quad N^a N_a = -1. \tag{17.4}$$

---

[2] In a two-dimensional world, the surfaces orthogonal to the integral curves are simply curves on the 2-surface. On the surface of a sphere the geodesics are the great circles. In the case when the great circles are lines of longitude, the 'surfaces' are the lines of latitude (which are not in general great circles) and they are orthogonal to every geodesic of the longitude-line congruence. In 3-dimensional Euclidean space the hypersurfaces orthogonal to the geodesics are 2-planes and it is not a priori clear what is the condition that they should mesh together in a simple way.

[3] Frobenius provided a necessary and sufficient condition for being able to find solutions to such equations. The necessity of the condition had been first proved by Alfred Clebsch and the sufficiency by Feodor Deahna. The way Frobenius approached the problem was through the use of Pfaffian forms, thereby opening the application of the theorem to differential geometry.

The tensor h_{ab} is generally referred to as the *projection tensor* or the *transverse metric*.[4] Contracting h_{ab} with a vector or a tensor finds the components of the vector or tensor that lie in the 3-surface Σ.

We can see this by projecting any vector with h^a_b and then finding the component in the N^a direction. This goes as follows. We note that in the particular case of the vector N^a, we have $N^a h^b_a = N^a(g_{ab} + N_a N_b) = N_b + (N^a N_a)N_b = 0$ since we normalised N^a to $N^a N_a = -1$. In this sense, N^a and h^a_b are orthogonal.

Now look at any arbitrary vector V^a: the projection with h_{ab}, i.e. $h^b_a V^a$ has a component $(V^a h^b_a)N^a$ which is parallel to N^a. But, as we just saw, $N_a h^b_a = 0$, and so the component of $h^b_a V^a$ in the direction of N^a is zero. The action of contracting a vector or tensor with h_{ab} is to find the component of the vector or tensor that lies in the space Σ.

The gradient of the vector field N^a plays an important role since it 'explores' the geometry of the local neighbourhood of normal vectors aound N^a. We can use h_{ab} and N^a to resolve the gradient tensor $N^a_{;b}$ into a time-like component parallel to N^a and a component orthogonal to N^a lying in the 3-surface Σ as follows. Define a vector $\dot{N}^a$:

$$\dot{N}^a \overset{\text{def}}{=} N^b N^a_{;b}. \tag{17.5}$$

Now we show that the tensor $\dot{N}^a N_b + N^a_{;b}$ lies in the 3-space Σ by calculating its component in the direction parallel to N^b. We get $(\dot{N}^a N_b + N^a_{;b})N^b = \dot{n}^a(N_b N^b) + N^b N^a_{;b} = 0$, as required. It is now easy to verify that

$$h^p_a h^q_b N_{p;q} = \dot{N}_a N_b + N_{a;b}. \tag{17.6}$$

Notice the acceleration term, $\dot{N}_a$.

We now know how vectors project, next we have to see how derivatives project. This inherits complications from the fact that the 3-space inherits a curvature from the parent 4-space: moving from the 4-space to the 3-space is technically non-trivial.

17.1.3 Calculus in the Orthogonal 3-space

From here we can go on to do calculations in the 3-space orthogonal to N^a: this is important because we live in such a 3-space and make all our observations of the Universe there. We shall call the 4-space in which the bundle of flow lines with tangents N^a are defined, $^{(4)}S$, and refer to the 3-space that is orthogonal to N^a at a point as $^{(3)}S$. In other words, $^{(3)}S$ is

---

[4] It is necessary that the vector field N^a be *hyper-surface orthogonal* for h_{ab} to be the metric of the surface (which it is in the current discussion, because we have defined $N^a = \alpha f^{,a}$). Nevertheless, the projection tensor can always be used to project tensors into some subspace that is orthogonal to N^a, and that can be done locally at all points of the underlying manifold.

The requirement of hyper-surface orthogonality of flow lines is guaranteed when $N^a = \alpha \partial_a f(x^m)$, which via the Frobenius Theorem (Poisson, 2007, Section 2.3.3) can be shown to be equivalent to N^a saying that the underlying flow is irrotational: $N_{a;b} - N_{b;a} = 0$. This is true even if the flow lines are not geodesics. This is the equivalent of the theorem in classical fluid mechanics that a potential flow is necessarily irrotational.

If the congruence is not hypersurface orthogonal we can only set up a hyper-surface that is orthogonal to a given geodesic, in which case the current discussion is valid for that geodesic and the orthogonal 3-spaces at each point of the geodesic.

the hyper-surface that is orthogonal to the vector N^a at a point in $^{(4)}S$. We can derive the properties of $^{(3)}S$ as seen from the point of view of $^{(4)}S$ using h^{ab}.

h_{ab} operating on a vector or tensor annihilates the components along the vector N^a: it decomposes tensors into a purely spatial part and a purely time-like part. The projection $T_{ab}^{\perp}$ of a tensor T_{ab} onto the space orthogonal to the vector N^a is

$$T_{ab}^{\perp} = h_a{}^c h_b{}^d T_{ab}. \tag{17.7}$$

Since h_{ab} is the metric of the 3-space $^{(3)}S$, we can define a Christoffel connection using h_{ab} in a manner that is directly analogous to the connection $\Gamma^a{}_{bc}$ defined by the metric g_{ab} of the 4-space $^{(4)}S$. In order to avoid confusion we shall use Greek indices $\alpha, \beta, \ldots = 1, 2, 3$ when working in the 3-space:

$$\gamma^{\eta}_{\mu\nu} = \tfrac{1}{2} h^{\eta\rho} \left(\partial_\mu h_{\rho\nu} + \partial_\nu h_{\mu\rho} - \partial_\rho h_{\mu\nu} \right). \tag{17.8}$$

As before, $\gamma^{\eta}_{\mu\nu}$ is not a tensor, but it allows us to define a covariant derivative for the 3-space, which we can denote by D_α when working in the 3-space, and D_a when working in the 4-space.[5] The 3-space derivative as viewed in the 4-space, i.e. D_a, is related to the familiar 4-space covariant derivative ∂_a by simple projection:

$$D_a \equiv h_a{}^p \partial_p : \quad D_a\phi = h_a{}^b (\partial_b\phi), \quad \text{and} \quad D_a T^b{}_c = h_a{}^m h_n{}^b h_c{}^p (\partial_m T^n{}_p), \tag{17.9}$$

with similar expressions for other types of tensor.[6] It is easy to check that $D_a h_{bc} = 0$ and $\dot{h}_{ab} = N_a \dot{N}_b + N_b \dot{N}_a$.

We can of course then define Riemann and Ricci tensors on the 3-space in the usual manner with the results:

$$D_a D_b V^c - D_b D_a V^c = r^c{}_{mab} V^m, \tag{17.10}$$

$$r_{db} = r^c{}_{dcb} \quad \text{and} \quad r = g^{ab} r_{ab}. \tag{17.11}$$

The 3-space Riemann tensor $r^c{}_{mab}$ is said to measure the *intrinsic curvature* of the 3-space.

Consider any 3-space: the geometry is completely determined by the Riemann tensor $r^\alpha{}_{\beta\gamma\delta}$, $\alpha, \beta, \ldots = 1, 2, 3$, which can easily be shown, on the basis of the symmetries on the indices, to have six independent components. Furthermore, the corresponding 3-space Ricci tensor $r_{\beta\delta}$, being symmetric, also has six components. Thus the 3-space geometry, which is specified by its Riemann tensor, is equally fully described by the Ricci tensor.[7]

We saw in Section 13.7 that the 4-space Riemann tensor $R^a{}_{bcd}$ had 20 independent components, so it is clear that the 3-spaces defined by the fluid paths are by themselves not enough to specify the global structure of the space-time.

---

[5] There is an important technical point here: it is necessary to distinguish the 4-space indices from the 3-space indices, even though the 3-space is a sub-manifold of the 4-space manifold. Formally, this is because of the difference between the tangent spaces of the manifold and the sub-manifold.
An often used alternative notation for D_a is $\bar{\partial}_a$.

[6] There are several alternate notations for D_a, for example $\bar{\partial}_a$. The notation for the 3-space covariant derivative, analogous to the 4-space semi-colon, is $D_a X_b \equiv X_{b\|a}$.

[7] Weinberg (1972, Ch.4 Section 7: Eq. (6.7.6)) shows that

$$r_{\alpha\beta\gamma\delta} = r_{\alpha\gamma} g_{\beta\delta} - r_{\alpha\delta} g_{\beta\gamma} + r_{\beta\delta} g_{\alpha\gamma} - r_{\alpha\delta} g_{\beta\gamma} - \tfrac{1}{2} r (g_{\alpha\gamma} g_{\beta\delta} - g_{\alpha\delta} g_{\beta\gamma}).$$

17.1.4 Fluid Flow: World Lines, Tangents and Projections

Now we go on to apply these principles to a particular case of N^a: the velocity field u^a of a fluid moving under gravity, and perhaps other forces such as internal pressure forces. Using the projection operator, $h_{ab} = u_a u_b + g_{ab}$ for u^a we can look at the flow in the 3-space about a given point and orthogonal to the flow, from the point of view of an observer moving with the fluid. In the absence of non-gravitational forces, these flow lines are geodesics of the space-time. With non-gravitational forces, the flow lines are no longer geodesics, but we can, as just shown, still split the space into a $3 + 1$ structure using h_{ab}.

It should be noted that this $3 + 1$ split can always be done locally, but there are circumstances where the local 3-spaces do not mesh together to give a global description of the flow. One such circumstance arises from the presence of vorticity, ω, in the flow: vorticity causes the flow lines in the 4-space to twist. So, for the sake of simplicity we shall at some point take $\omega = 0$ in order to make progress.

The fluid elements move on their 4-space *pathlines*,[8] each having a time-like unit tangent vector u^a normalised so that

$$u^a u_a = -1. \tag{17.12}$$

We will study the structure of flow lines having tangent vector u_a by resolving the 4-velocity gradient $u_{a;b}$ field into separate physically meaningful components.[9]

Suppose the geometry of the space-time is described by a metric tensor g_{ab}. Consider the projection tensor h_{ab} defined by the vector field u^a:

Projection tensor onto hyper-surface orthogonal to the flow lines:

$$h_{ab} = g_{ab} + u_a u_b, \quad u^a u_a = -1, \quad h_{ab} u^b = 0. \tag{17.14}$$

Also referred to as the *projection operator orthogonal to flow lines*.

Other properties of h_{ab} are[10]

$$h_{ab} u^a = 0, \quad h_{ap} h^p{}_c = h_{ac}, \quad h^{ap} h_{pb} = {}^{(3)}\delta^a_b, \quad h^a_a = 3, \tag{17.15}$$

where the (3) superscript in ${}^{(3)}\delta^a_b$ informs us that this is the 3-space delta symbol. So h_{ab} can be viewed as a metric tensor that acts in the 3-space that is orthogonal to the flow lines. The 3-space described by h_{ab} is the instantaneous rest space of an observer moving with 4-velocity u^a.

---

[8] As opposed to the *streamlines* which, in non-relativistic flows, are a snapshot of the lines whose tangents are the flow velocity field at that given time. The streamlines are a 3-space projection of the pathlines.

[9] A reminder about notation: any tensor T_{ab} can be split into its symmetric and anti-symmetric parts, S_{ab}, A_{ab}:

$$S_{ab} = T_{(ab)} = \tfrac{1}{2}(T_{ab} + T_{ba}), \quad A_{ab} = T_{[ab]} = \tfrac{1}{2}(T_{ba} - T_{ab}). \tag{17.13}$$

A_{ab} is trace-free: $A^a_a = 0$, and the trace of S_{ab} is just the trace of T_{ab}: $S^a{}_a = T^a_a$.

[10] We can calculate $u^a h_{ab} = u^a g_{ab} + u^a u_a u_b = u_b - u_b = 0$. Likewise $h^a{}_b u^b = 0$, $h^{ab} u_b = 0$ and so on. The inverse of h_{ab} is easily verified to be $h^{ab} = g^{ab} + u^a u^b$.

17.2 Describing a Cosmological Flow

We now go on to show how the geometry of the local bundle of geodesics provides information about the local gravitational field and other forces that may act on a fluid element. The result is a set of equations for a general relativistic cosmological flow described locally by three parameters: the expansion rate, θ, the shear tensor, σ_{ab}, and the vorticity, ω_{ab}.

17.2.1 Acceleration

We shall now see the condition that the observer is freely falling in a gravitational field. We define the *acceleration* of a particle as the projection of the 4-velocity gradient, $u_{a;b}$, in the direction of the velocity field: $\dot{u}_b = u^a u_{b;a} = u^a \partial_b u_a$. Using (17.12),

$$\dot{u}_b = u^a u_{b;a} \quad \text{and} \quad u^b \dot{u}_b = 0. \tag{17.16}$$

This shows that the acceleration is orthogonal to the velocity, i.e. the acceleration lies entirely in the tangent space to the geodesic. In classical fluid mechanics, $\dot{u}_b$ would be the time derivative following the motion of a fluid element. Technically, $\dot{u}_b$ is related to the curvature of the path to which u^a is tangential.

Consider for a moment the motion under gravity with no external forces. If we write out $\dot{u}^a$ in terms of coordinates we have

$$\dot{u}^a = u^b u^a_{;b} = u^b \frac{\partial u^a}{\partial x^b} + \Gamma^a_{kb} u^k u^b = 0, \qquad \text{dust}, \tag{17.17}$$

which describes a particle in free fall moving along a geodesic.[11] We have already seen this in Section 13.5.2.

With $\dot{u}_b = u^a u_{b;a} = 0$, we can show that

$$\dot{h}_{ab} = 0, \qquad \text{(dust)}. \tag{17.18}$$

This tells us that in the absence of non-gravitational forces, i.e. geodesic motion, the projection tensor is constant along a geodesic.

17.2.2 The 3-space Velocity Gradient, V_{ab}

Now we turn to the velocity field V^a as seen in the 3-space orthogonal to u^a. The character of the velocity field V^a is described by the 3-space gradient tensor, V_{ab}, of the fluid 4-velocity, u^a. Following Equation (17.7), this is defined by the equation

$$V_{ab} = h_a^c h_b^d u_{c;d}. \tag{17.19}$$

Here h_{ab} is the metric of the 3-space orthogonal to the 4-velocity, u_a: $h_{ab} = g_{ab} + u_a u_b$. It acts to project components of a vector or tensor into the 3-space orthogonal to the flow

---

[11] If we evaluate this, for $a = \alpha = 1, 2, 3$ in an instantaneous freely falling local reference frame, where $u^a = (1, 0)$, we recover Equation (13.47): $\dfrac{d^2 x^\alpha}{dt^2} = -\Gamma^\alpha_{00}$. This is how we identified the Christoffel symbol Γ^α_{00} with the acceleration due to gravity. See Sections 13.4, 13.4.1 and 14.4.1.

described by the flow-line tangents u^a. Hence, V_{ab} describes the flow as seen in the local neighbourhood of a comoving observer. It describes the gradient of the velocity field we measure in the Universe of galaxies. If a nearby galaxy is observed at position η^a relative to a comoving observer on the pathline defined by u^a, its velocity is $V_a = V_{ab}\eta^b$.

The tensor V_{ab} can be decomposed into its symmetric and skew-symmetric parts:[9]

$$V_{ab} = \underset{\text{expansion}}{\theta_{ab}} + \underset{\text{vorticity}}{\omega_{ab}} , \tag{17.20}$$

$$\theta_{ab} = V_{(ab)} = \tfrac{1}{2}(V_{ab} + V_{ba}), \quad \omega_{ab} = V_{[ab]} = \tfrac{1}{2}(V_{ba} - V_{ab}). \tag{17.21}$$

The tensor θ_{ab} is called the *expansion tensor* and ω_{ab} is the *vorticity tensor*; θ_{ab} is in general not trace-free, and can further be decomposed into its trace and trace-free parts:

$$\theta_{ab} = \sigma_{ab} + \tfrac{1}{3}\theta\, h_{ab}, \tag{17.22}$$

(remembering that θ_{ab} lies in the 3-space with metric h_{ab}). Here σ_{ab} is referred to as the (trace-free) *shear tensor*, and θ as the *expansion rate*.

V_{ab} as defined in (17.20) is constructed from tensors that lie in the 3-space. We can relate σ_{ab}, ω_{ab} and θ directly to projections of the 4-space velocity gradients: The components $\sigma_{ab}, \omega_{ab}, \theta, \dot{u}_a$ are defined as

$$\sigma_{ab} = \tfrac{1}{2}\, h^p{}_a h^q{}_b \left(u_{p;q} + u_{q;p} - \tfrac{2}{3}u^r{}_r h_{pq}\right), \tag{17.23}$$

$$\omega_{ab} = \tfrac{1}{2}\, h^p{}_a h^q{}_b \left(u_{p;q} - u_{q;p}\right), \tag{17.24}$$

$$\theta = u^p{}_{;p}, \tag{17.25}$$

$$\dot{u}_a = u^p u_{a;p}. \tag{17.26}$$

This can easily be verified by direct substitution into (17.20) and (17.22).

The vorticity tensor ω_{ab}, also called the *twist of the congruence*, is anti-symmetric, while the shear tensor σ_{ab} is symmetric. We can define a *vorticity vector*, ω^a, using ω_{ab}:

$$\omega^a = \tfrac{1}{2}\eta^{abcd} u_b \omega_{cd} = \tfrac{1}{2}\eta^{abcd} u_b u_{c;d}, \tag{17.27}$$

where η^{abcd} is the contravariant permutation tensor that takes the value $|\det g_{ab}|^{-1/2}$ whenever $(abcd)$ is an even permutation of (1234) and $\eta^{abcd} = -\eta^{1234}$ if $(abcd)$ is an odd permutation of (1234). η^{abcd} is zero otherwise.[12] The 4-vector ω^a lies in the hypersurface orthogonal to u_a: this follows from $u_a\omega^a = \tfrac{1}{2}\eta^{abcd} u_a u_b u_{c;d} = 0$ because $u_a u_b$ is symmetric in (ab). As we saw in Section 17.2.1, the acceleration vector $\dot{u}_a$ also lies in the 3-space.

The magnitudes σ and ω of the shear and vorticity tensors are defined as:

$$2\sigma^2 = \sigma_{ab}\sigma^{ab}, \qquad 2\omega^2 = \omega_{ab}\omega^{ab}. \tag{17.28}$$

The notation $\dot{u}_a$ for the component along the flow-line tangent, u_a, is clearly appropriate: in special relativity this corresponds to the time derivative of the velocity. The pre-factor

---

[12] Without the $|\det g_{ab}|^{-1/2}$ factor this would be a permutation symbol analogous to the familiar ϵ^{ijk}, but it would not transform as a tensor. The indices of η^{abcd} can be raised and lowered using the metric tensor, so $\eta_{abcd} = \pm|\det g_{ab}|^{-1/2}$ according as to whether $(abcd)$ is an even, $(+1)$, or odd, (-1) permutation of (1234). See Weinberg (1972, p.98 *et seq.*).

$h^p_a h^q_b$ appearing in the definition of σ_{ab} and ω_{ab} ensures that these are 3-tensors lying entirely in the hyper-surface orthogonal to u_a (see Equation 17.7).

Importantly, the relationship between the 3-space velocity gradient V_{ab} and the 4-space gradient $u_{a;b}$ is given by (17.6):

$$V_{ab} = h^c_a h^d_b \, u_{c;d} = \dot{u}_a u_b + u_{a;b}. \tag{17.29}$$

From here we have:

$$u_{a;b} = \underset{\text{shear}}{\sigma_{ab}} + \underset{\text{vorticity}}{\omega_{ab}} + \underset{\text{expansion}}{\tfrac{1}{3}\theta h_{ab}} - \underset{\text{acceleration}}{\dot{u}_a u_b} \,, \tag{17.30}$$

where $\sigma_{ab} = \sigma_{ba}$ is symmetric and trace-free and $\omega_{ab} = -\omega_{ba}$.

This equation resolves the 4-gradient $u_{a;b}$ into the sum of the 3-space shear, vorticity and expansion, and a term depending on the acceleration, $\dot{u}_a$.

17.3 The Raychaudhuri Equation

17.3.1 Some History

The Raychaudhuri equation is one of the most important equations in modern cosmology: in its modern form it provides a coordinate-free description of flows in space-time. However, the form in which the equation was given in those original papers would hardly be recognised today.[13]

The form in which the Raychaudhuri equations are presented nowadays was introduced in the 1961 publication of a set of lectures from the Hamburg General Relativity Group. The lectures are commonly known as the *Academy of Mainz Lectures* since they were published, in German, in the *Proceedings of the Academy of Mainz*.[14] One of the lecturers, Ray Sachs, published an English language paper on gravitational radiation in the same

---

[13] Raychaudhuri's paper is set in a specific coordinate system which tends to obscure the generality and interpretation of the result. A year after the appearance of Raychaudhuri's paper, Komar (1956) independently came to a somewhat clearer and stronger statement about singularities. Komar had not cited Raychaudhuri and in a subsequent short note Raychaudhuri (1957) refers to his own, now widely cited paper, as … *a previous paper of which Komar is apparently unaware.*

[14] The original German language articles by P. Jordan, J. Ehlers, W. Kundt and R.K. Sachs, were published in *Proceedings of the Mathematical Natural Sciences Section of the Mainz Academy of Sciences and Literature*, Nr. 11, 792 (1961). Ehlers, Kundt and Sachs, along with M. Trumper, worked in the influential research group of Pascual Jordan in Hamburg. See Ellis (2009a) for a historical perspective.

The *Mainz Lectures* changed the way we think about general relativity. They de-emphasised the choosing and interpretation of coordinate systems with which to do calculations and dealt instead with scalars, vectors and tensors, viewing them as physical objects and using concepts of differential geometry. One of their prime goals was to discover further exact solutions to the Einstein equations. This approach did a great deal to make the physical concepts underlying general relativity more transparent.

Of course, in the final analysis, to get numbers out of Einstein's theory it is necessary to invoke a coordinate choice, write down a metric and its Christoffel symbols and solve some equations.

year with a version of the Raychaudhuri equation (Sachs, 1961, Equation 4.11 *et seq.*). However, none of the Academy of Mainz Lectures were published in the English language until 1993 when an English language version of Lecture VI, *Exact Solutions of the Field Equations of General Relativity Theory* finally appeared (Ehlers, 1993, translated by G.F.R. Ellis (with P. Dunsby)). Fortunately, appreciation of the Hamburg group's approach had become available to non-German speakers through the outstanding lectures of George Ellis, delivered at the XLVIIth Enrico Fermi Summer School held in Varenna in July 1969 (Ellis, 1971). Those lectures were based around the Mainz Lectures and were a key factor in changing the face of relativity.[15]

The study of gravitational radiation was a prime subject of research during the late 1950s and 1960s, and this brought with it new, coordinate-free, techniques to describe vacuum solutions to the Einstein equations. Notable were the introduction to spinors by Penrose (1960) and the development of general relativity using spinors by Newman and Penrose (1962, 1963). There were also papers advocating the use of differential geometry to describe space-time, notably Misner and Wheeler (1957, Sections III.B, III.C). The *1964 Brandeis Lecture* of Trautman (1964) was a key step in introducing concepts of differential geometry into the description of space-times.[16] This in turn led to research on the boundaries of manifolds and eventually to the fundamental singularity theorems of Hawking and Penrose (see Hawking and Ellis, 1975, and references therein).

There has been further evolution in the way we think about general relativity since the change brought about by the Mainz lectures, though it had always been implicit in those lectures and in lectures and article written by others around the same time. This was the move towards using the concepts of *differential geometry* as the means for describing the space-time and the motions within it.

17.3.2 The Geodesic Flow Equations

We have described the salient features of u_{ab}: the expansion and vorticity tensors (Equations 17.20, 17.22). Now we need to see how these evolve. That requires dealing with the derivative of $u_{a;b}$, which in a curved geometry, depends on the order in which the

---

[15] Many non-German speaking GR newbies learned general relativity from George Ellis' *Varenna Lectures*. Until 2009, his lectures were only available in the relatively hard-to-find Summer School Proceedings, or in a typed preprint version having hand-written equations that was circulated to some institutes. The lectures have now been republished as a 'Golden Oldie' (Ellis, 2009b).

[16] For a pedagogic modern view of differential geometry in physics see Poisson (2007) and Frankel (2011), and, in particular Schutz (1980) for a fine presentation in terms of differential forms and Lie derivatives. From the mathematical point of view, the presentation of the geometry of Riemannian spaces in the classic book by Eisenhart (1926) is hard to beat, though it has no figures to guide the intuition. The excellent on-line lectures of Blau (2014, in particular Part B) deal with the geometry of space-time as seen from the point of view of classical tensor calculus (i.e. using indices on tensors) rather than the differential forms notation of differential geometry.

All these writings use the same metric signature, $-+++$ as is used here, and indeed that was one of the main reasons for choosing to use this signature in this book.

derivatives are taken. This derivative ordering is expressed through the Riemann tensor, R_{abcd} (see Equation 13.104), which for the underlying 4-velocity u^a is:[17]

$$u_{a;dc} - u_{a;cd} = R_{abcd}u^b.$$ (17.31)

Starting with Equation (17.31) we can derive equations for the rate of change of θ, σ, ω. The derivations are somewhat tedious, so attention will be focused on deriving $\dot{\theta}$.

First we need to contract (17.31) with u^c and rearrange the terms:

$$u^c u_{a;cd} = u^c u_{a;dc} + R_{abcd}u^b u^c.$$ (17.32)

To deal with the left-hand side we need the identity

$$(u^c u_{a;c})_{;d} = u^c{}_{;d}u_{a;c} + u^c u_{a;cd},$$ (17.33)

and we note that the second term on the right is just the left side of (17.32). Moreover $(u^c u_{a;c})_{;d} = \dot{u}_{a;d}$, so now we have

$$\dot{u}_{a;d} - u^c{}_{;d}u_{a;c} = u^c u_{a;dc} + R_{abcd}u^b u^c.$$ (17.34)

Raise the index a and contract on the indices a, d:

$$\dot{u}^a{}_{;a} - u^c{}_{;a}u^a{}_{;c} = u^c u^a{}_{;ac} - R_{bc}u^b u^c,$$ (17.35)

where we have used the definition of the Ricci tensor: $R_{bc} = R^a{}_{bac}$ (see Equation 13.113 and note the permutation of the indices).

Now it remains to write the terms $u^c u^a{}_{;ac}$ and $u^c{}_{;a}u^a{}_{;c}$ in terms of θ, σ and ω. The first of these is the simpler:

$$u^c u^a{}_{;ac} = u^c \theta_{;c} = \dot{\theta}.$$ (17.36)

For the second we can substitute for $u^c{}_{;a}$ using Equation (17.30) to give

$$u^c{}_{;a}u^a{}_{;c} = u_{c;a}u^{a;c} = (\sigma_{ac} + \tfrac{1}{3}\theta g_{ac} + \omega_{ac})(\sigma^{ac} + \tfrac{1}{3}\theta g^{ac} + \omega^{ac}),$$ (17.37)

$$= \tfrac{1}{3}\theta^2 + \sigma_{ac}\sigma^{ac} - \omega_{ac}\omega^{ac} = \tfrac{1}{3}\theta^2 + 2\sigma^2 - 2\omega^2.$$ (17.38)

So finally, with a bit of rearrangement, putting Equations (17.36) and (17.38) into (17.35), we get the Raychaudhuri equation for dust:

Raychaudhuri equation for geodesic flow:

$$\dot{\theta} + \tfrac{1}{3}\theta^2 + 2(\sigma^2 - \omega^2) + R_{ac}u^a u^c = 0.$$ (17.39)

For geodesic flow the term $\dot{u}^a$ is zero.
We can write the 3-space curvature $R_{ac}u^a u^c$ as $^{(3)}R$.
Note: at this point we have not used the Einstein equations, this reflects the geometric structure of space-time. The Einstein equations will determine $R_{ac}u^a u^c$ in terms of the energy density.

---

[17] The notation is understood to mean $u_{a;dc} = u_{a;d;c} = (u_{a;d})_{;c}$, so first we do the d-derivative and then the c-derivative. Some introduce $u_{a;d} = B_{ad}$ so that $u_{a;dc} = B_{ad;c}$ and work in terms of the gradient tensor B_{ad} to help resolve any ordering ambiguity.

If we view Equation (17.39) as an equation for $\dot{\theta}$, we see that the expansion and shear terms $\frac{1}{3}\theta^2 + \sigma^2$ tend to focus the congruence ($\dot{\theta} < 0$) while the ω^2 term tends to make it diverge ($\dot{\theta} > 0$). The issue of the focusing of the congruence then rests with the sign of the *Raychaudhuri scalar*, $R_{ac}u^a u^c$. Geometrically, this term is the projection of the Ricci tensor onto the tangent of the local world line and represents a mean curvature. It could have either sign depending on the direction of the normal to the surface.

17.3.3 Relating Ricci Curvature to Material Content

The task is now to evaluate the term $R_{ac}u^a u^c$ in Equation (17.39) in terms of the material content of the Universe. The Einstein field equations provide the solution since they relate the Ricci tensor, R_{ab}, directly to the energy momentum tensor.[18] The Einstein equations for a simple fluid having density ρ and pressure p can be written in the form:[19]

$$R_{ab} = \kappa(T_{ab} - \tfrac{1}{2}Tg_{ab}) + \Lambda g_{ab}, \tag{17.40}$$

$$R = -\kappa T + 4\Lambda, \tag{17.41}$$

$$R_{ab}u^a u^b = \kappa(T_{ab}u^a u^b + \tfrac{1}{2}T) - \Lambda, \tag{17.42}$$

$$T_{ab} = (\rho + p)u_a u_b + pg_{ab}, \tag{17.43}$$

$$T_{ab}u^a u^b = \rho, \tag{17.44}$$

$$T = -\rho + 3p, \tag{17.45}$$

$$T_{ab}u^a u^b + \tfrac{1}{2}T = \tfrac{1}{2}(\rho + 3p), \tag{17.46}$$

where we have put $\kappa = 8\pi G/c^2$ to handle diverse conventions on the choice of units.

Putting these together we get

$$^{(3)}R \equiv R_{ab}u^a u^b = \tfrac{1}{2}\kappa(\rho + 3p) - \Lambda. \tag{17.47}$$

This can be substituted into the Raychaudhuri Equation (17.39), but for a perfect fluid with pressure we must restore the acceleration term $\dot{u}^a_{;a}$ which we had set to zero in the dust case.

Projecting R_{ab} into the 3-space similarly gives:

$$R_{ab}h^a_c h^b_d = \left[\tfrac{1}{2}(\rho - p) + \Lambda\right]h_{cd}. \tag{17.48}$$

The term $R_{ab}u^a u^b$ in the Raychaudhuri Equation (17.39) and in (17.47) is the projection of the Ricci tensor in the direction of the time-like vector u^a (Eisenhart, 1926, Section 25). It is the component of curvature that is not accessible to the 3-space orthogonal to u^a, and is the extrinsic curvature of the surface that we shall be discussing in §17.4.

The Raychaudhuri Equation (17.39) explicitly shows the term $^{(3)}R = R_{ac}u^a u^c$ as a driver of the divergence of the flow lines as expressed through the rate of change of the expansion parameter θ. The fact that gravity curves the space-time geometry and is responsible

---

[18] Up to this point, we have not used the Einstein field equations, everything has been purely geometric, describing the relationship of geodesics to the curvature of the space-time.

[19] It may help to think of the energy momentum tensor for a simple fluid having density ρ and pressure p as $T^a_b = \text{diag}(-\rho, p, p, p)$ and the 4-space fluid velocity as $u^a = (1, 0, 0, 0)$.

for driving the divergence is seen in Equation (17.47). However, we notice that it is the curvature of the 3-geometry $^{(3)}R$ that is involved, not the curvature of the 4-geometry R.

17.3.4 Simple Fluid Flow

Using (17.47), the Raychaudhuri equation for a simple fluid flow is then:

Raychaudhuri–Ehlers equation for simple fluid flow:

$$\dot{\theta} + \tfrac{1}{3}\theta^2 + 2(\sigma^2 - \omega^2) - \dot{u}^a_{;a} + \tfrac{1}{2}\kappa(\rho + 3p) - \Lambda = 0. \qquad (17.49)$$

For geodesic flow the term $\dot{u}^a_{;a}$ is zero.
ρ is the density of the fluid and p is its pressure, related through an equation of state $p = p(\rho)$.
This is the first time that we have brought in the Einstein equations to describe a flow.

The Raychaudhuri equation was first given in this form in 1961 by Ehlers (1993, *Academy of Mainz Lecture, 1961*, Equation (81)). Consequently Ellis *et al.* (2012, Section 6.1) decided to add Ehlers' name to this form of the equation.

It is useful to introduce another parameter in place of $\dot{\theta}$: the *expansion scale*, or *scale factor*, ℓ, defined by

$$\theta = 3\frac{\dot{\ell}}{\ell} = 3H, \qquad (17.50)$$

where H denotes the familiar local 'Hubble expansion rate'. With this, we have

$$\dot{\theta} + \frac{1}{3}\theta^2 = 3\frac{\ddot{\ell}}{\ell}. \qquad (17.51)$$

In a homogeneous and isotropic expanding cosmological model, $\ell(t)$ is what we normally refer to as the cosmic expansion factor, or, simply, the scale factor. In that context it is generally normalised to have the value $\ell_0 = 1$ at the present epoch.

The local acceleration of the expansion in cosmology has been characterised by the *deceleration parameter*, $q(t)$, where t denotes parameter along the fluid world line, which in the case of a homogeneous isotropic solution of the Einstein equations is the cosmic time. Here we can derive a parameter q from θ and $\dot{\theta}$:

$$q = -\frac{\ell\ddot{\ell}}{\dot{\ell}^2} = -1 - 3\frac{\dot{\theta}}{\theta^2} = \frac{3}{\theta^2}\left[\frac{1}{2}(\rho + 3p) - \Lambda + 2\sigma^2\right]. \qquad (17.52)$$

$q > 0$ indicates deceleration of the local expansion.

17.4 Intrinsic and Extrinsic Curvature

Gauss (1827)[20] defined the notion of curvature for the first time. Gauss defined curvature two ways. The first way: by considering the normal at a point, P, on the plane and drawing

---

[20] This work of Gauss was the founding paper of modern geometry containing the famous *Theorema Egregium* and its proof. It also contained the first version of the *Gauss–Bonnet* theorem relating the sum of the angles

a plane containing that normal, as in a flag. The plane cuts the surface in a curve having radius of curvature r and curvature $\kappa = 1/r$. As the plane rotates about the normal there is a point where the radius is a maximum, r_1, and another point where the radius is a minimum, r_2. These two directions will be orthogonal to one another and the radii of curvature are referred to as the *principal radii of curvature* at the point P. The second way was quite different: the *Gauss map* in which a point P on the surface is associated with a point Q on a sphere such that the normal at P on the surface is parallel to the normal at the point Q on the sphere. All points in a neighbourhood of P can be similarly mapped into points in a neighbourhood of Q. In this way, a circle of area G in the neighbourhood of Q maps into a closed curve about P enclosing an area F on the surface. Gauss thereby defined a single number to characterise the curvature of the surface: the *Gaussian curvature* $|K_{\text{Gauss}}| = \lim_{F \to 0}(G/F)$. The relationship between the principal curvatures κ_1 and κ_2 and the Gaussian curvature K_{Gauss} is[21]

$$K_{\text{Gauss}} = \kappa_1 \kappa_2. \tag{17.54}$$

It is possible to define another scalar measure of curvature, *mean curvature*:

$$K_{\text{mean}} = \kappa_1 + \kappa_2. \tag{17.55}$$

These two scalar curvature measurements are fundamentally different: K_{Gauss} can be measured from observations made from within the surface and without reference to the larger Euclidean space in which the surface is embedded, but this is not true of K_{mean}, which is a property of the sheet within the context of the embedding space.[22]

To understand the difference consider the very simple example of a flat sheet of paper. The two principal curvatures are $\kappa_1 = \kappa_2 = 0$. If the sheet is rolled up into a cylinder of radius r the principal curvatures are $\kappa_1 = 0$ and $\kappa_2 = r$, whence the Gaussian curvature is still $K_{\text{Gauss}} = 0$. However, the mean curvature is $K_{\text{mean}} = 1/2r$. The Gaussian curvature is a constant under such deformations of the sheet, but the mean curvature is not. We conclude that K_{Gauss} is an intrinsic property of the sheet, while K_{mean} is a property of the way it was rolled up within the larger embedding space, $\mathbb{R}^3$. The theorem, discovered by Gauss, that the Gaussian curvature of an embedded surface does not depend on how the

---

of a spherical triangle drawn on a sphere to the radius of the sphere. Gauss used this to determine the radius of the Earth using accurate measurements of a triangle having a side of ~ 25 km. He thereby introduced the remarkable notion that we could know how curved a surface is simply by making observations on the surface and without reference to the 3-dimensional space in which it was embedded. Thus he defined *intrinsic curvature*.

This work was the seed for Riemann's work on non-Euclidean geometry (his Habilitationsschrift *Uber die Hypothesen welche die Geometrie zu Grunde liegen*, 1854).

[21] This generalises to a 3-dimensional surface embedded in a 4-dimensional manifold. Consider a point P in a 4-dimensional manifold, $^{(4)}\mathcal{M}$ and a time-like unit vector n^a at some point P. There is a 3-dimensional surface, $^{(3)}\mathcal{V}$ at P in $^{(4)}\mathcal{M}$ that is orthogonal n^a. There are three 'flagpoles' at P, each defining a Gaussian curvature for sections of $^{(3)}\mathcal{V}$: we get three Gaussian curvatures $\kappa_1, \kappa_2, \kappa_3$, one from each flagpole. The quantity

$$^{(3)}K = \kappa_1 \kappa_2 + \kappa_2 \kappa_3 + \kappa_3 \kappa_1 \tag{17.53}$$

is a measure of the extrinsic curvature of $^{(3)}\mathcal{V}$ in $^{(4)}\mathcal{M}$. In general relativity $^{(3)}K$ is directly related to the Ricci tensor R_{ab} and the Ricci scalar R of $^{(4)}\mathcal{M}$ at P.

[22] It is merely a point of curiosity that K_{Gauss} and K_{mean} have different physical dimensions: $[L^{-2}]$ and $[L^{-1}]$.

space is embedded, the *Theorema Egregium*, is quite remarkable: the Gaussian curvature depends only on the metric in the surface and not the embedding space.[23]

Intrinsic curvature is what the inhabitants of a space can discover through experiment. Extrinsic curvature can only be defined for a space by embedding it in a higher dimensional space. There is no measurement that the inhabitants can make to reveal extrinsic curvature.[24]

As a final remark we should note that in our rolled paper example we would use different coordinates to describe geometry in the embedding space (Cartesian coordinates, $\{x^a\}$, $a = 1, 2, 3$) and to describe geometry in the rolled-paper space (cylindrical coordinates, $\{x^\alpha\}$, $\alpha = 1, 2$). The relationship between these coordinate systems is not simple. This makes it cumbersome to relate the embedding space and the embedded space.

17.4.1 The Extrinsic Curvature Tensor

Consider a vector $u^a(P)$ that is normal to the 3-dimensional hyper-surface, Σ, at point P in Σ. The normals at points of Σ that are in a neighbourhood of P tell us about the curvature of Σ at P as seen from within the embedding 4-space. The tensor $u^a_{;b}$ defines the neighbourhood, and the projections $h_a^p h_b^q u_{a;b}$ of this tensor into the hyper-surface Σ define the neighbourhood of the points around P that lie in the surface Σ. We define the *extrinsic curvature tensor*, K_{ab}, by

$$K_{ab} = -h_a^p h_b^q u_{(p;q)}, \tag{17.56}$$

where $u_{(a;b)} = \frac{1}{2}(u_{a;b} + u_{a;b})$ is the shear of the congruence defined by the vectors u^a.

In the absence of vorticity $u_{(a;b)} = u_{a;b}$, in which case we can expand the right-hand side and using Equation (17.6) gives

$$K_{ab} = -u_{b;a} - u_a \dot{u}_b, \quad \omega = 0. \tag{17.57}$$

If we take the trace of this tensor we have the *mean curvature* (Eisenhart, 1926, see Section 50):

$$K = -u^a_{;a}. \tag{17.58}$$

---

[23] Nonetheless, embedding a 4-space in suitable higher dimensional spaces can reveal important properties of the 4-space, as had been done by Robertson (1928) when he embedded de Sitter's solution in a 5-dimensional Minkowski space, thereby showing de Sitter space as a surface containing the time dimension.

There is another way of looking at this embedding process: we can postulate that our 4-dimensional world is in fact a subspace, i.e. a $3 + 1$-dimensional hyper-surface (a *brane*), inside a higher dimensional Universe (the *bulk*). Any physically relevant higher dimensional theory must have a 4-dimensional subspace that looks like our Robertson–Walker space. It must also induce sub-space gravitational field equations that are the Einstein equations, possibly with a correction factor that would be regarded as physics leaking into our Universe from higher dimensions. This is one of the fundamental tenets of *Braneworld cosmology* (Clifton *et al.*, 2012).

One of the main links between our observed Universe and the higher dimensions are the Gauss–Codazzi equations which provide constraints on which higher dimensional space may be used. Romero *et al.* (1996) exploited a theorem of Campbell (1926, Ch.XII, p.220) showing how to locally embed any solution of the general relativity equations in a Ricci-flat 5-dimensional space. Subsequently Mashhoon and Wesson (2007) showed how this simple procedure could generate a Λ-term in the 4-space equations.

[24] Even though we cannot determine the extrinsic curvature of our Universe, we can, in principle, determine the topology of the Universe (Buddelmeijer, 2006).

Using Gaussian normal coordinates (i.e. synchronous coordinates) at the point P we can find an expression for K_{ab} in terms of the metric tensor which takes the form

$$ds^2 = -dt^2 + h_{\alpha\beta}\,(t, x^\mu)\,dx^\alpha dx^\beta, \quad \alpha, \beta = 1, 2, 3. \tag{17.59}$$

In these comoving coordinates, $u^a = (1, 0, 0, 0)$ and we can calculate $u^a_{;b}$ directly from the expression for the covariant derivative in this coordinate system (see Equation 13.68):

$$u_{a;b} = u_{a,b} - \Gamma^k_{ab} u_k, \tag{17.60}$$

$$= \tfrac{1}{2} g^{0m} \left[g_{am,b} + g_{mb,a} - g_{ab,m} \right]. \tag{17.61}$$

Here, Γ^d_{ab} is the usual Christoffel symbol. Summation over the index k is implied in the first line and summation over m in the second. Since, in this coordinate system, $u_a = (1, 0, 0, 0)$ and does not depend explicitly on the coordinates $\{x^\mu\}$, the term $u_{a,b}$ is zero. Moreover, in the sum $\Gamma^k_{ab} u_k$, only the $k = 0$ term survives the sum $\Gamma^k_{ab} u_k$, leaving $\Gamma^0_{ab} u_0$, which controls the indices in the second line. In the second line, the only non-zero m in g^{0m} is for $m = 0$ (the advantage of working in Gaussian normal coordinates!) and that leaves only the term $g_{ab,0}$. Putting all this together with (17.57) we are left with

$$K_{ab} = \frac{1}{2} \frac{\partial g_{ab}}{\partial t}. \tag{17.62}$$

The only time dependent components of the metric (17.59) are the spatial ones and so we can also write this as:

Extrinsic curvature of the $t = $ const hyper-surfaces of the embedded 3-space Σ:

$$K_{\alpha\beta} = \frac{1}{2} \frac{\partial h_{\alpha\beta}}{\partial t} \quad \text{or} \quad K^\gamma_\beta = \frac{1}{2} h^{\gamma\alpha} \frac{\partial h_{\alpha\beta}}{\partial t}. \tag{17.63}$$

Geometrically, this is known as the second fundamental form of the $t = $ const hyper-surfaces Σ.

($h_{\alpha\beta}$ is the metric on the 3-space orthogonal to u^a and can be used for raising and lowering indices in that space. See Section 22.2.2 for the relevance of this to the Sachs–Wolfe effect.)

In a more general case, if the metric is given in the form

$$ds^2 = g_{00}(t, x^\mu)\,dt^2 + h_{\alpha\beta}\,(t, x^\mu)\,dx^\alpha dx^\beta, \quad \alpha, \beta, \mu = 1, 2, 3, \tag{17.64}$$

then the extrinsic curvature of the surface Σ is given by

$$K_{\alpha\beta} = \frac{1}{2\sqrt{-g_{00}}} \frac{\partial g_{\alpha\beta}}{\partial t}. \tag{17.65}$$

This a slightly more general variant of (17.63). The $\sqrt{-g_{00}}$ factor arises because $K_{\alpha\beta}$ is a tensor in Σ, not in the embedding manifold.

17.4.2 Gaussian Curvature and the Riemann Tensor

The relationship between the extrinsic curvature K_{ab} and the Riemann tensor comes through the Gauss–Codacci equations. Wald (1984, Section 10.2, p.258) writes these as:

$$^{(3)}R_{abc}{}^{d} = h_a{}^{f}h_b{}^{g}h_c{}^{h}h^d{}_{j}R_{fgk}{}^{j} - K_{ac}K_b{}^{d} + K_{bc}K_a{}^{d}, \tag{17.66}$$

$$D_a K^a{}_b - D_b K^b{}\ b = h^c{}_b n^d R_{cd}. \tag{17.67}$$

The 3-space derivative D_a is defined in Section 17.1.3 and footnote 6. The derivation of these equations is somewhat intricate and not particularly edifying, see Wald *loc. cit.* or Hawking and Ellis (1975, Section 2.7 p.47).

18

Observing and Measuring the Universe

In the Newtonian perspective, light simply moves in straight lines with constant velocity. The propagation of light through space was not an issue that Newtonian theory could address without making additional assumptions. In the curved geometry of Einstein's space-times, light responds to the ever changing curvature of the space-time.

In special relativity the light rays are defined by the fact that the proper distance moved by light is zero. This is enshrined in saying that light rays are the curves on which the line element is $ds^2 = 0$. Given a coordinate system and a metric expressed in those coordinates, we can, in principle, solve the resulting equations. There is no reason to believe that our Minkowskian intuition about the propagation of light would be of help, and some of the results can at first be quite surprising. Like the result that the angular diameter subtended by an object of fixed size can increase with distance!

Aside from this counter intuitive behaviour there is another problem: the problem of understanding and interpreting the coordinates in terms of what a person sitting in an inertial frame observes. This is fundamental to the astronomer observing the Universe, and so we develop the relevant equation for an FLRW Universe.

18.1 Light Propagation in Curved Space-Time

Understanding the physics of light propagation is a vital step in interpreting the observed cosmological redshift. Getting to understand the redshift was not at all straightforward even though it is a consequence of the cosmological models constructed within the context of general relativity.[1] Some of this early story with regard to de Sitter's Universe was retold in Section 15.6.4.

As late as the 1960s and 1970s there was still an active, and often heated, debate as to whether the observed redshifts of galaxies and quasars were of cosmological origin. That was not simply a question of whether or not general relativity could provide the explanation for the phenomenon. The debate was more about whether there was an additional contribution from phenomena intrinsic to the source or from something that might happen along the path of the light rays (as in 'tired light'). Some of the debate even suggested that

---

[1] The story is told with a detailed historical perspective in the fine article by Duerbeck and Seitter (2001).

the constants of nature might be a function of time,[2] an idea that originated with Dirac (1937, 1938) following up on earlier ideas of Milne and Eddington.

Accepting the notion that the measured quasar redshifts were indeed cosmological opened up the possibility of using them as high redshift cosmological probes. However, they exhibited such a huge variation in intrinsic properties that they would never be suitable for plotting a Hubble diagram to determine the 'two numbers', σ_0 and q_0, that defined our cosmology at that time. Their cosmological impact would come from measurements of the absorption lines, mainly Ly-α, in their spectra which traced the state of the cosmic medium between us and the quasar. Up until that time, low-z approximations to the closed-form relativistic redshift formulae of Mattig (1958) and Terrell (1977) were used to interpret magnitudes, diameters and separations of distant objects (see Section 5.6.3).[3] However, these formulae were valid only for $\Lambda = 0$ models, there were no $\Lambda \neq 0$ counterparts. Clarity between the various general relativistic interpretations of 'distance' was needed.

18.1.1 The Cosmological Redshift

While most cosmologists today would embrace the theoretical conclusions derived from studying general relativistic cosmological models as providing the correct explanation of the phenomenon of the cosmic expansion, as evidenced by the redshift, there is still a debate about the *interpretation* of this. We ask ourselves how we are to understand this result in physical terms? What is in fact happening? [4] The interpretation, or understanding, of the cosmological redshift has long been a subject of controversial discussion, and to a certain extent it still is today (see, for example Bunn and Hogg (2009)).

We assign a redshift z to a galaxy when we observe a spectral line that was emitted at wavelength λ_E with frequency ν_E to be received with wavelength λ_O and frequency ν_O:

$$1 + z = \frac{\lambda_O}{\lambda_E} = \frac{\nu_E}{\nu_O} \quad \text{or} \quad z = \frac{\lambda_O - \lambda_E}{\lambda_E} = \frac{\delta\lambda}{\lambda_E}. \tag{18.1}$$

Within the framework of known physics there are only two ways to change the frequency of a light beam: the Doppler shift and the gravitational redshift.[5] The latter is a consequence

---

[2] In recent times the emphasis in this direction has been on studying possible spatial and temporal variations of the fine structure constant. See, for example, the papers of Barrow and Webb (2005) and Webb *et al.* (2011).

[3] Prior to 1970, hand-held scientific calculators were not available, and even after that they were costly 'must-have' items for scientists. Computing at that time was not a straightforward process: it consisted mainly of writing Fortran programs onto punched cards and submitting them as a batch job, waiting a day or so to get back the answers (if successfully executed) printed out on a line printer. Diagrams were plotted on special flat-bed printers, and submission of artwork for diagrams had to be prepared by a professional drafting person. That changed with the advent of the personal computer in the mid 1980s, though it was at first a relatively slow change.

[4] I might argue that while such questions are interesting, they are not really germane to the argument of why the redshift is what it is. Once we adopt Einstein's gravitational theory for the geometry of space-time the result follows whether or not we can rationalise it. We perform experiments to verify the result and while everything works out we are fine. If something goes wrong, we rethink the situation. The situation is little different from attempts to rationalise the orbits of the planets prior to time of Galileo and Kepler, or to rationalise quantum mechanics.

[5] It is difficult to change the frequency of a monochromatic light beam in the laboratory. When passing a monochromatic light beam though a refracting medium, the speed of light decreases and the wavelength

of the curvature of space and thus an effect that can only be properly discussed within the framework of general relativity.

The result we are aiming for is to show that the frequency shift, $1+z$, as defined in (18.1) is inversely proportional to the scale factor of the cosmic expansion:

$$1+z = \frac{1}{a(t)}. \tag{18.2}$$

This is the classical redshift *vs.* expansion factor formula connecting the scale of the Universe $a(t)$ at an earlier time t with the redshift we expect to measure for objects being seen at that time.

The redshift z of a galaxy is an observable and measurable quantity. Unfortunately we cannot know the local cosmic time, the age of the Universe, at which we observe the galaxy except indirectly through an understanding of how galaxies evolve. So, in order to interpret cosmological data we have to use the redshift as a surrogate for the time.

The observed value of z can be used as a cosmic clock. Of course, the clock has to be calibrated, we need to know the age of the Universe in years. This has to be done by fitting models to cosmological data and knowing the $z(t)$ or $t(z)$ relationships for those models.

There are many approaches to explaining the cosmological redshift phenomenon, some of them wrong or, at best, misleading. The simplest, frequently used for didactic purposes, is to 'explain' this as being due to the stretching of light waves by the cosmic expansion.

Redshifts in General Relativity

The general situation involves an emitter of a pulse of light, **E**, and an observer, **O**, moving on time-like paths Γ_E and Γ_O in space-time. This is depicted in Figure 18.1 where the 4-velocities of each observer, the tangents to their world lines, are denoted by u^a_E and u^a_O. The light ray sent from **E** is sent in direction k^a_E towards **O** who receives it travelling in direction k^a_O.

Emitter **E** emits two pulses of light at neighbouring positions A_E and B_E separated by proper time interval ds_E. These are received by **O** at A_O and B_O separated by proper time interval ds_O. If we think of these two events as marking successive crests in a light ray, then ds_E and ds_O define the period of the light at the time of emission and reception, respectively, i.e. the periods of the light wave. Hence the frequencies ν_E and ν_O assigned to the light by the emitter and observer are related by

$$\frac{\nu_E}{\nu_O} = \frac{ds_O}{ds_E} = \frac{d\tau_O}{d\tau_E}, \tag{18.3}$$

where τ_E and τ_O denote the proper time as measured by the emitter and the observer on their path. This is a quite general result, it makes no assumptions about the underlying space-time or the motions of the emitter and observer. The ratio of the frequencies at emission and observation is simply a function of the local proper times on each of the world lines. We cannot separate the contributions from the Doppler shift and the gravitational shift in frequency.

increases, but the frequency remains the same. There are some exotic solid state and atomic phenomena wherein frequency does change. See, for example, Padgett (2014).

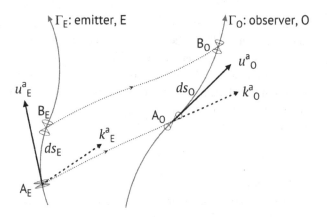

Fig. 18.1 Propagation of wavefront from an emitter Γ_e to an observer Γ_0. The light rays (dotted) are the lines of constant phase for the light which propagates along them, so the phase difference between A_e and B_e is the same as the phase difference between A_0 and B_0. The period of the light will be proportional to the elapsed time, ds_e between A_e and B_e, and and the time ds_0 between A_0 and B_0. Hence the frequencies ν_e and ν_0 measured by each observer will have $\nu_0/\nu_e = ds_e/ds_0$. ©

We can go one step further without needing to specify the space-time geometry, or the particle paths. In Figure 18.1 we have denoted the tangents to the light ray path as 4-vectors k_E^a at emission and k_O^a at observation. These are the frequency 4-vectors at **E** and **0**, having components $(\nu_E/c, \mathbf{k})_E$ and $(\nu_O/c, \mathbf{k})_O$ in the local inertial frames. The emitter and observer can be assigned 4-velocities u_E^a and u_0^a, which in the same inertial frames have components $u_E^a = \gamma_E(\mathbf{u}_E/c, 1)$ and $u_O^a = \gamma_O(\mathbf{u}_O/c, 1)$. In the local rest frames of **E** and **0**, the 3-velocities are zero and so the scalar product $(u^a k_a)_E = \nu_E$ and $(u^a k_a)_O = \nu_O$. Invoking the principle of equivalence, these scalar values in the local rest frames must have the same values in all frames. With this we can rewrite (18.3) as

$$\frac{\nu_E}{\nu_O} = \frac{(u^a k_a)_E}{(u^a k_a)_O},\qquad (18.4)$$

While (18.3) and (18.4) are quite general, a metric has to be specified in order to evaluate them. For the simple homogeneous and isotropic metrics we can proceed in a more direct manner, as exemplified by the calculation of the redshift shown earlier for the de Sitter Universe (see Section 15.6.4).

A Simple 'Explanation'

One way in which this is frequently presented is to consider two observers moving with the cosmic expansion, i.e. comoving observers, separated by a small distance dr. By the Hubble law their relative velocity is $dv = H(t)dr$. From (18.1) the redshift dz between the observers is the fractional change in wavelength, $d\lambda/\lambda$ of the light sent from one to the other. If this is due to a Doppler shift, we have $dz = dv/c = Hdr/c$. The light travel time, dt, between the observers is given by $cdt = dr$ and so $dz = Hdt$.

Next we bring the cosmic expansion into the picture via the scale factor $R(t)$. The functional form of $R(t)$ is determined by the Friedmann equations and the expansion rate is given by the Hubble parameter $H = \dot{R}/R$. This gives $dz = d\lambda/\lambda = dR/R$.

This says that $\lambda \propto R$ and so between being emitted at t_E and observed at t_O the light changes in wavelength from λ_E to λ_O given by $\lambda_O/\lambda_E = R(t_O)/R(t_E)$. If we normalise the radii $R(t)$ to the present value $R(t_O)$, i.e. write $a(t_O) = R(t_E)/R(t_O)$, we have $\lambda_E/\lambda_O = a(t_E)$ whence

$$1 + z = \frac{1}{a},\tag{18.5}$$

which is our result (18.2).

This explanation can be criticised on a number of grounds. It assumes that the Doppler shift is the cause of the redshift (it starts with $dv = H(t)dr$ and then assumes that H is given by the Friedmann equations, or can be obtained from Newtonian theory). There is no explicit role for any general relativistic contribution, and we do not need to know what the function $R(t)$ is. So, while such motivational thinking may be of some help, it does not take us any further towards dealing with a more complex situation like de Sitter's empty space, or Universes that are not homogeneous.

The General Relativistic Explanation

Now we turn to the general relativistic derivation. The redshift in a de Sitter space-time was already discussed in Section 15.6.4 and we can take a similar route here using the Robertson–Walker metric in the form (16.1).

The path of a radial light pulse emitted from a galaxy at time t_E is given by $ds^2 = 0$:

$$\frac{cdt}{R(t)} = \frac{dr}{\sqrt{1 - kr^2}}.\tag{18.6}$$

Note that the left-hand side depends only on t and the right-hand side depends only on the radial coordinate r. We can follow the path of the light towards an observer at O located at radial coordinate $r = 0$ who receives the light pulse at t_O:

$$\int_{t=t_E}^{t_O} \frac{cdt}{R(t)} = \int_{r=0}^{r_E} \frac{dr}{\sqrt{1 - kr^2}}.\tag{18.7}$$

We can write the same equation for a pulse emitted a little later at $T_E + \Delta t_E$ and received a little later, at $T_O + \Delta t_O$:

$$\int_{t=t_E+\Delta t_E}^{t_O+\Delta t_O} \frac{cdt}{R(t)} = \int_{r=0}^{r_E} \frac{dr}{\sqrt{1 - kr^2}}.\tag{18.8}$$

The integral on the right-hand side here is the same as before (i.e. in 18.7) since it only depends on the comoving coordinates of the emitter and receiver, which are fixed.

A bit of elementary calculus then gives that

$$\int_{t=t_E+\Delta t_E}^{t_O+\Delta t_O} \frac{cdt}{R(t)} = \int_{t=t_E}^{t_O} \frac{cdt}{R(t)} \quad \Rightarrow \quad \frac{c\Delta t_E}{R(t_E)} = \frac{c\Delta t_O}{R(t_O)}.\tag{18.9}$$

We can identify $c\Delta t_E$ with the wavelength, λ_E, of the pulse at t_E and $c\Delta t_O$ with the wavelength, λ_O, of the pulse at t_O to give

$$\frac{\lambda_O}{\lambda_E} = \frac{R(t_O)}{R(t_E)} = 1 + z, \qquad (18.10)$$

by definition of the redshift. Note that the Hubble expansion law does not figure in this: the cosmology comes in with the scale factor $R(t)$ and the metric.

18.2 Age, Distances and Angles

We need to discover which, if any, of the theoretical models for the Universe most closely represents the Universe in which we live. Whereas only 50 years ago the major data available for this task was the Hubble diagram relating distance to recession velocity, that did not probe far enough in distance or in accuracy to act as a strong discriminator among the various models. The measured recession velocities probed were generally only a small fraction of the speed of light.[6] No redshift even close to $z = 1$ was measured until the discovery of the quasars. While at the time of their discovery they cast strong doubt on non-evolving Steady State models, they were manifestly not the 'standard candles' that would be required to produce a useful Hubble diagram.

That of course changed with the discovery of the cosmic background radiation which saw the Universe as it was at a redshift of $z \sim 1000$, but several decades would have to elapse before that could be used as a cosmological probe that would establish the parameters of our Universe with high precision. That would only become possible through a detailed understanding of the physics of the Universe around the time that the CMB was being observed, i.e. a redshift of $z \sim 1000$.

Work on the Hubble diagram moved away from galaxies and onto observing very distant supernovae. This effort from Perlmutter (1999) and Riess and Filippenko (1998) mapped the Hubble diagram at redshifts $z \sim 0.5$ and beyond and provided the evidence that our Universe was geometrically flat and dynamically driven by a form of dark energy that was identified with Einstein's cosmological constant. An essential ingredient of this programme was to understand the physics of the supernova light curves well enough to recognise the stage of the explosion and to assign an absolute magnitude to the exploding object. This work won Perlmutter, Riess, and Schmidt the 2011 Nobel Prize in Physics.

18.2.1 Light Propagation in Curved Spaces

The arena of cosmology is Einstein's theory of general relativity and so, in order to discuss observations of phenomena outside our local volume, we must know how to deal with the propagation of light in cosmological models. So what is difficult about that? Nothing except that in studying this aspect of cosmology we move into an arena that

---

[6] Several other tests were proposed and tried, but faced the problem that they could not probe deep enough into space to be truly effective, or, if they did, they fell foul of poorly understood evolutionary problems.

is outside our common experience. It is natural to think of measuring distance in terms of using one or more rulers and pacing out the distance between two points. If that is impractical we have several alternatives that work in a variety of circumstances: parallax, apparent angular size of a measuring rod, or apparent brightness of a source of known brightness. In our Euclidean world, these give the same answer to within the accuracy that is afforded by the equipment. This is not so in general relativity. Light rays in general relativity do not behave the same way as light rays in a Euclidean background space.

We cannot attribute a direct operational meaning to the notion of distance in curved geometries. Distance, as we normally think of it, depends on a specific choice of coordinates. We nevertheless find it useful to define a variety of measures of distance for specific purposes. Thus we use *luminosity distance* when we want to relate the intrinsic brightness of an object to its apparent brightness when seen at a redshift z. The redshift is directly observable and so in these terms is a meaningful quantity. Likewise we may use the angular diameter subtended by an object of known intrinsic size to determine an *angular diameter distance*. In principle we could also use *parallax distance* to measure a distance. In the Euclidean geometry of the Newtonian world all these methods would give the same answer.

In cosmology these three measures of distance differ, but they are simply related via the redshift to the object whose 'distance' is required. The value of a distance depends on the details of the cosmological model: we shall give these equations below.

There is a further complication in the general relativistic theory of cosmological models. If the model is inhomogeneous the gravitational deflection of the light rays by the inhomogeneities affects the distance measurements. This is simply the phenomenon of gravitational lensing which is absent from Newtonian cosmology.

18.2.2 Distances

We can make a connection between the mathematical description of the Friedmann–Robertson–Walker ('FLRW') cosmology, with its various abstract coordinates, and the world of observational cosmology where we have only measurements of positions, redshifts, angular sizes and brightness of diverse observed targets. Notice that the fundamental concept of 'distance' is not in this list. Cosmological distances are not directly observed except for relatively nearby objects, and then only indirectly via proxies (like bright stars or supernovae) that have prior calibration.

In general relativity there is more than one definition of 'distance'. This is not a matter of choice of coordinate system, these different definitions are coordinate independent (as they should be). They present different geometrical aspects of the concept of distance that are not necessary in the Euclidean world of our experience.[7]

---

[7] The subject of distance is further confused by a variation in the names attached to these distances. The following articles go some way to clarifying the redshift law and the issues surrounding distance measurement: Harrison (1993), Kiang (1995, 2004), Hogg (1999) and Davis and Lineweaver (2004, see Appendix A). The article by Hogg was never published except as an arXiv e-Print.

Some Reminders about the Hubble Parameter

The solutions of the Friedmann equations depend on the constituents of the cosmic medium with their equations of state, and on the curvature of the space-time. The relative contributions of the constituents to the total cosmic density are parameterised by their fractional contribution to the present day total cosmic density. These contributions are the Ω-parameters.

The simplest model that could reasonably represent the post-recombination Universe contains zero-pressure (cold) matter of density ρ_m and has an Einstein cosmological term Λ that has no time dependence. The Ω-parameters for these two constituents are $\Omega_m = 8\pi G\rho_{m0}/3H_0^2$, where ρ_{m0} is the current matter density, and $\Omega_\Lambda = \Lambda/3H_0^2$, where H_0 is the present Hubble constant. We designate the contribution from the curvature by Ω_k, and if there are no other constituents

$$\Omega_m + \Omega_k + \Omega_\Lambda = 1. \tag{18.11}$$

We can therefore write the curvature contribution as $\Omega_k = (1 - \Omega_m - \Omega_\Lambda)$.

With this model we can write the expansion rate at redshift z as

$$H(z)^2 = H_0^2 \left[\underbrace{\Omega_m(1+z)^3}_{\text{matter density}} + \underbrace{(1 - \Omega_m - \Omega_\Lambda)(1+z)^2}_{\text{curvature}} + \underbrace{\Omega_\Lambda}_{\text{cosmological constant}} \right]. \tag{18.12}$$

In the case of zero curvature, i.e. a flat cosmology, $\Omega_k = 0$ and $\Omega_m + \Omega_\Lambda = 1$ and this is simply

$$H(z)^2 \stackrel{\text{flat}}{=} H_0^2 \left[\Omega_m(1+z)^3 + \Omega_\Lambda \right]. \tag{18.13}$$

If instead of the cosmological constant we have dark energy with equation of state $p = w\rho c^2$ in place of the cosmological constant Λ, Equation (18.12) becomes:

$$H(z) \stackrel{\text{flat}}{=} H_0 \left[\Omega_m(1+z)^3 + (1 - \Omega_m)(1+z)^{3(1+w)} \right]^{\frac{1}{2}}. \tag{18.14}$$

This follows because, with this equation of state, the dark energy density varies with radius as $a(t)^{-3(1+w)}$, $a(t)$ being the usual expansion scale factor.

Some Remarks about the Metric

The development that follows is done in terms of the FLRW metric written in 'hyperspherical' coordinates (see Equation 16.6):[8]

$$ds^2 = -c^2 dt^2 + R(t)^2 \left[d\psi^2 + S_k(\psi)^2 d\Omega^2 \right], \quad S_k(\psi) = (\sin\sqrt{k}\psi)/\sqrt{k}. \tag{18.15}$$

Here, t is the local proper time of a comoving observer and ψ is the radial coordinate the meaning of which was explained in Section 16.1 (see Figures 16.1 and 18.2).

---

[8] Chapters 15 and 16 describe the more mathematical aspects of a variety of solutions to the Einstein equations and provide a mathematical basis for much of what is described here.

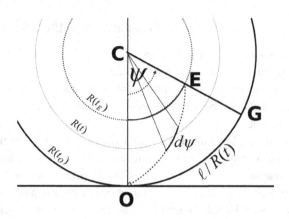

Fig. 18.2 Path of light ray from emitter **E**, to observer **O**. The light ray **EO** was emitted when the radius of the Universe was $R(t_E)$ and received when it was $R(t_0)$. The angular coordinate ψ changes through a sequence of elements $d\psi$: one is shown at a time t when the radius of the Universe was $R(t)$. The coordinate length of the light path is the sum of elements $Rd\psi = cdt$. The emitting galaxy will continue to move away from **C** as the Universe expands and at time t_0 will reach **G**. The length of the arc **OG** is the proper distance between the observed galaxy and the observer. The arc **OE** is a generator on **O**'s past light-cone.

The sketch depicts a sequence of snapshots of a $k = +1$ FLRW model, described in hyper-spherical coordinates, at three different stages of its expansion: this is the 'expanding balloon model'. See the caption to Figure 16.1 for more details. (cc)

In these coordinates the equation for the spatial motion of radial light rays is $cdt = \pm R(t)d\psi$. The element of radial distance between two neighbouring points separated by radial coordinate increment $d\psi$ is

$$d\ell = cdt = R(t)d\psi. \tag{18.16}$$

This equation plays an important role in defining distances in general relativity.[9]

A simple use of this last equation occurs when converting a small redshift interval $[z, z+dz]$ into a local depth, $d\ell$ (measured in Mpc). From $d\ell = cdt$ we get[10]

$$d\ell = \frac{cdz}{(1+z)\,H(z)}. \tag{18.17}$$

---

[9] If we write the integral of this for a constant value of ψ we have $\ell = R(t)\psi$. This gives $d\ell = \dot{R}(t)\psi\,dt$ and so, with $H = \dot{R}/R$, ℓ satisfies the Hubble law, $\dot{\ell} = H\ell$, for all FLRW cosmologies. However, ℓ is not a physically measurable quantity. It has to be transformed into other variables that can be measured.

Carroll (2003, Section 8.3) refers to ℓ as the *instantaneous physical distance*.

[10] This comes from the chain of substitutions

$$d\ell = cdt = \frac{cda}{\dot{a}} = \frac{cda}{aH(a)} = -\frac{cdz}{(1+z)H(z)},$$

with $H = \dot{a}/a$ and $(1+z) = 1/a$. These are useful when transforming the independent variable in derivatives.

In a flat ($k = 0$) Universe where $\Omega_m + \Omega_\Lambda = 1$ this is

$$d\ell = cH_0^{-1} \frac{dz}{(1+z) \left[\Omega_m(1+z)^3 + \Omega_\Lambda\right]^{1/2}}, \tag{18.18}$$

on using Equation 18.13.

18.2.3 Connecting with observables

We take the point of view of an observer **0** sitting at the origin of the (t, ψ, ϕ, θ) coordinate system. The observer's cosmic time is denoted by t_O. The observer observes galaxies that are taken to be fixed in the cosmic expansion, i.e. co-moving galaxies, they have no non-cosmological (peculiar) motions relative to the cosmic expansion. A galaxy that is being observed is being seen at cosmic time t_E and has a fixed value of ψ during the cosmic expansion. The physical distance to the galaxy, whichever distance measure we wish to use, depends only on the product $R(t)\psi$.

We are going to need to convert the radial coordinate ψ to redshift, z, an observable quantity. To do that it is useful to write $R(t) = a(t)R(t_O)$, where $a(t)$ is the expansion scale factor as given by the Friedmann equations. Then (18.16) is $cdt = R(t)d\psi = a(t)R(t_O)d\psi$.

We can write the important quantity $R_0 d\psi$ in a number of ways:

$$\frac{da}{a} = -\frac{1}{1+z}, \quad R_0\psi = \int_t^{t_O} \frac{cdt}{a(t)} = c\int_a^1 \frac{da}{a^2 H(a)} = c\int_0^z \frac{dz}{H(z)}, \tag{18.19}$$

where we have abbreviated $R(t_O)$ to R_O. Here, $R_O\psi$ is evaluated for a galaxy observed at redshift z when the cosmic time was t and the cosmic scale factor was $a(t)$. R_O has dimensions of length, while ψ is a dimensionless radial coordinate. $a(t)$ is a function giving the stage which the expansion has reached in terms of its current value, i.e. $R(t) = a(t)R(t_O)$. The actual value of $R(t_0)$ is never needed since it only appears as the ratio $a(t) = R(t)/R(t_O) = (1+z)^{-1}$.

Comoving, or 'Ruler', Distance

First we need to establish the 'comoving distance', or 'ruler distance', within the framework of this FLRW metric. This is simply the calculation of the length of a null geodesic line between two points of space-time and is analogous to calculating how far an ant travels along a geodesic drawn on an expanding spherical balloon. This is depicted in Figure 18.2.

Consider the distance cdt the light propagates in successive intervals dt: at each step from the emitter **E** to the observer **0**. Owing to the expansion, it has further to go at each step than at the previous step. To do the calculation we therefore write the distance cdt in terms of the radial coordinate ψ as $d\ell = cdt = R(t)d\psi$. In the depiction of the coordinates in Figure 16.1, the light ray moves through an angle $d\psi = d\ell/R(t)$. We add up all these elemental angles between emitting the light ray at t_E and receiving it at t_O, and then rescale the answer to a real length (as opposed to a physically meaningless angle):

Comoving distance:

$$D_p(t_O) = R(t_O) \int_{t_E}^{t_O} d\psi = R(t_O) \int_{t_E}^{t_O} \frac{c\,dt'}{R(t')} = \int_{t_E}^{t_O} \frac{c\,dt'}{a(t')} = c \int_0^z \frac{dz}{H(z)}. \qquad (18.20)$$

This is the distance between the galaxies when they are both on the time t_O hyper-surface of the FLRW space-time.

Here $a(t)$ is the scale factor $R(t)/R(t_O)$, given by the Friedmann equations, and the last equality comes from Equation (18.19). The first of these integrals is the total ψ-coordinate distance, $R(t)\psi$, traversed by the light ray going from the galaxy at t_E to the galaxy at t_O. The factor $R(t_O)$ converts this ψ-distance moved by the light to a physical distance, *scaled to the present time*.[11]

The *comoving distance* is the distance between the galaxies when they are both on the time t_O hyper-surface of the space-time. It is the distance between points in a snapshot of a simulation of the expansion of a perfectly uniform Universe. The observer at **0** cannot measure that, since the observer galaxy is only seen at time t_E and has yet to reach the t_O hyper-surface, see Figure 18.2.

With this $R(t_O)$ normalisation, $D_p(t_O)$ is the physical distance from the observer to the place where the emitting galaxy is at the current cosmic time, t_O. The observer cannot see that place, he/she only sees the galaxy where it was at t_E. So $D_p(t_O)$ is not a quantity that can be measured by our observer at **0**: we are only observing the galaxy at a time before it has arrived at t_O.

The nomenclature for this distance measure is somewhat variable. The distance $D_p(t_O)$ here is referred to as the *comoving distance* because it is a measure of distance between two points that are comoving with the expansion. However, it is sometimes referred to as the *proper distance* because it is a measurement of ds for two events with $dt = 0$.

Expressing $S_k(\psi)$ in Terms of Redshift

The ψ-distance from the emitter to the observer can be written, using the second and last terms in (18.20), as

$$\psi = \frac{c}{R_0} c \int_0^z \frac{dz}{H(z)} = \frac{cH_0^{-1}}{R_0} \int_0^z \frac{dz}{E(z)}, \qquad (18.22)$$

$$E(z) = H(z)/H_0 = \left[\Omega_m(1+z)^3 + (1 - \Omega_m - \Omega_\Lambda) + \Omega_\Lambda \right]^{1/2}. \qquad (18.23)$$

---

[11] We can derive the expression for the proper distance using the 'standard' form of the FLRW metric (16.1). The proper distance is then given by

$$D_p(r) = R(t) \int_0^r \frac{dr}{\sqrt{1 - kr^2}} = \begin{cases} \arcsin r & k = +1 \\ r & k = 0 \\ \operatorname{arcsinh} r, & k = -1 \end{cases} = R(t)\,\psi \qquad (18.21)$$

which is consistent with the definition (16.5) of ψ and with Equation (18.20).
Proper distance is independent of the choice of the metric used to describe the space-time.

This uses the form (18.12) for $H(z)$, but works equally for other mixes of cosmic constituents, such as (18.13) or (18.14). Doing this leaves an integral over the Hubble parameter independent quantity, $E(z)^{-1}$. This integral cannot be done analytically except in a few special cases.

The term cH_0^{-1}/R_0 is the ratio of two important lengths: cH_0^{-1} is the distance a light ray travels in the cosmic expansion time, *i.e.* the scale of the Universe. The quantity R_0 is the present day value of the scale factor $R(t)$ appearing in the metric, and represents the geometric radius of curvature of the space. We can make this explicit by noting the following.

From the Friedmann equations written down for the metric (18.15) we get

$$\dot{R}^2 = \frac{8\pi G\rho}{3}R^2 + \frac{1}{3}\Lambda c^2 - c^2 k \quad \Rightarrow \quad k = -\left(\frac{H_0 R_0}{c}\right)^2 \Omega_{k,0}, \quad k = 0, \pm 1. \quad (18.24)$$

The second equation follows from the first by dividing throughout by R^2 and recognising that $H = \dot{R}/R$, and evaluating the resultant expression at the present time. We will drop the 0-subscript on the $\Omega_{k,0}$ and understand Ω_k to mean the today-value unless otherwise stated.

This enables us to remove the term cH_0^{-1}/R_0 appearing in the expression (18.22) for ψ to get a neat formula for ψ using only Ω_k and k, so that:

$$R_0 S_k(\psi) = \frac{cH_0^{-1}}{\sqrt{|\Omega_k|}} \sin\left(\sqrt{\Omega_k} \int_0^z \frac{dz}{E(z)}\right). \quad (18.25)$$

Note that this has units of length, as would be expected.

Comoving Distance Evaluated in FLRW Models

We can give a simple example of this for a matter dominated Einstein de Sitter cosmological model with the metric in the form (15.130). For this model $R(t) = R_O(t/t_O)^{2/3}$ and so

$$D_p = c \int_{t_E}^{t_O} \frac{R(t_O)}{R(t)} dt = 3ct_O \left[1 - \left(\frac{t_E}{t_O}\right)^{1/3}\right]. \quad (18.26)$$

Using the $z(t)$ relationship from Section 18.1.1 for this model, $(1 + z)^{-1} = R(t)/R(t_0) = (t/t_O)^{2/3}$, we get the relationship between proper distance and redshift:

$$D_p \stackrel{\text{EdS}}{=} 2cH_0^{-1}\left[1 - (1 + z)^{-1/2}\right]. \quad (18.27)$$

This model has $\Lambda = 0$.

In the more general case of matter and non-zero cosmological constant we can write down the expression for D_p but we cannot do the integral analytically:

$$D_p(z) \stackrel{\text{flat}}{=} cH_0^{-1} \int_0^z \frac{dz}{\sqrt{\Omega_m(1 + z)^3 + \Omega_\Lambda}}, \quad \Omega_m + \Omega_\Lambda = 1, \quad (18.28)$$

where Ω_m is the usual fractional contribution of cold matter (*i.e.* baryons and dark matter) to the total cosmic density and Ω_Λ is the contribution of the dark energy (Equation 18.13). This applies to the post-recombination flat Universe, when dark energy and matter dominate the expansion rate.

It is straightforward to get a high-z approximation to Equation (18.28):

$$D_p(z) \simeq \frac{2cH_0^{-1}}{z\,\Omega_m^{1/2}}, \quad z \to \infty. \tag{18.29}$$

In effect this neglects the small extra contribution from the recent accelerated phase of the cosmic expansion in the integral $c \int_0^z dz/H(z)$ of Equation (18.20).

When we look at the CMB radiation we are looking at a surface of the sky at which all points are at the same redshift, the redshift of the last scattering surface at $z \sim 1100$. The physical distance of a point on the CMB sky from its position at the origin of the Universe is

$$D_H(z) = c \int_z^\infty \frac{dz}{H(z)}. \tag{18.30}$$

This defines the proper or co-moving *horizon size* at the epoch z. In this case the expression for $H(z)$ should contain a contribution from the cosmic radiation field which becomes dominant at very high z. However, this is a relatively small contribution to the integral, and a high-z approximation to this that is applicable after the Universe recombines (i.e. $z \lesssim 1500$) is

$$D_H(z) \simeq \frac{2cH_0^{-1}}{z^{1/2}\,\Omega_m^{1/2}}, \quad z \to \infty. \tag{18.31}$$

Points separated by more than this distance at redshift z are not causally connected.

18.2.4 Angular Diameter and Apparent Brightness

We now turn attention to relating the relativistic proper distance to something that astronomers can measure. There are several possible measurements that can be made on a galaxy: among which are the diameter, luminosity and surface brightness.[12] In a relativistic cosmological model we choose to define a *luminosity distance*, D_L, and an *angular diameter distance*, D_A, that agree with our naïve notion that apparent brightness should vary inversely as the square of the defined distance, and that angular size should vary inversely with the defined distance. These two distance measures are not the same and, because of the geometry of space time, they depend on redshift and on the cosmological parameters.

The outcome of this is shown in Figure 18.3, where the panels shows the behaviour as a function of redshift, z, of the luminosity distance (on the left) and the angular diameter distance (on the right). The assigned luminosity distance increases with redshift in a cosmology-dependent way, as would be expected. However, the angular diameter distance, in some models, increases with redshift, but then starts to decrease because of

---

[12] The paper of Kayser *et al.* (1997) is a clear discussion on this topic: they provide formulae for angular diameter distance in a wide variety of $\Lambda \neq 0$ models with corrections for inhomogeneity. The treatments by Ryden (2003, Sections 7.2,7.3) and by Ellis *et al.* (2012, Section 7.4) are didactically excellent.

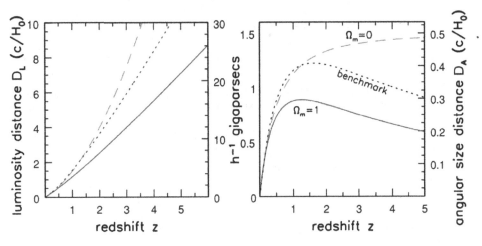

Fig. 18.3   Luminosity distance D_L (left panel) and angular diameter distance, D_A (right panel) as a function of redshift for three cosmological models. The models are as displayed on the right-hand panel. These diagrams show the relationship between distance and redshift. The surprise is that, in the right-hand diagram, the distance reaches a maximum in both the $\Omega_m = 1$ and the benchmark models. This is because in both these models a standard ruler subtends a smaller angle as the distance to the ruler increases, but at some point it reaches a minimum apparent size ($z = 5/4$ in the $\Omega_m = 1$ model) beyond which it appears to get larger.
(From Sparke and Gallagher (2007, Fig. 8.11), with permission.)

the focusing of light by the geometry of space-time. Since, by construction, the apparent angular diameter of an object varies as $\theta \propto D_L^{-1}$, this behaviour means that there comes a redshift beyond which the apparent angular diameter of the object starts to increase.

Angular Diameter Distance

It is convenient to measure the apparent angular size subtended by a distant object in steradians, Ω. If the physical area is A, in our Euclidean world, we assign a distance D_A given by

$$D_A^2 = \frac{A}{\Omega}, \tag{18.32}$$

and so the subtended solid angle is $\Omega = A/D_A^2$.

In an FLRW Universe with metric (16.6) the element of area is $S_k(\psi)d\Omega$ and so the solid angle subtended is in fact

$$\Omega = \frac{A}{R(t_e)^2 S_k(\psi)^2}, \tag{18.33}$$

where ψ_e is the radial coordinate of the object at the time t_E it is observed. We have Equation (18.25) for $R(t_0)S_k(\psi)$ with which to bring this into a more practical form:

The *angular diameter distance* of the object at redshift z:

$$D_A = R(t_e)S_k(\psi) = \frac{cH_0^{-1}}{(1+z)} \frac{1}{\sqrt{\Omega_k}} \sin\left(\sqrt{\Omega_k} \int_0^z \frac{dz}{E(z)}\right). \tag{18.34}$$

$E(z)$ is the normalised expansion rate as defined in Equation (18.23). This is the distance which we would assign to an object of known intrinsic size on the basis of its subtended angular diameter.

The redshift factor enters through $R(t_E)/R(t_0)$. Remember also that the $\sqrt{\Omega_k}$ term acts as a switch between the $\sin(x)$ or $\sinh(x)$ functions.

In a flat space where $\Omega_k = 0$ this reduces to

$$D_A \stackrel{\text{flat}}{=} \frac{cH_0^{-1}}{1+z} \int_0^z \frac{dz}{E(z)}. \tag{18.35}$$

High-z Approximation and Angular Scales

The integral in (18.35) can be done for the case of a spatially flat Universe with ordinary matter and a cosmological constant. In that case, the angular diameter distance at redshift z is given by

$$D_A(z) = \frac{c}{H_0(1+z)} \int_0^z \frac{dz}{\sqrt{\Omega_m(1+z)^3 + \Omega_\Lambda}}. \tag{18.36}$$

This follows directly from Equation (18.13). We often consider cosmological models made up of a wider variety of constituents than we have been considering here. In some special cases the functions $D_A(z)$ and $D_L(z)$ have closed form analytic expressions (Linder, 1988a,b), but otherwise they must be computed numerically.

In the large z limit (*e.g.* $z \sim 1000$) of this, and for a flat $k = 0$ model, we can neglect the Ω_Λ-term in the expression for $H(z)$ and do the integral to give

$$D_A(z) \simeq \frac{2cH_0^{-1}}{(1+z)} \frac{1}{\Omega_m^{1/2}} \left(1 - \frac{1}{\sqrt{1+z}}\right) \sim \frac{2cH_0^{-1}}{z\,\Omega_m^{1/2}}. \tag{18.37}$$

Notice that the distance scales as $(\Omega_m h^2)^{-1/2}$.

Angular Scales at Recombination

When we look at the surface of last scattering we are looking at a surface of constant time. In this case it makes sense to think of the comoving, or proper, distance as the physical distance separating pairs of points. The angular diameter subtended by the horizon at last scattering, i.e. Equation (18.31) evaluated at $z = 1100$ is

$$\theta_H = \frac{D_H(z_{rec})}{D_A(z_{rec})} \simeq z_{rec}^{-1/2} \simeq 1.73°. \tag{18.38}$$

This shows that the CMB sky is effectively covered with small patches having this angular size that have evolved independently from the Big Bang. The extraordinary degree of isotropy of the CMB shows that these patches are statistically identical, without there being any clear mechanism for such close synchronisation of the cosmic expansion in those different regions. This dilemma was the source of the idea that the Universe went through a very early inflationary phase during which such synchronisation could be achieved.

The angular diameter of a patch that has size L Mpc at a redshift z is then

$$\theta(L) \sim z^{1/2} \, (\Omega_m h^2)^{1/2} \, \frac{L}{8600 \, \text{Mpc}} \quad \text{radians.} \tag{18.39}$$

We notice that if the scale L is very large, then it is roughly co-expanding with the Universe, in which case, $(1+z)L = L_0$, the present day scale. With that, the angle subtended today by a feature of comoving size L_0 at a redshift of $z \sim 1000$, is

$$\theta_{1000}(L_0) \simeq 0.66° \, (\Omega_m h^2)^{1/2} \left(\frac{L_0}{100 \, \text{Mpc}} \right). \tag{18.40}$$

At the epoch of last scattering, we have $2ct \sim 0.24$ Mpc, which would subtend an angle of $\sim 1.6°$ on our sky.

Sound Horizon at Recombination

An important scale is the *sound horizon* at the time of recombination, $\sim c_S(t_{\text{rec}}) \, t_{\text{rec}}$, where $c_S(t)$ is the *adiabatic sound speed*. This is the furthest an adiabatic sound wave of wavelength $\lambda \ll c_S(t) t$ can travel in an expansion timescale. Adiabatic density perturbations of wavelength $\lambda \lesssim c_S(t) t$ do not have time to even perform one acoustic oscillation.[13] Observationally, the scale $\sim c_S(t_{\text{rec}}) \, t_{\text{rec}}$ corresponds roughly to the first peak in the CMB temperature power spectrum.

The adiabatic sound speed in the early Universe depends only on the baryon density and the radiation density. Dark matter and neutrinos do not couple to the baryons or photons:

$$c_S = \frac{c}{\sqrt{3}} (1+R)^{-1/2}, \quad R = \frac{3\rho_B}{4\rho_\gamma}. \tag{18.41}$$

At recombination the value of $R \sim 0.8h^2$ (using values from Table 6.1) and so $c_S(t_{\text{rec}}) \sim 0.45c$. Note that $R \propto (\Omega_B h^2)^{-1}$.

Eisenstein and White (2004) have given a nice expression for the sound horizon at recombination using this expression for the sound speed:

$$s = \int_0^{t_{\text{rec}}} c_S(1+z)dt = \int_{t_{\text{rec}}}^\infty \frac{c_S dz}{H(z)}, \tag{18.42}$$

$$= (\Omega_m H_0^2)^{-1/2} \frac{2c}{\sqrt{3z_{\text{eq}} R_{\text{eq}}}} \ln \frac{\sqrt{1+R_{\text{rec}}} + \sqrt{R_{\text{rec}} + R_{\text{eq}}}}{1 + \sqrt{R_{\text{eq}}}}. \tag{18.43}$$

Here $z_{\text{eq}} = \Omega_m / \Omega_r$ is the redshift at which the matter and radiation components have equal densities.

---

[13] Dodelson (2003, Fig. 8.1, p.218) illustrates the fact that the acoustic fluctuation that generated the second peak of the CMB power spectrum was the longest wavelength to perform one complete acoustic oscillation since entering the horizon. The first, and largest amplitude peak, is due to an over-density on scale $\sim c_S(t) t$ that just reaches its maximum density at the time the Universe becomes neutral.

The subscripts 'eq' and 'rec' on R refer to its value at z_{eq} and z_{rec}. Anderson (2014a) gives the value[14]

$$s = 149.28\,\text{Mpc}. \tag{18.44}$$

The sound horizon is only very weakly dependent on the baryon density, Ω_B, varying as $(\Omega_B h^2)^{-0.08}$, and so this number is almost independent of the present day cosmological parameters.

At the epoch of recombination the densities of the baryons and photons are roughly the same, but the Universe is still matter dominated by a factor of about ~ 7. The redshift as a function of time is approximated at very large z by $z \simeq \frac{3}{2}\left[(\Omega_m h^2)^{1/2} H_0 t\right]^{-2/3}$, which gives the horizon size

$$ct(z_{rec}) \simeq \frac{2}{3}(\Omega_m h^2)^{-1/2} c H_0^{-1} z_{rec}^{-3/2}. \tag{18.45}$$

We previously showed that the Jeans length-scale is a factor ~ 0.4 of ct_{rec}, with only weak h-dependence, i.e. the comoving Jeans length, measured at the present time. is

$$L_{J,0} \simeq 80(\Omega_m h^2)^{-1/2}\,\text{Mpc}. \tag{18.46}$$

With the angular diameter distance $D_{A,rec}$ given in (18.37) we see this scale at recombination subtends an angle of about

$$\theta_{LJ} \sim 0.5^\circ, \tag{18.47}$$

and that the cosmology dependent factor $(\Omega_m h^2)^{-1/2}$ has cancelled. This remark makes this scale particularly effective for cosmography since it is almost independent of the cosmological parameters: it is a standard ruler.

Luminosity Distance

What we mean by *luminosity distance* is the distance that we would assign to an object given knowledge of its intrinsic luminosity L and its apparent brightness. It is measured as the received flux of energy per unit area, F, on a detector. So if, in a Euclidean world, we know a source has luminosity L and we receive a flux F we *assign* a distance D_L such that

$$F = \frac{L}{4\pi D_L^2}. \tag{18.48}$$

In optical astronomy it is conventional to convert F into the logarithmic Pogson magnitude-scale.[15] It must be stressed that this is merely a matter of defining D_L.

Now consider this from the physical point of view of a bundle of photons that is emitted from the source isotropically over the 4π steradians of the source's sky. We need to identify the factors that dilute the flux of radiation as the radiation spreads out from the source.

---

[14] This value is determined using the CAMB (Code for Anisotropies in the Microwave Background) by Lewis and Challinor (`http://camb.info/readme.html`), using fiducial values for the cosmological parameters. See Anderson (2014a, Section 8.2) for an explanation of how this works.

[15] We might specify a wavelength for the flux, but since the Doppler and gravitational redshifts are colour-blind we do not need to do this.

There are three such factors: (a) the ever increasing area over which the radiation is spread; (b) the decrease in energy $E = h\nu$ of the photons due to the cosmological redshift ($E \propto R(t)^{-1}$); and (c) the rate at which the photons are falling onto the detector is decreased due to time dilatation.

First consider (a). We work in our hyper-spherical FLRW metric having radial coordinate ψ, where the spatial geometry is described by the function $S_k(\psi)$. Suppose the source **E** is at $\psi = 0$ and that radiation emitted at t_E is received by an observer **O** at time t_O. From the metric (16.6) the area of the sky at t_O is $4\pi R(t_0)^2 S_k(\psi)^2$ steradians,[16] and so the flux received at t_0 is diluted by a factor $S_k(\psi)^2$ due to the geometry of the Universe.[17]

The two further dilution factors, (b) and (c), are simple: the photon frequency falls off as $(1+z)^{-1}$ and the rate at which photons fall onto the detector also declines as $(1+z)^{-1}$.

Putting these together the received flux is

$$F = \frac{L}{4\pi R(t_0)^2 S_k(\psi)^2}_{(a)} \frac{1}{(1+z)^2}_{(b)(c)}. \qquad (18.49)$$

This must be the value of F which we assign the distance D_L in Equation (18.48) and so we assign a *luminosity distance*

$$D_L = (1+z)\, R(t_0)\, S_k(\psi) \qquad (18.50)$$

to the source. Using an argument like that leading to (18.34) we can write this as a function of redshift, giving the *luminosity distance* to a source at redshift z as:

The *luminosity distance* of the object at redshift z

$$D_L = (1+z)\frac{cH_0^{-1}}{\sqrt{\Omega_k}}\sin\left(\sqrt{\Omega_k}\int_0^z \frac{dz}{E(z)}\right). \qquad (18.51)$$

This is the distance which we would assign to an object of known intrinsic brightness when we want to calculate its apparent brightness.

The same remark applies about what happens when $\Omega_k < 0$.

In the flat cosmology case, $\Omega_k = 0$, this reduces to

$$D_L(z) \overset{\text{flat}}{=} cH_0^{-1}(1+z)\int_0^z \frac{ds}{H(s)}. \qquad (18.52)$$

Relationship between Distance Measures and Magnitudes

The *angular diameter distance $D_A(z)$*, Equation (18.34), and *luminosity distance, $D_L(z)$*, Equation (18.50), are defined as

$$D_A(z) = R(t_O)\, S_k(\psi)(1+z)^{-1}, \qquad (18.53)$$

---

[16] The factor $R(t_0)$ is the current radius of the Universe, and the area would be $4\pi R(t_0)^2$ in a Euclidean Universe. This is modified by the factor $S_k^2(\psi)$ if the metric is a non-flat FLRW.

[17] This can be rigorously derived from the geometric optics approximation to light propagation. See Ellis (1971, 2009b, Section 6.4.) and Ellis (2011, Section 7.4.).

$$D_L(z) = R(t_0)\, S_k(\psi)(1 + z). \tag{18.54}$$

We see that these two measures of distance are related through

$$D_L(z) = (1 + z)^2 D_A(z). \tag{18.55}$$

These distance definitions behave as one would expect. Given an object of intrinsic size D seen at redshift z, it subtends an angle

$$\theta = D/D_A(z). \tag{18.56}$$

Likewise, the flux of radiation, $F(z)$, received from an object of luminosity L, varies inversely as the square of the luminosity distance,

$$F(z) = \frac{L}{4\pi D_L(z)^2}. \tag{18.57}$$

Consider two objects, both with intrinsic luminosity L, seen at redshifts z_1 and z_2. The ratio of the received fluxes is

$$\frac{F(z_1)}{F(z_2)} = \frac{D_L(z_2)^2}{D_L(z_1)^2}. \tag{18.58}$$

We can convert this to an astronomical unit of measure, i.e. magnitude.[18] The measured (bolometric) magnitude difference is

$$m_1 - m_2 = -2.5 \log_{10}(F_1/F_2) = -5.0 \log_{10}(D_L(z_2)/D_L(z_1)). \tag{18.60}$$

This is what is plotted in the magnitude redshift diagram, the z-dependence and the cosmology enter through the expression (18.52) we have developed for $D_L(z)$.

Likewise, if an object of intrinsic (absolute) magnitude M_0 is observed at apparent magnitude m then the object's *distance modulus* $\mu = m - M_0$ is related to the luminosity distance by

$$\mu = 5 \log_{10}\left(\frac{D_L(z)}{1\mathrm{Mpc}}\right) + 25. \tag{18.61}$$

18.2.5 Cosmic Time

Ambiguities also arise when talking about time since there are numerous definitions of 'time' each depending on the coordinate system used to describe the geometry. We have, for example, proper time, conformal time, cosmic time and so on. As with distance, the one to be used depends on the issue that is being addressed. Thus questions like 'how long since the Big Bang was redshift z?', 'how long ago was recombination?' have answers. These are addressed using *cosmic time*.

---

[18] The total brightness of a source integrated over some waveband is measured by optical astronomers in units of $\mathrm{ergs\,cm^{-2}sec^{-1}}$ and is translated into *magnitudes*. The notion of a magnitude was formalised by Pogson (1856). Two sources of brightness, f_1 and f_2, are said to have magnitudes m_1 and m_2 related by

$$m_1 - m_2 = -2.5 \log_{10}(f_1/f_2). \tag{18.59}$$

Thus the magnitude scale is established once a standard has been set. Note that as an object gets fainter its magnitude increases.

The cosmic time that has elapsed since the scale factor was a is defined as

$$t = \int_a^1 \frac{da}{aH(a)}. \qquad (18.62)$$

The right-hand side is time measured in units of the present day Hubble time H_0^{-1}. In using this we are expected to substitute Equation (18.12) for $H(a)$ in the integrand of (18.62). Specific choices of the coefficients Ω_m, Ω_Λ etc., can provide useful analytic solutions or limiting cases, as we shall see next.

18.2.6 Cosmic Time in the Recent Past

Our Universe today is dominated by the Λ dark energy term and cold dark non-baryon matter. Our Universe also appears to be geometrically flat. It is well described by Ω_m and Ω_Λ with $\Omega_m = \Omega_\Lambda = 1$. In this case Equation (18.62) with (18.13) reduces to

$$H_0^{-1} t = \int_0^{a(t)} \frac{da}{a\left(\Omega_m a^{-3} + \Omega_\Lambda\right)^{\frac{1}{2}}}. \qquad (18.63)$$

With this the integral in (18.62) can be done analytically after a few well-chosen transformations (transform to a variable u such that $\sinh^2 u = a^3(\Omega_\Lambda/\Omega_m)$). The result is

$$H_0 t = \frac{2}{3\sqrt{\Omega_\Lambda}} \ln\left[\left(\frac{a}{a_\dagger}\right)^{3/2} + \sqrt{1 + \left(\frac{a}{a_\dagger}\right)^3} \right], \qquad (18.64)$$

where

$$a_\dagger = \left(\frac{\Omega_m}{\Omega_\Lambda}\right)^{\frac{1}{3}} = \left(\frac{\Omega_m}{1 - \Omega_m}\right)^{\frac{1}{3}} \qquad (18.65)$$

is the value of the scale factor at which the Λ-term takes control of the expansion.

The age of the Universe is given by putting $a = 1$ in Equation (18.64) and using Equation (18.65):

$$t_0 = H_0^{-1} \frac{2}{3\sqrt{\Omega_\Lambda}} \ln\left[\frac{1 + \sqrt{\Omega_\Lambda}}{\sqrt{\Omega_m}} \right], \qquad \Omega_m + \Omega_\Lambda = 1. \qquad (18.66)$$

It can be verified that t_0 increases with increasing Λ, and that $t_0 \to \frac{2}{3} H_0^{-1}$ for $\Omega_\Lambda \to 0$.

With the cosmological benchmark values for the parameters, $\Omega_m = 0.3$ and $\Omega_\Lambda = 0.7$ the age of the Universe is $t_0 = 0.94 H_0^{-1}$. With $H_0 = 100h\,\mathrm{km s^{-1} Mpc^{-1}}$ we have $H_0^{-1} = 3.086 \times 10^{17} h^{-1} \mathrm{s} = 9.42 h^{-1}\,\mathrm{Gyr}$. The flat benchmark model with $\Lambda = 0.7, H_0 = 70\,\mathrm{km s^{-1} Mpc^{-1}}$ therefore has an age of 13.47 Gyr. For the latter model the transition to accelerated expansion took place at $a_\dagger = 0.65$, corresponding to a redshift $z_\dagger = 0.53$.

This is the essence of the coincidence that we just happen to be living at a time that is remarkably close to the transition to accelerated expansion. In practical terms this means that by observing the Universe out to redshifts of $z \sim 2$ or more we are probing the switchover to accelerated expansion.

PART IV

THE PHYSICS OF MATTER AND RADIATION

19

Physics of the CMB Radiation

The Cosmic Microwave Background Radiation not only provided the key evidence for what has become our basic view of the Universe, but it has also provided a powerful tool with which to study the Big Bang itself. The radiation we see comes from a redshift of around 1000 and is the result of some complex physical processes that take place in the preceding 'fireball' phase of the cosmic expansion. The tools for studying the early Universe come from an understanding of those physical processes. The goal of this chapter is to explain the physics of the fireball from the point of view of the radiation field, and to present the means by which we analyse the measurements to extract information about the Universe.

These measurements are now being made with extraordinary accuracy over a wide range of angular scales and frequencies. It is that accuracy that allows us to use the CMB as a probe of the physics of the fireball and to determine the values of cosmological parameters with a precision of a mere few percent. In this chapter we present the theory of the radiation field, what we expect from our observations and how we interpret what is seen.

19.1 The Radiation Field

The cosmic fireball is a somewhat unusual environment, far from any laboratory experience.[1] The principal difference lies in the intensity of the cosmic radiation field: in the Universe photons outnumber baryons by some eight orders of magnitude.

Some of the material in this chapter is based on the seminal paper of Sunyaev and Zel'dovich (1970). This is one of the most important papers written on the subject of the CMB.

19.1.1 The Equilibrium Distribution of Photons

The radiation field can be most simply defined by telling how many photons there are of a given frequency moving in a particular direction. However, things are complicated

---

[1] It is perhaps no coincidence that the closest terrestrial experience we have is in the study of nuclear explosions and that among the early contributors to cosmology were scientists working in nuclear research: Gamow, Teller, Zel'dovich and Sakharov to name but a few. The ground-breaking book of Zel'dovich and Raizer (1967) is largely about the physics of nuclear fireballs and is a superb treatise on the issues presented here.

somewhat because the radiation field is measured in terms of the flux of energy carried across a unit surface by photons of a given frequency moving into a given solid angle about a direction normal to the surface. For radiation in thermodynamic equilibrium at some temperature, this flux follows the famous Planck law. In this section we discuss the concept of flux and then relate it to the equilibrium distribution function for the photons and so derive the Planck law from basic principles.

Of course the derivation assumes that there is a fundamental distribution function for particles like photons, the Bose–Einstein distribution. The details of this can be found in many standard textbooks on statistical physics.

19.1.2 Flux

Denote the number of photons in the frequency band ν to $\nu + \delta\nu$ contained in the volume element δV and having a direction of motion within an element of solid angle $\delta\Omega$ about a unit vector $\mathbf{\Omega}$ by

$$f(\nu, \mathbf{\Omega})\delta V \delta\Omega \delta\nu. \tag{19.1}$$

Consider a time interval δt and cone of length $c\delta t$ having solid angle $\delta\Omega$ about the direction $\mathbf{\Omega}$. In the time interval δt, all the photons having direction of motion in the cone, which has volume $\delta V = (c\delta t)^2\delta\Omega.c\delta t$, pass through the cap, which has area $(c\delta t)^2\delta\Omega$. The number of photons having frequencies in the range ν to $\nu + \delta\nu$ and lying in the cone with direction of motion into the solid angle $\delta\Omega$ is $f(\nu, \mathbf{\Omega})\delta V \delta\Omega \delta\nu$ by definition of f.

19.1.3 Spectral Radiation Intensity

Since each of these photons has energy $h\nu$, h being Planck's constant, the flux of energy through the cap from the photons travelling in this set of directions is

$$h\nu f (c\delta t)^2\delta\Omega \, c\delta t \, \delta\Omega \delta\nu. \tag{19.2}$$

So the flux of energy per unit area and per unit time is obtained by dividing by the area of the cap of the cone, $(c\delta t)^2\delta\Omega$, and by the time interval δt:

$$I_\nu(\mathbf{\Omega}) \, \delta\nu\delta\Omega = h\nu f(\nu, \mathbf{\Omega}) \, \delta\Omega \delta\nu. \tag{19.3}$$

The quantity I_ν is called the *spectral radiation intensity*, and the radiation field is completely specified by either I_ν or f. These quantities can of course be functions of position $\mathbf{r}$ and time t, as well as frequency, ν, and direction, $\mathbf{\Omega}$. Our goal is to calculate this for a distribution of photons in equilibrium at temperature T. The unit of solid angle is the *steradian*, abbreviated to 'sr'. The units in which I_v is measured are $\text{erg s}^{-1} \text{cm}^{-2} \text{Hz}^{-1} \text{sr}^{-1}$.

19.1.4 Spectral Radiant Energy Density

In general I_v will be a function of direction, and we shall discuss cases later where this is so. We will then find that the angular moments of the distribution of the radiation are useful quantities. The isotropic component of I_v is of most importance and this is obtained by averaging I_v over all solid angles:

$$U(\mathbf{r}, t) = h\nu \int_{4\pi} f d\Omega = \frac{1}{c} \int_{4\pi} I_v d\Omega.$$

The first of these shows readily that U is to be interpreted as the *spectral radiant energy density*, or more simply *spectral radiance*. The spectral radiant energy density is measured in *Jansky*, where $1 \, \text{Jy} = 10^{-23} \, \text{erg s}^{-1} \, \text{cm}^{-2} \, \text{Hz}^{-1} = 10^{-26} \, \text{W m}^{-2} \, \text{Hz}^{-1}$.

19.1.5 The Planck Distribution

Photons are mass zero bosons and there are two polarisations or spin states (left-handed or right-handed). This means that in a gas of photons in thermodynamic equilibrium at temperature T the number density of photon modes $n(\nu)d\nu$ having energy in the range $h\nu$ to $h(\nu + \delta\nu)$ is given by the Bose–Einstein distribution for zero chemical potential:

$$n_{th}(\nu) = \frac{1}{\left(e^{h\nu/kT} - 1\right)}, \qquad (19.4)$$

where T is the temperature of the distribution. This is sometimes referred to as the *Planck radiation law*.[2] It is usual to introduce a dimensionless variable for the frequency, normalising it by the temperature T:

$$x = \frac{h\nu}{kT}, \quad \eta(x) = \frac{1}{e^x - 1}. \qquad (19.5)$$

---

[2] The first appearance of the Planck radiation law was in 1900 (Planck, 1900a,b): in essence this was simply a fitting formula to the high and low frequency limits that were already known. The introduction of Planck's constant, h, came in Planck (1901, see Section 10 and Equation (16) where Planck gives the value of h). The notion that light was made up of discrete quanta came not with Planck, but with Einstein and his paper on the photoelectric effect (Einstein, 1905). However, a full derivation of the Planck law only came with the paper of Einstein (1917b). This very important paper is discussed in detail by Lewis (1973), and in the context of modern quantum many body theory by Derlet and Choy (1996). See Kragh (2000) for a detailed history of the Planck law.

Information 19.1.5

When dealing with black-body radiation, the function $\eta(x)$ appears in a variety of integrals that are of the form

$$\int_0^\infty \frac{x^{n-1}dx}{e^x - 1} = \zeta(n)\Gamma(n), \tag{19.6}$$

where $\zeta(x)$ is the Riemann zeta function and $\Gamma(x)$ is the gamma function. These are well-tabulated and well-known functions. For integer n, $\Gamma(n+1) = n!$. Relevant values, in the present context, for $\zeta(x)$ are $\zeta(3) = 1.202057$ (Apery's constant[a]), $\zeta(4) = \pi^4/90 = 1.0962$.

[a] The Riemann ζ-function is of central importance in number theory and in statistical physics. The probability that three (uniformly) randomly chosen positive integers less than some N are relatively coprime tends to $1/\zeta(3)$ as $N \to \infty$.

Now we need to calculate the fraction of these that are moving in a given direction, per unit time. This will be the photon flux. Consider a cone of length $c\delta t$ subtending a solid angle $\delta\Omega$ about a given direction. The area of the cap of the cone is $(c\delta t)^2\delta\Omega$ and the volume of the cone is $V = (c\delta t)^2\delta\Omega\, c\delta t$. The number of modes of frequency ν that can fit in the volume[3] is $2V\nu^2\delta\nu/c^3$, the number 2 accounting for the fact that there are two possible spin states for the photon. Thus the number of photons in the volume that will cross the cap of the cone in time δt is

$$\text{photons per mode} \times \text{modes in cone} = n(\nu) \times 2V\frac{\nu^2\delta\nu}{c^3}, \quad V = (c\delta t)^2\,\delta\Omega.c\delta t. \tag{19.7}$$

The number of photons crossing the surface per unit area and per unit time is thus

$$n(\nu)\frac{2\nu^2\delta\nu}{c^3}c. \tag{19.8}$$

Each photon carries energy $h\nu$ and so the flux of energy across the surface is

$$I_\nu = \frac{2h\nu^3}{c^2}\frac{1}{e^{h\nu/kT} - 1}. \tag{19.9}$$

This is the standard *Planck spectrum* for a radiation field in thermodynamic equilibrium at temperature T. It is also referred to as *Planck's law*. Note that in some texts, I_ν is denoted by $B_\nu(T)$ and is called the *brightness temperature*.

Sometimes it is useful to express the Planck spectrum as a function of wavelength, λ, rather than frequency:

[3] This follows from the quantum mechanics argument that the number of photon states in energy interval $d(h\nu)$ and with momentum $\mathbf{p}$ directed along the unit vector $\mathbf{k}$ into the interval $d\Omega$ of solid angle is

$$dN = \frac{2V}{h^3}d^3\mathbf{p} = 2V\frac{(h\nu)^2}{(hc)^3}d(h\nu)d\Omega.$$

The density $\rho(\nu, \mathbf{k}) = 2(h\nu)^2/(hc)^3$ will in general depend both on frequency and direction.

$$I_\lambda = \frac{2hc^2}{\lambda^5} \frac{1}{e^{hc/kT\lambda} - 1}. \tag{19.10}$$

The frequency for which I_ν (the *brightness*) reaches a maximum is

$$\nu_m \approx \frac{3kT}{h} \approx 6 \times 10^{10} T \ \text{Hz}, \tag{19.11}$$

corresponding to a wavelength

$$\lambda_m \approx 0.51 T^{-1} \ \text{cm}. \tag{19.12}$$

If I_ν is expressed as a function of wavelength, the maximum of the Planck curve lies at a wavelength of

$$\lambda_m \approx 0.29 T^{-1} \ \text{cm}. \tag{19.13}$$

Black-body Intensity

The *total black body intensity* is obtained by integrating the brightness over all angles and all frequencies:

$$I = \frac{1}{\pi}\sigma T^4 = 1.8046 \times 10^{-5} \ T^4 \ \text{erg s}^{-1}\text{cm}^{-2}\,\text{steradian}^{-1}, \tag{19.14}$$

$$\sigma = \frac{2\pi^5 k^4}{15 c^2 h^3} = 5.669 \times 10^{-5} \text{erg s}^{-1}\text{cm}^{-2}\,\text{K}^{-4}, \tag{19.15}$$

where σ is the Stefan–Boltzman constant.

Radiation Energy Density

The *radiation energy density* (integrated over all frequencies) is then

$$U = \frac{8\pi h}{c^3} \int_0^\infty \frac{\nu^3 d\nu}{\exp(h\nu/kT) - 1} = \frac{8\pi k^4 T^4}{h^3 c^3} \int_0^\infty \frac{x^3 dx}{e^x - 1} = aT^4, \quad a = \frac{4\sigma}{c}. \tag{19.16}$$

a is simply called the radiation density constant.[4] The value of a is variously expressed depending as to whether temperature is measure in Kelvin or electron volts,

$$a = 7.564 \times 10^{-15} \text{erg cm}^{-3}\text{K}^{-4} = 137 \ \text{erg cm}^{-3} \ \text{eV}^{-4},$$

with other variations for other systems of units.

---

[4] It is interesting to note that for neutrinos the calculation is identical except for two points: (a) there are three types of neutrino and their antiparticles; and (b) neutrinos obey Fermi–Dirac statistics rather than Bose–Einstein statistics. The first of these replaces the 2 counting the number of independent particle types in the above by a 6, and replaces the $-$ sign in $e^x - 1$ by a $+$ sign. With those changes the contribution to the energy density from the neutrinos when in equilibrium at temperature T is $\frac{21}{8}T^4$ (see Section 7.1.3).

Equivalent Antenna Temperature

The equivalent *antenna temperature* T_a is related to the radiation brightness I_v by

$$T_a \equiv \frac{c^2 I_v}{2k v^2}, \tag{19.17}$$

and hence to the thermodynamic temperature T via

$$kT_a = \frac{hv}{\left(e^{hv/kT} - 1\right)}. \tag{19.18}$$

At low frequencies $hv \ll kT$ the antenna temperature is a constant independent of frequency for a true Planck spectrum:

$$T_a = T\left(1 - \frac{1}{2}\frac{hv}{kT}\right). \tag{19.19}$$

This relationship holds for frequencies lower than that of the peak of the spectrum where the observations are easiest. Fitting this relationship to the data enabled early observers to check the expected deviation from the Rayleigh–Jeans law $T_a = T$ and thus establish the Planck law. It was not until many years later that observations were performed at frequencies higher than the peak of the spectrum.

Rayleigh–Jeans and Wein

There are two special limits for the Planck law: the low frequency limit where

$$I_v = 2\frac{v^2 kT}{c^2}, \qquad hv \ll kT, \tag{19.20}$$

and the high frequency limit where

$$I_v = \frac{2hv^3}{c^2}e^{-hv/kT}. \tag{19.21}$$

The low frequency limit is referred to as the *Rayleigh–Jeans law* and the high frequency limit as the *Wien law*.

19.1.6 Temperatures

Consider a distribution of photons, and for simplicity suppose that the directions of motion of the photons are randomly and isotropically distributed. Suppose further that the number of photons per unit frequency is given by the function $\eta(v)$. We define the *thermodynamic temperature* of this set of photons by the equation

$$T_r = \frac{1}{4}T_0 = \frac{\int_0^\infty x^4 \eta(1 + \eta)dx)}{\int_0^\infty x^3 \eta dx}. \tag{19.22}$$

This somewhat intimidating equation describes the temperature of a distribution of test electrons in equilibrium with the radiation field. The importance of this is that it is the temperature that an electron 'sees' in the sense that thermodynamic equilibrium between photons and a thermal gas of electrons can only be achieved when $T_r = T_0$. With that condition there is no net heat transfer between the electrons and photons.

For the specific case where the photon distribution is the *Planck law*,

$$\eta(\nu) = \frac{1}{e^{h\nu/kT_0} - 1},$$

(19.23)

and it can be verified by direct substitution and integration that

$$T_r = T_0.$$

(19.24)

In other words, the temperature T_0 appearing in the Planck law (19.23) is just the equilibrium temperature of the radiation field.[5] A well-known result seen from a slightly different point of view.

19.1.7 Equilibrium Photon Gas in an Expanding Universe

During some phases of the cosmic expansion, cosmic background photons are neither created nor destroyed. They are simply scattered if there are any free electrons around, and redshifted by the cosmic expansion. The frequencies change like $\nu \propto (1 + z)$, and so from Equation (19.23) the temperature assigned to the photon distribution must change like

$$T_0 \propto (1 + z).$$

(19.25)

If physical processes intervene to change $\eta(\nu)$, or to create photons, the Planck law is not preserved and this result is no longer necessarily true.

There is direct observational evidence that the temperature does have this dependence on the redshift. This is illustrated in Figure 19.1 which is a compilation of results in Muller *et al.* (2013). The best fit line for the data shown gives $T(z) \propto (1 + z)^{1-\alpha}$ with $\alpha = +0.009 \pm 0.019$.[6]

19.1.8 Photon Density and Photon to Baryon Ratio

The spatial density of photons in a Planckian distribution at temperature T_0 is

$$n_\gamma = \frac{8\pi}{c^3} \int \frac{\nu^2 d\nu}{e^{h\nu/kT_0} - 1}.$$

(19.26)

(This follows because there are $\nu^2 d\nu d\Omega/c^3$ modes of a given polarisation per unit volume moving into a solid angle $d\Omega$. There are 4π steradians in a sphere and there are two polarisations.) Doing the integral (see Equation 19.6), and putting in the numbers we have

$$n_\gamma = \Gamma(3)\zeta(3)\frac{1}{\pi^2}\left(\frac{kT}{\hbar c}\right)^3 = 0.2436\left(\frac{kT}{hc}\right)^3 = 20.284\left(\frac{T}{1\,\mathrm{K}}\right)^3.$$

(19.27)

---

[5] We shall see later, in Section 19.3, why this particular distribution arises in a plasma of photons and electrons.

[6] Saro *et al.* (2014), using the South Pole Telescope (SPT) and a more extensive set of S-Z data, derive an even tighter limit of $\alpha = 0.005 \pm 0.012$ using a combined data set.

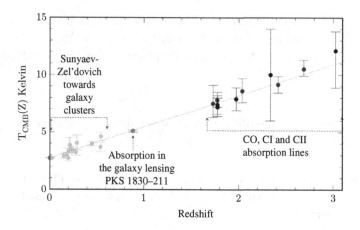

Fig. 19.1   CMB temperature measured at different redshifts. Three different methods have been used in the different redshift bands. The dotted line is the line that would be expected if the temperature varied with redshift as $T \propto (1 + z)$. The figure is an edited version of Muller *et al.* (2013, Fig. 5).
(Reproduced with permission from A&A.)

If we evaluate this at the present time when $T = 2.725$ K we have

$$n_{\gamma,0} \simeq 410 \, \text{cm}^{-3}, \quad T_0 = 2.725 \text{ K}. \tag{19.28}$$

We notice that

$$n_\gamma \propto T_0^3 \propto (1 + z)^3, \tag{19.29}$$

and so since a co-expanding volume scales as $(1 + z)^{-3}$, the number of photons per unit co-expanding volume is conserved with the expansion. This is of course just as it should be, since scattering neither creates nor destroys photons.

Since the number of baryons in a co-expanding volume is also conserved ($n_B \propto (1 + z)^{-3}$), the photon to baryon ratio is a constant. This is expressed through either of the quantities:

Photon to baryon ratio:

$$\frac{n_\gamma}{n_B} \simeq 3.63 \times 10^7 (\Omega h^2)^{-1}, \quad \eta_{10} = 10^{10} \frac{n_B}{n_\gamma} = 275(\Omega_b h^2), \tag{19.30}$$

for $T_\gamma = 2.725$ K. For the WMAP value of $\Omega_b h^2 = 0.02223$ we have $\eta_{10} = 6.11$ There are far more photons in any volume than there are baryons![7]

At the present epoch the Universe is transparent, the photons do not interact significantly with the matter and the fact that there are so many photons is of little importance. At earlier epoch, however, when the electron density was very high, the photons and electrons collided rapidly and thermal equilibrium was established. The fact that there was a far larger number of photons than electrons means that it is the matter that adjusts itself to the

---

[7] This is a significant aspect of the cosmic plasma that differs from the familiar terrestrial plasmas.

thermal history of the radiation rather than vice versa. This is best expressed in terms of the heat capacities of the electron and photon gases, which we will come to in the next section.

19.2 The Fireball Phase

In this section we will concentrate on the interaction between the baryonic and photon components of the Universe: the collisionless neutrinos and dark matter play no role in this interaction except insofar as they influence the cosmic expansion rate.

Once the light elements have formed at a redshift $z \sim 10^9$ the Universe expands and cools until it becomes largely neutral at a redshift $z \sim 10^3$. The cosmic plasma during this period of time is characterised by being completely ionised, and the pressure is dominated by Thomson scattering of the cosmic background photons off the free electrons:

$$p = \frac{1}{3}aT^4 + n_B kT. \qquad (19.31)$$

The first term is the radiation pressure (a is the radiation constant) and the second is the gas pressure (k is the Boltzmann constant). The ratio of these terms is $aT^3/3kn_B$, and during this period is independent of time and large: it is on the order of the number of photons per baryon, which we saw in Equation (19.30) was $\sim 10^8$.

The scattering time for photons is

$$t_{\text{Thomson}} = \sigma_T n_e c, \qquad (19.32)$$

where

$$\sigma_T = \frac{8\pi}{3}\left(\frac{e^2}{m_e c^2}\right)^2 = 6.65 \times 10^{-26}\,\text{cm}^2 \qquad (19.33)$$

is the Thomson cross-section, and n_e is the free electron number density.[8] During the fireball phase the ionisation is complete and so this timescale is very short in comparison with the cosmic expansion timescale.

In general, the spectrum of the photons is governed by two processes: the Thomson scattering between electrons and photons and the free-free absorption (Bremsstrahlung)

---

[8] The Thomson cross-section describes the scattering of light off electrons in the limit where the wavelength of the light, c/ν, is longer than the Compton wavelength $h/m_e c$ of the electron (see Equation 19.56). At shorter wavelengths, i.e. higher frequencies, the cross-section falls below the Thomson cross-section because of quantum effects (see Equation 19.54). We then speak of *Compton scattering*. The high frequency condition is equivalent to $h\nu > m_e c^2$, i.e. the energy of the photon is greater than the rest mass of the electron. One way of describing Compton scattering is to say that it is a generalisation of Thomson scattering, taking account of recoil in the case that the the photon is relativistic.

The usage of the names Compton and Thomson scattering is frequently mixed up, which is not surprising since they are different aspects of the same phenomenon. Peebles (1993, p.581 *et seq.*) decided to refer to the process as Compton–Thomson scattering.

and emission of photons by electron pair scattering. The equation describing this process is the Kompaneets equation, which we can write down symbolically as

$$\frac{\partial \eta}{\partial t} = \left(\frac{\partial \eta}{\partial t}\right)_{\text{Thomson}} + \left(\frac{\partial \eta}{\partial t}\right)_{\text{Bremsstrahlung}} . \tag{19.34}$$

In equilibrium, these processes balance to produce the Planck spectrum (19.23). In the fireball phase of the Universe these processes are very rapid when compared with the cosmic expansion timescale. This means that the photons and electrons can rapidly come into thermal equilibrium and can relax rapidly towards the Planck distribution.

Thus the fireball provides the *mechanism* for establishing the accurately Planckian distribution of the microwave background radiation that is observed today (see Figure 19.2). This is a very important point: the accuracy of the Planckian fit to the data demands that the Universe was far denser in the past than it is now. Any theory of the Universe that sought to revive ideas that the Universe did not start from a hot dense phase would have to face the problem of explaining why the cosmic background radiation is so accurately Planckian.

The rate of free-free absorption and emission can be calculated to be

$$t_{\text{ff}} \simeq 1.4 \times 10^{24} (\Omega_B h^2) T^{-5/2} \text{s}, \tag{19.35}$$

and so the Planck law must be established during periods when this is less than the cosmic expansion timescale, t_{exp}. This happens when $t < 10^{20} T^{-2}$ s, i.e. when

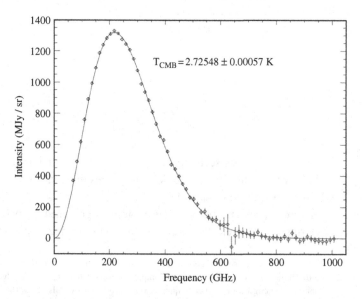

Fig. 19.2 The spectrum of the Cosmic Microwave Background Radiation (Fixsen, 2009). The points are the data from the FIRAS experiment on board the COBE satellite, but recalibrated using data from the WMAP experiment. The curve is not a fit to the data but the Planck curve at 2.725 K derived from the earlier FIRAS data calibrations. The result of $T_{CMB} = 2.72548 \pm 0.00057$ K is the weighted average of previous data, most of that weight coming from FIRAS itself.
Adapted from Fixsen (2009, Fig. 3). *(Reproduced by permission of the AAS.)*

$$T_{\text{Planck Law}} > 2 \times 10^8 \text{ K}. \tag{19.36}$$

If for some reason the spectrum of the radiation were perturbed substantially after that time, we would see today deviations from the Planck spectrum. This argument can be turned around and we can use the observed accuracy with which the Planck law fits the data to constrain the thermal history of the Universe.

The best data on the accuracy of the Planck law comes from the COBE satellite FIRAS experiment (Fixsen (2009), see Figure 19.2[9]).

19.2.1 A Universe with Gas and Radiation

In relativistic gas dynamics, there are some important technicalities that arise when dealing with a gas under pressure. The pressure in a gas is due to the random motions of particles relative to the frame of reference in which the fluid is locally at rest. This kinetic energy in random motions gives rise to relativistic corrections to the equations of motion, and these are best expressed in terms of the energy–momentum tensor for the matter, since that is the source of the Einstein field equations.

When talking about the 'density' of a parcel of gas we thus recognise two contributions: one from the rest mass energy density that is simply a matter of counting n particles of rest mass m so that $\rho = nm$, and the other due to the random motions whose mean square velocity is proportional to kT. We define a quantity μ, the *energy density*[10] which is the sum of these two contributions:

$$\mu = \rho + \epsilon. \tag{19.37}$$

For a perfect gas of adiabatic index γ,

$$\epsilon = \frac{1}{\gamma - 1} kT, \tag{19.38}$$

$$= \frac{3}{2} nkT, \qquad \gamma = \frac{5}{3}. \tag{19.39}$$

This last formula is familiar from undergraduate courses in the thermodynamics of the perfect gas. In relativity, the equation of state is given as a relationship of the form

$$p = p(\mu), \tag{19.40}$$

and for the dust case this is simply $p = 0$.

Now consider matter which is a mixture of baryons and radiation. If the mixture is at one temperature T, and the baryons have number density n, we have

$$p = nkT + \frac{1}{3} aT^4, \tag{19.41}$$

(the total pressure is the sum of the partial pressures of the constituents).

---

[9] The units on the vertical axis, MJy/sr, are Mega-Jansky per steradian. The Jansky, Jy, is defined in Section 19.1.4

[10] Technically, μ just is the 00-component, T_{00}, of the energy momentum tensor T_{ab}. The way in which the rest mass density ρ and the internal energy contribute is a result from relativistic kinetic theory.

Remark: The equation for the conservation of energy in general relativity takes the form

$$\frac{d\mu}{dt} + 3\frac{a}{a}\left(\mu + \frac{p}{c^2}\right) = 0. \tag{19.42}$$

In the case of a zero pressure gas this reduces to the equation of conservation of baryons in the volume under consideration.

This equation can be understood by noting that the total energy within a sphere of radius ℓ and mass M is Mc^2 (that is the relativistic bit). As the sphere expands into the surrounding medium it must do some 'pdV' work and hence the energy budget is expressed as

$$\frac{d}{dt}Mc^2 = -p\frac{dV}{dt}. \tag{19.43}$$

Relating M and V to ℓ via $M = (4\pi/3)\mu\ell^3$ and $V = (4\pi/3)\ell^3$ we get

$$\frac{d}{dt}\frac{4\pi}{3}\mu\ell^3 c^2 = -4\pi l^2 p\frac{d\ell}{dt}, \tag{19.44}$$

after which

$$\frac{d\mu}{dt} + 3\frac{\dot{\ell}}{\ell}\left(\mu + \frac{p}{c^2}\right) = 0. \tag{19.45}$$

Replacing ℓ with the scale factor reproduces the relativistic cosmic energy equation.

Now let us look at μ, the total mass energy density. If the particle mass is m then the density of particles is mn and the density of radiation is aT^4. The total energy density must, however, contain a kinetic energy contribution from the random (thermal) motion of the particles. The mean square random velocity of the particles at temperature T is approximately nkT/m, so the kinetic energy density is roughly nkT. Collecting these up, the total mass-energy density is just

$$\mu = nm + \frac{nkT}{\gamma - 1} + aT^4. \tag{19.46}$$

The quantity γ is the ratio of the specific heats of the gas particles, and is just $\gamma = 5/3$ for a simple monatomic gas. This factor follows from the theory of gases.

Remark: The specific heat at constant pressure, C_P, and the specific heat at constant volume, C_V, are related by $C_P = C_V + Nk$ where N is the number of particles per unit mass and $k = 1.38 \times 10^{-16}$ erg K^{-1} is Boltzmann's constant. If f is the number of degrees of freedom of a molecule of the gas, then it can be shown that

$$\gamma = \frac{C_P}{C_V} = \frac{2}{f} + 1. \tag{19.47}$$

For a monatomic gas, $f = 3$ giving $\gamma = 5/3$, whereas for a gas of rigid diatomic molecules in which the vibrational degrees of freedom are not excited, $f = 5$, giving $\gamma = 7/5$. Allowing the vibrational degrees of freedom to be fully excited gives $f = 7$

and $\gamma = 9/7$. For a gas of relativistic particles, or of photons, $\gamma = 4/3$. (It is useful to know that the specific heat of a perfect gas at constant volume is just $C_V = R/(\gamma - 1)$, where $R = k/m_p$ is called the *gas constant*.)

These are ideal cases. In the real world the specific heats of a gas need not be constants (such as would happen when a molecular gas is dissociating) and γ has to be defined via $pV = (\gamma - 1)U$ and a detailed knowledge of the equation of state.

Now we simply substitute for μ and p in the energy balance equation and get

$$\frac{a}{T}\frac{dT}{da} = -\left[\frac{\sigma + 1}{\sigma + \frac{1}{3}(\gamma - 1)^{-1}}\right], \tag{19.48}$$

where

$$\sigma = \frac{4aT^3}{3nk} \tag{19.49}$$

is called the *photon entropy per gas particle*. Note that it depends on the temperature T, but we shall see that for most of the period of cosmic evolution that concerns us $T \propto (1 + z)$ and $n \propto (1 + z)^3$ and hence σ is a constant. We have approximately

$$\sigma = \frac{4aT_0^3}{3n_0 k} \simeq 10^8 (\Omega h^2)^{-1}, \tag{19.50}$$

where the last equality refers to a cosmological context in which the Ω value is the contribution of the gas particles to the total cosmic mass density, given the rate h of cosmic expansion. σ is indeed a big number and as we shall see later it is roughly the ratio of the number of photons to baryons per unit volume.

The temperature history $T_r(z)$ of the radiation component changes with the expansion according to

$$T_r \propto (1 + z). \tag{19.51}$$

The temperature history of the matter $T_m(z)$ depends on whether or not the matter is coupled to the radiation field. If there is a strong interaction between the two components, as when the Universe was ionised during the fireball phase, then the matter temperature will follow the radiation temperature exactly. The matter particles are overwhelmed by the number of photons: the matter is in a heat bath. (The ratio of the specific heats of the matter and the radiation is approximately $nk/aT_0^3 \sim 10^{-8}\Omega_B$.) Once there is no longer any interaction between photons and matter, the matter temperature can evolve as a gas with $\gamma = 5/3$ and $T_m \propto (1 + z)^{3(1+\gamma)}$, independently of the radiation, which is cooling down as $T_r \propto (1 + z)$.

19.2.2 Electron–Photon Scattering

When the Universe was hot and ionised, the interactions between the electrons and photons played a major role in establishing the Planck equilibrium spectrum of the radiation field. There are a number of atomic processes contributing to this, but the dominant one, as we shall see, is the simple scattering of photons off free electrons. The details of this scattering process are worked out in some detail in Section 21.2.4.

Thomson Scattering

The scattering of unpolarised photons of energy $h\nu \ll m_e c^2$ off non-relativistic free electrons is described by the cross-section:

$$d\sigma = \frac{1}{2}\left(\frac{e^2}{m_e c^2}\right)(1 + \cos^2\theta)d\Omega. \tag{19.52}$$

This measures the probability that a low-energy photon ($h\nu \ll m_e c^2$) interacting with a stationary electron will be scattered into a solid angle $d\Omega$ about a direction making an angle θ with its initial direction.

Remark: The differential cross-section at high energies is given by the *Klein–Nishina* formula:

$$d\sigma = \frac{1}{2}\left(\frac{e^2}{m_e c^2}\right)\left(\frac{\nu}{\nu'}\right)\left[\frac{\nu}{\nu'} + \frac{\nu'}{\nu - \sin^2\theta}\right]. \tag{19.53}$$

The corresponding total cross-section is then

$$\sigma_{KN} = \sigma_T\left[1 - 2\left(\frac{h\nu}{m_e c^2}\right) + \frac{26}{5}\left(\frac{h\nu}{m_e c^2}\right)^2 + \cdots\right],$$

for $h\nu \ll m_e c^2$.

The total cross-section is obtained by integrating (19.52) over all solid angles:

Thomson cross-section:

$$\sigma_T = \frac{8\pi}{3}\left(\frac{e^2}{m_e c^2}\right)^2 = 6.65 \times 10^{-25}\,\text{cm}^2. \tag{19.54}$$

It is referred to as the *Thomson cross-section*. Note that σ_T is, in this low frequency limit, independent of the photon frequency.

Energy is transferred to the electron in such collisions. Applying simple conservation of energy and momentum shows that the emerging photon has frequency:

Compton formula for frequency shift of the scattered radiation

$$\nu' = \nu\frac{m_e c^2}{m_e c^2 + h\nu(1 - \cos\theta)} \approx \nu + \frac{h\nu^2}{m_e c^2}(1 - \cos\theta), \quad h\nu \ll m_e c^2. \tag{19.55}$$

This corresponds to a wavelength change by

$$\delta\lambda = \lambda_c(1 - \cos\theta) \quad \text{where} \quad \lambda_c = \frac{h}{m_e c} = 2.426 \times 10^{-10}\,\text{cm}. \tag{19.56}$$

λ_c is called the *Compton wavelength* of the electron. For a non-relativistic photon ($h\nu \ll m_e c^2$), we have approximately $\nu' = \nu$.

Note that the cross-section for scattering of photons off protons is a factor $(m_e/m_p)^2$ smaller, and so is entirely negligible in comparison. Note also that if the electrons are not free, but bound to atoms, the Compton formula still holds for photons whose energy is substantially larger than the binding energy of the electron in its atom.

> **Remark:** Things get a little more complicated if the electron is not at rest. When the electron is moving relativistically the process is referred to as *inverse Compton scattering*, because the photon gains energy from the electron. If an electron of energy $\gamma m_e c^2$ collides with a photon of energy $h\nu$, then the photon is boosted to an energy:
>
> $$h\nu' \simeq \gamma^2 h\nu, \qquad \gamma h\nu \ll m_e c^2,$$
> $$\simeq \gamma m_e c^2, \qquad \gamma h\nu \gg m_e c^2.$$
>
> Note that in the second limit, all the electron energy is transferred to the scattered radiation, regardless of the incident photon energy. The relevant cross-section is the Klein–Nishina cross-section.

The importance of electron–photon scattering in the Universe can be gauged by seeing how often a photon is likely to scatter during the time it takes the Universe to expand. The Thomson interaction between electrons and photons is a phenomenon of the early Universe.

Radiation Drag

Now consider the scattering process described in the previous section from the point of view of an electron moving at velocity v relative to an isotropic radiation field of temperature T. (The radiation field determines a preferred frame of reference.) From the point of view of the electron, the temperature distribution is

$$T(\theta) = T \left(1 + \frac{v}{c} \cos\theta \right). \tag{19.57}$$

The radiation field ahead of the electron motion is brighter than behind it, and so the electron experiences a retarding force. We can calculate the force on the electron. This force acts in the direction of motion of the electron and opposes the motion.

The momentum flux per steradian carried by photons coming from a direction θ is just

$$p(\theta) = \frac{aT(\theta)^4}{4\pi}, \tag{19.58}$$

(integrated over all frequencies). The cross-section for collisions is just σ_T, and so the force exerted on the electron is

$$F = -\int \sigma_T \cdot \frac{aT(\theta)^4}{4\pi} \cdot \cos\theta \, d\Omega = -\frac{4}{3}\sigma_T aT^4 \frac{v}{c}, \tag{19.59}$$

where we have used the previous expression for $T(\theta)$. This last formula describes what is referred to as the *radiation drag* on the electron.

The Eddington Limit

As a simple illustration of Thomson scattering, consider a plasma consisting of electrons and protons surrounding a star that is radiating with luminosity L. The radiation energy density is

$$U = \frac{L}{4\pi r^2 c},$$

(19.60)

and the force exerted on an electron is just $\sigma_T U$.[11] This is the radiation pressure.

The forces acting on an electron are the gravitational field of the star, the radiation pressure, and the electrostatic force due to the electric field of the protons:

$$F_e = \frac{GMm_e}{r^2} - \sigma_T U - eE_{ep}.$$

(19.61)

Similarly we can write a proton equation:

$$F_p = \frac{GMm_p}{r^2} - \sigma_T U \left(\frac{m_e}{m_p}\right)^2 + eE_{ep}.$$

(19.62)

If we set $F_e = F_p$, we can eliminate E_{ep}, and hence find the force on an electron or proton:

$$F_e = F_p = \frac{1}{2}\left(\frac{GMm_p}{r^2} - \sigma_T U\right),$$

(19.63)

where terms of order $(m_e/m_p)^2$ have been neglected. There is an interesting limit where

$$U = \frac{GMm_p}{\sigma_T r^2}.$$

(19.64)

Turning U back into luminosity with (19.60), we see that there is a maximum luminosity that an accreting star can have:

$$L_{Edd} = \frac{4\pi GMm_p c}{\sigma_T} \approx 10^{38} \frac{M}{M_\odot} M_\odot,$$

(19.65)

($M_\odot$ is the mass of the Sun). This is the maximum luminosity for stable stars.

Compton Cooling

If the electrons have temperature T_e, then the typical electron velocity v is given by $m_e v^2 = 3kT_e$. These electrons are slowing down due to the drag force and giving up their energy to the radiation field. The rate at which the electrons lose energy is

$$\frac{d}{dt}\frac{1}{2}v^2 = vF = -\frac{4}{3}\sigma_T ac T^4 \left(\frac{v}{c}\right)^2 = 4\sigma_T ca T^4 \left(\frac{kT_e}{m_e c^2}\right).$$

(19.66)

Since the typical electron has energy $3kT_e$, the timescale to lose its energy by virtue of the radiation drag is:

---

[11] The electron presents an area σ_T to the stream of radiation, and so a volume $\sigma_T c$ of radiation is effectively absorbed by the radiation per unit time. The momentum of that column of radiation is just U/c, and so the rate of absorption of momentum is $(U/c)\,\sigma_T c$.

Compton radiation drag timescale:

$$t_C = \frac{3m_e c}{4\sigma_T a T^4}.$$ (19.67)

This is the timescale on which electrons relax from a temperature T_e to come into equilibrium with the radiation field at temperature T. If the electrons are hotter than the radiation, this is effectively a cooling time.

Of course, not only do the electrons cool to the radiation temperature (if $T_e > T$), but the photon energy distribution is changed by virtue of that same interaction. The photons gain in energy. In each collision, a photon changes its energy by an amount corresponding to a relative frequency shift,

$$\frac{\delta v}{v} \simeq \frac{kT_e}{m_e c^2}.$$ (19.68)

A low frequency photon scattering off a non-relativistic electron travelling at velocity v suffers a frequency change $\delta v / v \simeq v/c$. However, if the photon scatters off a Maxwellian distribution of electrons with root mean square velocity v, calculation of the energy shift shows that the term of order v/c vanishes, leaving the $(v/c)^2$ term. Since $m_e v^2 \simeq kT_e$ we have the above result.

The number of collisions suffered in time t is $t\sigma_t n_e c$, since $(\sigma_t n_e c)^{-1}$ is the mean free time between collisions. The total energy change in this time is thus $t\sigma_t n_e c(\delta v / v)$, and this is on the order of unity on a timescale:

Compton timescale:

$$t_C \simeq \frac{1}{\sigma_T n_e c}\left(\frac{m_e c^2}{kT_e}\right).$$ (19.69)

This is the timescale for photons to change their energy due to collisions with the field of electrons.

Hot electrons, at temperature T_E, that are injected into a gas of photons at temperature $T_\gamma < T_e$ will cool down while transferring heat to the photons. During the process the frequency spectrum of the photon gas will change. The seemingly simple process is described by a fairly complicated-looking equation: the Kompaneets equation.

19.3 The Kompaneets Equation

The evolution of a distribution of photons interacting with a hot ionised gas was first discussed by Kompaneets (1957) and by Weymann (1965, 1966).[12] Sunyaev and Zel'dovich

---

[12] Weymann did not know of Kompaneets' work at that time since the paper was not translated into English until several years later. Weymann based his work on that of Dreicer (1964) who had independently derived

applied the Kompaneets equation to the question of the origin of the Planck spectrum in their seminal papers, Zel'dovich and Sunyaev (1969); Sunyaev and Zel'dovich (1970).[13]

In the cosmological context the Kompaneets equation describes the evolution of an ensemble of photons at some temperature T_0 interacting with electrons at higher temperature T_e (we think of energy being injected into the cosmic radiation field). The number of photons η in each frequency bin evolves under the influence of the Compton scattering of photons off electrons, and the production and absorption of photons by the bremsstrahlung, or free-free, process. The Compton process boosts the energy of the photons towards higher frequencies without creating any new photons, while the Bremsstrahlung process creates photons that can make up the Compton losses at lower frequencies.

There are two topics of interest: how the electrons change the radiation spectrum and what is the final equilibrium state between the electrons and photons. In the cosmological context a detailed study of the spectrum of the CMB might reveal deviations from the Planck law that indicate early phases of energy injection. This also relates to the effect in which the hot electrons in the gas in clusters of galaxies affect the radiation passing through it: the *Sunyaev–Zel'dovich effect*.

We start by introducing the Compton and bremsstrahlung processes and the Kompaneets equation and present the physical ramifications in the following section. The discussion of the Kompaneets equation here is non-relativistic, see Challinor and Lasenby (1998) for the relativistic version.

19.3.1 Thomson / Compton Scattering

The quantity α_0 defined by

$$\alpha_0(z) = \sigma_T n_e c \left(\frac{k T_\gamma}{m_e c^2} \right) \simeq 1.1 \times 10^{-28} (1+z)^4 \left(\Omega_b h^2 \right) \text{s}^{-1} \qquad (19.70)$$

characterises the rate at which the Compton process, $e^- + \gamma \leftrightharpoons e^- + \gamma$, transfers energy between electrons and photons. Note that the definition of α_0 involves T_γ, not T_e. This is merely a matter of convenience that simplifies some equations that will come in later. The z^4 dependence is the combination of z^3 from the density n_e and z from the temperature. At the temperature when the densities of non-relativistic matter and radiation are equal, i.e. z_{eq} (Equation 6.57), we have

$$\alpha_0(z_{eq}) = 1.62 \times 10^{-14} \Omega_b h^2 \text{ s}^{-1}, \quad \text{at } z_{eq} \text{ in the benchmark model.} \qquad (19.71)$$

If the radiation field is heated via injection of hot electrons, the Compton process transfers the excess energy from those electrons to the photons simply by boosting their energy. This cools the hot electrons and heats the photon gas. The Compton process conserves the number of photons. The hot electrons will come into equilibrium with the photon gas at a temperature given by Equation (19.22). However, the photon spectrum need no longer

---

Kompaneets' equation. One of the key remarks made in his 1966 paper was that if there were any contributions from high redshift free-free radiation, they should be visible now in the 20–30 cm wavelength range.

[13] The discussion of this topic from a physics viewpoint in Peebles (1971, Ch. VII Sections iv, v, pp.233 *et seq.*) and Peebles (1993, p.581 *et seq.*) are still among the clearest.

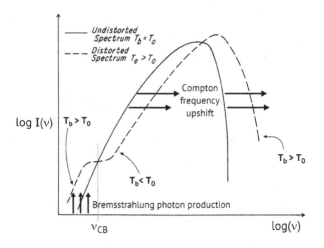

Fig. 19.3 Representation of a distorted Planck spectrum. The solid curve is an undistorted Planck spectrum at temperature T_0. The dashed curve depicts a distorted spectrum which results from heating the electrons to temperature $T_b > T_0$. The ordinate ν_{CB} represents the frequency at which the Compton frequency up-shifting of photons balances the rate of creation of photons through the Bremsstrahlung process. At frequencies $\nu < \nu_{CB}$ the Bremsstrahlung process creates enough photons to fill that part of a Planck curve at temperature T_b. At frequencies $\nu > \nu_{CB}$ the Compton process boosts the energy of the photons without creating any new photons.
From Jones and Steigman (1978, Fig. 1). *(Permission granted.)*

follow a Planck law since it will be deficient in photons at this higher temperature, unless there is a mechanism to create the necessary photons, see Figure 19.3.

There is another process that is important at the earliest times soon after the the $e^- e^+$ pairs annihilate to create the photon field: *double Compton scattering* (Gould, 1984). This process, represented as $e^- + \gamma_0 \leftrightharpoons e^- + \gamma + \gamma$, is for a short time the main source of any new photons that are required to create the cosmic Planck equilibrium distribution (Danese and de Zotti, 1982). At later times photons are mainly created through Bremsstrahlung.

19.3.2 Free-free / Bremsstrahlung

The acceleration of free electrons due to their interaction among themselves creates and absorbs photons via the free-free process (also known as Bremsstrahlung).[14] The amount of energy emitted per unit volume per second, the emissivity, by a gas of freely interacting electrons of density n_e and temperature T_e is

$$j(\nu, T_e) = 6.8 \times 10^{-38} G_{\text{ff}}(\nu, T_e)\, n_e^2 T_e^{-1/2} e^{-h\nu/kT_e} \quad \text{erg s}^{-1}\,\text{cm}^{-3}\,\text{Hz}^{-1}. \tag{19.72}$$

The exponential cut-off in the frequency distribution of the emitted radiation is of considerable importance. It means that this process acts mainly at lower frequencies.

---

[14] Bremsstrahlung is an important source of X-ray emission from the hot gas in galaxy clusters and is responsible for most of the cooling of the cluster gas.

The function $G_{ff}(\nu, T)$ is the infamous 'Gaunt factor'. This function provides a good approximation to the quantum aspects of Bremsstrahlung radiation and is extensively tabulated. For the present purposes of considering emission from electrons at temperature T_e sitting in a sea of photons at temperature T_0, we may take

$$G_{ff}(x) = \frac{\sqrt{3}}{\pi} \ln\left(\frac{2.35}{x}\frac{T_e}{T_0}\right), \quad x = \frac{h\nu}{kT_0}. \tag{19.73}$$

The emissivity integrated over all frequencies is

$$j(T) = 2.4 \times 10^{-27} \overline{G}_{ff}\, n_e^2 T_e^{1/2} \ \text{erg s}^{-1} \ \text{cm}^{-3}, \tag{19.74}$$

where $\overline{G}_{ff}$ denotes a frequency-averaged Gaunt factor.

In the inverse process, free-free absorption, electrons gain energy from the photons. The rate for free-free absorption is given by

$$\mu(\nu, T_e) = 3.7 \times 10^8 G_{ff}(\nu, T_e)\, \nu^{-3}\left(1 - e^{-h\nu/kT_e}\right) n_e^2 T_e^{-1/2} cm^{-1}. \tag{19.75}$$

This equation is made up from two components: the factor $n_e^2 T_e^{-1/2} G_{ff}(\nu, T)$ from the emissivity (19.72) and the factor $\nu^{-3}(1 - e^{-h\nu/kT_e})$, which is the inverse of the brightness I_ν (19.9). This relationship relating absorption to the emissivity and the brightness is called Kirchoff's law.

19.3.3 y-time

Consider only Compton scattering of CMB photons off free electrons in a gas having density n_e and temperature T_e. The average energy gain by a photon of the CMB having energy $h\nu$ in a collision with an electron in the gas is of the order of $\Delta h\nu/h\nu \sim 4kT_e/m_ec^2$. For non-relativistic electrons this is a small number, so it takes many such scatterings to make a substantial change in the photon energy.

The number of scatterings in a time interval Δt is $n \sim t/t_C$, where the collision time, t_C is given by the Thompson cross-section and electron density: $t_C = (\sigma n_e c)^{-1}$. In the time interval Δt we have $n \sim \sigma n_e c \Delta t$ collisions and the energy gain in time Δt is

$$\frac{\Delta h\nu}{h\nu} \sim \frac{4kT_e}{m_ec^2}\sigma_T n_e c\Delta t \quad \Rightarrow \quad \frac{dU}{dt} = \frac{4}{3}\sigma_T n_e c\frac{\langle v^2\rangle}{c^2}\,U, \tag{19.76}$$

where $U = aT^4$ is the radiation energy density (Equation (19.16)).

The hotter electrons cannot continue cooling below the radiation temperature. We should replace the kT_e in this expression by $kT_e - kT_0$, the excess energy the electrons have over what they would have if they were in equilibrium with the photons.

Consider an interval of time from t_h when the hot electrons are injected into the radiation field, to a later time t_0 when the radiation spectrum is observed. The parameter Y is given by:

The Sunyaev–Zel'dovich y-parameter:

$$Y(t_0) = \int_{t_h}^{t_0} \frac{k(T_e - T_0)}{m_ec^2}\sigma_T n_e c\,dt, \tag{19.77}$$

Sunyaev and Zel'dovich (1980, Section 2.4).

This is a measure of the relative boost in the photon energy over the time span t_h to t_0 due to Compton collisions with electrons at temperature T_e. So $Y \sim 1$ would indicate a significant change in the photon energy distribution.

Instead of using time as a parameter in the equations it will be convenient to use another parameter telling how much Thomson scattering has been taking place:

y-time:

$$y_{t_h}(t) = \int_{t_h}^{t} \alpha_0 \, dt, \qquad (19.78)$$

where α_0 was given in Equation (19.70). The convenience of using y is that it depends only on the ambient temperature, T_0, of the photon gas and not on the specifics of the heating mechanism, i.e. the electron temperature.[15]

Here t_h is a starting time, usually taken to be the time when the heating of the photon gas by the electrons started, and t is some subsequent time during the subsequent evolution of the spectrum. Generally, we are interested in the value of this parameter at the present time, t_0, and in that case it is convenient to think of $y_{t_h}(t_0)$ as function of t_h: i.e. $y_0(t_h)$. Notice that although $y(t)$ is an increasing function of time, most of the integral comes from the highest redshifts, i.e. from t_h. This provides a way to estimate the value of y without doing the integral.

Since most of the contribution to the integral in (19.78) comes from the higher redshift, we have, as a rough approximation for the benchmark model,

$$y_0(t_h = t_{eq}) = \left[\sigma_T n_e ct(kT_0/m_e c^2) \right]_{t=t_{eq}} = 0.024 \Omega_b h^2 = 5.35 \times 10^{-5}. \qquad (19.79)$$

Since $y \propto z^4$, as we go to earlier times we get to $y \sim 1$ at $z \sim 41\,000$. We shall see that injecting hot electrons at times when $y \gg 1$, when Compton collisions are very frequent, can lead to an equilibrium non-Planckian photon distribution.

19.3.4 The Terms of the Kompaneets equation

In what follows we shall normalise the photon frequency ν relative to the temperature T_0 in the obvious way:

$$x = \frac{h\nu}{kT_0}. \qquad (19.80)$$

This is particularly convenient since both T_0 and ν vary as $a(t)^{-1}$ with the expansion, and so x remains invariant. We shall also measure the time in terms of the number of Thomson scatterings that have taken place, i.e. the y-time of Equation (19.78).

The occupation number of photons, $\eta(\nu)$, at frequency ν evolves as:

---

[15] We most frequently see the definitions $y = \int (kT_e/m_e c^2) \sigma_T n_e c \, dt$, as originally given by Sunyaev and Zel'dovich *loc. cit.*. As recognised by Chan and Jones (1975a,d), that definition introduces further terms in the Kompaneets equation which result in numerical complications. See also Burigana *et al.* (1991, Appendix Eq.A1) who discuss this in some detail.

The Kompaneets equation

$$\frac{\partial \eta}{\partial y} = \frac{1}{x^2}\frac{\partial}{\partial x}x^4\left[\frac{T_e}{T_0}\frac{\partial \eta}{\partial x} + \eta(1+\eta)\right]$$

Compton diffusion shifting photons to higher energy

$$+ \frac{\kappa_0}{\alpha_0}\left(\frac{T_e}{T_0}\right)^{-1/2}\frac{G_{ff}(x)}{x^3}\left\{\underbrace{\exp\left(-x\frac{T_e}{T_0}\right)}_{\text{emission}} - \eta\underbrace{\left[1 - \exp\left(-x\frac{T_e}{T_0}\right)\right]}_{\text{absorption}}\right\}. \quad (19.81)$$

Bremsstrahlung emission and absorption of photons at lower energies

It is easy to see where the Bremsstrahlung emission and absorption rates (19.72, 19.75) come in. The quantity κ_0 is a convenient measure of the rate associated with the free-free process:

$$\kappa_0 = \frac{8\pi e^6 h^2}{2m_e(6\pi m_e kT_0)^{1/2}}\frac{n_e^2}{(kT_0)^3} \simeq 4.7 \times 10^{-25}(1+z)^{5/2}(\Omega_b h^2)^2 \text{ s}^{-1}. \quad (19.82)$$

Note the dependence on the *square* of the electron density n_e, and the $T^{-7/2}$ temperature dependence. These give rise to the $(\Omega_b h^2)^2$ and $(1+z)^{5/2}$ dependencies of κ_0.[16] The rate constants κ_0 and α_0 enter only as a ratio in the Bremsstrahlung term. Thus the evolution is governed by one number. That is the advantage of using y-time (19.78).

Any deviations from the Planck law are expected to be very small, we can make substantial progress towards understanding this equation by perturbing it about the Planck occupancy $\eta(x)$. We look at this in Section 19.3.5. However, a full solution can only be numerical. This was first done by Chan and Jones (1975a,b,c,d) and repeated within a wider cosmological context by Burigana *et al.* (1991).

19.3.5 Understanding the Kompaneets Equation

It is helpful to look at the effects of the Compton and bremsstrahlung terms separately in the low frequency limit ($x < 1$) of the Kompaneets equation:[17]

$$\left(\frac{\partial \eta}{\partial y}\right)_C \simeq \frac{2}{x}\left(1 - \frac{T_e}{T_0}\right), \quad (19.83)$$

$$\left(\frac{\partial \eta}{\partial y}\right)_B \simeq \frac{\kappa_0}{\alpha_0}G_{ff}\left(\frac{T_e}{T_0}\right)^{-3/2}\left(\frac{T_e}{T_0} - 1\right)\frac{1}{x^3}. \quad (19.84)$$

The terms have opposite signs and there is an important characteristic frequency, $x_{CB} \sim (\kappa_0/\alpha_0)^{1/2}$, at which the Compton and Bremsstrahlung processes balance (see Figure 19.4). More precisely:

---

[16] See Sunyaev and Zel'dovich (1970, Eq.3). The word 'Bremsstrahlung' is a translation of the German for 'braking radiation'. The timescale κ_0^{-1} is sometimes referred to as the 'braking time'.

[17] See Chan and Jones (1975a, Section II) where this is derived and compared with numerical integration of the Kompaneets equation.

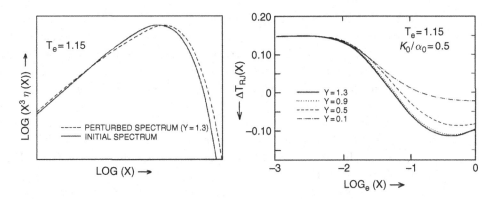

Fig. 19.4 The effects of Thomson scattering and bremsstrahlung, showing how a photon spectrum interacts with a hot electron gas at the time of maximum distortion (Chan and Jones, 1975a, Figs. 7,9). *Left:* shows a distorted spectrum relative to the original Planck spectrum for an unrealistically large distortion parameter value $y = 1.3$. *Right:* The deviations from the Planck law as a function of frequency $x = h\nu/kT_0 < 1$ for a variety of Y-parameters. The parameters used for both panels are $\kappa_0/\alpha_0 = 0.5$, and $T_e/T_0 = 1.15$. This was the first numerical integration of the Kompaneets equation.
(Reproduced by permission of the AAS.)

$$x_{CB} \simeq \frac{\kappa_0}{\alpha_0} \left[\frac{G_{\mathrm{ff}}}{2} \left(\frac{T_e}{T_0} \right)^{-3/2} \right]^{1/2} \simeq 2.2 \times 10^2 z^{-3/4} (\Omega_b h^2)^{1/2} \left[\frac{G_{\mathrm{ff}}}{2} \left(\frac{T_e}{T_0} \right)^{-3/2} \right]^{1/2}. \quad (19.85)$$

Note that the frequency x_{CB} increases with time: photon production 'eats' its way into the blue-shifted Rayleigh–Jeans part of the spectrum. It can be shown, however, that once $y = 1$, the frequency x_{CB} is essentially frozen in. Its march towards higher frequencies then takes longer than the expansion time. We denote the redshift at which $y = 1$ as $z_{y=1}$. The value of x_{CB} at $z_{y=1}$ is of particular importance. We come to that in the next section.

At still lower frequencies the Bremsstrahlung process can produce enough photons to fill out a Planck law. This frequency x_B is given by

$$x_B = (\kappa_0 t_{\mathrm{exp}})^{1/2} \left[G_{\mathrm{ff}} \left(\frac{T_e}{T_0} \right)^{-3/2} \right]^{1/2} \simeq 9.7 \times 10^{-4} z^{1/2} (\Omega_b h^2)^{3/4} \left[G_{\mathrm{ff}} \left(\frac{T_e}{T_0} \right)^{-3/2} \right]^{1/2}. \quad (19.86)$$

x_B decreases as a function of time and so the Bremsstrahlung part of the spectrum creeps to ever lower frequencies.

19.4 Spectral Distortions

The kind of spectral distortions that we might expect depend on the redshift at which the main energy injection takes place. The first division comes at values of the y-parameter

$y \sim 1$. This occurs at redshifts around $z_{y=1} \sim 10 z_{eq} \sim 35\,000$. For $y \gg 1$ collisions are frequent enough that we can expect an equilibrium situation to be established. At extremely large y we will get a Planck spectrum because the free-free and double Compton processes are capable of creating photons at a rate that is sufficient to counter the frequency up-shifting from the Compton effect. The Planck spectrum is maintained at redshifts higher than $z_{PL} \sim 2 \times 10^7$.

At lower redshifts, but while $y \gg 1$, there is a regime of frequent Compton collisions where the free-free and double Compton processes can no longer keep up with creating photons. In this regime the number of photons is conserved and the resulting photon distribution looks like a Bose–Einstein distribution. The redshift range associated with this kinetic equilibrium is $z_{BE} < z < z_{PL}$, with $z_{BL} \sim 2 \times 10^5$, see Section 19.4.1.

The opposite limit is $y \ll 1$ when, in effect, the Compton process is the only agent causing the spectrum change. There are two stages: pre- and post-recombination. In the pre-recombination phase the entire spectrum is up-shifted in frequency and no photons are created. This is simply referred to as the *y-distortion* as a result of which we have a *Comptonised spectrum*.

After recombination the Universe becomes effectively neutral and n_e becomes very small. However, at very recent times, we can have patches of fully ionised plasma at very high electron temperature. High electron density and high temperature result in a high value of y and the generation of localised spectral distortions. This happens in rich clusters of galaxies (Sunyaev and Zel'dovich, 1970).[18] This distortion from hot gas in clusters is referred to as the *Sunyaev–Zel'dovich effect* or '*SZE*' for short. There are two types of SZE: the *thermal SZE*, or *tSZE*, and the *kinetic SZE*, or *kSZE*, see Section 22.6.

19.4.1 Setting up the Bose–Einstein Distribution

The characteristic photon frequencies of the Kompaneets equation are x_{CB}, where the bremsstrahlung and Compton processes compete, and the frequency x_B, where photon production by Bremsstrahlung has the time to fill out a Planck distribution (*cf.* Equations 19.85, 19.86). If $x_{CB} < x_B$ then we have a situation where the photons that are created by the electrons are passed up the spectral frequencies by the Compton process, and it can be shown that under these circumstances the spectrum of the photons takes on the classical Bose–Einstein form:

$$\eta = \left[\exp\left(x\frac{T_0}{T_e} + \mu(x) \right) - 1 \right]^{-1}, \tag{19.87}$$

where the effective chemical potential $\mu(x)$ is given by

$$\mu(x) = \mu_0 \exp\left(-\frac{x_{CB}}{x} \right). \tag{19.88}$$

We found above that

$$x_{CB} \sim 220 (\Omega_b h^2)^{1/2} z^{-3/4}, \tag{19.89}$$

and commented that it was 'frozen in' at the epoch z_a when $y = 1$.

---

[18] This little paper was written shortly after the confirmation of X-rays from the Perseus cluster of galaxies (Gursky *et al.*, 1971, 1972).

At that time

$$x_{CB}(z_a) \simeq 220(\Omega_b h^2)^{1/2} z_{y=1}^{-3/4} \simeq 0.22(\Omega_b h^2)^{7/8}, \quad \Omega_b h^2 \lesssim 0.6. \tag{19.90}$$

For the benchmark values of $\Omega_b h^2$, this lies at very low frequencies such that we would not be able to select out the CMB signal from the foregrounds. The Bremsstrahlung term could be ignored when calculating the evolution of the spectrum (Zel'dovich *et al.*, 1972; Chan and Jones, 1975b).

19.4.2 Energy Injected at z such that $y \ll 1$

Zel'dovich *et al.* (1972) discovered a simple solution to the Kompaneets equation in the case $y \ll 1$.[19] The solution is the following.

The perturbed spectrum at time y is written as

$$\eta(x, y) = \eta_0(x) + Y_{\text{eff}} f(x), \quad Y_{\text{eff}} = y\left(\frac{T_e}{T_0} - 1\right), \tag{19.91}$$

where Y_{eff} is essentially (19.77). In this way the approximation works for $T_e \sim T_0$. It can then be shown that the perturbation $f(x)$ in the shape of the spectrum is

$$f(x) = \frac{xe^x}{(e^x - 1)^2} \left\{ \frac{x}{\tanh(x/2)} - 4 \right\}, \quad x^2 y \ll 1. \tag{19.92}$$

Note that in this approximation we use Y_{eff} rather than Y, but in the case $(T_e/T_0) \gg 1$ they are the same thing.

The form of the distortion $f(x)$ is somewhat unedifying, but in the small Y_{eff} limit it is easy to show that

$$\frac{T_{RJ}}{T_0} \equiv x\eta(x) = 1 - 2Y_{\text{eff}}. \tag{19.93}$$

---

[19] The solution is, however, not strictly valid at the high frequency ($\lambda < 1$mm) part of the spectrum, the sub-millimetre regime that FIRAS observed.

20

Recombination of the Primeval Plasma

The transition the Universe makes from a fully ionised plasma to a neutral gas happens very quickly when the age of the Universe is around 380 000 years, i.e. a redshift of $z \sim 1500$. From our point of view, looking out into the distant Universe, and hence into the past, we see this as a cosmic photosphere. This is referred to as the *surface of last scattering* since this corresponds to the time after which most of the CMB photons travel freely towards us. The event of the Universe becoming neutral is referred to as the *epoch of recombination*.[1]

The image of this surface we create with our radiometers is a snapshot of the Universe as it was when it was still only slightly inhomogeneous: we see the initial conditions for the birth of cosmic structure.

20.1 Recombination

The photons of the CMB that we observe come from the epoch at a redshift $z \sim 1090$ when the cosmic plasma was almost neutral and the timescale for electron–photon collisions became longer than the cosmic expansion timescale. The Universe was then some $\sim 380\,000$ years old. Prior to that time, electrons and photons had been held in thermodynamic equilibrium by the Thomson scattering process, while after that time the baryonic component of the cosmic plasma could evolve as an independent component of the plasma.

This was the time of the decoupling of matter from the radiation field. Shortly after that time most of the CMB photons were able to travel directly to us now without being scattered by free electrons. This gave us a direct view of the early conditions that led to the formation of the cosmic structures we see today, the galaxies, clusters of galaxies and their organisation into what is now known as the 'cosmic web'.

The Universe did not suddenly become neutral as it emerged from the fully ionised fireball phase of the cosmic expansion. The neutralisation, or 'recombination', of the cosmic plasma took place over several tens of thousands of years. That is enough to slightly blur the details of the structure that might be observed on the last scattering surface: we are looking into the cosmic photosphere. If we are to understand the nature of the primeval

---

[1] A remark is occasionally made about the 're' when talking of the epoch of recombination, it is not as though the Universe were ever neutral before that time. Technically, however, 'recombination' refers to the nature of the atomic processes that took place at that time. In the past this time was often referred to as the *epoch of decoupling*, because that was when the matter and radiation decoupled from one another.

structure, we must understand the recombination process well enough to de-blur what we see. Hence the detailed modelling of the decoupling process occupies a central role in our understanding of the Universe we observe.

Understanding the cosmic photosphere, the process of *cosmic recombination*, is of great importance. The pioneering work was done in papers by Peebles (1968), Zel'dovich *et al.* (1969) and Matsuda *et al.* (1971),[2] who considered the recombination using relatively simple models. Later, Jones and Wyse (1985), also using a simple 3-level model for the hydrogen atom, repeated these calculations, deriving convenient analytic approximations for the ionisation history, $x(z)$, in the presence of both baryonic and non-baryonic dark matter. It is now a highly technical domain of atomic physics and radiation gas-dynamics. The subject of the recombination and its aftermath is thoroughly reviewed by Glover *et al.* (2014).

20.1.1 The Saha Formula

As the Universe expands and cools, it remains fully ionised while there are enough photons that have sufficient energy to ionise a neutral atom. However, the cosmological situation is unusual in that there are several hundred photons for each baryon. Thus long after the radiation temperature has dropped below $T_r \sim 10^4$ K there are still enough energetic photons around in the high frequency Wein part of the CMB spectrum to maintain the ionisation level above what would be expected on a simple equilibrium model. The rate of recombination is governed by the rate at which Ly α photons are red-shifted out of the Ly α line by the cosmic expansion. Consequently, the bulk of the neutralization takes place at temperatures around 3500 K.

In the following discussion we confine ourselves for didactic reasons to the physics of a gas of hydrogen at temperature T.

The fundamental atomic process describing the mixture of protons, electrons and neutral hydrogen atoms is

$$p + e^- \rightleftharpoons H + h\nu. \tag{20.1}$$

The binding energy of the hydrogen atom in its ground state is[3]

$$I = 13.6\,\text{eV} \simeq 160\,000\,\text{K}, \tag{20.2}$$

and if the mixture were in local thermodynamic equilibrium (LTE), the densities of protons (ions), N_i, electrons, N_e and neutrals, N_n, would be related by the Saha equilibrium formula:

$$\frac{N_e N_i}{N_n} = \frac{(2\pi m_e kT)^{3/2}}{h^3} \exp\left(-\frac{I}{kT}\right) = 2.4 \times 10^{15} T^{3/2} \exp(-160\,000/T). \tag{20.3}$$

The total number of hydrogen atoms, ionised and neutral, is $N = N_i + N_n$. This, in our hydrogen-only model, is the baryonic number density. We can write the fractional ionisation as

$$x = \frac{N_e}{N} = \frac{N_i}{N_i + N_H}, \tag{20.4}$$

---

[2] The paper of Matsuda *et al.* (*loc. cit.*) focuses on the ionisation history in the presence of rapidly decaying cosmic turbulence. The turbulence becomes supersonic during the recombination process and reheats the cosmic plasma.

[3] 1 eV translates to 11 600 K, so 13.6 eV = 157 760 K.

since the number of free electrons must equal the number of ions: $N_i = N_e$. This reduces (20.3) to

$$\frac{x^2}{1-x} = \frac{1}{N} \frac{(2\pi m_e kT)^{3/2}}{h^3} \exp\left(-\frac{160\,000}{T}\right) \overset{\text{def}}{=} S(T), \qquad (20.5)$$

which is a simple quadratic equation $x^2 + Sx - S = 0$ for the ionisation history, $x(T)$.

We can write the baryonic number density, N, in terms of the temperature T as $N(T) = (T/T_0)^3 N_0$, where N_0 is the current baryon number density.[4] In our assumed pure hydrogen Universe, N_0 corresponds to a present-day mass density $\rho_{b,0} = m_p N_0$, which is a fraction, Ω_b of the total cosmic density. N is also small in the sense that it is far smaller than the number of photons. This is where the baryonic density parameter, Ω_b, of the cosmological model influences the recombination process.[5]

The Saha equation works reasonably well only in the higher temperature regime. In the lower temperature regime the Saha recombination is much too fast.

The Saha formula only gives a very rough description of what in fact happens. The presupposed equilibrium can only be maintained while the electrons and photons of the cosmic medium can interact. Equilibrium cannot be maintained when the timescale for photon–electron collisions, $(\sigma_T n_e(z)c)^{-1}$ is longer than the cosmic expansion timescale $\sim H^{-1}$ ($\sigma_T \simeq 6.65 \times 10^{-25}$ cm^2 is the Thomson cross-section).

It shows that the recombination is not instantaneous but it fails badly when the temperature falls much below 3000 K. The situation is unexpectedly complicated because there are some 10^9 cosmic radiation photons for every baryon. There are enough energetic photons to modify the populations of the various levels of the neutral atoms and, in particular, to inhibit the recombination to the ground state. This is a situation that does not occur in other environments, like the interstellar medium or a stellar atmosphere, where the Saha equation might be considered an adequate approximation. The equilibrium premise upon which the Saha formula is based is not correct in the cosmological environment.[6] We need a more detailed model of the atomic physics and of the role of the CMB photons.

---

[4] When considering the recombination as a function of redshift it is important to note that the temperature of the electrons is still falling off as $T_e \propto (1 + z)$ for $z \gtrsim 300$ since, despite the low ionisation, the remaining free electrons are still tightly coupled to the cosmic radiation field via Compton scattering.

[5] We could solve (20.5) for $x(T)$ by successive approximations, but it is sufficient to note that there are two distinct regimes. The ionisation x will be close to $x \simeq 1$ at high temperatures, and close to $x \simeq 0$ at low temperatures. Referring to the right-hand side of (20.5) as $S(T)$, then in the regime $S(T) \gg 1$, $x \simeq 1 - S(T)^{-1}$, while in the regime $S(T) \ll 1$, $x \simeq S(T)^{1/2}$. These approximations could easily be taken to higher order. The dividing line between the two limiting solutions occurs around $T \sim 4280$ K, at $z \sim 1560$ when the ionisation is $x \sim 0.8$. It is the presence of the factor $1/N$ on the right-hand side of (20.5) that causes the recombination process to happen at a lower temperature than seen in typical *HII* regions.

[6] Strömgren (1948, Eq. (29) *et seq.*) developed a modified version of the Saha equation for a non-equilibrium situation arising in the formation of absorption and emission lines in the interstellar medium. In his model he had allowed the electron and ion temperatures to be different, and for departures in the populations of the atomic levels from the standard Boltzmann distribution. Strömgren's equation was used by Michie (1969, Eq. (A11)) when discussing the passage of density perturbations through recombination, but he gave no details as to how this was implemented.

20.2 Recombination: the Physics

The computation of the ionisation history is a matter of atomic physics and, even in the simplest of models is a complex and tedious calculation.[7] The standard framework for the calculation was to consider a cosmological model in which the baryonic component was hydrogen. The assumption was that the $\sim 10\%$ of He atoms would have recombined at higher temperatures and so would not play a role as electron donors during the later period of hydrogen recombination. This was entirely reasonable in the days before precision cosmology. This simple model is depicted in Figure 20.1, where the left panel recalls the structure of the energy levels in the standard model of the hydrogen atom and the associated spectral lines. The right-hand panel adds information about the angular momentum states of those energy levels and, in particular, splits the $n = 2$ level into its two angular momentum states, referred to as the $2s, (l = 0)$, and $2p, (l = 1)$, states, l being the angular momentum quantum number for the electron orbital.[8] This is important because the allowed transitions must conserve energy and angular momentum.

20.2.1 Simple Model for Hydrogen – Radiative Processes

Figure 20.1 also depicts the transitions that have to be considered. At the temperatures considered here ($T < 5000\,\text{K}$) we can safely ignore collisional processes in which free electrons knock out bound electrons. The only processes to be considered involve photons, i.e. radiative processes.

The transitions between the continuum and the $n = 2$ level in which the electrons get captured to (*radiative recombination*) and knocked out of (*photoionisation*) the first excited state have rates that are denoted by R_{c2} and R_{2c}:[9]

$$H^* + \gamma \rightleftharpoons H^+ + e, \qquad \begin{cases} \leftarrow & \text{rate: } R_{c2} = 2.84 \times 10^{-11} T_e^{-1/2}, \\ \rightarrow & \text{rate: } R_{2c} \propto R_{c2} \, \dfrac{(2m_e kT)^{3/2}}{h^3} \, e^{-B_2/kT}. \end{cases} \tag{20.6}$$

Here, H^* denotes a neutral hydrogen atom in its first excited state ($n = 2$), and H^+ denotes an ionised, charged, hydrogen atom. $B_2 = 3.4\,\text{eV}$ is the binding energy of the $n = 2$ state.

---

[7] See, for example, Equations (1)–(20) and Appendix A of Matsuda *et al.* (1971).

[8] The nth level is allowed to have $l = n - 1$ discrete angular momentum states, and the angular momentum can take on any of the $2l + 1$ values $m = -l \cdots + l$. Thus n, l, m are three quantum numbers describing the state of the atom. They arise as eigenvalues of the solution of the Schrödinger equation for the motion of an electron about a proton. There is a fourth quantum number which describes the orientation of the electron spin in a state of given n, l, m.

For historical reasons the $l = 0, 1, 2, 3$ angular momentum states of hydrogen are referred to by the labels s, p, d, f. This strange nomenclature is a throw-back to the oldest days of atomic spectroscopy. It arose in describing the quality of the spectral lines of alkali metals, Li, Na, K, ... as 'sharp', 'principal', 'diffuse', 'fine', *etc.* . The s, p, d classifications of lines arose in the 1870s, while the f classification only came during the first decade of the 20th century. See Jensen (2007) for a brief history of this.

[9] This is the notation used by Matsuda *et al.* (1971). The more usual, but less clear, notation is to label R_{c2} and R_{2c} as α and β respectively (and often with a variety of subscripts).

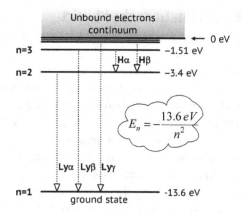
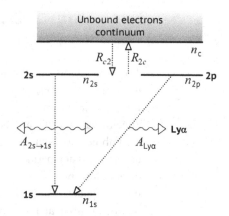

Fig. 20.1   Physics of the hydrogen recombination process. On the left the energy levels of the electron orbitals of the
Bohr–Schrödinger model of the hydrogen atom are shown. The transitions between these levels give rise to the
familiar lines of the hydrogen spectrum. On the right we depict the $n = 1$ ground state and the first $n = 2$ excited
state, with the transitions that can take place between these levels and the continuum. The limit $n \to \infty$ is the
continuum in which the electrons are not bound to the nucleus. This lies some 13.6 eV above the ground state.
The $n = 2$ state has two angular momentum states having the same energy, these are labelled by the angular
momentum quantum number $l = 0$, (the '2s' state) and $l = 1$ (the '2p' state). The occupancies of these states are
denoted by n_{1s}, n_{2s}, n_{2p} and the continuum by n_c.
A fairly accurate computation of the recombination can be achieved using this three-level model. The important
processes are transitions between the first excited state ($n = 2$) and the continuum, and the decays of the first
excited state to the ground state ($n = 1$). The symbols for the rates for key transitions are shown on the figure. ©

The rates for these processes, R_{c2}, R_{2c}, are the ones used by Peebles (1968), where more
details can be found.[10] The units of these coefficients are cm^{3}s^{-1}.

Clearly the rate R_{c2} will depend on the mean velocity of the free electrons, which can
be assumed to be given by $v_e \simeq (kT_e/m_e)^{1/2}$. Hence we have $R_{c2} \sim T_e^{-1/2}$. The reverse
process, photoionisation of the $n = 2$ state, will have a rate depending on the population of
the $n = 2$ level and will be proportional to the Boltzmann-like factor $(2m_e kT)^{3/2}e^{-B_2/kT}$,
where $B_2 = 3.4$ eV is the binding energy of the $n = 2$ state. A full calculation would
require taking account of higher n states, but here we can get away with using the Saha
equation for the relative occupancy of the $n \geq 2$ states. This assumes that electrons
captured to higher states will rapidly cascade their way down to the $n = 2$ level.

The population of the $n = 2$ level also depends on spontaneous radiative transitions from
the $n = 2$ to the $n = 1$ levels (*spontaneous emission*):

$$H_{2p} \to H_1 + \gamma, \quad \text{rate: } A_{Ly\alpha} \simeq 6.25 \times 10^8 \text{ s}^{-1}. \tag{20.7}$$

---

[10] The paper by Boardman (1964) was generally cited for the recombination rate R_{c2}, though it did not in fact
contain this fitting formula. The clearest discussion of this appears in the book of Tucker (1975, Section 6.1).
More recently, Pequignot *et al.* (1991, Eq. (1), Table 1, see the coefficient α_C) have updated this rate, giving
the approximation $R_{c2} = 10^{-13} a T_4^b/(1 + c T_4^d)$ with $a = 4.309, b = -0.6166, c = 0.6703, d = 0.5300$, where
T_4 is the electron temperature measured in units of 10^4 K. Such fitting formulae are often approximations that
are valid to within 1% or so, but some are less accurate. See the discussion in Tucker's book *loc. cit.* .

This leads to the emission of a Lyman-α photon (see the left panel of Figure 20.1). However, there is another process by which an electron in the $n = 2$ level can get to the ground state: two photon emission from the $2s$ state,[11]

$$H_{2s} \to H_{1s} + \gamma + \gamma, \quad \text{rate: } A_{2s \to 1s} \simeq 8.22\,\text{s}^{-1}. \qquad (20.8)$$

Some $\sim 57\%$ of all hydrogen atoms in the Universe become neutral via this $2s \to 1s$ two-photon decay channel. The decay rate, $A_{2s \to 1s}$, of the transition $2s \to 1s$ via two-photon emission is the key reaction (Spitzer and Greenstein, 1951; Kipper, 1952).[12]

20.2.2 Calculating Populations of Hydrogen Energy Levels

We now come to discuss the physical process that takes us away from the Saha equilibrium formula.[13] We shall see where the number of Lyman-α photons provided by the cosmic background radiation comes into play, and where the expansion of the Universe redshifts the Lyman-α photons down in energy, thereby controlling the rate at which the ionisation happens.

An important trick is to suppose that every electron radiatively captured into the $n = 2$ level finds its way down to the $n = 1$ level, resulting in a neutral ground state hydrogen atom. Then the rate at which the plasma is becoming neutral is just the rate at which the $n = 2$ level is being populated:

$$\text{rate}(c \to 2) = \frac{d}{dt}\left(\frac{n_c}{n}\right) = -\frac{1}{n}\left[R_{c2}n_e n_p - R_{2c}n_{2s}\right]. \qquad (20.9)$$

Here, n_c is the number density of free electrons and $n = n_e + n_p + n_H$ is the total particle number density, and so n_c/n is the ionisation. This would solve the problem for the ionisation history if we knew the populations n_e, n_p, n_{2s} and n_{2p}. This is where the Lyman-α issue comes in, and that in turn brings the two-photon decay of the $2s$ state into the equations.

One of the key factors in controlling the relative population of the $n = 1$ and $n = 2$ levels is the availability of Lyman-α photons to excite a ground state $n = 1$ atom into the $n = 2$ first excited state. This has to compete with the decay of the $2p$ excited state to the ground state by emission of a Lyman-α photon. The other factor is the rate at which the two-photon decay channel can depopulate the $2s$ state of the $n = 2$ level.

The contribution of these processes to the population of the ground state is described by the equation

$$\text{rate}(2 \to 1) = \frac{d}{dt}\left(\frac{n_1}{n}\right) = A_{2s \to 1s}\left[n_{2s} - n_1 e^{-h\nu_\alpha/kT_r}\right] + L. \qquad (20.10)$$

[11] Both the $2s$ and $1s$ have zero angular momentum, so this cannot happen by emission of one photon. Two photons are required for conservation of angular momentum.

[12] As a measure of the capability of precision cosmology the analysis for the 2015 Planck data release was able to provide an estimate for this transition rate, which they label as $A_{2s \to 1s}$, from within the CMB data itself (Planck Collaboration, 2015, Section 6.7.2). Combining data from measurements of the *TT*, *TE*, *EE* modes and the low-ℓ ($\ell < 30$) polarisation analysis they obtained $A_{2s \to 1s} = 7.72 \pm 0.60\,\text{s}^{-1}$, which is consistent with the theoretically determined value of $8.2202\,\text{s}^{-1}$.

[13] This is a fairly complex process and so the discussion presents the physics highlights without all the details. The discussion follows closely the detailed treatment of Matsuda *et al.* (1971).

The first term on the right describes the two-photon decay to the ground state (20.8). The second term, L, denotes the production rate per unit volume of Lyman-α photons from the $2p \rightarrow 1s$ decay to the ground state, and is given by

$$L = A_{Ly\alpha} \left(\tfrac{3}{4} n_2(1 + N_\alpha) - 3n_1 N_\alpha \right), \tag{20.11}$$

where N_α is the number of Lyman-α photons per mode.[14]

We need to bear in mind that being a 'Lyman-α' photon means only that its frequency lies within the natural line width $\Delta\nu_\alpha$ of the Lyman-α line. Since the Universe is redshifting the photons, a photon can only be within the Lyman-α line for as long as the cosmic expansion takes to redshift it though the range of frequency $\Delta\nu_\alpha$:

$$\Delta t = \frac{a}{\dot{a}} \frac{\Delta\nu_\alpha}{\nu_\alpha}, \tag{20.12}$$

where $a(t)$ is the cosmological scale factor, and $\dot{a}/a$ is the Hubble expansion rate.

We now focus on the photons that lie within the Lyman-α line width, $\Delta\nu_\alpha$. Some of those Lyman-α photons come from the Planck law background radiation and some are the result of allowed transitions:

$$N_\alpha \Delta\nu_\alpha = \frac{1}{e^{h\nu_\alpha/kT_r} - 1} \Delta\nu_\alpha + L \frac{c^3}{8\pi\nu_\alpha^3} \Delta t = \frac{1}{e^{h\nu_\alpha/kT_r} - 1} \Delta\nu_\alpha + L \frac{c^3}{8\pi\nu_\alpha^3} \frac{a}{\dot{a}} \frac{\Delta\nu_\alpha}{\nu_\alpha}, \tag{20.13}$$

where L is given in (20.11). Hence we can eliminate N_α from (20.11) to give us the approximation[15]

$$L = \frac{8\pi\nu_\alpha^3}{c^3} \frac{\dot{a}}{a} \left[\frac{n_2}{4n_1} - e^{-h\nu_\alpha/kT_r} \right], \quad N_\alpha \simeq \frac{n_{2s}}{n_1}. \tag{20.14}$$

(This uses the result $n_{2s} = \tfrac{1}{4} n_2$ that we noted in footnote 14.)

With this we have enough information to eliminate the densities appearing in the recombination rate Equation (20.15) in favour of the ionisation $x(T)$. The result, given all the approximations that are made, is remarkably simple.

20.2.3 The Recombination Rate Equation

The upshot of all this is that, given the many simplifying assumptions, the rate of change of the ionisation is given by the equation

$$\dot{x} = \frac{1 + A}{1 + A + C} \left[R_{2c}(1 - x) e^{(B_2 - B_1)/kT} - R_{c2}nx^2 \right], \tag{20.15}$$

$$A = KA_{2s \rightarrow 1s}n(1 - x), \quad C = Kn(1 - x)R_{2c}, \quad K = \frac{c^3}{8\pi\nu_\alpha} \frac{a}{\dot{a}}. \tag{20.16}$$

$B_1 = 13.6\,\text{eV}$ and $B_2 = 3.4\,\text{eV}$ are the energies of the ground state and first excited state of the hydrogen atom, and n is the hydrogen number density, whether ionised or

---

[14] The $2p$ level($l = 1$) has three spin states, while the $2s$ level ($l = 0$) has only one. Hence $\tfrac{3}{4}$ of the electrons lie in the $2p$ level which is associated with the Lyman-α transitions. This also tells us that in equilibrium $n_{2s} = \tfrac{1}{4} n_2$.

[15] In making this approximation, L is zero to the first two orders in n_2/n_1.

not. Note that the terms in square brackets are the terms that appear in the Saha equilibrium equation for the first excited state. Compare this with the ground-state Saha equilibrium of equation (20.5). This term would be zero if such an equilibrium were realised.

The factor $(1 + A)/(1 + A + C)$ is the only place the expansion of the Universe comes into this and is referred to as the *inhibition factor*.[16] It is this term, which is less than unity unless $C = 0$, that is responsible for the slow-down of the recombination relative to the Saha equilibrium formula. This is seen in Figure 20.2. We note that when $A \gg C$, the inhibition factor is unity and the expansion does not play any direct role in controlling the ionisation state.

20.2.4 Convenient Approximations

The equations for the recombination history are sets of linear differential equations that are to be solved numerically. As the number of physical processes included in the analysis increases the complexity of these equations increases and there is substantial increase in the demand for computing resources. Analytic solutions are therefore useful but limited to the simplest of models, or to producing fitting functions.

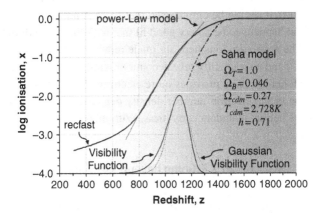

Fig. 20.2 Computed history of the ionisation of the cosmic medium for the standard cosmological parameter set.
The heavy curve is the **recfast** computation of Wong *et al.* (2008) which takes account of H and He recombination and includes a 'fudge factor' for additional uncertainties. The dot-dash curve is the Saha recombination model (Equation 20.5)
The simple power-law model is Equation (20.19): this is made to fit the redshift range. $800 < z < 1200$.
The Thomson visibility function and its approximation by a Gaussian are inset with an arbitrary but linear vertical axis. The Gaussian has its peak at $z_m = 1109$ and $\sigma = 83$. This is computed using Equation (20.30), which is itself based on Equation (20.18) which provides an excellent fit to the recombination history over the limited range $1000 < z < 1100$. ⓒ

---

[16] This appears in a slightly different form in Equations (30) and (31) of Peebles' paper, *loc. cit.*, in Equations (18), (19) (20) of Matsuda *et al.*, *loc. cit.* and in Equations (3) and (4) of Jones and Wyse.

Jones and Wyse (1985) provided analytic solutions to Equation (20.15) in the Saha regime, $2000 > z > 1300$ and in the non-Saha regime $1500 > z > 800$ using the WKBJ methodology. The somewhat unedifying solution in the latter range is

$$x_e(z) = 206.1 \frac{(\Omega_0 h^2)^{1/2}}{\Omega_B h^2} \frac{1}{z^2} \left[1 + 3.6 \times 10^{-5}(\Omega_0 h^2)^{-1}z\right]^{1/2} \left[1 + 2.26 \times 10^4 z e^{-14620/z}\right],$$
$$1300 > z > 800. \tag{20.17}$$

This is shown on Figure 20.2 using the standard model parameters. Over its range of validity it compares remarkably well with the numerical solution of the equations.

Note that the curve $x_e(z)$ has relatively low curvature over this range and so we can make approximations of the form $x_e(z) \simeq az^b$ over short segments of this range or, less accurately, over the whole range. One of the most important applications of the ionisation history $x_e(z)$ is the calculation of the optical depth of the Universe, $\tau(z)$, and the associated visibility function, i.e. the details of the cosmic photosphere. The cosmic photosphere covers a relatively narrow range of redshifts around $z \sim 1100$ where we can use the approximation[17]

$$x_e(z) \simeq 1.38 \times 10^{-3} \frac{(\Omega_0 h^2)^{1/2}}{\Omega_B h^2} \left(\frac{z}{1000}\right)^{12.75}, \quad 1200 > z > 800. \tag{20.18}$$

The approximation works fairly well even over a wider range than the region $z = 1000 - 1100$.[18] It should be emphasised that there is no physical basis for this simple power law, it just happens to be a very good fit that is useful when doing some analytic calculations.

The normalisations here are made relative to the Wong *et al.* (2008) computation using recfast, and so implicitly take account of additional effects such as helium recombination.[19] Each helium atom provides more electrons than a hydrogen atom, and so this changes the value of the electron number density per nucleon density: $n_e = n_c^0/(1 - Y_p/2)$, where Y_p is the helium abundance, by mass. With this convention, the ionisation can be greater than unity.

20.2.5 Going beyond the Simple Recombination Model

As precision cosmology gets ever more precise, we need to achieve an accuracy of around 0.1% in the recombination history. We have only dealt with a simple 3-level model of the

---

[17] The normalisation factor here is slightly different from the original Jones and Wyse (1985) value to account for other factors, such as He recombination, that are needed to fit the RECFAST curve of Wong *et al.* (2008).

[18] A similar fitting function with a wider range is shown in Figure 20.2:

$$x_e(z) \simeq 1.38 \times 10^{-3} \frac{(\Omega_0 h^2)^{1/2}}{\Omega_B h^2} \left(\frac{z}{1000}\right)^{11.3}, \quad 1300 > z > 800 \tag{20.19}$$

though this is less accurate than (20.18) over the range of the visibility function.

[19] The cosmic helium, which was ionised at early epochs, recombines at a higher temperature than does hydrogen, and so has by this time already become neutral (Kholupenko *et al.*, 2011). Helium recombines from a doubly ionised state to a singly ionised state and finally becomes neutral. At the start of hydrogen recombination, the He is all neutral: $He^{++} \rightarrow He^+$ took place at $z_{He++} \simeq 6000$, while $He^+ \rightarrow He$ took place at $z_{He+} \simeq 2500$. At recombination the main effect of the Helium is that the hydrogen number density, n_H, is smaller than the baryon number density, n_B.

hydrogen atom which seems to describe reasonably well the main features of the recombination process. But there are many other atomic processes to consider, none of which by themselves are especially significant, but which taken together will account for small but significant changes to the simple model. These changes will become observable in future experiments. Ultimately, we will see spectral lines associated with these various transitions, and we will observe small but significant deviations from the Planck spectrum due to these processes. The progress will not be unlike the progress that was made in understanding stellar atmospheres since the seminal work of Cecilia Payne-Gaposhkin a century ago.

There is the question of identifying the plethora of atomic processes that might play a significant role. This brings with it the question of how accurately we know the transition probabilities and cross-sections. There are also issues of radiative transfer in this recombining medium: in the previous discussion we did not consider the interactions between the electrons and photons in a regime where the optical depth is changing more rapidly than the expansion timescale.

This brings with it a greater number of nonlinear equations and it is clear that the simple approach just described will increase vastly in complexity. There are public domain programs which will do the computations, and these are often closely integrated or are a part of programs that compute the spectra of temperature fluctuations.[20]

20.3 Optical Depth to Thomson Scattering

The probability that a photon emitted at redshift z and coming towards us will encounter an electron and get scattered by Thomson scattering is determined by the free electron density, n_e, and the Thomson cross-section, $\sigma_T = 6.65 \times 10^{-25}\,\mathrm{cm}^2$. These are combined to give the *mean free path* to Thomson scattering, $\lambda_T = 1/\sigma_T n_e$.[21]

We start by considering a uniform medium of free electrons. The probability dP that a photon travelling a small distance $d\ell$ through such a medium will encounter an electron is

$$dP = d\ell/\lambda_T = \sigma_T n_e d\ell. \tag{20.20}$$

The mean distance travelled by any photon before it is scattered is therefore

$$\langle x \rangle = \int x dP = \lambda_T = 1/(\sigma_T n_e). \tag{20.21}$$

In a bundle of N photons travelling through the medium, $dN = -N n_e \sigma_T d\ell$ will be scattered out of the bundle in a distance $d\ell$. So, if a bundle starts with N_0 photons and travels a distance x, there will be only

---

[20] Examples of programs that explicitly compute the recombination are RICO (Fendt *et al.*, 2009), RECFAST and its descendants (Chluba *et al.*, 2010), and HYREC (Ali-Haïmoud and Hirata, 2011). These are discussed and compared within the framework of COSMOREC (Chluba and Thomas, 2011; Shaw and Chluba, 2011; Li *et al.*, 2014). Diverse programming languages are used in the implementations, which tend to be in a continual state of evolution and thereby continually out of date.

[21] This assumes that the electrons are at rest, or that any random velocity they have is much smaller than the photon velocity, i.e. the electrons are non-relativistic.

$$N(x) = N_0\, e^{-\sigma_T n_e x} = e^{-x/\lambda_T} \tag{20.22}$$

photons left not having been scattered. This describes the *attenuation* of the photon bundle. Considering a path of length L between a source and an observer in our medium of free electrons, we call $\tau = L/\lambda_T = \sigma_T n_e L$ the *optical depth* in the path. In the case of an inhomogeneous medium we will need to sum contributions $d\tau = \sigma_T n_e(\ell) d\ell$ from each of the path elements $d\ell$ at distance ℓ to get the total optical depth.

We expect a fraction $e^{-\tau}$ of the photons travelling along that path to have survived without being scattered. If $\tau < 1$ the path is said to be *optically thin*, whereas if $\tau > 1$, it is *optically thick*.[22]

20.3.1 Transparency of the Universe to Thomson Scattering

Now we apply this to the question of the transparency of the Universe in which the density of free electrons, $n_e(z)$, depends on distance (as measured by z). To calculate the total optical depth from an observer to a point at redshift z we will need to express $dP = \sigma_T n_e(z) d\ell$ as a function of redshift and so we need to relate $d\ell$ to dz. This is where the details of the cosmological model come in.

We change the element of path length, $d\ell$, to an element of redshift dz using the fact that in an interval of time dt the photon travels a distance $d\ell = cdt$. The interval of time dt corresponds to a redshift increment dz given by[23]

$$d\ell = cdt = c\frac{dt}{dz}dz = -c\frac{dz}{(1+z)H(z)}. \tag{20.24}$$

With this, the optical depth $\tau(z)$ out to redshift z, is given by

$$\tau(z) = -\int_0^z \sigma_T n_e(s) c\frac{ds}{(1+s)H(s)}, \tag{20.25}$$

$$\stackrel{\text{flat}}{=} -\frac{c}{H_0}\int_0^z \sigma_T n_e(s)\frac{ds}{(1+s)\sqrt{\Omega_m(1+s)^3 + \Omega_\Lambda}}. \tag{20.26}$$

The second of these equations is for the special case of a matter-dominated, $\Omega_r \ll \Omega_m$, flat space, where $\Omega_m + \Omega_\Lambda = 1$ and for which $H(z)^2 = H_0^2[\Omega_m(1+z)^3 + \Omega_\Lambda]$.[24] We shall see

---

[22] Thomson scattering is independent of the wavelength of the photons. However, for most scattering processes there is a wavelength dependence which means that a gas might be optically thick at some wavelengths and not at others.

[23] This follows simply from the definition $H(a) = \dot{a}/a$, which says $dt = da/aH(a)$. Since $a = 1/(1+z)$, transforming to z gives us that $dt = -dz/(1+z)H(z)$.

It is then useful to recall that the Friedmann equations gave

$$(\dot{a}/a)^2 = H(t)^2 = H_0^2\left[a^{-4}\Omega_r + a^{-3}\Omega_m + a^{-2}\Omega_k + \Omega_\Lambda\right], \tag{20.23}$$

where $\Omega_k = (1 - \Omega_r - \Omega_m - \Omega_\Lambda)$ is the contribution of the space-time curvature to the expansion rate, and is zero for a flat model.

The matter contribution Ω_m is made up of a baryonic contribution, Ω_B and a dark matter contribution, Ω_{dm}. The electron density, $n_e(z)$, is given by $n_e = \Omega_B x_e$, for a Universe in which the contribution to Ω_B is due only to hydrogen, a fraction $x_e(z)$ of which is ionised. In a more realistic model where the number fraction of singly ionised helium ions is y, then $n_e = n_H(1 + y)$, where $y \simeq 0.08$.

[24] This is a good approximation to the dynamics of the cosmic expansion following recombination. Noting that $n_e(z)$ tends rapidly to zero during the later stages of recombination, we can even ignore the Ω_Λ term when

that the explicit appearance of the $1/H_0$ factor conspires with the implicit Hubble constant dependence of $n_e(z)$ to make $\tau(z)$ independent of the Hubble constant.

The redshift z_* at which optical depth is $\tau_* = \tau(z_*) = 1$, in the absence of any intervening re-ionisation between us and the epoch of recombination, is called the *last scattering surface*. This, in effect, is where we are looking when we observe the cosmic background radiation. Because of the nature of the recombination, most of the photons we see come from a narrow range of redshifts around z_*, the last scattering surface is in fact a last scattering layer.

We shall explore this below when we discuss the visibility function. For simplicity, we restrict the discussion to a Universe containing only hydrogen with present day density n_B^0, and avoid the complications of taking the helium into account. $\sigma_T n_B^0 c$ is the Thomson collision rate if the present day Universe were completely ionised, and H_0 is the cosmic expansion rate. The leading term $\sigma_T n_B^0 (c/H_0)$ is interpreted as the ratio of the Thomson collision rate to the expansion rate at the present time.

We can reorganise the multiplying factors in Equation (20.26) by writing the coefficients in terms of present day values wherever possible.[25] Then $\sigma_T n_B^0 c H_0^{-1} = 0.0696\, \Omega_B h$ which we interpret as saying that the present day Thomson collision time is some two orders of magnitude longer than the expansion time. From Equation (20.25), ignoring Λ and putting $z \gg 1$, the optical depth due to Thomson scattering from the present to a redshift z is given by

$$\tau(z) = \sigma_T n_c^0\, c H_0^{-1} \frac{\Omega_B h^2}{(\Omega_0 h^2)^{1/2}} \int_0^z s^{1/2} x_e(s)\, ds = 0.0696 \frac{\Omega_B h}{(\Omega_0 h^2)^{1/2}} \int_0^z s^{1/2} x_e(s)\, ds. \tag{20.27}$$

We simply have to feed in an ionisation history $x_e(z)$ and do the integral, which is very straightforward using an approximation like (20.18). Using the approximation (20.18), we get

$$\tau(z) = 0.213 \left(\frac{z}{1000} \right)^{14.25}. \tag{20.28}$$

We notice the remarkable cancellation of the Hubble parameter h, Ω_B and Ω_T: the function $\tau(z)$ is independent of the cosmological parameters. This means that the last scattering surface redshift, defined to be where $\tau(z_*) = 1$, is $z_* \simeq 1146$ to this order of approximation, and the same in all cosmological models.[26]

---

considering the contribution of the optical depth from the cosmic decoupling of matter and radiation. This makes the integrals very easy.

[25] We have, for example, $n_e(z) = n_B(z) x_e(z) = n_B^0 (1+z)^3 x_e(z) = n_c^0 \Omega_B ((1+z)^3 x_e(z)$, where $n_c^0 = 3H_0'/8\pi G = 1.13 \times 10^{-5} h^2$ cm^{-3} is the present critical density, and Ω_B the density parameter for the baryonic component of the cosmic material.

[26] Hu and Sugiyama (1995, 1996, Appendices C and E respectively) show that in a higher order approximation there is a weak dependence on Ω_B such that $z_*/1000 = \Omega_B^{-0.027/(1.0+0.11\,\ln \Omega_B)}$. This gives $z_* = 1134$ for $\Omega_B = 0.0446$.

20.3.2 The Visibility Function

The *visibility function* reflects the probability that a photon received from the CMB came from a redshift z. If the recombination has been instantaneous then all photons would come from the redshift corresponding to that epoch. However, the decline in ionisation depicted in Figure 20.2 shows that we will receive photons from a variety of redshifts.

The visibility function is defined in terms of the optical depth $\tau(z)$ by the equation

$$\psi(z) = e^{-\tau} \frac{d\tau}{dz}. \tag{20.29}$$

$e^{-\tau} d\tau$ is the probability that a photon comes from where the optical depth lies between τ and $\tau + d\tau$, and so $e^{-\tau} \frac{d\tau}{dz} dz$ is the probability that an observed photon came from the redshift interval z to $z + dz$. The visibility function tells us the distribution of redshifts at which the photons received today were last scattered.

For the simple model of ionisation given in Equation (20.18) we have

$$\psi(z) = 3.06 \times 10^{-3} \left(\frac{z}{1000} \right)^{13.25} \exp \left[-0.213 \left(\frac{z}{1000} \right)^{14.25} \right]. \tag{20.30}$$

This has a maximum at $z_v \simeq 1109$.

This is a function that is well approximated by a Gaussian situated at redshift $z_M = 1109$ with dispersion $\sigma = 78.6$. Not surprisingly, this approximation holds for all values of Ω_0, Ω_B and h^2. So the microwave background photons carry information to us directly from a rather narrow band of redshifts centred on $z = 1109$.

The visibility function tells us two things. Firstly, that when we look back with the microwave background radiation, we are looking to a redshift of $z \sim 1100$ when most of the recombination of the cosmic plasma is over and the Universe is almost neutral. Secondly, the photons we see do not come from a 'wall' at one redshift, but rather from a narrow range of redshifts. This range is in fact wide enough to have important consequences for what we see.

The visibility function mixes information from different redshifts, but since the function is narrow this amounts to a blurring and attenuation of the signal from small scale fluctuations. We can estimate this as follows. The visibility function is peaked at $z \sim 1090$ with a half-height width of $\Delta z \sim 195$. This width corresponds to a scale of size $\Delta r \sim 16.2 (\Omega_m h^2)^{-1/2} \sim 42 \, \text{Mpc}$. The mass within a sphere of this radius is similar to a big galaxy cluster. This scale subtends an angle on the sky of $3'$. Temperature fluctuations on scales smaller than this will get superposed, and there will be statistical cancellation between the positive and negative fluctuations, reducing the observed fluctuation amplitude.

20.4 Re-ionisation

There are two main contributions to $\tau(z)$: the electrons that are left over from the recombination period and the electrons that reappear when the cosmic medium is re-ionised by the astrophysical processes occurring during galaxy formation.

Not all photons from the CMB reach us, some are scattered by subsequent re-ionisation of the cosmic medium during the early phases of galaxy formation when cosmic structure is evolving into its nonlinear phase of development. In rough terms about 93% of the photons we receive from the CMB come straight from the surface of last scattering, and the other 7% come from the period when the Universe has been re-ionised.

The evidence that there has been a period of re-ionisation comes from observations of the *Gunn–Peterson* effect,[27] wherein the light in the spectra of very distant quasars, $z \gtrsim 6$, is seen to be absorbed shortward of the Lyman-α (Ly-α) emission line at rest wavelength $\lambda = 121.5\,\text{nm}$ (1215Å). This happens because the light emitted at higher frequencies than the Ly-α line is redshifted on its way towards the observer and reaches a point where its wavelength is the wavelength of the Ly-α line. At that point, if there are any neutral hydrogen atoms around, it gets absorbed. The amount of absorption depends on the local density of neutral hydrogen at that point.

The *Gunn–Peterson trough* for a high redshift QSO ($z = 6.28$) is shown in Figure 20.3. The trough is seen in QSOs out to redshifts beyond 7. Mortlock and Warren (2011), in their discovery paper announcing the first redshift 7 QSO, remark that 'the numerous transmission spikes at $z \sim 5.5$ give way to increasingly long Gunn–Peterson troughs

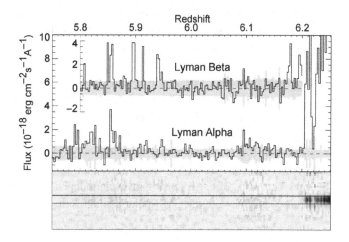

The Gunn–Peterson effect observed in QSOs. The top two plots show spectra of the $z = 6.28$ quasar QSO SDSS 1030+0524. There is a lack of flux shortwards of the Ly-α and Ly-β lines. This absorption shortwards of the lines is the Gunn–Peterson trough. The shaded regions show the 1σ noise levels. The bottom panel shows the sky-subtracted 2-dimensional spectrum in the Ly-α region, this emphasises the absence of any significant spectral signal in the trough.
From Becker *et al.* (2001, Fig. 3). The bottom panel has been processed for publication in monochrome. *(Reproduced by permission of the AAS.)*

[27] Several people had recognised that very distant quasars would provide information about the state of ionisation of the intergalactic medium at high redshift. (Gunn and Peterson, 1965) discussed the effect within weeks following the report by Schmidt (1965) of the observation of the Lyα emission line in the quasar 3C9. Of course, the fact that all three authors were at the California Institute of Technology at the time would have been a factor in developing this idea.

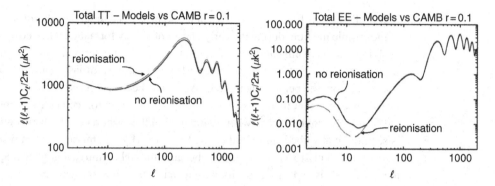

Fig. 20.4 Effect of various re-ionisation histories on the *TT* and *EE* CMB power spectra. The re-ionisation effects scales $\lesssim 20$, and has relatively more impact on the *EE* spectrum. The model sets the ratio r of the tensor to scalar modes to $r = 0.1$ Contamination from foregrounds and gravitational lensing are not taken into account. Adapted from Trombetti and Burigana (2012a, Fig. 6).
(Permission granted.)

at $z \sim 6$ and, finally, to almost complete absorption beyond $z \sim 6.3$'. The Universe is 3.15 billion years old at that time.[28] The discovery of gamma ray bursts out to redshifts of $z \sim 10$ and beyond will clearly add to this. The status of the ionisation of the cosmic plasma from the point of view of QSO observations is well reviewed by Fan (2012).

Observations of the CMB anisotropy provide an estimate of the redshift at which the re-ionisation of the Universe must have taken place. The primary parameter determined in this way is τ, the optical depth of the Universe to Thomson scattering due to the re-ionisation of the cosmic plasma. The value assigned by the Planck Team (2015, Section 10.2) is $\tau = 0.066 \pm 0.016$.[29] This comes from the low-ℓ polarisation modes of the temperature power spectrum, which implies that re-ionisation took place at around $z_{\rm re} = 8.8^{+1.7}_{-1.4}$ (see Section 22.5). Use of CMB lensing gives an independent estimate of $\tau = 0.071 \pm 0.016$, which is consistent with the CMB polarisation result.

---

[28] In a flat Universe, for redshifts $(1 + z) > (\Omega_\Lambda / \Omega_m)^{\frac{1}{3}} = 1.31$ with the Planck Collaboration I (2015a, Table 9) parameters, the age of the Universe is $t(z) \simeq (1 + z)^{-3/2} \times 17.38\,{\rm Gyr}$. Putting $z = 1089$ into this equation gives $t = 0.48$ Myr for the time of the last scattering surface. Of course, putting $z = 0$ in this over-estimates the age by a considerable factor, the difference between this and the computed value of 13.8 Gyr is due to the considerable Λ-acceleration.

[29] Prior to this, WMAP had returned an estimate of $\tau = 0.091 \pm 0.019$ (Dunkley and Spergel, 2009). With the assumption of instantaneous re-ionisation this corresponds to a redshift of $z_{\rm re} \simeq 11.0 \pm 1.4$.

21

CMB Polarisation

The polarisation of the CMB provides one of the most important tools we have for exploring the Universe. We will see that there are two distinct modes of polarisation, the E-mode and the B-mode, from which we can produce E-mode and B-mode maps. The modes are sensitive to different physical phenomena and so we can separate out different effects such as the gravitational lensing of the CMB and the contributions from late re-ionisation of the cosmic plasma. These would otherwise be inextricably mixed in with the usual temperature maps. The cosmological B-mode is only generated by primordial gravitational waves, so their discovery would be one of the most important advances in understanding the very earliest stages of the cosmic expansion.

21.1 A Bit of History

In 1968, Rees (1968) pointed out that in an anisotropic model Universe, the microwave background radiation emerging from the recombination period would be polarised. The observed polarisation would reflect the last few scatterings of the photons just prior to the recombination. The first attempt to detect polarisation was reported by Nanos (1979) on the basis of his thesis work in 1973–1974 under the supervision of David Wilkinson. Perhaps not surprisingly, he could only put upper limits on any anisotropy, but it was an important first step. Subsequently, Caderni *et al.* (1978); Lubin and Smoot (1981) attempted to improve on the limits set by Nanos. However, it was not until 2002 when Kovac *et al.* (2002) reported the first positive detection of polarisation by the DASI ground-based experiment team. Shortly afterwards a positive detection was reported from the WMAP experiment (Kogut *et al.*, 2003; Page, 2003).

The 2003 flight of the BOOMERANG long-duration balloon-borne experiment, launched from the South Pole, was one of the great CMB mapping experiments. The 1998 flight[1] had already revealed, for the first time, the first two peaks of the temperature power spectrum (de Bernardis, 2000; Ruhl, 2003), while the 2003 flight yielded the first polarisation power spectrum. The data and analysis papers of the 2003 flight did not appear until 2006 (Jones, 2006; MacTavish, 2006).

---

[1] The 1998 flight flew 16 unpolarised detectors operating in four wavebands.

21.2 The Background Physics

Here we present the physics of how electromagnetic waves in a vacuum scatter off electrons. The main process of interest in the cosmological context is Thomson scattering, the scattering of light off a free electron. We shall also discuss Rayleigh scattering, the scattering of light off a bound electron that is free to oscillate. Rayleigh scattering is considerably less important in the cosmological context (Peebles and Yu, 1970), but it makes enough of a contribution to the interpretation of precision observations of the cosmic background radiation that it has to be taken into account (Takahara and Sasaki, 1991).

But there is another important reason to discuss Rayleigh scattering here. Rayleigh scattering of sunlight by our atmosphere is responsible for the colour and polarisation of our sky on a clear day. That familiarity helps in coming to understand the phenomenon of polarisation that arises from the electron scattering of light.

21.2.1 The Polarisation of Light – Discovery

The polarisation of light was discovered by Huygens (1690, the work had been completed by 1678, but not published until later). The discovery was enabled by the report of Erasmus Bartholinus in 1669 that light passing through crystals of Iceland spar[2] would show two images. Bartholinus had performed an extensive series of experiments on the phenomenon of double refraction the results of which were reported in 1670 (Bartholinus, 1670),[3] but he did not present any explanation for what he saw.

It was much later, in 1809, that Malus discovered the polarisation of light by reflection. The story goes that he observed reflections in a window through a crystal of calcite and found that the intensity varied with the angle of the crystal: this became what is now called *Malus' law*. The intensity was found to vary as

$$I(\theta) = I_0 \cos^2 \theta, \tag{21.1}$$

where θ is the angle through which the crystal is turned, $\theta = 0$ corresponding to the angle at which the intensity is a maximum, I_0. That same year, Arago reported for the first time that the daylight sky is polarised[4] and that the maximum darkening of the sky as seen through a calcite crystal occurred at 90° from the Sun.

The first two decades of the 19th century were marked in physics by a series of experiments in France by Arago, Biot and Fresnel and in England by Brewster and Young that served to establish the wave theory of light for the next 100 years. By 1803, Young had

---

[2] Iceland spar is a form of Calcite, which is still used in optical polarimeters. However, all the Icelandic sources have been depleted and most of it now comes from Mexico.

[3] Rasmus Bartholin (1625–1698) was a Danish mathematician who was a professor of the University of Copenhagen. Although he conducted many experiments on the phenomenon, he did not know about polarisation. That was to come shortly afterwards from Huygens. Interestingly, Newton had also experimented with Iceland spar and used the results as one of the reasons to reject the idea of Hooke and Huygens that light was a wave phenomenon.

[4] There is a contentious suggestion that 1000 years earlier the Norwegian Vikings had used 'Sunstone' to navigate across the Atlantic using the polarisation of the sky much as birds and bees do.

put forward important evidence, based on the interference of light, that light was a wave phenomenon. The 'Young's slits' experiment is one of the iconic experiments in physics (although Young did not use slits, he used two holes). Young's argument for the wave nature of light was made by analogy with water waves. On the basis of that experiment, and with the analogy, he was able to estimate the wavelength of violet light, giving a value of around 400 nm, remarkably close to the right answer given that he was working with sunlight and not monochromatic light.

But the Newtonian view was so strongly held that the argument was not finally settled until 1817 with the work of Fresnel. Fresnel had been working on the phenomenon of interference using monochromatic light. He discovered that beams of monochromatic light that were polarised at right angles to each other did show interference. This enabled Fresnel to argue convincingly that light could not be a longitudinal wave phenomenon, but was in fact a transverse wave phenomenon.[5]

> **Short biography:** **François Arago** (1786–1853) was from the Catalan region of South-West France, and became an important scientific and political figure. He is well known among astronomers as a long time director of the Paris Observatory (1830–1853). He was elected to the Paris Academy of Sciences at the age of 23, and at the same time became a professor at the prestigious École Polytechnique. His political career started in 1830 when he was elected to the Chamber of Deputies as the member for Perpignan. By 1848, with the forced abdication of 'King of the French' Louis Phillippe I, Arago became Minister of War, Navy and the Colonies and shortly thereafter was made President of the Executive Power Commission (11 May 1848). He served in this capacity as provisional head of state until 24 June 1848. During that period he managed to abolish slavery, despite severe opposition from wealthy and powerful lobby groups (Monthly Notices Obituary, 1854).

Arago (1820) reported on the first observation of polarised light in an astronomical object: he observed the tail of a comet with a twin prism system that would split light into two polarised components and saw two images. His deduction was that comet tails contained dust.

If light really was a wave phenomenon, the issue to be faced was the nature of the medium in which the wave propagated. The complete wave theory of light was eventually produced by Maxwell (1865) by building on the work of Faraday. Specifically, what Maxwell did was to modify Ampère's law in the situation that the electromagnetic fields were time dependent. Maxwell's stroke of genius was to add a time derivative of the electric field to the current driving the magnetic field. In a vacuum and in Gaussian cgs units (see Appendix A.4) this is:

$$\nabla \times \mathbf{B} = \frac{4\pi}{c}\mathbf{J} \quad \xrightarrow{\text{Maxwell}} \quad \nabla \times \mathbf{B} = \frac{1}{c}\frac{\partial \mathbf{E}}{\partial t} + \frac{4\pi}{c}\mathbf{J}, \tag{21.2}$$

thereby creating an important symmetry with the Faraday Law:

$$\nabla \times \mathbf{E} = -\frac{1}{c}\frac{\partial \mathbf{B}}{\partial t}. \tag{21.3}$$

---

[5] There was very strong opposition to Fresnel's claims from Poisson who had studied Fresnel's theory in meticulous detail. Arago, who had up to that time also been an adherent of the corpuscular theory of light, was instrumental in helping the Fresnel view of the transverse wave nature of light gain acceptance.

Without this additional term, the field equations are in fact inconsistent. This can be seen by noting that, taking the divergence of the left-hand version of the Ampère law in Equation (21.2), we get $\nabla \cdot \mathbf{J} = 0$, which is correct for steady state systems, but not in a time dependent situation. Taking the divergence of Maxwell's version on the right-hand side yields a proper charge conservation law:

$$\frac{\partial \rho}{\partial t} + \nabla \cdot \mathbf{J} = 0. \tag{21.4}$$

See Appendix A (Section A.4) for a brief resumé of our units for Maxwell's equations.

But the Maxwell theory was not without controversy. Both Kelvin and Helmholtz, for example, objected strenuously to Maxwell's theory. The matter was not settled until 1886, with the detection of visible electromagnetic waves by Hertz.[6] But even then, the issue of the medium in which these waves propagated remained open and was a point of much speculation, argument and experiment until Einstein came up with special relativity and a set of equations that did not require an aether.[7]

21.2.2 Maxwell's Equations in vacuo: the Theory of Light

Maxwell's theory of classical electromagnetism is certainly one of the great achievements of 19th century physics, and, many would argue, ranks among the greatest fundamental theories of modern physics, alongside those advanced by Newton and Einstein. The Maxwell equations form the basis of all electromagnetic phenomena and in combination with the Lorentz force law are the basis of the subject that is referred to as *Classical Electrodynamics*.

After a decade of work, the 20 equations appeared in published form in 1865 (Maxwell, 1865, Part V1 p.497 is on the *Theory of Light*). It must be appreciated that, at that time, vectors were not known and so the 20 equations had to be written out in terms of the components of the field variables. It is, in fact, not easy to recognise these equations in the original paper. Maxwell himself published several works on the subject (Maxwell, 1954).

Short biography: **James Clerk Maxwell** (1831–1879) was born James Clerk and somewhat later adopted the surname Maxwell on inheriting a large estate in south-west Scotland. He went to Edinburgh University, in 1847, aged 16, during which period he conducted experiments with polarised light about which he published a paper when 18 years of age: the scene was set. Leaving Edinburgh in 1850, he went to Trinity College, Cambridge, until 1856, during which period he started working on what were to become the Maxwell equations. After a 5 year period in Aberdeen he went to King's College London from 1860–1865 where he completed and published the series of papers on electromagnetism for which he is known. In 1871 he went back to Cambridge where he became the first Cavendish Professor of Physics and died there, from abdominal cancer, at the age of 48.

---

[6] The first radio transmission was made by David Hughes in 1879 and demonstrated to the Royal Society in 1880. However, the transmission was at the time believed to be due to induction and not to propagation of waves. Hughes is also known as the inventor of, among other things, the crystal radio. He became a Fellow of the Royal Society in 1880 and won the Gold Medal of the Royal Society in 1885.

[7] Einstein did point out that while the equations of special relativity did not require an aether, they did not refute the existence of an aether.

The detection of electromagnetic waves in 1881 by Hertz set the seal on the Maxwell theory and the notion that light consisted of transverse oscillations of an electromagnetic field.

It is worth going back to the Maxwell wave equations for the electric and magnetic components, $\mathbf{E}(\mathbf{x}, t)$, $\mathbf{B}(\mathbf{x}, t)$, of a propagating electromagnetic wave:

$$
\begin{aligned}
\nabla^2 \mathbf{E} - \frac{1}{c^2} \frac{\partial^2 \mathbf{E}}{\partial t^2} &= 0, \\
\nabla^2 \mathbf{B} - \frac{1}{c^2} \frac{\partial^2 \mathbf{B}}{\partial t^2} &= 0.
\end{aligned}
\tag{21.5}
$$

There is a particularly important, yet simple, *plane wave* solution to these equations which can be written in the form

$$
\begin{aligned}
\mathbf{E} &= \mathbf{E}_0 \, e^{i(\mathbf{k} \cdot \mathbf{r}) - \omega t}, \\
\mathbf{B} &= \mathbf{B}_0 \, e^{i(\mathbf{k} \cdot \mathbf{r}) - \omega t},
\end{aligned}
\tag{21.6}
$$

where the vector $\mathbf{k}$ is in the direction of propagation of the wave and is perpendicular to the wave front. By substituting Equations (21.6) into the Maxwell equations it can be shown that the only components of $\mathbf{E}(\mathbf{x}, t)$ and $\mathbf{B}(\mathbf{x}, t)$ that propagate are those perpendicular to the direction of motion $\mathbf{k}/|\mathbf{k}|$: electromagnetic waves are *transverse waves*. Moreover the vectors $\mathbf{E}(\mathbf{x}, t)$ and $\mathbf{B}(\mathbf{x}, t)$ are mutually perpendicular. Writing $\mathbf{n} = \mathbf{k}/|\mathbf{k}|$ these statements can be expressed as the vector product

$$
\mathbf{B} = \mathbf{n} \wedge \mathbf{E}.
\tag{21.7}
$$

$\mathbf{E}(\mathbf{x}, t)$, $\mathbf{B}(\mathbf{x}, t)$ and $\mathbf{k}$ form a (right-handed) orthogonal triad. The direction of the electric field is referred to as the *direction of polarisation*. The plane perpendicular to the electric field is, by convention, called the *plane of polarisation* and contains the magnetic field of the wave and the direction of propagation.

21.2.3 Scattering of Light by Electrons, Atoms and Molecules

Light scatters off matter through a variety of mechanisms. Here we shall discuss two such mechanisms, Rayleigh scattering and Thomson scattering. Rayleigh scattering involves scattering of light off bound electrons. Thomson scattering involves scattering off free electrons. The comparison is useful in coming to an understanding of the phenomenon of polarisation and the mechanisms that cause it. This section generally follows the discussion of Panofsky and Phillips (2005, Sections 22–2, 22–3), which is one of the clearest discussions of this problem.

When light hits an electron, the motion of the electron is affected by the electric field of the light, and by the fact that the electron itself radiates when caused to accelerate. The latter causes a back-reaction on the motion of the electron. So there are two elements to this problem: the radiation of the electron under the influence of the light, and the back reaction this has on the motion of the electron. The former is solved by an equation due to Larmor, the *Larmor power equation*, and the second is addressed classically by the *Abraham–Lorentz model* for the radiative reaction of the electron.

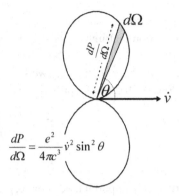

$$\frac{dP}{d\Omega} = \frac{e^2}{4\pi c^3} \dot{v}^2 \sin^2 \theta$$

Fig. 21.1   Radiation pattern from accelerating electron moving with instantaneous non-relativistic velocity v. This is to be seen as the cross-section of a doughnut shape about the acceleration vector $\dot{v}$. If the source is moving relativistically the radiation is thrown forwards into a beam. ⓒ

Larmor's Power Equation

Electrons radiate when accelerated, see Figure 21.1. The physical understanding of that was a key problem that was addressed during the latter part of the 19th century by some of the greatest physicists and mathematicians of the time. The story is told and details presented in most of the advanced textbooks on electromagnetism. Here we need highlight only a few of the main results. Perhaps the most important in the present context is the *Larmor power formula*[8] which describes the total power, P, radiated by the accelerating electron. This comes in various forms depending on the choice of units and whether we use velocity or momentum:

$$P \underset{\text{SI}}{=} \frac{2}{3} \frac{e^2}{6\pi\epsilon_0 c^3} \dot{\mathbf{v}}^2 \underset{\text{cgs}}{=} \frac{2}{3} \frac{e^2}{c^3} \dot{\mathbf{v}}^2 \underset{\text{cgs}}{=} \frac{2}{3} \frac{e^2}{m^2 c^3} (\dot{\mathbf{p}} \cdot \dot{\mathbf{p}}), \tag{21.8}$$

where $\mathbf{p} = m\mathbf{v}$ is the classical momentum of the electron. The last form involving the momentum allows generalisation to the relativistic version of Larmor's formula by making the substitution $\dot{\mathbf{p}} \cdot \dot{\mathbf{p}} \to \dot{p}'_\mu \dot{p}'_\mu$, where now p_μ is the four momentum and the dot derivative is with respect to proper time.

Short biography: Sir Joseph Larmor (1857–1942) was arguably the last great 19th century classical physicist. He was the Lucasian Professor at Cambridge in 1903 to 1932 and a member of Parliament for the University from 1911 to 1922. He had published the Lorentz transformation in 1897, two years before Lorentz did (Larmor, 1897, see p.229). Besides the *Larmor power equation* he is famous for the *Larmor Precession* which lies at the heart of modern magnetic resonance imaging. His book *Aether and Matter* (Larmor, 2012) was highly influential and marked the transition from the classical physics of the 19th century to the relativistic and quantum physics of the 20th.

He was somewhat unfortunate in that his work was eclipsed by that of Lorentz and Einstein, though in fact he never came to terms with either relativity or the new quantum principles of Pauli and Einstein.

---

[8]   The translation between SI and cgs units in such cases is simply a replacement of the squared charge: $e^2_{\text{cgs}} \Leftrightarrow e^2_{\text{SI}}/4\pi\epsilon_0$.

We need to go a bit beyond the Larmor formula, we need to know the angular distribution of the radiation relative to the instantaneous direction of motion of the electron. In other words, how much of the radiation goes in which direction? We characterise this by setting up a spherical polar coordinate system in which the direction of motion of the electron is the z-axis. By symmetry the radiation intensity in a given direction cannot depend on the azimuthal angle ϕ, it can only depend on the angle θ the direction makes relative to the z-axis.

The result, while deceptively simple, is technically complex and involves a considerable amount of work to derive. Most texts on advanced electrodynamics cover this important subject, e.g. Panofsky and Phillips (2005, Ch.20). So we simply quote the result for the power radiated into a small cone of solid angle $d\Omega$:

Power radiated into a solid angle $d\Omega$ by an accelerating electron:

$$\frac{dP}{d\Omega} = \frac{e^2}{4\pi c^3}\dot{\mathbf{v}}^2 \, \sin^2\theta. \tag{21.9}$$

The $\sin^2\theta$ is the signature of dipole radiation. We notice the absence of a velocity term in this equation, the velocity only comes into play when the electron motion becomes relativistic.

It should be stressed that Equation (21.9) is only correct for non-relativistic motion of the electron. A relativistic electron that is accelerating is seen to throw the radiation in a forward direction relative to its direction of motion. This is simply a consequence of the Lorentz transformation between the electron's instantaneous rest frame and the frame of the observer.[9]

We now move on to determining the acceleration $\dot{\mathbf{v}}$ experienced by an electron due to an incident electric field. First we tackle the simplest problem of a free electron moving in the field of an electromagnetic plane wave.

Electron Accelerated by an Electromagnetic Wave

When examining the scattered radiation field of an electron that is executing periodic oscillations with some frequency ω we have to deal with Equation (21.9) in a slightly different manner. We are generally not interested in the periodic fluctuations of the emitted power $P(t)$ which arise from the oscillations of frequency ω, we want to know the time averaged power over longer periods of time.

If the position $\mathbf{r}$ and velocity $\mathbf{v}$ are periodic with period $2\pi/\omega$ we can average the power given by (21.9) over a cycle of one oscillation (all cycles are the same in this case). We get

$$\left\langle \frac{dP}{d\Omega} \right\rangle = \frac{e^2}{4\pi c^3} \langle \dot{\mathbf{v}}^2 \rangle \sin^2\theta, \tag{21.11}$$

---

[9] For motion in a straight line at relativistic velocities, the analogue of this equation is

$$\frac{dP}{d\Omega} = \frac{e^2}{4\pi c^3}\dot{\mathbf{v}}^2 \, \frac{\sin^2\theta}{(1-\beta\cos\theta)^5}, \quad \beta = v/c. \tag{21.10}$$

The equation is considerably more complicated for a particle not moving in a straight line.

where the brackets $\langle \cdot \rangle$ denote an average over one cycle of the quantity concerned.[10] In the case that the oscillations are driven by a time varying electric field $\mathbf{E} = \mathbf{E}_0 \cos \omega t$ we can write

$$m\dot{\mathbf{v}} = m\ddot{\mathbf{r}} = e\,\mathbf{E}_0 \cos \omega t, \tag{21.13}$$

with which

$$\langle \dot{\mathbf{v}}^2 \rangle = \frac{e^2 E_0^2}{m^2} \langle \cos^2 \omega t \rangle = \frac{e^2 E_0^2}{m^2} \frac{\omega}{2\pi} \int_0^{2\pi/\omega} \cos^2 \omega t \, dt = \frac{e^2 E_0^2}{2m^2}. \tag{21.14}$$

With this we have an expression for the scattered energy per unit solid angle at an angle θ per unit time:

Angular distribution of power radiated when the electron oscillates in the electric field $E(y) = E_0 \cos \omega t$ of a light ray:

$$\left\langle \frac{dP}{d\Omega} \right\rangle = \frac{c}{8\pi} \left(\frac{e^2}{mc^2} \right)^2 E_0^2 \sin^2 \theta. \tag{21.15}$$

This approximation is valid in the non-relativistic limit and provided the displacement of the charge during one cycle is much less than the wavelength of the oscillating electric field. In other words, this is a non-relativistic long wavelength approximation.

Differential and Total Cross-section

The most general way of characterising the scattering of an incident beam of light by electrons is to look at the *differential cross-section*, which is defined as:

Definition 21.1
Differential scattering cross section:

$$\left\langle \frac{d\sigma}{d\Omega} \right\rangle = \frac{\left\langle \dfrac{dP}{d\Omega} \right\rangle}{\langle \text{Incident Flux} \rangle}. \tag{21.16}$$

In cgs units, the units of the differential cross section are $cm^2 \, str^{-1}$.

The numerator is the scattered energy per unit solid angle at an angle θ per unit time, and the denominator is the incident energy per unit area per unit time. So this quantity measures the part of the incident energy that is scattered into a direction θ, per unit solid angle.[11] An alternative way of writing this last equation is

---

[10] The time-average of a varying quantity $f(t)$ over a window of length T is defined as

$$\langle f \rangle = \frac{1}{T} \int_0^T f(t) \, dt. \tag{21.12}$$

Here we are going to put $F(t) = \cos^2 \omega t$ and $T = 2\pi/\omega$ and get $\langle \cos^2 \omega t \rangle = 1/2$.

[11] This corresponds with the familiar notions of cross-section and differential cross-section that we encounter when particles hit a target particle: we compare the number of incoming particles with the number scattered into a given solid angle in a particular direction.

$$\frac{d\sigma(\theta)}{d\Omega} = \frac{\text{scattered energy per unit solid angle at angle } \theta \text{ per unit time}}{\text{incident energy per unit area per unit time}}. \tag{21.17}$$

The total cross-section is simply the integral of the differential cross-section over all solid angles:

Definition 21.2
Total cross-section:

$$\sigma = \int_{4\pi} \left\langle \frac{d\sigma}{d\Omega} \right\rangle d\Omega. \tag{21.18}$$

In cgs units, the units of the total cross-section are cm^2.

The total cross-section is an estimate of the target size offered by the electron to the scattering process. The differential cross-section breaks this down to provide the angular dependence of the scattered light. If the density of target electrons is n_e particles per unit volume, the length scale $l_\gamma = (\sigma_T n_e)^{-1}$ is called the *mean free path* of the photons to the process of electron scattering described by cross-section σ_T. A low electron density n_e gives a longer mean free path than a high electron density, when there are, in effect, more electrons around with which to collide. A volume containing electrons that is smaller in size than the mean free path is effectively almost transparent: only a small amount of light is scattered out of the incident beam. As an example, the sky is transparent while clouds are not.[12]

21.2.4 Thomson Scattering

Scattering of a Monochromatic Linearly Polarised Wave

One of the simplest solutions to the Maxwell equations is the monochromatic linearly polarised electromagnetic plane wave, propagating in a vacuum. In this solution (see Equation 21.6), the electric field is given by the equation

$$\mathbf{E} = \mathbf{E}_0 e^{i(\mathbf{k} \cdot \mathbf{r} - \omega t)}. \tag{21.19}$$

Here the electric field of the wave always lies in the plane defined by $\mathbf{E}_0$, and the wave propagates in the direction of the vector $\mathbf{n} = \mathbf{k}/|\mathbf{k}|$ with frequency ω. The wavelength of the wave is $\lambda = 2\pi/|\mathbf{k}|$ and since the wave propagates at the speed of light, we have $\omega = |\mathbf{k}|c$.

The energy carried by a time-dependent electromagnetic field is given by the *Poynting vector*:

---

[12] As we shall see later (Section 21.2.5), in the ideal Earth's atmosphere the dominant scattering process is Rayleigh scattering, while in the clouds the dominant scattering process is Mie scattering. In the cloud-free and pollution-free atmosphere the scattering particles, molecules, are far smaller than the wavelength of visible light, so the scattering mechanism is Rayleigh scattering of the light. The ice particles in the clouds are bigger than the wavelength of visible light, so the dominant form of scattering there is Mie scattering. The sky is almost transparent to Rayleigh scattering (we can see stars in the daytime), while the clouds are opaque to Mie scattering.

$$\mathbf{S} = \frac{c}{4\pi}\mathbf{E} \times \mathbf{B}. \tag{21.20}$$

Here $\mathbf{S}$ is the flux of electromagnetic energy passing through a unit area in a unit time and has units of energy per unit area per unit time. It is expressed in Gaussian cgs units (see Appendix A, Section A.1).

In a plane electromagnetic wave $\mathbf{E}$ and $\mathbf{B}$ are orthogonal to each other and to the direction of propagation $\mathbf{n}$ of the wave. The $\mathbf{B}$-field is related to the electric field $\mathbf{E}$ via Equation (21.7), and so it is not necessary to describe them individually. Given $\mathbf{E}$, we can simply calculate $\mathbf{B}$.

By convention, the plane containing $\mathbf{n}$ and $\mathbf{B}$ is called *the plane of polarisation*: it is normal to $\mathbf{E}$. Occasionally, a *polarisation vector* $\mathbf{e}$ is defined to be the unit vector in the direction of the electric field: $\mathbf{e} = \mathbf{E}/|\mathbf{E}|$. The amplitudes and phases of plane waves may not be constants: they may fluctuate in time. In that case the amplitudes E_x^0 and E_y^0 can be replaced by time averages.

The Thomson Cross-section

Consider a small surface element of area dA having its normal along a unit vector $\mathbf{n}$. The energy flux across the surface element $\mathbf{n}\, dA$ in an interval of time dt is $dE = \mathbf{S} \cdot \mathbf{n}\, dA\, dt$. In the case of a light beam the time averaged power is the time average of the Poynting vector:

$$\langle S \rangle = \frac{c}{8\pi}|E_0^2|, \quad \mathbf{E} = \mathbf{E}_0 e^{i(\mathbf{k}\cdot\mathbf{r}-\omega t)}. \tag{21.21}$$

With this we get the famous expression for the differential cross-section:

Differential cross-section for electron scattering of a light beam:

$$\left\langle \frac{d\sigma}{d\Omega} \right\rangle = \left(\frac{e^2}{mc^2} \right)^2 \sin^2\theta = r_0^2 \sin^2\theta, \quad r_0 = \frac{e^2}{mc^2} = 2.8 \times 10^{-13}\text{ cm}. \tag{21.22}$$

This is independent of frequency, and depends only on the fundamental constants of nature e, c, m and on the angle θ. This combination of constants $r_0 = e^2/mc^2 = 2.8 \times 10^{-13}$cm appearing in (21.22) is called the *classical radius of the electron*.

It is trivial to integrate this over all solid angles to get:

Thomson cross-section:

$$\sigma_T = \frac{8\pi}{3} r_0^2 = 6.65 \times 10^{-25}\text{ cm}^2. \tag{21.23}$$

We can then express the differential cross-section for Thomson scattering (21.22) as

$$\frac{d\sigma}{d\Omega} = \frac{3\sigma_T}{8\pi}|\mathbf{n}_1 \cdot \mathbf{n}_2|^2, \tag{21.24}$$

where $\mathbf{n}_1$ and $\mathbf{n}_2$ are the directions of polarisation of the incident and scattered photons respectively.

Scattering of Unpolarised Light

When the incident beam of radiation is a mixture of waves having arbitrary direction of polarisation, the beam is said to be unpolarised. Because Maxwell's theory is linear in the fields we can simply view the beam as a superposition of waves, each having the electric field lying in the plane perpendicular to the direction of propagation of the beam. It is simply necessary to average the contributions from all of these field orientations; in other words, sum up the contributions to the light scattered in a direction **n** relative to the direction **k** of the incoming beam.

To do this we need to set up some coordinates and angles as shown in Figure 21.2. By symmetry the scattering into a solid angle $d\Omega$ can only be a function of the angle θ the element $d\Omega$ makes with the beam direction. The scattering of polarised light (Equation 21.22) is defined in terms of the angle Θ that the emergent radiation makes with the direction of the electric field **E**. Θ corresponds to the angle θ in Equation (21.9): it sets the angle the scattered component makes with the incoming electric field.

We have
$$\cos\Theta = \sin\theta\cos\phi, \tag{21.25}$$

with which

$$\sin^2\Theta = 1 - \cos^2\phi\sin^2\theta. \tag{21.26}$$

We need to replace $\sin^2\Theta$ (as in Equation 21.22) with its averaged value:

$$\langle\sin^2\Theta\rangle_0^{2\pi} = 1 - \langle\sin^2\phi\rangle_0^{2\pi}\sin^2\theta, \tag{21.27}$$

$$= 1 - \frac{1}{2}\sin^2\theta, \tag{21.28}$$

$$= \frac{1}{2}(1 + \cos^2\theta), \tag{21.29}$$

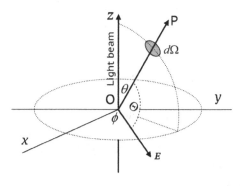

Fig. 21.2   Definition of angles for calculating the differential cross-section for scattering of unpolarised light. The Oz-axis is the direction of propagation of the beam, with the xy-plane set up arbitrarily perpendicular to Oz. The electric field **E** lies in the xy-plane making an angle ϕ with the Ox-axis. Light is scattered into a solid angle $d\Omega$ making an angle θ with the beam direction and Θ with the electric field. The differential cross-section is to be expressed as a function of the angle θ by averaging over all angles ϕ.
(Note the slight change in notation from that used in Equation (21.9): we have replaced the angle θ in (21.9) with Θ since we wish to express our result in terms of the angle denoted by θ, as is standard.) ©

where we have used the notation $\langle . \rangle_0^{2\pi}$ to denote the average of the quantity taken over the range $[0, 2\pi]$. Substituting this into Equation (21.22) we get the famous equations:

Thomson scattering of unpolarised light
Differential cross-section:

$$\left\langle \frac{d\sigma}{d\Omega} \right\rangle = \left(\frac{e^2}{mc^2} \right)^2 \frac{1 + \cos^2 \theta}{2} = \frac{1}{2} r_0^2 (1 + \cos^2 \theta). \tag{21.30}$$

Total cross-section:

$$\langle \sigma_T \rangle = \frac{8\pi}{3} \left(\frac{e^2}{mc^2} \right)^2 = \frac{8\pi}{3} r_0^2, \tag{21.31}$$

which is just the total Thomson cross-section of Equation (21.23).

We can eliminate r_0 between these equations to give the differential cross-section in terms of the Thomson cross-section:

$$\left\langle \frac{d\sigma}{d\Omega} \right\rangle = \frac{3\sigma_T}{16\pi} (1 + \cos^2 \theta). \tag{21.32}$$

With an incoming unpolarised monochromatic beam of intensity, i.e. incident flux, I_0, the amount of radiation scattered into direction θ is therefore

$$I(\theta) = \frac{3\sigma_T}{16\pi} I_0 (1 + \cos^2 \theta). \tag{21.33}$$

It should be noted that these results are independent of the photon frequency ω provided $\hbar\omega \ll m_e c^2$, where $\hbar = h/2\pi$, h being Planck's constant.

The Compton wavelength and frequency of the electron:

$$\lambda_{\text{Compton}} = \frac{h}{m_e c}, \quad \omega_{\text{Compton}} = \frac{m_e c^2}{\hbar}. \tag{21.34}$$

The general subject of scattering of photons off electrons is usually called *Compton scattering*, in honour of Arthur Compton who was awarded the Nobel Prize in 1927 for his work on scattering of X-Rays. Compton discovered that the wavelength of the scattered X-rays was increased in scattering off electrons, an effect that is now called the *Compton effect*. This was clear evidence for the particle nature of electromagnetic radiation and so reinforced the wave–particle duality of the electromagnetic field that had arisen out of the then-new quantum mechanics. At frequencies on the order of or higher than the Compton frequency, quantum effects come into play. The scattering cross-section for that case is called the *Klein–Nishina cross-section* and the low energy limit of the Klein–Nishina formula is the Thomson cross-section.

21.2.5 Rayleigh Scattering

The sky is blue because of Rayleigh scattering, a much stated aphorism. If there were no atmosphere our sky would look black at all times of day, as it does from space. We know

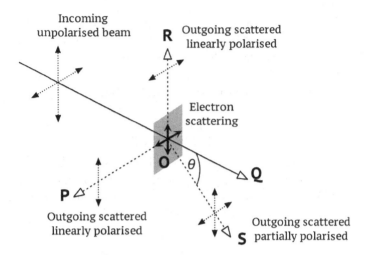

Incoming
unpolarised beam

R Outgoing scattered
linearly polarised

Electron
scattering

P Outgoing scattered
linearly polarised

Q

θ

Outgoing scattered
S partially polarised

Fig. 21.3 An unpolarised beam of light propagating in the direction **OQ** interacts with a bound electron at **O**. The electric field of the incident beam lies in a plane perpendicular to **OQ** when it strikes the electron at **O**. The observer looking along **PO** towards the electron only sees the electric field from components of the motion of the electron along **OR**. Likewise an observer looking along **RO** at the electron sees only the field from electron motion along **PO**. The observer at **S** is in the plane **ORQ** situated where the angle ∠ **QOS** is θ. He sees all of the radiation due to components of the electron motion along **OP** but only a part of the radiation due to electron motion along **OR**. Thus observers at **P** and **R** see linearly polarised scattered light, while observers at **S** see partially polarised light. ©

several things about the appearance of the sky in the daytime. Towards sunset the sky is red near the horizon, while still blue overhead. We know that clouds do not look blue, they tend to be white and various shades of grey. Importantly, from wearing polarising sun-glasses, we know that the blue sky is polarised, whereas sunlight itself is not polarised.

When sunlight strikes an atom or molecule the electric field in the light ray causes the electrons to vibrate (Figure 21.3). That vibration is transverse to the direction of propagation of the light ray since that is the direction of the electric field. The acceleration of the electron by the field of the light ray causes the electron to radiate like a dipole which is aligned perpendicular to the direction of propagation. Now we develop the equations of motion of this electron taking account of the fact that the electron radiates when it is under the influence of the incoming electromagnetic wave.

Scattering of Light off Bound Electrons

The equation of motion of a bound electron having mass m_e that is under the influence of an oscillating electric field $\mathbf{E} = \mathbf{E_0}e^{i\omega t}$ is[13]

$$\ddot{\mathbf{r}} + \gamma\dot{\mathbf{r}} + \omega_0^2\mathbf{r} = -\frac{e}{m_e}\mathbf{E_0}\,e^{i\omega t}. \tag{21.35}$$

[13] This equation derives from the *Abraham–Lorentz equation of motion* of a charged particle of mass m moving under the influence of an external force $\mathbf{F}_{ext}$ when there is a radiative back-reaction from the accelerated particle:

$$m(\dot{\mathbf{v}} - \tau\ddot{\mathbf{v}}) = \mathbf{F}_{ext}.$$

The frequency ω_0 is the natural frequency of vibration of the electron in its bound state. This is determined by the atom or molecule which hosts the electron. The parameter γ is a damping parameter which, in the Abraham–Lorentz model reflects the back-reaction on the electron due to the fact that it is radiating. We shall assume that $\gamma \ll \omega_0$.

This is a linear equation that is easily solved:

$$\mathbf{r}(t) = -\frac{1}{\omega_0^2 - \omega^2 - i\gamma\omega} \frac{e}{m_e} \mathbf{E}_0 \, e^{i\omega t}. \qquad (21.36)$$

We see the essence of this solution in the resonance at $\omega = \omega_0$: that is why it is good to have the γ damping term in the equation. The damping term also results in a small phase shift of the electron oscillation relative to the oscillation of the driving wave.

The acceleration experienced by the electron is

$$\ddot{\mathbf{r}} = -\frac{-\omega^2}{\omega_0^2 - \omega^2 - i\gamma\omega} \frac{e}{m_e} \mathbf{E}. \qquad (21.37)$$

The cross-section for scattering is then determined by the time average of the squared acceleration, $\langle \ddot{\mathbf{r}} \cdot \ddot{\mathbf{r}} \rangle$ (see Equation 21.8). This results in a total cross-section for Rayleigh scattering given by:

$$\sigma_{\text{Rayleigh}} = \frac{8\pi}{3} \left(\frac{e^2}{m_e c^2} \right)^2 \frac{\omega^4}{(\omega_0^2 - \omega^2)^2 + \gamma^2 \omega^2}. \qquad (21.38)$$

For a strongly bound electron $\omega_0 \gg \omega$ and we have

$$\sigma_{\text{Rayleigh}} \simeq \sigma_T \left(\frac{\omega}{\omega_0} \right)^4, \qquad \omega_0 \gg \omega, \qquad (21.39)$$

which is Rayleigh's important result saying that the scattering cross-section for scattering off a tightly bound electron varies as the fourth power of the frequency.

If we expand Equation (21.38) about $\omega \simeq \omega_0$ we get[14]

$$\sigma_R(\omega) \propto \frac{(\gamma/2)^2}{(\omega - \omega_0)^2 + (\gamma/2)^2}, \qquad \omega \simeq \omega_0, \qquad (21.40)$$

which is called the *Lorentz line profile*. We see that the radiation damping constant, γ, determines the line width.

This is derived using the Larmor power formula and by demanding overall conservation of energy. Since the Larmor formula involves the rate of change of acceleration (which engineers call the *jerk*, though a more pleasant and descriptive word might be *surge*) the energy balance and hence the equation of motion also involve the rate of change of acceleration. Note that this is a non-relativistic equation, the relativistic version is called the *Abraham–Lorentz–Dirac equation*. In the quasi-stationary limit where the forcing term is zero and the system is oscillating at its natural frequency ω_0 the rate of change of acceleration can be approximated by $\dddot{\mathbf{v}} = -\omega_0^2 \mathbf{v}$, which is used to remove the $\dddot{\mathbf{v}}$ term. The parameter τ reflects the natural line width of the radiation spectrum.

The $\dddot{\mathbf{v}}$ back-reaction term brings along with it many problems, not the least of which is an acausal behaviour in which the particle reacts to the force before the force is applied. For a recent discussion on this and classical alternatives see Griffiths *et al.* (2010).

[14] To make this expansion we write $(\omega_0^2 - \omega^2)^2 = (\omega_0 - \omega)^2(\omega_0 + \omega)^2 \simeq 4\omega_0^2(\omega_0 - \omega)^2$.

Polarised Blue Sky

Figure 21.3 and Equation (21.39) provide a simple explanation of what we see on a clear cloudless day. If, in the figure, **OQ** represents the line of sight to the Sun, the observer looking at 90° to that line of sight, as looking along **OR**, will see linearly polarised light. A general line of sight, as **SO**, will contain a portion of linearly polarised light, depending on the angle θ with the line of sight to the Sun. In other words, the maximum linear polarisation occurs 90° away from the line of sight to the Sun.

There is another effect arising out of Rayleigh scattering due to the ω^4 frequency dependence of Equation (21.39). This shows that more blue light is scattered out of the atmosphere than red light. If we regard 'red' light as having a wavelength of ~ 650 nm and 'blue' as having a wavelength of ~ 450 nm we have $(650/450)^4 \sim 4.35$ times as much blue light getting scattered than red.

This has two consequences. Firstly, more blue light is scattered than red, and hence the sky generally looks blue to our visual system. Secondly, when the Sun is low on the horizon, or just below the horizon, the path length through the atmosphere to the observer is long and so we see sunsets dominated by the red component of sunlight: the blue component has been scattered out of the sunbeams.

So why are clouds a variety of shades of grey ranging from almost white to the ominous dark greys that foretell storms? Clouds are made up of ice crystals whose size is substantially larger than the wavelengths of visible light. Scattering off these ice crystals is not described by the Rayleigh process which assumes that wavelengths are greater than the size of the scattering particles. The scattering off ice crystals and large particles (dust and pollutants) is described by a process called *Mie scattering*, which is not a very sensitive function of wavelength and so scatters all wavelengths roughly equally. Clouds look white, though very dense clouds will look darker. For the same reason, vehicle headlights when seen in fog are surrounded by a halo of Mie scattered light.

Mie scattering is also responsible for the varying degree of blueness, i.e. the *saturation*, of the sky comparing the deeper blue overhead and the milky blue nearer the horizon: the sky is more saturated when looking further from the Sun.

Thomson Scattering Revisited

We can carry out a similar analysis of scattering off a free electron by starting with the free particle analogue of Equation (21.35):

$$\ddot{\mathbf{r}} = -\frac{e}{m_e} \mathbf{E_0}\, e^{i\omega t}. \tag{21.41}$$

This equation is entirely trivial to solve, but it is worth noticing that it is just Equation (21.35) with $\omega_0 = 0, \gamma = 0$ and the solution for the total cross-section can be taken straight from Equation (21.38):

$$\sigma_T = \frac{8\pi}{3} \left(\frac{e^2}{m_e c^2} \right)^2. \tag{21.42}$$

The polarisation arises because the cross-section for Thomson scattering of photons off free electrons depends on the angle between the direction of the incident and scattered light.

21.3 Plane Waves and Polarisation

We now return to the description of plane wave solutions to the Maxwell equations that was begun in Section 21.2.2, where the simplest monochromatic, linearly-polarised, plane wave was discussed.

Now we focus on a fixed plane perpendicular to the direction of propagation. To describe what happens in this plane as the wave travels through it we need to fix a pair of orthogonal axes in the plane. If $\mathbf{l}$ and $\mathbf{m}$ are unit vectors along these axes, then $\{\mathbf{l}, \mathbf{m}, \mathbf{n}\}$ is an orthogonal triad: all vectors can be expressed as sums of components in these directions. We can set up a coordinate system on these axes: (x, y, z). Thus the electric field $\mathbf{E}$, which lies in the plane of $\mathbf{l}$ and $\mathbf{m}$, can be written as the sum of two components

$$\mathbf{E} = (E_1 \mathbf{l} + E_2 \mathbf{m})\, e^{-i(\omega t - kz)}, \tag{21.43}$$

for some constants E_1 and E_2 which will, in general, be complex numbers since we are using the complex number representation of wave motion. We can write

$$E_1 = E_1^0\, e^{i\alpha}, \quad E_2 = E_2^0\, e^{i\beta}, \tag{21.44}$$

where the constants E_1^0 and E_2^0 are real numbers. With this the plane wave solution (21.43) can be written out as:

$$\mathbf{E} = (e^{i\alpha}\, E_1^0\, \mathbf{l} + e^{i\beta}\, E_2^0\, \mathbf{m})\, e^{-i(\omega t - kz)}. \tag{21.45}$$

This is perhaps intimidating but it is all that is needed to describe what is meant by the polarisation of the wave.

There are two important special cases corresponding to specific choices for α and β: *linear polarisation* and *circular polarisation*. We discuss these cases first.

21.3.1 Linear and Circular Polarisation

Suppose that

$$\beta = \alpha \pm m\pi, \quad m = 0, 1, \dots \tag{21.46}$$

then

$$\mathbf{E} = (E_1^0 \mathbf{l} \pm E_2^0 \mathbf{m})\, e^{-i(\omega t - kz - \alpha)}. \tag{21.47}$$

The term $(E_1^0 \mathbf{l} \pm E_2^0 \mathbf{m})$ describes the electric field in the plane of the wave: it is a real constant independent of time.

The vector $\mathbf{E}$ performs a linear oscillation in the plane of the wave front. The linearly polarised wave is described by the two parameters E_1^0 and E_2^0 as in Equation (21.47), but it is physically more useful to think of the orientation of a line in the wavefront plane along which the electric field vector executes oscillations with amplitude $(E_1^{0^2} + E_2^{0^2})^{1/2}$. We shall see below that this amplitude and orientation are related to some more general parameters describing general polarisation: the Stokes parameters.

It should be noted that isotropic unpolarised radiation scattering off electrons will not result in polarisation. This is illustrated schematically in Figure 21.4. However, if the radiation field has an anisotropic distribution with at least a quadrupole component, the scattering off electrons will result in a polarised scattered component. A variety of mechanisms can give rise to a quadrupole distribution in the radiation field among which are simple acoustic oscillations and gravitational waves.

The next case is

$$E_1^0 = E_2^0 \equiv E^0, \tag{21.48}$$

$$\beta = \alpha \pm \pi/2, \tag{21.49}$$

which says that the amplitudes of the two components are equal and have phases that differ by $\pi/2$. Then we have

$$\mathbf{E} = E^0(\mathbf{l} \pm i\mathbf{m})\,e^{-i(\omega t - kz - \alpha)}. \tag{21.50}$$

In order to interpret this it is best to separate it into its $\mathbf{l}$ and $\mathbf{m}$ direction components and revert to the world of real numbers:

$$E_x = E^0 e^{-i(\omega t - kz - \alpha)}, \tag{21.51}$$

$$E_y = \pm i E^0 e^{-i(\omega t - kz - \alpha)}, \tag{21.52}$$

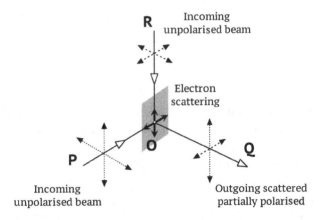

Fig. 21.4 Thomson scattering of unpolarised radiation off electrons. Unpolarised radiation comes in along mutually perpendicular directions **PO** and **RO** and scattered radiation emerges along **OQ**. The outgoing radiation is a blend of the components of the incoming electric fields that are transverse to **OQ**. If the radiation fields coming in along **PO** and **RO** are the same, the outgoing radiation is not polarised. ⓒⓒ

which in real space (taking the real part of each of these) is

$$E_x = E^0 \cos(\omega t - kz - \alpha), \tag{21.53}$$

$$E_y = \pm E^0 \sin(\omega t - kz - \alpha). \tag{21.54}$$

Thus, in the plane of the wave, the end of the electric field vector, $\mathbf{E} = (E_x, E_y)$ traces out a circle as t varies. The direction in which the circle is traced depends on which sign is chosen in Equation (21.54).

If the plus sign is chosen, e.g. $\beta = \alpha + \pi/2$, the end of this vector traces out a circle in the counter-clockwise direction as seen by looking at the on-coming wave from the positive z-axis. This is referred to as *left circular polarisation*. If the opposite sign is chosen, e.g. $\beta = \alpha - \pi/2$, then the circle is traced in the opposite direction and we describe this as *right circular polarisation*.

21.4 Characterising Polarisation

21.4.1 The Stokes Parameters

In the most general case we have

$$E_x = E_x^0 e^{-i(\omega t - kz - \alpha)}, \tag{21.55}$$

$$E_y = E_y^0 e^{-i(\omega t - kz - \beta)}. \tag{21.56}$$

Following the previous route by taking the real part of these equations, we have, as in (21.54),

$$E_x = E_x^0 \cos(\omega t - kz - \alpha), \tag{21.57}$$

$$E_y = E_y^0 \cos(\omega t - kz - \beta). \tag{21.58}$$

Replacing β with $\gamma = \beta - \pi/2$, these are

$$E_x = E_x^0 \cos(\omega t - kz - \alpha), \tag{21.59}$$

$$E_y = E_y^0 \sin(\omega t - kz - \gamma). \tag{21.60}$$

This form reveals that the $\mathbf{E}$ vector traces an ellipse in the plane of the wave front. The cases (21.50) and (21.54) are limiting cases wherein the ellipse degenerates to either a straight line or a circle.

A more convenient way of describing this elliptic pattern is via the *Stokes parameters*:

Definition 21.3
Stokes parameters:

$$I = E_x^{0\,2} + E_y^{0\,2},$$

$$Q = E_x^{0\,2} - E_y^{0\,2},$$

$$U = 2E_x^0 E_y^0 \cos(\alpha - \gamma),$$
$$V = 2E_x^0 E_y^0 \sin(\alpha - \gamma). \qquad (21.61)$$

With these, linear polarisation corresponds to $U = V = 0$, while circular polarisation is $Q = U = 0$. The quantity I is the intensity of the radiation, and so the Q, U, V describe the polarisation: $Q = U = V = 0$ describes unpolarised light. Q and U describe linear polarisation, and V describes circular polarisation. In effect, I measures the intensity of the wave, V measures the ratio of the axes of the ellipse and Q and U describe the orientation of the ellipse.

The *degree of elliptical polarisation* is defined as:

Definition 21.4
Degree of elliptical polarisation:

$$\Pi = \frac{1}{I}(Q^2 + U^2 + V^2)^{1/2}. \qquad (21.62)$$

For circular polarisation $\Pi = V/I$ and for linear polarisation $\Pi = Q/I$.

If the Oxy-axes in the plane of the wave front are rotated through an angle θ, I and V remain unchanged (we are merely re-orienting the ellipse), but Q and U change to new values Q' and U' given by

$$\begin{pmatrix} Q' \\ U' \end{pmatrix} = \begin{pmatrix} \cos 2\theta & \sin 2\theta \\ -\sin 2\theta & \cos 2\theta \end{pmatrix} \begin{pmatrix} Q \\ U \end{pmatrix}. \qquad (21.63)$$

This transformation[15] (21.63) is more often written as

$$\begin{pmatrix} Q' \\ U' \end{pmatrix} = \begin{pmatrix} \cos \theta & \sin \theta \\ -\sin \theta & \cos \theta \end{pmatrix} \begin{pmatrix} Q \\ U \end{pmatrix} \begin{pmatrix} \cos \theta & -\sin \theta \\ \sin \theta & \cos \theta \end{pmatrix}. \qquad (21.64)$$

Another way of writing this is to combine Q and U into a complex number $Q \pm iU$. The transformation (21.64) is then equivalent to

$$Q' + iU' = e^{-2i\theta}(Q + iU), \qquad (21.65)$$
$$Q' - iU' = e^{2i\theta}(Q - iU). \qquad (21.66)$$

---

[15] The $\sin 2\theta$ and $\cos 2\theta$ terms are perhaps surprising since the coordinates (x, y) themselves transform under a rotation θ according to

$$\begin{pmatrix} x' \\ y' \end{pmatrix} = \begin{pmatrix} \cos \theta & \sin \theta \\ -\sin \theta & \cos \theta \end{pmatrix} \begin{pmatrix} x \\ y \end{pmatrix}.$$

This equation describes the rotational transformation, not only of the position vector $\mathbf{r} = (x, y)$, but of any vector: it is part of the definition of what a *vector* quantity is. What Equation (21.63) tells us is that the pair of quantities Q and U are not components of some vector quantity (Q, U). The object whose components are (Q, U) is an example of a quantity known as a *spinor* (see Section 21.4.3).

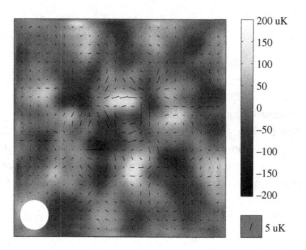

Fig. 21.5 The first map of the microwave background showing the detection of CMB polarisation (Kovac *et al.*, 2002, The DASI interferometer). The polarisation at each point is represented by a black line, whose orientation and length correspond to the direction and intensity of polarisation, respectively. The level of polarisation is about $\sim 10^{-6}$. The background image is the temperature map coded as per the bar on the right. The white spot is the beam size. The image has been processed for publication in greyscale.
From Kovac (2002), *(with permission from John Kovac for the DASI team).*

Dealing with this complex valued quantity is more straightforward than dealing with matrices. We shall return to these quantities later when discussing the distribution of the data over a sphere.

The most direct way of presenting linear polarisation data is to display maps of the measured values of Q and U. However, these quantities are not invariant under different choices of coordinates. Equations (21.63–21.66) show that we can define a *length*

$$P = \sqrt{Q^2 + U^2} \tag{21.67}$$

and an *orientation angle*

$$\gamma = \frac{1}{2} \arctan \frac{U}{Q} \tag{21.68}$$

that are invariant under rotation of the coordinate θ. Polarisation maps are presented as maps showing the value of P and γ, either singly or both on the same sky plot (see, for example, Figure 21.5).

In the WMAP data, γ is measured relative to the galactic meridian and is taken as positive in the directions north through west.

21.4.2 Thomson Scattering: the Stokes Parameters

If we write $I_x = E_x^{0^2}$ and $I_y = E_y^{0^2}$ in Equations (21.61) we can express the Stokes parameters for Thomson scattering as

$$I = I_x + I_y = \frac{3\sigma_T}{16\pi} I_0(1 + \cos^2\theta), \tag{21.69}$$

$$Q = I_x - I_y = \frac{3\sigma_T}{16\pi} I_0 \sin^2\theta, \tag{21.70}$$

$$U = 0, \tag{21.71}$$

$$V = 0. \tag{21.72}$$

That the circular polarisation $V = 0$ follows from symmetry. Because the cosmic background radiation is polarised by Thomson scattering, the microwave background radiation cannot be circularly polarised: it too must have $V = 0$.

21.4.3 Spinor Representations of Polarisation

The fundamental quantities Q and U described in Equation (21.66) can themselves be combined into a single object,

$$\mathbf{P} = \begin{pmatrix} Q & U \\ U & -Q \end{pmatrix}, \tag{21.73}$$

which is referred to as the *polarisation matrix*. Mathematically, $\mathbf{P}$ has the properties of the mathematical object called a *spinor* and so $\mathbf{P}$ might be referred to as the *polarisation spinor*.

A spinor is a mathematical object that describes the properties of a particular physical quantity much as the more familiar scalar, vector and tensor constructs describe mass, velocity and moment of inertia in dynamics.[16] Q and U are themselves not scalars, and nor can they be combined in any simple way into a vector or tensor that has real components. The polarisation of light has some additional fundamental properties that are not reflected in the mathematical structure of scalars, vectors or tensors.

The intuitive way to think of a spinor is as an axial vector that has a rigid flag attached (a 'flagpole'), see Figure 21.6. The series of articles by Payne (1952, 1955, 1959) are an intuitive introduction to spinors in the context of rotations and of special relativity (the flagpole is an 'axe' there). The vector aspect endows it with a direction, while the flag endows the object with the additional attribute of a plane containing the direction. Thus two spinors may have the same direction but different flag orientations. There is a fine description of spinors from the point of view of rigid body dynamics in Goldstein *et al.* (2001), where the three Euler angles for the orientation of a solid body are mapped onto four *Cayley–Klein parameters*.

Any matrix of the form

$$\mathbf{P} = \begin{pmatrix} z & x - iy \\ x + iy & -z \end{pmatrix} \tag{21.74}$$

can be written as the sum[17]

$$\mathbf{P} = x\boldsymbol{\sigma}_1 + y\boldsymbol{\sigma}_2 + z\boldsymbol{\sigma}_3, \tag{21.77}$$

---

[16] Spinors were introduced into general relativity by Penrose (1960).

[17] Any 3-vector, $\mathbf{v} = (v_1, v_2, v_3)$, can be associated with a matrix $\mathbf{V} = \mathbf{v} \cdot \boldsymbol{\sigma} = v_1\boldsymbol{\sigma}_1 + v_2\boldsymbol{\sigma}_2 + v_3\boldsymbol{\sigma}_3$. The length of the vector is given by $|v|^2 = \det(\mathbf{V}) = \det(\mathbf{v} \cdot \boldsymbol{\sigma})$. If we have two 3-vectors $\mathbf{a}, \mathbf{b}$ associated in this way with

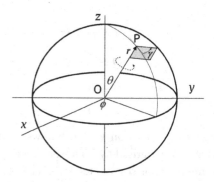

A *spinor* viewed as an axial vector with the additional attribute of having a rigid flag attached. The vector **OP** is described by the standard spherical polar coordinates (r, θ, ϕ), while the plane associated with the flag has an orientation angle γ measured from the meridian *ZP*. The angle γ is the same as the 'roll' angle, usually denoted by ψ, of the Euler angles (see Payne (1952)). ⓒⓒ

where the matrices $\sigma_1, \sigma_2, \sigma_3$ are the *Pauli spin matrices*

$$\sigma_1 = \begin{pmatrix} 0 & 1 \\ 1 & 0 \end{pmatrix}, \quad \sigma_2 = \begin{pmatrix} 0 & -i \\ i & 0 \end{pmatrix}, \quad \sigma_3 = \begin{pmatrix} 1 & 0 \\ 0 & -1 \end{pmatrix}, \quad \mathbf{1} = \begin{pmatrix} 1 & 0 \\ 0 & 1 \end{pmatrix}. \tag{21.78}$$

These four matrices are mutually independent (no one can be expressed as a sum of the other three) and so any 2×2 matrix can be expressed as a linear sum of these.

The polarisation matrix **P** of Equation (21.73) is a function of position on the sky, so we could formally write it as $\mathbf{P}(\mathbf{n})$ to explicitly refer to the direction **n** on the sky. Like any vector or tensor field, this spinor field can be decomposed into independent expansion, shear and vorticity components.

It is possible to define derivatives of this spinor field in the plane of the sky that are analogous to the covariant derivatives of tensor fields.

This can be illustrated in terms of steady irrotational incompressible flows in 2-dimensions for which we can define a velocity potential φ and a stream function ψ (Landau and Lifshitz, 1987, Section 10). The analogy is by no means perfect, but it does serve to display some of the basic ingredients that go into describing a 2-dimensional vector field in terms of complex numbers and functions.

the matrices $\mathbf{A} = \mathbf{a} \cdot \boldsymbol{\sigma}$ and $\mathbf{B} = \mathbf{b} \cdot \boldsymbol{\sigma}$, the matrix product **AB** is easily verified to satisfy

$$\mathbf{AB} = (\mathbf{a} \cdot \boldsymbol{\sigma})(\mathbf{b} \cdot \boldsymbol{\sigma}) = (\mathbf{a} \cdot \mathbf{b})\mathbf{1} + i(\mathbf{a} \times \mathbf{b}) \cdot \boldsymbol{\sigma}. \tag{21.75}$$

A noteworthy result is that if we define a matrix $\mathbf{R} = ct\mathbf{1} + \mathbf{v} \cdot \boldsymbol{\sigma}$, then

$$\det \mathbf{R} = c^2 t^2 - \mathbf{r} \cdot \mathbf{r} = c^2 t^2 - r^2. \tag{21.76}$$

In doing this we have a 2×2 matrix representation of real Lorentz 4-vectors. We can likewise define a matrix derivative operator $\partial = \mathbf{1}\partial_t - \boldsymbol{\sigma} \cdot \boldsymbol{\nabla}$ and, for example, a matrix electromagnetic potential $\mathbf{A} = \Phi\mathbf{1} + \mathbf{A} \cdot \boldsymbol{\sigma}$ from the components of the electromagnetic 4-potential $A_\mu = (\Phi, \mathbf{A})$. The exploitation of this and its relationship with spinors and quaternions is well described in Baylis (1980), where there are many interesting examples.

21.5 Polarisation: E-modes and B-modes

The Stokes parameters are the usual way of describing the polarisation of electromagnetic radiation. They are defined at a point on the sky and they are convenient in experimental situations. However, in cosmology we are interested in patterns of polarisation on the sky since these are generated by structure at or around the time of last scattering. We need to be able to describe and derive the power spectra of the various polarisation modes in the CMB sky.

This involves the use of the less familiar spin harmonics on the sphere. These are complex-valued functions of the direction $\mathbf{n}$, denoted by the symbol $^{(\pm 2)}Y_{lm}(\mathbf{n})$ and were described in the astrophysical context by Kamionkowski *et al.* (1997) and by Zaldarriaga and Seljak (1997, Section II).[18] The subject of representations of data on the sphere is reviewed in depth in the On-line Supplement *Functions on a Sphere*.

21.5.1 Spin-weighted Harmonic Transforms of Q and U

Consider the (difficult) problem of describing Q, U in terms of spherical harmonics. While Q and U look like scalars, they are components of a more complex object, a spinor. The usual familiar Y_{lm} spherical harmonics can be used to describe a scalar field, like a potential or charge distribution on a sphere, but they cannot describe vector and tensor fields. For this we need to use an appropriate generalisation of the Y_{lm}, which are called *spin-weighted harmonics*.

The first few of these, the spin-1 and spin-2 harmonics, are

spin-1: $s = 1, l = 1$ **spin-2**: $s = 2, l = 2$

$$^{1}Y_{1,0} = \sqrt{\frac{3}{8\pi}}\sin\theta, \qquad\qquad ^{(2)}Y_{2,0} = \sqrt{\frac{45}{96\pi}}\sin^2\theta,$$

$$^{1}Y_{1,\pm 1} = -\sqrt{\frac{3}{16\pi}}(1 \mp \cos\theta)\, e^{\pm i\phi}, \qquad ^{(2)}Y_{2,\pm 1} = \sqrt{\frac{5}{16\pi}}\sin\theta(1 \mp \cos\theta)\, e^{\pm i\phi},$$

$$^{(2)}Y_{2,\pm 2} = \sqrt{\frac{5}{64\pi}}(1 \mp \cos\theta)^2\, e^{\pm 2i\phi},$$

$$(21.79)$$

where we have written $^{s}Y_{l,m}$ for $^{s}Y_{lm}$ and likewise $Y_{l,m}$ for Y_{lm}. This is simply yet another set of useful orthogonal functions. There is nothing obscure about them other than explaining where they came from and their inevitably intimidating mathematical notation.

We have three parameters describing the CMB in directions $\mathbf{n}$ on the sky: the temperature $T(\mathbf{n})$, and the two Stokes parameters $Q(\mathbf{n})$, $U(\mathbf{n})$. The last two are combined into the complex quantities $(Q \pm iU)(\mathbf{n})$, which transform as spin ± 2 functions. So their representation in terms of spherical harmonics is

$$T(\mathbf{n}) = \sum_{lm} a_{T,lm}\, Y_{lm}(\mathbf{n}), \qquad\qquad (21.80)$$

---

[18] The following discussion follows closely the discussion of Zaldarriaga and Seljak *loc. cit.*

$$(Q + iU)(\mathbf{n}) = \sum_{lm} {}^{(+2)}a_{lm} \, {}^{(+2)}Y_{lm}(\mathbf{n}), \qquad (21.81)$$

$$(Q - iU)(\mathbf{n}) = \sum_{lm} {}^{(-2)}a_{lm} \, {}^{(-2)}Y_{lm}(\mathbf{n}). \qquad (21.82)$$

The inverse of Equations (21.80, 21.81, 21.82) gives the spherical harmonic representation of the CMB signals T, Q, U:

$$a_{T,lm} = \int T(\mathbf{n})Y_{lm}^*(\mathbf{n}) \, d\Omega, \qquad (21.83)$$

$${}^{(+2)}a_{lm} = \int (Q + iU) \, {}^{(+2)}Y_{lm}^*(\mathbf{n}) \, d\Omega, \qquad {}^{(-2)}a_{lm} = \int (Q - iU) \, {}^{(-2)}Y_{lm}^*(\mathbf{n}) \, d\Omega. \quad (21.84)$$

The last two equations express the spherical harmonic transform of the Q, U maps in terms of the spin-weighted spherical harmonics. The statistics of the coefficients $a_{T,lm}$, ${}^{(+2)}a_{lm}$ and ${}^{(-2)}a_{lm}$ provide a complete Fourier-style description of the polarisation.

The weight-2 coefficients ${}^{(\pm 2)}a_{lm}$ can be combined to produce the even and odd parity coefficients[19]

$$a_{E,lm} = -[{}^{(+2)}a_{lm} + {}^{(-2)}a_{lm}]/2, \qquad (21.85)$$

$$a_{B,lm} = i \, [{}^{(+2)}a_{lm} - {}^{(-2)}a_{lm}]/2. \qquad (21.86)$$

These describe the distribution of the even, 'electric', and odd, 'magnetic', parity components of the polarisation field,[20] Under a parity transformation in which the direction $\mathbf{n}$ is reversed, $\mathbf{n} \to -\mathbf{n}$, the E-mode component of the polarisation field does not change sign, while the B-mode component does change sign. This is precisely the same behaviour as the $\mathbf{E}, \mathbf{B}$ components of the electromagnetic field and this is what gives rise to referring to these as the *E-mode and B-mode components of the CMB polarisation*, see Figure 21.7.

With the amplitudes (21.86) we can define the E and B fields[21] by

$$E = \sqrt{\tfrac{l-1}{l+2}} \sum_{lm} a_{E,lm} Y_{lm}(\mathbf{n}), \quad B = \sqrt{\tfrac{l-1}{l+2}} \sum_{lm} a_{B,lm} Y_{lm}(\mathbf{n}). \qquad (21.87)$$

Since the transformations to get to this point are linear, these are exotic weighted combinations (21.81, 21.82) of the $Q \pm iU$. The analysis of polarisation is conducted entirely in terms of the amplitudes $a_{E,lm}, a_{B,lm}$, the fields E and B are not used for anything other than display and so the normalisations given here are quite arbitrary.

---

[19] The coefficients ${}^{(\pm 2)}a_{lm}$ have the symmetries

$$a_{T,lm}^* = a_{T,l-m}, \qquad {}^{(-2)}a_{lm}^* = {}^{(+2)}a_{l-m},$$

where the superscript * denotes complex conjugation.

[20] The *parity transformation*, P, involves changing the sign of one of the space coordinates: $P : (x, y, z) \to (-x, y, z)$. In effect this is 'physics in a mirror world': addressing the question of what happens to the laws of physics when a coordinate is flipped. Some vectors flip under parity transformation, others do not. Vectors that do not flip under parity transform, like the angular momentum and magnetic field vectors, are called *axial vectors* or *pseudo-vectors*.

[21] Despite the notation, not to be confused with the electric, $\mathbf{E}$ and magnetic, $\mathbf{B}$, fields. The only similarity, which probably gave rise to this nomenclature, is the fact the $\mathbf{E}$ is curl-free, while $\mathbf{B}$ is divergence-free.

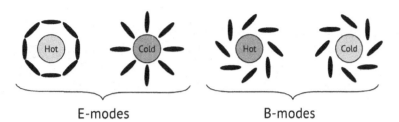

E-modes   B-modes

Fig. 21.7 Polarisation patterns defining *E*-modes and *B*-modes. If the *E*-mode patterns are looked at in a mirror, the pattern and its mirror image are the same. If the *B*-mode patterns are looked at in a mirror, one pattern will be clockwise and the other anti-clockwise. This defines the 'handedness', or 'parity', of the polarisation field pattern. The *E*-modes are the non-local divergence of the polarisation field and the *B*-modes are the non-local curl: this provides the mechanism for separating the two in a map.
See Figure 21.5 for an image of the polarisation field. ©

21.5.2 The Power Spectra

There are six possible spectra and cross-spectra involving T, E and B maps: these are referred to as TT, TE, TB, EE, EB and BB. Of these, because of parity considerations, the cross spectra, C_{TB} and C_{EB} must be zero in a complete noise-free sky map. So there are four power spectra and cross-spectra that can be calculated for each value of ℓ:

Observable power spectra of CMB sky:

$$C_{TT}(\ell) = \frac{1}{2\ell+1}\sum_m \langle a^*_{T,\ell m} a_{T,\ell m}\rangle, \qquad C_{BB}(\ell) = \frac{1}{2\ell+1}\sum_m \langle a^*_{B,\ell m} a_{B,\ell m}\rangle, \quad (21.88)$$

$$C_{EE}(\ell) = \frac{1}{2\ell+1}\sum_m \langle a^*_{E,\ell m} a_{E,\ell m}\rangle, \qquad C_{TE}(\ell) = \frac{1}{2\ell+1}\sum_m \langle a^*_{T,\ell m} a_{E,\ell m}\rangle. \quad (21.89)$$

The coefficients $a_{T,lm}, a_{E,lm}, a_{B,lm}$ are statistically orthogonal:

$$\langle a^*_{T,\ell'm'} a_{T,\ell m}\rangle = C_{TT}(\ell)\delta_{\ell'\ell}\delta_{m'm}, \qquad \langle a^*_{E,\ell'm'} a_{E,\ell m}\rangle = C_{EE}(\ell)\delta_{\ell'\ell}\delta_{m'm},$$

$$\langle a^*_{B,\ell'm'} a_{B,\ell m}\rangle = C_{BB}(\ell)\delta_{\ell'\ell}\delta_{m'm}, \qquad \langle a^*_{T,\ell'm'} a_{E,\ell m}\rangle = C_{TE}(\ell)\delta_{\ell'\ell}\delta_{m'm}, \qquad (21.90)$$

$$\langle a^*_{T,\ell'm'} a_{B,\ell m}\rangle = \langle a^*_{E,\ell'm'} a_{B,\ell m}\rangle = 0.$$

The expectations for the three power spectra values were first computed by Hu *et al.* (2003, Fig. 2). This was an important paper because it clearly sets out the goals for the future (see Section 22.5.1).

22   CMB Anisotropy

> The anisotropy of the Cosmic Microwave Background (CMB) has become a central tool in 21st century cosmology with prospects for a remarkable future.
>
> Some 95% of the photons we observe in the CMB come from the Universe at the time of recombination, carrying with them details of the physical processes acting in the first 380 000 years of cosmic evolution. The remaining 5% come to us after being scattered by reheated gas in more recent times, and they carry information to us about the past few billion years. In those photons we see the clues about the formation of cosmic structures. The main problem is to isolate all of the effects that take place to shape the observed power spectrum of the CMB.

22.1 Cosmological Sources of Anisotropy

22.1.1 The Sachs–Wolfe Effect

There are number of sources of anisotropy in the cosmic background radiation temperature distribution, and these act at various times during the expansion. Names are attached to these sources of anisotropy according to who first identified and wrote about them.[1] The main ones are as follows.

A   **The Sachs–Wolfe effect** which divides into two parts:

    1   *The recombination Sachs–Wolfe effect* This is the original effect as presented by Sachs–Wolfe:1967. The anisotropies are due to the scattering of photons off inhomogeneities in or at the last scattering surface. The relevant inhomogeneities are fluctuations in the gravitational potential or fluctuations in the peculiar velocity of matter at that time.[2] This is generally summarised through the equation

$$\frac{\Delta T}{T} = \frac{1}{3}\left(\frac{\Delta\phi}{c^2}\right)_E,\tag{22.1}$$

      relating the observed temperature fluctuation, $\Delta T/T_E$, to the fluctuating value of the gravitational potential at the point of last scattering, $\Delta\phi$. The factor '$\frac{1}{3}$' accounts

---

[1] The names are more or less consistently applied.

[2] This has also been referred to as the 'non-integrated Sachs–Wolfe effect'.

for a combination of the photons scattering off electrons that are at a different temperature than the mean background temperature, and a redshift contribution that is due to the gravitational potential well of the perturbation. We shall evaluate that later (*cf.* Equation 22.3).

There will also be a Doppler contribution from any systemic velocities, v_E, that the last scattering electrons may have relative to the overall expansion during the recombination.

2 *The integrated Sachs–Wolfe effect* Inhomogeneities that can scatter electrons and cause anisotropies also exist between the last scattering surface and ourselves. The effect of integrating these effects along lines of sight is referred to as the 'integrated Sachs–Wolfe effect' or simply 'the **ISW** effect'. The effect is important when there is a substantial amount of dark matter present, since then the potential wells through which the photons are passing on their way to Earth are evolving with time.

The ISW effect is sometimes divided into two further regimes: 'the early-time ISW effect', dealing with effects that happen near to the time of recombination and the 'late-time ISW effect', dealing with the relatively recent past.

B **The Rees–Sciama effect** This effect was first by discussed by Rees and Sciama (1968), who remarked that a photon falling into an evolving galaxy cluster or super-cluster would climb out of a different potential well than the well it fell into and so cause an anisotropy. The Rees–Sciama effect is a nonlinear SW effect in the sense that it vanishes in linear theory of perturbation growth. It is a late-time effect.

C **The Sunyaev–Zel'dovich effect**, or 'SZ effect'. This is an effect taking place in the hot X-ray emitting gas in galaxy clusters, wherein hot electrons scatter CMB photons out of the beam coming from the last scattering surface towards the observer. This is quite independent of the Sachs–Wolfe effect. It is usual to distinguish two aspects of the SZ effect:

1 *The thermal Sunyaev–Zel'dovich effect*, often referred to as 'the tSZ effect', in which the anisotropy is caused simply by the random motions of electrons in a cluster that is otherwise stationary relative to the background Universe.

2 *The kinetic Sunyaev–Zel'dovich effect*, often referred to as 'the kSZ effect', which is due to any systemic motion of the cluster relative to the microwave background radiation.

D **The Vishniac effect** This is an important 'second-order' effect that can under certain circumstances be bigger than the first order effects. This comes from the correlation of bulk motions of the electrons and the electron density. In order of magnitude, the size of the temperature fluctuation induced by bulk motions of velocity **v** is

$$\frac{\Delta T}{T} = - \int n_e \sigma_T \mathbf{n} \cdot \mathbf{v} e^{-\tau} dt \qquad (22.2)$$

where **n** is the direction along the line of sight and τ is the optical depth.

E **Gravitational lensing of the CMBR** distorts the observed character of the inhomogeneities as a result of the irregular deflection of light by intervening mass distributions. This has to be factored out when interpreting the spectrum of fluctuations, and in doing so we obtain a characterisation of the intervening dark matter.

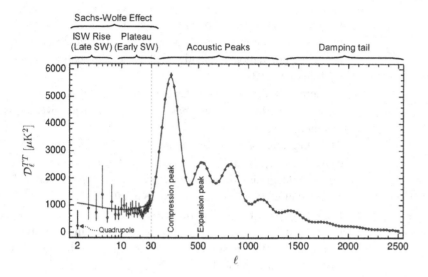

Fig. 22.1 Planck 2015 TT power spectrum. The quantity plotted on the vertical axis is $\mathcal{D}_\ell^{TT} = \ell(\ell + 1)C_\ell/2\pi$. The horizontal scale is logarithmic for $\ell < 30$ and linear for $\ell > 30$. The curve is the theoretical expectation for the standard Planck model.

At least five prominent peaks can be seen, the highest at $\ell \simeq 220$ is due to the sound horizon at the time of recombination. At higher ℓ-values the shape of the spectrum is dominated by viscous damping processes that attenuate the primordial waves. The lower ℓ-values are dominated by different aspects of what is known as the Sachs–Wolfe (SW) effect. Most of the temperature fluctuations arising from the last stages of the recombination create the Sachs–Wolfe Plateau, while the after-effects of late time re-ionisation of the cosmic medium are responsible for the integrated Sachs–Wolfe rise.

The point in the lower left corner of the plot at $\ell = 2$ is the Planck quadrupole. The error bars show $\pm 1\sigma$ uncertainty. While the deviation from the model at $\ell = 2$ looks marginally significant, it does not follow the trend suggested by the $\ell > 2$ data. See also Figures 3.10 and 22.2.

Based on Planck Collaboration (2015, Fig. 1). *(Reproduced with permission from A&A. © ESA.)*

These names are sometimes used by different authors under different circumstances. The reason for this subdivision is perhaps observational: if you see a dip in the temperature and there is an X-ray cluster there, this is the SZ effect. On the other hand a nearby large void might be associated with a temperature decrement because of the Rees–Sciama effect.

The recombination Sachs–Wolfe effect, (A), is regarded as a *primary source* of the observed temperature fluctuations. The Rees–Sciama effect, (B), the thermal and kinetic Sunyaev–Zel'dovich effects, (C), the Vishniac effect, (D), and the effects of gravitational lensing, (E), together comprise what are referred to as *secondary anisotropies*.

The physical processes acting during the pre-recombination Universe and the above effects conspire to mould the spectrum of the radiation field. The power spectrum resulting from these diverse effects is shown in Figure 22.1. The understanding of these physical processes enables us to decode this curve and accurately determine the parameters that characterise our Universe.

22.1.2 The Sachs–Wolfe Equation

Sachs and Wolfe (1967) wrote the first paper describing in detail how primeval structure would give rise to potentially observable temperature fluctuations in the CMB map.[3] At that time, no small scale anisotropies had been observed despite searches at ever increasing levels of sensitivity. That changed with the launch of the COBE satellite which was, for the first time, able to detect structure, albeit only through the angular correlation function of the noisy signal.[4] Later advances in the technology for measuring the temperature structure of the last scattering surface with increased sensitivity and angular resolution demanded technically more detailed models to be developed. It became necessary to understand both the recombination process and the evolution of structure with a precision of better than 1% and perhaps even 0.1%.

While the original paper of Sachs and Wolfe *loc. cit.* established the phenomenon of temperature anisotropies from the epoch of recombination, the best way to look at it is perhaps through the equation[5] of Martínez-González *et al.* (1990), who give the equation for the Sachs–Wolfe effect in the form:

Sachs–Wolfe effect:

$$\left(\frac{\Delta T}{T}\right)_O = \underbrace{\left(\frac{\Delta T}{T}\right)_E + \frac{1}{3}\phi_E}_{\substack{\text{Acoustic peaks +} \\ \text{Sachs–Wolfe}}} + \underbrace{\mathbf{n}\cdot(\mathbf{v_O} - \mathbf{v_E})}_{\substack{\text{Doppler} \\ \text{Shift}}} + \underbrace{2\int_E^O \frac{\partial\phi}{\partial t}\, dt}_{\substack{\text{Integrated} \\ \text{Sachs–Wolfe}}}, \qquad (22.3)$$

where $\mathbf{n}$ is the direction on the sky in which $\Delta T/T$ is being measured.

This is a perfectly general formula and involves no symmetry assumptions for the potential. We shall see how it comes about in the following sections. Note that this equation does not describe the angular distortion of the pattern of fluctuations by gravitational lensing.

---

[3] The realisation that fluctuations would result in small temperature variations was widely appreciated among the small community working in this area at the time, though it was not clear that these would ever be observable. The argument was simply that density fluctuations $\delta\rho \sim 0.1\%$ or larger would be needed on galaxy scales in order that galaxies should form by the present epoch. The scale and amplitude were inconceivably small. Nonetheless, Sachs and Wolfe *loc. cit.* went ahead and did the full relativistic calculation, after which it was only necessary to add in the extra physics needed to deal with what happens during and after the recombination process. The publication of the Sachs–Wolfe work is surely one of the defining papers of modern cosmology. The two contemporaneous papers of Silk (1967, 1968) provided a great leap forward in our understanding of the physics of the radiation field in an inhomogeneous Universe. This was also the first time we saw the famous equation $\Delta T/T = \frac{1}{3}\delta\rho/\rho$.

[4] Sachs and Wolfe (1967, Equation 49) had suggested a relatively high fluctuation amplitude for the relative temperature fluctuation $\Delta T/T$ of around one percent. This seemed like an attainable goal, but searches failed to reveal any evidence for fluctuations. A succession of improvements to the Sachs–Wolfe estimate continually lowered the expectations, while further experiments continually put ever more stringent upper limits on the anisotropy. There was a palpable sense of relief with the COBE team announcement that they had finally detected evidence for fluctuations.

[5] This form of the equation is from Martínez-González *et al.* (1990); Martínez-González and Sanz (1990). See also the fine review of Martínez-González (2009) and the lectures of Sanz (1992, 1997) which give a more relativistic treatment.

Of the four terms on the right hand side the first three describe the conditions at the point, E, where the photon was last scattered. The first reflects the temperature fluctuation value at the point of scattering, E.[6] The second is a general relativistic term describing the change in temperature due to the fact that the photon has to climb out of the gravitational potential at E. A simple, heuristic, argument can be given as to where the $\frac{1}{3}$ comes from (Section 22.2.1). The third term is the Doppler shift due to the resultant bulk motion $\mathbf{v}_E$ of the scattering medium at E, relative to any motion that the observer at O has relative to the frame in which the CMB is isotropic.

The fourth term describes what happens to the photons as they travel between E on the last scattering surface and the observer at O, it does not depend on the conditions at recombination. Since it is an integral it reflects the variations in the potential all along the line of sight, and that is why it is referred to as the *integrated Sachs–Wolfe effect*, or ISW for short. This adds information about the Universe between the last scattering surface and the observer and so, if it could be separated out from the $\Delta T/T$ total signal, it would provide information on the evolution of cosmic structure.

The light from the last scattering surface may pass through regions of partially re-ionised gas on its way to the observer. This gas, if dense enough, will scatter light out of the beam diminishing the amplitude of the signal $\Delta T/T$ and possibly adding spectral features from interaction with any neutral gas. This is likely to happen during and after the re-ionisation of the cosmic medium during the period when stars and galaxies are forming. Modelling the ISW effect presents a potentially powerful tool for learning about those processes (Cai *et al.*, 2009).

There is an additional, important, feature that is not contained in Equation (22.3): the gravitational deflection of the light path as it passes through the evolving inhomogeneities on its way to the observer (Martínez-González *et al.*, 1990; Martínez-González and Sanz, 1990). The angular deflection α depends on the local potential gradient, $\nabla\phi$ at all points along the photon path:

$$(\delta\alpha)^2 - \boldsymbol{\gamma}^2 = (\boldsymbol{\gamma}\cdot\mathbf{n})^2, \quad \boldsymbol{\gamma} = -2\int_E^O \nabla\phi\,\frac{dt}{a(t)}, \tag{22.4}$$

where, again, $\mathbf{n}$ is the direction on the sky where the observer O is looking. This is the *gravitational lensing distortion* of the $\Delta T/T$ pattern on the sky and it serves to modify the character of the observed angular distribution. Such a distortion may at first sight seem like a potential problem in interpreting the $\Delta T/T$ map, but separating it out and mapping the deflection displays the gravitational potential of all the gravitating material in the Universe. This is an important tool in locating the dark matter in the Universe.

These physical processes that contribute to the temperature fluctuations depend on the details of the cosmology and in particular the values of the various density parameters. Interestingly, different features respond differently to variations in those cosmic parameters. This is shown in Figure 22.2.

---

[6] The second term is sometimes written as $\frac{1}{3}(\phi_E - \phi_O)$ to stress that the observer O may lie in a special place.

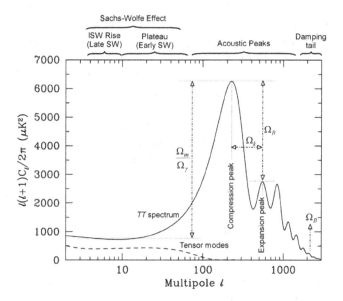

Fig. 22.2 Effect of cosmology on the shape of the power spectrum. The first two peaks shown here are the compression and rarefaction of the same acoustic mode. That ratio is determined mainly by the baryon density Ω_B. The distance between the first and second peaks is the key to the curvature, Ω_k. The baryon to photon density ratio, which is the factor controlling the sound speed determines the height of the first peak over the Sachs–Wolfe plateau at low ℓ. ©

22.2 Deriving the Sachs–Wolfe Result

We give two 'derivations' of the Sachs–Wolfe formula showing how the relative temperature fluctuation relates to the gravitational potential along the path of a light ray. See Martínez-González *et al.* (1990); Sanz (1992, 1997).

22.2.1 A Heuristic Argument

The Sachs–Wolfe effect is a general relativistic effect: it cannot be derived within Newtonian theory any more than can the gravitational redshift of a photon escaping from Earth. However, we can make the following heuristic argument.

The anisotropies we shall be discussing are associated with inhomogeneities in the temperature field, ΔT, that are associated with fluctuations, $\Delta\phi$, in the gravitational potential, ϕ. CMB photons emanating from, or travelling through, gravitational potential hills and valleys will be redshifted or blueshifted as they respond to the gravitational field. The frequency shift, $\Delta\nu$ in the photons will be associated with a temperature fluctuation[7]

$$(\Delta T/T)_{\text{grav}} \sim \Delta\phi/c^2, \quad \text{redshift.} \tag{22.5}$$

---

[7] This is just the relationship $kT \propto h\nu$ between frequency and temperature.

Generally speaking, the gravitational potential fluctuations will be due to inhomogeneities, $\Delta\rho$ in the density field of one or more constituents of the cosmic medium.[8]

From the point of view of an observer looking at the last scattering surface in the cosmic photosphere there are two contributions to the measured temperature fluctuation. One will be due to the fact that the photon was last scattered in a place where the gravitational potential differed from the average value. The other contribution will be from the potential fluctuations that the photon passes through on its way to us. Such fluctuations will be associated with forming or formed cosmic structures such as clusters of galaxies and cosmic voids. The observed temperature fluctuations are tracing such inhomogeneities right back until the time of recombination.

There is an additional complication over and above the gravitational redshift described by (22.5): the perturbation is not stationary, it is expanding with the Universe at a slightly different rate than the background Universe. Hence clocks within the perturbation are seen by an outside observer as ticking at a slightly different rate than clocks in the ambient Universe. Consider a photon climbing out of the gravitational potential well where it was last scattered. In addition to the redshift contribution (22.5) there will be a time dilatation as viewed from afar: $\Delta t/t = -\Delta\phi/c^2$. If the region is a small perturbation, then it will be growing in radius R as the photon propagates through it, with $R \propto t^{2/3}$. The temperature falls off as $T \propto R^{-1}$, and so $\Delta T/T = -\Delta R/R \propto -\frac{2}{3}(\Delta t/t)$, from which we see that the time dilatation contribution is

$$(\Delta T/T)_{\text{time}} = \tfrac{2}{3}\Delta\phi/c^2, \quad \text{time dilatation.} \tag{22.6}$$

The sum of the two contributions (22.5) and (22.6) is

$$\left(\frac{\Delta T}{T}\right)_{\text{SW}} = \frac{1}{3}\frac{\Delta\phi}{c^2}, \tag{22.7}$$

which is the Sachs–Wolfe contribution to the temperature fluctuation at the place of last scattering.

There is an important heuristic argument concerning the time dependence of $\Delta\phi$ during the post-recombination linear growth stage of the density fluctuations. Consider a spherical co-expanding region of radius $R(t)$ having a small excess of density, $\Delta\rho$, relative to the background Universe density ρ. If we denote the mass fluctuation in this region by ΔM, then the fluctuation in the gravitational potential on scale R is

$$\Delta\phi = \frac{G\Delta M}{R} = \frac{4\pi}{3}G\rho R^3 \frac{1}{R}\frac{\Delta\rho}{\rho}. \tag{22.8}$$

The quantity $\frac{4\pi}{3}\rho R^3$ is the mass contained in the co-expanding sphere, and so is constant ($\rho \propto R^{-3}$). From small perturbation theory we found that $\Delta\rho/\rho \propto R^{-1}$, and so, in linear theory, $\Delta\phi$ does not change with time.

---

[8] There will also be Doppler terms arising from variations in the velocity of the photon gas at the time of emission, but for the moment we will assume that any non-Hubble velocities are driven by the fluctuating gravitational potential. There may also be fluctuations in the gravitational potential that are not associated with matter, these are gravitational waves.

This means that the last term in Equation (22.3) vanishes while the cosmic structure is in its earliest stages of post-recombination evolution towards higher density contrast. Early-on, there are no extra contributions to the temperature fluctuations and so

$$(\Delta T / T)_{\text{time}} = \text{constant}, \quad \text{during the linear structure growth phase.} \tag{22.9}$$

Once we come out of that phase at relatively recent redshifts the Rees–Sciama contribution kicks in and we get contributions to the temperature fluctuations on top of the recombination-time fluctuations. Since those scales are relatively close to us, they manifest themselves on the larger angular scales.

22.2.2 The Motion of a Photon

To properly describe the motion of photons in a gravitational field requires the general theory of relativity. Consider the metric of the space-time written in the form

$$ds^2 = -c^2 dt^2 + a(t)^2 [\delta_{\alpha\beta} - h_{\alpha\beta}(\mathbf{x}, t)] dx^\alpha dx^\beta, \tag{22.10}$$

where the $h_{\alpha\beta}$ is considered a small perturbation to the background Robertson–Walker metric. The metric written in this form, with $c^2 dt^2$ as the only explicit dt dependence, is said to be in synchronous coordinates.

Consider a null geodesic (light ray) that passes two observers at times t and $t + \delta t$. The proper separation of these two events is

$$\mathbf{r} = \mathbf{x} a(t)[1 - h_{\alpha\beta} \gamma^\alpha \gamma^\beta], \tag{22.11}$$

where the γ^α are the direction cosines of the 3-space connecting vector $\mathbf{x}$. If the observers are comoving, the connecting vector $\mathbf{x}$ is fixed. If the observers have a motion relative to the cosmic frame this term will depend on time and give rise to the Doppler terms in the temperature variation.

Restricting ourselves to the case where the observers are comoving, we can easily show that

$$\dot{r} = \frac{a}{a} r - \frac{1}{2} a \dot{h}_{\alpha\beta} \gamma^\alpha \gamma^\beta, \tag{22.12}$$

whence, assuming $\dot{h}$ to be small,

$$\frac{\dot{r}}{r} = \frac{a}{a} - \frac{1}{2} \dot{h}_{\alpha\beta} \gamma^\alpha \gamma^\beta. \tag{22.13}$$

This is the contribution to the redshift of the photon, per unit time. The first term on the right represents the contribution to the redshift from the expansion of the Universe, while the second is the contribution from the perturbation itself.

The fluctuation in the radiation brightness will thus be

$$\frac{\delta T}{T} = \frac{1}{2} \gamma^\alpha \gamma^\beta \int_E^O \frac{\partial h_{\alpha\beta}}{\partial t} dt, \tag{22.14}$$

since this is the integral of all the redshift contributions from fluctuations along the line of sight.[9]

Now we need to know how the metric fluctuation $h_{\alpha\beta}$ evolves with time, and in particular we need to relate it to some quantity we can understand, like the relative density fluctuation $\delta\rho/\rho$. This is a classical problem of cosmological perturbation theory. The answer for an Einstein de Sitter background Universe is

$$\frac{\partial h_{\alpha\beta}}{\partial t} = \frac{1}{3\pi t}\frac{\partial^2}{\partial x^2}\int \frac{\delta(\mathbf{x}',t)}{|\mathbf{x}-\mathbf{x}'|}d^3\mathbf{x}', \qquad (22.16)$$

where $\delta \equiv \delta\rho/\rho$. For a Universe with $\Omega \neq 1$, the time dependence is the only thing that changes.

The fluctuating gravitational potential is

$$\phi(\mathbf{x}) = -Ga^2 \int \frac{\delta\rho(\mathbf{x}',t)}{|\mathbf{x}-\mathbf{x}'|}d^3\mathbf{x}'. \qquad (22.17)$$

The conversion from comoving coordinates $\mathbf{x}$ to physical coordinates is $\mathbf{r} = a(t)\mathbf{x}$.

The temperature fluctuation is thus

$$\frac{\delta T}{T} = \frac{1}{3}\phi(\mathbf{x}). \qquad (22.18)$$

Note that in the right-hand side of the equation for ϕ, $\delta \propto t^{2/3} \propto a(t)$ in the linear regime for gravitational growth of perturbations. Since $\rho \propto a^{-3}$, we see that ϕ is in fact independent of time.

22.3 Analysis of the Temperature Fluctuations

The analysis of data on the sky is generally done in terms of spherical harmonic functions, $Y_{l,m}(\theta,\phi)$ where (θ,ϕ) are coordinates on the sky (we take ϕ to be the azimuthal angle, as in right ascension or in longitude). The basic mathematical results are briefly described in Appendix F, the full mathematical details can be found in the On-line Supplement *Functions on a Sphere*.

22.3.1 Analysing the Sky Map

Let us write $\Delta T(\mathbf{n})/T_0$ for the fractional temperature fluctuation in direction $\mathbf{n}$, where T_0 is the mean temperature of the background. This map of the temperature fluctuations

---

[9] In a space-time having metric in the form $ds^2 = -dt^2 + h_{\alpha\beta}(t,x^\mu)\,dx^\alpha dx^\beta$, $\alpha,\beta = 1,2,3$, the quantity

$$K_{\alpha\beta} = \frac{1}{2}\frac{\partial h_{\alpha\beta}}{\partial t} \qquad (22.15)$$

is the *extrinsic curvature* of the spatial sections $t = constant$. This is formally known as the *second fundamental form* of the $t = $ const. hyper-surfaces of the space-time. This is an expression of the relationship between the curvature of space and light rays, and the consequences in terms of observable quantities. This is discussed in greater depth in Sections 17.4 and 17.4.1. See also Peebles (1980, Section 82 *et seq.*and Section 93B *et seq.*) for the detailed derivation of the Einstein equations for $h_{\alpha\beta}$.

can be resolved into independent spherical harmonic components using the Legendre functions:

$$\Delta T(\mathbf{q}) = T(\mathbf{q}) - \bar{T} = \sum_{l,m} a_{lm} Y_{lm}(\mathbf{q}). \tag{22.19}$$

The a_{lm} are random variables taken from some underlying distribution. Because of the definition of $\bar{T}$ being the mean temperature averaged over the sky, the mean of each of the a_{lm} is zero. When mapping the CMB sky we observe only one realisation of this process.

We can speak of the statistical properties of the a_{lm} appearing in Equation (22.19). The means (i.e.: the expectation values) of the a_{lm} are zero. The expectation values of the mean square a_{lm} are:

$$C_l = \langle |a_{lm}|^2 \rangle. \tag{22.20}$$

Given a theory, we can calculate the expected C_l, but when analysing a sky map the values merely correspond to the realisation we have observed.

If we prefer to talk about fluctuations on particular angular scales instead of the spherical harmonic content of fluctuations, we can use the transformation

$$C(\theta) = \frac{1}{4\pi} \sum_l (2l + 1) C_l P_l(\cos\theta), \tag{22.21}$$

where $P_l(\cos\theta)$ is the usual Legendre polynomial of order l (see Equation F.17).

Now we need to put the detector into the equation. The simplest thing to do is to characterise an experiment by its *instrumental response function*, F_l, which tells us how sensitive the experiment is to each value of l, and in principle we know the F_l if we have the details of the instrumental sensitivity as a function of angular scale.[10] As a specific example, for COBE, we had $F_l(\mathrm{COBE}) \sim 0$ for $l > 20$: COBE was not sensitive to fluctuations on an angular scale smaller than about $2.5°$. WMAP offered some 30 times this resolution and so was able to produce anisotropy spectra out to $l \sim 1000$, or $10'$.

The mean square temperature fluctuation reported by an instrument that has measured the C_l is then:

$$\left\langle (\Delta T)^2_{\mathrm{expt}} \right\rangle = \sum_l \frac{2l + 1}{4\pi} F_l C_l. \tag{22.22}$$

This sums up the contributions from all modes that can be measured given F_l.

22.3.2 Multipoles

We can be a bit more specific and calculate the amplitude of the lth spherical harmonic

$$\Delta T_l^2 = \sum_m \frac{|a_{lm}|^2}{4\pi}. \tag{22.23}$$

The $l = 2$ harmonic is referred to as the r.m.s. quadrupole and is often denoted by

$$Q_{\mathrm{rms}} \equiv \Delta T_2. \tag{22.24}$$

---

[10] It is often useful to note that the characteristic angular scale corresponding to a particular value of l is $\theta_l \sim 10^4/l$ arc minutes.

On smaller angular scales, where the curvature can be neglected, we have the useful approximation that the angular scale, θ, corresponding to mode l is

$$\theta = \pi/l, \tag{22.25}$$

so that $l \sim 100$ corresponds to degree scales on the sky. For large l the power per logarithmic interval is

$$\Delta_T^2 = \frac{l(l+1)}{2\pi} C_l T^2, \tag{22.26}$$

and this is what is usually plotted when displaying power spectra.

22.3.3 Correlation Function of Temperature Fluctuations

If $\Delta T(\mathbf{n})/T_0$ is the fractional temperature fluctuation in direction $\mathbf{n}$, where T_0 is the mean temperature of the background, we can define the correlation function between temperature fluctuations in two directions $\mathbf{m}$ and $\mathbf{n}$ by

$$C(\theta) = \langle \Delta T(\mathbf{m}) \Delta T(\mathbf{n}) \rangle, \quad \mathbf{m} \cdot \mathbf{n} = \cos\theta. \tag{22.27}$$

The angular brackets denote the ensemble average over all pairs of directions separated by an angle θ.

$C(\theta)$ can be decomposed into a series of Legendre polynomials:

$$C(\theta) = \frac{1}{4\pi} \sum_{l=2}^{\infty} (2l+1) C_l P_l(\cos\theta), \tag{22.28}$$

where the $P_l(\cos\theta)$ are the Legendre polynomials. This last result is far from trivial to prove, (see Appendix F.1.3, footnote 5).

There is no $l = 1$ dipole term here since that is removed prior to doing the analysis: it is assumed to be due to the motion of the observatory relative to the frame in which the microwave background is isotropic. If the Universe did indeed have an intrinsic dipole component, this would be indistinguishable since the sum of two dipoles is another dipole. Corroboration using other tracers of our motion relative to observed galaxies and systems of galaxies is an important part of putting limits on any intrinsic cosmic component of the dipole.

22.3.4 Power Spectrum of Temperature Fluctuations

The power spectrum of the fluctuations is usually presented as a plot of $l(l+1)C_l$ against l. Figure 22.3 shows the observed spectrum of fluctuations for l-values up to $l \sim 1500$. The rule of thumb translation between l and θ is $\theta \sim \pi/l$ radians, or $180/l$ degrees.

The C_l in Equation (22.28) constitute what is referred to as the *power spectrum* of the temperature fluctuations. The equation describes how each of the l-values contributes to the

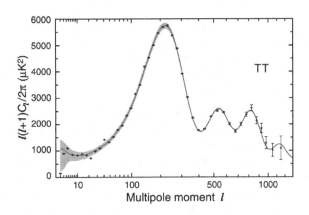

Fig. 22.3   The 9-year WMAP TT angular power spectrum. The WMAP data are shown with error bars. The curve is the best fit
model. The shaded region is the smoothed binned cosmic variance curve. The first three acoustic peaks are
well-determined. Note the data point at the lower left, $\ell = 2$. This is the quadrupole measurement. It lies just on the
edge of the greyed area which indicates the bounds of cosmic variance, so from a statistical point of view this should
occasion no concern. However, it does buck the manifest trend set by the points $2 < \ell < 30$. This low value for the
quadrupole has since been confirmed by the Planck data.
From Bennett (2013, Fig. 32). *(Reproduced by permission of the AAS.)*

total variance on an angle θ. For an ideal temperature map completely covering the sky,
the theoretical C_l are statistically independent. Figure 22.3 displays the power spectrum
measured by the WMAP satellite. The resolution into the components C_l shows distinct
features.

The power spectrum is dominated by a large peak around $l \sim 200$, and it dies down
towards larger l. The decline towards large l is not monotonic, it shows a number of peaks
in addition to the main one at $l \sim 200$, at least three of which in Figure 22.3 appear to be
significant. These additional peaks are roughly at $l \sim 500$, $l \sim 800$, $l \sim 1100$. Planck data
reveals other peaks at $l \sim 1400$ and beyond (see Figure 22.1).

These features depend on the details of the cosmological model: the density of the
Universe as characterised by Ω, the baryon density as characterised by Ω_b, the Hubble
constant H_0, and the cosmological constant, Λ. Other parameters such as the number
of light neutrino species will also play a role. Of course details of the statistical distribution of the primordial density and velocity fluctuations that are responsible for the
temperature fluctuations also come into this, primarily through the shape of their power
spectrum.

The main peak is the *sound horizon* at the time of recombination: the distance a sound
wave travels in the expansion time at that epoch. The sound speed depends on the baryon
density and so reflects Ω_b and H_0. Λ has an overall cosmological effect that serves to
make the peaks shift to larger angular scales as Λ is increased. The second peak tends to
be sensitive to the amount of baryonic dark matter, while the third peak is sensitive to dark
matter in general.

Working out the details of how the fluctuations emerge from the fireball is a complex
task involving many physical processes. With careful modelling we can predict not only

the location of the peaks, but also their relative amplitudes. We are entering the era of *precision cosmology*.

The power spectrum shown in Figure 22.3 is mostly created at the epoch of last scattering of the photons: some 95% of the photons making up that curve come from that epoch, while the remaining few percent come from the time when the cosmic plasma is re-ionised by the processes of galaxy formation.

There are many physical effects that combine to produce the power spectrum. The basic phenomena are depicted in Figure 22.4. The main peaks are due to photons coming from hotter places on the sky, but there is also a Doppler contribution from the residual velocities that the matter in these inhomogeneities has relative to the background cosmic expansion. The peaks due to the Doppler effect are 90° out of phase with the temperature peaks. As we go to smaller length-scales, higher ℓ, we see the effects of earlier attenuation of the higher frequency waves by dissipative processes. This is the Sachs–Wolfe effect which was discussed earlier in Section 22.1.2 and Equation 22.3.

22.3.5 Cosmic Variance

Because we have errors in the measured values of the temperature fluctuations, and because we have only one sample of the sky from which to form our estimate $\bar{C}_l$ of C_l, we will only achieve an approximation $\bar{C}_l \simeq C_l$. A normalised chi-squared distribution with ν degrees of freedom has mean $\mu = \nu$ and variance $\sigma^2 = 2\nu$. Using σ/μ as a normalised measure of the variance we have $\sigma/\mu = \sqrt{2/\nu}$. Whatever the situation might be as regards measurement

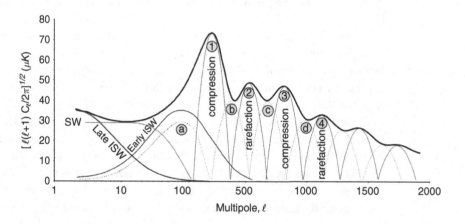

Fig. 22.4 Contributions to the CMB power spectrum. This cartoon depicts the contribution to each angular scale, $\theta \sim 2\pi/\ell$, from the various components that make up Equation (22.3). The components are shown to the right of the picture. The solid curve with the peaks labelled '1,'2,'3,'4', is the temperature contribution to the amplitude spectrum while the dotted curve with peaks labelled 'a,'b,'c,'d' is the Doppler contribution due to the non-Hubble motion of the perturbations. The Doppler peaks are 90° out of phase with the temperature. The part of the curve labelled SW is the Sachs–Wolfe plateau, but at large scale that becomes dominated by the late re-ionisation of the cosmic plasma. The thick curve is a model that is fits the Planck data, while the other curves are merely schematic, illustrating the ℓ-values at whch we see the various sources of fluctuations. ⓒⓒ

errors, we cannot overcome the fact that we only have data from one realisation of the sky. So in the case of the C_l the observed mean deviates from the ensemble average by

$$\frac{\Delta C_l}{C_l} = \sqrt{\frac{2}{2l+1}}. \tag{22.29}$$

Even with perfect measurements we would expect such a difference between our estimate and the real thing: this is referred to as *cosmic variance*.

There are other potential issues: because of measurement problems, the difficulty of accurately subtracting contaminating backgrounds to get at T_0, incomplete sky sampling and so on, the a_{lm} will not in fact be independently distributed random variables. Moreover, they may not in fact be Gaussian distributed.

22.3.6 Correlations vs. the Power Spectrum

It is worth asking the question as to the relative merits of working in real angular space, where correlations tell us what is happening in terms that we can easily relate to, or whether we learn more from the statistical independence of the a_{lm}, despite their lack of intuitive appeal. The sky map of $\Delta T/T$ appears to be a random distribution, displaying structure on all scales. How are we to analyse this in order to extract information about the conditions at the last scattering surface and about the inhomogeneity of the Universe between that point and us? The answer is to use the 2-point correlation function of the values of $\Delta T/T$ at all points of the sky map:

$$\zeta(\alpha) = \left\langle \frac{\Delta T}{T}(\mathbf{m}) \frac{\Delta T}{T}(\mathbf{n}) \right\rangle, \quad \cos\alpha = \mathbf{m} \cdot \mathbf{n}, \tag{22.30}$$

where the average is taken over all pairs of lines of sight $\mathbf{m}, \mathbf{n}$ separated by an angle α. The existence of primordial temperature fluctuations was established by the COBE-DMR experiment (Smoot *et al.*, 1991, 1992). Although the sky maps produced from that experiment showed structure on all scales, the existence of an underlying cosmic contribution could only be demonstrated by means of this correlation function, which is reproduced in Figure 3.9. COBE's angular resolution was around $7°$, which corresponds to a cosmic scale that is some 5 times the scale of the particle horizon at the epoch of recombination.

The correlation function (22.30) has the problem that it mixes the data coming from different angular scales: what is measured at one angle α is not independent of what happens on all other scales. The way of getting better scale separation of the sky map is to do the spherical equivalent of Fourier analysis, which is to analyse the map of

$$\Theta = \Delta T(\theta, \phi)/T \tag{22.31}$$

in terms of its spherical harmonic components, and look at the mean squared amplitudes of the individual components.

The mean squared amplitudes, expressed as a function of angular scale, constitute what is called the *power spectrum* of Θ. We cover this in more detail later, but it is useful to say something about this here since the discussion of the fluctuations is generally expressed in terms of the power spectrum.[11] If the statistically random pattern on the sky, $\Theta(\theta, \phi)$, has correlation function $\zeta(\alpha)$ on scale α as in Equation (22.30), we can define a set of coefficients C_ℓ by the equations

$$C_\ell = 2\pi \int_0^\pi \zeta(\psi) P_\ell(\cos\psi) \sin\psi \, d\psi, \quad \zeta(\alpha) = \frac{1}{4\pi} \sum_{\ell=0}^\infty (2\ell+1) C_\ell P_\ell(\cos\alpha), \quad (22.32)$$

where $P_\ell(\cos\theta)$ is the Legendre polynomial of order ℓ. The first equation is the Legendre transform of $\zeta(\alpha)$ and the second is the inverse transform reconstructing the correlation function $\zeta(\alpha)$ from the coefficients C_ℓ.[12] A simulation of this relationship by Szapudi $et\ al.$ (2000) is displayed in Figure 22.5, taken from their paper.[13]

From the expression for $\zeta(\alpha)$ in (22.32), we see that the total variance of the fluctuations, $\zeta(0)$, is given by

$$\zeta(0) = \frac{1}{4\pi} \sum_\ell (2\ell+1) C_\ell. \quad (22.33)$$

If we approximate this expression as an integral we have

$$\zeta(0) = \frac{1}{4\pi} \sum_\ell (2\ell+1) C_\ell \simeq \frac{1}{2\pi} \int \ell(\ell+1) C_\ell \, d\ln l. \quad (22.34)$$

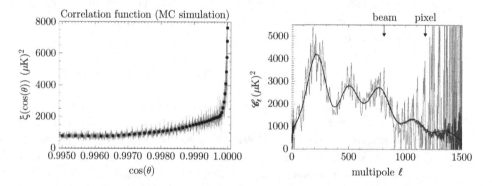

Fig. 22.5 Correlation function (left) and power spectrum (right) derived for simulated data. The left panel is the correlation function for the simulated data, together with a Gaussian resampled version shown as large dots. This resampled version is used to construct the power spectrum, shown on the right. Note that the horizontal scale of the correlation function covers only angles $0° \leq \theta \leq 5.7°$ and most of the correlation appears at angles $\theta < 1°$, $\cos\theta > 0.99985$. The horizontal axis is the order of the multipole $\ell(\sim 180/\theta)$) and we see that the small scales are now better resolved. This provides an efficient and fast way of estimating power spectra from a sky map. From Szapudi $et\ al.$ (2001, Fig. 1). $Reproduced\ by\ permission\ of\ the\ AAS.$

---

[11] The importance of power spectrum analysis is depicted in the inset to Figure 3.10, which displays the result of Fourier analysis of vocal vowel sounds: each sound is characterised by a particular mix of frequencies called a 'formant'.

[12] This is the spherical equivalent of Parseval's theorem. For more details on this and on spherical harmonic analysis in general see On-line Supplement $Functions\ on\ a\ Sphere.$

[13] The C_ℓ in the paper are normalised differently from Equation (22.32).

This shows that $\ell(\ell + 1)C_l/2\pi$ is the contribution of the mode ℓ to the total variance of the field, per unit logarithmic interval $d\ln\ell$. Accordingly it is usual to plot the values of $\ell(\ell+1)C_\ell/2\pi$ versus $\log\ell$ when presenting power spectrum data. This is true regardless of the form of the power spectrum.

22.3.7 Going from 3-d to 2-d

Now we turn to considering the simple case of a power-law power spectrum. Despite the fact that a power law is not even a reasonable approximation to the distribution there is a lot to be learned by understanding this trivial case.

There is a technical question to address. The fluctuating density field, $\Delta\rho/\rho$, or equivalently, the fluctuating potential field $\Delta\phi$ are 3-dimensional random fields whose randomness will be characterised at any instant by a power spectrum $P(k)$. The power spectrum $P(k)$ is the contribution of the mode having wavenumber k to the total variance of the field. However, what we are observing when we look at the CMB is a 2-dimensional spherical slice of the field, i.e. the sky. So how do the C_ℓ describing the distribution on the sphere relate to $P(k)$?

Equation (F.11) of Appendix F provides the direct relationship between the Fourier and correlation descriptors of the inhomogeneity:

Angular power spectrum derived from 3-dimensional power spectrum:

$$C_l = \langle a_{lm}a_{lm}^* \rangle = 16\pi^2 \int_0^\infty j_l(kr)^2 \, P_g(k) \, k^2 dk. \tag{22.35}$$

There is a simple and relevant case that can be treated analytically arising when the 3-dimensional field has a power law power spectrum:

$$P(k) = Ak^n. \tag{22.36}$$

When the field is a field of density fluctuations, $n = 1$ corresponds to the Harrison–Zel'dovich initial spectrum. Going through the calculation it is found that

$$C_l = |a_l|^2 = \frac{4A}{2^{3-n}} \frac{\Gamma(3-n)}{\Gamma^2(2 - \frac{1}{2}n)} \frac{\Gamma\left(l + \frac{1}{2}(n-1)\right)}{\Gamma\left(l + \frac{1}{2}(5-n)\right)}. \tag{22.37}$$

If we normalise with respect to C_2, we have

$$C_l = C_2 \frac{\Gamma\left(\frac{1}{2}(9-n)\right)}{\Gamma\left(\frac{1}{2}(3+n)\right)} \frac{\Gamma\left(l + \frac{1}{2}(n-1)\right)}{\Gamma\left(l + \frac{1}{2}(5-n)\right)}. \tag{22.38}$$

Equation (22.37) is somewhat intimidating: the Γ-functions are not very intuitive! However, there are some useful and important limiting cases.[14]

---

[14] To derive these limiting cases we need the identity $\Gamma(z + 1) = z\,\Gamma(z)$ (Abramowitz and Stegun, 1965, Eq. 6.1.15) and Stirling's large z approximation $\Gamma(z) \sim (2\pi)^{\frac{1}{2}}e^{-z}z^{z-\frac{1}{2}}(1 + O(z^{-1}))$ (Abramowitz and Stegun, 1965, Eq. 6.1.37) to zeroth order.

For $n = 0, 1$:

$$n = 0: \quad C_l = 8A \frac{1}{(4l^2 - 1)(2l + 3)},$$

$$n = 1: \quad C_l = \frac{4A}{\pi} \frac{1}{l(l+1)}, \tag{22.39}$$

and for large l:

$$C_l \propto l^{n-3}, \quad l \gg 1. \tag{22.40}$$

As expected, the shape of the power spectrum C_l depends on the spectrum of fluctuations. In the large l limit for $n = 0, 1$ we have

$$l^3 C_l \to \text{constant}, \quad l \to \infty, \quad n = 0, \tag{22.41}$$

$$l(l+1)C_l = \text{constant}, \qquad n = 1. \tag{22.42}$$

We note from Equation (F.22) that $l(l + 1)C_l$ is proportional to the contribution of mode l to the total variance per unit logarithmic interval $d \ln l$ in l. Hence the 'flat' $n = 1$ spectrum contributes equal power per unit logarithmic interval, $d \ln l$, at all l-values.

Turning back to the case of general values of n, we see that a plot of the power spectrum using $\ell(\ell + 1)C_\ell$ instead of C_ℓ, versus $\ln k$ (or $\log_{10} k$), shows the relative contribution to the total power from the modes of wavenumber k. The plot will be flat in the case of a spectrum with $n = 1$. However, as just remarked, the power spectrum that emerges from the physical processes occurring before and during the period of recombination is dominated by prominent peaks, and falls off at high ℓ due to dissipative processes. At values of ℓ below the values for the largest of these peaks, other more recent physical effects take place to modify the large angular scale appearance of the spectrum.

22.4 Model Fitting – the Angular Power Spectrum

Maps of the CMB have huge numbers of pixels and so handling pixel-based data can be a problem, especially given the complexity of computing all the multipole coefficients. This is handled by binning the multipoles into groups and doing the analysis with these groups.

The signal from the CMB is viewed through a detector and so the detector beam has to be taken into account. For a circular beam the spherical harmonic transform will depend only on the l-value and so the measured signal will be

$$s(\mathbf{n}) = \sum_{lm} B_l a_{lm} Y_{lm}(\mathbf{n}), \tag{22.43}$$

where B_l is the harmonic transform of the beam pattern. This influences the correlations of the sky signal. The correlation between pixel pairs separated by an angle ψ is

$$S(\psi) = \sum_l \frac{2l + 1}{4\pi} b_l^2 C_l P_l(\psi), \tag{22.44}$$

where C_l is the power spectrum, and the P_l are the Legendre polynomials of order l. The situation is further complicated by the fact that we do not observe the entire sky, so there has to be a sky mask imposed on this, and this means that the C_l are not a complete orthonormal basis.

For a given set of cosmological parameters **p** we can compute the expected, i.e. theoretical, components $C_l^{\text{th}}(\mathbf{p})$ of the power spectrum. If we are to fit binned C_l measurements, we can also compute the theoretical binned version of the $C_b^{\text{th}}(\mathbf{p})$, where b is the bin identifier.

If we consider the binned data with measured values $\hat{C}_b$, the deviation between theory and measurement is

$$\Delta C_b(\mathbf{p}) = \hat{C}_b - C_b^{\text{th}}(\mathbf{p}). \tag{22.45}$$

We can model the observations assuming that the $\Delta C_b(\mathbf{p})$ are Gaussian distributed (see Equation 26.52):

$$\mathbb{P}[\hat{C}_b \,|\, C_b^{th}\,] = \exp\;-\frac{1}{2}\left\{\Delta C_b^T \mathbf{M}^{-1} \Delta C_{b'}\right\}, \quad \mathbf{M} = \langle \Delta C_b^T \Delta C_b \rangle, \tag{22.46}$$

with the corresponding likelihood function:

Log-likelihood using band-power covariance matrix:

$$2\ln\mathcal{L} = -\Delta C_b^T \mathbf{M}^{-1} \Delta C_{b'} - \text{tr}\,\mathbf{ln}\,\mathbf{M}. \tag{22.47}$$

The matrix **M** is referred to as the *band-power covariance matrix* and is determined from simulations of the CMB sky. In principle **M** will depend on the cosmological parameters used for generating the simulation, but in practice only one set of parameters is used since this dependence is of a higher order.

22.5 Polarisation

The radiation we see comes from the time of recombination of the cosmic plasma around a redshift of $z \sim 1100$. This was the last time the radiation was scattered before we received it. At that time the matter in the Universe was in motion and partially ionised. The photons scattered off moving electrons: that resulted in polarisation of the radiation.

The physics of the CMB polarisation is discussed at some length in Section 21.4.1 *et seq.* To recapitulate: the polarisation pattern, described by the Stokes parameters Q and U, can be described as a superposition of two independent components: E-modes and B-modes. The polarisation patterns associated with these modes have opposite parity, odd or even. This is somewhat like splitting the velocity gradient tensor of a chaotic flow into its shear and vorticity components.

Temperature is a scalar quantity, while polarisation is a second rank tensor. Scalar perturbations produce only E-modes, while tensor mode perturbations produce both E-mode and B-mode patterns, in roughly equal strength. As far as cosmology goes, The B-modes, if they exist, would be a consequence of tensor perturbations in the Universe, i.e. gravitational waves. However, gravitational lensing of the CMB also produces a B-mode component, which at large ℓ can dominate the gravitational wave component. In addition, B-mode polarisation patterns are also generated by non-cosmological foregrounds such as dust in the galaxy. Gravitational waves produce both E-modes and B-modes.

The most important tool we have for analysing the patterns of polarisation are the cross correlations, or the cross power spectra, between the patterns generated by the corresponding modes. Of the six possible correlations, TT, EE, BB, TE, TB and EB, only four are relevant: TT, EE, BB, TE (see Figure 22.6). The TB, and EB cross-correlations must be zero, which provides a useful check on the consistency of the data acquisition and processing.

Figure 22.7 shows the expectations for the TT, EE and BB power spectra. The TT spectrum has been well determined and dominates the others. What we are seeing there are the various contributions to the recombination Sachs–Wolfe effect and from the integrated Sachs–Wolfe effect. The E modes are associated with the T modes: any contribution to the E-mode pattern from gravitational waves is swamped by the TT spectrum. So the EE and TE spectra serve mainly to confirm what we have seen in the TT spectrum and contribute to tighter limits on the cosmological parameters. However, the existence of a significant correlation on larger scales $l < 10$ indicates that the microwave background radiation has been scattered since the epoch of recombination. This is therefore evidence of re-ionisation of the primeval plasma since the time of recombination.

22.5.1 Expectations for Future Measurements

The future grail of cosmology is the detection of the cosmological B-mode, which would provide evidence for the existence of cosmological gravitational waves, and hence strong support for the theory of an inflationary origin for the Universe. However, separating out the very weak cosmological B-modes from the non-cosmological contributions is very difficult. The situation is depicted in Figure 22.7.

The peak contribution to the B-mode from primordial gravitational waves occurs at recombination on angular scale $\ell \sim 80$. This corresponds to the horizon scale for gravitational waves just entering the horizon at the epoch of last scattering of the CMB radiation.

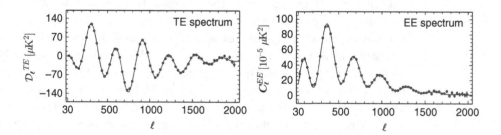

Fig. 22.6 *TE (left)* and *EE (right)* spectra from the Planck satellite data, 2015. The quantity plotted on the vertical axis is $\mathcal{D}_\ell = \ell(\ell+1)C_\ell/2\pi$. The line passing through the data points is derived from the best fit to the TT spectrum shown in Figure 22.1. This shows a remarkable degree of consistency between the temperature and the polarisation data.

From Planck Collaboration I (2015a, Fig. 10). *Reproduced with permission from A&A. © ESA.*

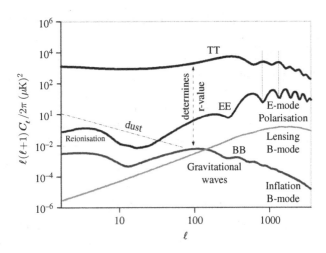

Fig. 22.7 Expected *TT*, *EE*, and *BB* power spectra for the temperature, *E*-mode and *B*-mode CMB signals on a log–log plot. The Planck satellite data has revealed the E-mode polarisation in the range $30 < \ell < 2000$ (see Figure 22.6). We see that the expected B-mode signal would be some 3 or more orders of magnitude below the *TT* signal, and that it would be swamped by gravitational lensing on small scales, $\ell \gtrsim 100$. The dot-dash line labelled 'dust' is merely schematic, it depicts the contribution from foreground dust which threatens to obscure the *B*-mode signal at low ℓ-values. The ratio of the heights of the *TT* and *BB* curves (squared) relates to the tensor to scalar ratio, r, of Equation (9.23). Measuring r tells us how much power was in primordial gravitational waves. At $\ell < 10$ we would be seeing the effects of the re-ionisation of the cosmic plasma so the optimal ℓ-range for discriminating the *B*-mode is around $\ell \sim 100$.
Based on plots from Trombetti and Burigana (2012b) *Permission granted.* ©

On smaller angular scales, the waves entered the horizon at earlier times since when they have been attenuated by the cosmic expansion. So the power in the *B*-modes falls off towards higher ℓ-values and the signal is lost among the *B*-modes generated by gravitational lensing. Towards low ℓ-values there is an ever-increasing contribution from the late re-ionisation phase of the cosmic plasma, and a contribution from polarising dust. The latter provides a difficult foreground to eliminate. That leaves a window of *l*-values centred around $\ell \sim 100$ as the best possibility of detecting cosmic gravitational waves. Whether they are detected depends on the nature of the inflation where they were generated.

22.6 The Sunyaev–Zel'dovich Effect (SZE)

The Kompaneets equation describing the interaction between electrons and photons at different temperatures was discussed in Section 19.3. An important aspect of the interaction was the distortion of the photon spectrum by Compton scattering. In Section 19.4 we touched on the distortion produced by hot electrons in rich galaxy clusters in our line of sight. The quasi-random distribution of such clusters on the sky would create anisotropies in the observed temperature.

If clusters at high redshift have hot (but non-relativistic) gas, microwave background photons passing through the cluster will inverse Compton scatter off the electrons that pervade the intracluster medium (ICM), receiving a boost of energy. On the order of $\sim 1\%$ of photons passing through a cluster undergo such scattering, while the electrons cool off by this process. This will cause a small change in the measured temperature of the CMB, as first discussed by Sunyaev and Zel'dovich (1972), after whom the effect is named.

Taking only the Compton term of the Kompaneets equation, but this time working in 'observer units' of frequency v and spectral energy density f_v, we have

$$\Delta f_v = \tau \frac{kT_e}{m_e c^2} v \frac{\partial}{\partial v} \left(v^4 \frac{\partial}{\partial v}(v^3 f_v) \right), \tag{22.48}$$

where n_e is the electron density through which the photons pass, T_e is the electron temperature and $\tau = \int n_e \sigma_T dl$ is the optical depth of the cluster gas to Compton scattering.

The magnitude of the effect depends on the *Zel'dovich–Sunyaev parameter, y*:

Zel'dovich–Sunyaev y-parameter:

$$y = \int \frac{kT_e}{m_e c^2} n_e \sigma_T dl = \frac{\sigma_T k}{m_e c^2} \int T_e n_e dl, \tag{22.49}$$

Also known as the Compton y-parameter or the Kompaneets y-parameter,

where n_e is the electron density in the cluster and T_e is the electron temperature of the gas.[15]

This can be understood as follows. In a time dt the photon travels a distance cdt. Since the mean path length between collisions is $(n_e \sigma_T)^{-1}$, the number of collisions in time dt is $n_e \sigma_T cdt = n_e \sigma_T dl$. So Equation (22.49) can be understood as measuring the energy change per collision, $\sim kT_e/m_e c^2$, multiplied by the number of collisions.

The major contribution to y from the integral in (22.49) comes from regions of high temperatures and electron densities on the photon trajectory.

For the hottest and densest X-ray clusters, $y \sim 10^{-4}$ and we expect a dip of around 1 mK. We notice that $n_e T_e$ is proportional to the polytropic gas pressure $p = nkT$. We can model our expectations by assuming a cluster density profile and using a polytropic equation of state from the hot gas.

If we focus attention on the Rayleigh–Jeans part of the spectrum where $f_v \propto v^2$, we have simply that

$$\Delta f_v \overset{\text{RJ}}{=} -2\tau \frac{kT_e}{m_e c^2} f_v. \tag{22.51}$$

There is a dip in the measured Rayleigh–Jeans temperature. This is the (thermal) Sunyaev–Zel'dovich effect (tSZE). Typically, the temperature dip is on the order of $\Delta T \sim 10^{-4}$ K.

---

[15] Since the effect is generally measured over an area of the cluster, another variant of the y-parameter is introduced:

$$Y = \int y \, dA \propto \int n_e T_e dV \propto E_{\text{thermal}}. \tag{22.50}$$

(The argument is $dydA \sim n_e T_e dldA \sim n_e T_e dV$.)

However, in the Wein law part of the spectrum, the antenna temperature increases, thereby providing the S–Z effect with an important signature. The change-over in sign occurs at a wavelength of $\lambda = 1.4$ mm. This signature is important because it enables us to distinguish the tSZE from a related effect, the *kinetic Sunyaev–Zel'dovich effect*, (kSZE).

The kinetic Sunyaev–Zel'dovich effect is due to the motion of clusters relative to the cosmic expansion. This motion results in a Doppler shift of the distortion and, since the Doppler shift is colour-blind, this does not change the shape of the spectrum. For a cluster moving with radial peculiar velocity v, the Doppler shift produces a temperature change in the CMB on the order of

$$\left(\frac{\Delta T}{T}\right)_{kSZE} \simeq -\tau_e \frac{v}{c} \qquad (22.52)$$

which is on the order of 10^{-5}, more than an order of magnitude below the thermal Sunyaev–Zel'dovich effect.

The South Pole Telescope (SPT) and the Atacama Cosmology Telescope (ACT) detect clusters out to redshifts of ~ 1.5 (see for example Bleem and Stalder (2015)). These are cosmologically important probes of the Universe.

PART V

PRECISION TOOLS FOR PRECISION COSMOLOGY

23

Likelihood

The concept of likelihood is fundamental in the statistical estimation of the parameters that enter into a parameterised model of a dataset. We write down an expression telling us how well the model represents the data, and choose the parameter values that do the best job. Despite the apparent simplicity of this statement, it hides a wealth of issues that have caused a great schism in thinking about how we should draw conclusions from data.

This is the great divide between the frequentist and the Bayesian schools of thought concerning the way likelihood should be implemented in practice. Should we simply take a mechanistic approach and find the most likely parameter set, with an estimate of how confident we are in asserting the answer? Should we somehow fold in our prior prejudices to take account of our past experience? How should we interpret the result of any process that purports to assign a confidence to an answer? What, indeed, would confidence mean in this sense, especially if we cannot analyse any further samples to confirm our result? [1]

23.1 The Great Schism

In this chapter we develop the theory of likelihood which is widely used in deriving parameters that fit models to data. Here, we regard the parameters of the model simply as numbers that are to be determined by some optimisation process. We shall generalise this in subsequent chapters when we come to discuss Bayesian inference, but along the way we shall point out salient differences between the two ways of thinking.

We are interested in the selection of parameters that provide the *best fit* to the given data. What we mean by best fit is generally described by a *cost function* that provides a quantitative measure of what we mean by *goodness of fit*. The usual criterion is to ask for the smallest set of parameters that provides the least total deviation of the model from the given data. That too involves an assumption regarding precisely what we mean by 'smallest total deviation', or indeed by a measure of the deviation.

[1] Examples of the use of likelihood will be found in the On-line Supplement *Likelihood in Action*. It is assumed in the following chapters that the reader is familiar with elementary statistics. An overview of elementary probability is given in the On-line Supplement: *Probability and Statistics Primer*. The latter supplement provides a basic overview of elementary concepts such as expectation, variance, statistical moments, etc. Appendix E provides a brief summary of some the fundamental statistical distributions which are used in this chapter.

Once we have made that fit we may be confronted by a new data set, or an augmentation of the first data set. In either case, we may repeat the fitting procedure only to find a different set of parameters than we found in the first place. This raises the important question about reliability of the model and the fitting criteria. It brings up the question of what is our degree of confidence in the values we have found, and indeed, what is our degree of confidence in the underlying model. This is the frequentist world of *hypothesis testing*, where we create a decision procedure that will decide in favour of one hypothesis or another.

There are two major schools of thought about the process of testing hypotheses that are made about data: the 'frequentists' and the 'Bayesians'. In a simplistic sense, fitting a particular model to a set of data results in the determination of a set of parameter values for the model. The parameters are determined by making a fit to that particular data set. If someone else were to independently generate a data set and test the same model, they would get a different data set and so determine different values for the parameters. The discussion would then centre around whether these determinations were consistent. There would also be the hope that getting a larger data set would lead to a better determination of the parameters. This is the essence of what is dubbed 'the frequentist approach'.

The alternative is to view the parameters of the model as being themselves random variables, on much the same level as the data. We would then need a model for the statistical distribution of the parameters, and that would have to come either from some prior information or from inspired guesswork. This is the essence of the 'Bayesian approach'. The distribution of a parameter that is to be determined expresses a degree of belief rather than what would normally be described as a probability distribution.[2]

23.1.1 Bayesian Inference: the Philosophy

Experimental physicists could be said to be natural Bayesians in the sense that when they design an experiment they do so with a prior knowledge of what they are looking for and the technological and financial limitations on what they can achieve. Very few, if any, experiments are designed to 'see what's out there' in the hope of making a serendipitous discovery.[3] The theorist is likewise led to interpret data in terms of current thinking, and in terms of current models if there are any. Sometimes, a given observation may have two or more opposing theories that are yet to be discriminated by future experiments. These experiments are designed knowing what these hypotheses are and what the expectations are for those hypotheses.

This leads us to think in terms of physical science as being driven by ideas that are modelled by the theories available at the time. Those models always have measurable parameters that are needed to explain given datasets. Generally speaking, it would be safe

---

[2] It is at this point that discussions get lost in philosophical discussion about the nature of probability. We will sidestep that and simply argue that this is yet another model assumption among the many assumptions that go into fitting models to data. There are even factions within the Bayesian camp ranging from the purists to the empiricists. Numerous well respected and influential philosophers (Popper, 2002), mathematicians (Doob, 1949), physicists (Jaynes, 2003), economists (Keynes, 1921) and statisticians (Fisher, 1922) have written about and debated Bayes' theorem, its use and its interpretation.

[3] That is not to imply that there are no serendipitous discoveries in science! But they are often a by-product of an experiment, or data analysis process, that was designed for a different investigation.

to say that, at any given time, the then acceptable models would be expected to reproduce the available data that they sought to explain or understand. The task of the theorist is to think of some datum that could be measured to discriminate between models.[4]

23.1.2 The Rise of Bayesians

Although there were murmurings of dissent during the post-Fisher period, it was not until the second half of the 20th century that the Bayesian view rose to prominence. One of the first papers in astronomy advocating the use of Bayesian methods was that of Sturrock (1973) in which, as an example, he didactically examined the then available evidence for the two hypotheses that pulsars were either white dwarfs or neutron stars.

Several factors contributed to the sudden rise in the use of Bayesian statistics. Not the least of these was the dramatic advance of computing facilities in the post-1980s period, and with this a corresponding rise in the acquisition of vast amounts of data. Implementing Bayesian methods for data analysis is not easy and it would be almost impossible to perform this kind of analysis on a calculating machine such as had existed prior to the computer revolution. The advent of Monte Carlo Markov Chain (MCMC) methods was certainly a part of that revolution since these methods are essential to implementing Bayesian methodology in real life data analysis. It has been argued (Robert and Casella, 2011) that Markov Chains were only brought into the world of mainstream statistical analysis after the publication of the paper by Gelfand and Smith (1990).

Bayesian analysis, with its use of prior information to filter out 'unreasonable' parameter values, is an appropriate part of the scientific process. Of course, 'unreasonable' can be interpreted as 'not conforming to our prior prejudices', but this is not a constraint on our conclusions. The advantage of a Bayesian approach is that if one physicist does not share the prior beliefs that are being applied to a data set, he/she is quite at liberty to apply different prejudices and see if the result is any better.

Another way of looking at this[5] is to say that in the frequentist methodology, one can invent a variety of estimators for a given quantity without providing a basis for saying which is optimal, while in the Bayesian view the basis for choice is made explicit. In the Bayesian view, the parameters describing a model are themselves statistical quantities so the art of model building in the Bayesian world consists in defining models not only for the data, but also for the parameters that enter into the models.

---

[4] In that context we can think of the schism of the 1950s and 1960s between the Steady State theory of Bondi, Gold and Hoyle and the Hot Big Bang theory of Gamow. Bondi *et al.* said, in modern parlance, that the density parameter $\Omega = 1.00$. Gamow's Hot Big Bang could accommodate an $\Omega = 1.0$ model (though there would be no 'natural' explanation for this), but the Steady State theory could not easily accommodate the cosmic radiation field. The CMB was the crucial factor.

[5] This is admittedly a Bayesian view: see MacKay (2003, Ch. 24). However, it is true that in the Bayesian world one can marginalise a distribution over any of the parameters of the model simply because these parameters are statistical quantities. The result of that marginalisation will often conform to what the frequentist night have written down in the first place. The result of marginalisation will of course depend on the assumed distribution of the parameters in question, but at least that assumption is laid bare and not hidden.

23.1.3 Frequentists versus Bayesians

At the base of the problem there are two different interpretations of what probability is:

1. Probability is a measure of our state of knowledge.
2. Probability is a measure of our degree of belief that an event will occur.

This choice divides the world into frequentists who hold to the first and Bayesians who hold to the second. Put another way, the frequentist wants to establish the statistics by repeated measurements alone, while Bayesians want to take account not only of the experimental data, but also our prior knowledge (or prejudice!) insofar as that may help bolster our expectations.

Historically, during the first half of the 20th century, the world of statistics was dominated by the frequentists following the dogma as laid down by Fisher (1922, 1925 (1st edition)). The arguments on the philosophical side questioned the very notion of 'probability'[6] (de Finetti, 1937, 1974; Ramsey, 1931), and on the mathematical side there were attempts to put the concept of probability on to a firm axiomatic footing. On the practical side there was the issue as to which was the correct way to go about understanding hypotheses in the light of data.

One might even say that the frequentists sensed that the Bayesians were prejudicing their analysis by appealing to some ill-defined notion of *prior information*, while the Bayesians felt that the frequentists were being unnecessarily restrictive in their view by not using all the information available to them.[7] The Bayesian process is not in any way 'unnatural' or 'irrational': it is used, often unconsiously, in many aspects of everyday life. A common example is the way in which computer programmers at all levels test their programs. The programmer writes a chunk of code, runs it and tests it. If it fails or does not behave as expected, the programmer modifies the code and goes through the test cycle again. If it meets expectations, then there may be further testing, or, perhaps optimistically, the programmer will declare it 'free of bugs'. This process of 'refinement by experience' is, in effect, a Bayesian process: the refinement continues until the programmer asserts a belief that the program will behave as expected and that nothing untoward will happen.

Having said that, we should note that both Bayesian analysis and frequentist analysis are based on constructing a *likelihood function*. The operational difference is that Bayesian

---

[6] Perhaps none more so than in de Finetti's aphorism that *Probability does not exist*. This encapsulated his notion that probability was merely a matter of the observer's degree of belief that an event might or might not happen. In that sense statistical inference is not an empirical process that delivers a verdict as to the meaning of data, but rather a measure of which among a variety of opinions about data might be favoured. Probability, according to de Finetti, does not exist independently of the human mind.

[7] An old story encapsulates the difference between the two philosophies. The story describes a hypothetical conversation between Harold Jeffries, a Bayesian, and Ronald Fisher, the noted frequentist. Harold looks up at the cloudy sky and says 'I think there is a 40% chance of rain tomorrow, Ronald, what do you think?'. To which Ronald replies 'Harold, how can you possibly know that when you have not determined the frequency of rain falling?'. Harold, the Bayesian, expresses a belief, while Ronald, the frequentist, expresses the need to determine a frequency before coming to a conclusion. Cynically, one might say that Ronald has to watch the next week's weather before he can say anything about tomorrow. In fact, what Ronald would do is to rely on his folkloric experience (that it always rains in April) and extrapolate that to the current situation: he would become an 'empirical Bayesian'.

analysis allows a rational modification of the likelihood, based on prior knowledge or experience. It should be made clear, however, that frequentists also allow modification to likelihood (Reid, 2010, 2013, for an overview) in order to constrain the model in terms of the number of parameters and in terms of additional information that might be present.

So we start by exploring likelihood, the factor common to both of these two approaches.

23.2 Likelihood of an Exponential Process

As an introduction to how likelihood works, let us consider a specific process: the exponential process as exemplified by the decay of a particle. This has the basic elements of a physical model: the decay law with a to-be-determined parameter, and a set of measurements from which we want to determine the parameter value indicated by the data.

Suppose we have a theoretical model suggesting an exponential decay law for which the statistical distribution of the decay times, t, should follow a probability distribution function,

$$f(t; \tau) = \tau^{-1} e^{-t/\tau}. \tag{23.1}$$

The experiment is to observe decays taking place within some intervals $(t, t + dt)$, which results in a sample t_i of $i = 1, \ldots, n$ decay time measurements. Our data set is $\mathcal{D} = \{t_i\}$, our model is (23.1) and we need to determine τ.

If the measurements are independent the probability of getting that particular data set is obtained by multiplying together the individual probabilities for any one measurement.

The probability that the measured decay time falls in the interval $(t, t + dt)$ is

$$p(t) \, dt = \frac{1}{\tau} e^{-t/\tau} \, dt. \tag{23.2}$$

This describes a model for the probability that a particle will be observed to decay in the time interval $[t, t + dt]$. The goal of the experiment is to determine the best value of τ, the lifetime. Note that this model is simply one in which the decay process is a statistical process following some rule mandated by physics. There is no accounting for measurement error in this model.

We are going to construct a function $\mathcal{L}(\tau)$ that reflects the probability of generating the given data set, i.e. the set of measurements $t_1, \ldots, t_n$ for a given value τ of the decay time we are trying to determine. The likelihood[8] for a set of n observations of decay time $t_1, \ldots, t_n$ will simply be taken as the product of the individual probabilities (23.1):

$$\mathcal{L}(\tau) = \frac{1}{\tau} e^{-t_1/\tau} \ldots \frac{1}{\tau} e^{-t_n/\tau} = \frac{1}{\tau^n} e^{-(\Sigma_{i=1}^{n} t_i/\tau)}. \tag{23.3}$$

While $\mathcal{L}(\tau)$ is not itself a probability (it is not normalised), its magnitude can be seen as a figure of merit: the closer the measured t_i are to the putative τ value, the bigger will be the number $\mathcal{L}$. So we will assert that the most likely value of τ that would reproduce this set of measurements is the one that maximises $\mathcal{L}(\tau)$. That maximum occurs where

---

[8] We formalise this later, see Equation (23.7).

$\mathcal{L}'(\tau_{\max}) = 0$. Finding that maximum is most easily done by taking the (natural) logarithm of this expression and differentiating $\ln \mathcal{L}(\tau)$:

$$\ln \mathcal{L}(\tau) = -\frac{1}{\tau} \sum_{i=1}^{n} t_i - n \ln \tau. \tag{23.4}$$

This is called the *log-likelihood* for this exponential process. Note that if we take another sample of t_i values with the same number n of measurements, then the log-likelihood function will be different if Σt_i is different in the two samples.[9] We can easily show how this affects the estimated value of the parameter τ, as follows.

The maximum of (23.4) occurs where

$$\frac{d}{dt} \ln \mathcal{L}(\tau) = \frac{1}{\tau^2} \sum_{i=1}^{n} t_i - \frac{n}{\tau} = 0, \tag{23.5}$$

so that the maximum likelihood estimator for the lifetime is

$$\tau_{\mathrm{ML}} = \frac{1}{n} \sum_{i=1}^{n} t_i. \tag{23.6}$$

This is just the mean of the observed lifetimes. Formally we should write this as $\tau_{\mathrm{ML}}(t_1 \ldots, t_n)$ since its value depends directly on all n data points t_i.

The details of this example are illustrated in Figure 23.1.

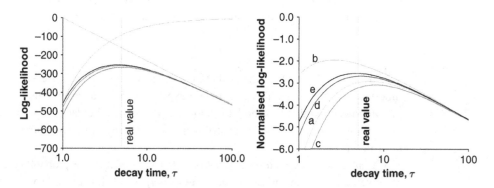

Fig. 23.1 The log-likelihood function for the particle decay model (23.1) with $\tau = 5.0$. The left panel shows the log-likelihood function of Equation (23.4) for each of three independent samples of $n = 100$ observations. The curves differ because Σt_i for the data values in the samples are different. The two faint lines are the two terms in (23.4). On the right, the log-likelihood is built up as the data is acquired. The curves labelled a, b, c, d, e correspond to the stages when $n = 1, 4, 10, 30, 100$ data values have been collected. The fluctuations in Σt_i as data is gathered shift the position of the maximum of the log-likelihood as per Equation (23.6). ©

[9] The fact that the single parameter $\bar{t} = \Sigma_{i=1}^{n} t_i$ is the only datum appearing the likelihood is discussed in Section 23.8.3. The statistic $\bar{t}$ built from the data in this way is referred to as a 'sufficient statistic'.

23.3 Formalising Likelihood

Following on from our motivational example of Section 23.2, we can put likelihood into a slightly more general context in which our model has m parameters and where we have n measurements with which to determine these parameters.

We will denote the parameters of the model by, $\boldsymbol{\theta} = \{\theta_1, \ldots, \theta_m\}$. These are to be determined from a set of measurements, the data, $\mathcal{D} : \{x_1, \ldots, x_n\}$. For convenience we use the notation $\mathbf{x} = \{x_1, \ldots, x_n\}$. We will assume the simplest case that the measurements are independent and identically distributed (*iid*) samples[10] drawn from a probability density $f_X(x_i; \boldsymbol{\theta})$ that describes the probability of getting the measurement x_i on the basis of the model having underlying parameters $\boldsymbol{\theta}$. [11] We put the subscript X on f_X to remind ourselves that it is the measurements x_i that are the random variables, not θ. In the spirit of classical statistics θ is merely a number that is to be determined by repeated (independent) sets of measurements $\{x_i\}$. It is because of this that we write the θ-dependence of the underlying probability density as $f_X(x_i; \theta)$ rather than $f_X(x_i \mid \theta)$.

The *likelihood function* is then:

Likelihood function:

$$\mathcal{L}(\boldsymbol{\theta}; \mathcal{D}) \stackrel{\text{def}}{=} f_X(\mathcal{D}; \boldsymbol{\theta}) = f_X(x_1, \ldots, x_n; \boldsymbol{\theta}) = \prod_{i=1}^{n} f_X(x_i; \boldsymbol{\theta}). \tag{23.7}$$

It should be emphasised that the likelihood function is not a probability density. It is a function of both the parameters of the model θ_i, and of the measurements x_i that the model is trying to emulate. In the frequentist world there is nothing random about the θ_i, they are simply numbers that we have to determine. The likelihood function itself is however a random function, but only because of its explicit dependence on the data.

The definition $\mathcal{L}(\boldsymbol{\theta}; \mathcal{D}) \stackrel{\text{def}}{=} f_X(\mathcal{D}; \boldsymbol{\theta})$ highlights an important distinction between likelihood and probability. The term likelihood is used in relation to an event that has already occurred or measurements that have already been made and reflects what we might know about the possible values of an underlying parameter set $\boldsymbol{\theta}$ that led to those particular measurements. This is expressed through the notation $\mathcal{L}(\boldsymbol{\theta}; \mathcal{D})$ indicating that it is the given data $D: \{x_1, \ldots, x_n\}$ that is telling us about the relative merit of parameters $\boldsymbol{\theta}$. $\mathcal{L}(\boldsymbol{\theta}; \mathcal{D})$ is merely a function of a collection of numbers $\{\theta_1, \ldots, \theta_m, x_i, \ldots, x_n\}$. [12]

---

[10] The usual statistical abbreviation for *independent and identically distributed* is '*iid*'. This is an important assumption that underlies much of model fitting.

[11] The variables x_i could have different scales. For example x_1 might be the maximum rotation speed of a galaxy and x_2 might be the intrinsic luminosity of a galaxy. On the other hand, the random variables x_i might themselves be vectors such as position in space or velocity components of a velocity field at each point, in which case the components of x_i will have the same scale (e.g. km s^{-1}). Put simply, each of the variables x_i could be a scalar, vector or tensor (objects which are invariant under coordinate transformations).

[12] As MacKay (2003, p.29) remarks, we should not talk of 'the likelihood of the data', but rather 'the likelihood of the parameters'. In the frequentist view θ are just numbers that we have to determine: they are not random variables.

The term 'probability' is used in relation to the distribution of possible values of a yet-to-be-made measurement that would occur, given particular values of parameters $\boldsymbol{\theta}$. This is expressed through the notation $f_X(\mathcal{D}; \boldsymbol{\theta})$.

The function $\mathcal{L}(\theta)$ varies from one measurement set to another, and so the estimator θ_{ML} of θ is itself a random variable, the expectation of which gets closer to the actual value θ^* of the parameter θ as the number of experiments and measurements increases. In the limit of large numbers of measurements the distribution of the estimates θ_{ML} becomes normal and the mean of that distribution tends to the required value of θ^*.

The last product term in (23.7) comes from the fact that the measurements are *iid*(compare with Equation 23.3). Because of the presence of this product it is often more convenient to work in terms of the log likelihood:

Log-likelihood of a dataset:

$$\ln \mathcal{L}(\boldsymbol{\theta}; \mathcal{D}) = \sum_{i=1}^{n} \ln f(x_i; \boldsymbol{\theta}). \tag{23.8}$$

Here, $f(x_i; \boldsymbol{\theta})$ is the probability of getting the datum x_i, given $\boldsymbol{\theta}$.

The quantity $\ln \mathcal{L}(\boldsymbol{\theta}; \mathcal{D})$ is nothing more than a function that is used to describe the merit of various parameter choices relative to some set of measurements, the maximum value being in some sense the most meritorious.

We can take derivatives of $\ln \mathcal{L}(\boldsymbol{\theta}; \mathcal{D})$ with respect to the parameters θ_i, and because it is a function of the data we can take expectation values of $\ln \mathcal{L}(\boldsymbol{\theta}; \mathcal{D})$ and its derivatives.[13] So, for example, in the case of a single parameter θ, we can consider the expectation value of the derivative of the log-likelihood with regard to the parameter θ:

$$\mathbb{E}\left[\frac{\partial \ln \mathcal{L}(\theta)}{\partial \theta}\right] = \int_{-\infty}^{\infty} \frac{\partial \ln f_X(\mathbf{x}; \theta)}{\partial \theta} f_X(\mathbf{x}; \theta) d\mathbf{x}, \tag{23.9}$$

$$= \int_{-\infty}^{\infty} \frac{\partial f_X(\mathbf{x}; \theta)}{\partial \theta} d\mathbf{x} = \frac{\partial}{\partial \theta} \int_{-\infty}^{\infty} f_X(\mathbf{x}; \theta) d\mathbf{x} = 0. \tag{23.10}$$

We shall use this technique several times in the forthcoming discussions.[14] This will lead us in Section 23.7 to the concept of the Fisher information.

---

[13] This distinction between data and parameters exposes one of the fundamental aspects of the frequentist–Bayesian schism. In the frequentist view the parameters $\boldsymbol{\theta}$ are nothing more than fitting parameters that enter into the model, they represent the 'true', albeit unknown, values that we seek to discover by experiment. In the Bayesian view the $\boldsymbol{\theta}$ are treated as random variables on an equal footing with the data: they are simply statistics that are subject to uncertainty.

[14] Equation (23.9), as written, may give the impression that we are taking the expectation value of a function of the parameter θ. However, $\mathcal{L}$ is not a probability density and θ is not a random variable. To make this clearer the expectation $\mathbb{E}\left[\partial \ln \mathcal{L}(\theta)/\partial \theta\right]$ could be written $\mathbb{E}_{\mathbf{x}}[\partial \ln \mathcal{L}(\theta)/\partial \theta]$, emphasising that it is the expectation value of a function of the data $\mathbf{x}$ that is involved. The right-hand side of (23.9) is, in effect, the definition of what we mean by the expectation on the left.

23.4 Maximum Likelihood Estimators

For didactic reasons, let us consider the simple single parameter model.

The basic tenet of the *maximum likelihood method* is that the maximum of the likelihood function $\mathcal{L}(\theta)$, viewed as a function of the model parameter θ gives the estimate θ_{ML} for the parameter θ, for the given dataset $\{x_1, \ldots, x_n\}$. For our single parameter discussion:

$$\frac{d}{d\theta} \ln \mathcal{L}(\theta) = \frac{1}{\mathcal{L}} \frac{d\mathcal{L}(\theta)}{d\theta} = 0 \quad \text{at } \theta = \theta_{\mathrm{ML}}, \tag{23.11}$$

and we see that the likelihood and log-likelihood functions have maxima at the same θ-value. The log-likelihood is particularly convenient to work with and is often given its own special symbol. Note also that the maximum of the likelihood function, the mode, provides the estimate, not the mean value.

There are issues to be aware of. It could be that the likelihood has more than one maximum. For complex models, it might be difficult to find an explicit solution for θ_{ML}, even if there is only one parameter to determine. In the worst case, the likelihood function may not even be differentiable. It should also be remarked that there are alternatives to maximising the likelihood such as maximising the entropy or information content of the dataset, and we should not forget *least squares* minimisation and its relatives.

23.4.1 Maximum Likelihood Estimation

Equation (23.11) forms the basis of one of the most important tools we have for estimating parameters in models of data, $\mathcal{D}$, – the method of *maximum likelihood*.

Suppose we are trying to determine a single parameter θ in a theory. We can make a plot of the likelihood function $\mathcal{L}(\boldsymbol{\theta} \mid \mathcal{D})$, viewed as a function of θ, that the data $\mathcal{D}$ would have turned out the way it did given different values of $\boldsymbol{\theta}$ (see Figure 23.2). Such a function will generally be bell-shaped, hopefully with a well-defined maximum at some value θ_{ML} of the parameter θ. θ_{ML} is the *maximum likelihood estimator* of θ.

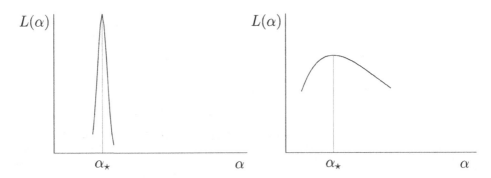

Fig. 23.2 Maximum likelihood estimation: on the left the parameter α is depicted as being sharply defined, whereas on the right the uncertainty is larger. ©

23.4.2 Errors in Parameter Determination

Up to now we have constructed a way of deriving the parameters of a model from measurements without broaching the issue of how well those parameters are determined, or even whether the model is at all appropriate. The intuitive way of thinking about how well determined the parameters are would be to gather more data, either as an extension of the current data set, or from a separate experiment. Either way we would at least expect consistency between the two sets of measurements, and as the data set grew in size we would expect the values of the parameters to approach the 'real world' values.[15]

If the likelihood function is sharp, we can argue that we have a good estimator of θ. The curvature of the function $L(\theta)$ at its maximum is a measure of the sharpness. We may also notice that for a Gaussian form of $\mathcal{L}(\theta)$,

$$\left[-\frac{d^2 \ln \mathcal{L}(\theta)}{d\theta^2} \right]_{\theta=\theta_{\mathrm{ML}}} = \frac{1}{\sigma^2}. \tag{23.12}$$

So, an alternative way of specifying how good is the estimator for θ is to characterise the width of $\mathcal{L}(\theta)$ by its second derivative at the peak:

$$\Delta\theta = \left[-\frac{d^2 \ln \mathcal{L}(\theta)}{d\theta^2} \right]_{\theta=\theta_{\mathrm{ML}}}^{-\frac{1}{2}}. \tag{23.13}$$

Thus the curvature of the likelihood function in the vicinity of its maximum provides a measure of the error of the estimated value. This readily generalises to situations involving many to-be-determined parameters where the likelihood is a function of many variables, $\mathcal{L}(\theta_1, \ldots, \theta_n)$: we can use the Hessian of the likelihood as a measure of the curvature of the likelihood function at a maximum.[16]

23.4.3 A Formal Error Estimate

The maximum value $\mathcal{L}(\theta_{\mathrm{ML}})$ of the likelihood function $\mathcal{L}(\theta)$ is not a useful indicator of how good that estimate might be: the likelihood is not a probability density. In other words,

---

[15] Of course they might converge, but would only converge on the 'real world' values if the model itself was a good physical model for the phenomenon under consideration.

[16] The Hessian matrix of a real valued function $f(x_1, \ldots, x_n)$ is defined as the matrix having components

$$H_{ij}(x_1, \ldots, x_n) = \frac{\partial^2 f}{\partial x_i \, \partial x_j}. \tag{23.14}$$

This is a function defined on the space $\mathbf{x}$ which describes the local behaviour of the function $f(x_1, \ldots, x_n)$. The Taylor expansion of $f(\mathbf{x})$ about a point is

$$f(\mathbf{x} + \delta\mathbf{x}) = f(\mathbf{x}) + (\delta\mathbf{x} \cdot \nabla)f(\mathbf{x}) + \tfrac{1}{2}\delta\mathbf{x}^T \mathbf{H}(\mathbf{x}) \, \delta\mathbf{x} + \text{higher order terms}, \tag{23.15}$$

where $\mathbf{H}$ is given by Equation (23.14). The eigenvalues of H_{ij} at a point $\mathbf{x}$ describe the n-dimensional curvature of the function $f(\mathbf{x})$ and so, in this case, the Hessian can be said to describe the geometric structure of the likelihood landscape. Note that the Hessian is a linear operator: if f and g are two functions $H[f + g] = H[f] + H[g]$, so we can easily build up a Hessian from its constituent components. This is used when defining the statistical information (see Section 23.8). The Hessian is a fundamental tool for finding structural elements in point sets in general and in particular for identifying structural elements in the galaxy distribution (Aragón-Calvo et al., 2007; Platen et al., 2007).

$\mathcal{L}(\theta_{\mathrm{ML}})$ does not tell us how likely θ_{ML} is to represent the true value of the parameter θ. However, for large samples, it can be shown that the likelihood estimates of parameters tend to a Gaussian distribution. That is a consequence of the central limit theorem. So, in that limit, the log-likelihood function is approximately parabolic in the vicinity of the maximum:

$$\mathcal{L}(\theta) = \mathcal{L}(\theta_{\mathrm{ML}}) + \frac{1}{2}\frac{d^2\mathcal{L}}{d\theta^2}\bigg|_{\theta_{\mathrm{ML}}}(\theta - \theta_{\mathrm{ML}})^2 + \cdots \tag{23.16}$$

In this limiting case the variance of the estimate θ_{ML} is then

$$\sigma^2(\theta_{\mathrm{ML}}) = \left(\mathbb{E}\left[\frac{d^2\mathcal{L}}{d\theta^2}\right]\right)^{-1}, \tag{23.17}$$

$$\simeq \left(\frac{d^2\mathcal{L}}{d\theta^2}\bigg|_{\theta=\theta_{\mathrm{ML}}}\right)^{-1}, \tag{23.18}$$

where $\mathbb{E}[X]$ denotes the statistical expectation value of the random variable X.[17] This last equation provides a basis for defining the variance of the estimate, though it is not without ambiguity when the log-likelihood function is not symmetric, as can be seen in the case of Equation (23.4). However, we do know that in the limit of large samples, the likelihood function tends towards having a parabolic shape, and so the issue of lack of symmetry disappears with ever larger data samples.

23.4.4 Alternative Error Estimates

There are several alternatives to using the curvature of the likelihood function at its maximum. The width of the function $\mathcal{L}(\theta)$ can alternatively be calculated by

$$\Delta\theta = \left[\frac{\int(\theta - \theta_{\mathrm{ML}})^2\mathcal{L}(\theta)d\theta}{\int\mathcal{L}(\theta)d\theta}\right]^{\frac{1}{2}}, \tag{23.19}$$

and this can be regarded as the standard error of the estimate. Note that the normalisation of $\mathcal{L}(\theta)$ does not come into this. In the particular case when $\mathcal{L}(\theta)$ is a Gaussian of width σ and mean θ_{ML} the integral can be evaluated. We find that

$$\Delta\theta = \sigma, \tag{23.20}$$

which agrees with the previous estimator.

23.4.5 Issues Arising when Using Maximum Likelihood

Maximum likelihood estimation is not without its difficulties and there are many examples in the literature of relatively simple (though often contrived) instances where it fails. Perhaps the most difficult situation arises where the maximum is not unique. It should also

---

[17] If the probability density of the random variable X is $f(x)$, then $\mathbb{E}[X] = \int xf(x)dx$, where the integral is taken over the range of value taken by X.

be noted that the more familiar least squares fitting is the same as the likelihood estimator of the parameters in the case when the measurement errors are independent and normally distributed. Least squares fitting is often done even when the errors are not Gaussian distributed, when results can be highly misleading. Computational packages that do least squares fitting are readily available in most data handling environments and it is advisable to verify the distribution of the residuals after the fit has been obtained.

23.5 Marginal and Conditional Likelihood

Typically, an experiment may be designed to measure some subset of the model parameter set. Alternatively, it may happen that the number of parameters θ_i is large and we wish to focus on fewer of these parameters. Or we may simply wish to use the data to constrain only one of the parameters. It is obviously advantageous to be able to ignore the model parameters that are not of direct interest.

It is at this point that we begin to see the advantages of supposing that the unknown parameters $\theta_1, \ldots, \theta_n$ were themselves taken from some statistical distribution, since then we could simply integrate out the unwanted parameters. Our likelihood would then consist of a product of marginalised or conditional probabilities. Attributing probability distributions to the model parameters is an aspect of Bayesian inference that does not exist in the frequentist world where the parameters $\theta_1, \ldots, \theta_n$ are simply numbers.

In the frequentist world it is necessary to rewrite the likelihood so as to refer specifically to the variable of interest. Suppose the model parameter set is $\boldsymbol{\theta} = (\phi, \eta)$, where ϕ is the parameter of interest from among the $\boldsymbol{\theta}$, and the η are now nuisance parameters, the ones we wish to not consider.

Take as a simple example the situation in which our model is that observations X of a quantity x are normally distributed, we write this as $X \sim N(\mu, \sigma^2)$. If all we want is the location μ, then σ is a *nuisance parameter*. μ is our *parameter of interest*.

We need, if possible, to transform our random variable X into a pair of random variables (Y, Z) such that in terms of (Y, Z) and (ϕ, η), our underlying model, which is now $f(Y, Z; \phi, \eta)$, is written as either of

$$f(Y, Z; \phi, \eta) = f_Y(Y; \phi) f_{Z \mid Y}(Z \mid Y; \phi, \eta), \quad f_Y(Y; \phi) : \text{marginal}, \tag{23.21}$$

$$f(Y, Z; \phi, \eta) = f_{Y \mid Z}(Y \mid Z; \phi) f_Z(Z; \phi, \eta), \quad f_{Y \mid Z}(Y \mid Z; \phi) : \text{conditional}. \tag{23.22}$$

In each of these equations there is a term that depends only on the parameter of interest ϕ and not on the η. This is the key factor in trying to select appropriate variables Y, Z. If such a decomposition is possible, the *marginal likelihood* is built using $f_Y(Y; \phi)$, or the *conditional likelihood* is built using $f_{Y \mid Z}(Y \mid Z; \phi)$.

23.6 Likelihood Invariance

Maximum likelihood has many attractive features, such as the important property which is referred to as *invariance*.

Suppose that we wish to estimate the maximum likelihood value of some parameter ξ which is a function of θ, $\xi(\theta)$, and suppose further that $\xi(\theta)$ is simple in the sense of having a singled valued inverse.[18]

> **Invariance of ML estimator:**
>
> If $\mathcal{L}(\theta)$ has a maximum at θ_{ML} then, provided the function $\xi(\theta)$ has a single-valued inverse, the maximum likelihood estimator, ξ_{ML}, of $\xi = \xi(\theta)$ is $\xi(\theta_{ML})$.

We can present a simple proof of the invariance property for the case that the function $\xi(\theta)$ has a single valued inverse. In that case we can write

$$\mathcal{L}(\theta) = \mathcal{L}(\xi^{-1}(\xi(\theta))). \tag{23.23}$$

The maxima of the left- and right-hand sides are obviously the same, so the value θ_{ML} that maximises the left side must be the same as the value that maximises $\xi^{-1}(\xi(\theta))$. If we denote $\xi(\theta_{ML})$ by ξ_{ML}, then

$$\theta_{ML} = \xi^{-1}(\xi_{ML}), \tag{23.24}$$

whence

$$\xi(\theta_{ML}) = \xi_{ML}. \tag{23.25}$$

This is one of several important and useful properties of maximum likelihood.[19]

23.7 Information Measure

The narrower the likelihood or log-likelihood function, the more precise information we have about the underlying parameter θ. This leads us to the important concept of *information*. In the case of a one-parameter system the information provided by the data is defined as:

> **Likelihood and information:**
>
> $$J(\theta) = -\frac{\partial^2 \ln \mathcal{L}(\theta)}{\partial \theta^2}, \qquad \sigma^2(\theta_{ML}) = -\frac{1}{\mathbb{E}\left[J(\theta)\right]}, \tag{23.26}$$

where the second relationship is just Equation (23.17).

Despite this notation, which is conventional, we should not lose sight of the fact that $\mathcal{L}(\theta)$, and hence $J(\theta)$ are functions that depend directly on the realisation of the data. That dependence is made explicit in Equation (23.8). To emphasise this we could write $J(\theta)$ as $J(\theta; \mathcal{D})$ or $J_{\mathcal{D}}(\theta)$, or even $J(\theta; x_1, \ldots, x_n)$.

---

[18] In an important paper, Berk (1967) shows that invariance holds even if the transformation is not one-to-one.

[19] Note that if the likelihood function had been a probability density then the transformation $\theta \to \xi(\theta)$ would have brought in a factor of the Jacobian of the transformation which would have shifted the location of the maximum. See Cox and Hinkley (1979, Section 9.2) for a more precise statement of invariance and some examples.

Because of this explicit data dependence it makes sense to talk about the expectation value of $J(\theta)$, i.e. the value of $J(\theta)$ that would be obtained from a large number of experiments for the true value of θ. (In this frequentist view the parameter θ is not itself a random variable). We can make this explicit by going back to the definition of likelihood in terms of the underlying data model. Then, for independent observations $\{x_1, \ldots, x_n\}$,

$$\log \mathcal{L}(\theta) = \sum \ln f(x_i, \theta) \quad \text{and} \quad J(\theta) = -\sum \frac{\partial^2}{\partial \theta^2} \ln f(x_i, \theta). \tag{23.27}$$

23.7.1 Fisher's Information Measure

The expectation of the information, $\mathbb{E}[J(\theta)]$, is referred to in the statistical literature as the *Fisher information*, which for our single parameter model is:

Fisher Information – I:

$$I(\theta) = \mathbb{E}[J(\theta)] = \mathbb{E}\left[-\frac{\partial^2 \ln \mathcal{L}(\theta)}{\partial \theta^2}\right]. \tag{23.28}$$

We can write this in terms of the underlying data model $f(x, \theta)$ for the usual case that the data points $\{x_i\}$ are mutually independent, as

$$I(\theta) = \mathbb{E}\left[-\sum_i \frac{\partial^2}{\partial \theta^2} \ln f(x_i, \theta)\right]. \tag{23.29}$$

Here we see the contribution to the Fisher information $I(\theta)$ that each data point makes.

There is another form for the Fisher information that exploits the fact that the curvature of the likelihood function is evaluated at a maximum of the log-likelihood:

Fisher Information – II:

$$I(\theta) = \mathbb{E}\left[\left(\frac{\partial \ln \mathcal{L}}{\partial \theta}\right)^2\right]. \tag{23.30}$$

We can show that (23.30) follows from (23.28) by differentiating Equation (23.9) with respect to θ. We get

$$0 = \frac{\partial}{\partial \theta} \int_{-\infty}^{\infty} f \frac{\partial \ln f}{\partial \theta} d\mathbf{x} = \int_{-\infty}^{\infty} \left[\frac{\partial f}{\partial \theta} \frac{\partial \ln f}{\partial \theta} + f \frac{\partial^2 \ln f}{\partial \theta^2}\right] d\mathbf{x}$$

$$= \int_{-\infty}^{\infty} \left(\frac{\partial \ln f}{\partial \theta}\right)^2 f d\mathbf{x} + \int_{-\infty}^{\infty} \frac{\partial^2 \ln f}{\partial \theta^2} f d\mathbf{x} = \mathbb{E}\left[\left(\frac{\partial \ln \mathcal{L}}{\partial \theta}\right)^2\right] + \mathbb{E}\left[\frac{\partial^2 \ln \mathcal{L}}{\partial \theta^2}\right], \tag{23.31}$$

which gives the desired result. The definition of the likelihood function (23.7) enables the switch from $f(\mathbf{x}; \theta)$ to $\mathcal{L}(\theta; \mathbf{x})$ in the last line.

The interpretation of this information measure is relatively straightforward. It is the mean squared gradient of the log-likelihood and so if the likelihood function varies rapidly with θ a single measurement will contain a lot of information about the parameter θ. In other words a bell-shaped likelihood function that is narrow specifies the value of θ more accurately than a wide bell-shaped function.

23.7.2 Efficiency

It is usual to talk in terms of the *efficiency* of the estimate θ_{ML}

$$\mathrm{Eff}(\theta_{\mathrm{ML}}) = \frac{I(\theta_{\mathrm{ML}})^{-1}}{\mathrm{Var}\,(\theta_{\mathrm{ML}})}. \tag{23.32}$$

This compares $\sigma^2(\theta_{\mathrm{ML}})$ of Equation (23.36) with the true variance, var(θ_{ML}), of the likelihood estimator (θ_{ML}). We recall from Equation (23.26) that $I(\theta_{\mathrm{ML}})^{-1}$ is an approximation for the variance based on the likelihood function. So Equation (23.32) is telling us how effective the curvature of the log-likelihood function is in estimating the variance. If the likelihood function is rather flat then our efficiency, as defined in (23.32) will be low. The ideal is to get an efficiency of unity. We shall see this again when discussing the Cramer–Rao–Frechet bound.

23.7.3 Fisher's Score Function

This last form (23.30) of the Fisher information is related to an important quantity called the *score function*:

$$S = \frac{\partial \ln \mathcal{L}(\theta)}{\partial \theta} = \frac{1}{\mathcal{L}} \frac{\partial \mathcal{L}(\theta)}{\partial \theta}, \tag{23.33}$$

which is just the gradient of the log-likelihood. The score vanishes at the maxima of the likelihood (or, equivalently, the log likelihood). The score S is a statistic. It is a function of the to-be-determined parameter θ, and of the data. In particular, if the log-likelihood is everywhere concave, the maximum likelihood estimator is the solution of the equation $S(\theta_{\mathrm{mle}}) = 0$.

We recall from Equation (23.10) that the expectation of the score is

$$\mathbb{E}\,[S] = \mathbb{E}\left[\frac{\partial \ln \mathcal{L}(\theta)}{\partial \theta}\right] = 0, \tag{23.34}$$

and the variance is the Fisher information:

$$I(\theta) = \mathbb{E}\left[\left(\frac{\partial \ln \mathcal{L}(\theta)}{\partial \theta}\right)^2\right] = \mathbb{E}\,[S^2]. \tag{23.35}$$

A number of statistical tests assessing the significance of a hypothesis about the data (i.e. the data model) are based on S.

The usual form of such hypotheses is that θ has a particular value, θ^*, and the question is to what extent the data supports that hypothesis.[20]

---

[20] We will touch on this in the simple case of testing the hypothesis that the variance of a zero-mean normal distribution underlying a data set $\{x_i\}$ had the value σ_0. We shall use the statistic $Z^2 = \Sigma x_i^2 / \sigma_0^2$ (see Equation 24.16) to set confidence limits on the hypothesis. In other words we will compare the ratio of what the data told us with an expected value.

The usual assumption is that the estimates θ_{ML} of θ^* derived from repeated experiments are normally distributed with mean θ^* and variance $I^{-1}(\theta^*)$.[21]

It turns out that the distribution of the score is asymptotically normal, $N(0, I(\theta))$ (which is of course consistent with Equations 23.34 and 23.35). In that case there are many known tests for the hypothesis $H^* : \theta = \theta^*$. A straightforward test using the score statistic is to make the comparison

$$W = \frac{S(\theta^*)^2}{I(\theta^*)},\tag{23.36}$$

where $S(\theta^*)$ is the score (23.33) and $I(\theta^*)$ is the Fisher information measure evaluated at the test value θ^*.[22] The statistic W is asymptotically χ^2 distributed and provides a significance level for the hypothesis H_0.

23.7.4 Maximising the Likelihood: Iterative Solutions

Unless the likelihood function has a particularly simple mathematical form, finding the maximum has to be done iteratively. For a likelihood that is a function of a single variable θ the simplest method is a simple generalisation of the Newton–Raphson method. This starts with a guess θ_0 for the value of θ_{ML} that maximises the likelihood $\mathcal{L}(\theta)$ or, equivalently, $L(\theta) = \ln \mathcal{L}(\theta))$. If we can calculate the likelihood gradient at θ_0 we simply march uphill. There are many algorithms for deciding how far we march: in the Newton–Raphson method that is determined by the second derivation of the likelihood.

Expanding the gradient $L(\theta)'$ of the log-likelihood function $\ln \mathcal{L}(\theta)$ about its maximum at θ_{ML}, we have

$$L(\theta_{\mathrm{ML}})' = L(\theta_0)' + (\theta_{\mathrm{ML}} - \theta_0)L(\theta_0)'' + \cdots = 0.\tag{23.37}$$

If we simply use the first two terms of the expansion, we have an approximation θ_1 to θ_{ML} given by

$$\theta_1 = \theta_0 - \left.\frac{L(\theta)'}{L(\theta)''}\right|_{\theta = \theta_0}.\tag{23.38}$$

Not surprisingly, how far we march is determined by the inverse of second derivative, i.e. the local curvature of the likelihood function at our starting point θ_0. Obviously we can iterate by replacing θ_0 with the value of θ_1 and repeating until two successive values are within a specified agreement.

A twist on this was added by Fisher. We note from Equation (23.28) that $\mathbb{E}\left[-L(\theta_0)''\right] = I(\theta_0)$ is the information at θ_0 , and Fisher proposed using $I(\theta)$ in place of $1/L(\theta)''$. With that we can write the $(k + 1)$th iteration in terms of the kth as

$$\theta_{k+1} = \theta_k + I(\theta_k)^{-1}L(\theta_k)', \quad k = 0, 1, \ldots\tag{23.39}$$

---

[21] This is an essentially frequentist view of hypothesis testing: you can only determine θ^* by doing repeated independent experiments. In cosmology we only have one Universe to observe.

[22] Since this result is asymptotic, it is possible to use $I(\theta_{\mathrm{ML}})$.

23.8 Information in Two or More Dimensions

The previous discussion relates to the situation where there is a single parameter, which we have denoted by θ, to be determined from n observations x_i, $i = 1, \ldots, n$ of a random variable X having probability density (pdf) $f_X(x \mid \theta)$. We need to progress towards the problem of estimating m parameters $\boldsymbol{\theta} = \{\theta_i\}$, $i = 1, \ldots, m$ from n observations of X where the underlying model for the pdf of X is $f(\mathbf{x} \mid \theta_1, \ldots, \theta_m)$. Towards this end it is of value to consider a worked example of the situation in two dimensions, i.e. to discuss the case wherein the model for the data has pdf $f(\mathbf{x} \mid \theta_1, \theta_2)$.

The log-likelihood for this case is (see Equation 23.8)

$$\ln \mathcal{L}(\theta_1, \theta_2) = \sum_{i=1}^{n} \ln f(x_i \mid \theta_1, \theta_2), \tag{23.40}$$

and the maxima of this function are determined from the *Hessian matrix*[23] of $\ln \mathcal{L}(\theta_1, \theta_2)$.

Consider a single measurement x_i in a sample of size n from a random variable X with distribution function $f_X(\theta_1, \theta_2)$. The Hessian of the log-likelihood is

$$\mathbf{H}(\theta_1, \theta_2 \mid x_i) = \begin{pmatrix} \dfrac{\partial^2}{\partial \theta_1^2} & \dfrac{\partial^2}{\partial \theta_1 \, \partial \theta_2} \\ \dfrac{\partial^2}{\partial \theta_1 \, \partial \theta_2} & \dfrac{\partial^2}{\partial \theta_2^2} \end{pmatrix} \log f(x_i \mid \theta_1, \theta_2). \tag{23.41}$$

23.8.1 Information in Higher Dimensions

The information $I(\theta)$ is proportional to the (negative) curvature of the log-likelihood function. If there is more than one parameter we replace the second derivative of the log-likelihood by the matrix of second derivatives:

$$\frac{\partial^2 \ln \mathcal{L}}{\partial \theta^2} \quad \longrightarrow \quad H_{jk}(\boldsymbol{\theta} \mid \mathbf{x}) = \frac{\partial^2 \ln f(\boldsymbol{\theta} \mid \mathbf{x})}{\partial \theta_j \, \partial \theta_k}. \tag{23.42}$$

With this we can define the *information matrix* as

$$I(\boldsymbol{\theta} \mid \mathbf{x}) = -\mathbb{E}[H(\boldsymbol{\theta} \mid \mathbf{x})], \tag{23.43}$$

which is the multi-dimensional analogue of (23.28). The information in more than one dimension becomes a matrix which is the expectation of the Hessian matrix of the log-likelihood.

For n independent random samples $\mathbf{x}_i$ we can write the Hessian as

$$\mathbf{H}(\boldsymbol{\theta} \mid \mathbf{x}) \equiv H_{jk}(\boldsymbol{\theta} \mid \mathbf{x}) = -\sum_{i=1}^{n} \frac{\partial^2 \ln f(\boldsymbol{\theta} \mid x_i)}{\partial \theta_j \, \partial \theta_k} = \sum_{i=1}^{n} H_{jk}(\boldsymbol{\theta} \mid x_i), \tag{23.44}$$

---

[23] See footnote 16.

which breaks the Hessian into a sum of contributions from each individual datum, x_i. With this the expectation of the Hessian, the information, becomes a sum over contributions from each of the x_i:

$$I(\theta \mid \mathbf{x}) = -\sum_{i=1}^{n} \mathbb{E}\left[H_{jk}(\theta \mid x_i)\right]. \tag{23.45}$$

Since the measurements x_i are independent, all the expectations in the sum are equal (they are merely independent samples of the same random variable), and so we can write this as

$$I(\theta \mid \mathbf{x}) = -n\mathbb{E}\left[H_{jk}(\theta \mid x)\right] = nI(\theta \mid x), \quad \text{for any } x \in \{x_i\}. \tag{23.46}$$

This shows that all points contribute equally to the information, and that the information increases with the number of points.

23.8.2 Example: the Univariate Normal Distribution

Suppose now that $f(x_i \mid \theta_1, \theta_2)$ is the density of a univariate normal distribution, $N(\mu, \sigma^2)$ (see Appendix E.1). Then

$$\log f(x) = -\frac{1}{2}\log(2\pi\sigma^2) - \frac{1}{2}\frac{(x-\mu)^2}{2\sigma^2}, \tag{23.47}$$

and it is straightforward to calculate the needed derivatives to get

$$\mathbf{H}(\mu, \sigma^2 \mid x_i) = \begin{pmatrix} -\dfrac{1}{\sigma^2} & -\dfrac{x_i - \mu}{\sigma^2} \\ -\dfrac{x_i - \mu}{\sigma^2} & \dfrac{1}{2\sigma^4} - \dfrac{(x_i - \mu)^2}{\sigma^6} \end{pmatrix}. \tag{23.48}$$

The information matrix is $\mathbf{I}(\mu, \sigma^2) = -\mathbb{E}\left[\mathbf{H}(\mu, \sigma^2 \mid x_i)\right]$, and in order to get the expectation values of each of the terms we need the expectation value of $(x_i - \mu)$ and of $(x_i - \mu)^2$. Clearly

$$\mathbb{E}[X] = \mu \quad \text{and} \quad \mathbb{E}[(X - \mu)^2] = \sigma^2, \tag{23.49}$$

whence the information matrix for a single observation is:

Information matrix for a single observation:

$$\mathbf{I}(\mu, \sigma^2) = \begin{pmatrix} \dfrac{1}{\sigma^2} & 0 \\ 0 & \dfrac{1}{2\sigma^4} \end{pmatrix}. \tag{23.50}$$

From Equation (23.46) the information matrix for n observations is then

$$\mathbf{I}(\mu, \sigma^2) = n\,\mathbf{I}(\mu, \sigma^2 \mid x_i) = \begin{pmatrix} \dfrac{n}{\sigma^2} & 0 \\ 0 & \dfrac{n}{2\sigma^4} \end{pmatrix}. \tag{23.51}$$

Note that in this case the estimators are uncorrelated.

We could have arrived here without appeal to the result (23.46) had we used the log-likelihood for *n* measurements in place of (23.47):

$$\log f(\mathbf{x}) = -\frac{n}{2}\log(2\pi\sigma^2) - \frac{1}{2}\sum_{i=1}^{n}\frac{(x_i - \mu)^2}{2\sigma^2}. \tag{23.52}$$

The maximum likelihood estimators of the mean and variance for the data set $\{x_i\}$ are then

$$\mu_{\mathrm{ML}} = \frac{1}{n}\sum_{i=1}^{n}x_i, \quad \sigma_{\mathrm{ML}} = \frac{1}{n}\sum_{i=1}^{n}(x_i - \mu_{\mathrm{ML}})^2. \tag{23.53}$$

Notice from this that μ_{ML} is unbiased, while σ_{ML} is biased.

23.8.3 Sufficient Statistics

The parameters μ, σ of the Gaussian distribution are determined from two quantities, or statistics, involving the data: the sample mean $\bar{x}$ and the sample variance $\bar{s}$:

$$\bar{x} = \frac{1}{n}\sum_{i=1}^{n}x_i \quad \text{and} \quad \bar{s} = \frac{1}{n}\sum_{i=1}^{n}(x_i - \bar{x})^2. \tag{23.54}$$

The log-likelihood can be written as

$$\ln \mathcal{L}(\mu, \sigma) = -n\ln(\sqrt{2\pi}\sigma) - n\frac{(\mu - \bar{x})^2 + \bar{s}^2}{2\sigma^2}. \tag{23.55}$$

The data only appears in this expression through the data-related quantities $\bar{x}$ and $\bar{s}$ and not through any other quantity involving the individual data values. Because of this, $\bar{x}$ and $\bar{s}$ are referred to as *sufficient statistics*. Moreover, they are both necessary for specifying the likelihood, and so they are *minimally sufficient statistics*.

23.9 Maximum Likelihood with *n* Variables

The preceding discussion generalises quite easily to the situation when *m* variables are to be determined by maximising the log-likelihood function. In this case the likelihood function for *n* data values $\mathbf{x} = \{x_1, \ldots, x_n\}$ is a function of *m* parameters $\boldsymbol{\theta} = \{\theta_1, \ldots, \theta_m\}$ and, as before the log-likelihood is

$$\ln \mathcal{L}(\boldsymbol{\theta}) = \sum_{i=1}^{n}\ln p(x_i \mid \boldsymbol{\theta}). \tag{23.56}$$

The maxima of this function occur where $\nabla \ln \mathcal{L} = 0$, which gives the set of *m* equations

$$\frac{\partial \ln \mathcal{L}(\boldsymbol{\theta})}{\partial \theta_j} = 0, \quad j = 1, \ldots, m. \tag{23.57}$$

Solving these equations, which may not be so easy, yields the maximum likelihood estimators $\theta_{i,\mathrm{ML}}$ of the parameters θ_i defining the model.

Formulating the errors for these estimates is a little more difficult than in the case of a single variable. We can likewise be guided by the expansion of the log-likelihood in the neighbourhood of a maximum in the limit of a large number of data points:

$$\ln \mathcal{L}(\boldsymbol{\theta}) = \ln \mathcal{L}(\boldsymbol{\theta}_{\mathrm{ML}}) + \frac{1}{2} \sum_{i,j=1}^{n} \frac{\partial^2 \ln \mathcal{L}(\boldsymbol{\theta})}{\partial \theta_i \, \partial \theta_j} \bigg|_{\boldsymbol{\theta}=\boldsymbol{\theta}_{\mathrm{ML}}} (\theta_i - \theta_{i,\mathrm{ML}})(\theta_j - \theta_{j,\mathrm{ML}}) + \cdots \quad (23.58)$$

In other words, the behaviour in the vicinity of a maximum of $\ln \mathcal{L}$ is a multi-dimensional quadratic with coefficients.

Hessian of log-likelihood:

$$H_{ij} = \frac{\partial^2 \ln \mathcal{L}(\boldsymbol{\theta})}{\partial \theta_i \, \partial \theta_j}, \qquad (23.59)$$

evaluated at the maximum of $\ln \mathcal{L}(\boldsymbol{\theta})$ with respect to the parameters θ_i.

The matrix H_{ij} is called the *Hessian* of the function $\ln \mathcal{L}(\boldsymbol{\theta})$. In matrix form equation (23.58) looks somewhat less intimidating:

$$\ln \mathcal{L}(\boldsymbol{\theta}) = \ln \mathcal{L}(\boldsymbol{\theta}_{\mathrm{ML}}) + \frac{1}{2}(\boldsymbol{\theta} - \boldsymbol{\theta}_{\mathrm{ML}})^T \mathbf{H}\,|_{\boldsymbol{\theta}=\boldsymbol{\theta}_{\mathrm{ML}}}\,(\boldsymbol{\theta} - \boldsymbol{\theta}_{\mathrm{ML}}). \qquad (23.60)$$

The expectation value of the Hessian is called the *Fisher information matrix*:

Fisher information matrix:

$$\mathbf{F} = F_{ij} = \mathbb{E}\left[\frac{\partial^2 \ln \mathcal{L}(\boldsymbol{\theta})}{\partial \theta_i \, \partial \theta_j}\right]. \qquad (23.61)$$

23.10 ML Linear Regression

As an important example of the maximum likelihood method we can look at the task of fitting a straight line through a set of points (x_i, y_i) in which the y_i values are subject to an uncertainty, or error, that is modelled as a Gaussian distribution of zero mean and variance σ^2. Note that the variance[24] does not depend on the value of x_i. Our model for the distribution of points is then

$$y_i = \alpha + \beta x_i + \epsilon_i, \qquad (23.62)$$

where β is the to-be-determined slope of the line and α is its y-axis intercept. The errors ϵ_i are $N(0, \sigma)$ distributed random variables. The parameters to be determined for the data

[24] We could write this as $\sigma_{y|x}$ to emphasise that this is the variance in y given x.

$\mathcal{D}: (\mathbf{x}, \mathbf{y}) = \{(x_i, y_i),\ i = 1, \ldots, n\}$ are α, β, σ^2. The likelihood $\mathcal{L}(\alpha, \beta, \sigma^2)$ is the product of the individual likelihoods:

$$\mathcal{L}(\alpha, \beta, \sigma^2) = \frac{1}{(2\pi\sigma^2)^{n/2}} \exp\left(\frac{1}{2\sigma^2} \sum_{i=1}^{n} [y_i - (\alpha + \beta x_i)]^2 \right), \tag{23.63}$$

since, according to our model, $\epsilon_i = y_i - (\alpha + \beta x_i)$. The log-likelihood is

$$\ln \mathcal{L}(\alpha, \beta, \sigma^2) = -\frac{n}{2} \ln(2\pi\sigma^2) - \frac{1}{2\sigma^2} \sum_{i=1}^{n} \left[y_i - (\alpha + \beta x_i) \right]^2. \tag{23.64}$$

Finding the maximum of $\ln \mathcal{L}(\alpha, \beta, \sigma^2)$ by differentiating with respect to each of the parameters α, β, σ^2 in turn yields the system of equations[25]

$$\frac{\partial}{\partial \alpha} \ln \mathcal{L}(\alpha, \beta, \sigma^2): \quad \alpha n + \beta \, \mathbf{\Sigma} x_i = \mathbf{\Sigma} y_i, \tag{23.65}$$

$$\frac{\partial}{\partial \beta} \ln \mathcal{L}(\alpha, \beta, \sigma^2): \quad \alpha \, \mathbf{\Sigma} x_i + \beta \, \mathbf{\Sigma} x_i^2 = \mathbf{\Sigma} x_i y_i, \tag{23.66}$$

$$\frac{\partial}{\partial \sigma^2} \ln \mathcal{L}(\alpha, \beta, \sigma^2): \quad \sigma^2 = \frac{1}{n} \sum (y_i - \hat{y}_i)^2. \tag{23.67}$$

The maximum likelihood fit to the value of y at x_i is

$$\hat{y}_i = (\alpha_{\mathrm{ML}} + \beta_{\mathrm{ML}} x_i), \tag{23.68}$$

where α_{ML} and β_{ML} are the solutions of (23.65) and (23.66).

There are two remarks to be made about this. Firstly, the maximum likelihood estimate (23.67) of the variance σ^2 is biased and underestimates the variance. The unbiased sample variance is

$$s^2 = \frac{1}{n-2} \sum (y_i - \hat{y}_i)^2. \tag{23.69}$$

The divisor is $n - 2$ because there are two other disposable constants, α, β, that we can choose so as to make s^2 as small as possible. The second remark concerns the matrix

$$\mathsf{I} = \begin{pmatrix} n & \mathbf{\Sigma} x_i \\ \mathbf{\Sigma} x_i & \mathbf{\Sigma} x_i^2 \end{pmatrix}. \tag{23.70}$$

The *Hessian* of the log-likelihood, i.e. the *information matrix* is

$$\mathbf{H} = \begin{pmatrix} \dfrac{\partial^2}{\partial \alpha^2} & \dfrac{\partial^2}{\partial \alpha \, \partial \beta} \\[2ex] \dfrac{\partial^2}{\partial \alpha \, \partial \beta} & \dfrac{\partial^2}{\partial \beta^2} \end{pmatrix} \ln \mathcal{L}(\alpha, \beta, \sigma^2) = -\frac{1}{\sigma^2} \begin{pmatrix} n & \mathbf{\Sigma} x_i \\ \mathbf{\Sigma} x_i & \mathbf{\Sigma} x_i^2 \end{pmatrix}. \tag{23.71}$$

---

[25] There is a complete discussion of this from the point of view of χ^2 fitting in Press *et al.* (2007, Section 15.2), where the discussion allows each measurement to have its own variance. Their matrix S in that section is the normalised information matrix of Equation 23.70 .

23.11 Likelihood Multivariate Gaussian

The multivariate Gaussian is certainly one of the most important of all the distributions. In matrix form the probability density of a p-dimensional random vector $\mathbf{X}$ with normally distributed components is

$$f_{\mathbf{X}}(\mathbf{x} \mid \boldsymbol{\mu}, \boldsymbol{\Sigma}) = \frac{1}{(2\pi)^{p/2}|\boldsymbol{\Sigma}|^{1/2}} \exp\left[-\frac{1}{2}(\mathbf{x} - \boldsymbol{\mu})^T \boldsymbol{\Sigma}^{-1}(\mathbf{x} - \boldsymbol{\mu})\right]. \tag{23.72}$$

The matrix $\boldsymbol{\Sigma} = \{\langle X_i X_j \rangle\}$ is the covariance matrix of the data. The log-likelihood of a data set $\mathcal{D} : \{\mathbf{x}_1, \ldots, \mathbf{x}_n\}$ of n samples of $\mathbf{X}$ is then

$$\ln \mathcal{L}(\boldsymbol{\mu}, \boldsymbol{\Sigma}) = \sum_{i=1}^{n} f_{\mathbf{X}}(\mathbf{x_i} \mid \boldsymbol{\mu}, \boldsymbol{\Sigma}), \tag{23.73}$$

$$= -\frac{np}{2}\ln(2\pi) - \frac{1}{2}\ln|\boldsymbol{\Sigma}| - \frac{1}{2}\sum_{i=1}^{n}(\mathbf{x}_i - \boldsymbol{\mu})^T \boldsymbol{\Sigma}^{-1}(\mathbf{x}_i - \boldsymbol{\mu}). \tag{23.74}$$

To find the maxima of this we need the help of some results in matrix theory (Anderson, 2003),[26] the main one being

$$\sum_{i=1}^{n}(\mathbf{x}_i - \boldsymbol{\mu})^T \boldsymbol{\Sigma}^{-1}(\mathbf{x}_i - \boldsymbol{\mu}) = \operatorname{tr}(\boldsymbol{\Sigma}^{-1}\mathbf{A}) + n(\bar{\mathbf{x}} - \boldsymbol{\mu})^T \boldsymbol{\Sigma}^{-1}(\bar{\mathbf{x}} - \boldsymbol{\mu}), \tag{23.75}$$

$$\text{where} \quad \mathbf{A} = \sum_{i=1}^{n}(\mathbf{x}_i - \bar{\mathbf{x}})(\mathbf{x}_i - \bar{\mathbf{x}})^T \quad \text{and} \quad \bar{\mathbf{x}} = \frac{1}{n}\sum_{i=1}^{n}\mathbf{x}_i. \tag{23.76}$$

The matrix $\mathbf{A}$ is positive definite. This allows us to do the summation:

$$\ln \mathcal{L}(\boldsymbol{\mu}, \boldsymbol{\Sigma}) = -\frac{np}{2}\ln(2\pi) - \frac{1}{2}\ln|\boldsymbol{\Sigma}| - \frac{1}{2}\operatorname{tr}(\boldsymbol{\Sigma}^{-1}\mathbf{A}) - \frac{n}{2}(\bar{\mathbf{x}} - \boldsymbol{\mu})^T \boldsymbol{\Sigma}^{-1}(\bar{\mathbf{x}} - \boldsymbol{\mu}). \tag{23.77}$$

We need to know how to maximise each of these terms.

The last term is the only one containing $\boldsymbol{\mu}$, and since $\boldsymbol{\Sigma}^{-1}$ is positive definite[27] this term is maximised at $\boldsymbol{\mu} = \bar{\mathbf{x}}$. Hence the maximum likelihood estimator of $\boldsymbol{\mu}$ is $\boldsymbol{\mu}_{\mathrm{ML}} = \bar{\mathbf{x}}$.

Once we set $\boldsymbol{\mu} = \bar{\mathbf{x}}$ the last term vanishes no matter what $\boldsymbol{\Sigma}$ is, so to find the variance we need only the middle two terms. These middle two terms need yet another theorem telling us that $-n\ln|\boldsymbol{\Sigma}| - \operatorname{tr}(\boldsymbol{\Sigma}^{-1}\mathbf{A})$ is maximised by $\boldsymbol{\Sigma} = \frac{1}{n}\mathbf{A}$ ((Anderson, 2003)). With that we see that the estimator of $\boldsymbol{\Sigma}$ is $\frac{1}{n}\mathbf{A}$.

In summary

$$\boldsymbol{\mu}_{\mathrm{ML}} = \frac{1}{n}\sum_{i=1}^{n}\mathbf{x}_i, \quad \boldsymbol{\Sigma}_{\mathrm{ML}} = \frac{1}{n}\sum_{i=1}^{n}(\mathbf{x}_i - \bar{\mathbf{x}})(\mathbf{x}_i - \bar{\mathbf{x}})^T, \tag{23.78}$$

which is pretty much what one might have expected.

---

[26] The following discussion is from Olive (2012, Ch.5, Example 5.4).

[27] This follows from a theorem on quadratic forms.

23.12 Cramér–Rao–Frechet Lower Bound

Suppose we have an estimator $T(\mathbf{x})$ for the value $\theta_\dagger$ of the parameter θ in the data model. Different data sets $\mathbf{x}$ will yield different values for the estimate $T(\mathbf{x})$, and so $T(\mathbf{x})$ is a random variable, and is independent of θ. Unless the data is biased, $\mathbb{E}[T(\mathbf{x})] = \theta_\dagger$. The difference between these two quantities is called *the bias*:

$$b(\theta) = \mathbb{E}[T(\mathbf{x})] - \theta_\dagger. \tag{23.79}$$

Sometimes it is appropriate to put a subscript n on the bias, $b_n(\theta)$, because the bias will depend on the dimension of $\mathbf{x}$. We can be formal and write $t(\mathbf{x})$ for the value of the random variable T in a given realisation.

The expectation of the data $\mathbf{x}$, given the value of a parameter θ in the data model, is

$$\mathbb{E}[T; \theta] = \int t(\mathbf{x}) f_X(\mathbf{x}; \theta) \, d\mathbf{x}. \tag{23.80}$$

In the frequentist world, θ is just a number that we are trying to determine, it is not a random variable.[28]

The variance of θ_{ML} can be regarded as a measure of the *precision* of the estimate θ_{ML}:

$$\mathrm{Var}(\theta_{\mathrm{ML}}) = \int_{D_X} (\theta - \theta_{\mathrm{ML}})^2 \, \mathcal{L}(\theta) \, dx_1 \ldots dx_n. \tag{23.81}$$

The Cramér–Rao–Frechet bound[29] on the variance of an unbiased estimator θ_{ML} is:

Cramér–Rao–Frechet lower bound for $\theta_{\mathbf{ML}}$ unbiased:

$$\mathrm{Var}(\theta_{\mathrm{ML}}) \geq \frac{1}{I(\theta_{\mathrm{ML}})}. \tag{23.82}$$

This bound was first written down by Fisher (1925) but not in the present form until the work of Rao (1945) and Cramér (1946). It was first so-named by Neyman and Scott (1948).

In the case that the estimator of θ is biased this is modified to:

Cramér–Rao–Frechet lower bound:

$$\mathrm{Var}(\theta_{\mathrm{ML}}) \geq \frac{\left(1 + \dfrac{\partial B}{\partial \theta}\right)^2}{I(\theta_{\mathrm{ML}})}, \tag{23.83}$$

for an estimator θ_{ML} based on n measurements having bias

$$B_n(\theta_{\mathrm{ML}}) = \mathbb{E}[\theta_{\mathrm{ML}}^{(n)}] - \theta_*. \tag{23.84}$$

---

[28] Since the likelihood is formally defined by $\mathcal{L}(\theta; \mathbf{x}) = f_X(\mathbf{x}; \theta)$ we sometimes see this integral (23.80) written as $\theta_\dagger = \int \theta \mathcal{L}(\theta; \mathbf{x}) d\mathbf{x}$. See the discussion about Equation (23.7) in Section 23.3.

[29] In astrophysics and in many statistics texts this is, for some reason, referred to as the Cramér–Rao bound without Frechet's name. Frechet had in fact derived the same result as Cramér and Rao a little before them.

23.12.1 Deriving the Cramér–Rao–Frechet Lower Bound

We can sketch a simple proof as follows. The argument depends on the Cauchy–Schwarz inequality, which for functions $u(x)$ and $v(x)$ states that:

$$\int u^2 dx \int v^2 dx \geq \left(\int uv \, dx \right)^2. \tag{23.85}$$

Equality is achieved if and only if u and v are linearly dependent, i.e. if there exist an α, independent of x, such that $u(x) = \alpha v(x)$.

As before we work with the probability distribution of the data $\mathbf{x}$: $f(\mathbf{x}; \theta)$ and consider the situation where the estimator of the parameter θ is some function of the given data: $T(\mathbf{x})$. The important quantity is the expectation value of the estimates $T(\mathbf{x})$, which of course does not depend on θ, taken over the pdf of the data:

$$\theta_\dagger = \mathbb{E}\left[T(\mathbf{x})\right] = \int t(\mathbf{x}) f(\mathbf{x}; \theta) d\mathbf{x}. \tag{23.86}$$

Crucially, we want the estimator $\theta_\dagger$ to be unbiased, which is expressed as $\mathbb{E}[\theta_\dagger - \theta] = 0$:

$$\int \left(t(\mathbf{x}) - \theta \right) f(\mathbf{x}, \theta) \, dx = 0. \tag{23.87}$$

This asserts that the mean of the sampling distribution is the parameter θ of the population from which the data is drawn. We can differentiate this with respect to the parameters θ:

$$\int (t - \theta) \frac{\partial f}{\partial \theta} \, d\mathbf{x} - \int f \, d\mathbf{x} = 0 \quad \Rightarrow \quad \int (t - \theta) \frac{\partial \ln f}{\partial \theta} f d\mathbf{x} = 1, \tag{23.88}$$

since $\int f \, d\mathbf{x} = 1$. Now we use the Schwarz inequality (23.85), identifying

$$u = \int (t - \theta) \sqrt{f} \, d\mathbf{x}, \quad \text{and} \quad v = \int \frac{\partial \ln f}{\partial \theta} \sqrt{f} \, d\mathbf{x}. \tag{23.89}$$

This leads to the inequality

$$\left[\int (t - \theta)^2 d\mathbf{x} \right] \left[\int \left(\frac{\partial \ln f}{\partial \theta} \right)^2 f d\mathbf{x} \right] \geq \int (t - \theta) \frac{\partial \ln f}{\partial \theta} f d\mathbf{x} = 1. \tag{23.90}$$

The term $\int (t - \theta)^2 f \, d\mathbf{x}$ is just the variance of the estimate $\theta_\dagger = \mathbb{E}\left[T(\mathbf{x})\right]$ about the mean θ, and so we can write this as

$$\text{Var}\,(\theta_\dagger) \, \mathbb{E}\left[\left(\frac{\partial \ln f(\mathbf{x}; \theta)}{\partial \theta} \right)^2 \right] \geq 1, \tag{23.91}$$

which is the result we wanted.[30]

---

[30] If the estimator $T(\mathbf{x})$ had been biased, then $\mathbb{E}\left[T(\mathbf{X})\right] = \theta_\dagger + B(\theta)$ and the right-hand side of Equation (23.87) would have been $B(\theta)$ instead of zero. Then the right-hand side of (23.88) would have been $1 + B'(\theta)$ which would have carried through to the right-hand side of the bound (23.91), which is Equation (23.83) for the bound on a biased estimator.

We noted that the Cauchy–Schwarz bound (23.85) was reached if and only if the integrands were linearly related. This is important because it tells us what is the best that can be achieved, i.e. it tells us when our estimator $T(\mathbf{x})$ is most efficient. Equality in Equation (23.90) is achieved when

$$(t - \theta) = \alpha(\theta)\left(\frac{\partial \ln f}{\partial \theta}\right) \qquad (23.92)$$

for some proportionality factor α which does not depend on the $\mathbf{x}$, but may depend on the value of θ since that is just a number and is independent of the data. By definition, $\ln f_X(\mathbf{x}; \theta)$ is the likelihood function which when viewed as a function of θ has a maximum at the estimator θ_{ML}. That is where the right-hand side of this equation vanishes, and so where $t(\mathbf{x}) = \theta_{ML}$.

> **Remark:** We come to the remarkable fact that the maximum likelihood estimator θ_{ML} is the one which achieves the optimal result of having the smallest variance as provided by the Cramér–Rao–Frechet bound.

23.12.2 The James–Stein Estimator

In 1956, the statistics community was shaken by a paper written by Stein (1956) and refined by James and Stein (1961). To quote the *Scientific American* article on the subject by Efron and Morris (1977): *His result undermined a century and a half of work on estimation theory, going back to Karl Friedrich Gauss and Adrien Marie Lagrange.* Dramatic perhaps, but it is certainly true that the papers led to a large amount of, often heated, debate. Papers explaining the paradox are still being published.

The standard way of presenting the paradox is to first consider the simple problem of estimating the mean of a population, and then generalising this to simultaneously determining the means of a number of independent populations. We would expect that determining the means collectively would yield no better an answer than determining them individually. That expectation is incorrect: that is the essence of *Stein's paradox*. Intuitively, we feel that each independent experiment should have no influence on another experiment, they are independent.

Suppose we have independent random variates $\{X_i\}$, $i = 1, \ldots, p$. Then, because they are independent the sample mean is the minimum variance unbiased estimator of the population mean: $\hat{\theta}_i = \mathbb{E}[X_i]$, for each of the p samples. In this case, the estimates $\hat{\theta}_i$ are also the maximum likelihood estimators for each population. Stein's estimator (James and Stein, 1961) is generally presented in the form

$$\hat{\boldsymbol{\theta}}(\mathbf{X}) = \boldsymbol{\theta}_0 + \left(1 - \frac{p-2}{(\mathbf{X} - \boldsymbol{\theta}_0)^T(\mathbf{X} - \boldsymbol{\theta}_0)}\right)(\mathbf{X} - \boldsymbol{\theta}_0), \quad p \geq 3. \qquad (23.93)$$

In this form we see that the estimate of the mean of any one of the populations is decreased by the factor in parentheses, which for $p \geq 3$ is smaller than unity. This factor 'shrinks' the estimator of $\mathbf{X}$ towards the value $\boldsymbol{\theta}_0$. The James–Stein estimator was the first of a number of such *shrinkage estimators*.[31] The quantity θ_0 in Equation (23.93) requires a comment.

The idea that the sample mean should be the best estimator of the population mean is one of the cornerstones of basic statistics: it is an unbiased minimum variance estimator and is the result of applying the maximum likelihood principle. The result seems totally bizarre if the populations $\{X_i\}$ are manifestly independent, as if we were to choose, say, the weight of a truckle of cheese made in Cheddar in the UK, the mean density of traffic on a US Interstate Highway, and the population of a species of 1 year old kangaroos in Australia. Yung (1999) provides a more realistic example of measuring the IQs of p children in a class. The value achieved in the test θ is the widely adopted estimator of the true value $\hat{\theta}_i$ of a child's IQ. However, a better estimator is to consider all of the children in the class together and to use Stein's shrinkage estimator.[32]

---

[31] The factor $(p-2)$ in Equation (23.93) is referred to as the *shrinkage factor*. In other shrinkage estimators, the shrinkage factor differs from Stein's risk-based factor $(p-2)$, but the mathematical form of the estimator is unchanged.

[32] If we take a Bayesian view on what is happening, the situation is slightly less surprising. In the world of the Bayesian estimation of the parameters the estimate depends on assigning some 'priors'. In the case of the children in a class undergoing an IQ test the children are not seen as being independent: they are all of the same age, in the same classroom, taking the same test for which we have prior knowledge of the spread of the values achieved by such a group. The (empirical) Bayesian analysis of Efron and Morris (1977) does result in a Stein-like shrinkage estimator of the same form as (23.93).

24

Frequentist Hypothesis Testing

In this chapter we cover one particular aspect of frequentist statistics in order to be able to compare and contrast the approach with the Bayesian approach that we shall be discussing shortly. In much of the world that uses statistics, the principle objective appears to be to test hypotheses about given data and come to some conclusion. The conclusion might be the answer to either of the questions: 'Is this supported by the data or not?', or 'Which is the better descriptor of the data?'. Control samples are commonly used as an alterrnative against which the data will be evaluated. Whatever the question, the questioner expects a quantitative answer.

In cosmology we only have the one Universe as a source of data, and we have no other Universe that can act as a control. Of course we do have numerical simulations that can play that role. Our goal is often to fit a parameterised model to data and we can reasonably ask how confident we should be that this is the 'best' answer.

24.1 Classical Inference

In generic terms, the goal of statistical analysis is to discover how some factor Y responds to changes in some input, or set of inputs X. The observations $(\mathbf{x}, \mathbf{y})$ are paired, and while the values of $\mathbf{x}$ are generally under control of the experimentalist, the response $\mathbf{y}$ is subject to measurement errors ϵ. The mechanism for doing this may take the form of establishing a model for the dependence of Y on X, or might be to discover which of the factors X, or which combination of the elements of X, are the main contributory factors to the response Y. Here we will consider only the first of these, fitting parameterised models.

24.1.1 Linear Regression

The simplest data model, linear regression is commonplace and involves determining parameters α and β in the relationship of the form

$$\mathbf{Y} = \mathbf{A}\mathbf{X} + \epsilon, \tag{24.1}$$

where the experiment consists of observing the values of $\mathbf{Y}$, the *response*, resulting from *treatments* $\mathbf{X}$. The observations $\mathbf{Y}$ are subject to errors ϵ.

This approach makes the simplifying assumptions that Y and ϵ are independent, and moreover that the ϵ are zero mean normally distributed errors with variance independent of X (referred to as *homoscedasticity*[1]). Formally, the error distribution ϵ is homoscedastic if $\mathrm{Var}(\epsilon \mid X)$ is independent of X. Classically, the X-independence of the errors ϵ is dealt with by adopting a model for the trend in the variance and removing it.

24.1.2 Hypothesis Testing and Fitting

One of the factors that dominates some sciences more than others is the notion of the *null hypothesis*, H_0, versus the *alternative hypothesis*, H_a. Associated with this is a *significance level*, denoted by α, for rejecting H_0. The question is how much of the variation in the response Y is due to any of the factors X, and the null hypothesis would be that none of the variation in Y is due to X. A related question is how to measure the relative merits of two hypotheses, H_A and H_B, given a set of observations.

χ^2 Least Squares Goodness of Fit

Minimising the χ^2 statistic is an essential part of the (frequentist) toolbox for determining 'best fit' parameters.[2] The basic idea is to 'fit' observed data points $\{x_i, y_i\}$ with a model of the form $y(x) = f(x; a_1, \ldots, a_M)$ for some specific functional form of f given in terms of parameters $a_1, \ldots, a_M$ that are to be determined. How we get to the function form of f, our *model*, and how we decide on our parameters is completely arbitrary: the straight line model, $y = mx + c$, and its generalisations to higher order polynomials are the most familiar. The order and type of such a fitting polynomial is generally pre-ordained by the way the data looks.

Given the model with M fitting parameters a_i, we define a *goodness of fit* statistic, χ^2 for N data points:

χ^2 'Goodness-of-fit' statistic for fitting N data points with M parameters:

$$\chi^2 \equiv \chi^2(a_1, \ldots, a_M; \epsilon_1, \ldots, \epsilon_N) = \sum_{i=1}^{N} \epsilon_i^2 \big(y_i - f(x_i; a_1, \ldots, a_M)\big)^2. \qquad (24.2)$$

The number of degrees of freedom is $N - M$.

Here the factors ϵ_i are the weights to be attached to each of the data points. A small ϵ_i means that a point makes a small contribution to the sum, reflecting a lower level of confidence in the value of that datum. That might perhaps be due to a larger measurement error in the experimental measurement of y_i.

The goal is to find the set of parameters a_i that minimise the value of χ^2 in (24.2). Differentiating (24.2) with respect to each of the a_i and setting this to zero to locate the minimum of χ^2 yields the set of M equations:

$$\sum_{i=1}^{N} \frac{y_i - f(x_i)}{\sigma_i^2} \frac{\partial}{\partial_k} f(x_i; a_1, \ldots, a_M), \quad k = 1, \ldots, M. \qquad (24.3)$$

---

[1] Also spelled 'homoskedasticity'. The opposite is *heteroscedasticity*.
[2] See Press *et al.* (2007, Ch. 15) and Papoulis and Pillai (2002, pp.361 *et seq.*).

Here we have replaced the weighting factor ϵ_i with $1/\sigma_i^2$, σ_i being a measure of the uncertainty of the measurement. These, possibly nonlinear, equations can, in principle, be solved to find the M values of a_i that yield the 'best fit'.

The equations are linear in the case that the fitting function is linear is each of the a_i:

$$f(x_i, \mathbf{a}) = \sum_{j=1}^{M} a_j P_j(x_i), \quad i = 1, \ldots, N, \tag{24.4}$$

where the $P_j(x)$ are linearly independent functions of x. In the case of polynomial models these could simply be the powers of x or even a suitable set of orthogonal polynomials such as Chebyshev or Legendre polynomials.

Remarks about χ^2 Fitting

The reason this statistic is labelled with the symbol χ^2 is that, under the assumption that the data errors are independent and normally distributed, the statistical distribution of the statistic is the χ_{N-M}^2 distribution.[3]

There are several general remarks to make about the χ^2 statistic. Firstly, we are minimising the weighted sum of squares of the deviations of the model from the data. That is why the resulting values of the parameters giving the minimum χ^2 are referred to as the *least squares fit* to the data. Secondly, it should be noted that we can still use the χ^2 statistic even when the errors are not Gaussian. However, in that case the statistic is no longer χ^2-distributed and it is harder to report confidence levels for the fit. Thirdly, if the errors are not independent then the χ^2 statistic has to be modified to account for their correlation.

There is no formal method prescribing the way in which the weights are to be assigned. The simplest option is to argue that the ϵ_i are random values taken from a Gaussian distribution having zero mean, $\mu = 0$, and to-be-determined variance σ^2. It is of course possible to assign each point an ϵ_i that is taken from a Gaussian distribution of variance σ_i (perhaps reported by the experimentalist, or inferred from the data itself).[4]

If the observations are not independent but are such that they have a covariance $\mathbf{A} = \text{Cov}(y_i, y_j)$, the least squares estimator is simply

$$\chi^2(\mathbf{a}) = \left(\mathbf{y} - \mathbf{f}(\mathbf{a})\right)^T \mathbf{A}^{-1} \left(\mathbf{y} - \mathbf{f}(\mathbf{a})\right), \quad \mathbf{f}(\mathbf{a}) = f(x_i; a_1, \ldots, a_M). \tag{24.5}$$

The notation here is that $\mathbf{y}$ is the column vector of the data values y_i observed at the ith data point, and $f(\mathbf{a})$ is the column vector of data values according to the model when using the parameter values $\mathbf{a}$.

---

[3] Formally, the statistical distribution of $Z = \Sigma_1^k X_i^2$, where the X_i are all independent and $N(0, 1)$ distributed, is the χ_k^2 distribution. k is referred to as the *number of degrees of freedom* of the distribution. The expectation of the sum Z is $\bar{Z} = k$.

[4] Peebles (1971, Table V-1, p.134 and p.139 *et seq.*) gives a nice χ^2 analysis of the first 14 temperature measurements to be made of the cosmic background radiation in the Rayleigh–Jeans part of the spectrum. He derives a weighted-mean temperature of $T_{RJ} = 2.72 \pm 0.08\,\text{K}$ with a $\chi^2 = 7.1$. The value of χ^2, on the basis of the quoted experimental error bars, is $\chi^2 = 7.1$, whereas the expected value was $\chi^2 = 13 \pm 5$. We conclude that there is no inconsistency in the data. If the experimental $\chi^2 = 7.1$ is viewed as being on the low side, that would most likely reflect the systematic effects that inevitably arise when making absolute measurements. That this is so close to the most accurately determined temperature $T_{CMB} = 2.725\,\text{K}$ is certainly a tribute to those early pioneers who made what at the time was a very difficult measurement.

24.2 Estimation: Risks and Decisions

It should be emphasised that χ^2 minimisation is not the only way of estimating parameters or arriving at a confidence estimate in their assigned values. An important alternative is, loosely speaking, to minimise the risk of making a bad decision. The game-changing paper of Stein (1956) showed that neither unbiased estimators nor maximum likelihood estimators have a low 'risk'. This is discussed at some length in Section 23.12.2.

The subject of evaluating 'risk' and making informed decisions is a major sub-branch of statistics. We can only touch on it here by considering a simple case of evaluating risk in adopting a particular parameter estimate, see Cox and Hinkley (1979, Section 11.6).

Consider the problem of fitting measurements $\mathbf{y}$ with a function that has a free parameter θ whose value we wish to determine. Suppose that $t(\mathbf{y}; \theta_1, \ldots, \theta_p)$ is an estimator of the value of θ, based on the given data $\mathbf{y}$ and a number of to-be-determined parameters θ_i.

In the simplest cases, $t(\mathbf{y})$ does not depend explicitly on the parameters θ_i. As an example, if θ represents the mean of the population from which the measurements y_i are drawn, then we might use $t(\mathbf{y}) = \sum_{i=1}^{N} y_i / N$, which does not itself involve the parameter θ. On the other hand, in standard least squares fitting, we find an estimate for the θ_i by minimising $(\theta - t(\mathbf{y}; \theta_1, \ldots, \theta_p))^2$ over the θ_i for the given data set.

24.2.1 The Loss Function

It is convenient and there is no loss of generality in considering the determination of a single parameter θ from the point of view of risk. The *risk* in using an estimate $t(\mathbf{y})$ for a parameter θ, based on data $\mathbf{y}$, is defined in terms of a *loss function* $\Lambda(\theta, t)$. Λ is a function of the data $\mathbf{Y}$ via the value $t(\mathbf{y})$ determined by the data $\mathbf{Y}$. So $(\theta - t)$ is the deviation of the data from the fit. There are various ways of determining t. If, for example, we are determining the mean of a set of measurements we might use either sample mean, $\bar{Y}$ or the sample median $\tilde{Y}$, or even take a wild guess independently of what the data says.

So, what motivates the choice for the loss function? These might be explicit functions involving an actual cost estimate for being wrong, or it might reflect a degree of belief that the data is telling us the right thing. Accepting this latter view brings in the Bayesian aspect of risk determination.

Expressing a 'degree of belief' is not the sole province of the Bayesian. When a frequentist accepts a hypothesis at a 92% level of confidence, the choice of 92% expresses the belief that this is an adequate margin for acceptance of a parameter value, albeit without making clear the basis for this decision.

A simple example of a loss function is $\Lambda(x) = x^2$, in which case we may say that the loss on the estimated value $t(Y)$, given the data Y is $\Lambda(\theta - t) = (\theta - t)^2$. This choice of loss function is of course directly related to the normal χ^2 statistic for goodness of fit and the log-likelihood for a normal distribution.

Frequently used examples of loss functions are

$$
\begin{array}{ll}
(a)\ \Lambda(x) = x^2, & (b)\ \Lambda(x) = |x|, \\[2mm]
(c)\ \Lambda(x) = \begin{cases} 1 & x \neq 0 \\ 0 & x = 0, \end{cases} & (d)\ \Lambda(x) = 1 - e^{-\alpha x^2},
\end{array}
\tag{24.6}
$$

where $x = \theta - t$ and $\Lambda(\theta - t)$ is the cost or loss incurred by using the estimator t when the actual value is θ. All these functions have $\Lambda(0) = 0$: there is no cost in getting the correct answer. Also, $x = 0$ is the minimum of the function: getting it wrong always costs something. Λ reflects the dependence of the cost on how wrong the estimate is.

The quadratic loss function (24.6a) assigns a disproportionate price for estimates that are further away from the true value, and therefore unfairly represents the effects of outliers in the data. The linear loss function (24.6b) is better in this sense. (24.6c) is the '0–1' cost function that expresses that any answer except the correct answer bears the same cost, while (24.6d) is a softer version of that: being almost correct is OK.

24.2.2 Estimating Risk

The risk R is estimated by taking the expectation value of $\Lambda(\theta - t)$ over all realisations of the data Y:

$$
R(\theta, t) = \mathbb{E}_Y[\Lambda(\theta - t(\mathbf{y}))] = \int \Lambda(\theta - t(\mathbf{y}))\, f_Y(\mathbf{y})d\mathbf{y}.
\tag{24.7}
$$

The integral would provide a mechanism for evaluating the risk if we knew the underlying distribution function $f_Y(\mathbf{y})$. In general that can only be obtained from the data itself, in which case the risk is sometimes referred to as the 'empirical risk'. Alternatively, a probability density with some assumed parameters can be used.

In the case where $\Lambda(x) = x^2$ we have $R = \mathbb{E}[(\theta - t(\mathbf{y})^2]$, which we recognise as the expectation of the mean squared error. This is minimised when $\theta = \mathbb{E}[t(\mathbf{y})]$. In the case that $R = \mathbb{E}[|(\theta - t(\mathbf{y})|]$, θ is the median value of the data.

24.3 Confidence Intervals

The subject of the meaning of *confidence intervals* in the estimation of statistical parameters is highly controversial. When we say 'the value of some parameter X is X_0 with 95% confidence' or 'we are 95% confident that ...', what, precisely, do we mean? The naïve understanding of this is that repetitions of the measurements will lie within the specified range 95% of the time. However, this is not correct, as remarked by Neyman (1937, Section II, p.347 and in particular p.349), the person who introduced this concept. The assigned probability refers to the reliability of the measurement process, i.e. the experiment and the data.[5]

---

[5] Indeed it was Neyman (1937) who in that paper emphasised the fundamental assertion of frequentist philosophy, that 'The parameter is an unknown constant and no probability statement concerning its value may be

Confidence, whatever that is, is generally given as a p-value. The p-value is a measure of the significance of a hypothesis given a particular data set. The (frequentist) probability value, p, assigned to a hypothesis is the probability that the data, or some more extreme set of data, would arise by chance when the hypothesis is true.[6] We shall discuss p-values in relation to 24.1.

The multi-variate version of the confidence interval is referred to as a *confidence region*. Whereas, when determining a single parameter, θ from data one can express the confidence level in the form $\theta_{5\%} < \theta^* < \theta_{95\%}$, in the case of simultaneously determining two parameters, (θ_1, θ_2) it is usual to present a contour plot in the (θ_1, θ_2)-plane. For larger numbers of variables we can only show contours in slices of the parameter space.

Understanding the shape of the contours displayed in two dimensions is important. While ideally they should be nice round elliptical shapes, they may in practice be extremely elongated or even not elliptic in shape.

24.3.1 Confidence Intervals in 1-dimension

The χ^2 distribution plays a major role in assigning a level of confidence to estimates of parameters. The reason for this is best explained in terms of an example.

Suppose we have a set of independent measurements $\mathbf{x} = \{x_1, \dots, x_n\}$ that are assumed to be taken from a normal distribution of known mean μ and unknown variance σ^2. In other words $(x_i - \mu) \in N(0, \sigma^2)$. In practice, the mean is determined using $\mu = \sum_i^n x_i/n$, and we want to know how 'accurate' this determination is.[7]

It is convenient to use the estimator s^2 of the variance σ^2 of the x_i given by

$$s_n^2 = \frac{1}{n-1} \sum_{i=1}^n (x_i - \mu)^2, \quad \mathbb{E}[s_n^2] = \sigma^2. \tag{24.8}$$

The mean squared error in the estimate of the variance[8] is the mean square deviation of the estimate s_n^2 from the actual value σ^2:

$$\mathbb{E}\left[(s_n^2 - \sigma^2)^2\right] = \text{Var}(s_n^2) = \frac{2\sigma^4}{n}. \tag{24.9}$$

We note that $\text{Var}(s_n^2) \to 0$ as $n \to \infty$. We can rewrite (24.8) as a sum of squares of $N(0, 1)$ normalised random variates, $Z_i = (x_i - \mu)/\sigma$:

$$Z^2 = \sum_{i=1}^n Z_i^2 = (n-1)\frac{s_n^2}{\sigma^2}, \quad Z_i = \frac{(x_i - \mu)}{\sigma}, \quad Z_i \in N(0, 1). \tag{24.10}$$

made'. Neyman's view of the use of Bayes Theorem *loc. cit.* (p.343) for inference is particularly interesting: he asserts that the to-be-determined parameters are not random variables, while nonetheless recognising that within the frameworks of Jeffreys' notion of probability, they could be regarded as random variables if one only knew how to assign their prior probabilities.

[6] Of course, that is a rather typical frequentist statement.

[7] The word 'accurate' is to be interpreted in the standard frequentist way here. If the experiment were repeated so as to produce a number of independent measurement sets $\mathbf{x} = \{x_1, \dots, x_n\}$, for each of which we determined a mean $\mu = \sum_i^n x_i/n$, what would be the dispersion of the μ-values?

[8] To prove this you need to know that for n measurements $\{x_i\}$ of an $N(\mu, \sigma^2)$ normally distributed random variate, $\mathbb{E}[(x_i - \mu)^4] = 3\sigma^4$.

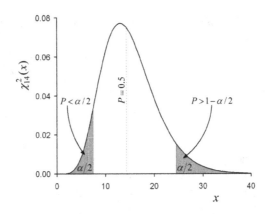

Fig. 24.1 A χ^2 distribution for 14 degrees of freedom (dof). The parameter α is the confidence level, set here to $\alpha = 0.1$ for purposes of illustration. The quantity P is the value of the cumulative probability distribution of χ^2 for 14 dof. The line marked $P = 0.5$ is the expectation value of x. This displays a 'two-sided' confidence region lying between the range of x-values bounded by the shaded areas $P < \alpha/2$ and $P > 1 - \alpha/2$. 90% of repeated estimates are expected to lie in the unshaded region. For larger dof the χ^2 curve is more symmetric and approaches a Gaussian form. ©

From (24.8) we see that $\mathbb{E}[Z^2] = n$. The statistical distribution of the sum of the squares of these n normalised and normally distributed zero-mean random variates Z_i is the χ^2 distribution with $(n-1)$ degrees of freedom. The χ^2 pdf for 14 degrees of freedom is illustrated in Figure 24.1.

Let us use Figure 24.1 to interpret this in practical terms. Suppose we wish to design an experiment such that 90% of the time the measured value lies within a given range.[9] As shown in the figure, this requirement defines a range of values covering 90% of the area under the χ^2 function. The remaining $\alpha = 10\%$ is divided equally between the left and right extremes of the distribution, and so each identifies a range of $\alpha/2 = 5\%$ of the extreme cases sampled from the distribution. If $x_< = \chi^2_{\alpha/2}$ denotes the value of the variable x such that a fraction $\alpha/2$ of the probability density lies to the left of $x_<$, and $x_> = \chi^2_{1-\alpha/2}$ the value of x such that an equal fraction $\alpha/2$ lies to the right of $x_>$ then a fraction α of measurements will lie in the range $x_< < x < x_>$. We can write this in terms of the n normalised measurements Z_i as

$$\chi^2_{\alpha/2} \le (n-1)\frac{s_n^2}{\sigma^2} \le \chi^2_{1-\alpha/2}, \tag{24.11}$$

from which we get the bounds on variance σ^2:

$$(n-1)\frac{s_n^2}{\chi^2_{1-\alpha/2}} \le \sigma^2 \le (n-1)\frac{s_n^2}{\chi^2_{\alpha/2}}. \tag{24.12}$$

---

[9] We might think, for example, of a machine part that will fit into a given slot of a given size. The manufacturer will specify how accurate the fit must be for the machine to work properly. This imposes a tolerance on the size distribution of parts so as to minimise the fraction of parts that will have to be rejected.

Using the definition (24.8) of s_n^2, this is expressed in terms of the data as

$$\frac{\Sigma(x_i - \mu)^2}{\chi_{1-\alpha/2}^2} \leq \sigma^2 \leq \frac{\Sigma(x_i - \mu)^2}{\chi_{\alpha/2}^2}. \tag{24.13}$$

The *confidence level* for σ is therefore

$$\frac{\sqrt{\Sigma(x_i - \mu)^2}}{\chi_{1-\alpha/2}} \leq \sigma \leq \frac{\sqrt{\Sigma(x_i - \mu)^2}}{\chi_{\alpha/2}}. \tag{24.14}$$

This is easily generalised to two or more dimensions, though in 3-dimensions and above, visualisation can only be achieved by looking at 2d sections of the space using the marginalised distribution obtained by integrating the distribution over the non-displayed variables (see Section 25.3.1 for more on this).

24.3.2 Confidence Intervals: a Simple Case Study

It is worth looking at a simple example clarifying what 'confidence levels' is about. We shall consider the case of determining the cosmological density parameters Ω_Λ and Ω_M simultaneously from the observations of high redshift supernovae. In particular we will look at the first paper (Schmidt, 1998) to suggest that $\Lambda = 0$ might not be likely. This was done on the basis of the observation of a single supernova.[10]

Figure 24.2 shows the result of the analysis of the first high redshift supernova, SN 1995K at $z = 0.479$, from the High-Z Supernova Search group (Schmidt, 1998). At that time the default cosmological model had $\Lambda = 0$ and they constructed a model of their expectations for an open Universe with matter density $\Omega_M = 0.4$. The two parameters to be determined were Ω_Λ and Ω_M. They produced a set of contours shown in the figure where, with a given probability, their data would lie if this were the correct model (i.e. their null hypothesis). An alternative $\Lambda = 0$ possibility was the Einstein de Sitter model which is shown on the diagram as a point marked by a star symbol.

Their data point defines a cosmological model with $\Omega_\Lambda = 0.6$ and $\Omega_M = 0.4$, albeit with the relatively large error bars shown in the figure. It clearly lies in the low-probability wings of their confidence contours for the $\Omega_M = 0.4, \Omega_\Lambda = 0.0$ model and they stated that it rejects the Einstein de Sitter model at the 80% confidence level.

24.4 Hypothesis Testing

A key application of this is to test hypotheses about the variance of a distribution. The previous discussion provides a test for the hypothesis that the variance of

---

[10] In 1997, the Supernova Cosmology Project group published a paper (Perlmutter, 1997) analysing seven supernovae with redshifts in the range $z = 0.35$–0.46, and demonstrated that Ω_Λ and Ω_M could be determined together. They obtained the result that $\Omega_M = 0.88^{+0.69}_{-0.60}$ for a $\Omega_\Lambda = 0$ cosmology, and $\Omega_M = 0.94^{+0.34}_{-0.28}$ for a flat cosmological model, thus giving $\Omega_\Lambda = 0.06^{+0.28}_{-0.34}$, or $\Omega_\Lambda < 0.51$ at the 95% confidence level. From this they concluded that 'The results for Ω_Λ-versus-Ω_M are inconsistent with Λ-dominated, low density, flat cosmologies ...'. Just before the publication of the paper of Schmidt (1998) using SN 1995K at $z = 0.479$, Perlmutter (1998) reported on the discovery of SN 1997ap at $z = 0.83$.

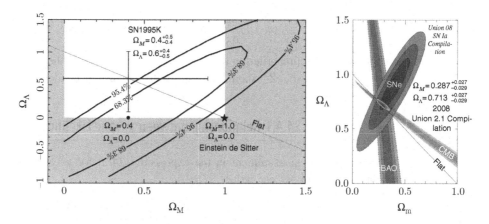

Fig. 24.2 First determination of the cosmological parameters with only a single type Ia Supernova (SN 1995K). The data on SN 1995K (Schmidt, 1998), interpreted in terms of a flat model, is shown with error bars. The contours represent the confidence contours for observations of a hypothetical sample of 30 high redshift ($z > 1$) supernovae in a $\Lambda = 0$ cosmological model having $\Omega = 0.4$ (the large dot). The broken line is the locus of $\Omega_M + \Omega_\Lambda = 1$ models, one of which with $\Lambda = 0$ is the Einstein de Sitter model (the star), which the authors concluded is ruled out at the 80% confidence level.

The plot on the right shows the result of 10 years of further data acquisition and analysis (Kowalski and Rubin, 2008, *Union 2.1 SN Ia Compilation*). Superposed on the left-hand figure, this plot would occupy only the area of the error cross in the left panel. The three contours in the supernova data are 99.7%, 95.4%, 68.3%. Also shown are the parameter determinations from BAOs and from the CMB as of 2008.

Left: Adapted from Schmidt (1998, Fig. 13). *Reproduced by permission of the AAS.*
Right: Kowalski and Rubin (2008, Fig. 15). *Reproduced by permission of the AAS.*

a distribution from which the data sample $\{x_i\}$ is drawn has some particular value, say, σ_0:

$$H_0: \quad \sigma^2 = \sigma_0^2. \tag{24.15}$$

We can measure the standard deviation s_n for the sample: $s_n^2 = (\Sigma(x_i - \bar{x}))/(n-1)$ where $\bar{x} = (\Sigma x_i)/n$ is the sample mean. The parameter of interest is

$$Z^2 = (n-1)\frac{s_n^2}{\sigma_0^2}, \tag{24.16}$$

since, according to (24.10), Z^2 would be χ^2-distributed if $\sigma = \sigma_0$. If we want to be, say, 90% confident that our hypothesis was correct, we would only accept our hypothesis at that level of confidence provided Z^2 fell inside the unshaded region of the function shown in Figure 24.1. This is the criterion provided by the bounds in Equation (24.11), with σ_0 in place of σ. We would reject the hypothesis if Z^2 lay outside of these limits, i.e. in either of the shaded regions of Figure 24.1.

We refer to the hypothesis (24.15) as a *two-sided* or *two-tailed* hypothesis, since we are prepared to accept values on either side. The hypothesis $H_>: \sigma^2 > \sigma_0^2$ is referred to as *upper one-tailed*, while $H_<: \sigma^2 < \sigma_0^2$ is referred to as *lower one-tailed*. We reject $H_>$ if

$$(n-1)\frac{s_n^2}{\sigma_0^2} > \chi_{1-\alpha}^2 \quad \text{reject } H_> : \sigma^2 > \sigma_0^2, \tag{24.17}$$

with a similar rejection criterion for $H_<$. The area of rejection is the area of the histogram where our test statistic, Z^2, should not fall. In the case of a one-sided test all of the area α falls on one side.

24.4.1 Relationship between χ^2 and Likelihood

Normalising each measurement in the expression (24.2) for χ^2 with the measurement variance $\epsilon_i = 1/\sigma_i^2$, we can write

$$\exp-\tfrac{1}{2}\chi^2(\mathbf{a}) = \prod_{i=1}^{N} \exp\left(-\frac{y_i - f(x_i; \mathbf{a})}{2\sigma_i^2}\right)^2, \tag{24.18}$$

where we have written $\epsilon_i = 1/\sigma_i^2$ and written the parameters of the model as the vector $\mathbf{a} = (a_1, \ldots, a_M)$. We recognise the product as being proportional to the product of the probabilities of getting the individual data values y_i, given a Gaussian distribution of zero-mean errors having variance σ_i^2. This is just the likelihood of the data, and so

$$\chi^2(\mathbf{a}) = -2\ln \mathcal{L}(\mathbf{a}, \mathbf{y}) + \text{constant.} \tag{24.19}$$

The minimum of χ^2 is the maximum of the likelihood.

24.5 Removing Variance Trends

The emphasis on Gaussian error distributions, and the associated use of the χ^2 statistic, are fundamental to much of statistics. This is in some sense justified by the central limit theorem, which assures us that, under quite general circumstances, the arithmetic mean of a large number of independent random variables will be approximately normally distributed, no matter what their underlying statistical distribution might be. It is normally required that the random variates come from the same underlying distribution, but that is not always necessary.

This gives rise to a number of *variance stabilising transformations* of the measurement values, X, of the data that are applied to the raw data.[11]

24.5.1 Linear Models

Suppose a random variable X has variance σ_X^2 that is a known (or assumed) function $\sigma_X^2 = \phi(\mu_X)$ of the mean μ_X of X. Transform to a new variable $Y = g(X)$, where $g(X)$ is related

---

[11] This may look like a heinous fix, but it is done quite unconsciously when people decide to plot variables in log-coordinates simply to make the data look like a power law prior to fitting. An extreme example occurred in astronomy when, in establishing the light profile of an elliptical galaxy, the angular distance from the centre of a galaxy, r, was plotted as $e^{r^{1/4}}$.

to the known function $\phi(\mu_X)$ in a model dependent way. The simplest model is the linear model,

$$Y = g(\mu_X) + g'(\mu_X)(X - \mu_X). \tag{24.20}$$

(This might be regarded as the first term in a Taylor series.) Then, since $\mathbb{E}\,[(X - \mu_X)] = 0$, we have

$$\mathbb{E}\,[Y] = g(\mu_X), \quad \mathbb{E}\,[(Y - g(\mu_X)^2)] = \{g'(\mu_X)\}^2\phi(\mu_X). \tag{24.21}$$

We need to make $\mathbb{E}\,[(Y - g(\mu_X)^2)] = \text{const.}$, and this is achieved by the indefinite integral

$$g(\mu) = \int \frac{d\mu}{\sqrt{\phi(\mu)}}, \quad \text{where } \sigma_X^2 = \phi(\mu_X). \tag{24.22}$$

Simple important cases occur when the distribution of X is Poisson, binomial or exponentially distributed:

Transformation	Density example	Variance σ^2	$g(\mu)$	Result
Square root	$e^{-m}m^s/m!$	m	$\sqrt{m}$	$Y = 2\sqrt{X}$
Arcsine	$\binom{n}{r}(1-p)^{n-r}$	$np(1-p)$	$\arcsin\sqrt{p}$	$Y = \arcsin X$
Logarithmic	$e^{-x/\beta}/\beta$	β^2	$\log\beta$	$Y = \log X$
Anscombe	Poisson			$Y = \sqrt{\left(X + \frac{3}{8}\right)}$

There are experience-driven refinements on this, such as in the square root transform using $2\sqrt{(X + 1)}$ or $\sqrt{X} + \sqrt{(X + 1)}$ in place of $2\sqrt{X}$ when the values of the random variable X are small. Likewise, the logarithmic transform can be replaced by $\log(X + 1)$. Akin to the logarithmic transformation is the *logit transformation*: $\text{logit}(X) = \frac{1}{2}\ln[X/(1 - X)]$, $0 < X < 1$. The logarithmic transforms are appropriate when the errors are proportional to the data values, rather than simply additive. The Anscombe transform has a somewhat better performance than the simple square root transform and, in astronomy, is used when analysing photon count statistics.

By Gaussianising the distribution in this way it is possible to use noise filtering algorithms that are predicated on the assumption of signal Gaussianity (but see Neyman and Scott (1960)). However, the most general method might be to resolve the observed distribution into a sum of Gaussians. This is a prevalent technique in image analysis and segmentation.

24.5.2 Fisher Z-transform

An important example of variance stabilisation is the *Fisher Z-transform*. This addresses an issue in estimating the population correlation ρ of pairs of variables X and Y having the bivariate normal distribution. As $|\rho| \to 1$, the variance of the sample correlation, r becomes smaller, which violates the precepts of linear regression analysis. If

r is the sample correlation of the joint bi-normal distribution of X and Y, then the variable Z:

$$Z = \frac{1}{2} \ln \left(\frac{1+r}{1-r} \right), \tag{24.23}$$

is approximately normally distributed as $\mathcal{N}(\mu_Z, \sigma_Z^2)$, where

$$\mu_Z = \mathbb{E}[Z] = \frac{1}{2} \ln \left(\frac{1+\rho}{1-\rho} \right) + \underbrace{\frac{\rho}{2(n-1)}}_{\text{Bias correction}}, \quad \sigma_Z^2 = \frac{1}{n-3}, \tag{24.24}$$

for samples of size $n > 10$ or so. The important result is that the variance of Z depends only on the sample size. The bias correction is generally omitted from this expression.

24.6 Histograms and Likelihood

The histogram is perhaps one of the most common ways of summarising data. The histogram was first introduced in scientific journals by Pearson (1895) (see Figure 24.3): in a footnote on page 399 of the paper Pearson writes that the term *histogram* was 'Introduced by the writer in his lectures on sciatics [*sic*] as a term for a common form of graphical representation, i.e. by columns marking as areas the frequency corresponding to the range of their base'. Pearson saw the concept not only as a way of visualising data, but also as a convenient way of compressing data for the complex computation of statistical moments.[12]

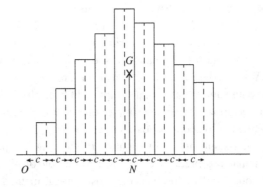

Fig. 24.3   Perhaps the first ever sketch of a histogram in a scientific journal, drawn by Pearson (1895, p. 346). There is neither figure number nor figure caption in the article. The figure is a plot of the binomial distribution illustrating how histograms binning works and how binning helps to calculate statistical moments. Much of the article concerns the derivation of statistical moments for a variety of binned data sets.

---

[12] Pearson's paper is well worth a look it develops a large number of concepts that today we take for granted. In particular, he devotes a significant effort to make the computation of statistical properties efficient. Perhaps that is not surprising given that his 'computer' was a hand driven Brunsviga calculating machine.

Interestingly, the machine had been invented in 1878 by the Swedish engineer W.T. Odhner, but was not manufactured until 1890 when F. Trinks bought the manufacturing rights and sold it under the brand name 'Brunsviga'. Given that Pearson's paper was published in 1895, it is clear that he was up to date with the very latest in computational technology.

As recognised by Pearson, the counts N_i in the ith histogram bin follow the multinomial distribution for which the probability density $p_i(\boldsymbol{\theta})$ depends on some set of parameters $\boldsymbol{\theta}$. If the total number of observations is N, then

$$N = \Sigma_{i=1}^{B} N_i, \tag{24.25}$$

where B is the number of bins and the expected count in the ith bin is

$$n_i(\boldsymbol{\theta}) = N p_i(\boldsymbol{\theta}). \tag{24.26}$$

The histogram is a frequently used tool. The possible range of values for a measurement of a physical quantity is divided into bins of smaller sub-ranges. The bins cover the entire range and do not overlap. The histogram is presented as the number of measurements O_i that fall into each of n bins. The typical question that is asked is whether the binned data, $\{O_i\}$, i.e. the histogram, conforms to our expectations E_i for the count in the ith bin. In other words, how well is the data fit by some parameterised statistical function.

If there are n bins in all, the degree of agreement between what is expected, E_i, and what is measured, O_i, is generally described by the number

$$\chi^2 = \sum_{i=1}^{n} \frac{(O_i - E_i)^2}{E_i}, \qquad \text{dof} = n - 3. \tag{24.27}$$

In the limit of large counts per cell this statistic is indeed χ^2-distributed, since the right hand side is a sum of squares of normalised normally distributed random variates $Z_i = (O_i - E_i)/\sqrt{E_i}$. The reason that the denominator is E_i is because the counts in a given cell are taken to be Poisson-distributed. Note that the expression (24.27) is invariant under a relabelling (re-ordering) of the bins. This asserts that the bins are themselves independent, the content of a particular bin does not depend on the values in the neighbouring bins.[13] A serious problem with this elementary test is that it depends on the choice of bin-size.

In the Poisson distribution the variance and mean are both equal, so if the expected count is E_i, the variance is also E_i. In the large n limit the Poisson distribution tends to the normal distribution, and so this can be expected to work. The number of degrees of freedom of the statistic χ^2 in (24.27) is $n - 3$ because in writing (24.27) there are two further constraints from assuming that we know the mean, E_i, and the variance of the counts, which is also E_i, in each of the cells.

In this context it is also interesting to note that the first commercially successful calculating machine was the 'Arithmometer' invented in France by Charles Xavier Thomas de Colmar in 1820, but not marketed until 1851. The Thomas Arithmometer was capable of doing additions, subtractions, multiplication and division. The user interface of the Brunsviga machine was the same as that used earlier in the Arithmometer. (Source: IBM Archive – Exhibits (2013))

[13] There is a distinction drawn between a *bar chart* and a *histogram*: bar charts represent categorical data while histograms represent numerical data. In software packages, bar charts are plotted with vertical bars that do not touch, while in histograms the bars are draw so as to touch their neighbours. However, the distinction is somewhat more subtle. Numerical data where, for example, the lifetime of a particle is measured in a series of independent measurements, can be tested via χ^2 (though it makes little sense to shuffle up the bins which are naturally sequential). The important point is that to use χ^2 as a statistic the bins must be independent, which for commonly used overlapping bins it is not. See the words of wisdom and alternative tests in Press *et al.* (2007, Section 14.3).

24.6.1 The Poisson Approximation

We can make the simplifying approximation that $n_i \ll N$ and then model the counts N_i in each bin as being Poisson distributed with mean n_i:

$$p_i(\boldsymbol{\theta}) = \frac{e^{-n_i} n_i^{N_i}}{N_i!}, \tag{24.28}$$

where, as per (24.26), the $\boldsymbol{\theta}$ dependence is implicit in the expected bin count n_i.

With this, the likelihood of getting the histogram $\{N_i\}$ of counts is

$$\mathcal{L}(\boldsymbol{\theta}) = \prod_{i=1}^{B} \frac{e^{-n_i} n_i^{N_i}}{N_i!}. \tag{24.29}$$

The log-likelihood for the entire histogram is then

$$\ln \mathcal{L}(\boldsymbol{\theta}) = \sum_{i=1}^{B} (-n_i + N_i \ln n_i), \tag{24.30}$$

plus terms that do not depend on $p_i(\boldsymbol{\theta})$ and which have been ignored. Note the important fact that the sum extends only over the bins, not over the individual measurements as would have been necessary for unbinned data.

If there is only one parameter θ we can find the maximum likelihood value of the bin count N_i to be

$$(N_i)_{\text{ML}} = n_i, \quad p_i(\boldsymbol{\theta}) \text{ depends on one parameter only}, \tag{24.31}$$

which is as expected.

A different example is to fit a model of the form

$$n_i = \lambda x_i, \tag{24.32}$$

where x_i represents the value attached to the ith bin. The data is now subject to a trend on top of the Poisson statistics of the bin count. Putting this into (24.30) yields

$$\ln \mathcal{L}(\boldsymbol{\lambda}) = \sum_{i=1}^{B} (-\lambda x_i + N_i \ln \lambda x_i). \tag{24.33}$$

24.6.2 The Normal Approximation

We can go one step further by recalling that the Poisson distribution, $p(n \mid \mu) = e^{-\mu} \mu^n / n!$, approaches a normal distribution with variance and mean n. So the probability density of the bin count in the ith bin is

$$p_i(N_i \mid \boldsymbol{\mu}) = \frac{1}{\sqrt{(2\pi \mu)}} \exp\left(-\frac{(N_i - \mu)^2}{2\mu}\right), \tag{24.34}$$

for which the log-likelihood for the histogram is

$$\ln \mathcal{L}(\mu) = -\frac{(N_i - \mu)^2}{2\mu} - \frac{1}{2} \ln 2\pi - \frac{1}{2} \ln \mu. \tag{24.35}$$

24.6.3 The classical view: χ^2

The question of how close two distributions are is a central point of classical statistical analysis, where the questions such as 'what is the probability that these points come from that distribution?' or 'are these two distributions (histograms) drawn from the same population?' are addressed.[14]

Data is often binned into histograms from which we wish to derive estimates of some quantity, θ, which could be a mean value, a slope, position of a spectral line, or simply a fit to a probability distribution. The binning might arise because of the way the experiment works, or because it is a convenient way of looking at large amounts of data. Within a histogram, each bin has an event count, d_i, for which there is an underlying probability density $p_i(\theta)$. If we make a total of N measurements the expected count in the ith bin will be $k_i = Np(\theta)$.

For the moment let us adopt the common view that two data samples are presented as binned histograms, $\{U_i\}$ and $\{V_i\}$ in which the data are binned in the same n bins. Suppose the ith bin of U contains a count u_i and the ith bin of V contains a count v_i.[15]

The statistic we shall focus on is the count differences in each bin:

$$\Delta_i = u_i - v_i. \tag{24.36}$$

When the counts are large and the bins independent, the joint distribution of the count differences Δ_i will be a product of Gaussians:

$$\mathbb{P}[\Delta] = \frac{1}{\sqrt{2\pi}\,\sigma_1} \exp\left(-\frac{1}{2}\frac{\Delta_1^2}{\sigma_1^2}\right) \cdots \frac{1}{\sqrt{2\pi}\,\sigma_n} \exp\left(-\frac{1}{2}\frac{\Delta_n^2}{\sigma_n^2}\right), \tag{24.37}$$

$$= \frac{1}{2\pi)^{n/2} \prod_{i=1}^n \sigma_i} \sum_{i=1}^n \exp\left(-\frac{1}{2}\frac{\Delta_i^2}{\sigma_i^2}\right). \tag{24.38}$$

Here we assume the mean of Δ_i is zero and σ_i denotes the variance of Δ_i in the ith bin.

If the bin counts are independent, then the sample variance of the difference of counts in the ith bin is to be estimated as the sum of the variances of the individual counts in the bin: $\sigma_i^2 = \sigma_{U_i}^2 + \sigma_{V_i}^2$. If the counts in the bins follow Poisson statistics, then the sample variances are $\sigma_{U_i}^2 = u_i$ and $\sigma_{U_i}^2 = v_i$. So we have the estimate for the sample variance of the count difference in the ith bin is

$$\hat{\sigma}_i^2 = u_i + v_i, \tag{24.39}$$

where we have put a 'hat' on the σ to indicate that it is an estimate based on the data. We could alternatlively have used the symbol s_i^2 in place of this.

---

[14] One might respond to such questions by something like 'what do you mean by different?'. Although it is interesting to know that there might be a difference, it would be more interesting to know the nature of the difference. Are the shapes of the distributions the same? Are the means or variances different?

[15] This problem is discussed in most textbooks on statistics, but let us refer to Press *et al.* (2007, Section 14.3) and Porter (2008) as useful references discussing the 'how-to' and the 'why' with great clarity. Here I follow the notation of the latter paper.

25

Statistical Inference: Bayesian

The second half of the 20th century saw several revolutions in statistical data analysis, among which was Bayesian analysis. What was particularly striking about the rise of Bayesian analysis was that Bayes did the work that bears his name in the early 18th century: the revolution took more than two centuries to happen. What Bayes had pointed out was that if you knew the chance of an event A happening, given that B had happened, i.e. $\mathbb{P}[A\,|\,B]$, you could calculate the inverse probability of B happening, given that A had happened, i.e. $\mathbb{P}[A\,|\,B]$.[1]

From that simple idea, that a probability could be pre-conditioned, an entire philosophy of statistical inference has emerged. That emergence started in the first half of the 20th century, but has only fully blossomed within the last few decades with the advent of vast datasets and computing power to handle them. The interpretation of 'Bayes theorem' has always been somewhat contentious. Its descendant, Bayesian inference, is still regarded as contentious, though not for any obvious, compelling, logical reasons. We are seeing a rapid growth in the acceptance of Bayesian methods in a wide variety of fields.

25.1 Bayes Theorem

Elementary probability[2] tells us that if A and B are events with probabilities $\mathbb{P}[A]$ and $\mathbb{P}[B]$, then

$$\mathbb{P}[AB] = \mathbb{P}[A\,|\,B]\,\mathbb{P}[B] = \mathbb{P}[B\,|\,A]\,\mathbb{P}[A]. \tag{25.1}$$

This is in fact a statement of what we mean by the conditional probability, $\mathbb{P}[A\,|\,B]$. From this we have:

Bayestheorem:

$$\mathbb{P}[B\,|\,A] = \mathbb{P}[A\,|\,B]\frac{\mathbb{P}[B]}{\mathbb{P}[A]}, \qquad \mathbb{P}[A\,|\,B] = \mathbb{P}[B\,|\,A]\frac{\mathbb{P}[A]}{\mathbb{P}[B]}. \tag{25.2}$$

---

[1] This phrasing is immediately contentious, since it suggests a violation of causality unless it applies to events that are themselves not causally related – like throws of a dice or the drawing of cards.

[2] There is an overview of elementary probability in the On-line Supplement: *Probability and Statistics Primer*.

This is one of the most famous equations in statistics: it is *Bayes equation* and has appeared in almost every book about the statistics of games and gambling.

Short biography: Thomas Bayes (1701?–1761) was an accomplished English mathematician from Sheffield in South Yorkshire. He received his university education in Edinburgh, Scotland. Bayes was a contemporary of Abraham de Moivre who wrote, among other things, *The Doctrine of Chances: a Method of Calculating the Probabilities of Events in Play* (1711), in which he discovered the Poisson distribution, *A Treatise of Annuities upon Lives* and an essay *On the Law of Normal Probability*. (de Moivre was a French Huguenot exile living in London where he became a Fellow of the Royal Society and was close friends with Halley and Newton.) Bayes was elected to the Royal Society in 1742.
  During his lifetime, Bayes published no scientific papers: the theorem that bears his name was edited and published posthumously by his friend Richard Price who, apparently, was uncertain about several parts of Bayes' work. (See Bellhouse (2004) for an interesting biography of Bayes.) There is even the suggestion by Stigler (1983) that the theorem was first derived by Nicolas Saunderson. Saunderson was blind from the age of one year, but was a formidable mathematician who became the fourth Lucasian Professor at Cambridge (Newton had been the second). However, it was Laplace who brought us Bayes theorem in the form we would recognise today in his *A Philosophical Essay on Probabilities* (1814).

Bayes theorem provides a relationship between $\mathbb{P}[A\,|\,B]$ and $\mathbb{P}[B\,|\,A]$. A simple example shows the power of this. Suppose that galaxies are either field galaxies (F) or cluster galaxies (C), and that the probability of being a field galaxy is $\mathbb{P}[F] = 0.9$ (and so $\mathbb{P}[C] = 0.1$). Suppose further that galaxies can be classified either as spirals (S) or ellipticals (E), and that 5% of the field galaxies are ellipticals, while 45% of the cluster galaxies are ellipticals. This can be phrased as $\mathbb{P}[E\,|\,F] = 0.05$ and $\mathbb{P}[E\,|\,C] = 0.45$.

If I observe an elliptical galaxy, what do I conclude? Using Bayes theorem (25.2) we have

$$\mathbb{P}[F\,|\,E] = \mathbb{P}[E\,|\,F]\,\mathbb{P}[F]/\mathbb{P}[E], \qquad (25.3)$$

$$\mathbb{P}[C\,|\,E] = \mathbb{P}[E\,|\,C]\,\mathbb{P}[C]/\mathbb{P}[E]. \qquad (25.4)$$

We know everything on the right-hand side, though we have not said anything about $P(E)$. In any case, with the given numbers

$$\frac{\mathbb{P}[F\,|\,E]}{\mathbb{P}[C\,|\,E]} = \frac{0.05 \times 0.9}{0.45 \times 0.1} = 1.0. \qquad (25.5)$$

So, given that you are observing an elliptical, it is as likely that this is a cluster elliptical as a field elliptical. (Of course, we did not really need Bayes to sort that out – it is a simple counting exercise!)

We had some prior knowledge of $\mathbb{P}[F]$ and $\mathbb{P}[C]$ and we were given the likelihood that a galaxy in a cluster was an elliptical or a galaxy in the field was an elliptical. In fact, in Bayes theorem as written in Equation (25.2), $\mathbb{P}[A]$ is the *prior distribution* of the parameter we are trying to estimate and $\mathbb{P}[A\,|\,B]$ is its *posterior distribution*.

Knowledge of the prior distribution is an important aspect of the Bayesian approach, see Figure 25.1. That knowledge corresponds to the fact that you can obviously make better estimates of a parameter if you have some prior knowledge about it. The question is where that knowledge comes from: usually it comes from previous examination of relevant data, so in the example of the field versus cluster elliptical galaxies, we derive the prior distribution by noting in our data what fractions of the field and cluster populations are in fact ellipticals.

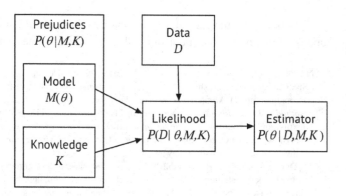

Fig. 25.1   A theory reflecting our state of knowledge, K, is based on a model $M(\theta)$ depending on some parameters θ. When the theory is confronted with data D we can provide a measure of our confidence via the likelihood $P(D|\theta, M, K)$ of getting that particular data, given the model $M(\theta)$ and a value of θ. In the frequentist view we select the value of θ that maximises $P(D|\theta, M, K)$. In the Bayesian point of view we can compute $P(\theta|M, K)$ quantifying in our degree of belief in the theory given our prior knowledge as described by $M(\theta)$. ⓒⓒ

25.1.1 Bayes: the Simple Form

The two versions are presented in Equation (25.2) to stress that we can determine both the probability of finding situation A given situation B, as we can the probability of situation B given situation A.

If A and B referred to events occurring in time, it might appear that we can determine the probability of the outcome B given that A occurs from knowledge of the probability of the outcome A given that B occurred. However, we should be careful since there is no implied sense of a causal link between A and B in (25.2). If A and B were in fact causally related events we would have to make a choice about causal order and use the appropriate version of Equations (25.2). There is nothing paradoxical about this: Equation (25.2) simply does not embody the notion of cause and effect: it is merely a statement about probabilities.

Priors, Posteriors, Likelihoods and Evidence

A bit of rephrasing of the above rendition of Bayes theorem explains why this is so important. For B write $\mathcal{H}$ for 'hypothesis'. For A write E for 'evidence' for the hypothesis, E is usually some set of measurements: the data with which we are going to test the hypothesis. Then we have

$$\underset{\text{Posterior}}{\mathbb{P}[\mathcal{H}|E]} = \underset{\text{Likelihood}}{\mathbb{P}[E|\mathcal{H}]}\,\underset{\text{Evidence}}{\frac{\overset{\text{Prior}}{\mathbb{P}[\mathcal{H}]}}{\mathbb{P}[E]}}. \tag{25.6}$$

The left-hand side is a measure of how strongly the hypothesis $\mathcal{H}$ is supported by the evidence E, which is a principal aim for most experiments. This is referred to as the *posterior*. The first term on the right-hand side, $\mathbb{P}[E|\mathcal{H}]$, tells us how likely the data or evidence is on the basis of the hypothesis $\mathcal{H}$ that we are testing. This term is referred to as the *likelihood*. It reflects the degree to which the data is likely to have arisen on the basis of our hypothesis $\mathcal{H}$. The importance of this formulation is that we can generally write an explicit mathematical expression for $\mathbb{P}[E \mid \mathcal{H}]$.

Another view of this equation puts these concepts in a different light:

$$\mathbb{P}[\mathcal{H} \mid E] = \underbrace{\frac{\mathbb{P}[E \mid \mathcal{H}]}{\mathbb{P}[E]}}_{\text{Impact}} \mathbb{P}[\mathcal{H}]. \tag{25.7}$$

In this version, the quantity $\mathbb{P}[E \mid \mathcal{H}] / \mathbb{P}[E]$ is referred to as the *impact* of the evidence E on our degree of belief in $\mathcal{H}$. However, in this form the quantities on the right-hand side are not manifestly calculable.

The Debate Surrounding Interpretation

A debate centres around the second term in (25.6), which is the ratio of the *prior*, $\mathbb{P}[\mathcal{H}]$, to the *evidence*, $\mathbb{P}[E]$. We might well ask what we mean by either of these probabilities. It is perhaps easier to make sense of $\mathbb{P}[\mathcal{H}]$ in the usual situation that the hypothesis $\mathcal{H}$ refers to a mathematical or physical model for a particular phenomenon. The model inevitably involves parameters that may themselves be experimentally determined, and in that way $\mathbb{P}[\mathcal{H}]$ is a reflection of how well we can determine those parameters. Nevertheless, there still remains the motivation for the model itself: what is our degree of certainty that this is correct? This is a reflection on the quality of our understanding or *knowledge*, $\mathbb{P}[\mathcal{K}]$.[3]

From the technical point of view the evidence, $\mathbb{P}[E]$, is merely a normalisation factor that ensures that the total probability of the left-hand side, i.e. the posterior, should be unity.

The Gambler's Concept of Odds

The amount of money a gambler is prepared to bet on an event is proportional to his/her expectation that the event will occur. A bet is always 'against' another opinion, be it another gambler or a casino. The bet may be a personal one of 'my opinion versus yours', or it may be a generic bet that 'my opinion is the correct one among all other possible outcomes'. In the first case it is a bet of one opinion $\mathcal{H}_1$ versus a specific alternative $\mathcal{H}_2$. in the second case it is a bet of the opinion $\mathcal{H}_1$ against its alternative $\overline{\mathcal{H}_1}$.[4]

Formally, the *odds* of one outcome $\mathcal{H}_1$ versus another $\mathcal{H}_2$ are defined:

---

[3] This takes us to the fascinating article of Patton and Wheeler (1975): *Is Physics Legislated by Cosmogony?*. In Footnote 18 of the article they refer to Føllesdal's aphorism that *Meaning is the joint product of all the evidence that is available to people who communicate*, and which they paraphrase by substituting 'those' for 'people'. Here we might adopt a version asserting that *Our understanding is the joint product of all the evidence that is available to those who communicate their research*. That might provide an interpretation for the quantity $\mathbb{P}[\mathcal{K}] = \mathbb{P}[\mathcal{H}] / \mathbb{P}[\mathcal{E}]$ as being a measure of our state of understanding, or our knowledge, $\mathcal{K}$. In the physical sciences our understanding usually expresses itself in terms of a theory which is expressed in the universal language of mathematics.

[4] The first corresponds to $\mathcal{H}_1 =$ horse '1' will win versus $\mathcal{H}_2 =$ horse '2' will win (neither might win), while the second is simply $\mathcal{H}_1 =$ horse '1' will win the race (it either wins or it doesn't).

Odds of outcome $\mathcal{H}_1$ versus the outcome $\mathcal{H}_2$:

$$O(\mathcal{H}_1, \mathcal{H}_2) = \frac{\mathbb{P}[\mathcal{H}_1]}{\mathbb{P}[\mathcal{H}_2]}. \tag{25.8}$$

Within the Bayesian framework, the odds expressed in this way as a ratio of probabilities express your relative belief in $\mathcal{H}_1$ over $\mathcal{H}_2$. The notation also indicates that this is an initial belief, i.e. a prior belief, since there is no indication that any data is taken into account. We come to the problem of assessing the influence of having a data set in the following section.

25.1.2 Comparing Hypotheses on the Basis of Data

If we have two hypotheses $\mathcal{H}_1$ and $\mathcal{H}_2$ we can write an Equation (25.6) for each of $\mathcal{H}_1$ and $\mathcal{H}_2$ and deduce that:

Bayes factor:

$$\frac{\mathbb{P}[\mathcal{H}_1 \mid E]}{\mathbb{P}[\mathcal{H}_2 \mid E]} = \frac{\mathbb{P}[E \mid \mathcal{H}_1]}{\mathbb{P}[E \mid \mathcal{H}_2]} \frac{\mathbb{P}[\mathcal{H}_1]}{\mathbb{P}[\mathcal{H}_2]}. \tag{25.9}$$

Shows the effect that new evidence E has on the odds that hypothesis $\mathcal{H}_1$ is preferred over $\mathcal{H}_2$.

We see that in doing this division we have eliminated the normalisation factor $\mathbb{P}[E]$. This is an important equation. The first term on the right expresses the relative performance of $\mathcal{H}_1$ and $\mathcal{H}_2$ in explaining the data E. We have encountered the second term in equation (25.8) where it was interpreted as the ratio of our prior beliefs.

To interpret this equation we shall assume that the hypotheses $\mathbb{P}[\mathcal{H}_1]$ and $\mathbb{P}[\mathcal{H}_1]$ are mutually exclusive, so that the evidence, or data, E, could have arisen from either of these hypotheses (see for example Kass and Raferty (1995)). With this assumption,

$$\mathbb{P}[\mathcal{H}_2] = 1 - \mathbb{P}[\mathcal{H}_1], \tag{25.10}$$

$$\mathbb{P}[\mathcal{H}_2 \mid E] = 1 - \mathbb{P}[\mathcal{H}_1 \mid E], \tag{25.11}$$

and so we see that Equation (25.9) can be rewritten as:

Odds for hypotheses:

$$O(\mathcal{H}_1 \mid E) = \frac{\mathbb{P}[E \mid \mathcal{H}_1]}{\mathbb{P}[E \mid \mathcal{H}_2]} \; O(\mathcal{H}_1), \tag{25.12}$$

Updated odds $=$ Bayes factor $\times$ initial odds.

Another way of putting this is to say that:

      Improved belief $\Longleftarrow$ initial belief $\times$ new data,

or, in modern terms relative to a given model or hypothesis,

      Posterior belief $\Longleftarrow$ prior belief $\times$ likelihood of new datum under hypothesis.

We now need to know what to do when new data becomes available, This brings us to the subject of the Bayes factor.

Bayes Factor

The ratio $\mathbb{P}[\mathcal{H}_1]/\mathbb{P}[\mathcal{H}_2]$ does not depend on the data or evidence E: it represents the ratio of the *prior odds* for each of the hypotheses $\mathcal{H}_1$ and $\mathcal{H}_2$. On the other hand the ratio $\mathbb{P}[\mathcal{H}_1\,|\,E]/\mathbb{P}[\mathcal{H}_2\,|\,E]$ represents the ratio of those odds in the light of the evidence E. We call $\mathbb{P}[\mathcal{H}_1\,|\,E]$ the *posterior odds* for the hypothesis $\mathcal{H}_1$ in the light of the evidence E. The factor:

Bayes factor:
$$B_{12} = \frac{\mathbb{P}[E\,|\,\mathcal{H}_1]}{\mathbb{P}[E\,|\,\mathcal{H}_2]}, \tag{25.13}$$

is called the *Bayes factor* for evaluating the hypotheses $\mathcal{H}_1$ against $\mathcal{H}_2$. If we had a set of hypotheses that we were comparing pairwise, the Bayes factor for hypothesis $\mathcal{H}_i$ against $\mathcal{H}_j$ would be denoted by B_{ij}.[5] The Bayes factor B_{12} represents the effect of the evidence on the prior odds $\mathcal{H}_1$. If the Bayes factor is zero then the hypothesis $\mathcal{H}_1$ is falsified by the evidence. We notice that the factors $\mathbb{P}[E\,|\,\mathcal{H}]$ are simply the likelihoods of getting the data E given the hypothesis $\mathcal{H}$.

Suppose now that we have performed an experiment which produces evidence E_1 and compares two hypotheses as per Equation (25.12). Then we perform a second *independent* experiment which produces additional evidence E_2. How does that affect our valuation of the hypotheses?

The calculation is straightforward: we use Equation (25.6) to bring $\mathbb{P}[\mathcal{H}_1\,|\,E_2E_1]$ into a form that separates the contributions from E_1 and E_2. It is the assumed independence of these experiments that allows us to make this separation. The derivation is a little tedious but it is worth seeing the point at which the requirement that the experiments E_1 and E_2 be independent is invoked.

Going back to Equation (25.9) we have

$$\frac{\mathbb{P}[\mathcal{H}_1\,|\,E_2E_1]}{\mathbb{P}[\mathcal{H}_2\,|\,E_2E_1]} = \frac{\mathbb{P}[E_2E_1\,|\,\mathcal{H}_1]}{\mathbb{P}[E_2E_1\,|\,\mathcal{H}_2]}\frac{\mathbb{P}[\mathcal{H}_1]}{\mathbb{P}[\mathcal{H}_2]}, \tag{25.14}$$

$$= \frac{\mathbb{P}[E_2\,|\,E_1\mathcal{H}_1]\,\mathbb{P}[E_1\,|\,\mathcal{H}_1]}{\mathbb{P}[E_2\,|\,E_1\mathcal{H}_2]\,\mathbb{P}[E_1\,|\,\mathcal{H}_2]}\frac{\mathbb{P}[\mathcal{H}_1]}{\mathbb{P}[\mathcal{H}_2]}. \tag{25.15}$$

Now we exploit the independence of the experiments and the evidence they provide. This tells us that

$$\mathbb{P}[E_2\,|\,E_1\mathcal{H}_1] = \mathbb{P}[E_2\,|\,\mathcal{H}_1], \tag{25.16}$$

---

[5] When the hypothesis under consideration, $\mathcal{H}_1$, is being tested against the null hypothesis $\mathcal{H}_0$, we might write the Bayes factor simply as B_1 to denote B_{10}. Note the order: the null hypothesis then appears in the denominator of Equation (25.13) in this notation.

since our knowledge of the results of the second experiment E_2 cannot depend on what happened in the first experiment, E_1. Here E_2 can only depend on the hypotheses $\mathcal{H}_1$ or $\mathcal{H}_2$. Hence we arrive at

$$\frac{\mathbb{P}[\mathcal{H}_1 \mid E_2 E_1]}{\mathbb{P}[\mathcal{H}_2 \mid E_2 E_1]} = \frac{\mathbb{P}[E_2 \mid \mathcal{H}_1] \, \mathbb{P}[E_1 \mid \mathcal{H}_1]}{\mathbb{P}[E_2 \mid \mathcal{H}_2] \, \mathbb{P}[E_1 \mid \mathcal{H}_2]} \frac{\mathbb{P}[\mathcal{H}_1]}{\mathbb{P}[\mathcal{H}_2]}, \tag{25.17}$$

$$= \frac{\mathbb{P}[E_2 \mid \mathcal{H}_1]}{\mathbb{P}[E_2 \mid \mathcal{H}_2]} \frac{\mathbb{P}[\mathcal{H}_1 \mid E_1]}{\mathbb{P}[\mathcal{H}_2 \mid E_1]}, \tag{25.18}$$

Odds on $\mathcal{H}_1$ given E_1 and E_2 = Bayes factor for $E_2 \times$ odds on $\mathcal{H}_1$ given E_1.

So the odds after the second experiment are changed from the odds after the first experiment by the Bayes factor for the second experiment. Each time an experiment is performed to test a hypothesis we just multiply the odds by the appropriate Bayes factor. The Bayes factors for independent evidence simply multiply.

To write Equation (25.18) in a form like (25.12) we need to enhance our notation for both the Bayes factor (25.13) and the odds: the dependencies have to be made explicit. In particular B_{12} must state explicitly which experiment is relevant, so we replace B_{12} by $B_{12}(E)$, where E represents the dataset from the experiment E. We can also drop the explicit reference to the hypothesis and write $\mathcal{O}(E)$ for $\mathcal{O}(\mathcal{H} \mid E)$. We then have:

Odds for hypotheses given independent experiments:

$$\mathcal{O}(E_1, E_2) = B_{12}(E_2) \, \mathcal{O}(E_1). \tag{25.19}$$

This can be generalised to a statement for multiple independent experiments to

$$\mathcal{O}(E_1, \ldots, E_n) = \prod_{i=1,\ldots,n} \mathcal{O}(E_i) \, \mathcal{O}(\mathcal{H}), \quad \mathcal{O}(\mathcal{H}) = \frac{\mathbb{P}[\mathcal{H}_1]}{\mathbb{P}[\mathcal{H}_2]}. \tag{25.20}$$

Taking logarithms base 10 changes the product into a sum:

$$\log_{10} \mathcal{O}(E_1, \ldots, E_n) = \sum_{i=1,\ldots,n} \log_{10} \mathcal{O}(E_i) + \log_{10} \mathcal{O}(\mathcal{H}). \tag{25.21}$$

The quantity $\log_{10} \mathcal{O}(E)$ is sometimes referred to as the *weight of evidence* provided by E. The additive nature of this equation suggests a balance achieved by combining individual pieces of evidence $\mathcal{O}(E_i)$: the weight of all the evidence $\log_{10} \mathcal{O}(E_1, \ldots, E_n)$ can be either for or against depending on the sign of $\log_{10} \mathcal{O}(E_1, \ldots, E_n)$.

25.1.3 Bayes Theorem: More Complex Forms

Multiple Mutually Exclusive Hypotheses

We now wish to consider using data to evaluate one particular hypothesis, H_i, among a number of competing hypotheses $\{H_j\}$.

To this end we can exploit the *total probability theorem*[6] that, for a complete set of mutually exclusive events A_i:

$$\mathbb{P}[X] = \sum_i \mathbb{P}[X \,|\, A_i]\,\mathbb{P}[A_i], \quad A_i \cap A_j = \varnothing, \ \forall i \neq j. \tag{25.22}$$

Using Bayes theorem in the form

$$\mathbb{P}[A_i \,|\, X] = \mathbb{P}[X \,|\, A_i]\frac{\mathbb{P}[A_i]}{\mathbb{P}[X]}, \tag{25.23}$$

we can now express $\mathbb{P}[X]$ in terms of the A_i using the total probability theorem (25.22):

$$\mathbb{P}[A_i \,|\, X] = \frac{\mathbb{P}[X \,|\, A_i]\,\mathbb{P}[A_i]}{\mathbb{P}[X \,|\, A_1]\,\mathbb{P}[A_1] + \cdots + \mathbb{P}[X \,|\, A_n]\,\mathbb{P}[A_n]}. \tag{25.24}$$

We can interpret the A_i as a set of mutually exclusive hypotheses, and X as the evidence available to examine the merits of particular hypothesis H_i. We should notice that $\mathbb{P}[X \,|\, A_i]$ is the likelihood of the situation X given the hypothesis A_i.

The card game bridge provides a good example of how this equation might be interpreted (Epstein, 1977). The game involves four players, each holding 13 of the 52 cards from a standard pack. The play consists of a sequence of verbal bids leading to a final contract that goes to the highest bidder. The bids from the ith player reveal information about the hand of cards h_i that player holds. Then, as the cards are played the players see more and more information about which cards the players were holding. The data, D, is thus the sequence of bids and card play. The goal of the players is to infer the likely hands H_i held by the others on the basis of the bids and the play: $\mathbb{P}[H_i \,|\, D]$. The good player knows from experience about $\mathbb{P}[D \,|\, H_i]$: this follows what a rational player would do in the given circumstances D when holding a particular hand H_i. Under these circumstances Bayes' theorem as expressed by (25.24) can be written

$$\mathbb{P}[H_i \,|\, D] = \mathbb{P}[H_i]\frac{\mathbb{P}[D \,|\, H_i]}{\sum_j \mathbb{P}[H_j]\,\mathbb{P}[D \,|\, H_j]}. \tag{25.25}$$

The 'likelihood' $\mathbb{P}[D \,|\, H_i]$ here is what the player understands about the game on the basis of experience.

Determination of Multiple Parameters

Generally speaking, theories are expressed in terms of parameters that are to be determined experimentally. Denote this set of parameters by the vector $\boldsymbol{\theta} = \{\theta_1, \ldots, \theta_n\}$. The goal is to determine the values of the θ_i on the basis of the data that is available. We can do this by using the Bayes theorem to write $\mathbb{P}[ABC]$ as

$$\mathbb{P}[A]\,\mathbb{P}[B \,|\, A]\,\mathbb{P}[C \,|\, AB] = \mathbb{P}[A]\,\mathbb{P}[C \,|\, A]\,\mathbb{P}[B \,|\, AC].$$

Cancelling the $\mathbb{P}[A]$s and substituting $A \rightarrow \mathcal{H}, B \rightarrow D$ and $C \rightarrow \boldsymbol{\theta}$ gives

$$\mathbb{P}[D \,|\, \mathcal{H}]\,\mathbb{P}[\boldsymbol{\theta} \,|\, \mathcal{H}D] = \mathbb{P}[\boldsymbol{\theta} \,|\, \mathcal{H}]\,\mathbb{P}[D \,|\, \mathcal{H}\theta], \tag{25.26}$$

---

[6] See the On-line Supplement *Probability and Statistics Primer*.

which leads to the form:[7]

$$\mathbb{P}[\boldsymbol{\theta} \mid D \; \mathcal{H}] = \mathbb{P}[D \mid \boldsymbol{\theta} \; \mathcal{H}] \frac{\mathbb{P}[\boldsymbol{\theta} \mid \mathcal{H}]}{\mathbb{P}[D \mid \mathcal{H}]}. \tag{25.27}$$

This is just a re-expression of Equation (25.6) that makes it clearer where the hypothesis comes into play. The quantity $\mathbb{P}[\mathcal{H}]$ that appeared there is replaced with $\mathbb{P}[\boldsymbol{\theta} \mid \mathcal{H}]$, the prior information about the parameterisation of the theory provided by the hypothesis itself. We refer to $\mathbb{P}[D \mid \boldsymbol{\theta} \; \mathcal{H}]$ as the *likelihood*, $\mathbb{P}[\boldsymbol{\theta} \mid \mathcal{H}]$ as the *prior* and $\mathbb{P}[D \mid \mathcal{H}]$ as the *evidence*. So we have, in summary,

$$\text{Posterior} = \text{likelihood} \times \frac{\text{prior}}{\text{evidence}}, \tag{25.28}$$

which brings us back to Equation (25.6).

25.1.4 Bayes with Continuous Probability Densities

We now wish to move towards the limit where the sets of events A and B described in the preceding are replaced by continuous random variables X and Y having probability densities that we shall denote by $f_X(x)$ and $f_Y(y)$. When there is no ambiguity about which random variable is being referred to we can simplify the notation for these pdfs to $f(x)$ and $f(y)$. If the joint distribution of x and y is $f(x, y)$, then

$$f(x) = \int f(x, y) \, dy, \qquad f(y) = \int f(x, y) \, dx, \tag{25.29}$$

where the integrals are over the range of their respective integrands. The definitions of the conditional probabilities $f(x \mid y)$ and $f(y \mid x)$ are then

$$f(y \mid x) = f(x, y)/f(x), \qquad f(x \mid y) = f(x, y)/f(y). \tag{25.30}$$

We can combine the second of (25.29) with the first of (25.30) to eliminate $f(x, y)$ and give

$$f(y) = \int f(y \mid x) f(x) \, dx. \tag{25.31}$$

The generalisation of Bayes theorem as given in Equation (25.2) is:

$$f(x \mid y) = \frac{f(y \mid x) f(x)}{f(y)}. \tag{25.32}$$

This follows directly from (25.30). We can then write the continuous random variables version of (25.24) as

$$f(x \mid y) = \frac{f(y \mid x) f(x)}{\int f(y \mid x) f(x) \, dx}. \tag{25.33}$$

---

[7] The paper by MacKay (1992) is thoroughly good reading. It is available through MIT Cognet at
`http://cognet.mit.edu/library/journals/`
but only with a subscription or through a subscribing library. Otherwise copies of the article can be found through Google Scholar. MacKay's book (MacKay, 2003) goes into considerably more detail.

25.1.5 Relative Merits of Different Models

Now let us consider some theoretical models $\mathcal{M}$ that predict data values D which we can measure experimentally. We propose to use the data to assess the relative merits of these models. We might interpret the symbols:

$\mathbb{P}[D \,|\, \mathcal{M}]$ probability that this particular data set D would arise on the basis of the model. This is often referred to as the *evidence*.

$\mathbb{P}[\mathcal{M} \,|\, D]$ probability that the model $\mathcal{M}$ is correct given that our experiment gave the data set D.

And so we write the *likelihood* of our theory (or hypothesis), given the data, as:

Likelihood of a data model or theory $\mathbb{P}[\mathcal{M}]$:

$$\mathcal{L} = \mathbb{P}[\mathcal{M} \,|\, D] = \mathbb{P}[D \,|\, \mathcal{M}]\frac{\mathbb{P}[\mathcal{M}]}{\mathbb{P}[D]}. \qquad (25.34)$$

So if we have a set of models or theories $\mathcal{M}_i$, given a dataset D we can in principle give relative weights for the acceptability of the theories given these data. Using this equation we can calculate the ratio of $P(\mathcal{M}_i|D)$ and $P(\mathcal{M}_j|D)$:

Likelihood ratio of two hypotheses:

$$\lambda_{\mathcal{M}_i \mathcal{M}_j} = \frac{\mathbb{P}[D \,|\, \mathcal{M}_i]}{\mathbb{P}[D \,|\, \mathcal{M}_j]}\,\frac{\mathbb{P}[\mathcal{M}_i]}{\mathbb{P}[\mathcal{M}_j]}. \qquad (25.35)$$

The ratio does not depend on $\mathbb{P}[D]$ since we are comparing on the basis of the same data. We appear to have a mechanism for deciding which is the best theory: all other things being equal, we just compare the probabilities of getting the observed data set D on the basis of each of the hypotheses $\mathcal{M}_i$ and $\mathcal{M}_j$. On this basis, the best model is the one that maximises the likelihood ratio.

The argument is, however, not quite as straightforward as has been stated. How are we to understand the quantity $\mathbb{P}[\mathcal{M}]$? How can we assign prior probabilities to models if not via an intuition gained from looking at data? After all, it is observation of patterns in data that motivates the creation of a model in the first place. Clearly, $\mathcal{M}$ and D do not have the same status *vis-à-vis* A and B in (25.2).

25.2 Bayesian Inference: the Basics

At this point we are ready to make the leap into Bayesian inference. We have already encountered in Chapter 23 the frequentist's use of likelihood in determining the parameters of models that purport to fit data. With likelihood analysis we create a cost function for fitting a data set with a parameterised model in which a good fit is the parameter choice that minimises the cost, or maximises the likelihood. The data by itself determines the parameter values and assigns confidence levels to that determination. There is no input to the determination from other sources of information about the parameters that might have existed prior to acquiring the data set. Moreover the method does not itself tell us how to choose a model nor how many parameters the model should have.

The Bayesian approach to inference is that prior knowledge or prior bias should enter into parameter determination in an explicit way. The goal is achieved by multiplying the likelihood by a function of the parameters that reflects prior knowledge or bias. In so modifying the likelihood we alter the optimal parameter values that reflect the existence of prior information.

In using this approach we have to adopt a somewhat different view of the parameters of the model. If we encapsulate our prior knowledge of a parameter θ in terms of a function $\pi(\theta)$, we are saying that the θ appearing in the model is not a simple number whose value is to be determined. The parameter adopts the role of a random variable to which Bayes Theorem can be applied.

> **Remark: Bayesians vs. frequentists**
> In the Bayesian world, parameters of models are random variables.
> In the frequentist world, parameters are fixed unknown quantities that are to be determined.

Along with this comes an interesting and somewhat controversial discussion about the very nature of probability. The Bayesian can meaningfully assign a value to a parameter in a model for an event given a single datum, while the frequentist would argue that this is meaningful only for events that are repeatable. The Bayesians re-interpret the meaning of probability as a measure of belief rather than as frequency of occurrence under repetition.

25.2.1 Going Bayesian, a First Look

Following our previous discrete variable discussion, we now look at the situation where x in the above equations is replaced by a continuous parameter θ that parameterises a continuous probability density function $f(y; \theta)$ of a random variable Y. So θ might, for example, be the mean of the distribution of Y, or some other statistical parameter. We then rewrite equation (25.33) in the more edifying form

$$f(\theta \mid y) = \frac{f(y \mid \theta)f(\theta)}{f(y)} = \frac{f(y \mid \theta)f(\theta)}{\int f(y \mid \theta)f(\theta)\, d\theta}. \tag{25.36}$$

The term $f(y \mid \theta)$ is the likelihood and is a function of θ, while $f(\theta)$ is our prior. The output, $f(\theta \mid y)$, is not a single value, but another distribution of the parameter θ: the posterior distribution. If our prior $f(\theta)$ does not depend on θ (in other words, it is uniform), then the posterior $f(\theta \mid y)$ is simply proportional to the likelihood $f(y \mid \theta)$, and the best guess θ is obtained by maximising the likelihood.

When applying Bayesian methods it is usual to write this equation in a way that makes explicit the fact that we are using the data to test a physical model, M. For greater clarity, replace the data label y with the symbol D and the prior $f(\theta \mid M)$ by the more suggestive symbol $\pi(\theta \mid M)$:

Generating a posterior from the likelihood and a prior:

$$\underset{\text{Posterior}}{f(\theta \mid D, M)} = \frac{f(D \mid \theta, M) f(\theta \mid M)}{\int f(D \mid \theta, M) f(\theta \mid M)\, d\theta} = \frac{\overset{\text{Likelihood}}{f(D \mid \theta, M)}\ \overset{\text{Prior}}{\pi(\theta \mid M)}}{\underset{\text{Evidence}}{E(D \mid M)}}. \tag{25.37}$$

The integral in the denominator of the fraction is the likelihood averaged over the prior $f(\theta)$: it is the *Bayesian evidence*, which we have denoted by $E(D \mid M)$.[8] The evidence has no manifest dependence on θ and so this is not relevant for evaluating the θ-dependency of the posterior $f(\theta \mid D, M)$. However, we cannot ignore it when it comes to assessing the relative merits of two different models relative to the same data D. We shall come to this later.

Before moving on we should notice two limiting cases of the prior $\pi(\theta \mid M)$. The first case is the case of a delta-function $\pi(\theta \mid M) = \delta(\theta - \theta_0)$. In that case our prior belief is that the value of the parameter θ is merely a number that we have to determine by using the likelihood of the data alone and determining θ_0 using maximum likelihood. In this way Bayesian analysis encompasses the frequentist methodology. The second case is the case of the *flat prior* $\pi(\theta \mid M) = $ constant, which is independent of θ. This prior asserts that we have no idea where the value of θ may lie, though, more realistically, the statement $\pi(\theta \mid M) = $ constant would be accompanied by a range $[\theta_{\min}, \theta_{\max}]$ in which we believe θ will be found. With this prior we are again relying on the likelihood of the data to tell us about θ, though now the specified range $[\theta_{\min}, \theta_{\max}]$ may not contain the maximum of the likelihood function.

25.2.2 Understanding Priors and Posteriors

A simple example of how Equation (25.37) modifies the likelihood to reflect our prior prejudices is shown in Figure 25.2 for two choices of a prior: a Gaussian and a Cauchy distribution.[9] A familiar example arises in political elections where we might take the previous election results as a prior for a likelihood made up of pre-election opinion polls. The posterior then reflects both of these sources.[10] Both diagrams depict the same likelihood function as a zero-mean unit variance normal distribution. The priors are two standard deviations away from this and so have a strong effect in shifting the maximum

---

[8] The term *marginal likelihood* is also used.

[9] The Cauchy distribution $f(x) = (1 + x^2)^{-1}/\pi$ has a well-defined mean at $x = 0$, but the variance is infinite. Using a Cauchy distribution as a prior is one way of asserting that we believe we know whereabouts the value of the parameter lies, but we cannot assign an uncertainty to that.

[10] The purist Bayesian might object that this is merely an empirical Bayesian attitude on the grounds that using previous elections is a repeatable (frequentist) formula for selecting a prior and thus represents more than a sense of belief. This is the philosophy that 'today's posterior is tomorrow's prior'.

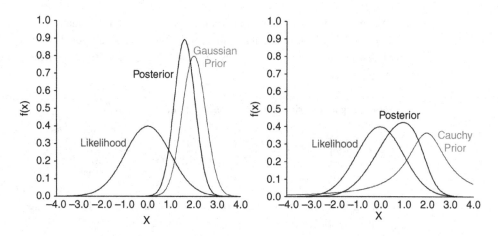

Fig. 25.2 Effect of Gaussian and Cauchy distributed priors on the likelihood function for a parameter X. In this sketch the likelihood has been represented as a Gaussian, and, in the case of a Gaussian prior, the posterior is then also a Gaussian. (For normally distributed data with unknown mean μ, the normal prior for μ is said to be *conjugate*. See Table 25.1 and Section 25.4.1.) In the case of the narrow Gaussian prior, we see that the posterior strongly discriminates against values $x < 0$. The Cauchy prior with its wide wings does not impose as strong a conclusion. In this sketch all functions have been normalised to unit area, but it should be stressed that in general they may not be proper probability densities. This kind of plot is often referred to as a *triplot*. (cc)

of the posterior curve. The Gaussian prior has a relatively small variance and so the posterior can exclude negative values of the parameter x. The Cauchy distribution does not have a finite variance, so while it results in a posterior with a shifted maximum relative to the maximum likelihood, it nevertheless leaves a broad range of values for the parameter x.

The constant, 'flat', prior reflects complete ignorance, but only for the given parameterisation of the model. If we choose a different set of parameters, such as a nonlinear function of the first set of parameters, the resultant parameterisation will no longer be flat. We clearly cannot make our prior invariant under all possible transformations of parameters, but we could at least demand that our prior be invariant under scale transformations. A function $f(x)$ is scale invariant when $f(\lambda x) = \lambda^{\alpha} f(x)$, for some scaling factor α. For example, the function $f(x) = 1/x$ has $f(\lambda x) = 1/(\lambda x) = (1/\lambda) f(x)$ and so $\alpha = -1$. For the case $\alpha = -1$ the transformation $y = \ln x$ makes $dy = dx/x$ and the rules for transformation of probabilities tell us $f(y) = f(x)dx/dy = $ constant. Hence the prior $\pi(\theta) \propto \theta^{-1}$ is scale invariant.[11] Of course, the range over which this holds has to be finite. Harold Jeffreys (1961, Section 3.10) was the first to argue[12] that priors should be invariant under scale transformations of the variables (see Jaynes, 2003, Ch. 12 for an extensive discussion).

---

[11] See the example of the gamma distribution as a prior in Equation (25.63).

[12] The first edition of Jeffrey's book was published in 1939.

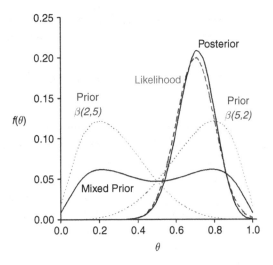

Priors can handle a diversity of differing beliefs about the outcome of an experiment. Here this is illustrated by two priors that are taken to be β-functions having different parameters, $\beta(2, 5)$ and $\beta(5, 2)$, which are combined with equal weight to give a relatively flat *mixed prior*. The likelihood here (dashed curve) is depicted as a Gaussian and the posterior is constructed from the mixed prior. We note that the posterior is rather similar to the likelihood, showing a slight shift of the likelihood towards the model parameterised by $\beta(5, 2)$.

The vertical axis is arbitrary, but all curves are normalised to the same area. ⓒⓒ

When there is more than one parameter, the situation gets more complicated. In that case, the *Jeffrey's prior* is given by

$$\pi(\boldsymbol{\theta}) \propto \sqrt{\det I(\boldsymbol{\theta})}, \qquad I_{ij} = -\mathbb{E}\left[\frac{\partial^2 \ln f(D \mid \boldsymbol{\theta}, M)}{\partial \theta_i \, \partial \theta_j}\right], \qquad (25.38)$$

where I_{ij} is the Fisher information matrix for the log-likelihood function (see Section 23.9 and Equation 23.61). Because it involves the likelihood of the very data that is being analysed, this prior is not generally regarded as representing a measure of the 'degree of belief' (it has a frequentist flavour because it uses the very data we are analysing to motivate the choice of prior).

An interesting situation arises when we have two sets of quite distinct priors, perhaps reflecting a difference of opinion about the outcome of the experiment. This is sketched in Figure 25.3, where the two prior opinions are modelled as β-functions favouring opinions of $\mu_1 = 0.2$ and $\mu_2 \sim 0.8$.[13] If multiple priors are denoted by π_i, $i = 1, \ldots, n_P$, then the mixed prior is simply the weighted sum

$$\pi_{\text{mixed}}(\theta) = \boldsymbol{\Sigma}_1^{n_P} \, w_i \, \pi_i(\theta). \qquad (25.39)$$

The appropriate determination of the relative weights is, of course, open to some debate. In Figure 25.3 the two priors are simply added with equal weights.

---

[13] The mode of the function $\beta(x; a, b)$ is at $\hat{x} = (a - 1)/(a + b - 2)$.

A similar issue arises when two data sets are presented for analysis with a single prior: how do we combine the data sets to compute a likelihood?

25.2.3 Classifying Priors: Two Philosophies

There is a diversity of interpretations of what 'Bayesian inference' should mean and how it should be correctly implemented. The differences of approach within the Bayesian community are as different from one another as they are from the opposing side, the frequentists, where there is at least a substantial degree of unity. The top-level schism among Bayesians is the rift between *subjective Bayesians* and *objective Bayesians*. In 2006 the open journal *Bayesian Analysis* featured a debate in which the subjective side was presented by Goldstein (2006) and the objective side by Berger (2006). Along with those two papers the journal published critiques of the papers by leading experts and rejoinders by the authors.[14] This distinction lies at the very base of how we are to interpret and use the theory of probability.

Without going into too much detail, a subjective prior reflects a personal feeling, perhaps based on experience and analogous situations, about the likely range taken on by a parameter. If we want to discuss the statistics of mountain ranges we need not consider mountains higher than 10 km because we have never seen a report of one such.

On the other hand an objective prior might arise from pertinent historical data, or, more likely, from small, exploratory, experiments performed to motivate and justify a larger study. Part of the notion of an objective prior seems to require that the motivational data be independent of the full study: it would not do to take out, say, 5% of the full data sample and use that as a prior for the remaining 95%. The objective prior is supposed to reflect knowledge obtained before any data had been taken for the current project.[15]

25.2.4 Classifying Priors: More Pragmatic Views

There are subclasses of Bayesian analysis that come down to more 'practical matters'.

Given the importance in Bayesian analysis of the prior $\pi(\theta \mid M)$ as an expression of prior degree of belief, it is useful to be able to state what the intent of using a particular prior might be. Different priors can be used to express different perspectives of what a data set is telling us.[16]

The prior might, for example, represent a strong opinion about the parameters that are being measured. This would generally be the case in an experiment designed to improve the

---

[14] The critiques are sometimes politely vitriolic and sometimes great fun to read. One starts with the sentence: *The dangerous heresy of so-called objective Bayesian methods is again propounded by Berger. These comments are my attempt to save Bayesian statistics.* The articles making up this debate, as published in the *Bayesian Analysis* journal, can be accessed from the Project Euclid website:
`http://projecteuclid.org/euclid.ba/1340371033`

[15] This description of an objective prior would seem to exclude what is done on quality control where one applies a precisely formulated procedure to extract samples as a 'control' for the rest of batch: the *empirical prior*.

[16] This is important, for example, in the analysis of clinical trials for drug treatments. Pharmaceutical companies will usually want to report enthusiastically about the results of a new treatment, while medical professionals might want to express a degree of scepticism about such claims. See for example O'Hagan and Luce (2003).

accuracy of previous measurements, or, less strongly, the range of values where an experiment should search to find a particular signal. Such priors are referred to as *informative priors*.

The converse, the *non-informative prior*, refers to a situation where one does not have any credible guidance as to where to look, except perhaps for a broad range of values covering most of the range of sensitivity of the experiment providing the data.

In addition to that there are priors that can be chosen to express scepticism about a given hypothesis by adding support for a null hypothesis. In that case the prior embraces parameter values that indicate 'no significant difference'. These, not surprisingly, are called *sceptical priors* and they are informative priors. The converse might be called the *enthusiastic prior*.

25.2.5 Computing the Bayes Factor

The posterior $f(\theta \mid D, M)$ in fact contains all the information that we have about the phenomenon that M represents. It is often convenient to encapsulate that information simply by giving the expectation and variance of the function $f(\theta \mid D, M)$, but some care is necessary when doing that. Not only does this over-simplify the true state of knowledge gained from the data, but quoting a mean or a mode of the distribution with percentile error bounds can only be interpreted relative to the nature of the posterior itself.

The Bayes factor is clearly an important part of analysing data. The issue to be addressed is how we compute $\mathbb{P}[E \mid \mathcal{H}]$, the likelihood of getting the data E on the basis of the hypothesis $\mathcal{H}$. The mathematical expression of the hypothesis $\mathcal{H}$ will contain parameters that are to be determined in the light of E, that is the posterior $\mathbb{P}[\mathcal{H} \mid E]$. So we let $\mathcal{H}$ depend on a set of to-be-determined parameters $\theta = \{\theta_1, \ldots, \theta_n\}$ and incorporate this fact into our notation. The parameterised hypothesis is referred to as a *model* and we shall denote that by $\{\theta; \mathcal{H}\}$. We write

$$\mathbb{P}[E \mid \mathcal{H}] = \int \mathbb{P}[E \mid \theta, \mathcal{H}] \, \mathbb{P}[\theta \mid \mathcal{H}] d\theta. \tag{25.40}$$

The quantity $\mathbb{P}[E \mid \mathcal{H}]$ is called the *marginal likelihood* for the model $\{\theta; \mathcal{H}\}$.[17] That is because in the integral we have integrated over the distribution of θ. In other words we have calculated the mean of $\mathbb{P}[E \mid \theta, \mathcal{H}]$ relative to the statistical distribution of θ under the hypothesis $\mathcal{H}$.

This should be contrasted with the maximum likelihood approach to parameter determination which uses the maximum of the function $\mathbb{P}[E \mid \mathcal{H}]$. One of the advantages the Bayesian approach has over the likelihood method is that the Bayesian method is robust to the introduction of too many parameters in the model, while likelihood methods are prone to over-fitting. However, there is the important issue of setting the function $\mathbb{P}[\theta \mid \mathcal{H}]$: the *prior distribution*.

---

[17] The more descriptive term *integrated likelihood* is also used for $\mathbb{P}[E \mid \mathcal{H}]$.

25.2.6 Doing Bayesian: the Integrals

In the case of a model with only one or two parameters θ it is possible to display a plot of the posterior $f(\theta \mid D, M)$. It may not be possible to analytically perform the integrals necessary to extract the means, median, variances and so on from this pdf, but it is straightforward to do this numerically. The problems arise as the dimensionality of the parameter space θ increases. It then becomes difficult to plot anything but sections of the function (the marginal distributions) and doing the integrals to extract key statistical parameters requires special algorithms.

Over the past few decades such algorithms have emerged based on sampling the parameter space. These are the *Monte Carlo* and the *Markov Chain Monte Carlo* methods.[18]

25.3 Using Bayes to Analyse Data

25.3.1 Marginalising over Parameters

Since in the Bayesian view the parameters of the model are assigned probability density functions, it is possible to 'marginalise' a model over some or all of the parameters. This is important when the model of the data has numerous parameters and we wish to focus on just one or two of them without committing ourselves to specific values of the other parameters.[19] How are we to deal with the extra parameters that we want to temporarily ignore? The process of ignoring parameters is called *marginalisation*. It is here that there is a big difference between the frequentist and Bayesian views.[20]

Consider a model for the data resulting from measurements of a random variable X having the probability density $f_X(x \mid \theta)$, where θ is a to-be-determined parameter of the model. In the frequentist world, θ is just a parameter and we simply work with maximising the likelihood,

$$\mathcal{L}_{\text{Freq}}(x_1, \ldots, x_n \mid \theta) = \prod_{i=1}^{n} f_X(x_i \mid \theta). \tag{25.41}$$

The likelihood is simply a function of the data we have available, expressed via the data model $f_X(x_i \mid \theta)$ for the individual data points. In this view, θ is a to-be-determined number that appears in the data model.

[18] Information about this can be found in the On-line Supplement: *MC and MCMC*.

[19] It might be that we simply want to plot a graph showing the inter-dependency of the two parameters, or the distribution of one parameter.

[20] Cox and Hinkley (1979, p.17.), when discussing 'extended definitions of likelihood', i.e. the ideas of *conditional likelihood* and *marginal likelihood*, say that ... *it is difficult to express this precisely and for that reason we shall not make extensive direct use of marginal and conditional likelihoods, although they will be implicit in much of the discussion of Chapter 5.* Their Chapter 5 is about significance tests.

In the world of the Bayesian, θ can have a prior distribution function that reflects any prior prejudices we may have. We would expect this to modify the likelihood. When that prior information is modelled by distribution $\pi(\theta)$ we have learned how to use Bayes theorem to modify the likelihood to reflect that additional information. The modified likelihood could be written:

$$\mathcal{L}_{\text{Bayes}}(\theta \mid x_1, \ldots, x_n, \pi) = \mathcal{L}(x_1, \ldots, x_n, \theta \mid f)\, \pi(\theta), \qquad (25.42)$$

where the dependence on the prior $\pi(\theta)$ has been included to emphasise that it is there. The data and θ are now on the same footing. We can contemplate marginalising the likelihood over the distribution of θ:

Marginalised likelihood:

$$\mathcal{L}(x_1, \ldots, x_n \mid f) = \int_{D_\theta} \mathcal{L}(x_1, \ldots, x_n, \theta \mid f)\pi(\theta)\, d\theta. \qquad (25.43)$$

D_θ is the domain of values of the parameter θ. This equation simply weights the contributions $\mathcal{L}(x_1, \ldots, x_n, \theta \mid f)$ to the overall likelihood $\mathcal{L}(x_1, \ldots, x_n \mid f)$ with the probability $p(\theta)$ of getting a particular value of θ. It is inevitable that $p(\theta)$ will depend on some further unknown parameters that reflect our prejudices in deciding what $p(\theta)$ might be.[21]

If we were to use a delta-function model $\pi(\theta) = \delta(\theta - \alpha)$ for some general nominal value α, in doing the integration (25.43) we would return to the frequentist view and end up finding the particular value of α that maximised $\mathcal{L}_{\text{Freq}}(x_1, \ldots, x_n \mid \alpha)$. Conversely we can choose $\pi(\theta)$ so as to exclude or give low weight to values of θ that are not consistent with our prior prejudices.[22]

We could integrate out all of the parameters and so end up with what was nominally a parameter-free likelihood $\mathcal{L}(x_1, \ldots, x_n)$ reflecting only what might be regarded as the objective data obtained from the experiment, unsullied by any manifest parametric prejudice. We would be back in the world of the frequentist.

The power of being able to marginalise a likelihood over one or more of the to-be-determined parameters, as in (25.43), comes into its own when the number of parameters θ gets large. We can then no longer visualise the likelihood surface, and so instead we can concentrate on a few parameters by marginalising out the rest of the parameters. Such visualisation is an important part of data analysis.

---

[21] The cynic might wonder why we do not assign probabilities to these too! In fact the Bayes purist might regard this approach to Bayesian statistics as paralleling what is termed the *empirical Bayes* approach ((Casella, 1989). The empirical Bayesian might, for example, try to get a handle on $p(\theta)$ from the data itself (which might seem somewhat circular). The key to 'empirical Bayes' is that the prior is a frequency distribution to be assigned on the basis of some stable physically motivated random mechanism. There is a strong element of 'frequentism' which Bayesian purists find unacceptable.[7] Empirical Bayesian methods do, however, provide a highly effective tool for statistical inference, particularly in quality control on production lines and in process control in general.

[22] It has already been remarked that physical theories do not appear from nowhere: they are generally based on a lot of prior work which at least indicates and motivates a need for further exploration, and also provides indications as to what the values of these parameters might be.

One of the key questions is how to evaluate integrals like (25.43). Even in one dimension, i.e. only one parameter θ, it is not necessarily the case that the integral can be evaluated analytically. When there are many dimensions, the only way forward is to do the integral numerically via a Monte Carlo integration.[18]

There are some one-dimensional situations which do allow analytic work. This happens when the underlying model $f_X(x \mid \theta)$ has one of several particular analytic forms. Then, in each case, there is a particular $p(\theta)$ that allows integration.

25.3.2 Mixed Priors

We mentioned the situation in which there might be two or more prior belief sets about the outcome of analysing a data sample, in which case the prior, $\pi(\theta)$, for data analysis would be a weighted sum of the individual priors $\pi_i(\theta)$:

$$\pi_{\text{mixed}}(\theta) = \Sigma_1^{n_P} w_i \pi_i(\theta), \tag{25.44}$$

where $w_i, i = 1, \ldots, n_P$ are the to-be-chosen relative weights of the belief sets $\pi_i(\theta)$ (see Figure 25.3 and Equation 25.39). If we normalise the priors $\pi_i(\theta)$ and $\pi_{\text{mixed}}(\theta)$ so that

$$\int \pi_{\text{mixed}}(\theta) \, d\theta = 1, \quad \int \pi_i(\theta) \, d\theta = 1, \quad i = 1, \ldots, n_P, \tag{25.45}$$

then the weights must sum to 1:

$$\Sigma_1^{n_P} w_i = 1. \tag{25.46}$$

If the likelihood of the data from the measurement set is $\mathcal{L}(\theta)$, then the posterior for any one of the priors is

$$f(\theta \mid \pi_i) = \frac{1}{E_i} \mathcal{L}(\theta) \pi_i(\theta), \quad E_i = \int \mathcal{L}(\theta) \pi_i(\theta) \, d\theta, \tag{25.47}$$

(see Equation 25.37). Putting everything together we find that using an overall prior that is the weighted sum of the individual priors gives a posterior,

$$f_{\text{mixed}}(\theta) = \sum_i^{n_P} W_i f(\theta \mid \pi_i), \quad W_i = \frac{w_i E_i}{\Sigma w_i E_i}. \tag{25.48}$$

This is the not very surprising result that the overall posterior, taking everyone's prior beliefs into account, is the weighted sum of their individual posteriors, and the weighting factors for this are simply the w_i-weighted fraction of the total evidence provided by each prior.

The question is how to select the weighting factors w_i for each prior. Democracy of opinion would dictate something like equal weights, but one could, in the true Bayesian spirit, take the weights themselves to be statistical quantities taken from some suitable prior. Such weights are described as *hyper-parameters* since they have nothing to do with the model that is taken to represent the data being analysed (Hobson *et al.*, 2002).

25.3.3 Hierarchical Bayes

Bayes theorem can be used hierarchically in the sense that instead of writing

$$f(\theta \mid D) \propto f(D \mid \theta)\pi(\theta), \qquad (25.49)$$

we introduce another parameter ϕ which itself underlies θ in such a way that the prior $\pi(\theta)$ can be written as $\pi(\theta \mid \phi)\pi(\phi)$. Then we have the 2-stage hierarchical expression

$$f(\theta \mid D) \propto f(D \mid \theta)\,\pi(\theta \mid \phi)\,\pi(\phi), \qquad (25.50)$$

which is clearly extensible to a 3-stage expression by expressing ϕ in terms of another hyper-parameter ψ, and so on. This hierarchical approach is used in a number of areas, such as Bayesian learning.

25.4 Conjugate Priors

Choice is a wonderful thing, but the issue remains that (Cox and Hinkley, 1979, p.376), given a dataset, how is the appropriate prior chosen for a parameter about which little or nothing is known? Simplicity is one factor to consider: there is generally a prior with a convenient mathematical form that simplifies the computation of integrals like (25.43).[23]

As it happens, there are a few pairs of distributions whose product can be integrated analytically, both of of which might serve as a likelihood–prior pair. These are listed in Table 25.1, where apart from the Dirichlet distribution these are all well-known frequently discussed distributions.[24] These are called *conjugate priors*. The fact that they can be integrated analytically means that we get explicit expressions for the posterior distribution which are then easy to apply in real situations. We shall show a couple of well-known examples below.

25.4.1 Single Realisation of One Variate

We can elucidate the somewhat uninspiring equations of the previous section by looking at the simple case of analysing a single sample drawn from a normally distributed random variable $X : N(\mu, \sigma)$. To simplify things further we shall suppose that the variance σ^2 is known and that we want to determine μ.

Following the Bayes approach, we shall regard the parameter μ as a random variable to which we will assign a probability density, $f(\mu)$. We shall discuss this assignment process later, it is the main bone of contention surrounding Bayesian inference.

Bayes theorem says that we can determine the distribution of μ, given the measurement x, from the equation

---

[23] Of course, the response to that might be 'why should we care when we can compute it?'.

[24] The Dirichlet distribution is a particularly versatile distribution. In its general form it depends on k parameters $\alpha_1, \ldots, \alpha_k$ and has the important limiting cases that $\alpha_1 = \alpha_2 = \cdots = 1$, which is the uniform distribution, and $\alpha_1 = \alpha_2, = \cdots = 1/2$, which is the least informative *Jeffreys prior*. In two dimensions the beta distribution is a special case of the Dirichlet distribution. See MacKay (2003, Section 23.5).

Distribution	Parameter	Conjugate prior
Table 25.1 Conjugate prior distributions.		
Binomial	$\mathbb{P}[\text{success}]$	beta
Multinomial	$\mathbb{P}[\text{success}]$	Dirichlet
Poisson	mean	gamma
Exponential	$1/\text{mean}$	gamma
Normal	mean μ (given variance)	normal
Normal	variance σ^2 (given mean)	inverse gamma

$$\overset{\text{posterior}}{f(\mu \,|\, x, \sigma^2)} \propto \overset{\text{likelihood}}{f(x \,|\, \mu, \sigma^2)} \; \overset{\text{prior on } \mu}{f(\mu),} \tag{25.51}$$

where $f(x \,|\, \mu, \sigma^2)$ is the probability that the dataset x would arise on the basis of this model with parameters μ and σ_μ^2. In other words $f(x \,|\, \mu, \sigma^2)$ is the likelihood of the data given the model. The denominator appearing in Bayes theorem (25.34) is merely a normalisation factor and in this case it is the probability density $f_X(x)$ of the data itself. $f_X(x)$ is unknown but at least fixed for a given experiment.

In this case of the single measurement,

$$\mathcal{L}(x \,|\, \mu, \sigma^2) \propto f(x \,|\, \mu, \sigma) \propto \frac{1}{\sigma^2} \exp\left[-\frac{1}{2\sigma^2}(x - \mu)^2 \right], \tag{25.52}$$

which is just (25.41) with $n = 1$.

We can choose whatever we like for the prior $f(\mu)$. It could be a uniform distribution on some interval, in which case we would be asserting that we believe that this is where the values of μ are, but we have no idea where on that interval the value might be. We could choose $f(\mu)$ to be a Gaussian, which says that we believe that we know roughly where μ will be.

We shall take μ to be normally distributed as $N(\mu_0, \sigma_\mu)$:

$$f(\mu \,|\, \mu_0, \sigma_\mu) \propto \frac{1}{\sigma_\mu^2} \exp\left[-\frac{1}{2\sigma_\mu^2}(\mu - \mu_0)^2 \right]. \tag{25.53}$$

We have all we need to build Equation (25.51):

$$f(\mu \,|\, x) \propto \frac{1}{\sigma^2} \exp\left[-\frac{1}{2\sigma^2}(x - \mu)^2 \right] \overset{\text{likelihood}}{} \times \frac{1}{\sigma_\mu^2} \exp\left[-\frac{1}{2\sigma_\mu^2}(\mu - \mu_0)^2 \right] \overset{\text{prior}}{}$$

$$\propto \frac{1}{\sigma^2 \sigma_\mu^2} \exp -\frac{1}{2}\left[\frac{(x - \mu)^2}{\sigma^2} - \frac{(\mu - \mu_0)^2}{\sigma_\mu^2} \right]. \tag{25.54}$$

We see that the argument of the exponential is a quadratic in x and μ, so this is in fact a normal distribution for two correlated normally distributed random variables x and μ. We can now derive the functional form of $f(\mu \mid x)$. Getting to the following expression involves a lot of 'book-keeping' work,[25]

$$f(\mu \mid x) \propto \exp\left[-\frac{1}{2}\left(\frac{1}{\sigma_\mu^2} + \frac{n}{\sigma^2}\right)(\mu - \bar{\mu})^2\right], \quad \bar{\mu} = \frac{\dfrac{\mu_0}{\sigma_\mu^2} + \dfrac{n\bar{x}}{\sigma^2}}{\dfrac{1}{\sigma_\mu^2} + \dfrac{n}{\sigma^2}}, \quad (25.56)$$

from which we can immediately see the parameters of the posterior distribution of μ. For convenience the number of observations has been restored to n. The expression describes the posterior when we have n measurements of X and we have put $\bar{x} = \sum x_i/n$.

This works because we assumed that the distribution of the parameter defining the prior is itself normal, and so we are multiplying and integrating the product of two normal distributions to yield a normally distributed posterior.

25.4.2 Priors for Parameters

In this example illustrating the use of conjugate priors we derive an estimator for the scale parameter of the exponential distribution (see Appendix E.2) in two ways: one directly from the likelihood and one using a Bayesian prior for the distribution of the parameter.

Likelihood Estimation for Exponential Distribution

Consider a specific example of estimating the parameter θ of an exponentially distributed random variable X from a set of independent measurements $\{x_1, \ldots, x_n\}$. The distribution of X is

$$f_X(x \mid \theta) = \theta e^{-\theta x}, \quad x > 0, \; \theta > 0. \quad (25.57)$$

The log-likelihood of the n measurements is

$$\ln \mathcal{L}(\theta \mid x_1, \ldots, x_n) = n \ln \theta - \theta \, \boldsymbol{\Sigma}_i x_i, \quad (25.58)$$

from which we can determine the maximum likelihood estimator of θ:

$$\hat{\theta}_{\mathrm{ML}} = \frac{n}{\boldsymbol{\Sigma}_i x_i}. \quad (25.59)$$

There are other estimators that can be constructed. For example the uniformly minimum variance unbiased estimator (UMVUE)[26] is

$$\hat{\theta}_{\mathrm{UMVUE}} = \frac{n-1}{\boldsymbol{\Sigma}_i x_i}, \quad (25.60)$$

---

[25] It helps to know the identity

$$a(y - x)^2 + b(x - z)^2 = (a + b)\left(x - \frac{ay + bz}{a + b}\right)^2 + \frac{ab}{a + b}(y - z)^2, \quad (25.55)$$

in which the first term on the right-hand side is a quadratic in x and the second term does not involve x.

[26] See Papoulis and Pillai (2002, Section 8.3, p334).

while the estimator that minimises the mean square error (MSE) $(\theta - c/\Sigma x_i)^2$ is

$$\hat{\theta}_{\text{MSE}} = \frac{n-2}{\Sigma_i x_i}. \tag{25.61}$$

For large n these converge to the same value, as they should.

Bayesian Estimation for the Exponential Distribution

Now we turn to what the Bayesian analysis gives.[27] According to the Bayes theorem, if we assume θ has a prior probability density $\pi(\theta)$,

$$f_\Theta(\theta \mid x_1, \ldots, x_n) = \frac{f_X(x_1, \ldots, x_n \mid \theta)\pi(\theta)}{f_X(x_1, \ldots, x_n)}. \tag{25.62}$$

The function $f_X(x_1, \ldots, x_n \mid \theta)$ is the likelihood $\mathcal{L}(\theta \mid x_1, \ldots, x_n)$ for data from the exponential distribution (25.57) (given as $\ln \mathcal{L}$ in Equation 25.58).

We can make progress analytically since the exponential distribution has a Bayesian conjugate prior in the form of the gamma distribution:

$$\pi(\theta; \alpha, \beta)) = \text{gamma}(\theta; \alpha, \beta) = \frac{\beta^\alpha}{\Gamma(\alpha)}\theta^{\alpha-1}\exp(-\beta\theta), \quad \alpha, \beta > 0. \tag{25.63}$$

Here α is referred to as the 'shape' and β as the 'rate'. The parameters α, β are freely chosen so as to define a prior that is appropriate to the problem being addressed by the data. This is a scale invariant prior for $\alpha \neq 1$ since under scale transformations $\theta \to \lambda\theta$ this scales as λ^{-1}. The case $\alpha = 1$ and $\beta = 0$ corresponds to a flat prior, $\pi(\theta) = \text{constant}$.

Collecting everything together, Equation (25.62) for the posterior distribution becomes

$$f_\Theta(\theta \mid \mathbf{x}) \propto \underbrace{\exp(n\ln\theta - \theta\,\Sigma_i x_i)}_{\text{likelihood}}\,\underbrace{\beta^\alpha\theta^{\alpha-1}\exp(-\beta\theta)}_{\text{prior}} \tag{25.64}$$

$$\propto \beta^\alpha\theta^{n+\alpha-1}\exp[-\theta(\beta + \Sigma_i x_i)], \tag{25.65}$$

$$\propto \text{Gamma}(\alpha + n, \beta + n\bar{x}), \qquad \bar{x} = \Sigma_i x_i/n. \tag{25.66}$$

Here we see the advantage of using conjugate priors: we know, or can look up, all the properties of the gamma-distribution (see Appendix E.4). Using either the mean or the mode of the posterior we get for general α, β:

Posterior mean and mode estimators for sample modelled by the exponential distribution:

$$\bar{\theta}_{\text{mean}} = \frac{\alpha + n}{\beta + \Sigma_i x_i}, \quad \hat{\theta}_{\text{mode}} = \frac{\alpha + n - 1}{\beta + \Sigma_i x_i}. \tag{25.67}$$

For particular choices of α, β we derive the Bayesian estimators using the mode $\hat{\theta}_{\text{mode}}$:

$$\bar{\theta}_{\text{flat}} = \frac{n}{\Sigma_i x_i}, \qquad \alpha = 1, \beta = 0, \quad \text{flat prior} \tag{25.68}$$

---

[27] This follows the article by Elfessi and Reineke (2001). See also the section on Bayesian inference at
http://en.wikipedia.org/wiki/Exponential_distribution

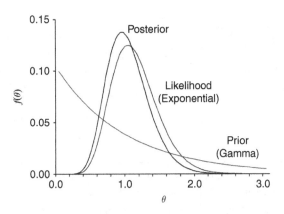

Parameter estimation for $n = 10$ samples taken from an exponential distribution $f(x) = \theta^{-1} \exp(-\theta x)$. The prior, taken to be the Bayesian conjugate to the exponential, i.e. the gamma distribution, is shown for parameters $\alpha = \beta = 1$. The posterior is simply the product of the likelihood and the prior and has a maximum that is shifted to smaller values of θ relative to the likelihood maximum, as expected from the solution (25.67). ⓒⓒ

$$\bar{\theta}_{\text{Jeffreys}} = \frac{n-1}{\Sigma_i x_i}, \qquad \alpha = 0, \beta = 0, \qquad \text{Jeffreys prior} \qquad (25.69)$$

$$\bar{\theta}_{\alpha=-1} = \frac{n-2}{\Sigma_i x_i}, \qquad \alpha = -1, \beta = 0. \qquad (25.70)$$

These correspond respectively to the likelihood derived estimates $\hat{\theta}_{\text{ML}}$, $\hat{\theta}_{\text{UMVUE}}$, and $\hat{\theta}_{\text{MSE}}$ of Equations (25.59–25.61). We notice that the cases $\alpha = 0$ and $\alpha = -1$ are formally excluded from the definition of the gamma distribution probability density function (25.63), which requires $\alpha > 0$. The procedure is nonetheless valid since there is no requirement that the prior $\pi(\theta)$ be a proper probability density function (it merely expresses a degree of belief). The likelihood, prior and posterior are shown in Figure 25.4 for the case $\alpha = 1$ and $\beta = 1$ when there are $n = 10$ data points. We see the expected shift in the maximum of the posterior following Equation (25.67).

26

CMB Data Processing

The CMB data comes from many different kinds of experiment, and in the competition between the small balloon-based missions and huge satellite-borne programs it is occasionally the former who win. But in the end, the high quality sky-wide data comes from space experiments. During this century, two have dominated cosmology so far: WMAP and Planck.

In this chapter we describe how the data is acquired, how it is stored in special format maps, and how it is treated to remove all the non-cosmological contributions. After this has been done we have our data, but it is the need to remove noise and foreground contamination that is the most challenging, and the most demanding of computing resources.

We treat the problem of noise and foreground removal in as generic a way as possible, there is not a single best way of doing this. The mathematical formalism is quite complex, as is the statistical data analysis that will follow.

26.1 Introduction

Some of the history of the observations of the Cosmic Microwave Background radiation (CMB) has been recounted in the first part of this book (see Chapter 3). The culmination of the early efforts, the 'first 30 years', was perhaps the COBE DMR/FIRAS mission. COBE DMR produced the first all-sky maps of the CMB (Smoot *et al.*, 1991), while the FIRAS experiment on the same satellite had established the remarkable accuracy of the Planckian form of the radiation spectrum (Mather and the COBE collaboration, 1990). Importantly, COBE had delivered a low resolution, albeit noisy, picture of the microwave sky that not only made substantial scientific advances, but also attracted widespread public attention and provided a vital scientific stimulus.

The manifest success of the COBE mission inspired a series of ground-based, balloon-based and and space observatories to look at the fluctuations with more sensitivity and at higher angular resolution. Satellite experiments are expensive and take a long time to build and so the competition to get the key results from lower cost ground based experiments before the space experiments could deliver was fierce, and in large part successful.

There was a considerable mixture of balloon-borne and ground-based experiments, each with its own label: Archeops, BOOMERang, MAXIMA, MSAM, Python QMAP, Sask,

Table 26.1 WMAP data acquisition.					
Parameter[a]	K	Ka	Q	V	W
Frequency range (GHz)	20–25	28–36	35–46	53–69	82–106
Wavelength (mm)	13.6	10.0	7.5	5.0	3.3
Detectors	2	2	4	4	8
Nominal beam size (deg)[b]	0.88	0.66	0.51	0.35	0.22
Noise sensitivity (mK s$^{1/2}$)[c]	0.65	0.78	0.92	1.13	1.48

[a] From Greason *et al.* (2012, Table 1.3.)

[b] Beam not Gaussian, this is square root of beam solid angle.

[c] Noise per radiometer and defined in terms of the Rayleigh–Jeans temperature.

Viper, VSA, to name but a few.[1] It was difficult for those not directly involved with the quest to recall which label did what and to sort out the differences between the results they published. Nevertheless, they delivered consistent and significant discoveries and results, and undoubtedly provided an important basis for the planning of future space-based missions.

26.2 WMAP and Planck Experiments

There have now been three space-borne experiments that have mapped the entire sky and reported detections of structure on degree or less angular scales: COBE, WMAP and Planck (see Figure 26.1).

The raw maps are dominated by emission from the galaxy, or from pollution by additional sources in the Universe that lie between us and the time of last scattering. Learning how to remove this interference in order to show a map of the CMB has been a long and complex process having to meet the demands of ever increasing sensitivity, resolution and wavelength coverage of the experiments. We can check on how well we are doing by comparing maps of the same part of the sky, either among the all-sky space-borne experiments, or in diverse patches observed from the ground or on balloon-borne missions.

26.2.1 Data Acquisition

The Wilkinson Microwave Anisotropy Probe (WMAP), named in honour of David Wilkinson, was launched on 30 June 2001 and gathered data for 9 years, see Table 26.1.[2] The release and analysis of that data is referred to as the *WMAP 9-year results release*. The

---

[1] There is a pre-WMAP review and complete compilation of these early experiments by Gawiser and Silk (2000). The NASA-LAMBDA website (2012) maintains a list of all known experiments to observe the Cosmic Background Radiation, and also holds any publicly available data from these experiments.

[2] David Wilkinson, one of the great pioneers of cosmic background radiation experiments, died a year after the launch of what was then called the MAP satellite. The satellite was renamed in his honour shortly thereafter.

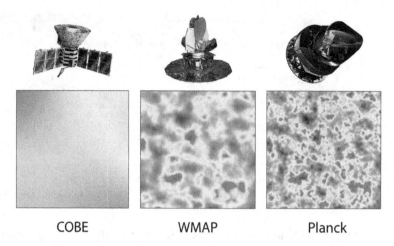

COBE   WMAP   Planck

Fig. 26.1 The same patch of sky mapped by three space-borne experiments. The three panels show 10-square-degree patches of all-sky maps created by space-based missions capable of detecting the CMB. The first spacecraft, launched in 1989, was NASA's Cosmic Background Explorer, or COBE (left panel) which had an angular resolution of $\sim 7°$. The anisotropies in the background radiation came into sharper focus with NASA's next-generation spacecraft, the Wilkinson Microwave Anisotropy Probe, or WMAP (middle panel). This mission, launched in 2001, found strong evidence for inflation, the very early epoch in our Universe when it expanded dramatically in size, and measured basic traits of our Universe better than ever before. Planck, launched in 2009, images the sky with more than 2.5 times greater resolution than WMAP, revealing patterns in the ancient cosmic light as small as one-twelfth of a degree on the sky. Planck has created the sharpest all-sky map ever made of the Universe's CMB, precisely fine-tuning what we know about the Universe.

Source: http://photojournal.jpl.nasa.gov/catalog/PIA16874
(NASA/JPL-Caltech/ESA)

results are presented in Bennett (2013). WMAP can be said to have established the science of 'precision cosmology' in which numerous parameters that describe our standard model of the Universe have been determined with what 50 years earlier would have been regarded as unforeseeable precision.[3]

As the satellite scans the sky in a pre-set pattern, data is sent back to Earth and assembled into maps. The COBE data was stored as maps on the faces of a face-centred cube. The COBE map had 6144 pixels, while the later BOOMERANG experiment scanned a region of the sky with a map of $\sim 200\,000$ pixels. For vastly larger data sets of WMAP (millions of pixels) and Planck (tens of millions of pixels) in multiple wavebands, this map projection would prove cumbersome. There were several alternatives available by the year 2000, but

---

[3] In June 2012, Charles Bennett and the WMAP team were awarded the notable *Gruber prize*. The citation states that 'The Gruber Foundation proudly presents the 2012 Cosmology Prize to Charles Bennett and the Wilkinson Microwave Anisotropy Probe team for their exquisite measurements of anisotropies in the relic radiation from the Big Bang – the Cosmic Microwave Background. These measurements have helped to secure rigorous constraints on the origin, content, age, and geometry of the Universe, transforming our current paradigm of structure formation from appealing scenario into precise science.'

Parameter[a]	LFI (Radiometers)			HFI (Bolometers)					
Centre frequency (GHz)	30	40	70	100	143	217	353	545	857
Effective frequency[b] (GHz)	28.4	44.1	70.4	100	143	217	353	545	857
Wavelength (mm)	10.6	6.8	4.25	3.0	2.1	1.4	0.85	0.55	0.35
Detectors	4	6	12	8	11	12	12	3	4
FWHM (arcmin)	33.16	28.09	13.06	9.59	7.18	4.87	4.7	4.73	4.51

Table 26.2 Planck instruments for data acquisition.

[a] From Planck Collaboration I (2013, Table 2) unless otherwise stated.
[b] From Planck Collaboration I (2013, Table 6).

it was the tree-structured HEALPix tessellation of the sky of Górski *et al.* (2005) that was adopted.[4]

The COBE successors, WMAP and Planck are now defunct, but have had a dramatic impact on attaining the goal of precision cosmology. An overview of the history and achievements of WMAP is presented in the on-line document of Greason *et al.* (2012) with the final data analysis in 2013 (Bennett, 2013). WMAP had achieved Dave Wilkinson's life long goal of producing all-sky polarisation maps of the CMB (Kogut *et al.*, 2003). These were to prove useful in the analysis of the Planck early data of 2013, until Planck was to release its own polarisation data and papers in 2015.

As regards Planck, much of the process, starting with the acquisition of the data (see Table 26.2) and going through to the final deliverables, CMB maps, covariance functions software and so on, is described in the *Planck Explanatory Supplement*. There are two versions of the Supplement (Public Release 1): an electronic version and a print version (Planck Collaboration ES, 2013).[5]

The Planck Collaboration released its first results on the temperature anisotropies derived from the first 15.5 months of data acquisition in 2013. This included the simultaneous release of 29 papers and the first data products. These are summarised in the first of the 29 papers, Planck Collaboration I (2013, *Overview of Products and Scientific Results*). The second data release with another batch of papers took place in 2015.

26.3 CMB Maps and Correlations

One of the goals of the space-based experiment is to make a full-sky map of the CMB, which inevitably involves dealing with contamination of the CMB data by the radiation from the galaxy and other extragalactic sources. A ground-based or balloon-based

---

[4] There were in the late 1990s a number of alternative suggestions for sky pixelations. These are reviewed by Górski *et al.* (2005), who themselves had put forward the first draft of HEALPix with a 'HEALPix primer' in 1999 (Górski *et al.*, 1999). The current version of the primer is Górski *et al.* (2010).

[5] Many space-based experiments tend to give every element of the project a TLA (three-letter acronym), and sometimes four-letter variants. The Glossary of the *PLA Explanatory Supplement* (Planck Collaboration ES, 2013, 328pp. in the printed version) defines over 130 acronyms ('PLA' = Planck Legacy Archive), while the Glossary of Greason *et al.* (2012, 189 pages) *WMAP 2012 Nine-year Explanatory Supplement* lists about 500 acronyms.

experiment will choose a limited area of sky so as to minimise those contributions. As a consequence we find space-based experiments making maps over a wide range of frequency bands and developing software systems to model and remove these unwanted foregrounds. That is not to say the non-space experiments did not have to deal with foregrounds: high Galactic latitude 'cirrus' clouds had already been discovered in the mid-1980s and had to be taken into account.

In what follows we shall focus mainly on the Wilkinson Microwave Anistropy Probe (WMAP), which published its final result after 9 years of data acquisition in 2013, and the Planck Satellite, which completed its data acquisition in 2013 when it published its first results. The data from WMAP and Planck has been combined with data from contemporary ground-based experiments, notably the Atacama Cosmology Telescope (ACT: Dünner et al. (2013); Sievers et al. (2013)) and the South Pole Telescope (SPT: Hou et al. (2012); Story and Reichardt (2013)).

26.3.1 Data Acquisition

We begin with an overview of the data acquisition and how it is processed to make maps of various kinds and how parameterised models are fitted to these maps. The ultimate goal is the determination of the cosmological parameters. The process is complex and the details of how things are done vary from experiment to experiment. We shall try to keep the discussion as generic as possible, leaving the details of individual procedures to the original papers. However, we shall frequently refer to the series of 'first data release' papers from the Planck mission since this provides a homogeneous and detailed view of one particular experiment as described by a single group: the Planck Collaboration.

The Data Pipeline

Most of the data in CMB experiments involves stacking measurements from a series of scans of regions of the sky. In the all-sky experiments the sky is not scanned uniformly: data is acquired as the satellite and its detectors spin about an axis that may or may not be fixed. This gives rise to an inhomogeneous almost circular pattern of sampling of the sky. A given point on the sky may be observed many times with intervals of time between measurements and, if such scans are relative to fixed poles, the poles will be sampled more often than the equatorial regions of the scan pattern (see Figure 26.2). Other experiments might do raster or boustrophedontic scan patterns.[6]

Cleaning Raw Data

The acquired 'raw' data will suffer from instrumental artefacts, perhaps the simplest of which is blurring by the optical system. To this must be added interference from instrumental noise and effects from drifts in detector sensitivity. The list of such data

---

[6] *Raster*: scanning an area in lines from left to right, as in writing.
Boustrophedonic: scanning an area from left to right and and back from from right to left, in the manner of ploughing a field.

Fig. 26.2   Planck sky coverage for data acquired in two phases of the mission: the 'nominal mission' (left) and the 'cryogenic mission' (right). The grey scale represents the total integration time (measured in units of seconds pe square degree). Processed for clarity in grey-scale.
From Planck Collaboration I (2013, Fig. 5). *(Reproduced with permission from A&A. © ESA.)*

pollutants is particular to each experiment, and much of this, but by no means all, can be calibrated in a laboratory prior to data acquisition.[7] Detailed understanding of this aspect of an experiment is what distinguishes a good experiment from a great experiment.

The resulting frequency maps are shown in Figure 26.3. This is the data we have to work with: it is necessary from this point to uncover the signal that is due solely to the cosmic background radiation.

It is apparent from the figure that there is substantial pollution of the CMB data by the galaxy and other non-CMB sources. The contributions from all non-CMB sources to the maps are referred to as *foregrounds*. It should be recognised that these foregrounds are valuable astronomical data in their own right.

Separating out Foregrounds

The data in such an experiment comes in as a time ordered sequence of measurements ('time ordered information', or 'TOI') as the data acquisition proceeds. There will be one set of time ordered information for each of the frequency channels. The result of cleaning this raw data is a stream of clean data ('time ordered data', or 'TOD'), the output of which is then assembled into maps which are stored for analysis. There will be one map for each of the frequency bands of observation, though there may be variants on each map resulting from different data models.

Eliminating contributions from these foregrounds is one of the most delicate tasks in the data reduction: failure to do this correctly would lead to systematic errors. The general approach is to use only the data itself without bringing in information for external data sets. However, the data could be combined with simulations to make maps of the individual components that make up the foregrounds and the CMB map. The creation of the maps of the individual components involves the difficult process of separating out the different astronomical components that contribute to the overall maps. As noted above, these are described as the *astronomical foregrounds* and come from a variety of sources.

---

[7]   See, for example, Planck Collaboration ES (2013, section on TOI processing) for an overview of what this involves in relation to the Planck mission.

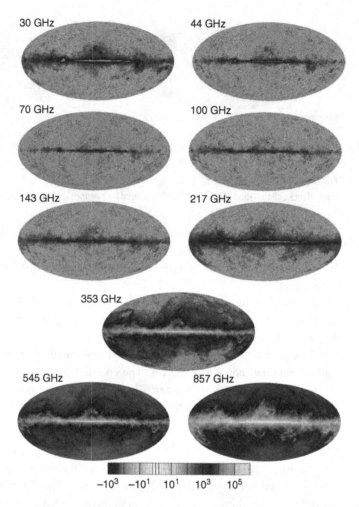

Fig. 26.3   Frequency maps of temperature relative to the mean. The 30—353 GHz. maps show δT in μK. The 545 and 857 GHz maps are in units of kJy str^{-1}.
The colours and labels have been changed for clarity in monochrome reproduction.
From Planck Collaboration I (2013, Fig.9). *(Reproduced with permission from A&A. © ESA.)*

The following diffuse foregrounds can be identified:

1 Dust: thermal emission.
2 Dust: tiny spinning grains.
3 CO emission lines (at high frequencies).
4 Free-free radiation.
5 Synchrotron radiation.
6 Cosmic infrared background (CIB).
7 Sunyaev–Zel'dovich (SZ) secondary CMB distortion.

In addition to this there are contributions from identified point sources and galaxy clusters.

There are several different ways of separating out these components and using different separation methodologies will lead to statistically consistent, but nonetheless different maps.

Data Products

When all this is done we emerge with a set of *data products*. The astronomical data is a set of maps of the cosmic microwave background and the separated foregrounds, and analyses of these. In addition to the astronomical data, there are the software products which were used to analyse the data, such as computer programs to generate simulations of the sky.

26.4 Reconstruction of the CMB Map

We have seen in the previous section some of the techniques that are available to process the data from one frequency channel to make a map at that frequency. The results are shown for the Planck data channels in Figure 26.3. Somewhere in those maps is buried the signal from the cosmic background radiation. The issue we address now is how to isolate the one component we want.

The photons we are primarily interested in for cosmology emanate from the last scattering surface at a redshift of ~ 1000. However, the flux along any line of sight consists of photons not only from the cosmic background radiation, but also from numerous other sources such as galaxies and clusters of galaxies, unresolved point sources, and a variety of contributions from our own galaxy such as emission from galactic dust, to name but a few. Furthermore, the light from the last scattering surface is deflected on its way by the gravitational influence of intervening mass inhomogeneities which serve to distort the actual distribution of temperature on the last scattering surface.

A casual look at the sky maps within each of the frequency bands shows the extent of the problem: the signal we are looking for is, in many parts of the sky, the smallest of all the contributions to the photon flux. To extract a credible map from this requires a formidable data processing pipeline, and one of the main tasks of the data analysis is to verify the accuracy and integrity of the pipeline. This requires access to models of the emission from the sky over a wide range of frequencies. Because the input of the models can be controlled it is possible to evaluate where the uncertainties may lie and identify potential biases in the analysis.

26.4.1 Cleaning the Maps

The all-sky maps derived from the Planck wavebands are shown in Figure 26.3: these show graphically what the problem is. The three highest frequency channels clearly show little or no manifest evidence of the CBR, and even the lower frequency maps show the imprints

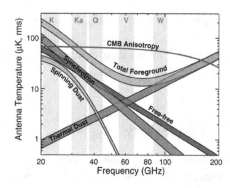 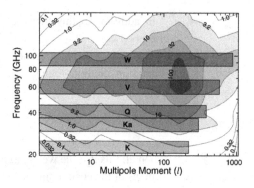

Fig. 26.4 Diagrams from the WMAP archive showing what the five different detector wavebands are measuring. The left figure shows how the foregrounds contribute to the signal received in each waveband. The CMB is most clearly distinguished in the V and W bands. The figure on the right shows which multipoles of the CMB power spectrum covariance function are being picked up by which detector spectral bands. The structure in the contour shapes is due to the acoustic oscillations at the last scattering surface. The darkest area corresponds to the first and dominant peak of the signal. The pictures have been enhanced for greater clarity.
Left: Bennett (2013, Fig. 22). *Reproduced by permission of the AAS.*
Right: http://map.gsfc.nasa.gov/mission/observatory_freq. html *(Courtesy NASA).*

of foregrounds even at high galactic latitudes. We can gain some understanding of this by looking at the left panel of Figure 26.4, which shows, for the WMAP wavebands, what is making the most contribution at which frequencies. It is clear that there is a dominant dust contribution on the highest four frequency bands (> 200 GHz) used by the Planck experiment. This, however, clearly gives a greater leverage in determining the contribution from the thermal dust to the overall Planck CMB signal.

The process of generating foreground-subtracted maps for the Planck mission is described in considerable detail in Planck Collaboration, XII *et al.* (2014), where there is a detailed comparison of several techniques for achieving the result.

The first Planck data release and its products were based on some 1000 simulated maps. The process of creating these maps involved two processes: a sky model comprising a set of maps for each of a number of backgrounds, based on pre-Planck data, and a simulation of the raw data acquisition process of the satellite, its instruments, detectors and their noise characteristics. The first part of this process is referred to as the *Planck sky model* (PSM), while the second set of simulations is referred to as the *full focal plane* simulations. The simulations were generated via Monte Carlo realisations of the models.

SMICA and Other Planck Map Reconstructions

The SMICA[8] reconstruction has been selected as the main 'Planck product', see Figure 26.5. Using maximum likelihood, the various Planck channels are combined with a

---

[8] Spectral matching independent component analysis

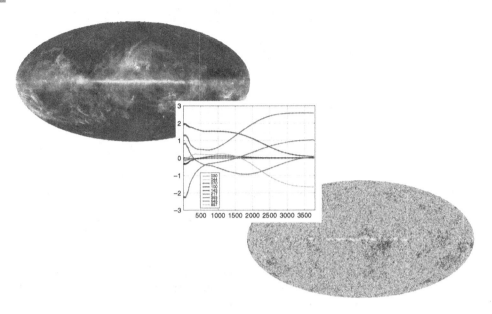

Fig. 26.5   Constructing the Planck SMICA map of the cosmic background radiation. The figure shows a composite
multi-frequency map of the with the contributions from the galaxy being the dominant feature. By using filters that
are both frequency dependent and scale-dependent, as shown in the small graph, the foreground shown in the
original map is suppressed, leaving behind a smooth map that is thought to be a good representation of the CMB
component.
Composite figure made using images from Planck Collaboration ES (2013, Section 6.7).
(Reproduced with permission from A&A. © ESA.)

weighting scheme that is l-dependent, though spatially uniform.[9] Two other maps were
produced using different background removal algorithms, the 'NILC' and 'SEVEM' maps,
see Figure 26.6. SEVEM [10] was used by the WMAP mission for component separation.
The comparison of the methods for reconstructing CMB temperature maps is shown in
Planck Collaboration, XII *et al.* (2014, Fig. 1). The SEVEM method works with the spherical harmonic components of the temperature distribution, while SMICA works on the
pixel data.

26.5 Map-making: HEALPix

A vast amount of software goes into a CMB project. Some is written by the project
teams and moulded to support the special characteristics of their experiments, while other
software has been written by groups outside the project.

Here we shall briefly describe the HEALpix program which, in the first place, serves as
a storage format for data on a sphere, see Figure 26.7. The HEALpix package also contains
a substantial suite of software to analyse and display maps.

---

[9] This process is described in detail in Planck Collaboration, XII *et al.* (2014, Appendix D).

[10] Spectral estimation via expectation maximisation.

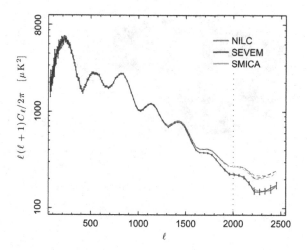

Fig. 26.6   Reconstruction of the temperature fluctuation power spectrum before and after foreground subtraction, using three different methods of dealing with the foregrounds. The broken curves are derived from the raw data, while the full curves show the effect of removal of unwanted foregrounds. Despite the great difference in the methods, the level of consistency is very good.
Adapted from Planck Collaboration, XII *et al.* (2014, Fig. 10.) *(Reproduced with permission from A&A. © ESA.)*

Finally, the map that has to be constructed, and all its by-products, have to be held in some format that can be accessed by users who wish to view and analyse the data. The classical longitude–latitude types of map provide a basic access to data values at a point, but they are not convenient to work with large quantites of data. It is best to divide the sky into a number of patches all having the same area and similar shape.[11] The COBE map was stored digitally as what is now referred to as a *COBE quadrilateralised spherical cube*, or 'COBE sky cube' for short. The celestial sphere was projected onto the six faces of an inscribed cube using a curvilinear area-conserving projection. This divided the sphere into six equal regions, each corresponding to a face of the cube. Each square face is then hierarchically subdivided into nested sub-squares.

The data acquired by COBE had an effective FWHM resolution of $\sim 7°$ and so produced around 6000 pixels in each of three sky maps. WMAP produced five maps, each of $\sim 3 \times 10^6$ pixels, and Planck produced nine sky maps each of $\sim 50 \times 10^6$ pixels. This growth in the volume of data and and the complexity of the analysis of the data required an efficient data representation. Many alternatives to the COBE sky cube were proposed but,

---

[11] It is only the regular Euclidean polyhedra that have identical faces, the one with the greatest number of faces being the icosahedron made up of 20 identical equilateral triangles. Each vertex is the meeting point of five triangles. Five triangles surround each of the poles, while the remaining ten form an equatorial band. There is a circumscribing sphere that touches each of the 12 vertices, and an inscribed sphere that touches each of the triangles in such a way that they are tangential to the sphere. If the 20 triangles were to be subdivided into four identical equilateral triangles we would lose this important contact property: the projection of the edges back onto the sphere would not be regular. This limits the usefulness of the icosahedral mesh. See Tegmark (1996); Teanby (2006).

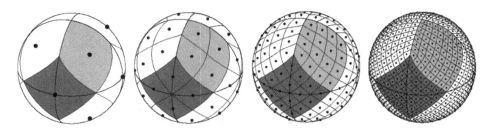

Fig. 26.7   Orthographic view of HEALPix partition of the sphere. The first panel shows the 12 base-tiles that make up the tessellation. The subsequent pictures show the subdivision of each base-tile into a 2 array of sub-tiles, and those sub-tiles themselves are subdivided. Each base-tile is thus the domain of an octree.
Based on Górski *et al.* (2010, Fig. 2.). *(Reproduced by permission of the AAS.)*

for a variety of good reasons, HEALPix[12] has emerged as the dominant standard and has now been adopted as an International Standard for data on a sphere.

Since the advent of HEALPix and its adoption as a standard many ground-based, balloon-based and space-based experiments have adopted the format.[13]

26.5.1 The HEALPix Projections

The HEALPix projection is now an International Standard recognised by NASA, the IAU and other international bodies and is a part of the FITS image format. It is best understood in terms of the final projected map: see Figure 12.2 for a picture of the Earth in HEALPix format, and Figure 26.8 for a description of the coordinate system. There are two fundamental choices in making such a projection: (a) the number of longitudinal subdivisions, N_ϕ and the number of rows, N_θ of base tiles; and (b) the number of subdivisions N_{side} into which the side of each tile is divided to make the sub-tiles. In HEALPix $N_\phi = 4$, $N_\theta = 3$ and $N_{\text{side}} = 4$, but there are many alternative ways of making such subdivisions onto cylinders and cones (Calabretta and Roukema, 2007).

Each of the 12 HEALPix base-tiles has a central point: there are four such points regularly distributed about the equator and four on each of the latitude circles $\theta = \pm\pi/4$. It should be noted that the boundaries between the tiles, when projected back onto the sphere, are not great circles, and so the HEALPix tiles are not spherical quadrilaterals in the strict mathematical sense.

The flattened-out map of Figure 26.8 shows the 12 base rectangles. The 'cap' of the sphere is represented by four triangles which together constitute 4/24s, i.e. 1/6 of the surface area of the sphere. So the latitude θ_* where the 'cap' starts is given by $2\pi(1-\sin\theta_*) = \pi/3$ from which $\theta_* = \arcsin(2/3)$, i.e. $\theta_* = 41.81°$.

The structure on each base-tile provides a *quadtree* on the 2-sphere. Successive levels of the quadtree have $N_{\text{pix}} = 12, 48, 192, 768, \ldots$ pixels in the overall tessellation of the sphere. For recent descriptions of the HEALPix tessellation see Górski *et al.* (2005, 2010).

[12] **H**ierarchical **E**qual **A**rea iso**L**atitude **Pix**elisation.

[13] HEALPix programs and resources are hosted at http://healpix.jpl.nasa.gov/.

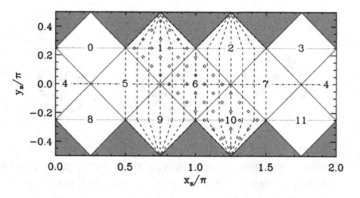

Fig. 26.8 The HEALPix map. The latitudes $| \theta | \leq \pi/2$ are projected onto a cylinder surrounding the sphere, while the corresponding polar caps $| \theta | \geq \pi/2$ are projected onto cones. This defines the enclosing capped cylinder onto which the lines of latitude, θ and longitude, ϕ are projected from the centre of the sphere. The lines of longitude and latitude project onto straight lines. The sphere is divided into four zones of longitude $0 \leq \phi \leq \pi/2$, $\pi/2 \leq \phi \leq \pi$, ... This map is obtained when the cones are cut on these boundaries and the enclosing capped cylinder is unrolled. The resulting 12 tiles are identical and, by convention, are numbered as shown. Each tile is divided into 2×2 sub-tiles and these can be further subdivided in the same way to form a quadtree within each parent tile. The sub-tiles are viewed as a hierarchy of different sized pixels in the map. A central feature of this regular tessellation is that the pixel centres are located on rings of constant latitude. Figure from Górski *et al.* (2005, Fig. 5.) *(Reproduced by permission of the AAS.)*

Working with HEALpix: Angular Power Spectra

Within each of the 12 base-tiles the data is treated like a regular square array of pixels at different resolutions, but it is crucial to note that the base-tile refers to an area on the celestial sphere, not in the projected space of the data description, and so the calculations must be done using spherical harmonics on the sphere. A function $f(\gamma)$ defined in a direction $\gamma = (\theta, \phi)$ on the HEALPix pixel array is represented in spherical harmonics as

$$f(\gamma) = \sum_{l=0}^{l_{\max}} \sum_{m=-l}^{l} a_{lm} Y_{lm}(\gamma). \tag{26.1}$$

Here $l_{\max}$ represents the highest angular frequency at which the function f is to be represented (see Appendix F.1).

The spherical harmonic transform of $f(\gamma)$ is, of course, achieved in the limit $l_{\max} \to \infty$. The estimate $\hat{a}_{lm}$ of the values of the coefficients a_{lm} is the inverse of (26.1), which written in terms of the HEALPix data arrays is

$$\hat{a}_{lm} = \frac{4\pi}{N_{\text{pix}}} \sum_{p=0}^{N_{\text{pix}}-1} Y_{lm}^*(\gamma_p) f(\gamma_p), \tag{26.2}$$

(Górski *et al.*, 2010, Appendices). The function Y_{lm}^* is the complex conjugate of Y_{lm}.[14] The estimator of the power spectrum from these coefficients is then (see Equation F.10)

$$\hat{C}_l = \sum_{m=-l}^{l} \hat{a}_{lm}^2. \tag{26.3}$$

26.6 Data Analysis

Data and image reconstruction is a vast field going back to the reconstruction of radar images during the Second World War. There was a rapid rise in image processing technology during the 1970s due to the advances in computer technology and the growth of the remote sensing industry. In addition, there have been substantial developments in the techniques of image analysis, some of which we shall describe in this chapter.[15]

According to Molina *et al.* (2001) the impetus for sophisticated image restoration methods in astronomy came in 1990 with the realisation that the mirror in the newly-launched Hubble space telescope suffered from serious spherical aberration. The mirror had been polished with a faulty device and was checked using another faulty device with which it passed the quality test (see Wilson (1990) for a complete early discussion of how the problem could have arisen). The high cost of acquiring images (stated to be $100 000 per minute) provided an extreme incentive to correct for the problem in software while a correction lens was built to deal with the aberration. See, for example Baade and Lucy (1990) for a discussion of the use of the Richardson–Lucy method of image enhancement.

26.6.1 Data Reconstruction

The process of data reconstruction is, in effect, one of optimisation of a cost function $\mathcal{C}(x)$ based on the reconstructed function values on some set of points points x. The choice of $\mathcal{C}(x)$ reflects our prejudices about what we consider to be a 'good reconstruction'. By 'optimisation', we simply mean that we shall seek the minimum of the cost $\mathcal{C}(x)$ over the range of values x where it is defined. The formal way of writing this is the rather unedifying:

Minimum of a function defined on a set of points $\{x\}$:

$$\arg\min_{x} \mathcal{C}(x) \overset{\text{def}}{=} \{x \mid \forall y : \mathcal{C}(y) \geq \mathcal{C}(x)\}. \tag{26.4}$$

---

[14] The definition of the spherical harmonics used in HEALPix is the Darwin convention, not the Condon–Shortley convention. The difference lies in the presence or absence of a $(-1)^m$ factor in the expressions for the Y_{lm}. The choice of convention does not affect the calculated power spectrum.

[15] The modern view on signal processing and image analysis is covered in the textbooks of Kay (1993, 1998, 2013) which form an indispensable trilogy of reading for what will be covered relatively briefly here. For a broad introduction to image processing see Solomon and Breckon (2010).

This is a frequently-used notation denoting the place where a function $\mathcal{C}(x)$ attains its minimum value over the support of the function $\mathcal{C}(x)$. This says nothing about the mathematical form of $\mathcal{C}(x)$. It could, for example, be the sum of squares of the deviations of data values from the reconstructed values, or it could be the sum of the absolute values of these deviations. Moreover this refers only to the global minimum of $\mathcal{C}(x)$ over the specified range. It says nothing about what happens outside that range, nor about any local minima that are not the smallest minimum occurring in the range. There is a similar definition for the maximum $\arg\max_x \mathcal{B}(x)$ of what might be called a 'benefit function', $\mathcal{B}(x)$.

The Generic Source-data Model

We shall begin with a simple generic model for cleaning data that underlies most of the data production and analysis that is being used. The model posits that there is a source $\mathbf{s}$ that is being observed and measured, that the measurements $\mathbf{d}$ are polluted by instrumental effects and extraneous influences.

In the case that the data $\mathbf{d}$ is a two-dimensional image of a scene $\mathbf{s}$, the data may suffer from lack of focus, it may be blurred or it may be polluted by noise from, say, cosmic rays if we think in terms of CCD images of astronomical objects. In the case of an image of a source, $\mathbf{s}$ and $\mathbf{d}$ are matrices, the rows and columns reflecting the discretisation of the image by the detector. Note that we do not assume that $\mathbf{s}$ and $\mathbf{d}$ have the same dimensions.

In the case of time ordered data the data measurements come in as a time ordered sequence that can be represented as a column vector $\mathbf{d}$, which is a noise-polluted version of the signal $\mathbf{s}$ that we wish to extract from $\mathbf{d}$. Although the data is time-ordered, it may, as in the case of many CMB experiments, be assembled after cleaning to make a 2-dimensional array which is our cleaned map. [16]

Whatever the dimensionality, the base model is simply expressed by the equation:

The basic data model

$$\underset{\text{measured data}}{\mathbf{d}} = \underset{\text{source}}{\mathbf{H}\ \mathbf{s}} + \underset{\text{noise}}{\mathbf{n}}, \quad \langle \mathbf{n} \rangle = 0. \qquad (26.5)$$

This is represented diagrammatically in Figure 26.9. Here $\mathbf{n}$ represents the noise, which will have zero mean and a calibrated covariance function $\langle \mathbf{n}\mathbf{n}^T \rangle$.

$\mathbf{H}$ is a matrix that describes the degradation of the signal $\mathbf{s}$, and will be made up of contributions from a variety of instrumental and acquisition-time effects that are presumed to have been modelled.[17]

---

[16] In medical imaging, CT scans (computed tomography) and PET scans (positron emission tomography) produce a 3-dimensional image either by scanning in slices or by acquiring data in three dimensions.

[17] The standard examples in imaging science are out of focus images and motion-blurring, where the image processing task is to find a matrix $\mathbf{H}$ that describes how these processes affect the original scene. We use knowledge of $\mathbf{H}$ to de-noise and de-blur the scene. In the simplest of cases these will apply uniformly to the entire scene (the blur happens because of camera-shake). However, in realistic every-day situations both these issues may apply differently to different parts of the scene and $\mathbf{H}$ has to be modelled by an anisotropic point spread function with parameters that depend on position.

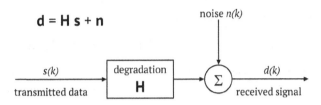

$$\mathbf{d = H\,s + n}$$

noise $n(k)$

$s(k)$ →   degradation **H**   → Σ → $d(k)$

transmitted data   received signal

Fig. 26.9 The basic signal degradation model. A sequential data signal $\boldsymbol{s}$, represented by a sequence of values $s(k)$, is degraded by processes represented by an operator **H**. Noise $\boldsymbol{n} = \{n(k)\}$ is added by external processes to yield a measured signal $\boldsymbol{d}$, represented by a sequence of measurements $d(k)$. ⓒⓒ

There is an important remark to make about the matrix **H**: it may not be square. In other words we may have fewer or more data measurements **d** than we have source points **s**:[18] the number of data samples per point may be more than one. In other words the dimension m of **d** could be larger than the dimension, n of the vector **s**, in which case **H** would be an $m \times n$ matrix and the noise vector would have dimension m: one element for each of m measurements.

In a scan where a map is built up of a sequence of scans that may cross each other, the crossing points will be sampled more often than the points where there is no crossing. Thus, in the Planck sky scan, the poles are scanned more frequently than the equatorial regions.[19] When acquiring time ordered data (i.e. a time sequence of measurements) we would not want to simply average the repeated readings for one patch on the sky since the noise characteristics may have differed between successive scans.

Reconstruction without Noise

Reconstructing **s** from **d** is a statistical process since the time ordered raw data stream is only one particular realisation of the noise. In the absence of noise,

$$\mathbf{d = H\,s,} \qquad (26.7)$$

---

[18] There is a simple, practical example of this in the world of correcting errors in audio signal transmission. Imagine a sequential input signal stream **s** which takes only values -1 or 1, and suppose that instrumental noise might occasionally and randomly turn a +1 into a -1 or vice versa. We can generate a toy model that presents two options for choosing the correct value:

$$\begin{bmatrix} d(k) \\ d(k-1) \end{bmatrix} = \begin{bmatrix} h_0 & h_1 & 0 \\ 0 & h_0 & h_1 \end{bmatrix} \begin{bmatrix} s(k) \\ s(k-1) \\ s(k-2) \end{bmatrix} = \begin{bmatrix} n(k) \\ n(k-1) \end{bmatrix}. \qquad (26.6)$$

The idea is to accept one of $d(k)$ and $d(k-1)$. Our two options, $d(k)$ and $d(k-1)$ are generated in an identical manner using either the pair $s(k)$ and $s(k-1)$ or the pair $s(k-1)$ and $s(k-2)$. Algorithmically, the choice would be made by assigning a probability to each of $d(k), d(k-1)$ that depended on experience or knowledge of the noise, i.e. based on calculating weights for eight possibilities of the triple $s(k), s(k-1), s(k-2)$. This could be done by a Bayesian decision process based on prior knowledge (the training set).

This is an example of a 'feed forward equaliser' for correcting noise in signals.

A banded matrix having the same values down each of the diagonals that are parallel to the main diagonal is referred to as a *Toeplitz matrix*.

[19] In an astronomical CCD image we may stack numerous frames to make a better image: in this case it may be that we have more than one data sample per sky point, but all points will have the same number of samples.

the noise-free reconstruction is given by

$$\mathbf{s} = \mathbf{H}^{-1}\,\mathbf{d}, \quad \mathbf{H} \text{ square and non-singular.} \tag{26.8}$$

This inversion assumes that the noise is zero and that the matrix $\mathbf{H}$ is square and invertible. What if the matrix were not square because some points have been observed more than once?

The inversion may still be possible since $\mathbf{H}$ is considered to be known. Let us suppose that there are M measurements $\mathbf{d}$ and that there are N source points $\mathbf{s}$. Then $\mathbf{H}$ is an $M \times N$ matrix and we can write (26.5) as

$$\underset{N \times M}{\mathbf{H}^T}\; \underset{M \times 1}{\mathbf{d}} \;=\; \underset{N \times M}{\mathbf{H}^T}\; \underset{M \times N}{\mathbf{H}}\; \underset{N \times 1}{\mathbf{s}}, \tag{26.9}$$

where the dimensionality, rows $\times$ cols, of each element is shown for clarity.

The matrix $\mathbf{H}^T\mathbf{H}$ is square $N \times N$ and may be invertible, in which case:

Reconstruction of noise-free signal:

$$\mathbf{s} = (\mathbf{H}^T\mathbf{H})^{-1}\,\mathbf{H}^T\,\mathbf{d}, \quad |\mathbf{H}^T\mathbf{H}| \neq 0. \tag{26.10}$$

(This is a fit to the data, there is no noise.)

This is the solution we are after.[20] This defines a *pseudo-inverse* $(\mathbf{H}^T\mathbf{H})^{-1}\mathbf{H}^T$ of $\mathbf{H}$ which reduces to the familiar form if $\mathbf{H}$ is square and non-singular. The existence of this pseudo-inverse requires that $\mathbf{H}$ be such that $\mathbf{H}^T\mathbf{H}$ be non-singular.

There is another pseudo-inverse that is related to the *singular value decomposition* of a matrix (SVD).

Reconstruction with Noise

If we have a noise contribution then traditionally we would handle the model (26.5) by minimising the quantity[21]

$$\|\,\mathbf{n}\,\|^2 = \|\,\mathbf{H}\hat{\mathbf{s}} - \mathbf{d}\,\|^2. \tag{26.11}$$

---

[20] As an example that has no particular physical significance,

$$\mathbf{H} = \begin{pmatrix} 1 & 0 & 0 \\ 1 & 1 & 0 \\ 0 & 3 & 1 \\ 0 & 0 & 1 \end{pmatrix}, \quad \mathbf{H}^T\mathbf{H} = \begin{pmatrix} 2 & 1 & 0 \\ 1 & 10 & 3 \\ 0 & 3 & 2 \end{pmatrix}, \quad (\mathbf{H}^T\mathbf{H})^{-1} = \frac{1}{20}\begin{pmatrix} 11 & -2 & 3 \\ -2 & 4 & -6 \\ 3 & -6 & 19 \end{pmatrix},$$

$$(\mathbf{H}^T\mathbf{H})^{-1}\mathbf{H}^T = \frac{1}{20}\begin{pmatrix} 11 & -9 & -3 & 3 \\ -2 & 2 & 6 & -6 \\ 3 & -3 & 1 & 19 \end{pmatrix},$$

from which we can quickly verify that $(\mathbf{H}^T\mathbf{H})^{-1}\mathbf{H}^T\mathbf{H} = \mathbf{I}$.

This form of inversion is known as the *Moore–Penrose inverse* and is formally equivalent to finding, for a given matrix $\mathbf{A}_{m \times n}$ the matrix $\mathbf{X}_{n \times m}$ that minimises $\|\mathbf{H}\mathbf{X} - \mathbf{I}_n\|_F$, where $\mathbf{I}_n$ is the $n \times n$ identity matrix, and $\|M\|_F = \Sigma_i \Sigma_j |m_{ij}|^2$ is the *Frobenius norm* of $\mathbf{M} = \{m_{ij}\}$ (Golub and van Loan, 1996, Section 5.5.4).

[21] The norm used here, unless otherwise stated, is understood to be the l_2 norm: $\|z\|^2 = \Sigma_i z_i^2$. To emphasise that this is the l_2 norm we could write it as $\|z\|_2$.

In the formal notation of the subject (see Equation 26.4) we write the result of this minimisation as

$$\hat{\mathbf{s}} = \arg\min_{\mathbf{s}} \| \mathbf{H}\mathbf{s} - \mathbf{d} \|^2. \tag{26.12}$$

Since this is a statistical estimate we denote the estimator of $\mathbf{s}$ by $\hat{\mathbf{s}}$. This can be done by minimising the cost function [22]

$$\mathcal{C}(\hat{\mathbf{s}}) = \| \mathbf{H}\hat{\mathbf{s}} - \mathbf{d} \|^2, \tag{26.13}$$

$$= (\mathbf{H}\hat{\mathbf{s}} - \mathbf{d})^T (\mathbf{H}\hat{\mathbf{s}} - \mathbf{d}), \tag{26.14}$$

which is, of course, just the *mean square error* characterising the deviation of the reconstruction relative to the data.

The minimisation is easily done by differentiating (26.14):[23]

$$\frac{dC(\hat{\mathbf{s}})}{d\hat{\mathbf{s}}} = 2\mathbf{H}^T(\mathbf{H}\hat{\mathbf{s}} - \mathbf{d}) = 0 \quad \Rightarrow \quad \mathbf{H}^T\mathbf{H}\hat{\mathbf{s}} = \mathbf{H}^T\mathbf{d}. \tag{26.15}$$

This gives us the least squares solution for the reconstruction of the signal $\mathbf{s}$:

> **Least squares reconstruction of noisy signal:**
>
> $$\hat{\mathbf{s}} = (\mathbf{H}^T\mathbf{H})^{-1}\mathbf{H}^T\mathbf{d}, \quad |\mathbf{H}^T\mathbf{H}| \neq 0. \tag{26.16}$$
>
> (The solution is unconstrained.)

We recognise this as being the same as the zero-noise Equation (26.10).

In the case of a square matrix $\mathbf{H}$ this reduces to

$$\hat{\mathbf{s}} = \mathbf{H}^{-1}\mathbf{d}, \quad \mathbf{H} \text{ square and } |\mathbf{H}| \neq 0. \tag{26.17}$$

This is an important underlying principle for reconstructing data: most least squares type reconstructions can be derived on this basis.

In practice things are not quite so simple: given this kind of linear model for the experimental degradation of the data, it might be that the matrix $\mathbf{H}$ is ill-conditioned even if $|\mathbf{H}| \neq 0$. We might also want to use a specific model for this 'noise'. Before we deal with that we should derive a few more fundamental results which will enable us to improve on this simplistic least squares optimisation model.

Reconstruction with Noise and Additional Constraints

Suppose we have some prior knowledge about the data, or some prejudice, how do we include that in the task of making a least squares reconstruction like (26.16)? We may, for

---

[22] This simple *least squares estimator* is a special case of the more general *weighted least squares estimator* that we shall deal with later. For the moment, we keep it simple for didactic reasons.

[23] The scalar $\alpha = \mathbf{x}^T\mathbf{A}\mathbf{x}$ written in component form is $\alpha = \sum_{i,j} A_{ij}x_ix_j$. Hence

$$\frac{\partial \alpha}{\partial x_k} = \sum_j A_{kj}x_j + \sum_i A_{ik}x_i, \quad \Rightarrow \quad \frac{\partial \alpha}{\partial \mathbf{x}} = (\mathbf{A} + \mathbf{A}^T)\mathbf{x}.$$

If $\mathbf{A}$ is square symmetric then $\partial \alpha / \partial \mathbf{x} = 2\mathbf{A}\mathbf{x}$. We get our result by putting $\mathbf{A} = \mathbf{H}^T\mathbf{H}$, which is symmetric.

example, know that the source **s** always takes on positive values, or we may believe that the source data is polluted by random noise spikes that distort the reconstruction (26.16). This is, in general, a difficult optimisation problem. However, if the constraint or prejudice can be expressed as a condition on some linear function of the data we can make some progress.

We shall, to be specific, consider constraints that can be expressed as an optimisation condition on some linear function of the source values $\mathbf{c} = \mathbf{G}\mathbf{s}$, where $\mathbf{G}$ is a constant matrix. This means that we will want to minimise a cost function like $\| c \|^2$ while maintaining a fit to the data.

Using constraints of the form $\mathbf{c} = \mathbf{G}\mathbf{s}$ may seem unduly restrictive on the type of constraints that we handle: we cannot for example, easily insist that the reconstructed function be strictly positive in this way. But there are nevertheless some useful constraints that have this form as illustrated by the following 1-dimensional example in which we impose a measure of smoothness on the reconstruction of the source.

If we reconstruct our source at a set of uniformly distributed points $\{x_i\}$, the curvature at the ith point, calculated on the basis of three consecutive points, is $c_i = (s_{i-1} - 2s_i + s_{i+1})$. We can thus write the vector of curvature values **c** schematically as

$$
\mathbf{c} = \begin{pmatrix} \cdot \\ c_{i-1} \\ c_i \\ c_{i+1} \\ \cdot \end{pmatrix} = \begin{pmatrix} -2 & 1 & 0 & 0 & 0 \\ 1 & -2 & 1 & 0 & 0 \\ 0 & 1 & -2 & 1 & 0 \\ 0 & 0 & 1 & -2 & 1 \\ 0 & 0 & 0 & 1 & -2 \end{pmatrix} \begin{pmatrix} \cdot \\ s_{i-1} \\ s_i \\ s_{i+1} \\ \cdot \end{pmatrix} = \begin{pmatrix} \cdot \\ s_{i-2} - 2s_{i-1} + s_i \\ s_{i-1} - 2s_i + s_{i+1} \\ s_i - 2s_{i+1} + s_{i+2} \\ \cdot \end{pmatrix}, \quad (26.18)
$$

where we show only a part of the scheme. We see that $\mathbf{G}$ has the entries $(1 \; -2 \; 1)$ down its principal diagonal and zeros elsewhere. Our basic requirement is that $\sum c_i^2$ should be minimised while at the same time our reconstruction should be a least squares fit to the data.[24] Constraining the curvature of the fitted solution serves an important mathematical purpose in the reconstruction process: it serves to suppress iregularities in the reconstructed data that might be due, for example, to anomalous noise spikes.

We can formalise the process of simultaneously satisfying two constraints through the use of a *Lagrange multiplier*, α, in which we minimise the cost function

$$
\mathcal{C}(\hat{\mathbf{s}}) = \| \mathbf{H}\hat{\mathbf{s}} - \mathbf{d} \|^2 + \alpha \| \mathbf{G}\hat{\mathbf{s}} \|^2. \quad (26.19)
$$

Here α is the Lagrange multiplier whose value is fixed by the need to satisfy the constraint. The first term assures the fidelity of the reconstruction $\hat{\mathbf{s}}$ to the data $\mathbf{d}$, while the second term, called the *regulariser*, captures any prior knowledge we may have about the expected behaviour of the data. The parameter α controls a trade-off between the effects of the two terms and assigns a relative level of importance between the two constraints.

Equation (26.19) can be written out explicitly as

$$
\mathcal{C}(\hat{\mathbf{s}}) = (\mathbf{H}\hat{\mathbf{s}} - \mathbf{d})^T (\mathbf{H}\hat{\mathbf{s}} - \mathbf{d}) + \alpha \, \hat{\mathbf{s}}^T (\mathbf{G}^T \mathbf{G}) \hat{\mathbf{s}}. \quad (26.20)
$$

---

[24] For a continuous function $s(x)$ this is equivalent to making a fit that minimises $\int | s''(x) |^2 dx$. So we are minimising the mean-square second derivative of the function: i.e. requiring a smoother fit. This is the basis of defining spline functions. See Press *et al.* (2007, Sections 18.4,18.5) for more on this.

The formal way of writing the minimisation of the cost function (26.19) is

$$\hat{\mathbf{s}}(\alpha) = \arg\min_{\mathbf{s}} \| \mathbf{H}\,\mathbf{s} - \mathbf{d} \|^2 + \alpha \| \mathbf{G}\,\mathbf{s} \|^2. \qquad (26.21)$$

We can do the same minimisation as we did in getting (26.16):

$$\frac{dC(\hat{\mathbf{s}})}{d\hat{\mathbf{s}}} = 2\mathbf{H}^T(\mathbf{H}\,\hat{\mathbf{s}} - \mathbf{d}) + 2\alpha\mathbf{G}^T\mathbf{G}\hat{\mathbf{s}} = 0, \qquad (26.22)$$

from which we deduce that:

> **Reconstruction in the presence of an additional constraint:**
>
> $$\hat{\mathbf{s}} = (\mathbf{H}^T\mathbf{H} + \alpha\mathbf{G}^T\mathbf{G})^{-1}\mathbf{H}^T\,\mathbf{d}, \qquad (26.23)$$
>
> where the parameter α must satisfy the imposed constraint.

This is subject to the existence of the inverse of the matrix in parentheses.

The Constraint Parameter α

Let us discuss the choice of the parameter α. The residual of the solution can be taken as the N-vector

$$\boldsymbol{\delta} = \mathbf{d} - \mathbf{H}\hat{\mathbf{s}}. \qquad (26.24)$$

With the solution (26.23) we can write this entirely in terms of the data $\mathbf{d}$:

$$\boldsymbol{\delta}(\alpha) = \mathbf{d} - \mathbf{H}\,(\mathbf{H}^T\mathbf{H} + \alpha\mathbf{G}^T\mathbf{G})^{-1}\,\mathbf{H}^T\,\mathbf{d}, \qquad (26.25)$$

where $\boldsymbol{\delta}(\alpha)$ is a vector that depends on the value of α. From this we can produce a scalar figure of overall closeness to satisfying the imposed constraints:

$$c(\alpha) = \boldsymbol{\delta}(\alpha)^T\boldsymbol{\delta}(\alpha). \qquad (26.26)$$

We aim to find a value of $\alpha = \alpha_0$ so that

$$\sqrt{c(\alpha_0)} = \| \mathbf{n} \| \pm \epsilon, \qquad (26.27)$$

where ϵ is the tolerance within which our solution is required. The value of α_0 can be found via any of a variety of methods, such as the Newton–Raphson method. One of the problems is that some methods of implementing this process may be subject to large oscillations. We shall discuss some of the main alternatives below, but before doing that we present the simplest iterative solution.

Iterative Solution

Consider first the case without any constraint. The mean square deviation of the reconstructed data $\hat{\mathbf{s}}$ from the data $\mathbf{d}$ is $\Delta\,(\hat{\mathbf{s}}) = \| \mathbf{d} - \mathbf{H}\hat{\mathbf{s}} \|^2$. The goal is to iterate successive estimates $\hat{\mathbf{s}}^k$ until $\Delta\,(\hat{\mathbf{s}}) = 0$, and then we have our solution.

The gradient of $\Delta(\hat{\mathbf{s}})$ with respect to $\hat{\mathbf{s}}$ is $\nabla_{\hat{\mathbf{s}}}\Delta = -2\mathbf{H}^T(\mathbf{d} - \mathbf{H}\hat{\mathbf{s}})$, and so if we head 'downhill' from the estimate $\hat{\mathbf{s}}^k$ we will, in principle, get an improved estimate:

$$\hat{\mathbf{s}}^{(k+1)} = \hat{\mathbf{s}}^{(k)} - \beta\, \mathbf{H}^T(\mathbf{d} - \mathbf{H}\hat{\mathbf{s}}^{(k)}), \tag{26.28}$$

$$= \hat{\mathbf{s}}^{(k)} - \beta\left[\mathbf{H}^T\mathbf{d} - \mathbf{H}^T\mathbf{H}\hat{\mathbf{s}}^{(k)}\right], \tag{26.29}$$

where β is a *stepping parameter* controlling how far we step downhill at each iteration.

We can do likewise with the constrained Equation (26.21) by writing the deviation of the solution (26.23) from the data as

$$\Delta\,(\hat{\mathbf{s}}) = \|\,\mathbf{d} - \mathbf{H}\hat{\mathbf{s}}\,\|^2 + \alpha\|\,\mathbf{G}\hat{\mathbf{s}}\,\|^2. \tag{26.30}$$

This gives the iteration process

$$\hat{\mathbf{s}}^{(0)} = 0, \tag{26.31}$$

$$\hat{\mathbf{s}}^{(k+1)} = \hat{\mathbf{s}}^{(k)} + \beta\left[\mathbf{H}^T\,\mathbf{d} - (\mathbf{H}^T\mathbf{H} + \alpha\mathbf{G}^T\mathbf{G})\,\mathbf{s}^{(k)}\right], \tag{26.32}$$

which is the obvious generalisation of (26.29). The stepping parameter β is referred to as the *relaxation parameter*.

26.6.2 Regularisation Methods

Producing clean image data from noisy images is a widespread need, be it in medical imaging, astronomical observation, military applications of imaging or in financial modelling where the data is not even an image. Consequently the information on data de-noising or data reconstruction is scattered over numerous journals with relatively little cross-fertilisation and considerable re-invention. This also means that what is essentially the same process often has several different names.

Here we take a glance at three well known methods of data de-noising (or data reconstruction): one of the oldest methods of image reconstruction: Tikhonov regularisation, a commonly used one; maximum entropy reconstruction; and one of the more recent ones, total variation regularisation.

Tikhonov Regularisation

When looking at Equation (26.21) and other forms of data reconstruction, we see that these reconstructions have the generic form

$$\hat{\mathbf{s}}(\alpha) = \arg\min_{f} \|\,\mathbf{d} - \mathbf{H}\mathbf{s}\,\|^2 + \alpha\|\,\mathbf{G}\mathbf{s}\,\|^2, \tag{26.33}$$

where we have written the additional constraint simply as $\mathbf{G}\mathbf{s}$. As we have seen in (26.23), this equation is equivalent to solving

$$\left(\mathbf{H}^T\mathbf{H} + \alpha\mathbf{G}^T\mathbf{G}\right)\hat{\mathbf{s}} = \mathbf{H}^T\,\mathbf{d} \tag{26.34}$$

for $\hat{\mathbf{s}}$ given $\mathbf{d}$.

In this situation $\mathbf{G}$ is referred to as the *Tikhonov matrix*. When the constraint term involving $\mathbf{G}$ is zero, i.e. when $\alpha = 0$, the solution reduces to the standard least squares solution without regularisation. As α gets larger $\mathbf{G}$ has a greater influence on the solution. Since $\mathbf{G}$ is usually a localised operation on the samples (as in the earlier example of controlling the local curvature to suppress instabilities in the solution), it tends to suppress the high frequency components of the noise, and, inevitably, of the signal.[25]

Tikhonov regularisation goes under a variety of different names and is closely related to *singular value decomposition* (SVD). See Press *et al.* (2007, Sections 18.4, 18.5) for a broader discussion of Tikhonov regularisation. An alternative to Tikhonov regularization is the Backus–Gilbert method (Press *et al.*, 2007, Section 18.6). While the Tikhonov method seeks to succeed by adding a smoothness constraint into the problem, the Backus–Gilbert method focuses on the stable solutions to the problem.

Other Regularisation Methods

Generally, the regularisation term $\mathbf{G}^T\mathbf{G}$ is chosen to be localised so as to minimise the adverse effects of high frequency noise in the reconstruction. However, the regularisation term does not have to be of this form nor does it have to be localised in this way.

Perhaps the most commonly used alternative is the *maximum entropy regularisation* in which the cost function is an entropy cost, a measure of the uncertainty in the image. The use of entropy for this purpose was due to Frieden (1972) and was promoted in astronomy by Gull (Gull and Daniell, 1978) and by Skilling (Skilling *et al.*, 1979; Skilling and Bryan, 1984).

If the image data is measured at N_s points, the entropy cost of the entire image is $\sum_{i=1}^{N_s} s_i \log s_i$ and the reconstruction is the solution of

$$\hat{\mathbf{s}}(\alpha) = \arg\min_f \| \mathbf{d} - \mathbf{H}\mathbf{s} \|^2 + \alpha \sum_{i=1}^{N_s} s_i \log s_i. \tag{26.35}$$

Unlike most other forms of regularisation, the maximum entropy reconstruction guarantees a solution that is everywhere positive. The downside is that it is a nonlinear optimisation problem and so has to be solved iteratively.

Another method that also takes account of the total dataset in the formulation of the cost function is *total variation regularisation* in which the cost function is the sum of the magnitudes of the pixel to pixel variations $\| \mathbf{D}\mathbf{s} \| = \sum_{i=1}^{N_s} | (\mathbf{D}\mathbf{s})_i |$:

$$\hat{\mathbf{s}}(\alpha) = \arg\min_f \| \mathbf{d} - \mathbf{H}\mathbf{s} \|^2 + \sum_{i=1}^{N_s} | (\mathbf{D}\mathbf{s})_i |. \tag{26.36}$$

There are various definitions for the variation operator $\mathbf{D}$, the simplest of which is the absolute value of the difference between a pixel and the mean of its neighbours, or, in a time series $Ds_i = | s_i - s_{i-1} |$.

This is also known as *total variation de-noising* and is just a measure of the total 'jiggliness' of the data. In recent years there have been some important developments in this technique both for noise removal and for image segmentation. As with maximum entropy, this is a nonlinear optimisation and so it has to be performed iteratively.

---

[25] The first use of Tikhonov regularisation to create a map of the CMB was probably that of Klypin *et al.* (1992) in their analysis of the data from the Relikt satellite. In that paper they simply chose $\mathbf{G} = \mathbf{I}$, the identity matrix.

26.6.3 Statistical Regularisation

There is a different, less mechanistic, view of data reconstruction that sees the process of reconstruction as a statistical one of determining the source **s** from the data **d**. The problem is phrased as asking what is, in a probabilistic sense, the most likely source **s** given the data **d**. In other words, we could seek to find the solution **s** that maximises the conditional probability density $\mathbb{P}(\mathbf{s}\,|\,\mathbf{d})$. This is the solution that maximises the likelihood that **s** is a consequence of having measurements **d**.[26]

Maximum a posteriori Reconstruction (MAP)

By Bayes theorem (Section 25.1) we can write

$$\mathbb{P}[\mathbf{s}\,|\,\mathbf{d}] = \frac{\mathbb{P}[\mathbf{d}\,|\,\mathbf{s}]\,\mathbb{P}[\mathbf{s}]}{\mathbb{P}[\mathbf{d}]}, \tag{26.37}$$

with which, disregarding the normalisation factor $\mathbb{P}(\mathbf{d})$,[27] we have

$$\underset{\text{posterior likelihood}}{\log \mathbb{P}[\mathbf{s}\,|\,\mathbf{d}]} = \underset{\text{likelihood of data given model}}{\log \mathbb{P}[\mathbf{d}\,|\,\mathbf{s}]} + \underset{\text{prior}}{\log \mathbb{P}[\mathbf{s}]}. \tag{26.38}$$

The first term on the right-hand side contains the data, while the second depends only on having a prior model for the source.[28]

The optimal solution to this equation is the one that maximises the left-hand side:

$$\hat{\mathbf{s}}_{MAP} = \arg\max_{\mathbf{s}} \ \log \mathbb{P}[\mathbf{s}\,|\,\mathbf{d}], \tag{26.39}$$

$$= \arg\max_{\mathbf{s}} \ \log \mathbb{P}[\mathbf{d}\,|\,\mathbf{s}] + \log \mathbb{P}[\mathbf{s}]. \tag{26.40}$$

The first term on the right, in maximising the likelihood $\log \mathbb{P}[\mathbf{d}\,|\,\mathbf{s}]$, ensures the fidelity of the reconstruction, while the second imposes the constraint on that maximisation arising from any prior knowledge we might have about what it is we are trying to reconstruct. The constraint has nothing to do with the data on which the reconstruction is based.

This is a Bayesian interpretation of the standard regularisation principle that we started with in Equation (26.21). This form of reconstruction is referred to as the *maximum a posteriori* (MAP) estimate of **s**, which is conventionally denoted by $\hat{\mathbf{s}}_{\mathrm{MAP}}$ which is the most probable source, given the data.

Making the Data Model Explicit

There is an aspect of Equation (26.38) that is not immediately apparent: when we do this kind of reconstruction there is an implicit underlying model which provides a parameter

---

[26] The restoration of images by Bayesian methods can be traced back to Richardson (1972) and to Hunt (1977, see his Appendix). The present discussion follows Hunt. See also Kay (1993, Ch. 11,12) for a more detailed exposition.

[27] $\mathbb{P}(\mathbf{d})$ is sometimes referred to as the *evidence* since it depends only on what knowledge you have: i.e. the data.

[28] The terms $\sum_{i=1}^{N_s} s_i \log s_i$ in Equation (26.35) and $\sum_{i=1}^{N_s} |(\mathbf{D}\mathbf{s})_i|$ in Equation (26.36) are, in effect, priors for their respective methods of reconstruction.

set that is to be determined. In fitting a line through a set of points our first choice model might be a straight line, which provides us with two parameters that are to be determined: the slope and an intercept. Alternatively, we might have chosen a quadratic fit, or some other different underlying model.

If we were to rewrite (26.38) so as to make this explicit we could write

$$\log \mathbb{P}[\mathbf{s} \mid \mathbf{d}, \mathcal{M}] = \log \mathbb{P}[\mathbf{d} \mid \mathbf{s}, \mathcal{M}] + \log \mathbb{P}[\mathbf{s} \mid \mathcal{M}], \qquad (26.41)$$

where $\mathcal{M}$ represents the model, or a set of parameters that reflect the model and that are to be determined.

In its original form (26.38) we are implicitly assuming that the model $\mathcal{M}$ is something that is fixed. The alternative is to suppose that the model has to-be-determined parameters. In other words that there exists an underlying probability density $\mathbb{P}[\mathbf{d}, \mathbf{s}, \mathcal{M}]$ to which we can apply Bayes theorem.

Starting from $\mathbb{P}[\mathbf{d}, \mathbf{s}, \mathcal{M}]$ we can write an expression for the probability $\mathbb{P}[\mathbf{s}, \mathcal{M} \mid \mathbf{d}]$ of the source $\mathbf{s}$ and model $\mathcal{M}$, given the data $\mathbf{d}$:

$$\mathbb{P}[\mathbf{s}, \mathcal{M} \mid \mathbf{d}] = \frac{\mathbb{P}[\mathbf{d} \mid \mathbf{s}, \mathcal{M}] \; \mathbb{P}[\mathbf{s} \mid \mathcal{M}] \; \mathbb{P}[\mathcal{M}]}{\mathbb{P}[\mathbf{d}]}. \qquad (26.42)$$

We argue, as before, that $\mathbb{P}[\mathbf{d}]$ is a constant, it is what we have, and we can argue that $\mathcal{M}$ has to be the same for all models in the absence of any source or data. So we can write

$$\log \mathbb{P}[\mathbf{s}, \mathcal{M} \mid \mathbf{d}] = \log \mathbb{P}[\mathbf{d} \mid \mathbf{s}, \mathcal{M}] + \mathbb{P}[\mathbf{s} \mid \mathcal{M}], \qquad (26.43)$$

where, now, we are going to maximise this with respect to both the source data and the model parameters: i.e. we are finding the most probable source and model (Frieden, 1972; Pina and Puetter, 1993).

26.6.4 The Simple Gaussian Model

At this point the relationship between Equation (26.40) and our starting point (26.19) is not manifestly evident. So we now show how to perform the maximisation in Equation (26.40) in the important special case that all statistical quantities are Gaussian with known covariances. We will first rewrite (26.40) for the Gaussian case and then we perform the maximisation following the procedure used in solving (26.20).

We can construct a simple Gaussian model to make that link as follows. Our model for the data $\mathbf{d}$ given the $\mathbf{s}$ is

$$\mathbf{d} = \mathbf{H}\mathbf{s} + \mathbf{n}. \qquad (26.44)$$

We suppose, for simplicity, that the noise and the source have Gaussian distributions having zero mean and covariances $\mathbf{C}_n = \langle \mathbf{n}\mathbf{n}^T \rangle$ and $\mathbf{C}_s = \langle \mathbf{s}\mathbf{s}^T \rangle$,

$$\mathbf{n} = \mathbf{d} - \mathbf{H}\mathbf{s} : \; N(\mathbf{0}, \mathbf{C}_n) \quad \text{and} \quad \mathbf{s} : \; N(\mathbf{0}, \mathbf{C}_s). \qquad (26.45)$$

Written out explicitly, these distributions are multi-variate normal distributions:

$$\mathbb{P}[\mathbf{d} \mid \mathbf{s}] \propto \exp \; -\tfrac{1}{2} (\mathbf{d} - \mathbf{H}\mathbf{s})^T \mathbf{C}_n^{-1} (\mathbf{d} - \mathbf{H}\mathbf{s}), \qquad (26.46)$$

$$\mathbb{P}[\mathbf{s}] \propto \exp \; -\tfrac{1}{2} \mathbf{s}^T \mathbf{C}_s^{-1} \mathbf{s}. \qquad (26.47)$$

With this (26.40) reads:

Maximum *a posteriori* reconstruction of source field:

$$\hat{\mathbf{s}}_{MAP} = \arg \max_{\mathbf{s}} (\mathbf{d} - \mathbf{H}\mathbf{s})^T \mathbf{C}_n^{-1} (\mathbf{d} - \mathbf{H}\mathbf{s}) + \mathbf{s}^T \mathbf{C}_s^{-1} \mathbf{s}. \qquad (26.48)$$

We recover our starting point (26.20) if we put $\mathbf{C}_s^{-1} = \alpha \mathbf{G}^T \mathbf{G}$ and $\mathbf{C}_n = \mathbf{I}$, the identity matrix.

Suppose that the joint distribution of k noise samples $\mathbf{n}$ can be modelled by a multivariate Gaussian distribution:

$$\mathbb{P}[\mathbf{n}] = \frac{1}{\sqrt{(2\pi)^k |\mathbf{C}_n|}} \exp -\frac{1}{2}\left\{\mathbf{n}^T \mathbf{C}_n \mathbf{n}\right\}, \quad \mathbf{C}_n = \langle \mathbf{n}\mathbf{n}^T \rangle. \qquad (26.49)$$

The noise is assumed to have zero mean: $\langle \mathbf{n} \rangle = 0$. Using the result from matrix theory:

$$|e^{\mathbf{X}}| = e^{\mathrm{tr}\,\mathbf{X}}, \qquad (26.50)$$

we can write[29]

$$|\mathbf{C}_n| = \exp(\mathrm{tr}\,\mathbf{ln}\mathbf{C}_n), \qquad (26.51)$$

$$\mathbb{P}[\mathbf{n}] = (2\pi)^{-k/2} \exp -\tfrac{1}{2}\left\{\mathbf{n}^T \mathbf{C}_n \mathbf{n} + \mathrm{tr}\,\mathbf{ln}\mathbf{C}_n\right\}. \qquad (26.52)$$

With our data model $\mathbf{d} = \mathbf{H}\mathbf{s} + \mathbf{n}$ we can rewrite this as

$$\mathbb{P}[\mathbf{d}\,|\,\mathbf{s}] = (2\pi)^{-k/2} \exp -\tfrac{1}{2}\left\{(\mathbf{d} - \mathbf{H}\mathbf{s})^T \mathbf{C}_n (\mathbf{d} - \mathbf{H}\mathbf{s}) + \mathrm{tr}\,\mathbf{ln}\mathbf{C}_n\right\}. \qquad (26.53)$$

Compare this with Equation (26.48).

With this, the maximum likelihood reconstruction of the map is obtained by maximising the likelihood, $\ln \mathcal{L} = \ln \mathbb{P}[\mathbf{d}\,|\,\mathbf{s}]$, with respect to $\mathbf{s}$ and is given by

$$\hat{\mathbf{d}} = (\mathbf{H}^T \mathbf{C}_n^{-1} \mathbf{H})^{-1} \mathbf{H}^T \mathbf{C}_n^{-1}\, \mathbf{d}. \qquad (26.54)$$

The MAP Solution

We can perform the maximisation (26.48) following the same path as when dealing with (26.20), which resulted in the estimator (26.23). The only difference is that now we have the covariance matrices appearing in the expression. The result is

$$\left(\mathbf{H}^T \mathbf{C}_n^{-1} \mathbf{H} + \mathbf{C}_s^{-1}\right) \hat{\mathbf{s}}_{MAP} = \mathbf{H}^T \mathbf{C}_n^{-1}\mathbf{d}. \qquad (26.55)$$

If we put $\mathbf{C}_n = \mathbf{I}$ and $\mathbf{C}_s^{-1} = \mathbf{G}^T \mathbf{G}$ we recover the simple Tikhonov regularisation equation (26.34). We note the presence of the inverse covariance matrices $\mathbf{C}_n^{-1}$ and $\mathbf{C}_s^{-1}$ in this equation.

Equation (26.55) can be rewritten in a slightly different form which does not require the inversion of the covariance matrices by using the *Woodbury formula* (see On-line Supplement *Matrices*). This gives:

---

[29] The logarithm of a matrix $\mathbf{A}$ is denoted by a boldface $\mathbf{ln}$ to emphasise that $\mathbf{ln}\mathbf{A}$ is itself a matrix. The result of the trace operator tr is a scalar, and so this is not boldfaced.

MAP reconstruction for Gaussian fields:

$$\hat{\mathbf{s}} = \mathbf{C}_s \mathbf{H}^T (\mathbf{H}\mathbf{C}_s\mathbf{H}^T + \mathbf{C}_n)^{-1} \mathbf{d}, \tag{26.56}$$

which is computationally more efficient.

Equation (26.55) can be put into yet another form, again by using the Woodbury formula:

$$\hat{\mathbf{s}} = (\mathbf{H}^T\mathbf{H})^{-1}\mathbf{H}^T \left[\mathbf{I} + (\mathbf{H}\mathbf{C}_s\mathbf{H}^T\mathbf{C}_n^{-1})^{-1} \right]^{-1} \mathbf{d}. \tag{26.57}$$

This looks more complicated than the previous version of the MAP reconstruction, but it opens up an important route to understanding what is happening in the reconstruction.

We recall from (26.16) that the least squares reconstruction with no additional constraints was $\hat{\mathbf{s}}_{\text{LSQ}} = (\mathbf{H}^T\mathbf{H})^{-1}\mathbf{H}^T\mathbf{d}$. Here we have the same term, but in the MAP reconstructions it is modified by the term in square brackets. This term involves the matrix product $(\mathbf{H}\mathbf{C}_s\mathbf{H}^T)\mathbf{C}_n^{-1}$, which describes the ratio of the signal $(\mathbf{H}\mathbf{C}_s\mathbf{H}^T)$, via the matrix $\mathbf{C}_s$, to the noise, via the matrix $\mathbf{C}_n$. Schematically we can think of this as

$$\text{source} = \frac{\text{least squares data reconstruction}}{1 + (S/N)^{-1}}, \tag{26.58}$$

which tells us that when the signal to noise is high, we recover the least squares reconstruction, but as the signal to noise deteriorates we move away from the least squares reconstruction. This is a direct reflection of the fact that in high signal to noise situations, the prior information is not of much use, while in the case of poor signal to noise, the prior information is vital.

We shall show later that, under some quite general conditions, this is the same as the parameterised spatial Weiner filter.

Weighted Least Squares Reconstruction

An important aspect of the MAP reconstruction is that it shows that the optimal weighting for the data is the inverse covariance function of the noise $\langle \mathbf{n}\mathbf{n}^T \rangle$. We have arrived at what is, in effect, a weighted least squares reconstruction of the sky signal using the presumably known noise covariance function. Of course, we do not know what the covariance function of the source model $\mathbf{C}_s$ is, so that has to be modelled and parameterised and the parameters determined as a part of this Bayesian model.

Preconditioned Conjugate Gradient Algorithms

If we put $\alpha = 0$ in Equation (26.55) we come back to the well-known unconstrained least squares fit to the data for which the solution is

$$\hat{\mathbf{s}}_{\text{MAP}} = \left(\mathbf{H}^T\mathbf{C}_n^{-1}\mathbf{H} \right)^{-1} \mathbf{H}^T\mathbf{C}_n^{-1}\mathbf{d}. \tag{26.59}$$

The difficulty with this, and with many of the previously discussed approaches that involve inverting matrices, is that the data might have 10^6 to 10^8 or more pixels, and so the computational effort of handling the inversion of $\mathbf{H}^T\mathbf{C}_n^{-1}\mathbf{H}$ is prohibitive. We have to use some sophisticated techniques to produce the map from the data while using reasonable computational resources. This is a widely discussed problem, which can be addressed by any of a number of iterative techniques. We shall outline one family of techniques involving the use of *preconditioning*. See Golub and van Loan (1996, Ch.10 and in particular Section 10.3) for a complete discussion of several approaches to iterative solutions of linear systems.

Let us go back to our starting point of the noise-free equation (26.7), $\mathbf{d} = \mathbf{Hs}$, and its solution (26.8), and consider what to do in the situation where the dimensionality of the vectors is so huge that it makes the direct inversion impractical.

Preconditioning involves transforming the coordinates from $\mathbf{s}$ to $\mathbf{y}$ with a suitable linear transformation $\mathbf{s} = \mathbf{Ty}$, where $\mathbf{T}$ is invertible. Typically, we might choose $\mathbf{T}$ to be an approximate inverse of $\mathbf{H}$, and use that as a starting point of an iteration. Once we have decided on $\mathbf{T}$, we solve $\mathbf{T}^T\mathbf{HTy} = \mathbf{T}^T\mathbf{d}$ to find a solution $\hat{\mathbf{y}}$ with which we can obtain $\hat{\mathbf{s}} = \mathbf{T}^{-1}\hat{\mathbf{y}}$. The matrix $\mathbf{TT}^T$ is called the *preconditioner* (van Loan, 1987).

One important strategy is to select $\mathbf{T}$ to be an approximate inverse of $\mathbf{H}$ and iterate the solution. The approximate inverse could, in some circumstances, simply be the diagonal matrix whose elements are the inverse of the corresponding diagonal elements of $\mathbf{H}$ (i.e. $T_{ii} = H_{ii}^{-1}$). On the other hand it might be a stage in the inversion of the matrix through Cholesky decomposition. There is an entire literature on preconditioned least squares discussing the choice of $\mathbf{T}$ and efficient strategies for doing the computation.

26.6.5 Spatial Weiner Filter

Here we shall start with the assumption that the number of observations m is the same as the number of sample points n, so all matrices are square $n \times n$.

We have seen in the MAP reconstruction (26.55) the role played by the covariances of the source $\mathbf{s}$ and the noise $\mathbf{n}$:

$$\mathbf{C}_n = \langle \mathbf{nn}^T \rangle, \tag{26.60}$$

$$\mathbf{C}_s = \langle \mathbf{ss}^T \rangle. \tag{26.61}$$

Typically, both these covariance matrices will have the dominant elements down their main diagonal and the off diagonal elements will fall off rapidly as we go further away from the main diagonal. The elements of these matrices are the expected products of the values at two points of the data sample and the covariance reflects the extent to which having a particular value at one point influences the value at another. Generally, at least in the simple cases, we can suppose that these functions are only functions of the distance between the places where the data samples are taken.

We can view $\mathbf{C}_n$ as the amplitude of the noise and $\mathbf{C}_s$ as a measure of the signal strength in the vicinity of a given point. A quantity of great importance is the *signal to noise ratio*, which indicates the level at which the signal can be distinguished within the data. In our restoration of the data we wish to optimise (i.e. minimise) the ratio of noise to signal. That suggests that the constraint matrix $\mathbf{G}$ be chosen so that

$$\mathbf{G}^T\mathbf{G} = \mathbf{C}_s^{-1}\mathbf{C}_n. \tag{26.62}$$

This is the *noise-to-signal matrix*, i.e. the inverse of the *signal to noise matrix*.

In this case our reconstruction becomes:

Parameterised Weiner filter:

$$\hat{\mathbf{s}} = (\mathbf{H}^T \mathbf{H} + \alpha \mathbf{C}_s^{-1} \mathbf{C}_n)^{-1} \mathbf{H}^T \mathbf{d}. \tag{26.63}$$

The case $\alpha = 1$ is the (spatial) Weiner filter.

See Equation 26.23: this is referred to as the *parametric Weiner filter*, and if we set $\alpha = 1$, this is the important *Weiner filter*. Usually this is presented in Fourier space, but starting from this formulation of the Weiner filter we can make transformations to Fourier space, wavelet spaces or any other basis we require.

We should note that setting $\alpha = 1$ does not in fact result in an optimal reconstruction: the reconstruction for $\alpha = 1$ does not satisfy the noise level requirement that $\| \mathbf{Hs} - \mathbf{d} \|^2 = \| \mathbf{n} \|^2$ (see Equation 26.27). On the other hand, $\alpha = 1$ does minimise $\mathbb{E}[(\mathbf{s} - \hat{\mathbf{s}})^2]$, albeit to some other value.

It will not have escaped the notice of the diligent reader that we do not know what $\mathbf{C}_s$ is until we have solved the problem, so, operationally, Equation (26.63) has an element that we do not know. There might be a temptation to deal with this iteratively: assume, for example, that $\mathbf{C}_s^{-1} \mathbf{C}_n = \mathbf{I}$, and pick a reasonable value for α in order to get the iteration started. However, as pointed out by Press *et al.* (2007, Section 13.3, p554), this naïve iteration does not converge to the required field. (See also Press *et al.* (2007, Section 18.5) for how to make an iterative scheme that works.)

27 Parameterising the Universe

It has been found that six parameters are sufficient to describe the physical properties of our Universe. All the other parameters can be derived from these. These are derived from the power spectra of the temperature and polarisation maps of the CMB sky. The characteristic shapes of these curves are the signature of the Universe.

We need to understand how we go from the CMB map to the power spectrum. This involves the spherical analogue of Fourier analysis, which is the description of the sky in terms of Legendre functions and their generalisations.

The power spectra have strong features that are sensitive to different combinations of the cosmological parameters. This means that we can determine the parameters as accurately as the noise in the power spectrum allows. Of the two polarisation modes, E and B, the discovery of the B-mode will yield information about a seventh parameter that will measure the gravitational waves that would have come from the earliest moments of the Universe, the period of inflation, if there was one.

27.1 The Cosmic Parameter Set

Cosmology was described by Sandage (1970)[1] as 'a search for two numbers', the Hubble constant, H_0, and the deceleration parameter, q_0. Neither was known to within a factor of two. It is astonishing that we now believe we can determine almost a dozen parameters, some with a precision of around 10%, and many to better than 1%. We have come to the era of *precision cosmology*, in which large groups of scientists handle vast data sets using sophisticated models and statistical techniques.

The possibility of accurately determining the values of the cosmological parameters that define our Universe merely by using the observations of the Cosmic Microwave Background (CMB) is quite extraordinary. We do not have to observe the expansion of the system of galaxies, we do not even have to know their distances: the CMB alone can tell us most of the things that define our Universe at large.

Having said that, we shall see that other entirely independent data sets, based on the distribution of galaxies and on the expansion of the Universe as measured by observations of Type Ia Supernovae, serve to provide quite independent and very accurate corroborating

---

[1] Sandage in this excellent article on the cosmological distance scale, quotes H_0 in the range $50 - 100 \ \mathrm{km \ s^{-1} \ Mpc^{-1}}$, and $q_0 = 1.0 \pm 0.4$. 40 years later Riess and Macri (2011) gave the value 69.32 ± 0.80.

measures of these parameters. Combining these datasets yields even greater accuracy. It is very satisfying that our knowledge and understanding of the Universe is not based on a specific interpretation of observations of the one phenomenon.

What makes this possible is our understanding of the physical processes that took place in the early Universe, and in particular those processes that were taking place just before and during the period, the *epoch of recombination*, when the cosmic plasma was becoming neutral. The light we see as the CMB radiation has passed through the emerging cosmic structure on its way to us. It falls into and climbs out of growing gravitational potential wells, and passes through the hot gas in galaxy clusters. We can learn about these phenomena by studying the CMB, and they have to be taken into account when analysing the CMB.

Sachs and Wolfe (1967) were the first to recognise the importance of observing temperature fluctuations and calculate the effect with a fully relativistic linearised theory describing a Universe with small deviations from homogeneity. This important phenomenon is now known as *the Sachs–Wolfe Effect*. This was followed by an article of Rees and Sciama (1968) who presented another mechanism for generating temperature fluctuations in which the time-dependent potential of forming large scale structures would change the frequency of cosmic background photons.

27.1.1 CMB Experiments

The Cosmic Microwave Background is a very smooth and isotropic temperature field with temperature T_0 on which are superposed small deviations in temperature ΔT. Once the noise in the signal has been removed together with the interfering sources of radiation such as dust from within our galaxy, the result is that the values of ΔT are closely Gaussian distributed. The fluctuations are, however, correlated: the measurements at two close-by points on the sky are not independent. The main tool used to describe the temperature fluctuation field is the power spectrum of the ΔT, the components of which are given by a set of coefficients C_i:

$$\Delta T(\mathbf{q}) = T(\mathbf{q}) - \bar{T} = \sum_{l,m} a_{lm}, Y_{lm}(\mathbf{q}), \quad C_l = \langle |a_{lm}|^2 \rangle. \tag{27.1}$$

The principal measurements of CMB analysis are the power spectra for the temperature and polarisation maps. These CMB power spectra are, in effect, 'signatures' of the Universe which encode the values of the basic parameter set. The details of the analysis of temperature fluctuations are found in Appendix F.1 and Section 22.3.

27.1.2 Features in the Power Spectrum

Figure 27.1 shows how the spectrum of the CMB fluctuations varies if one of the parameters of a reference cosmological model is changed. In the following we look at the physical effects associated with such changes.

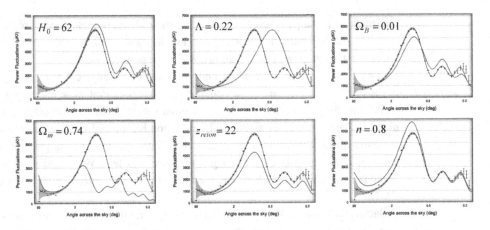

Fig. 27.1 These figures illustrate the results of varying one of the parameters of a flat reference model while the others are kept fixed. The model has $H_0 = 73$ km s^{-1}Mpc^{-1}, $\Omega_B = 0.04$, $\Omega_{CDM} = 0.22$, and $\Omega_\Lambda = 0.74$, a power law spectral index with $n = 0.95$ and a re-ionisation redshift of $z_{reion} = 11$. The reference CMB spectrum is shown as the solid line fitted through the observed data points (shown as small dots). The shaded area represents the possible error due to cosmic variance. The solid line is the spectrum that results from changing a different one of the reference parameters in each panel. There are no constraints on the viability of the parameter set. Thus the change in Ω_B results in a very closed model. The variety and size of the changes means that it is relatively easy to fit these different parameters with considerable precision. However all parameters must be adjusted simultaneously, which means that the fitting procedure has to be more sophisticated.

(Courtesy NASA) These images are snapshots from a demonstration on the NASA website
`http://map.gsfc.nasa.gov/resources/camb_tool/index.html`.
This is part of the teacher resources, *"Build a Universe"*, for WMAP outreach.

The Peaks

The values of C_ℓ for a specific data set (Planck Collaboration XIII, 2015) are shown in Figure 22.1 where $\ell(\ell + 1)C_\ell/2\pi$ is plotted against ℓ. Several features of this plot are of note. We notice a set of almost uniformly spaced peaks in the power spectrum: we can easily see seven or eight peaks. Planck Collaboration I (2015a, Table 7) gives the positions of eight peaks at the following ℓ-values:

#1 : 220.0 ± 0.5	#2 : 537.5 ± 0.7	#3 : 810.8 ± 0.7	#4 : 1120.9 ± 1.0
#5 : 1444.2 ± 1.1	#6 : 1776 ± 5	#7 : 2081 ± 25	#8 : 2395 ± 24

The peaks are almost equally spaced, the mean spacing between them being $\delta\ell \sim 310$.

The location and relative amplitudes of the peaks are one of the most important indicators of the parameters of the underlying cosmological model. The expected power spectrum was first computed in the framework of a variety of $\Lambda = 0$ cosmological models, including the Brans–Dicke theory, by Peebles and Yu (1970, Figs. 4,5).

The observed angular scale of the first and largest peak in the power spectrum reflects the material content of the Universe at the time of last scattering of the photons (see Section 20.1.4). We shall denote the redshift of the last scattering surface by z_*.

The first peak is due to the longest wavelength modes that reach their maximum compression just at the time of last scattering of the photons. The pre-recombination behaviour of adiabatic perturbations is determined by the *sound horizon*, which is determined by the adiabatic sound speed in the cosmic medium at that time. The sound speed depends on the relative densities of matter and radiation, which in turn, depends on the redshift of the last scattering surface and on the present ratio of the baryonic matter and radiation densities:[2]

$$c_s(z_*) = \frac{c}{\sqrt{3}} \left(1 + \frac{3\Omega_b}{4\Omega_\gamma}(1 + z_*)^{-1} \right)^{-1/2}. \tag{27.2}$$

The redshift z_* of the last scattering surface is, for any plausible model, almost independent of the cosmological parameters we are trying to determine and, in effect, depends only on the constants determining the atomic transitions that govern the recombination process. Of course, now we are in the regime of *precision cosmology* such details have to be taken into account. The physical size of the sound horizon $r_h(z)$ at a redshift z is given by

$$r_h(z) = \frac{1}{1+z} \int_{1+z}^{\infty} \frac{c_s(u)}{H(u)} du, \tag{27.3}$$

where $H(z)$ is the expansion rate at redshift z (see Section 18.2.4). The angular diameter distance $D_A(z_*)$ to the last scattering surface translates this distance into an observed angular scale.

We can make some simple order of magnitude estimates of the length scale associated with this first peak. Since this is a rough estimate we may as well take the Universe to be spatially flat. If the recombination takes place $400\,000$ yr after the Big Bang, then the scale of the Universe, $cH^{-1} = \frac{3t}{2}$ at that time is ~ 0.18 Mpc (1pc $= 3.26$ ly). The speed of sound in a radiation dominated situation is $c_S \sim c/\sqrt{3}$ and so the sound horizon at recombination is $\lambda_J \sim 0.10$ Mpc. If we set the redshift at the end of recombination to be $z \sim 1000$ this corresponds to a scale of ~ 100 Mpc at the present time.

The look-back time to the recombination is $t \sim 13 \times 10^9$ yr, which translates to a distance of $\frac{3}{2}ct \sim 6000$ Mpc. The relativistic equation for angular diameters then gives a scale of 100 Mpc$/6000$ Mpc radians which is just $1°$.

The measured value for this angle from the Planck satellite data is (Planck Collaboration XVI, 2013, Section 3.1):

Observed angular size of the sound horizon:

$$\text{Planck:} \quad \theta_* = (1.04148 \pm 0.00066) \times 10^{-2} = 0.596724° \pm 0.00038°. \tag{27.4}$$

This is the most precise measurement in cosmology.

This important measurement is highly robust to many side-effects that might blur the data (e.g. gravitational lensing). It is here that we see almost incontrovertible evidence for the existence of a non-zero cosmological constant.[3]

---

[2] The sound horizon is an attribute of adiabatic fluctuations in which the perturbations in the baryonic and radiation components are tightly coupled so as to keep the entropy per baryon unchanged. Only photons and baryons are relevant, dark matter does not affect this.

[3] The possibility of using the angular distribution of the CMB fluctuations to determine cosmological parameters, and in particular to put limits on the value of Λ, was first discussed by Blanchard (1984). However, it is not

The Damping Tail

Prior to the epoch of recombination, adiabatic perturbations smaller than a critical size are damped by photon diffusion (Silk, 1968). The mechanism is simple to understand. While the photon mean free path is smaller than this critical scale, the photons and electrons are tightly bound by Thomson scattering and behave as a single fluid. Adiabatic density perturbations on such scales preserve the entropy per baryon and so the quantity T^3/n_b is a constant. Hence the fluctuation in the temperature field, δT is related to the fluctuation in the baryon density δn_b via $\delta T \propto \frac{1}{3}\delta n_b$. However, if the photon mean free path becomes long enough that photons can random walk their way out of the fluctuation by repeated scatterings, this relationship no longer holds. The photons carry away heat and momentum and this causes the pressure oscillations perturbation to attenuate.

CMB Lensing

When we think of the spectrum of fluctuations we tend to think only of the temperature fluctuations that are revealed at the surface of last scattering and that are due to primordial fluctuations, the scenario discussed by Peebles and Yu (1970). There are other effects that distort the fluctuation spectrum even after the time when the photons were last scattered. One is immediately visible in Figure 27.2: the high-l part of the damping tail lies above the pure CMB contribution.

Parameter Degeneracy

The shape of the correlation function of the temperature fluctuations depends on the cosmological parameters. However, it happens that it is possible to vary two parameters in such a way that each approximately compensates the shape change caused by the other. This is referred to as a *parameter degeneracy*: to determine either of these parameters requires that the degeneracy be somehow broken. See Figure 27.3 which illustrates the degeneracy between Ω_Λ and Ω_m when trying to determine them using the CMB data alone. Use of other data will break the degeneracy. Generally speaking, this degeneracy breaking data will come from some other direction, such as data on the expansion of the Universe from observations of supernovae or baryonic acoustic oscillations. Some degeneracies may be broken by considering other data on the CMB such as polarisation data.

27.1.3 External Data Sources

Several non-CMB related data are also used to determine cosmological parameters. Most of these are local, i.e. they concern data closer than redshifts $z \sim 2$. Some of these methods provide excellent estimates of some of the parameters:

1. Large scale structure: power spectrum and BAO;

only Λ that enters into determining the position of the peaks, so obtaining the value of Λ is not quite as simple as reading a number off a graph.

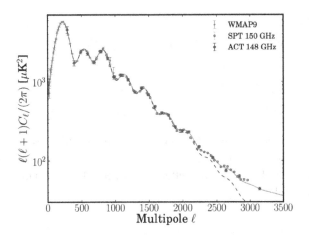

Composite power spectrum using data from the WMAP9, ACT and SPT temperature maps. The angular frequency on the sky is measured by the multipole number l which corresponds roughly to an angle of $\theta_l \sim 180°/l$ on the sky. Note that the horizontal axis is plotted linearly here. The power spectrum amplitude $l(l+1)C_l$ measures the contribution of the mode of frequency l to the total variance of the sky signal.

The solid line shows the best fit model to the ACT 148 GHz data combined with WMAP 7-year data, and the dashed line shows the CMB-only component of the same best fit model. (Note that the vertical axis is logarithmic). From Das *et al.* (2014, Fig. 12). *(Reproduced by permission of the AAS.)*

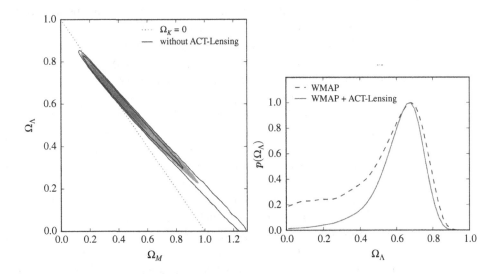

Degeneracy in determining Ω_Λ and Ω_m using CMB alone. The degeneracy must be broken by analysis of another data source. In the left panel, the shaded area is the 95% confidence limit from analysis of the WMAP and ACT lensing data taken together. The lines are the same using the WMAP data alone. Combining the data sets has improved the acceptance area. This is shown in a different way in the right panel which shows the posterior probability distributions for the estimate of Λ from WMAP alone (dashed curve) and for the combined WMAP + ACT lensing data. We see that WMAP by itself would allow $\Omega_\Lambda \neq 0$, while this is strongly disfavoured by the combination data.

(Reproduced from Sherwin et al. (2011, Figure 3) with permission. ©2011 APS, see p.696.)

2. Ly-α forest;

3. Weak and strong gravitational lensing;

4. Structure growth;

5. Integrated Sachs–Wolfe effect (ISW);

6. Supernovae.

We can combine the CMB observations with these observations, though some of the parameters. $\Omega_b h^2, n_s$ and τ are hardly changed by doing so. However, Ω_m and the fluctuation amplitude $\Delta_{\mathcal{R}}$ are better determined. The systematics of such data have to be handled carefully.

27.2 The Basic Parameters

The parameters we seek to determine are summarised in Table 27.1.[4] The table gives what would be considered the best values for these various parameters derived from the 9 years of the WMAP experiment. Since the parameters are determined from combining data from many large observational projects, the model they describe is often referred to as the *concordance model*. The alternative term *benchmark model*, first used by Barbara Ryden in her book (Ryden, 2003), seems somewhat more appropriate and so will be used here.

The table is organised into three sections. The first six values are the basic parameters of the so-called 6-parameter model: $\Omega_b h^2, \Omega_c h^2, \Omega_\Lambda, 10^9 \Delta_{\mathcal{R}}^2, n_s$ and τ. These establish the model from which all the other parameters can be derived.[5] These are sufficient to fit a model to the level that we can say that there is no evidence or need to add further cosmological parameters into the analysis. When the B-mode polarisation is detected there will be a need to add a 7th parameter, the ratio r of the tensor to scalar fluctuation amplitudes. The other cosmological quantities are derived from these six.

The three Ωs need no explanation but it is worth passing comment on the other three: the re-ionisation optical depth, τ, and the parameters defining the primordial spectrum A_s, n_s.

The re-ionisation optical depth is the contribution of the total optical depth between us and the surface of last scattering. By definition, if there were no re-ionisation, then $\tau = 1$.

The re-ionisation optical depth is the fraction of photons that are scattered after they have left the last scattering surface. This $\tau = 0.08$ means that 92% of the photons from the surface of last scattering reach us directly. and 8% have been scattered on their way here.

---

[4] Since there are many ways of analysing a given dataset and many combinations of data from different sources there can be no unique authority giving definitive values of these parameters. Table 27.1 is the table based on the 9 year, final, WMAP data release combined with selected information from a number of other astronomical sources that break degeneracies in or refine the WMAP parameter determination. We can easily understand the logic behind doing that: we want to use the best data available for any of the parameters, and, moreover, bringing in a parameter determination from another source means that there is one less parameter to determine from within the CMB data itself.

[5] The Planck Collaboration I (2015a, Table 9) uses θ_{MC}, the angle subtended by the first peak of the temperature fluctuation power spectrum, in place of Ω_Λ. This angle is the most precisely measured quantity in cosmology.

Table 27.1 Basic cosmological parameter set
(WMAP 9-year observations from Hinshaw and Larson (2013, Table 4)).

Parameter	Name	Approximate value[a]
$\Omega_b h^2$ [b]	baryonic matter density[c]	0.02223 ± 0.00033
$\Omega_c h^2$	cold dark matter density [d]	0.1153 ± 0.0019
Ω_Λ	dark energy density (cosmological constant)	$0.7135^{+0.0095}_{-0.0096}$
$10^9 \Delta_{\mathcal{R}}^2$	spectrum amplitude at $k_0 = 0.002 \, \mathrm{Mpc}^{-1}$	2.464 ± 0.072
n_s	spectrum slope at $k_0 = 0.002 \, \mathrm{Mpc}^{-1}$	0.9608 ± 0.0080
τ	re-ionisation optical depth	0.081 ± 0.012
t_0	age of Universe (Gyr)	13.772 ± 0.059
H_0	Hubble constant ($\mathrm{km s^{-1} Mpc^{-1}}$)	69.32 ± 0.80
σ_8	density fluctuation amplitude at $8h^{-1}$ Mpc	$0.820^{+0.013}_{-0.014}$
Ω_γ	photon density	5.0×10^{-5}
Ω_ν	neutrino density	3.4×10^{-5}
Ω_r	radiation density	8.2×10^{-5}
Ω_k	curvature	0
z_{reion}	epoch of re-ionisation	10.1 ± 1.0
z_{eq}	epoch when densities of matter and radiation were equal	3293 ± 47
w	dark energy equation of state index	-1.12 ± 0.12

[a] Values given here include priors from BAO data and determinations of H_0 by Riess and Macri (2011).
[b] $h = H_0/100 \, \mathrm{km s^{-1} Mpc^{-1}}$.
[c] Sometimes denoted by $\omega_b = \Omega_b h^2$.
[d] Sometimes denoted by $\omega_c = \Omega_c h^2$.

The matter power spectrum (Section 8.2.5) is characterised by an amplitude, A_s (called $\Delta_{\mathcal{R}}$ in Table 27.1), and a slope n_s:

$$P_m(k) = A_s \left(\frac{k}{k_0} \right)^{n_s - 1}. \tag{27.5}$$

This is related to the quantity σ_8, which is the rms fluctuation in the total matter density within a sphere of $R = 8h^{-1}$ Mpc:

$$\sigma_8^2 = \frac{1}{2\pi^2} \int k^3 P_m(k) \left[\frac{3 j_1(kR)}{kR} \right]^2 \frac{dk}{k}, \tag{27.6}$$

where $j_1(x)$ is the spherical Bessel function of order 1. The matter comprises baryons, cold dark matter and massive neutrinos.

There is a degeneracy between A_s and τ which is broken by using the polarisation data at low l.

The Confirmation of Dominant 'Dark Energy'

We see here the dominance of Ω_Λ over all other contributions to the energy density of the Universe. The first demonstration of this, by Perlmutter (1998); Riess and Filippenko

(1998); Schmidt (1998), came with the Hubble diagram for distant Type Ia Supernovae (see Section 28.2). Subsequent data from supernova surveys and data from other sources has only strengthened that conclusion. Importantly, the value of the cosmological constant derived from the supernova Hubble diagram is consistent with the values determined from the analysis of the CMB.

The problem with simply adding in a constant Λ-term to the Einstein equations is that its introduction is entirely ad hoc. On any argument from physics, the required value for this constant is vastly greater than would be considered reasonable in the cosmological context. Moreover, the apparent value when expressed as a dimensionless density parameter, Ω_Λ, is coincidentally close to one. So it is not surprising that there have been many attempts to 'legitimise' the need for an additional component in the Universe in terms of a 'reasonable' modification of some existing theory.

However, with the lack of any viable theory for dark energy, we can only produce a variety of cosmological models in which we parameterise the deviation from the standard theory and use the data to put bounds on the parameters. The hope is that we can put strong enough constraints on models for the 'dark energy' to point a way forward for other experiments and theories.

27.2.1 The Derived Parameters

The data from the first three acoustic peaks is sufficient to determine the main parameters of the ΛCDM model, but measurements at smaller angular scales serve to refine the measurements. On smaller angular scales we are dominated by the process of Silk damping which occurs during and before the last scattering at $z \sim 1100$, after which the Universe became transparent to present day observers. These small scales also provide information about the slope of the power spectrum of these primordial fluctuations. The polarisation measurements are particularly important in establishing that our understanding of the physics at the time of recombination is correct since they establish a relationship between the density fluctuation amplitude and velocity field.

Beyond the 6-parameters there are the so-called *derived parameters*, σ_8, Ω_Λ, Ω_m and the Hubble constant H_0, which describe the expansion and the material content.

H_0: the Hubble Constant

The determination of the Hubble constant cited in Table 27.1 is from Riess and Macri (2009). This result comes from distance determinations of 240 Cepheid variables in galaxies hosting type Ia supernovae. It is a direct determination without reference to the WMAP data. The improvement in the accuracy with which the Hubble constant is determined improves the accuracy of the determination of other cosmological parameters.

CMB observations do not directly determine either the Hubble constant or the age of the Universe. The determination comes from the determination of $\Omega_m h^2$ and the use of the positions of the peaks to find the distance to the surface of last scattering. The age is quite well determined, but the Hubble constant determination depends on assuming a flat ΛCDM model.

27.2.2 The Extended Parameters

Other derived parameters, sometimes referred to as *extended parameters*, impose limits on parameters that underlie the benchmark model, and that, if varied, would give what would currently be regarded as non-standard models. Such non-standard models might have a varying fine structure constant, they might not have a flat space-time geometry, they might have a detectable component of primordial gravitational waves, and so on. It is important to use the CMB data not only to define our benchmark model, but also to impose constraints on models that differ in some fundamental way.

Among these 'extended parameters' are the number of neutrino species, N_{eff}, constraints on the sum of the neutrino masses, σm_ν, and various parameters describing the dark energy. In addition, constraints can be imposed on any curvature of the spectrum, i.e. $n_s/d \ln k$, the primordial helium abundance Y_p, and the so-called *tensor to scalar ratio* $r = \Delta_h^2(k_0)/\Delta_{\mathcal{R}}^2(k_0)$ describing the ratio of the amplitudes of the primordial gravitational wave and matter density power spectra at the fiducial scale $k_0 = 0.002\,\text{Mpc}^{-1}$. These are also 'extended parameters'.

Ω_k: the Curvature of the Universe

Historically, the most important cosmological parameter is the curvature of the Universe: is the Universe flat? This is quantified by the parameter Ω_k that, on the basis of the supernova data is consistent with $\Omega_k = 0$. This parameter constrains models of inflation, and if $\Omega_k < 10^{-4}$ the inflationary paradigm would be falsified (Guth and Nomura, 2012; Bull and Kamionkowski, 2013).

Internal Parameters

There are also a number of non-cosmological parameters that arise from fitting procedures and internal calibrations. While these have no direct physical significance, they are nonetheless of some importance since they influence the reliability of the model that is fitted.

27.3 Fitting Models for the Dark Energy

As we shall see, most of the cosmological tests that are used to determine the parameters describing the Universe are based on a model for the expansion rate, H, expressed as a function of time, $H(t)$, or of redshift, $H(z)$. From H alone we can calculate many of the various observables. The explicit form of $H(z)$ will depend on the various assumptions underlying the model, like the existence of dark energy of a specific kind. Those underlying assumptions will in turn be parameterised and so $H(z)$ inherits those parameters.

Take for example a model of the dark energy having equation of state $p = w\rho$, relating the pressure p to the density ρ. The model will postulate what the functional form of

w might be and from that it might be possible to predict how w will vary with redshift: $w = w(z)$. Since $H(z)$ depends on $w(z)$ it is then possible to ask how the postulated $w(z)$ can match the data.

However, there are many theories for what the dark energy might be and so there are many $w(z)$s. It would be somewhat tedious to test each possible one against the ever changing data. A more economical approach is to parameterise $w(z)$ and use the observations to constrain what forms of $w(z)$ might be allowable. For example, we might fit any of the models

$$w(z) = \begin{cases} w_0 = \text{constant} & \text{like } \Lambda \\ w_0 + w_z z & \text{simplest linear change} \\ w_0 + w_a \dfrac{z}{1+z} & \text{varies between } w_0 \text{ and } w_0 + w_a. \end{cases} \tag{27.7}$$

When $H(z)$ is derived from such a model it will contain explicit reference to those of the parameters w_0, w_a or w_z that are part of this simple dark energy evolution model.

The procedure then is to derive some quantity based on $H(z)$ and fit those parameters by nonlinear least squares fitting or any other of a number of available fitting methods. The following sections describe this process.

For what we later derive to be our *plausible model* (equation (6.106)) we have

$$H(z)^2 \stackrel{\text{flat}}{=} H_0^2 \left[\Omega_m (1+z)^3 + (1 - \Omega_m)(1+z)^{3(1+w_0+w_a)} \exp\left(-3w_a \frac{z}{1+z} \right) \right]. \tag{27.8}$$

It might be useful to bear this in mind when reading the following sections.

We see that, in this particular model, the parameters to be determined from the data are Ω_m, w_0 and w_a. If we make a set of observations $O_i(z_i)$ of objects having measured redshifts z_i, and if our model predicts a value $P_i(z_i \mid \Omega_m, w_0, w_a)$ we can seek to determine the best fitting values of Ω_m, w_0 and w_a by choosing the set that minimise

$$\chi^2(\Omega_m, w_0, w_a) = \sum_i \frac{[O_i(z_i) - P_i(z_i \mid \Omega_m, w_0, w_a)]^2}{\sigma_i(z_i)^2}, \tag{27.9}$$

where σ_i is the error associated with the ith observation. Such a procedure is rather standard, but we can see some practical difficulties here.

The first is to know what the values of the σ_i are for each observation. The second is that with three parameter-values to be determined the minimisation process might be somewhat lengthy unless we use a smart algorithm. This is particularly true if getting the $P_i(z_i \mid \Omega_m, w_0, w_a)$ requires the numerical evaluation of integrals.

28

Precision Cosmology

While the CMB experiments have played a central role in setting precise values for many cosmological parameters, there are other astronomical datasets which can, by themselves, produce remarkably precise results, in particular the Type Ia supernovae (SN Ia) and the baryon acoustic oscillations (BAO). Bringing these together will both serve as mutual confirmation and generally reduce the errors in the estimates of the parameters. This may also highlight areas of tension or discord between estimates of some parameters.

It is important to realise that, had the CMB data not been taken, we would almost certainly have measured the values of the cosmological parameters, possibly with comparable accuracy in many cases, by studying distant objects, e.g. supernovae, and huge catalogues of galaxy positions and redshifts. What is scientifically important is that the CMB data analysis largely rests on a foundation that is based in the physics of the early Universe.

With CMB data, we calibrate fundamental cosmic-length scales by understanding the physics of the power spectrum of the temperature fluctuations, which arise mostly in a narrow redshift band around the time when the Universe was becoming neutral. In contrast, the astronomical determinations tend to be based on calibrations of distant data by detailed understanding of relatively nearer objects, and rarely, if ever, make reference to the CMB.

This duality of approaches is powerful evidence that we are not being somehow fooled by something unknown when coming to our conclusions: the approaches are mutually corroborative. So far there is compelling evidence from both approaches for the existence of copious amounts of both dark matter and for dark energy. Both approaches will play a part in coming to terms with these two great unknowns.

28.1 Observing the Universe

The increased level of precision with which we can determine the parameters that define our Universe has grown as a result of research by thousands of astronomers and physicists. Although a few lines of research have grown to dominate our thinking in regard to cosmological parameters, we have only arrived at that point due to efforts in a large number of different explorations. The culmination of that effort was perhaps the discovery of the acceleration of the cosmic expansion, and the subsequent efforts to pin down the properties of the expansion using high-precision techniques.

Pinning down the parameter values in models for the dark energy is more likely to tell us what the dark energy is rather than to enlighten us as to its nature. But, at least, that process of narrowing down options points in directions where we should go next.

Discovering the nature of the dark energy is regarded as one of the prime targets for future research in cosmology. In 2006 The *Dark Energy Task Force* produced a document (Albrecht *et al.* 2006, 2009)[1] describing what, at the time, could be done to explore the nature of dark energy and set out recommendations for future research. The methods available for increasing the precision of our model were presented and much of what follows is related to that list. However, the focus here will be on the supernovae surveys and the BAO surveys (see Figure 28.1).

28.2 Supernovae

The quest to map the expansion of the Universe via the Hubble's redshift versus distance diagram occupied much of the 20th century. At the start of this quest was Vesto Slipher's compilation of the first two dozen recession velocities of 'nebulae' (Slipher, 1915, 1917) which, when combined with Henrietta Leavitt's establishment of a distance scale using Cepheid variables (Leavitt, 1908; Leavitt and Pickering, 1912), allowed Hubble to plot his epochal velocity–distance diagram. Over the following decades, Milton Humason and others strove to extend the Hubble diagram to greater distances, culminating in the great *HMS Catalogue* of Humason, Mayall and Sandage (Humason *et al.*, 1956). The main problem with interpreting the resulting diagram for cosmology was how to select galaxies that were standard candles whose relative distance could be estimated simply from their 'corrected' relative magnitudes.

It had long been realised that supernovae could outshine their parent galaxy and so be seen at equally great distances, but again the variation in intrinsic brightness prevented them from being used as standard candles. That situation persisted until Branch and Miller (1993) and Phillips (1993) recognised that supernovae of different intrinsic brightness decayed with different time-scales that were strongly correlated with the intrinsic brightness. This opened the door for two great supernova cosmology projects to determine the rate of expansion of the Universe as a function of redshift (Riess and Filippenko, 1998; Schmidt, 1998), thereby establishing the remarkable fact that the cosmic expansion is accelerating.

The cause of this acceleration, be it a cosmological constant or some other agency, is one of the prime targets for future research.

---

[1] To quote from the document: *The Dark Energy Task Force (DETF) was established by the Astronomy and Astrophysics Advisory Committee (AAAC) and the High Energy Physics Advisory Panel (HEPAP) as a joint sub-committee to advise the Department of Energy, the National Aeronautics and Space Administration, and the National Science Foundation on future dark energy research.* See also Kolb (2006) and the later informal update in Kolb (2012).

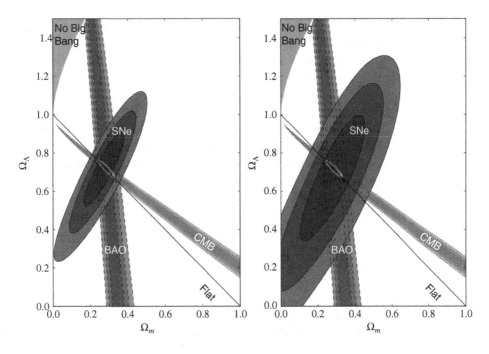

Fig. 28.1   Resolving the degeneracy in determining Ω_m and Ω_Λ separately. Confidence contours are shown for the analysis of
three quite different data sets: CMB, SN Ia and BAOs. While each one has this degeneracy, the nature of the
degeneracy in each case is rather different, allowing an overlap region in which the degeneracy is broken: the joint
analysis yields the small confidence ellipses in the region of intersection. (See also Figure 27.3). The figure on the left
takes account of statistical errors for the supernova data, while the figure on the right also includes the systematic
errors and so has bigger confidence contours.
From the analysis of the supernova 'Union' datasets by Suzuki *et al.* (2012, Fig. 5).
Reproduced by permission of the AAS.

During the first part of the 20th century, after the establishment of the homogeneous
and isotropic relativistic cosmological models and prior to the discovery of the cosmic
background radiation, the key question in cosmology was to decide whether our Universe
would expand forever or simply collapse back to a singularity. Was our space-time open
or closed, or did it lie on the boundary between the two? The only means available at the
time was the Hubble diagram for distant galaxies that could be identified in some way as
'standard candles', and whose distance could be determined independently of their redshift.
The hope was to measure the curvature of the Hubble redshift–distance relationship and
thereby determine the rate of deceleration (or acceleration) of the cosmic expansion. This
would provide a direct measure of the space-time curvature.

The quest to discover the nature of the cosmic expansion via observations of distant
galaxies reached its peak at around the time of the publication of the HMS catalogue
(Humason *et al.*, 1956) of galaxy magnitudes and redshifts. The final conclusion of that
study, based on the cluster data, had been that there was marginal evidence for a deceleration. The authors realised that it would be necessary to understand more about the

evolution of the stellar populations in the galaxies in order to come to any more definitive conclusions.

However, the realisation around 1970 that the interpretation of the Hubble diagram for galaxies would be obfuscated by the ill-understood effects of galaxy evolution made this seem like a depressingly unachievable goal (Gunn and Tinsley, 1975; Tinsley, 1977b, Section 2.7.1).

The break-through in addressing this question came from the use of Type Ia supernovae as distance indicators. These were not standard candles, but a study of their light curves allowed for the calibration of their intrinsic brightness in a relatively straightforward way. With this, it was possible to determine not only the rate of deceleration or acceleration of the cosmic expansion, but also to see how that rate varied with time. The execution of that task was a formidable problem of data acquisition and analysis that would eventually lead to the award of the Nobel Prize in Physics for the two groups who undertook this project. It also provided the first direct evidence for the existence of a non-zero cosmological constant.

2011 Nobel Prize in Physics

The Nobel Prize in Physics 2011 was divided, one half awarded to Saul Perlmutter, the other half jointly to Brian P. Schmidt and Adam G. Riess *for the discovery of the accelerating expansion of the Universe through observations of distant supernovae.*

The review article of Perlmutter (2012) is a fine overview of the subject.[2]

This was certainly one of the great experimental achievements of the 20th century. It was the culmination of the work of many people over the preceding decades and it is worth recounting some of the history of how this came about. The current status of what has become known as 'supernova cosmology' is presented in the compendium of articles edited by Ruiz-Lapuente (2010),[3] where the history is eloquently reviewed by Kirshner (2010).[4] The present chapter is in part based on the Kirshner's detailed account of the history of supernovae cosmology.

28.2.1 Supernovae as Distance Indicators

The potential of supernovae as distance indicators was recognised as long ago as 1939 by Wilson (1939). In 1968 Charles Kowal produced the first Hubble diagram using archival data on Type I supernovae[5] going back to 1895 (Kowal, 1968). At the time of Kowal's

---

[2] Not only does the article talk of the scientific issues, it also gives a wonderful insight into the teamwork aspect of the project.

[3] I should express my gratitude to Pilar Ruiz-Lapuente for her help with several aspects of this chapter.

[4] The spectroscopy of supernovae, which is important for dividing them into different types, is reviewed in detail by Filippenko (1997). The history of supernova research is engagingly reviewed by Trimble (1982, 1983), which also reviews the state of understanding of the supernova phenomenon up until the early 1980s.

[5] Since the work of Minkowski (1941), supernovae have been broadly classified, on the basis of spectroscopy, into Type I, which have no hydrogen lines in their spectra, and Type II, which do have hydrogen lines. However, it later became clear that the Type Is should be further subdivided into three classes: Ia, Ib and Ic. The Type Ias showed lines of singly ionised silicon (SiII), which were absent in Type Ib and Ic spectra. The difference between Ib and Ic was that the Ib spectra showed helium lines, while the Ics did not. This spectroscopic botany

work, the most distant supernovae were a pair of objects in the Coma cluster of galaxies, so there was no chance of discovering any curvature in the Hubble plot. However, Kowal was able to establish that the dispersion in the absolute magnitude estimates from this rather heterogeneous sample was around 0.6^m, despite his being unable to discriminate the Type Ia s from the Type Ib s.

Shortly after that Rust (1974a,b,c) modelled the light curves of Type I supernovae[6] and recognised that the number of days for the brightness of a supernova that had attained a maximum apparent magnitude m_0 to decay in brightness from $m_0 + 0.5$ to $m_0 + 2.5$ was an indicator of its intrinsic brightness. de Vaucouleurs *et al.* (1976), building on earlier work by Rust (1974a), attempted to calibrate and apply this to a sample of 36 nearby ($z < 0.06$) supernovae of types Ia and Ib, but were unable to improve on the scatter in their Hubble diagram. However, they did conclude that . . . *even though available data are insufficient for a conclusive test, the initial rate of decay of Type I supernovae appears to offer a promising tool for testing the cosmological time dilatation effect, but significant results will require additional observations of faint supernovae at $z \geq 0.05$ ($m_0 > 15$).*

28.2.2 Early Work Post-1980

The science methodology that led to the 2011 Nobel Prize did not suddenly appear from nowhere. As with most large projects it was built on a foundation of previous experience from groups working on supernovae as distance indicators. The incentive to use supernovae had originated in the late 1970s as a possible project for the then to-be launched Hubble Space Telescope (HST). In 1986, a group of Danish astronomers initiated a project using CCD detectors on the Danish 1.54 m telescope at La Silla in Chile (Hansen *et al.*, 1987), and using other large telescopes to acquire redshifts of any objects that might be discovered. They devised the strategy for cycling through the search fields on a monthly basis, and developed a real time data analysis pipeline, concepts that were to become standard in the supernova searches that followed during the 1990s.

While they did discover some distant supernovae (Hansen *et al.*, 1989; Norgaard-Nielsen *et al.*, 1989), their rate of discovery of about one supernova per year was far too low to be of any long-term use.[7] The Danish attempt was undoubtedly ahead of its time. They were using a small telescope equipped with small CCDs that had little R-band

---

reflects the mechanism by which the star explodes to form a supernova. Types II, Ib and Ic all explode via core collapse of massive stars ($M > 8\,M_\odot$, while the Ias are simply thermonuclear explosions of a carbon/oxygen white dwarf. Type Ias are generally extremely bright and are seen in both spiral and elliptical galaxies.

[6] Rust (1975) was perhaps the first person to suggest that the Type I supernovae fell into two groups: Type Ia and Type Ib. This was on the basis of his fitting and modelling of supernova light curves: the light curves split into two distinct classes having different rates of decay. Shortly thereafter, de Vaucouleurs *et al.* (1976) used Rust's classification when calibrating Type I supernova absolute magnitudes with a view to determine redshift independent distances.

[7] Their low detection rate was somewhat surprising in the light of the supernova detection rate suggested by Tammann (1979) when discussing the future use of the HST in establishing distances. Tammann's article is interesting in a number of respects, as is the ensuing discussion (*loc. cit.* p285 *et seq.*). In particular, he emphasises the use of HST to study supernovae for cosmology: *The independent determination of Ω/Ω_{crit} and q_0 shall give an estimate of the smoothly distributed, invisible matter in the Universe and test the assumption* $\Lambda = 0$.

response. A few years later, advances in CCD technology would make it possible for others to realise their experimental strategy (Perlmutter, 1995, The Supernova Cosmology Project).[8]

28.2.3 Calibration of Type Ia Supernovae Light Curves

Not all Type Ia supernovae[9] are identical: this is one reason why the scatter in Kowal's early use of a supernova-based Hubble diagram was on the order of ~ 0.6 mag. The key to using supernovae as accurate distance estimators lies in understanding the relationship between supernovae of the same type having different explosion strengths, i.e. differing maximum brightness. As a result of the work done by a number of astronomers investigating the nature of different Type Ia events and their light curves, we can use these supernovae as standard candles having, after calibration, a very small dispersion in intrinsic brightness, ~ 0.1 mag.

As our understanding of the nature of the explosions improves, we develop newer techniques for increasing the accuracy of the distance calibration. Here we shall look at three methods that were particularly important in the Supernova Cosmology projects that went on to win the 2011 Physics Nobel Prize.

Some general remarks should be made about these renormalisations. Firstly, they are purely empirical: there is little or no physical basis for the procedures. They are chosen simply so as to minimise the dispersion of the fit to the Hubble diagram. Secondly, the methods outlined here are simply the ones that provided the first strong evidence that the cosmic expansion was accelerating. The ensuing decades have seen a considerable improvement of these techniques, particularly by moving to the near infrared where the absorption corrections are less of an issue. Finally, these methods only involve comparing the light curves in multiple wavebands, no spectroscopic data is used except to determine the redshift.

Phillips' Calibration

The first steps in re-normalising supernova light curves came with the claim by Pskovskii (1977, 1984) that the rate of decline in the brightness of a supernova depended on its absolute magnitude. Doubts had been expressed about the accuracy of Pskovskii's photographic photometry and so Mark Phillips at the Cerro Tololo Inter-American Observatory (CTIO) took on a crucial re-assessment of this claim using multi-waveband (B, V, I -bands) CCD photometry (Phillips, 1993).

---

[8] The Supernova Cosmology Project team demonstrated in 1995 that it was feasible using such techniques and a megapixel CCD to detect supernovae at redshifts $z \sim 0.5$ (Perlmutter, 1995). It is interesting to note the impact of the CCD detector. The observing run acquired deep images of 54 fields, each covering $\sim 100\square'$ and containing on average some 200 galaxies in the redshift range $z \sim 0.25-0.50$. Two exposures, separated by a couple of days, were made for each field. With good seeing it took a mere 10 minutes of exposure to reach magnitude $R \simeq 23$. The redshift of their supernova, $z = 0.458$, beat the previous record of $z = 0.31$ set by Norgaard-Nielsen et al. (1989).

[9] In what follows we shall use the term 'supernova' to mean specifically a 'Type Ia supernova'. No confusion can arise since these are the only types of current relevance to cosmology.

Pskovskii had simply used the slope of the light curve, β, as measured between the maximum brightness and a recognisable point where the light curve has an elbow. Typically, this covers a period of some 25–30 days after maximum light. Phillips instead used the fall-off in B-band brightness, $\Delta m_{15}(B)$, during a specified period of time from the maximum, which he empirically decided was best set to 15 days.[10] There were two requirements of Phillips' sample, Firstly, the supernova should, for purposes of calibration, be in a galaxy that had a known Fisher–Tully distance. Secondly, the supernova photometry had to be available before maximum light is reached. These requirements reduced his sample to nine events. Figure 28.2 shows his plots of the maximum absolute magnitude of his nine sample events in each of the B, V, I-bands, as a function of $\Delta m_{15}(B)$. The trend is clear, and later use of better distance indicators for the parent galaxy served to provide a better calibration of the light curve.[11]

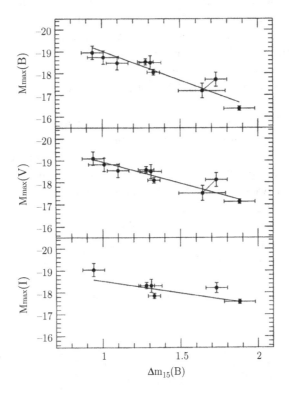

Fig. 28.2 Peak luminosity versus rate of brightness decline for the small supernova sample of Phillips (1993, Fig. 1). The diagram shows the absolute magnitude as a function of decline in brightness, measured 15 days after the maximum brightness, in three wavebands: B, V, and I. To create such a diagram, the supernovae must be observed before they achieve their maximum luminosity.
Reproduced by permission of the AAS.

[10] In his paper, Phillips acknowledges George Jacoby for the suggestion of using $\Delta m_{15}(B)$.
[11] Phillip's B-band graph looks similar to Pskovskii's plots from which he determined the slope β. However, this apparent agreement is now considered to be fortuitous.

There was another advantage of Phillips' approach: the $B-V$ colour of the supernova at maximum light also correlated with its $\Delta m_{15}(B)$. This would allow corrections to be made for extinction by dust in the parent galaxy, as later developed by Phillips *et al.* (1999).

MLCS and Stretch Calibration

The fundamental realisation was due to Phillips (1993), who recognised that the intrinsically fainter supernovae declined in brightness at a faster rate than the brighter ones. This allowed an estimate of the relative absolute magnitudes of supernovae and thus provided a renormalisation of the data to a standard absolute magnitude, see Figure 28.3.

However, cosmological time dilatation is also a major factor in understanding supernova light curves. Even if supernovae all had the same absolute magnitude, the more distant supernovae would appear to evolve more slowly than their nearby counterparts. Thus, for a Type Ia supernova viewed before it reaches maximum light, the rise time at $z \sim 0.5$ will be around 20 days, while at $z \sim 1$ it will be around 30 days. This is important for the scheduling of follow-up observations confirming the discovery of each object.

This suggests that, to a first approximation, simply stretching the time axis of a supernova light curve would help the renormalisation of the light-curves to a frame local to the supernova (Perlmutter, 1997). Putting this together with the Phillips relationship is the basis of the *stretch method* for renormalising supernova light curves. A combination of these two will cause the light curves to sit on top of one another (see Figure 28.3).

The Phillips (1993) $\Delta m_{15}(B)$ method suffered from two simple disadvantages: the supernova had to have been observed prior to the maximum, and only 15 days of the light curve, post maximum, are used, whereas light curves can be tracked for as much as 100 days. An alternative approach was due to Riess *et al.* (1995) who came up with a method that involves matching light curves to templates, the *multi-colour light curve shape*, (MLCS), method. This first version of the method used

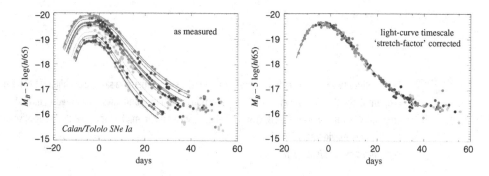

Fig. 28.3 Renormalising Type Ia supernova light curves using the 'stretch' methodology. Presented in various articles by Perlmutter (1999, 2003, 2012). *Permission granted.*

data from one waveband only, but was soon generalised to multiple waveband light curves (Riess *et al.*, 1996), 'MLCSk2'. The parameterised templates were constructed applying a learning algorithm, a classifier, to a wide variety of supernova light curves.[12]

One of the main problems was dealing with the possibility that dust obscuration was playing a role in dimming the light from the supernovae, and, worse still, dimming it in a wavelength dependent way. Introducing observations in more colour bands produced a modification of the basic methods, and there were important proposals for full spectrum modelling (see for example Guy and The SALT2 collaboration (2007)). Nonetheless, as the sample of supernovae has increased it has become apparent that there are still variations among the supernovae that cannot be accounted for by a simple dust model.

28.3 Cosmology with Supernovae

28.3.1 Fitting a Cosmological Model

The analysis of light curves is a complex process, and there are a number of alternative strategies. Here we focus on one particular fairly generic mechanism that is used in the analysis of the Union2.1 data sample (Suzuki *et al.*, 2012), which itself stems from the stretch mechanism of normalising the objects to a standard candle. This is a multi-step process

The first step is to do the stretching by one means or another and get, for each supernova, a stretch measure s. From there we move to getting the absolute magnitude and the distance modulus. This is done by fitting a model relating the measured apparent magnitude, m, absolute magnitude $\mathcal{M}$ and relativistic distance in a relationship of the form

$$m = 5\log_{10}[\mathcal{D}_L(z, \Omega_m, \Omega_\Lambda)] + \mathcal{M} - \alpha(s-1), \quad \mathcal{D}_L = H_0 D_L(z, \Omega_m, \Omega_\Lambda). \tag{28.1}$$

Here, D_L is the luminosity distance for a redshift z in a given cosmological model, and $\mathcal{D}_L$ is the luminosity distance with the present Hubble parameter H_0 factored out (H_0 is the same for the entire sample and so is just a normalisation factor). The Hubble constant is absorbed in $\mathcal{M} = M - 5\log_{10} H_0 + 25$, so in the absence of fitting parameters (i.e. if all supernovae had the same absolute magnitude, $\alpha = 0$) this last equation would be the familiar $m(z) - \mathcal{M} = 5\log_{10} \mathcal{D}_L$ (Perlmutter, 1997, Eq. 3).

The term $\alpha(s-1)$ is the term that brings all supernovae in the sample objects to the same absolute magnitude. The parameter s is the *stretch* which maps the light curve shape onto a standard shape by stretching the time evolution of the light curve. For each supernova in the sample we have measured m, z, s and so the tunable parameters of the model are $(\Omega_m, \Omega_\Lambda, \mathcal{M}, \alpha)$. If it is assumed that the geometry of the Universe is flat, then $\Omega_\Lambda + \Omega_m = 1$, reducing the number of parameters by one. Otherwise the curvature $\Omega_k = 1 - \Omega_\Lambda - \Omega_m$.

---

[12] The article of Riess *et al.* (1996) has complete details of the method.

The parameters $\mathcal{M}$ and α are examples of *nuisance parameters* that enter into tuning the fit, and have to be determined as part of the fitting procedure. Nuisance parameters do not necessarily have any obvious physical interpretation.

If we also wanted to fit an equation of state parameterised by the equation of state parameter w, an additional parameter would enter into the expression for $\mathcal{D}_L$. If w were a function of redshift, a simple model, $w = w_0 + w_a z/(1 + z)$, could be used (Linder, 2003). The simplest case requires a 4-dimensional space in which to search for solutions: $(\Omega_m, \Omega_\Lambda, \mathcal{M}, \alpha)$.

28.3.2 Parameter Determination

The goal is to find the cosmological parameters that lead to a supernova Hubble diagram, such as that shown in Figure 28.4. The technical issue is that the to-be-determined cosmological parameters are needed for every step of the fitting procedure. Establishing the absolute magnitude of the supernova requires knowledge of the cosmological parameters, but this is precisely what we are trying to find out. This is not atypical of many 'hard' optimisation, i.e. 'global extremisation' problems. There is a vast literature on that discussing many ways of doing this and, as usual, Press *et al.* (2007, Ch. 10) has words of wisdom about the subject.

We start with the model shown in Equation (28.3) and interpret it as a prediction m_{mod} for the visual magnitude, given the parameters and the pseudo-absolute magnitude $\mathcal{M}$:[13]

$$m_{\mathrm{mod}} = 5\log_{10}[\mathcal{D}_L(z, \Omega_m, \Omega_\Lambda)] + \mathcal{M} - \alpha(s - 1), \quad \mathcal{D}_L = H_0 D_L(z, \Omega_m, \Omega_\Lambda). \quad (28.3)$$

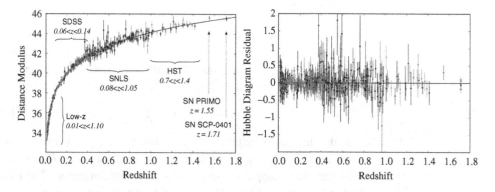

Fig. 28.4 The Supernova Ia Hubble diagram in 2013. The figure is based on a compilation of data by Suzuki *et al.* (2012), with the addition of two subsequently analysed objects having $z > 1.5$. The surveys used in the compilation, with the redshift range they cover, are marked. The fitted model is from Suzuki et al. *loc. cit.* and corresponds to a value $\Omega_\Lambda = 0.729 \pm 0.014$ (68% *confidence level including systematics*).
From Rubin *et al.* (2013, Fig. 5). *Reproduced by permission of the AAS.*
The references to the data points that appeared on the diagram have been removed since they serve no purpose in a monochrome image. These appear in the original paper.

[13] Conley *et al.* (2011) introduce an additional colour related term and write

$$m_{\mathrm{mod}} = 5\log_{10}[\mathcal{D}_L(z, \Omega_m, \Omega_\Lambda)] + \mathcal{M} - \alpha(s - 1) + \beta\mathcal{C}, \quad (28.2)$$

Naïvely, we might write the goodness of fit as a χ^2 value involving all of the data:

$$\chi^2 = \sum_{\text{sample}} \frac{(m_{\text{obs}} - m_{\text{mod}})^2}{\sigma^2}, \qquad (28.4)$$

where m_{obs} is the observed magnitude. The value assigned to the weighting factor σ for each supernova is in itself a major subject of discussion. Amanullah *et al.* (2010) split σ into three components: $\sigma^2 = \sigma_{\text{lc}}^2 + \sigma_{\text{ext}}^2 + \sigma_{\text{sys}}^2$: σ_{lc} is the contribution due to issues that arise from fitting light curves, notably colour corrections; σ_{ext} is the contribution from sources not directly related to the supernova itself, e.g. Galactic extinction, gravitational lensing effects; and σ_{sys}^2 is an estimate of the errors due to systematic effects from differences in the samples.[14]

We seek the parameter set $(\Omega_m, \Omega_\Lambda, \mathcal{M}, \alpha)$ that minimises this. Since all methods of finding the minimum of χ^2 need values of m_{mod} for arbitrary cosmological parameters, it might be as well to evaluate m_{mod} on a grid and interpolate as necessary. This look-up table approach has advantages in that the table need only be calculated once, though it helps to know what parts of the parameter space should be in the grid. It also tends to be inefficient when there are more than three dimensions to the grid.

There is a further difficulty in that (28.4) would only be a good measure of the deviation of the model from the data if the individual data points, i.e. the supernovae, were independent. There are, inevitably, systematic effects that are common to all objects in a given survey, and so mixing data from multiple sources will introduce artificial correlations. There may also be redshift dependent effects.

To take account of correlations the generalisation of Equation (28.4) introduces a *covariance matrix* **C** so that

$$\chi^2 = \Delta\mathbf{m}^T \mathbf{C}^{-1} \Delta\mathbf{m}, \qquad (28.5)$$

where $\Delta\mathbf{m}$ is the column matrix of the values of $(m_{\text{obs}} - m_{\text{mod}})$ for each of the supernovae in the sample. Conley *et al.* (2011, Table 7), discuss the data systematics that go into building the matrix **C**. This includes uncertainties in calibration, contamination by objects that are not in fact of type SNIa, the relationship between the host galaxy and the supernova, and Malmquist bias. Having done that the large covariance matrix has to be inverted and the optimisation proceeds by whichever method one prefers.

There is a final important step, the determination of how good the fit is. The value of χ^2 in $(\Omega_m, \Omega_\Lambda, \mathcal{M}, \alpha)$-space varies around the minimum point that is the proposed 'best fit'. The (frequentist) idea behind using the χ^2 statistic is that the values of the function $\chi^2(\Omega_m, \Omega_\Lambda, \mathcal{M}, \alpha)$ reflect the statistical distribution of the values that one might obtain for the parameter estimates, given a large number of 'trials'. To a first approximation, i.e. near to the minimum, the surfaces of constant χ^2 are ellipsoids, in this case in 4-dimensions.

---

where $\mathcal{C}$ is a measure of the colour of the supernova. This introduces an extra hyper-parameter, β, into the problem.

[14] This is calculated by setting the reduced χ^2 for each sample to unity and has the effect of giving the samples equal weight.

With the probability interpretation these surfaces are surfaces of constant probability in $(\Omega_m, \Omega_\Lambda, \mathcal{M}, \alpha)$-space.[15]

However, we are not interested in assigning confidence intervals to the hyper-parameters $\alpha, \mathcal{M}$. So the χ^2 probability distribution has to be integrated over all possible values of $\alpha, \mathcal{M}$ to get the *marginalised* probability distributions of the parameters we want: Ω_Λ, Ω_m.

28.3.3 'Blind' Parameter Determination

The method used by Suzuki *et al.* (2012) in generating Figure 28.4 is a fitting procedure referred to as a *blindness analysis*. This is claimed to be a way of eliminating systematic bias on the part of the researcher when selecting the starting point for parameter searches.[16] In blindness analysis the researcher cannot tune or guide the analysis by imposing some preconceived values for the results. He/she has to commit to publication of the result when it is delivered.[17] The method is explained in some detail in Amanullah *et al.* (2010); Conley *et al.* (2006).[18] Basically it involves finding the minimum of χ^2 by iterating from some almost arbitrary starting point. At each iteration towards the minimum the values of the two parameters Ω_m, Ω_Λ are updated but not exposed to the researcher so there can be no researcher bias. The researcher can however monitor the hyper-parameters to gauge whether convergence is being achieved. Then, in theory, once convergence is achieved the result is published, no matter how different it is from previous estimates.[19]

28.3.4 Acceleration of the Cosmic Expansion

The *Union* catalogue of Type Ia supernovae, (started by Kowalski and Rubin, 2008) and updated by Suzuki *et al.* (2012) is a compilation of a number of catalogues which have the

---

[15] The aim is to answer the question as to 'what is the probability, P, that this minimum value of χ^2, or a larger one, could arise by chance?'. P is interpreted as the chance of obtaining measurements at least this discrepant from the 'true' model. In the case that the model was the 'true model', we would take P to be the chance of mistakenly rejecting the model we have found. This is the so-called P-value on the basis of which a particular hypothesis is ruled in or ruled out.

[16] A Bayesian prior would achieve a similar goal, perhaps at the cost of a more computationally intensive fitting procedure. The Bayesian procedure attempts to embrace researcher-bias by making it explicit in the form of the prior rather than to try avoiding it. The fact that the prior is explicitly given in the Bayesian analysis invites those who do not like the given prior to re-do the analysis using their own prior (if they have the resources to do that).

[17] Perhaps the most common form of (honest) researcher-bias occurs when making an absolute measurement of some physical parameter. The CMB was a case in point when, shortly after the discovery, a number of groups made measurements of the Rayleigh–Jeans temperature of the spectrum. The process was one of successively subtracting sources of noise that would cause a systematic error in the measurement. The question was when to stop looking for additional things to subtract. See Section 3.1.6, footnote 18.

[18] See also Heinrich (2003b,a) for a wider discussion from a statistics point of view. My own opinion is that researcher bias depends as much on the researcher as anything else, particularly when it comes to sample selection and culling the data. Some people are simply not very good at dealing with data. The way to deal with this is to make the data accessible (in a usable form), with the tools that were used in the analysis.

[19] An example of researcher bias is the researcher who analyses data using a number of statistical tests, and then chooses to publish the result of the one test that is most favourable to his/her point of view. This happened in the past when fewer details of the data and the analysis were made available for scrutiny, and probably still happens today.

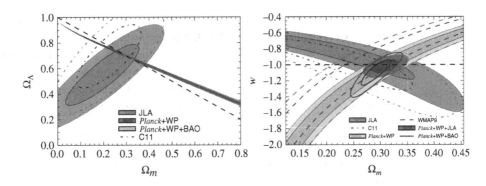

Fig. 28.5 Confidence areas for cosmological parameters from combining SN Ia data, Planck data combined with WMAP polarisation data and the results of BAO analysis. The supernova data, an elongated ellipse, breaks the degeneracy of the Planck+Polarisation+BAO data. *Left* The $\Omega_m : \Omega_\Lambda$ plot. The dashed line is the locus of models for which $\Omega_\Lambda + \Omega_m = 1$. *Right* The $\Omega_m : w$ plot. The horizontal line is the classical cosmological constant, $w = -1.0$. The best fit is at $\Omega_m = 0.303 \pm 0.012$ and $w = -1.027 \pm 0.055$ with $h = 0.685 \pm 0.0127$. From Betoule and Kessler (2014, Figs. 15,16., see Section 7.1 for details). *Reproduced by permission of the AAS.*

same light curve fitting methodology and which have the same data rejection criteria. It includes estimates for systematic errors. Figure 28.4 shows the best fit cosmological model fit from that paper, giving a value for the dark energy contribution to the present cosmic energy density, $\Omega_\Lambda = 0.705^{+0.040}_{-0.043}$ using the SNe Ia only.

Taking into consideration data from other sources, BAO, CMB and H_0, and, under the assumption of a flat cosmological model, yields $\Omega_\Lambda = 0.729^{+0.014}_{-0.014}$. The densities of the dark energy and non-relativistic matter components were equal at the redshift $z_{m\Lambda}$ when $\Omega_\Lambda(z_{m\Lambda}) = \Omega_m(z_{m\Lambda})$. With $\Omega_m(z) \propto (1+z)^3$ and $\Omega_\Lambda(z) = const.$, we have $z_\Lambda = 0.39$. The transition to undecelerated expansion would have taken place at redshift $z \sim 0.75$, corresponding to a look-back time of $\sim 6.6\,\mathrm{Gyr}$. The galaxy redshift surveys of the mid-20th century, only covering redshifts $z < 0.5$, could not have easily detected a non-zero value for Λ, even if the standard evolution-free candles had been available. This has been a triumph for the study of supernova cosmology.

The confidence contours for these determinations of Ω_m and Ω_Λ from this supernova sample and from BAOs amd the CMB are shown in Figure 28.5.

28.3.5 Cosmological Parameters and the Dark Energy

The 'dark energy' is taken to be the stuff that causes the acceleration of the cosmic expansion, and in its simplest form it is the cosmological constant Λ. In the absence of any idea what the dark energy might be we can only explore the evidence for the assumption that Λ is a form of energy whose density remains constant throughout the expansion: it is indeed a constant of nature. If we model the dark energy as a perfect gas with equation of state relating its effective pressure p and energy density ρc^2,

$$p = w\rho c^2, \tag{28.6}$$

then Λ is a gas with $w_\Lambda = -1$. It behaves as though it has negative pressure, not a situation that we a familiar with, nor for that matter something we can easily come to terms with. However, putting $w = -1$ in either the Einstein or Newtonian cosmological equations yields the same result as putting $\Lambda = 0.7$, a result that is in accord with what we observe. When using data to impose constraints on the dark energy, w is simply taken to be a function of redshift, $w(z)$, with a simple parameterisation,

$$w(z) = w_0 + w_a \frac{z}{1+z}, \tag{28.7}$$

where w_0, the present value of w, and w_a are two parameters that are to be determined by a fit to the data.[20]

The *equation of state parameter*, w, is detected through its influence on the history, $H(z)$ of the Hubble expansion rate:

$$H(z)^2 = H_0^2 \left[\Omega_r(1+z)^4 + \Omega_m((1+z)^3 + \Omega_k(1+z)^2 + \Omega_w f(z) \right]. \tag{28.8}$$

Here the dark energy contribution is denoted by Ω_w, and $f(z)$ is a function describing how it evolves during the cosmic expansion. In the case of the cosmological constant Λ, where $w = -1$, $\Omega_w(w = -1) = \Omega_\Lambda$ and $f(z) = 1$. For the simple model (28.7) of $w(z)$ we have

$$f(z) = (1+z)^{3(1+w_0+w_a)} \exp\left(-3w_a \frac{z}{1+z}\right). \tag{28.9}$$

The parameters w_0 and w_a enter into the fitting function (28.3) via the distance function $\mathcal{D}_L = H_0 D_L(z, \Omega_m, w_0, w_a)$, because

$$D_L = \int_0^z \frac{dz'}{H(z')}. \tag{28.10}$$

Given the form of $H(z)$, this adds another level of complexity to the process of getting the best fit model.

28.4 Gamma-ray Bursts (GRB)

For the moment and for the foreseeable future, supernova searches are limited to a redshift of $z \sim 1.7$ with a mean value of only $z \sim 0.5$. To get data beyond this redshift requires the use of samples of alternative redshift-independent distance indicators. Two candidates come to mind in this regard: quasars and gamma-ray bursts, both of which are visible out to redshifts of $z \sim 7$ (Tanvir *et al.*, 2009; Mortlock and Warren, 2011). Neither GRBs nor quasars are standard candles: the energies of GRBs span some seven orders of magnitude.[21]

For such high redshift objects to be useful in probing the cosmological parameters at high redshifts, it is necessary, as with supernovae, to use relationships among their observed

---

[20] The term $z/(1+z)$ is simply the inverse of the scale factor $a(t)$.

[21] The relatively nearby GRB 080319B at a redshift $z = 0.937$, discovered by the *Swift* burst alert telescope at 06:12 UTC on 19, March 2008 (Perley, 2009), attained an estimated visual magnitude $V \simeq 5.5$ some 18 days after the burst first notification, thus making it a naked eye object.

spectral and luminosity/energy characteristics to minimise the dispersion in their Hubble diagrams. However, there is an important distinction between renormalising supernovae to become standard candles and doing the same thing for these more distant and energetic objects. The supernovae can be calibrated using local supernova events and distance indicators that are independent of the cosmology. The calibration of distant objects for which there are no local counterparts is inevitably intertwined with the cosmological model that we are trying to fit.

There are several plausible solutions for the renormalisation of GRBs to a fiducial standard candle, two of which we will discuss briefly below (but see the salutary warning of Petrosian *et al.* (2009)). However, quasars appear to be so diverse that there is, as yet, no strong suggestions as to how one might select a subclass that is renormalisable, and so we focus on the GRB data.

28.4.1 What are Gamma-ray Bursts?

Gamma-ray bursts have been known since the 1960s, but it was only in 1997 that they were shown to be extragalactic objects (Metzger *et al.*, 1997).[22]

The gamma-ray bursts (GRBs) are the most energetic phenomenon in the Universe and are thought to be the consequence of relativistic explosions associated with exceptional phenomena such as the death of stars or with the collision of two neutron stars. The radiation from the explosion is thought to be beamed into a relativistic jet that occupies only 0.1–10% of the sphere, and so we only see those bursts that happen to be pointing roughly in our direction. The relativistic particles moving along the main axis of the jet are moving faster than the off-axis particles, and so the beaming of the radiation is enhanced by special relativistic effects. The interaction between the beam and the medium surrounding the explosion produces an after-glow at optical wavelengths. The observer does not need to be on the beam axis to see the after-glow.

In general terms, we can say that most of the total energy budget for the explosion comes out during the burst in the form of γ-rays (energies > 25 kev), with somewhat less than 10% coming out as 1–10 kev X-rays, and only 0.1% coming out in the optical bands. The after-glow only has less than $\sim$10% in γ-rays and about the same in X-rays. The optical output in the after-glow is, however, much enhanced and is typically a few percent, together with a fraction of a percent as radio emission. What the observer sees depends on the precise alignment of the beam and the observer: whether we are on-axis or slightly off-axis, results in our seeing a variety of types of burst.

Gamma-ray bursts are divided into two classes depending on the duration of the after-glow: the *short-burst* GRBs have durations ranging from milliseconds to $\sim$ 2 s with a mean of around $\sim$0.2 s, while the *long bursts* have durations in the range 2−100 s. There are also rare bursts lasting longer than $\sim$10 000 s, and in one case, up to two days, the so-called *ultra-long bursts*.

---

[22] First reported in an IAU Circular (Galama *et al.*, 1997). During the preceding decade there had been a growing body of opinion, influenced mainly by their apparently isotropic distribution on the sky, that they were indeed extragalactic. But this was the key discovery.

The duration of the short burst GRB suggests a size on the order of ~ 0.2 light-seconds, which is on the order of a neutron star size. One idea is that we are seeing the disintegration of a neutron star, possibly a collision, an idea supported by the flickering of the X-ray flux in the aftermath of the main explosion. The long burst GRBs are associated with the deaths of massive stars, an idea supported by the fact that most such events appear to be associated with galaxies that have a very high rate of star formation. There are currently more ideas on the origin of the ultra-long bursts than there are objects of that kind.

Despite the apparent diversity of properties, there have been numerous suggestions as to how these may all be manifestations of the same basic jet mechanism (see, for example, Sari *et al.* (1998); Mészáros (2006); Hakkila and Preece (2011)). In that case there is some hope of producing a standard candle from this apparently heterogeneous population of events.

28.4.2 Making Standard Candles out of GRBs

Discovering quantities among the observed properties of the GRBs that correlate with intrinsic (rest frame) luminosity provides a set of estimators for the intrinsic luminosity. The precise nature of the correlation will inevitably depend on the cosmological parameters which are used since time scales and frequencies have to be referred back to the local rest frame of the source.

In a study of 69 γ-ray bursts having redshifts between $z = 0.17$ and $z = 6.60$, Schaefer (2007) created a Hubble diagram for the bursts (see Figure 28.6) by exploiting correlations between five parameters that serve to characterise the bursts and the luminosity, L, of the source. Most of these relationships, had been found in the early 2000s, notably the so-called Amati and Yonetoku relations.

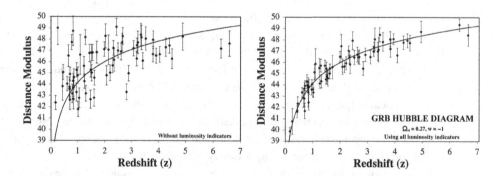

Fig. 28.6 GRB Hubble diagram before (left) and after (right) applying renormalisation of luminosity to standard candle for 69 GRBs. The renormalisation is achieved using a combination of luminosity indicators. The distance on the vertical axis is measured logarithmically as the distance modulus to source, expressed in magnitudes. The curve is what would be expected from the standard Λ-model of dark energy in the standard $\Omega_M = 0.27$, $\Omega_\Lambda = 0.73$ cosmological model. As impressive as this is, it should be remembered that this Hubble diagram may, in part, reflect a systemic variation of intrinsic brightness with redshift.

Adapted from Schaefer (2007, Figs. 7,15). *Reproduced by permission of the AAS.*

Amati *et al.* (2002) found a tight correlation between the spectral peak energy, E_{peak} of the νF_ν spectrum (see right-hand panel of Figure 28.7 and Equation 28.11) and the isotropic equivalent energy, E_{iso}. The isotropic energy E_{iso} is determined from the K-corrected bolometric flux, S_{bolo}, using the luminosity distance D_L (Equation 18.51): $E_{iso} = 4\pi D_L^2 S_{Bolo}(1 + z)^{-1}$. Yonetoku *et al.* (2004) found a relationship between E_{peak} and the isotropic *peak* luminosity, $L = 4\pi D_L^2 P_{bolo}$ where P_{bolo} is determined from the peak rest frame photon flux. Aside from the parameters just mentioned, L, E_{iso}, E_{peak}, the other parameters are a measure, V, of the variability of the source, the rise-time of the burst, τ_{RT}, and the lag time between the peak luminosity being achieved in two distinct wavebands, τ_{lag}. All quantities are reduced to the local rest frame, using a K-correction which is important at such high redshift. For a detailed review see Petitjean et al. (2016).

Combining the linear relationships between these variables allows a renormalisation of the intrinsic brightness to create what is, in effect, a standard γ-ray burst candle. Shaefer *loc.cit.* found that the tightest Hubble diagram, shown here in Figure 28.6, could best be generated by using all of these parameters. The result is quite impressive, notwithstanding that the procedure is somewhat model dependent. His Hubble diagram is consistent with values of $\Omega_m = 0.73$ and $w = -1$. Lin *et al.* (2016) presented a more model-independent formal determination $\Omega_m = 0.302 \pm 0.142(1\sigma)$. γ-ray bursts may well represent one of the best hopes for probing w to very high redshifts.

28.4.3 GRB Spectra

The entire energy spectrum, and its frequency-dependent time variation, is fundamental to understanding the physics of the gamma-ray burst. Models for the bursts were proposed long before there was proof that they were extragalactic events. A somewhat typical spectrum is shown in Figure 28.7 of the 1999 $z = 1.6$ burst GRB 990123, which was the first burst to be observed simultaneously at both optical and gamma-ray wavelengths.[23]

The observables are the X-ray and γ-ray spectrum of the burst and after-glow as a function of time, with optical data on the luminosity and redshift.

In high energy astrophysics it is usual to count the rate of arrival of photons of a given energy, over a given area of detector. This is converted to a count $dn(E)$ of photons of energy arriving in an energy bin $(E, E + dE)$, per unit area, so that $N_E = dn(E)/dE$ is the photon energy flux per energy interval. The photon energy flux $dn(E)/dE$ conventionally is given in units of photon per second, per square centimetre, per kev, $(s^{-1} cm^{-2} kev^{-1})$.

Since each of the photons arriving in that bin has energy E, $E dn(E)/dE$ is the energy flux arriving in the energy bin $(E, E + dE)$. Hence the energy flux per unit (natural) logarithmic interval of energy $d \ln E = dE/E$ is

$$P(E) = E^2 \frac{dn(E)}{dE} = E^2 N_E, \quad E_{peak} = \max P(E). \tag{28.11}$$

---

[23] GRB 990123 reached magnitude $V \simeq 9$ and so was, in principle, visible using modest amateur telescopes.

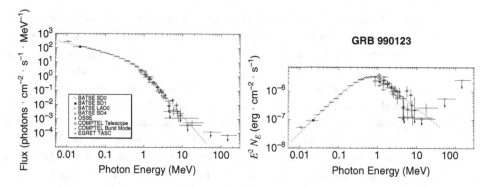

Fig. 28.7   Spectrum of the γ-ray burst GRB 990123 (left) and the radiation flux per unit logarithmic frequency interval (right). The latter is referred to as a 'νF_ν-plot' by analogy with the corresponding visual-band plot. The νF_ν-plot shows the energy output per unit logarithmic interval, $d \ln E$, of energy and, in this case, shows clearly where most of the radiated energy is coming from.
Adapted from Briggs (1999, Fig. 2). *Reproduced by permission of the AAS.*

$P(E)$ is referred to as the power per logarithmic bandwidth, in the cgs units of $\mathrm{erg\,cm^{-2}s^{-1}}$ or in the SI units of $\mathrm{watt\,m^{-2}\,Hz^{-1}}$. $P(E)$ peaks at the energy, E_peak, where the power output $P(E)$ is a maximum.[24]

Figure 28.7 shows the spectrum of the gamma-ray burst GRB 990123 displayed on a logarithmic energy axis. The left panel shows $dn(E)/dE$ while the right-hand panel shows $P(E) = E^2 N(E)$ as defined in Equation (28.11).

28.5 Baryonic Acoustic Oscillations (BAOs)

28.5.1 Density Fluctuations on the Largest Scales

We see structure in the CMB sky and, with current technology, this can be mapped and analysed to reveal key structures in the power spectrum of the distribution of the baryonic matter. The temperature map of the sky reflects the way the photons last interacted with the baryonic component of the Universe. Those baryonic density fluctuations then interact with the dark matter component and become more prominent, finally evolving into the structures we see today: galaxies, clusters of galaxies and the large scale cosmic web.

That structure is a direct consequence of the initial conditions that we are seeing on the last scattering surface. However, it has been modified since the time of last scattering by a string of nonlinear evolutionary processes such as galaxy and structure formation. The processes are generally sufficient to have destroyed all but traces of what the structure looked like at recombination. The way of tracking this is through simulations of the process of structure formation in which the physical processes taking place are modelled numerically.

---

[24] The plot of $P(E)$ versus $\log E$ is conventionally referred to as a 'νf_ν plot'. See Gehrels (1997) for more on this.

Some of these processes are simple, such as the gravitational aggregation of matter, which is very costly to compute. Other processes such as star formation are so complex that they must be simulated by simple parameterised models.

Finding relict structure in the matter distribution requires working on such a large scale that these physical processes do not influence the evolution. However, these processes may nonetheless affect what we see: we can only see stars and hot gas, the baryonic component that radiates. The relationship between the unseen dark matter and the luminous baryonic component is barely understood, but it can be modelled. This is the subject of *biasing*.

Nevertheless, there is nothing to stop us from looking into ultra-large scale structure, a process initiated in a quantitative manner by Peebles (1973) and Totsuji and Kihara (1969) using the few catalogues of galaxies that were available at the time. In those early days the catalogues were simply catalogues of positions on the sky and apparent brightness (measured in magnitude). Now, we have redshift catalogues having over a million objects, so this opens up the door to detecting weaker signals on the largest scales that might be visible relics of the initial conditions.

The development of the subject of large scale galaxy clustering dominated observational cosmology during the 1970s to early 1990s, stimulating ever larger large scale redshift surveys.[25]

28.5.2 2-point Correlation Function and Power Spectrum

The appearance of dominant peaks in the power spectrum of the CMB shows that the initial conditions for structure formation were such that certain peaks should be picked out in the present day galaxy distribution. The most prominent peak at $\ell \sim 200$ (see Figure 22.1) should certainly be visible in structure analysis, given a sufficiently large galaxy redshift sample.

The key tools in the analysis of clustered density fields are the computation of the 2-point galaxy correlation function, $\xi(r)$, and the clustering power spectrum, $P(k)$ (see Section 8.2.5). These functions, applied to galaxy catalogues, give information about the pairwise distribution of galaxies.[26] For more about this see Section 8.2 where this was discussed in relation to a continuous density field $\delta(\mathbf{x})$ describing a continuous random field of small amplitude density fluctuations. We need a few words to orientate what this means in terms of a distribution of discrete particle, i.e. galaxies.

Suppose we have two elemental volumes δV_1 and δV_2 located at $\mathbf{r}_1$ and $\mathbf{r}_2$. The probability δP_1 of finding a galaxy in δV_1 is simply

$$\delta P_1 = n\delta V_1, \tag{28.12}$$

where n is the expectation value of the galaxy density, i.e. the mean density.

---

[25] By the beginning of the 21st century there were some million galaxy redshifts and magnitudes. The history of this development is beautifully reviewed by the personal reminiscences of Einasto (2014), himself one of the great contributors and drivers for the probing of structure on the largest scales.

[26] These are described in more formal detail in the On-line Supplement on *Random Fields*. See also Bertschinger (1992, Section 5.8.1) for a fine discussion of this subject.

If the galaxies were Poisson-distributed, then the event of finding a galaxy in δV_1 would be independent of the event of finding a galaxy in δV_2 and the probability of finding a galaxy in both of δV_1 and δV_2 would be the product of the elemental probabilities (28.12): $n^2 \delta V_1 \delta V_2$. However, this would not be correct if the galaxy distribution were not uniform: if galaxies were clustered the events would not be independent. In the general case we can write

$$\delta^2 P_{12} = n^2 \delta V_1 \delta V_2 [1 + \xi(\mathbf{r}_1, \mathbf{r}_2)], \tag{28.13}$$

where the function $\xi(\mathbf{r}_1, \mathbf{r}_2)$ measures the deviation from the Poissonian result due to clustering.[27]

In cosmology, the function $\xi(\mathbf{r}_1, \mathbf{r}_2)$ is called *the galaxy 2-point correlation function*. Since $\delta^2 P_{12}$ is a probability, we must have

$$\xi(\mathbf{r}_1, \mathbf{r}_2) > -1, \tag{28.14}$$

and if we require that the volumes become statistically independent at very large separations, then

$$\xi(\mathbf{r}_1, \mathbf{r}_2) \to 0, \qquad |\mathbf{r}_1 - \mathbf{r}_2| \to \infty. \tag{28.15}$$

If the galaxy distribution is statistically homogeneous and isotropic the galaxy correlation function will depend only on the magnitude of the vector joining the volumes:

$$\xi(\mathbf{r}_1, \mathbf{r}_2) = \xi(r), \qquad r = |\mathbf{r}_1 - \mathbf{r}_2|. \tag{28.16}$$

The *power spectrum* of the fluctuation part of the density distribution is the Fourier transform of the correlation function. If the galaxy distribution is statistically isotropic we have

$$P(k) = \frac{1}{(2\pi)^3} \int \xi(r) e^{-i\mathbf{k} \cdot \mathbf{r}} d^3 \mathbf{x}. \tag{28.17}$$

This describes the contribution of Fourier modes of a given frequency l to the total variance of the 1-point function.[28] The inverse transform is simply

$$\xi(r) = \int P(k) e^{i\mathbf{k} \cdot \mathbf{x}} d^3 \mathbf{k}. \tag{28.18}$$

The power spectrum is the Fourier transform of $\xi(r)$.

In the early days when redshifts were not widely available, the sky-projected two point angular correlation function was measured and transformed into the 3-dimensional $\xi(r)$ using an equation developed long before by Limber (1953, 1954, 1957).

The galaxy distribution is treated as statistically isotropic and so treating the redshifts as radial distances would inevitably cause problems. Because the radial velocities were

---

[27] What we in cosmology call the galaxy 2- point correlation function is in fact the normalised autocovariance function of the underlying continuous process from which the sample is constructed.

The fact that $\xi(r)$ is a renormalised autocovariance function leads to some semantic difficulties, especially when talking with scientists who do not work in cosmology. It is therefore perhaps fortunate that we call $\xi(r)$ the 'galaxy 2-point correlation function' and refer to its Fourier transform as the 'cosmic power spectrum'. That way we can keep the terms 'autocovariance function' and 'spectral density' for what they are generally taken to represent in the literature of stochastic processes.

[28] See Chapter 8, footnote 9, for a remark about where to put the 2πs.

polluted by large scale systemic non-Hubble motions of galaxies in groups and clusters, this induced a bias in the radial distance that would have to be removed when computing a 'true' $\xi(r)$. However, this radial distortion is in itself an important cosmological, dynamical, probe of large scale cosmic structure.

28.5.3 Observations of Oscillations

The baryonic acoustic oscillations, BAOs, are relics of the largest scale inhomogeneities at the time of recombination, $z \sim 1100$ (see Section 27.1.2). They are detected as peaks in the angular distribution of the cosmic microwave background fluctuations on angular scales corresponding to C^l values having $100 < l < 1500$. Measuring the peaks is a matter of fundamental importance in the understanding of our Universe. Our observations of the acoustic peaks in the spectrum imply that most of the ions and electrons in the Universe combined to make neutral hydrogen and helium at $z \sim 1100$.

The third peak position constrains the value of $\Omega_m^{0.275}h$, while the third peak height strongly constrains the matter density, $\Omega_m h^2$ (Hu & White 1996; Hu & Sugiyama 1995).

During the late 1990s there had been several reports of detections of structure on a scale $100h^{-1} - 200h^{-1}$ Mpc. Einasto *et al.* (1997), for example, reported finding a $120h^{-1}$ Mpc periodicity in the redshift distribution of galaxy clusters and super-clusters, describing the distribution as resembling a 3-dimensional chessboard.

The first hint that the galaxy distribution counterpart of the oscillations seen in the CMB power spectrum came with Percival (2001, Fig. 3) through Fourier analysis of the 2dF galaxy survey. Detecting ultra-large scale structure is difficult because the galaxy distribution becomes dominated by shot noise on such scales, but there was a definite hint of structure. Confirmation that there was a significant feature on a scale of around $105h^{-1}$Mpc came almost simultaneously from three directions: Eisenstein *et al.* (2005), using the correlation function, Cole and Percival (2005), using Fourier analysis, and Hütsi (2005, 2006a,b), using power spectrum analysis. Figure 28.8 shows the significant peak in the correlation functions obtained by Eisenstein *et al.* and by Hütsi.

Since that time several special purpose BAO surveys have been undertaken using a variety of redshift catalogues probing the galaxy distribution to ever greater depths.

28.5.4 What is Being Measured

The surveys are, in effect, measuring overlapping, randomly placed, spherical over-densities in the galaxy distribution having a scale $r_S \sim 150$ Mpc. This scale is now our standard cosmic ruler. The resultant density enhancement is seen in both the spatial 2-point correlation function and in the power spectrum. The galaxy samples are large enough that they can be detected in sub-samples defined by redshift ranges. If we consider a redshift z we can calculate two numbers for our idealised spherical bumps: the angular diameter distance $D_A(z)$ corresponding to the angle subtended by the standard ruler; and a redshift distance $D_H = cH(z)^{-1}$, where $H(z)$ is the local Hubble expansion rate at that redshift.

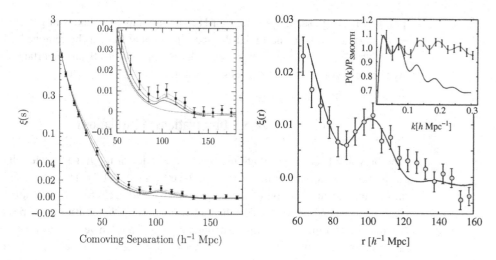

Fig. 28.8 Two analyses of the redshift space correlation function, $\xi(s)$, for SDSS luminous red galaxies (LRGs) sample showing the BAO bump at $\sim 100h^{-1}$Mpc. Both show the BAO hump in $\xi(s)$.

On the left is the analysis from Eisenstein *et al.* (2005, Fig. 2). The continuous lines are different models: the bottom one, without a bump, being a pure CDM model, $\Omega_{CDM}h^2 = 0.105$ with no baryons and hence no baryonic feature. The other models have $\Omega_m h^2 = 0.12, 0.13, 0.14$ (from top to bottom on the diagram), and the same $\Omega_B h^2 = 0.024$. The underlying power spectrum has a slope of $n = 0.98$. The vertical axis is linear for $\xi(s) \leq 0.04$ and logarithmic above that. The $\xi(s) \leq 0.04$ data is replotted on an expanded vertical scale in the inset. *(Reproduced by permission of the AAS.)*

On the right is the analysis of Hütsi (2005, Figs. 6 & 8). The inset shows the corresponding power spectrum, revealing the presence of other smaller scale features in the power spectrum. The curves show an evolved benchmark model. Hütsi's evaluation of $\xi(s)$ is consistent with that of Eisenstein *et al.* *(Permission granted.)*

The power of the BAO measurement is that because we know the absolute size of the object we are looking at, it is possible to estimate both $H(z)$ and D_A separately (Anderson, 2014b). $H(z)$ comes from the radial velocity spread across the feature and D_A comes from its apparent angular diameter. These are absolute measurements because we know the length of our standard ruler.[29]

We need to relate the bump in the correlation function to the apparent size of the feature at a given redshift. The correlation function reflects the mean density excess in a volume of scale r_S, and so it is essentially a measure of the volume occupied by one of these spheres. The volume measure is proportional to the square of D_A and to the distance D_H: these two numbers measure the geometric distortion of the sphere, as seen from our point of view. This leads us to think in terms of a *volume distance* D_V defined by

$$D_V = \left(1 + z)^2 D_A^2 \frac{cz}{H(z)}\right]^{1/3}. \tag{28.19}$$

---

[29] If we did not know the length of our ruler all we could determine would be the ratio of the sizes of the feature in the transverse (D_A) and line of sight, ($H(z)$) direction. This is the Alcock–Paczynski test (Alcock and Paczynski, 1979) which in effect measures $\Delta z / \Delta \theta$: the ratio of the radial size to the angular size.

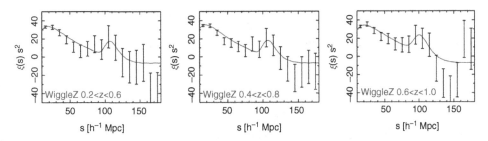

Fig. 28.9   The spatial correlation function $\xi\,(s)$, where s is the *comoving separation*, for three redshift bands of the WiggleZ survey. The vertical axis is $s^2\xi\,(s)$, this keeps the curves within a small vertical range on the plot. The plotted curves are models fitted to each redshift bin, each with its own cosmological parameter $\Omega_m h^2$ and rescaling factors. From Blake and Kazin (2011, Fig. 1), edited to fit horizontally. *(Permission granted.)*

By looking at BAO correlations at different redshift we can track the variation of D_V with redshift.

Figure 28.9 shows the correlation functions in three rather wide and overlapping redshift slices. The fitted model is taken as a distortion of a fiducial correlation function ξ_{fid}:

$$\xi_{\mathrm{model}}(s) = b^2 \xi_{\mathrm{fid}}(\Omega_m h^2, \sigma_v, \alpha s). \tag{28.20}$$

The factor α is a rescaling factor, b is a renormalisation parameter, and σ_v characterises the deviations of the galaxies from their original position due to bulk flows. b is thought of as modelling the effects of linear galaxy biasing and redshift space distortions (Blake and Kazin, 2011, Section 3). The rescaling parameter accounts for any difference in the comoving BAO scale via the distance parameter D_V:

$$D_V(z_{\mathrm{eff}}) = \alpha D_{V,\mathrm{fid}}(z_{\mathrm{eff}}). \tag{28.21}$$

We would expect $\alpha \simeq 1.0$ for all redshift slices if the model were correct, as indeed was found in the paper.

28.5.5 Cosmology with BAOs

The Hubble Parameter as a Function of Redshift

Using the BAO signatures at different redshifts makes it possible to derive the Hubble expansion parameter $H(z)$ at those redshifts (see Figure 28.10). There are two reasons why this should work. Firstly, the BAOs measure a feature that is a standard ruler whose length is, in principle, known at all redshifts. Secondly, the facilities are available to generate redshifts for enormous numbers of galaxies, so the statistical errors can be beaten down. However, there are numerous side-effects to be taken care of, not the least of which is the redshift distortion which produces a systematic bias in the radial direction. This can be calibrated out using models, but there are still effects due to nonlinear evolution that need correction.

Figure 28.11 shows a real BAO Hubble diagram based on this data: the angular diameter distance, $D_A(z)$ is plotted against the redshift. $D_A(z)$ has been calculated using the distance

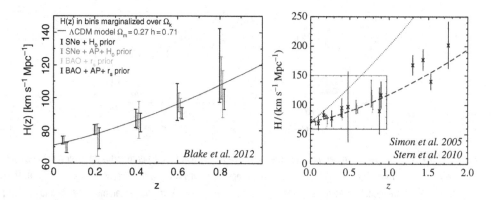

Fig. 28.10 *Left:* The Hubble expansion rate, $H(z)$, measured in five redshift bins using supernova and BAO data. The BAO galaxy sample uses 158 741 galaxies with measured redshifts $0.2 < z < 1.0$ from the WiggleZ Dark Energy Survey, and covers an area of $800\square°$. The supernova data is the Amanullah *et al.* (2010) compilation of 557 supernovae. $H(z)$ is estimated from the SNe and BAO datasets individually. The scale of the acoustic horizon, $r_S(z_d)$, at recombination is used as a prior in the analysis of the BAO data, and the present day Hubble constant, H_0, is used as a prior for the analysis of supernova data. The curve passing through the estimates of $H(z)$ in the five bins is not a fit to the data, it is the benchmark ΛCDM model having $\Omega_m = 0.27$ and $h = 0.71$.
From Blake (2012, the WiggleZ Dark Energy Survey). *(Permission granted.)*

Right: The Hubble expansion rate, $H(z)$, as determined from the ages of passively evolving galaxies in a sample of 24 rich clusters at $z < 1$ by Stern *et al.* (2010). The galaxies for $z > 1$ are from the earlier study by Simon *et al.* (2005). The left hand figure has been inset for comparison. The dotted line is the Einstein de Sitter model.
From Stern *et al.* (2010, Fig. 9). *(Permission granted.)*
Inset: from Blake (2012, Fig. 9). *(Permission granted.)*

$D_V(z)$ of Equation (28.19). This brings in the cosmological parameters because we need $H(z)$ to make the transformation to $D_A(z)$. The data is shown for $\Omega_m = 0.31, h = 0.67$. These are the values given in the Planck data release (Planck Collaboration XIII, 2015, Eq. 28) using the Planck *TT* and low frequency polarisation data, together with BAO analysis from Anderson (2014a).

Dark Energy

One of the prime aims for the BAO surveys is to probe the dark energy w-parameter relating the energy and pressure of whatever it is that is causing the accelerated expansion. BAOs have already probed a similar depth to supernovae and so are effective probes of the cosmic expansion up to and beyond redshifts of $z \sim 1$. We see this in Figure 28.11.

The key is the role of the Hubble parameter $H(z)$, which defines $D_V(z)$, and has the general form

$$H(z)^2 = H_0^2\left[\Omega_m(1+z)^3 + \Omega_r(1+z)^4 + \Omega_k(1+z)^2 + \Omega_w(1+z)^{3(1+w)}\right], \quad (28.22)$$

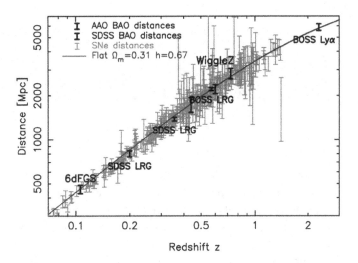

Fig. 28.11 Modern version of the Hubble diagram based on a number of BAO surveys and SN Ia observations. The fitted model has cosmological parameters $\Omega_m = 0.31$, $h = 0.67$. The BAO distances, $D_V(z)$ have been converted to angular diameter distances, $D_A(z)$ using $H(z)$ from this cosmological model. For the SN Ia sample, the conversion $D_A(z) = (1 + z)^2 D_L(z)$ has been used.
From Lahav and Liddle (2014, Fig. 24.2), based on Blake and Kazin (2011, Fig.12).
Credit: The Particle Data Group, Olive and Particle Data Group (2014) *(Permission granted.)*
`http://pdg.lbl.gov/2014/figures/figures.html`

where w could be an arbitrary function of redshift. The right-hand panel of Figure 28.4 shows the confidence contours for the distribution of Ω_m and w on the assumption that w is a constant. The confidence contours for both supernovae and BAOs are very elongated, but, remarkably, they are almost orthogonal to one another and so conspire to make a definitive measurement of Ω_m and w.

29

Epilogue

During the course of the last 100 years, our understanding of the Universe has increased dramatically from a point of ignorance about its extragalactic nature to a depth which enables us to model its behaviour and history to a high level of precision. This great change has been brought about through a series of key discoveries, theories and calculations, each one leading to another step in the ladder of detail about how we understand the Universe: another decimal point in the key parameters that determine our cosmic models.

'Top 10' lists are always contentious. This list is much abbreviated; it is not intended to be comprehensive. Inevitably, many names and results are left out, and the dates are indicative rather than accurate. I give the list so that people may reflect on the great achievements we have seen during the last 100 years, on the steps we have gone through to get to this point, and to appreciate what lies behind the extraordinary precision we can now achieve in our models of the Universe and its history.

Date	Author(s) + key observation/theory	Key realisation
1908–12	Leavitt Cepheid variable distances	extragalactic nature of nebulae
1913–1917	Slipher Systemic redshift of galaxy spectra	expansion of the Universe
1917–	Einstein; Friedmann; Lemaître relativity & gravitational physics	models for an expanding Universe
1926	Eddington stellar nucleosynthesis	theory of origin of elements
1929–36	Hubble and Humason the redshift distance relation	evidence for expanding Universe
1929–36	Robertson and Walker establish relativistic cosmology	show that the RW metric is the metric of a homogeneous isotropic space-time
1946–56	Gamow calculation for hot cosmic origin	model for Hot Big Bang Universe
1955+	Cambridge/Sydney groups radio source counts	evidence of cosmic evolution
1957	Burbidge, Burbidge, Fowler & Hoyle stellar nucleosynthesis	origin of most heavy elements
1961–64	Osterbrock, Rogerson; Hoyle, Taylor cosmic helium abundance	support for cosmic fireball

1965	Penzias & Wilson discovery of CMB	establishes the Big Bang theory
1965–68	Peebles; Silk; Doroshkevich; Sunyaev; Zel'dovich	Hot Big Bang structure formation
1967	Wagoner, Fowler & Hoyle calculation of cosmic nucleo-synthesis	cosmic origin of light elements
1967–68	Sachs & Wolfe; Silk physics of CMB	details in the CMB power spectrum
1969–72	Sunyaev & Zel'dovich distortions of the CMB spectrum	information from the Big Bang and after re-ionisation epoch
1970	Peebles, Yu; Sunyaev, Zel'dovich the radiation power spectrum	peaks in CMB power spectrum BAOs and the cosmic ruler
1974	Peebles clustering correlation functions	understand processes behind large scale structure in the Universe
1977	Smoot (U2) CMB dipole	motion relative to distant matter
1979	Bosma extended flat rotation curves	detailed study of galaxy rotation maps reveals extensive unseen matter
1979	Aarseth, Gott & Turner cosmology N-body simulation	provides multiple realisations of our otherwise unique Universe
1980	Rubin dark matter	presence of dark matter in the Universe on galactic scales
1981	Guth; Steinhardt; Linde inflation theory	why the Universe is flat and homogeneous
1984	Bardeen, Bond, Kaiser, Szalay transfer function at recombination	evolution of initial power law spectrum of inhomogeneities
1984–	Bond & Efstathiou CDM predictions of CMB polarisation	methodology for analysing CMB power spectrum
1985–9	Davis, Efstathiou, Frenk & White simulations with/without dark matter	cold dark matter in big quantities plays big role in galaxy formation
1990	Efstathiou, Sutherland, & Maddox evidence for cosmological constant	there is dark energy in Universe first claim of detection
1991–4	Kaiser & Squires; Schnieder theory of weak lensing	enable detection of distant matter in the Universe
1995	Perlmutter, Schmidt & Riess supernovae Hubble diagram	Universe is accelerating
1996	Bond & van de Weygaert cosmic web dominated by voids	Analysis of structure allows a new measurement of the Universe.
2000	Bacon;Kaiser;van Waerbeke;Wittmann detected cosmic shear	map dark matter through distortions of CMB inhomogeneity
2000–	2dF, SDSS complete galaxy redshift surveys	cosmological parameters Universe is a cosmic web

2002	DASI	confirmation of prediction according
	polarisation of CMB	to Big Bang theory
2003	BOOMERANG and WMAP	polarisation and 6-parameter fits
	power spectrum of CMB (BAOs)	precision CMB cosmology
2005–6	Eisenstein; Hutsi; Cole; Percival	cosmic measuring rod
	baryonic accoustic oscillations	
2013	Planck Collaboration	map of distribution of dark matter
	Planck satellite observations	
2015	LIGO Scientific Collaboration with	gravitational wave detected
	VIRGO Collaboration	from binary black hole merger
		GW1509014

Appendix A SI, CGS and Planck Units

Systems of units are always a problem, despite international agreement to standardise on the SI system. The old cgs system and its multitude of mutually inconsistent variants (Gauss, Lorentz–Heaviside, esu, emu, etc.) is still in use. This is particularly an issue when it comes to using the Maxwell equations. In astrophysics we generally see a mixture of cgs units and subject-specific units, like the magnitude scale for brightness, parsecs for distance, and flux units in the older radio astronomy literature. This is what we might call the *Astrophysical System of Units* and, like many contemporary authors, I have adhered to that in this text. However, certain issues must be clarified, and in particular the conventions used in the Maxwell equations.

This appendix provides translation between Gaussian cgs units and the SI system for several of the quantities that are used in the text, and provides some update on the values of the fundamental physical constants that have occurred since the decision to fix the speed of light at a given numerical value. The Maxwell equations have a small section to themselves.

A.1 SI, MKS and cgs

There has been a slow move in astronomy away from the traditional *centimetre-gram-second* (CGS) based system of units that evolved in the 19th century towards the metric *metre-kilogram-second* (MKS) based system of units and the subsequent *International System* (SI),[1] which is also known as the MKSA system. However, the change has been slow or, as in the case of astronomy, only partial. By the 1950s many fields had adopted their own subject-specific units, chosen largely because they were particularly convenient or simply entrenched in the culture of the discipline. There are perhaps just two reasons for the slowness of this change: teachers today were often taught in cgs or hybrid systems, and used textbooks with which they were familiar. Moving from cgs to SI units takes people out of their comfort zone.

The full details of the system of physical units are maintained and documented by the *National Institute of Standards and Technology*, NIST, which is maintained by the US

---

[1] The cgs system was proposed by Gauss in 1832, and formalised in 1874 by the British Association for the Advancement of Science under the influence of Maxwell and Thomson. cgs was rapidly adopted by all communities, whereas the MKS system, formalised in 1954 by the 10th Conference on Weights and Measures has taken much longer. The name 'SI' was adopted in 1960 by the 11th Conference.

Department of Commerce. The information presented here is for convenience and is largely abstracted from the NIST web site.[2]

The SI system is based on seven independent base quantities:

Base systems of units				
	SI base unit		cgs equivalent	
Item	Name	Symbol	Name	Symbol
Length	metre	m	centimetre	cm
Mass	kilogram	kg	gram	g
Time	second	s	second	s
Electric current	Ampère	A	abAmpère (or Biot)	abA (Bi)
Thermodynamic temperature	Kelvin	K	Celsius	°C
Amount of substance	mole	mol	mole	mol
Luminous intensity	candela	cd	stilb	sb

Adapted from the NIST website.

In the cgs system the only fundamental quantities are the centimetre, the gram and the second. Definitions of other quantities like temperature and units of electromagnetism lead to a diversity of cgs variants. The cgs unit for electric current is referred to as either the 'abAmpère' or the 'Biot', abbreviated 'Bi'.

A.2 Common cgs to SI Conversions

Conversion between SI and cgs				
Name	cgs unit	is	SI	symbol
Length	cm	10^{-2}	meter	m
Mass	g	10^{-3}	kilogram	kg
Time	second	$=$	second	s
Force	dyne	10^{-5}	Newton	N
Energy	erg	10^{-7}	Joule	J
Power	erg/s	10^{-7}	Watt	W or J s^{-1}
Energy flux	erg/cm^2/s	10^{-3}	W m^{-2}	J m^{-2} s^{-1}
Charge	esu	$10\,c^{-1}$	Coulomb	C
Electric current	statA	$10\,c^{-1}$	Ampère	A
Potential	statV	$10^{-7}c$	Volt	V
Electric field	statV/cm	$10^{-7}c$	Volt/metre	V m^{-1}
Magnetic field	gauss	10^{-4}	Tesla	T
Magnetic flux	Maxwell	10^{-8}	Weber	Wb
Luminance	stil	10^4	Candela/m^2	cd m^{-2}
Radiative flux	erg cm^{-2} s^{-1}			W m^{-2}

---

[2] Their web site is at
http://physics.nist.gov/cuu/index.html

In this table, $c = 299\,792\,458\,cms^{-1}$ (in cgs since the table converts cgs values to SI values). Note that the velocity of light is now fixed at this value (see Section A.3). Some of the cgs units have been given new names. The cgs unit of charge, the esu, can be referred to as the 'Franklin' with the abbreviation 'Fr'. The unit of magnetic flux has been named the 'Maxwell' with abbreviation 'Mx', which was previously a gauss.cm^2.

A.3 Fundamental Physical Constants

This list is provided for convenience. There are more extensive lists elsewhere, but the following is an extract from the recent recommendations of Mohr *et al.* (2012, Tables XL, XLI).

Physical constants (Mohr *et al.*, 2012)			
Item	symbol	value	units
Speed of light in vacuum	c	299 792 458	$m\,s^{-1}$
Newtonian constant of gravitation	G	$6.673\,84(80) \times 10^{-11}$	$m^3\,kg^{-1}\,s^{-2}$
Planck constant	h	$6.626\,069\,57(29) \times 10^{-34}$	$J\,s$
Elementary charge	e	$1.602\,176\,565(35) \times 10^{-19}$	C
Electron mass	m_e	$9.109\,382\,91(40) \times 10^{-31}$	kg
Proton mass	m_p	$1.672\,621\,777(74) \times 10^{-27}$	kg
Fine structure constant	α	$7.297\,352\,5698(24) \times 10^{-3}$	
Boltzmann constant	k	$1.380\,6488(13) \times 10^{-23}$	$J\,K^{-1}$
Stefan–Boltzmann constant	σ	$5.670\,373(21) \times 10^{-23}$	$W\,m^{-2}\,K^{-4}$

The speed of light has no error, this is what it is now defined to be. The meter is defined as the distance light travels in $1/299\,792\,458$ of a second. With this, any measurement of the speed of light is, in effect, measuring the meter. The number of seconds it takes light to travel one astronomical unit is determined to be 499.004783836(10)s. This comes from timing radio signals to various spacecraft in the Solar System whose positions are known by accurate orbital calculations involving the gravitational influence of all planets with general relativistic corrections.[3] Given that the speed of light is now fixed, this is a determination of the astronomical unit in meters.

See in particular the definitions of the base quantities at
`http://physics.nist.gov/cuu/Units/current.html`
See also `http://www.unitconversion.org/` which was used to make some of these conversions.

[3] See the IAU 2009 System of Astronomical Constants from the IAU Working Group on Numerical Standards for Fundamental Astronomy:
`http://maia.usno.navy.mil/NSFA/IAU2009_consts.html`
and Luzum (2011), which has open access at
`http://link.springer.com/article/10.1007%2Fs10569-011-9352-4`

In 2018 there will be a change in the definition of mass. The International Prototype Kilogram (IPK), defined in 1879, will be replaced by a base unit defined in terms of a fixed value of Planck's constant.

A.4 Electromagnetic Fields in vacuo

When it comes to electromagnetism, there are several sub-versions of the cgs system: electrostatic units (esu), electromagnetic units (emu), Lorentz–Heaviside units (HLU), and Gaussian units (cgs). In the Lorentz–Heaviside system the Maxwell equations have no factors of 4π and the only dimensional quantity appearing in Maxwell's equations is the speed of light, c. The key advantage of the Gaussian version of the cgs system is that the units of electric and magnetic field are the same.

A.4.1 Lorentz–Heaviside Units

Lorentz–Heaviside units are used in quantum field theory. In those units the Maxwell equations and Lorentz force law in free space conveniently coincide with the SI system if we set $\epsilon = \mu$:

$$\nabla \cdot \mathbf{E} = \rho, \tag{A.1}$$

$$\nabla \cdot \mathbf{B} = 0, \tag{A.2}$$

$$\nabla \times \mathbf{E} = -\frac{1}{c}\frac{\partial \mathbf{B}}{\partial t}, \tag{A.3}$$

$$\nabla \times \mathbf{B} = \frac{1}{c}\frac{\partial \mathbf{E}}{\partial t} + \frac{1}{c}\mathbf{J}, \tag{A.4}$$

$$\mathbf{F} = q(\mathbf{E} + \frac{1}{c}\mathbf{v} \times \mathbf{B}), \tag{A.5}$$

Note the absence of 4π. We can further set $c = 1$ without invoking any inconsistency, since that merely changes the unit of time.

These are the units used, for example, by Kolb and Turner (1990) and by Carroll (2003).

A.4.2 Gaussian Units

In Gaussian units these equations would be:

$$\nabla \cdot \mathbf{E} = 4\pi\rho, \qquad \text{Coulomb's law} \tag{A.6}$$

$$\nabla \cdot \mathbf{B} = 0, \qquad \text{no magnetic monopoles} \tag{A.7}$$

$$\nabla \times \mathbf{E} = -\frac{1}{c}\frac{\partial \mathbf{B}}{\partial t}, \qquad \text{Faraday's law} \tag{A.8}$$

$$\nabla \times \mathbf{B} = \frac{1}{c}\frac{\partial \mathbf{E}}{\partial t} + \frac{4\pi}{c}\mathbf{J}, \qquad \text{Ampère's law} \tag{A.9}$$

$$\mathbf{F} = q(\mathbf{E} + \frac{1}{c}\mathbf{v} \times \mathbf{B}), \qquad \text{Lorentz force law} \tag{A.10}$$

in free space. The sources ρ and $\mathbf{J}$ have acquired the 4π factors. So in vacuo, with no sources, the Lorentz–Heaviside version and the cgs version of Maxwell's equations coincide.

The cgs unit of $\mathbf{E}$ is statvolts. cm^{-1} and the cgs unit of $\mathbf{B}$ is the gauss. Common names for $\mathbf{E}$ and $\mathbf{B}$ are the *electric field vector* and *magnetic field vector*, respectively, although we also see *electric intensity vector* and *magnetic induction vector*.

It should be noted that both cgs and the SI system involve other physical constants which enter into the electromagnetic field equations when there are charge or current sources, or when the field is present in a medium.

A.5 Planck Units

When working with physics at the earliest cosmological epochs the formulae can be made to look less cluttered if we choose a system of units based on Planck units.[4]

The Planck length, time and mass are defined in Section 9.4.1. Planck units are based on setting the Planck length $l_P = 1$, the Planck time $t_P = 1$ and the Planck mass $M_P = 1$.[5] We still have a degree of freedom with regard to temperature scales and so we can put the Boltzmann constant $k_B = 1$, which means that the reference temperature is the Planck temperature, $T_P = 1$.

To recover 'normal' units, we merely rescale the units by the factors listed in Section 9.4.1. In Planck units the present day cosmological parameters take on the values below.

Cosmological parameters in Planck units			
Unit	Value	Unit	Value
Age, t_0	$8.1 \times 10^{60}\ t_P$	Radius, ct_0	$5.4 \times 10^{61}\ l_P$
Total cosmic density, ρ_0	$1.8 \times 10^{-123}\ \rho_P$	Cosmological constant, Λ	$1.3 \times 10^{-123}\ t_P^{-2}$
Hubble constant, H_0	$1.24 \times 10^{-61}\ t_P^{-1}$	Temperature, T_0	$1.9 \times 10^{-32}\ T_P$

Of course these are not independent (so the Hubble constant H_0 in these units, $c = 1$, is the inverse of the diameter ct_0). That three of the independent ones all have dimensionless factors of $\sim 10^{60}$ appearing is somewhat striking, but generally taken as a pure coincidence. What is a coincidence is that the present day total cosmic density ρ_0 and Λ are so similar: this restates, in different units, that Λ makes up $\Omega_\Lambda \simeq 0.7$ of the total mass density.

---

[4] The Planck system of units was in fact introduced by Planck in 1899 after his discovery of the constant h that is now named after him.

[5] We often see the use of units such that $\hbar = 1, G = 1$ and $c = 1$. Another option with regard to gravitation is to put $8\pi G = 1$, thereby eliminating the $8\pi G$ from the Poisson equation and from the Einstein equation. Such units are called *reduced Planck units*. Just to confuse the issue we also see the setting $4\pi G = 1$.

A.6 Gev Units

The other option is to work entirely in Gev. The conversions to Gev are as follows:

GeV units			
Unit	Value	Unit	Value
1 s	$= 1.5 \times 10^{24}$ GeV	1 cm	$= 5.1 \times 10^{13}$ GeV
1 K	$= 8.6 \times 10^{-14}$ GeV	M_P	$= 1.2 \times 10^{19}$ GeV

For reference, the proton mass is $m_p = 0.908 \, \text{GeV}$.

Appendix B   Magnitudes and Distances

B.1 Distances

The distance to an astronomical source is a fundamental astronomical datum about the source. While astronomers measure sizes of objects in centimeters, meters or kilometers, distances to distant objects are measured in light years or parsecs. Large objects like galaxies and clusters of galaxies are also measured in light years or parsecs.

The light year is simply the distance travelled by light in one year and it is a fairly convenient unit of measure for the distances to stars. However, distances to stars are not measured using the travel time of light rays: nearby stellar distances are measured using parallax. The parallax is the maximum apparent angular shift in position of a star on the sky as the Earth goes from one side of its orbit to the other. This is the same as the angle subtended by the Earth's orbit as viewed from the star.

The parsec is the distance at which the parallax of an object would be 1 arc second. The parallax of the nearest star, 4.2 light years away, is 0.772 arcsec. The relationship between light years and parsecs is 1 parsec = 3.26 light years. Distances to cosmological objects are to be measured in Megaparsecs (Mpc).

Because of the curvature of space the distance as measured by parallax is not the same as the distances measured by brightness, or by diameter.

B.2 Magnitudes and all that

It is difficult to avoid discussing the apparent brightness of objects such as stars and galaxies without encountering one of astronomy's major idiosyncrasies – the *magnitude scale*. Ever since the time of Hipparchus of Niacea (c190BC–120BC), the brightnesses of stars have been measured relative to one another on a logarithmic scale. Around 150BC, Hipparchus had classified the brightest stars he could see as being of 'magnitude 1' and the faintest as being of 'magnitude 6': a change of 5 magnitudes represented a factor of 100 in brightness. Because the brain perceives increments in brightness logarithmically, each increment of 1 magnitude corresponded to a change in brightness by a factor of $10^{1/5} = 2.512$.[1]

---

[1] Gustav Fechner, regarded as the founder of *psychophysics*, discovered a general relationship between a stimulus of intensity S and the perception P of that stimulus: $P \propto \ln(S/S_0)$. The increments in the perception were logarithmically dependent on the stimulus. He applied this to stimuli of all kinds. His work on this started in

The magnitude scale, as we know it today, was introduced by Pogson (1856). Nowadays we might decide to base a logarithmic brightness scale on powers of two, so that each 'magnitude' was twice as bright as its predecessor. Pogson, presumably following Hipparchus, decided to keep an increment in scale of 2.512.

Following Pogson's dictum, two sources of brightness f_1 and f_2 are said to have magnitudes m_1 and m_2 related by

$$m_1 - m_2 = -2.5 \log_{10} \frac{f_1}{f_2}, \quad \text{i.e.} \quad \frac{f_1}{f_2} = 10^{-0.4(m_1 - m_2)}. \tag{B.1}$$

In this way, two sources differing in brightness by 'one magnitude' would differ in brightness by $10^{0.4} = 2.512$. Moreover, the source with the numerically smaller magnitude is the brighter![2]

The total flux over all wavelengths is referred to as the 'bolometric magnitude'. However, life is a little more complicated than that: we do not in fact observe the flux integrated over all wavelengths. Observations are always restricted to some specific waveband determined by the observing instrument and its detector.

The total brightness of a source integrated over some waveband is measured by optical astronomers in units of $\mathrm{ergs\,cm^{-2}s^{-1}}$ and is translated into magnitudes. Thus the magnitude scale is established once a standard reference has been set in each waveband or for each wavelength filter that is used.

If $f(\lambda)$ denotes the energy flux measured in a filter centred at wavelength λ and having a passband of width $\Delta\lambda$, then the *flux density* is given by

$$S(\lambda) = \frac{f(\lambda)}{\Delta\lambda} \; \mathrm{ergs\,cm^{-2}s^{-1}Hz^{-1}}, \tag{B.2}$$

when, for example, the wavelength is given in Hz. The total flux is the integral of this flux density over all wavelengths. The zero points of the magnitude scale have been defined for specific filter sets. For example, in a particular filter known as the 'visual band' centred on 5500 A we have

$$\log_{10} S(\lambda) = -0.4m_V - 8.42, \tag{B.3}$$

and in this system the Sun has apparent magnitude $m_V(\odot) = -26.77$.

The choice of different filter bands leads to different magnitude systems. Data papers will always say precisely what their pass-bands are. Of course, this reflects a waveband at a range of given terrestrial frequencies. When the object is far away and its light is redshifted that pass-band does not reflect the same part of the spectrum as it would locally.

1850 and was published in 1860 in his *Elemente der Psychophysik*. In 1957, the American psychophysicist, Stanley Stevens, published a revised version of Fechner's relationship, giving a power-law relationship $P \propto S^\alpha$ where the value of α depended on the type of stimulus.

[2] We could have adopted the widely used measure of relative power, the *decibel*, abbreviated to 'dB'. Each decibel corresponds to a factor of $10^{0.1} = 1.259$, which corresponds in the world of audio to a 'just noticeable difference' (JND). A change of 10dB corresponds to a factor 10 in brightness (or loudness). The dB is formally defined in the International System of Quantities (ISQ) Standard ISO/IEC 80000 (2009). The dB is widely used in acoustics and electronics, and in particular in radio astronomy as a measure of signal to noise, or signal attenuation.

A correction has to be made that depends on redshift and on details of the spectrum, this is the infamous *K-correction*.

The SI unti of power, or luminosity, is the Watt, $1\text{W} = 1$ Joule $\text{s}^{-1} = 10^7$ erg s^{-1}. Radio astronomers measure flux density per unit frequency in units of $\text{Wm}^{-2}\,\text{Hz}^{-1}$. For numerical convenience, they give their flux density measurements in *Jansky* (Jy), where $1\text{Jy} = 10^{-26}\,\text{Wm}^{-2}\,\text{Hz}^{-1}$.

There are two standards for absolute calibration of fluxes and magnitudes: the *Vega-based* magnitudes and *AB-magnitudes*. The latter are commonly used in extragalactic work. The AB magnitude system defines brightness relative of a flat reference spectrum, and the zero point is defined to be 3631 Jy in all bands. Thus the magnitude m_{AB} corresponding to a flux F_ν is

$$m_{AB} = -2.5\log_{10}\left(\frac{F_\nu}{3631\,\text{Jy}}\right) = -2.5\log_{10}\left(\frac{F_\nu}{1\text{Jy}}\right) + 8.90. \qquad (B.4)$$

AB magnitudes convert to micro-Jansky according to $F(\mu\text{Jy}) = 10^{0.4(23.9-\text{AB})}$.

B.3 Absolute Magnitude and Distance Modulus

The luminosity of a body is the amount of energy the body radiates per unit time, measured in Watts. The Sun has a luminosity of 3.846×10^{26} W. For historical reasons, in optical astronomy this is measured on a magnitude scale: it is the apparent magnitude the object would have if placed at a distance of 10 parsecs.

Astronomers use another measure of distance based on the magnitude scale: the *distance modulus*. This is defined as the difference between the apparent magnitude m and absolute magnitude M of an object:

$$\mu = m - M. \qquad (B.5)$$

M is independent of distance and m increases with distance, so the distance modulus of an object increases with increasing distance.

The distance modulus is related to the distance in parsec by the equation

$$m - M = 5\log_{10}(d/10\text{pc}), \qquad (B.6)$$

and conversely

$$d = 10^{0.2(m-M+5)} = 10^{0.2\mu+1}\text{pc} \qquad (B.7)$$

relates the distance in parsecs to the distance modulus.

There are many refinements of these definitions that take account of the waveband used to define the magnitudes and to account for any absorption of light along the path between the source and the observer that affects the apparent magnitude m.

Appendix C Representing Vectors and Tensors

C.1 Some Ground Rules

It is convenient, for didactic purposes, to be able to model vectors and tensors as matrices. Indeed there is a close connection via vectors and dyadics, which were introduced into physics by Willard Gibbs in 1884 as a part of his goal to simplify the appearance of equations for teaching purposes. While vectors gained a permanent place within physics, the dyadics never made it into regular use and are now considered obsolete. Vectors and dyadics have given way to Cartan's differential forms, which provide a powerful way of handling physical quantities in a coordinate independent way (Schutz, 1980).

The goal here is considerably less ambitious, it is simply to provide a visualisation of vectors and tensors and simple mathematical operations involving them. Like vectors and dyadics, we confine ourselves to vectors and rank 2 tensors which, within a given coordinate system, have simple representations in component form within the theory of matrices. However, there is an important difference arising from the fact that we wish to work with 4-vectors and tensors in a space where the metric signature is the Minkowski $(-+++)$. We also wish to differentiate between covariant and contravariant representations of these objects within the given coordinate system.

The representations of a vector and a tensor in 4-matrix form are

$$u_a = \mathbf{u} = \begin{pmatrix} u_0 \\ \mathbf{u} \end{pmatrix}, \quad M_{ab} = \begin{pmatrix} m_0 & \mathbf{m}^T \\ \mathbf{m} & \mathbf{M} \end{pmatrix}. \tag{C.1}$$

Bold Latin letters denote 3-vectors and bold sans-serif symbols like $\mathbf{M}$ denote 3×3 matrices. The superfix ^{T} denotes the transpose of the vector or matrix. So here $\mathbf{u}$ denotes a 3-vector (the spatial components of u_a), u_0 is the time-like component of u_a. u_0 and m_0 are scalars.

We want to differentiate between the covariant and contravariant matrix representations of 4-vectors, u_a and u^a, and of tensors $M_{ab}, M^a{}_b, M_a{}^b$ and M^{ab}. The transformation between representations is achieved via a given matrix g_{ab} and its inverse g^{ab}:

$$g_{ab}g^{bc} = \delta_a^c, \quad \text{summation implied over index } b. \tag{C.2}$$

With this we define

$$u^a = g^{ab}u_b \Rightarrow \mathbf{u}^{\cdot} = \mathbf{g}^{\cdot\cdot}\mathbf{u}, \tag{C.3}$$

where the dots on the 4-matrices $\mathbf{u}$ and $\mathbf{g}$ indicate where the indices would be.

In this notation the Minkowski space metric η is

$$\eta^{cd} \equiv \boldsymbol{\eta}^{\cdot\cdot} = \begin{pmatrix} -1 & \mathbf{0}^T \\ \mathbf{0} & {}^{(3)}\mathbf{I} \end{pmatrix}, \quad \eta_{cd} \equiv \boldsymbol{\eta}_{\cdot\cdot} = \begin{pmatrix} -1 & \mathbf{0}^T \\ \mathbf{0} & {}^{(3)}\mathbf{I} \end{pmatrix}, \tag{C.4}$$

where ${}^{(3)}\mathbf{I}$ is the 3-identity matrix $\mathrm{diag}(1, 1, 1)$, and $\mathbf{0}$ is the null 3-vector whose components are zeros. Note that

$$\boldsymbol{\eta}^{\cdot\cdot}\boldsymbol{\eta}_{\cdot\cdot} = \eta^{ap}\eta_{pb} = \begin{pmatrix} 1 & \mathbf{0}^T \\ \mathbf{0} & {}^{(3)}\mathbf{I} \end{pmatrix} = \delta^a_b. \tag{C.5}$$

This conforms to (C.2) and suggests that it may be easier to work in terms of mixed up and down indices.

We need one more convention which distinguishes a product $M^{ab}u_a$ from $M^{ab}u_b$, i.e. we need to show whether the summation is over the first of two indices or the second. For this we need to adopt the convention that in M^{ab} the index a labels the rows and the index b labels the columns. So, as with all matrix multiplication, we have the identities

$$v^a = M^{ab}u_b \Leftrightarrow \mathbf{v}^{\cdot} = \mathbf{M}^{\cdot\cdot}\mathbf{u}, \quad \text{and} \quad w^b = M^{ab}u_a \Leftrightarrow \mathbf{w}^{\cdot T} = \mathbf{u}^{\cdot T}\mathbf{M}^{\cdot\cdot}. \tag{C.6}$$

The vectors $\mathbf{v}$ and $\mathbf{u}$ on the left are column 4-vectors, while the vectors on the right $\mathbf{w}^T$ and $\mathbf{u}^T$ are row 4-vectors. In this way we have for the product of two matrices the correspondences $A_{am}B^{mb} = \mathbf{AB}$ and $A_{ma}B^{bm} = \mathbf{BA}$. So, whereas in matrix notation the order of the matrices is important, in index notation it is the order of the indices that is important. Indeed, a good rule of thumb is to keep the summed indices adjacent as in $A_{am}B^{mb}$ rather than $P_{am}Q^{bm}$. The latter should be written $P_{am}(\mathbf{Q}^T)^{mb}$, so $\mathbf{Q}$ has to be transposed before multiplying.

We want to represent the covariant tensor F_{ab}, or its contravariant counterpart F^{ab}, in a given coordinate system as a matrix $\mathbf{F}$. [1] The matrix $\mathbf{F}$ comes in various forms:

$$F^{ab} \equiv \mathbf{F}^{\cdot\cdot} \quad F_{ab} \equiv \mathbf{F}_{\cdot\cdot} \quad F^a{}_b \equiv \mathbf{F}^{\cdot}{}_{\cdot} \quad F_a{}^b \equiv \mathbf{F}_{\cdot}{}^{\cdot} \tag{C.7}$$

It is necessary to indicate which of the indices is being raised or lowered because $F^a{}_b$ and $F_a{}^b$ are not necessarily the same. We indicate which has been raised or lowered via the position of the index. The order of the indices is important. We will retain the convention that the first index a of F_{ab} labels the rows of the matrix $\mathbf{F}$, while the second index b labels the columns. The covariant representation (F_{ab}) is generated from (F^{ab}) using the Minkowski metric η_{cd} (Equation C.4),

$$F_{ab} = \eta_{am}\eta_{nb}F^{mn} = \eta_{am}F^{mn}\eta_{nb}, \quad \text{summation over } m, n \text{ implied}, \tag{C.8}$$

which becomes

$$\mathbf{F}_{\cdot\cdot} = \boldsymbol{\eta}^T_{\cdot\cdot}\mathbf{F}^{\cdot\cdot}\boldsymbol{\eta}_{\cdot\cdot}. \tag{C.9}$$

We make use of this in Section D.3.

---

[1] Representing these index-related operations in matrix form is a little subtle, since matrix notation does not itself distinguish between representing the elements of a matrix as either upper or lower indices. As can be seen from these equations, the tensor-related operation of raising and lowering indices via η changes the representation of the matrix $\mathbf{F}$.

C.2 Lorentz Transformation: Matrix Form

The Lorentz transformation **L** of a 4-vector between two frames moving with relative velocity v along their shared x-axis is also represented by a 4×4 matrix, but **L** is not a tensor object: it merely expresses a relationship between two inertial observers. As a square matrix its elements are labelled by two indices, the row and column in which each element appears. But there is no sense in which these indices are to be regarded as being labels of either a covariant or contravariant object. Nevertheless, the Lorentz transformation acts on vector and tensor objects and so we have expressions which involve matrix multiplication in the sense of a linear transformation of variables. So we shall label the elements of the **L** matrix with indices as appropriate for the consistent application of the summation convention.

For the transformation of the coordinates $(ct, x, y, z) \rightarrow (ct', x', y'z')$ we have

$$\begin{pmatrix} ct' \\ x' \\ y' \\ z' \end{pmatrix} = \begin{pmatrix} \gamma & -\beta\gamma & 0 & 0 \\ -\beta\gamma & \gamma & 0 & 0 \\ 0 & 0 & 1 & 0 \\ 0 & 0 & 0 & 1 \end{pmatrix} \begin{pmatrix} ct \\ x \\ y \\ z \end{pmatrix}, \quad \beta = \frac{v}{c}, \quad \gamma = (1 - \beta^2)^{-1/2}, \quad (C.10)$$

which, when multiplied out gives the familiar transformation

$$t' = \gamma \left(t - \frac{vx}{c^2} \right), \quad x' = \gamma(x - vt), \quad y' = y, \quad z' = z. \quad (C.11)$$

For two frames of reference moving with relative velocity **v** in an arbitrary direction

$$\begin{pmatrix} ct' \\ \mathbf{r}' \end{pmatrix} = \begin{pmatrix} \gamma & -\gamma\boldsymbol{\beta}^T \\ -\gamma\boldsymbol{\beta} & {}^{(3)}\mathbf{1} + (\gamma - 1)\frac{\boldsymbol{\beta}\boldsymbol{\beta}^T}{\beta^2} \end{pmatrix} \begin{pmatrix} ct \\ \mathbf{r} \end{pmatrix}, \quad \boldsymbol{\beta} = \frac{\mathbf{v}}{c}, \quad \beta = |\boldsymbol{\beta}|. \quad (C.12)$$

In the same way the transformation of any 4-vector **q** can be written

$$\mathbf{q}' = \mathbf{L}\,\mathbf{q} \quad \Leftrightarrow \quad (q^a)' = L^a{}_b q^b, \quad (C.13)$$

where **L** is the 4-matrix in Equation (C.12).

The generalisation of this to the Lorentz transform of a tensor quantity T_{ab} is

$$\mathbf{T}'_{..} = \mathbf{L}^T \mathbf{T}_{..} \mathbf{L} \quad \Leftrightarrow \quad T'_{ab} = L^m{}_a T_{mn} L^n{}_b, \quad (C.14)$$

where the transpose $\mathbf{L}^T$ arises because of our convention on index ordering.

Appendix D The Electromagnetic Field

D.1 4-vectors and Tensors

We denote the coordinates of a point in Minkowski space by the 4-vector, x^a,

$$x^a = (ct, \mathbf{x}),$$ (D.1)

and so the 4-velocity is

$$u^a = (c, \mathbf{u}).$$ (D.2)

(Note that this is not normalised like the 4-velocity used in general relativistic fluids where $u^a u_a = -1$.)

The raising and lowering of indices is done with the Minkowski metric, which with our adopted conventions, is

$$\eta_{ab} \equiv \eta = \begin{pmatrix} -1 & \mathbf{0}^T \\ \mathbf{0} & {}^{(3)}\mathbf{I} \end{pmatrix}, \quad \eta^{cd} \equiv \eta = \begin{pmatrix} -1 & \mathbf{0}^T \\ \mathbf{0} & {}^{(3)}\mathbf{I} \end{pmatrix},$$ (D.3)

as per Equation (C.4). We note that $\eta^a{}_b = \eta^{am}\eta_{mb} = \delta^a_b = \mathrm{diag}(1111)$, the Kronecker delta. Note the difference between the representations of the 4-vector W with components W W^a and $W_a = \eta_{ab}W^b$:

$$W: \quad W^a = (w, \mathbf{W}), \quad W_a = (-w, \mathbf{W}).$$ (D.4)

The 4-gradient is variously expressed as

$$\partial_a \equiv \frac{\partial}{\partial x^a} \equiv \left(\frac{\partial}{\partial(ct)}, \mathbf{\nabla} \right) \equiv (\partial_{(ct)}, \mathbf{\nabla}), \quad \text{where } \partial_{(ct)} \equiv \frac{1}{c}\frac{\partial}{\partial t}.$$ (D.5)

Note that $\partial^a = \eta^{ab}\partial_b = (-\partial_{ct}, \mathbf{\nabla})$. The double derivative $\partial_a\partial_b$ will sometimes be denoted by ∂_{ab}, this makes the notation more compact.[1]

The 4-dimensional Minkowski space Laplacian is $\Box \equiv \partial_a\partial^a$:

$$\Box \equiv \partial_a\partial^a = \eta^{ab}\partial_a\partial_b = -\frac{\partial^2}{\partial t^2} + \nabla^2,$$ (D.6)

as can be verified by evaluating $\eta^{ab}\partial_a\partial_b$ via its matrix representation. Note that if we had used the $(+---)$ signature for the Minkowski metric the sign of this operator would have been reversed.

---

[1] But $\partial_{(ct)}$ should not be interpreted as $\partial_c\partial_t$! When we set $c = 1$ then the notation ∂_t should be unambiguous.

D.2 Electromagnetic Potentials

The vector fields $\mathbf{E}$ and $\mathbf{B}$ are the components of the *Faraday field tensor*, which is itself expressed in terms of derivatives of the field potentials ϕ and $\mathbf{A}$. The relationships between the $\mathbf{E}$ and $\mathbf{B}$ fields of Maxwell's equations and the electromagnetic scalar potential ϕ and vector potential $\mathbf{A}$ are

$$\mathbf{E} = -\nabla\phi - \frac{1}{c}\dot{\mathbf{A}}, \quad \mathbf{B} = \nabla \times \mathbf{A}, \quad \frac{1}{c}\frac{\partial\phi}{\partial t} + \nabla \cdot \mathbf{A} = 0. \tag{D.7}$$

The last of these equations is the *Lorenz gauge condition* which conveniently exploits the gauge freedom inherent in the definitions of the potentials ϕ and $\mathbf{A}$. These combine into a 4-potential with components

$$A^a = (\phi, \mathbf{A}), \quad A_a = (-\phi, \mathbf{A}), \quad \partial_a A^a = 0. \tag{D.8}$$

The last equation is just the Lorenz gauge condition.

It is straightforward to verify that these are consistent with the Maxwell equations.[2] With the choice of the Lorenz gauge condition the potentials satisfy the equations

$$\frac{1}{c^2}\frac{\partial^2\phi}{\partial t^2} - \nabla^2\phi = 4\pi\rho, \quad \frac{1}{c^2}\frac{\partial^2\mathbf{A}}{\partial t^2} - \nabla^2\mathbf{A} = \frac{4\pi}{c}\mathbf{J}. \tag{D.11}$$

In terms of the 4-potential, A^a, the Maxwell equations are simply

$$\Box A^a = -\frac{4\pi}{c}J^a, \quad \Box \equiv -\frac{1}{c^2}\frac{\partial}{\partial t} + \nabla^2. \tag{D.12}$$

Here, the 4-current J^a is the 4-vector having components

$$J^a = (c\rho, \mathbf{J}), \tag{D.13}$$

where ρ and $\mathbf{J}$ are the charge and current densities appearing on the right-hand side of the Maxwell equations (A.6–A.9).

D.3 The Faraday Tensor

The Faraday tensor is the tensor that unifies the electric and magnetic fields of Maxwell's theory into a single object $\mathbf{F} = (F^{ab})$ whose components transform correctly under Lorentz transformations.

---

[2]　This is derived as follows. Take the divergence and curl of $\mathbf{E}$ and $\mathbf{B}$ respectively in (D.7):

$$\nabla \cdot \mathbf{E} = -\nabla \cdot \nabla\phi - \frac{1}{c}\frac{\partial}{\partial t}\nabla \cdot \mathbf{A} = -\nabla^2\phi + \frac{1}{c^2}\frac{\partial^2\phi}{\partial t^2} = 4\pi\rho, \tag{D.9}$$

$$\nabla \times \mathbf{B} - \frac{1}{c}\frac{\partial\mathbf{E}}{\partial t} = \nabla \times (\nabla \times \mathbf{A}) - \frac{1}{c}\frac{\partial}{\partial t}\left(-\nabla\phi - \frac{1}{c}\frac{\partial\mathbf{A}}{\partial t}\right) = \nabla(\nabla \cdot \mathbf{A}) - \nabla^2\mathbf{A} + \nabla\frac{1}{c}\frac{\partial\phi}{\partial t} + \frac{1}{c^2}\frac{\partial^2\mathbf{A}}{\partial t^2} = \frac{4\pi}{c}\mathbf{J}. \tag{D.10}$$

In Equation (D.10) we have used the identity $\nabla \times (\nabla \times \mathbf{A}) = \nabla(\nabla \cdot \mathbf{A}) - \nabla^2\mathbf{A}$ and the Lorenz gauge condition.

The components of the Faraday tensor F_{ab}, or its contravariant form F^{ab} are

$$F_{ab} = A_{b,a} - A_{a,b} \quad F^{ab} = A^{b,a} - A^{a,b}, \tag{D.14}$$

which, by direct calculation using $A_a = (-\phi, \mathbf{A})$ (Equation D.8), leads to the expression for the Faraday tensor expressed in terms of the field components,

$$F_{0\mu} = \frac{\partial A_\mu}{\partial ct} - \frac{\partial A_0}{\partial x^\mu} = \frac{1}{c}\frac{\partial A_\mu}{\partial t} - \nabla\phi = -E_\mu, \tag{D.15}$$

$$F_{\mu\nu} = \frac{\partial A_\nu}{\partial x^\mu} - \frac{\partial A_\mu}{\partial x^\nu} = \epsilon^{\mu\nu\alpha} B_\alpha, \quad \text{no summation on } \alpha. \tag{D.16}$$

Here, as elsewhere in this book, the Latin indices, $a, b, \ldots$ range over $0, 1, 2, 3$ while Greek indices $\alpha, \beta, \ldots$ run over $1, 2, 3$, and the summation convention over repeated indices is assumed unless otherwise stated. The tensor $\epsilon^{\mu\nu\alpha}$ is the antisymmetric permutation symbol over the indices $\{1, 2, 3\}$, and so we have $F_{12} = B_3$, $F_{13} = -B_2$, $F_{23} = B_1$.

Let us define a column vector $\mathbf{e}$ and a 3×3 matrix $\mathbf{B}$ by

$$\mathbf{e} = \begin{pmatrix} E_1 \\ E_2 \\ E_3 \end{pmatrix}, \quad \mathbf{B} = \begin{pmatrix} 0 & B_3 & -B_2 \\ -B_3 & 0 & B_1 \\ B_2 & -B_1 & 0 \end{pmatrix}, \tag{D.17}$$

where (E_1, E_2, E_3) are the components of the electric field and (B_1, B_2, B_3) are the components of the magnetic field as they appear in Maxwell's equations. Then we have

$$F_{ab} \equiv \mathbf{F} = \begin{pmatrix} 0 & -\mathbf{e}^T \\ \mathbf{e} & \mathbf{B} \end{pmatrix}. \tag{D.18}$$

We can raise both indices using the metric $\eta^{\cdot\cdot} = (\eta^{cd})$:

$$\mathbf{F}^{\cdot\cdot} = \eta^{\cdot\cdot T}\mathbf{F}_{\cdot\cdot}\eta^{\cdot\cdot} = \eta^{ca}F_{am}\eta^{mb} = F^{cb} = \begin{pmatrix} -1 & \mathbf{0}^T \\ \mathbf{0} & {}^{(3)}\mathbf{I} \end{pmatrix}\begin{pmatrix} 0 & -\mathbf{e}^T \\ \mathbf{e} & \mathbf{B} \end{pmatrix}\begin{pmatrix} -1 & \mathbf{0}^T \\ \mathbf{0} & {}^{(3)}\mathbf{I} \end{pmatrix} = \begin{pmatrix} 0 & \mathbf{e}^T \\ -\mathbf{e} & \mathbf{B} \end{pmatrix}, \tag{D.19}$$

where the dot-notation for indices is as in Equations (C.7).[3] We shall also require the mixed forms

$$\mathbf{F}\eta \equiv F^{am}\eta_{mb} = F^a{}_b = \begin{pmatrix} 0 & \mathbf{e}^T \\ -\mathbf{e} & \mathbf{B} \end{pmatrix}\begin{pmatrix} -1 & \mathbf{0}^T \\ \mathbf{0} & {}^{(3)}\mathbf{I} \end{pmatrix} = \begin{pmatrix} 0 & \mathbf{e}^T \\ \mathbf{e} & \mathbf{B} \end{pmatrix}, \tag{D.20}$$

$$\eta^T\mathbf{F} \equiv \eta_{am}F^{mb} = F_a{}^b = \begin{pmatrix} -1 & \mathbf{0}^T \\ \mathbf{0} & {}^{(3)}\mathbf{I} \end{pmatrix}\begin{pmatrix} 0 & \mathbf{e}^T \\ -\mathbf{e} & \mathbf{B} \end{pmatrix} = \begin{pmatrix} 0 & -\mathbf{e}^T \\ -\mathbf{e} & \mathbf{B} \end{pmatrix}. \tag{D.21}$$

Because the Faraday matrix is skew-symmetric, it matters which of the indices is raised.

We can now write down the Maxwell equations in terms of the Faraday tensor:

$$\frac{\partial F^{ab}}{\partial x^b} = \frac{4\pi}{c}J^a, \tag{D.22}$$

---

[3] The difference between the products $\mathbf{AB}$ and $\mathbf{BA}$ lies in the summation: $\mathbf{AB} = A^{am}B_{mb}$ versus $\mathbf{BA} = B_{am}A^{mb}$. The summed index is the second index in the pre-multiplier and the first index of the post-multiplier so that we are fixing a row, a, of the pre-multiplier and convolve it with the column, b, of the post-multiplier.

and it is easily verified that the relativistic version of the Lorentz force law follows from

$$f^a = \frac{e}{c} F^a{}_b u^b, \tag{D.23}$$

where $u^a = (c, \mathbf{u})$ is the 4-velocity of the particle having charge e. The spatial part of this is

$$\frac{d\mathbf{p}}{dt} = \frac{d}{dt}\left(\frac{m_0 \mathbf{u}}{(1 - u^2/c^2)^{-1/2}}\right) = e\left(\mathbf{E} + \frac{1}{c}\mathbf{u} \times \mathbf{B}\right), \tag{D.24}$$

where $\mathbf{p} = m_0 \gamma \mathbf{u}$ is the relativistic momentum of the particle. This is the Lorentz force law (Equation 10.115). We can write equation (D.23) in terms of the current associated with the moving electron: $J^a = eu^a$, and then use the Maxwell Equation (D.22) to get:

$$f^a = \frac{1}{4\pi} F^a{}_b \frac{\partial F^{bc}}{\partial x^c}. \tag{D.25}$$

D.4 Lagrangian for the Electromagnetic Field

The Lagrangian density for the free electromagnetic field is

$$\mathcal{L}_{\text{EM}} = -\frac{1}{16\pi} F_{ab} F^{ab} = -\frac{1}{16\pi} \operatorname{tr}(F_{ab} F^{cb}), \tag{D.26}$$

$$= -\frac{1}{8\pi}\left(A_{a,b} A^{a,b} - A_{a,b} A^{b,a}\right), \tag{D.27}$$

where for compactness we have written $A_{a,b} = \partial_b A_a$. $\operatorname{tr} \mathbf{M} = \mathbf{M}^a{}_a$ is the trace of the matrix $\mathbf{M}$, i.e. the sum of the diagonal components of $\mathbf{M}^a{}_b$. Obtaining the Maxwell equations from a variational principle must be done via the second form (D.27): the potentials A^a are mutually independent whereas the fields $\mathbf{E}$ and $\mathbf{B}$ in the first form are not.

The summation in the last expression is over the second index, and so following our convention on matrix multiplication, this is the product $\mathbf{F}..\mathbf{F}^{..T}$. From Equations (D.18) and (D.19) we have

$$\mathbf{F}..\mathbf{F}^{..T} = \begin{pmatrix} 0 & -\mathbf{e}^T \\ \mathbf{e} & \mathbf{B} \end{pmatrix} \begin{pmatrix} 0 & \mathbf{e}^T \\ \mathbf{e} & \mathbf{B}^T \end{pmatrix} = \begin{pmatrix} -\mathbf{e}^T\mathbf{e} & -\mathbf{e}^T\mathbf{B}^T \\ \mathbf{B}\mathbf{e} & -\mathbf{e}\mathbf{e}^T + \mathbf{B}\mathbf{B}^T \end{pmatrix}. \tag{D.28}$$

Hence

$$\mathcal{L}_{\text{EM}} = -\frac{1}{16\pi}\left[\mathbf{e}^T\mathbf{e} + \operatorname{tr}(\mathbf{e}\mathbf{e}^T + \mathbf{B}\mathbf{B})\right], \tag{D.29}$$

$$= -\frac{1}{8\pi}(E^2 - B^2), \tag{D.30}$$

on using Equations (D.17) for the field components.

D.5 Energy–Momentum Tensor for the Electromagnetic Field

It is worth writing out these matrices to see what the energy–momentum tensor looks like. The key equations we shall use are the components $\mathbf{e}$ and $\mathbf{B}$ as given in (D.17) that make up the Faraday tensor (D.18) and its transpose (D.19). These are

$$\mathbf{F}_{..} = \begin{pmatrix} 0 & -E_1 & -E_2 & -E_3 \\ E_1 & 0 & B_3 & -B_2 \\ E_2 & -B_3 & 0 & B_1 \\ E_3 & B_2 & -B_1 & 0 \end{pmatrix}, \quad \mathbf{F}^{..} = \begin{pmatrix} 0 & E_1 & E_2 & E_3 \\ -E_1 & 0 & B_3 & -B_2 \\ -E_2 & -B_3 & 0 & B_1 \\ -E_3 & B_2 & -B_1 & 0 \end{pmatrix}. \quad (D.31)$$

We note that the sign of $\mathbf{B}$ has not changed. This is because the transpose is done using the raising and lowering operators η (Equations D.3).

We can now write out the product $\mathbf{F}_{..}\mathbf{F}^{..}$ as given in (D.28):

$$\mathbf{F}_{..}\mathbf{F}^{..T} = \begin{pmatrix} \mathbf{E}^2 & B_3E_2 - B_2E_3 & B_1E_3 - B_3E_1 & B_2E_1 - B_1E_2 \\ B_3E_2 - B_2E_3 & E_1^2 - B_3^2 - B_2^2 & E_1E_2 + B_1B_2 & E_3E_1 + B_3B_1 \\ B_1E_3 - B_3E_1 & E_1E_2 + B_1B_2 & E_2^2 - B_1^2 - B_3^2 & E_2E_3 + B_2B_3 \\ B_2E_1 - B_1E_2 & E_3E_1 + B_3B_1 & E_2E_3 + B_2B_3 & E_3^2 - B_2^2 - B_1^2 \end{pmatrix}. \quad (D.32)$$

The trace of this, which is the term in square brackets of (D.29) is

$$\mathrm{tr}(\mathbf{F}_{..}\mathbf{F}^{..T}) = 2(E^2 - B^2), \quad (D.33)$$

which was used to get to (D.30).

We can now do some notational simplification, first by introducing the vectors

$$\mathbf{s} = \mathbf{E} \times \mathbf{B} = (E_2B_3 - E_3B_2, E_3B_1 - E_1B_3, E_1B_2 - E_2B_1), \quad (D.34)$$

$$\mathbf{S} = \frac{c}{4\pi}\mathbf{E} \times \mathbf{B}, \quad (D.35)$$

where $\mathbf{S}$ is the *Poynting vector*, in our units. The energy–momentum tensor is

$$T_a{}^b = \frac{1}{4\pi}\left[F_{am}F^{mb} + \frac{1}{4}\delta_a^b F_{pq}F^{pq} \right]. \quad (D.36)$$

When written out as a matrix, the term in square brackets in the previous equation is

$$F_{am}F^{mb} + \tfrac{1}{4}\delta_a^b F_{pq}F^{pq} =$$

$$\begin{bmatrix} \frac{1}{2}(\mathbf{E}^2 + \mathbf{B}^2) & s_1 & s_2 & s_3 \\ s_1 & -\frac{1}{2}(\mathbf{E}^2 + \mathbf{B}^2) + E_1^2 + B_1^2 & E_1E_2 + B_1B_2 & E_3E_1 + B_3B_1 \\ s_2 & E_1E_2 + B_1B_2 & -\frac{1}{2}(\mathbf{E}^2 + \mathbf{B}^2) + E_2^2 + B_2^2 & E_2E_3 + B_2B_3 \\ s_3 & E_3E_1 + B_3B_1 & E_2E_3 + B_2B_3 & -\frac{1}{2}(\mathbf{E}^2 + \mathbf{B}^2) + E_3^2 + B_3^2 \end{bmatrix}.$$

$$(D.37)$$

With this we can write the energy–momentum tensor (D.36) in the shortened form

$$
\mathbf{T} = \left(
\begin{array}{c:cc}
\mathcal{E} & \frac{1}{c}\mathbf{S} \\
\hdashline
\frac{1}{c}\mathbf{S} & \mathbf{T}^M
\end{array}
\right), \tag{D.38}
$$

where

$$
\mathcal{E} = \frac{1}{8\pi}(\mathbf{E}^2 + \mathbf{B}^2) \tag{D.39}
$$

is the *electromagnetic field energy density* and $\mathbf{T}^M$ is the 3×3 *Maxwell stress tensor*:

$$
T^M_{\mu\nu} = \frac{1}{4\pi}\left[E_\mu E_\nu + B_\mu B_\nu - \tfrac{1}{2}\delta_{\mu\nu}(\mathbf{E}^2 + \mathbf{B}^2)\right], \qquad \mu, \nu = 1, 2, 3. \tag{D.40}
$$

We notice that the energy momentum 4-tensor $\mathbf{T}$ is trace-free.

As a trivial example let us suppose that there is no electric field, $\mathbf{E} = \mathbf{0}$, and that the $\mathbf{B}$-field is aligned along the 3-axis, $\mathbf{b} = (0, 0, B)$, then

$$
T_{ab} = \tfrac{1}{2}\mathrm{diag}(B^2, B^2, B^2, -B^2), \tag{D.41}
$$

$$
T^a_b = \tfrac{1}{2}\mathrm{diag}(-B^2, B^2, B^2, -B^2). \tag{D.42}
$$

The second of these comes by raising the first index using η. There are no off-diagonal terms since two of the B_ν are zero.

Appendix E Statistical Distributions

E.1 Gaussian, or Normal, Distribution

The Gaussian, or 'normal', distribution is arguably the most important of all continuous distributions. The distribution is specified via its probability density function (pdf) $f(x)$ since its cumulative distribution function (cdf) can only be expressed in terms of a somewhat unfamiliar function, the 'error function' $\text{erf}(x)$.[1] The Gaussian pdf is specified by two numbers: the mean μ and the variance σ and is generally denoted as $N(\mu, \sigma^2)$:

Definition E.1

Normal distribution, $N(\mu, \sigma)$: The probability density function is

$$f(x\,;\mu,\sigma) = \frac{1}{\sqrt{2\pi}\,\sigma}\exp\left(-\frac{(x-\mu)^2}{2\sigma^2}\right), \tag{E.1}$$

for which the distribution function can be written

$$F(x\,;\mu,\sigma) = \Phi\left(\frac{x-\mu}{\sigma}\right) = \frac{1}{2}\left[1 + \text{erf}\left(\frac{x-\mu}{\sigma\sqrt{2}}\right)\right]. \tag{E.2}$$

A random variable that is Gaussian distributed is also said the be 'Normally distributed as $N(\mu, \sigma^2)$'.

The special case $\mu = 0, \sigma = 1$ is the *standard normal distribution* and has probability density denoted by $\phi(x)$ and distribution function $\Phi(x)$ given by:

Definition E.2

Standard normal distribution, $N(0, 1)$:

$$\phi(x) = \frac{1}{\sqrt{2\pi}}e^{-x^2/2}, \quad \Phi(x) = \frac{1}{2}\left[1 + \text{erf}\left(\frac{x}{\sqrt{2}}\right)\right]. \tag{E.3}$$

---

[1] The error function is formally defined as

$$\text{erf}(x) = \frac{2}{\sqrt{\pi}}\int_0^x e^{-t^2}dt, \quad \text{erf}(\infty) = 1.$$

Note the lower limit of the integral. The complementary error function is simply $\text{erfc}(x) = 1 - \text{erf}(x)$.

The mean and variance are given by

$$\text{mean} = \mathbb{E}[X] = \mu, \tag{E.4}$$

$$\text{variance} = \mathbb{E}[(X - \mu)^2] = \sigma^2. \tag{E.5}$$

If random variables $X_1, X_2, \ldots, X_n$ are independently and identically distributed (iid) with distribution $N(\mu, \sigma)$, their mean $\bar{X} = (\sum_i X_i)/n$ is normally distributed as $N(\mu, \sigma/\sqrt{n})$.

If the random variables $X_1, X_2, \ldots, X_n$ are independently and identically distributed with finite mean μ and finite variance σ^2, then the sampling distribution of $\bar{X} = (\sum_i X_i)/n$ tends to $N(\mu, \sigma/\sqrt{n})$ as $n \to \infty$. This is a version of the central limit theorem which assures us that, in large samples, the distribution of estimates for sample means tends to a Gaussian, even if the underlying distribution is non-Gaussian.

E.2 Exponential Distribution

The exponential distribution is a single parameter distribution for a random variable X defined on the positive real line:

$$f_X(x) = \begin{cases} \lambda e^{-\lambda x} & x \geq 0\, x \in [0, \infty) \\ 0 & x < 0 \end{cases}. \tag{E.6}$$

The scale parameter λ is referred to as the *rate parameter*. The moments of the exponential distributed random variable X are

$$\mathbb{E}[X] = \lambda^{-1}, \quad \mathbb{E}[X^2] = \lambda^{-2}, \ldots, \mathbb{E}[X^k] = k!\,\lambda^{-k}. \tag{E.7}$$

E.3 Poisson Distribution

The Poisson distribution, discovered in 1711 by de Moivre, plays a key part in the understanding and modelling of point distributions and the theory of random walks. The Poisson distribution concerns a random variable X that can take on positive integer values with probability given by:

Definition E.3
Poisson distribution, $Pois(\lambda)$**:**

$$f(k; \lambda) = P[x = k] = e^{-\lambda}\frac{\lambda^k}{k!}, \quad k = 0, 1, \ldots, \infty. \tag{E.8}$$

The distribution describes the occurrence of rare events in a large number of trials. The mean and variance are given by

$$\text{mean, } \mu = \mathbb{E}\,[X] = \lambda, \tag{E.9}$$

$$\text{variance, } \text{Var}\,(X) = \mathbb{E}\,[(X - \mu)^2] = \lambda. \tag{E.10}$$

This distribution is the limit of the binomial distribution for a large number of trials.

We can exploit this to derive an important result about random *Poisson points*. Consider an interval $[-T/2, T/2]$ on the real line. Place n points at random in this interval and ask what is the probability that k of these points will lie in a subinterval of length τ. The probability that any one of the n points falls in τ is $p = \tau/T$ and so the probability that k out of n will fall in the interval τ is

$$\mathbb{P}\,[k \text{ points in } \tau\,] = \binom{n}{k} p^k (1 - p)^{n-k}, \quad p = \tau/T. \tag{E.11}$$

Then for $n \gg 1$ and $p \ll 1$ we have

$$\mathbb{P}\,[k \text{ points in } \tau\,] = e^{-n\tau/T} \frac{(n\tau/T)^k}{k!}. \tag{E.12}$$

Now we wish to make T arbitrarily large while keeping the density of points constant. We do this by writing the density of points as $\lambda = n/T$, and then

$$\mathbb{P}\,[k \text{ points in } \tau\,] = e^{-\lambda\tau} \frac{(\lambda\tau)^k}{k!}, \tag{E.13}$$

which is the continuous exponential distribution.

E.4 Gamma Distribution

The gamma distribution is defined in terms of its probability density function:

Definition E.4
Gamma distribution: shape parameter k:

$$f_X(x; k, \theta) = \frac{1}{\Gamma(k)} \frac{x^{k-1} e^{-x/\theta}}{\theta^k}, \quad x > 0, \quad k, \theta > 0, \tag{E.14}$$

where k is its *shape function* and θ is its *scale function*. $\Gamma(x)$ is the familiar gamma function and for integer values of the argument $\Gamma(n) = (n - 1)!$ for $n \in \mathbb{Z}^+$. Alternatively we can replace the scale θ by the *rate*, $\beta = \theta^{-1}$ and write:

Definition E.5
Gamma distribution: rate parameter β:

$$\text{Gamma}(x; \alpha, \beta) = \frac{1}{\Gamma(\alpha)} \beta^\alpha x^{\alpha-1} e^{-\beta x}. \tag{E.15}$$

The two versions differ only in that (E.14) uses the scale to characterise the exponential part of the function, while (E.15) uses the rate. Trivial as this may be, it is important to keep the distinction between the versions (E.14) and (E.15) clear and not confuse the two.

In the form (E.14) the mean and variance are

$$\mu_{\text{Gamma}} = \frac{\alpha}{\beta}, \quad \sigma^2 = \frac{\alpha}{\beta^2}, \quad \text{Skew} = \frac{2}{\sqrt{\alpha}}, \quad \text{Kurt} = \frac{6}{\alpha}, \quad \text{mode} = \frac{\alpha - 1}{\beta}. \quad (E.16)$$

The case $\alpha = 1$ is the exponential distribution.

E.5 χ^2 Distribution

This distribution depends on one free parameter, ν, describing the number of degrees of freedom of the distribution:

Definition E.6

χ-squared pdf with ν degrees of freedom:

$$f_X(x \mid \nu) = \frac{1}{2^{\nu/2}\Gamma(\frac{\nu}{2})} x^{\frac{\nu}{2}-1} e^{-\frac{x}{2}}, \quad x \geq 0. \quad (E.17)$$

This is also variously denoted by $\chi_\nu^2(x)$ or $\chi^2(x \mid \nu)$.

The mean and variance of this distribution are

$$\mu = \nu, \qquad \sigma^2 = 2\nu. \quad (E.18)$$

The cumulative distribution has been one of the most important functions in statistics for determining confidence bounds on estimators. The probability that the value of a χ^2 distributed random variable will be less than $x_{\text{max}} = \chi^2$ is given by the integral of the probability density (E.17):

$$F_X(x < \chi^2 \mid \nu) = \frac{1}{2^{\nu/2}\Gamma(\frac{\nu}{2})} \int_0^{\chi^2} t^{\frac{\nu}{2}-1} e^{-\frac{t}{2}} dt, \quad 0 \leq \chi^2 < \infty. \quad (E.19)$$

There is no analytic expression for this integral, and so most statistics texts and software packages have tables of its values for various values of ν and χ^2.

Appendix F Functions on a Sphere

F.1 Analysis of a Scalar Field on the Sphere

Here we summarise the main aspects of spherical harmonic analysis of functions on a sphere. This is used to analyse the CMB sky. Proofs and extensions to vector and tensor fields are found in the On-line Supplement *Functions on a Sphere*.

The mathematical description of a scalar field on a sphere begins with the specification of the field values, generally in spherical polar coordinates: ϕ, the azimuthal coordinate (i.e. longitude) and θ, the altitude coordinate (i.e. the latitude).[1] The conventional ranges for these coordinates are $0 \leq \phi < 2\pi$ and $0 \leq \theta \leq \pi$. (θ, ϕ) is a coordinate grid on the sphere.

The equivalent of Fourier analysis on a Cartesian grid is to represent the function $f(\theta, \phi)$ in terms of spherical harmonics, $Y_{l,m}(\theta, \phi)$. These are orthogonal eigenfunctions of the inhomogeneous Poisson equation

$$\nabla^2 Y_{l,m} = -\frac{l(l+1)}{r^2} Y_{l,m}. \tag{F.1}$$

The first few of these functions are

monopole: $l = 0$

$$Y_{0,0} = \sqrt{\frac{1}{4\pi}},$$

dipole: $l = 1$

$$Y_{1,0} = \sqrt{\frac{3}{4\pi}} \cos \theta,$$

$$Y_{1,\pm 1} = \mp \sqrt{\frac{3}{8\pi}} \sin \theta \, e^{\pm i\phi}.$$

quadrupole: $l = 2$

$$Y_{2,0} = \sqrt{\frac{5}{16\pi}} \left(3 \cos^2 \theta - 1\right),$$

$$Y_{2,\pm 1} = \mp \sqrt{\frac{15}{8\pi}} \sin \theta \cos \theta \, e^{\pm i\phi},$$

$$Y_{2,\pm 2} = \sqrt{\frac{15}{32\pi}} \sin^2 \theta \, e^{\pm 2i\phi}, \tag{F.2}$$

The range of l-values is $l \geq 0$ and for each l there are $2l + 1$ m-values, $-l, \ldots, +l$. The Y_{lm} are closely related to the associated Legendre functions $P_l^m(x)$ and the Legendre polynomials $P_l(\cos \theta)$. The functions Y_{lm} satisfy a variety of orthogonality conditions.

---

[1] There is no standard convention as to which of θ or ϕ should be the azimuthal angle, nor in which order they are specified, as in $f(\theta, \phi)$.

The way these are used is as follows. Suppose we wish to represent a function $f(\theta, \phi)$ defined on a sphere by its spherical harmonic expansion:

$$f(\theta, \phi) = \sum_{l=0}^{\infty} \sum_{m=-l}^{l} a_{lm}\, Y_{l,m}(\theta, \phi), \tag{F.3}$$

then the coefficients a_{lm} are given by

$$a_{lm} = \int_0^{2\pi} d\phi \int_0^{\pi} d\theta \sin\theta\, f(\theta, \phi)\, Y_{l,m}^*(\theta, \phi). \tag{F.4}$$

For each l, there are $2l + 1$ complex coefficients a_{lm}, corresponding to $m = -l, \ldots, +l$.

The Fourier and spherical harmonic representations of a function are related simply by knowing how to express plane wave $e^{i\mathbf{k}\cdot\mathbf{x}}$ in spherical harmonics:

Plane wave expressed in Legendre polynomials:

$$e^{i\mathbf{k}\cdot\mathbf{x}} = \sum_{l=0}^{\infty} i^l (2l+1) j_l(kr) P_l(\cos\gamma), \tag{F.5}$$

where γ is the angle between $\mathbf{k}$ and $\mathbf{x}$. The function $j_l(kr)$ is a *spherical Bessel function of the first kind* of order l.

If a function $f(\mathbf{r})$, is defined in 3-volume and has Fourier transform $g(\mathbf{k})$ then, with our normalisation of the Fourier transform:[2]

$$f(\mathbf{r}) = \int_{\mathbf{k}} g(\mathbf{k}) e^{i\mathbf{k}\cdot\mathbf{r}} d^3\mathbf{k}. \tag{F.6}$$

If we now write the spherical harmonic transform of $f(\mathbf{r})$ as

$$f(\mathbf{r}) = f(r, \hat{\mathbf{r}}) = \sum_{l=0}^{\infty} \sum_{m=-l}^{+l} a_{lm}\, Y_{lm}(\hat{\mathbf{r}}), \tag{F.7}$$

and compare this with the previous equation, we get the important result that:

Relating Fourier plane wave expansion to spherical harmonic expansion:

$$a_{lm} = 4\pi \int_{\mathbf{k}} i^l j_l(kr) Y_{lm}^*(\hat{\mathbf{k}})\, g(\mathbf{k}) d^3\mathbf{k}. \tag{F.8}$$

This equation will enable us to get the spherical harmonic representation, i.e. the coefficients a_{lm}, that represent a function whose Fourier transform is $g(\mathbf{k})$. In particular we shall apply this to the case when $g(\mathbf{k})$ is a random field with power spectrum $P(k)$. If $g(\mathbf{k})$ is the distribution of cosmic density fluctuations, the a_{lm} provide the spherical representation of that density field on a sphere of radius r from which we can calculate the power spectrum of measurements of the field on the surface of the sphere.

---

[2] Note that this equation, and hence the subsequent analysis, applies only to a function $f(\mathbf{r})$ defined on a flat space in which the functions $e^{i\mathbf{k}\cdot\mathbf{r}}$ form a complete orthonormal basis on the Cartesian frame of reference $\{x_i\}$.

In the cosmological context of observing temperature fluctuations on the surface of last scattering of CMB photons, we have $r = 2cH_0^{-1}$.

F.1.1 From 3-dimensions to 2-dimensions on a Sphere

The key to this is Equation (F.8) which tells us how to relate the 3-dimensional spatial Fourier amplitudes of a field to the spherical harmonic representation of the field on the surface of a sphere.

Consider a random function $f(\mathbf{x})$ having Fourier transform $g(\mathbf{k})$ and power spectrum $P_f(k)$ defined by $\langle g(\mathbf{k}) g(\mathbf{k}') \rangle = P_g(k) \delta^D(\mathbf{k} - \mathbf{k}')$ ($\delta^D(\mathbf{x})$ here being the Dirac delta function). Collecting up earlier results, $f(\mathbf{x})$ has Fourier and spherical harmonic representations:

$$f(\mathbf{x}) = \int g(\mathbf{k}) \, e^{i\mathbf{k}\cdot\mathbf{x}} d^3\mathbf{k}, \quad f(\mathbf{x}) = \sum_{l=0}^{\infty} \sum_{m=-l}^{+l} a_{lm} Y_{lm}(\hat{\mathbf{x}}), \tag{F.9}$$

where $\hat{\mathbf{x}}$ is the unit vector in the direction of $\mathbf{x}$. The coefficients a_{lm} are related to the $g(\mathbf{k})$ via the relationship (F.8).

We define the *angular power spectrum* of the function $f(\theta, \phi)$ as a function of the multipole index, l:

Angular power spectrum of $f(\theta, \phi)$:

$$c_l = \sum_{m=-l}^{l} |a_{lm}^2|. \tag{F.10}$$

This is just the sum of the squares of the $(2l+1)$ coefficients a_{lm} belonging to a given value of l. In this way l plays a role analogous to the frequency in Fourier analysis.

If $f(\theta, \phi)$ is a realisation of a zero-mean random process then different realisations have different $f(\theta, \phi)$ and yield different c_l. The coefficients c_l of Equation (F.10) are specific to a particular realisation. The statistical average of the c_l over many realisations, if it converges, is the expected power spectrum and is denoted by C_l. Parameterised theoretical models for $f(\theta, \phi)$ provide parameterised models for the expected power spectrum. Determining the model parameters from an observation of a particular realisation $f(\theta, \phi)$ is a problem in statistical inference.

From this, after some work, we can show that:

Angular power spectrum derived from 3-dimensional power spectrum:

$$c_l = \langle a_{lm} a_{lm}^* \rangle = 16\pi^2 \int_0^{\infty} j_l(kr)^2 P_g(k) k^2 dk. \tag{F.11}$$

The variable r is set to the radius of the sphere on which the C_l are defined. Although $j_l(kr)$ has no neat expression in terms of simple functions, the integral can be done analytically for some specific $P_g(k)$.

F.1.2 The Power Spectrum of the Sky Map

Now we turn to analysing the distribution on the sky of the fluctuations, $\Delta T(\theta, \phi)/T$ in the cosmic background radiation temperature, $T(\theta, \phi)$. It is conventional to normalise these fluctuations relative to the mean all sky temperature, T_0:

$$\frac{\Delta T(\theta, \phi)}{T} = \frac{T(\theta, \phi) - T_0}{T_0}, \tag{F.12}$$

where on the left side we drop the subscript on T_0 for convenience.[3] $\Delta T/T$ is a random function of position on the sky. It has, by construction, zero mean. Given the data we can calculate its spherical harmonic representation:

$$a_{lm} = \int_0^{2\pi} d\phi \int_0^{\pi} d\theta \sin\theta \, \frac{\Delta T(\theta, \phi)}{T} \, Y_{l,m}^*(\theta, \phi). \tag{F.13}$$

The a_{lm} are independently-distributed complex-valued random variables having zero mean and variance,

$$C_l = \langle |a_{lm}^2| \rangle, \tag{F.14}$$

where $\langle X \rangle$ denotes the statistical expectation of the random variable X. Note that if the temperature fluctuations, ΔT, are Gaussian distributed, then so are the a_{lm}, and hence the C_l are chi-square distributed with $2l + 1$ degrees of freedom.[4]

In practice we have to estimate C_l from the given data. This can be done by averaging the measured a_{lm} for each l over its $2l + 1$ m-values:

$$\bar{C}_l = \frac{1}{2l+1} \sum_{m=-l}^{l} |a_{lm}^2|. \tag{F.15}$$

This is an unbiased estimator. We see from these last two equations that the contribution to the variance from mode l is then

$$\langle |a_{lm}^2| \rangle = (2l+1)\bar{C}_l. \tag{F.16}$$

We understand this because there are $2l + 1$ modes that contribute to the variance at each l.

In the cosmological context we are interested in relating the observed power spectrum of the temperature fluctuations to the 3-dimensional power spectrum of the underlying temperature field and relating that to the power spectrum of the density fluctuations.

F.1.3 Correlation Function of the Sky Map

The correlation between the temperature variations $\Delta T(\theta, \phi)/T$ in directions on the sky separated by some angle ψ is defined as the average,

$$C(\psi) = \left\langle \frac{\Delta T(\mathbf{n_1})}{T} \frac{\Delta T(\mathbf{n_2})}{T} \right\rangle, \quad \mathbf{n_1} \cdot \mathbf{n_2} = \cos\psi \, , \tag{F.17}$$

where the average is taken over all pairs of directions $\mathbf{n_1}, \mathbf{n_2}$ separated by an angle ψ.

[3] We often see the notation $\Theta = \Delta T/T$ in the literature.
[4] See also Peebles (1993, Ch. 21, Equations 21.78 *et seq.*) where C_l here is denoted by a_l^2.

The spherical coordinate analogue of the well-known statement that the correlation function is the Fourier transform of the power spectrum then reads:[5]

$$C(\psi) = \frac{1}{4\pi} \sum_{l=0}^{\infty} (2l+1) C_l P_l(\cos\psi), \tag{F.19}$$

$$C_l = 2\pi \int_0^\pi C(\psi) P_l(\cos\psi) \sin\psi \, d\psi. \tag{F.20}$$

The fact that the correlation function $C(\psi)$ depends only on the angle ψ between the two directions $\mathbf{n}_1, \mathbf{n}_2$ is a reflection of the assumed statistical isotropy of the underlying random process $\Delta T(\mathbf{n})/T$.

The variance of the fluctuations is just $C(0)$ and is given by

$$C(0) = \frac{1}{4\pi} \sum_l (2l+1) C_l. \tag{F.21}$$

If we approximate this expression as an integral we have

$$C(0) = \frac{1}{4\pi} \sum_l (2l+1) C_l \simeq \frac{1}{2\pi} \int l(l+1) C_l \, d\ln l. \tag{F.22}$$

This shows that $l(l+1)C_l/2\pi$ is the contribution of the mode l to the total variance of the field, per unit logarithmic interval $d\ln l$. Accordingly it is usual to plot the values of $l(l+1)C_l/2\pi$ versus $\log l$ when presenting power spectrum data.

---

[5] We derive this as follows:

$$C(\psi) = \langle \Delta T(\mathbf{n}_1) \Delta T(\mathbf{n}_2) \rangle = \sum_{lm} \sum_{l'm'} \underbrace{\langle a_{lm} a^*_{l'm'} \rangle}_{C_l \delta_{ll'} \delta_{mm'}} Y_{lm}(\mathbf{n}_1) Y^*_{l'm'}(\mathbf{n}_2)$$

$$= \sum_l C_l \underbrace{\sum_m Y_{lm}(\mathbf{n}_1) Y^*_{lm}(\mathbf{n}_2)}_{\frac{1}{4\pi}(2l+1)P(\cos\psi)} = \frac{1}{4\pi} \sum_{l=0}^{\infty} (2l+1) C_l P_l(\cos\psi). \tag{F.18}$$

Appendix G Acknowledgements

Use of Figures

The publications for which a licence to re-use the material had to be obtained from the publisher, or their agent, are as follows:

- Figure 1.6 reproduced from Gamow (1956, Fig. 1. p.1731) with permission (©Elsevier 1956).
- Figure 3.2 reproduced from Bortolot *et al.* (1969, Fig. 1 p.308) with permission, ©1969 American Physical Society.
- Figure 3.7 (right) reproduced from Sunyaev and Zel'dovich (1970a, Fig. 1b p.9) with permission from the publishers via CCC 37001813366238.
- Figure 3.11 (left) reproduced from Gawiser and Silk (2000, Fig. 4 p.17) with permission from the publishers via CCC 3700450644616.
- Figure 4.3 adapted by permission from Macmillan Publishers Ltd: Massey *et al.* (2007, Nature) ©2007. CCC licence 3730161227530, article #6.
- Figure 27.3 reproduced from Sherwin *et al.* (2011, Figure 3) with permission from the publishers via CCC 3721150136987. ©2011 by the American Physical Society.

In all cases the full title of the article and other bibliographic data are available from the corresponding entry in the References section. 'CCC' refers to permissions gained through the Copyright Clearance Centre[1] and the number following is the granted licence number. These cover situations where the authors' permission was required but was not available.

NASA copyright policy states that 'NASA material is not protected by copyright unless noted'. Thus Figure 3.1[2] is in the public domain.

Unless otherwise noted, images and video on Laser Interferometer Gravitational-wave Observatory (LIGO) public web sites (public sites ending with a ligo.caltech.edu or ligo.mit.edu address) may be used for any purpose without prior permission.[3] See Figure 4.6.

[1] http://www.copyright.com/rightsholders/rightslink-permissions/
[2] https://en.wikipedia.org/wiki/Holmdel_Horn_Antenna
[3] https://www.ligo.caltech.edu/WA/page/image-use-policy

Figure 3.10: The *Particle Data Group* publishes annual *Reports on Particle Physics (RPP* which are published in the journal *Chinese Physics, C.* Since 2014, the figures from RPP are in the public domain (Olive and Particle Data Group, 2014)[4] and author permission is automatically granted. This concerns Figure 27.1 of Scott and Smoot (2014).

---

[4] http://pdg.lbl.gov/2014/figures/figures.html

References

Abbe, C. 1867. On the distribution of the nebulæ in space. *MNRAS*, **27**(May), 257–263.

Abbott, B. P., Abbott, R., Abbott, T. D. *et al.* 2016. Observation of gravitational waves from a binary black hole merger. *Physical Review Letters*, **116**(6), 061102.

Abbott, E.A. 2015. *Flatland: A Romance of Many Dimensions*. 2015 edn. CreateSpace Independent Publishing Platform.

Abbott, L. F., and Wise, M. B. 1984. Constraints on generalized inflationary cosmologies. *Nuclear Physics B*, **244**(Oct.), 541–548.

Abramowitz, M., and Stegun, I.A. 1965. *Handbook of Mathematical Functions*. 1965 edn. Dover Publications Inc.

Adams, W. S. 1941. Some results with the COUDÉ spectrograph of the Mount Wilson Observatory. *ApJ*, **93**(Jan.), 11.

Adler, R.J. 1981. *The Geometry of Random Fields*. John Wiley & Son.

Adler, R., Bazin, M., and Schiffer, M. 1975. *Introduction to General Relativity*. McGraw-Hill.

Albrecht, A., and Steinhardt, P. J. 1982. Cosmology for grand unified theories with radiatively induced symmetry breaking. *Physical Review Letters*, **48**(Apr.), 1220–1223.

Albrecht, A., Bernstein, G., Cahn, R. *et al.* 2006. Report of the Dark Energy Task Force. *arXiv astro-ph,* 0609591.

Albrecht, A., Amendola, L., Bernstein, G. *et al.* 2009. Findings of the Joint Dark Energy Mission Figure of Merit Science Working Group. *arXiv astro-ph.* 0901.0721.

Alcock, C., and Paczynski, B. 1979. An evolution free test for non-zero cosmological constant. *Nature*, **281**(Oct.), 358.

Aldrovandi, R., Caser, S., Omnes, R., and Puget, J. L. 1973. Matter-antimatter cosmology. *A&A*, **28**(Oct.), 253.

Alfvén, H., and Klein, O. 1962. Matter-antimatter annihilation and cosmology. *Arkiv för Fysik*, **23**, 187–194.

Ali-Haïmoud, Y., and Hirata, C. M. 2011. HyRec: A fast and highly accurate primordial hydrogen and helium recombination code. *Phys. Rev. D*, **83**(4), 043513.

Alpher, R. A. 1948. A neutron-capture theory of the formation and relative abundance of the elements. *Physical Review*, **74**(Dec.), 1577–1589.

Alpher, R. A., and Herman, R. C. 1950. Theory of the origin and relative abundance distribution of the elements. *Reviews of Modern Physics*, **22**(Apr.), 153–212.

Alpher, R. A., and Herman, R. C. 1953. The origin and abundance distribution of the elements. *Annual Review of Nuclear and Particle Science*, **2**, 1–40.

Alpher, R. A., Follin, J. W., and Herman, R. C. 1953. Physical conditions in the initial stages of the expanding Universe. *Physical Review*, **92**(Dec.), 1347–1361.

Amanullah, R., Lidman, C., Rubin, D. *et al.* 2010. Spectra and Hubble Space Telescope light curves of six type Ia supernovae at $0.511 < z < 1.12$ and the Union2 compilation. *ApJ*, **716**(June), 712–738.

Amendola, L., Kunz, M., and Sapone, D. 2008. Measuring the dark side (with weak lensing). *Journal of Cosmology and Astro-Particle Physics*, **4**(Apr.), 13.

Ames, A. 1950. A catalogue of 2778 nebulae, including the Coma-Virgo group. *Annals of Harvard College Observatory*, **88**, 1–40.

Anderson, L. *et al.* 2014a. The clustering of galaxies in the SDSS-III Baryon Oscillation Spectroscopic Survey: baryon acoustic oscillations in the Data Releases 10 and 11 galaxy samples. *MNRAS*, **441**(June), 24–62.

Anderson, L. *et al.* 2014b. The clustering of galaxies in the SDSS-III Baryon Oscillation Spectroscopic Survey: measuring D_A and H at z = 0.57 from the baryon acoustic peak in the Data Release 9 spectroscopic galaxy sample. *MNRAS*, **439**(Mar.), 83–101.

Anderson, T.W. 2003. *An Introduction to Multivariate Statistical Analysis.* Wiley–Blackwell; 3rd Edition.

Arago, F.J.D. 1820. Quelques nouveaux details sur le passage de la comte découverte dans le mois de Juillet 1819, devant le disque du soleil. *Annales de Chimie et de Physique, serie 2*, **13**, 104–110.

Aragón-Calvo, M. A., Jones, B. J. T., van de Weygaert, R., and van der Hulst, J. M. 2007. The multiscale morphology filter: identifying and extracting spatial patterns in the galaxy distribution. *A&A*, **474**(Oct.), 315–338.

Arnold, V.I. 1980. *Mathematical Methods of Classical Mechanics.* Graduate Texts in Mathematics, Vol. 60. Springer.

Ashby, N. 2002. Relativity and the Global Positioning System. *Physics Today*, **55**(5), 41–47.

Ashby, N. 2003. *Relativity in the Global Positioning System.* [Online]. Available: . http://www.livingreviews.org/lrr-2003-1.

Audouze, J., Pelletan, M.-C., Szalay, A., Zel'dovich, Y. B., and Peebles, P. J. E. (eds). 1988. *Large Scale Structures of the Universe. Proceedings of the 130th Symposium of the International Astronomical Union,* dedicated to the memoryof Marc A. Aaronson (1950 – 1987), held in Balatonfured, Hungary, 15–20 June 1987.

Baade, D., and Lucy, L. B. 1990. HST images: what can image processing do? *The Messenger*, **61**(Sept.), 24–27.

Baade, W., and Minkowski, R. 1954. Identification of the radio sources in Cassiopeia, Cygnus A, and Puppis A. *ApJ*, **119**(Jan.), 206.

Bade, W. L. 1953. Relativistic rocket theory. *American Journal of Physics*, **21**(Apr.), 310–312.

Baker, J. G., Centrella, J., Choi, D.-I., Koppitz, M., and van Meter, J. 2006a. Binary black hole merger dynamics and waveforms. *Phys. Rev. D*, **73**(10), 104002.

Baker, J. G., Centrella, J., Choi, D.-I., Koppitz, M., and van Meter, J. 2006b. Gravitational-wave extraction from an inspiraling configuration of merging black holes. *Physical Review Letters*, **96**(11), 111102.

Baldwin, O. R., and Jeffery, G. B. 1926. The relativity theory of plane waves. *Proceedings of the Royal Society of London Series A*, **111**(May), 95–104.

Bardeen, J. M. 1980. Gauge-invariant cosmological perturbations. *Phys. Rev. D*, **22**(Oct.), 1882–1905.

Barnes, L., Francis, M. J., Lewis, G. F., and Linder, E. V. 2005. The influence of evolving dark energy on cosmology. *Publications of the Astronomical Society of Australia*, **22**, 315–325.

Barrow, J. D., Juszkiewicz, R., and Sonoda, D. H. 1985. Universal rotation – How large can it be? *MNRAS*, **213**(Apr.), 917–943.

Barrow, J.D., and Webb, J.K. 2005. Inconstant constants. *Sci. Am.*, 70–77.

Bartholinus, E. 1670. *Experimenta Crystalli Islandici Disdiaclastici Quibus Mia & Infolita Refractio detegitur*. Hafniae.

Bateman, H. 1908. The conformal transformations of a space of four dimensions and their applications to geometrical optics. *Proc. London Math. Soc.*, **7**, *(2)*, 49–69.

Bateman, H. 1910. The transformation of the electrodynamical equations. *Proc. London Math. Soc.* **8**, *(2)*, 277–294.

Baumann, D. 2009. *TASI Lectures on Inflation. arXiv astro-ph*, 0907.5424

Baur, J., Blanchard, A., and Von Ballmoos, P. 2015. Is a symmetric matter-antimatter Universe excluded? *arXiv astro-ph*, 1512.08482

Baxter, Stephen. 2007. *Last Contact*. Solaris Science Fiction.

Baylis, W. E. 1980. Special relativity with 2x2 matrices. *American Journal of Physics*, **48**(Nov.), 918–925.

Becker, R. H., Fan, X., and White, R. L. *et al.* 2001. Evidence for reionization at $z \sim 6$: Detection of a Gunn-Peterson trough in a $z = 6.28$ quasar. *AJ*, **122**(Dec.), 2850–2857.

Begeman, K. G. 1989. H I rotation curves of spiral galaxies. I – NGC 3198. *A&A*, **223**(Oct.), 47–60.

Bekenstein, J. D. 2004. Relativistic gravitation theory for the modified Newtonian dynamics paradigm. *Phys. Rev. D*, **70**(8), 083509.

Bekenstein, J. D. 2010. Alternatives to dark matter: Modified gravity as an alternative to dark matter. *arXiv astro-ph*, 1001.3876

Belenkiy, A. 2013. 'The waters I am entering no one yet has crossed': Alexander Friedman and the origins of modern cosmology. Page 71 of: Way, M. J., and Hunter, D. (eds), *Origins of the Expanding Universe: 1912–1932*. Astronomical Society of the Pacific Conference Series, vol. 471.

Belinfante, F. 1939. On the spin angular momentum of mesons. *Physica*, **6**(July), 887–898.

Belinfante, F. J. 1940. On the current and the density of the electric charge, the energy, the linear momentum and the angular momentum of arbitrary fields. *Physica*, **7**(May), 449–474.

Belinskiĭ, V. A., and Khalatnikov, I. M. 1969. On the nature of the singularities in the general solutions of the gravitational equations. *Soviet Journal of Experimental and Theoretical Physics*, **29**, 911.

Belinskiĭ, V. A., Khalatnikov, I. M., and Lifshitz, E. M. 1970. Oscillatory approach to a singular point in the relativistic cosmology. *Advances in Physics*, **19**, 525–573.

Bellhouse, D.R. 2004. The Reverend Thomas Bayes, FRS: A biography to celebrate the tercentenary of his birth. *Statistical Science*, **19**, 3–43.

Bender, C.M., and Orzag, S.A. 1999. *Advanced Mathematical Methods for Scientists and Engineers: Asymptotic Methods and Perturbation Theory (v. 1)*. Springer.

Bennett, A. S. 1962. The preparation of the revised 3C catalogue of radio sources. *MNRAS*, **125**, 75.

Bennett, C. L. *et al.* 2013. Nine-year Wilkinson microwave anisotropy probe (WMAP) observations: Final maps and results. *Astrophys. J. Suppl.*, **208**(Oct.), 20.

Benson, A. J. 2010. Galaxy formation theory. *Phys. Rep.*, **495**(Oct.), 33–86.

Berger, J. 2006. The case for objective Bayesian analysis. *Bayesian Analysis*, **1**(*3*), 385–402.

Berk, R. 1967. Review 1922 of 'invariance of maximum likelihood estimators' by Peter W. Zehna. *Mathematical Reviews*, **30**, 343.

Bernoulli, D. 1738. *Hydrodynamica*. Reinhold, Basle.

Bernstein, J. 1988. *Kinetic Theory in the Expanding Universe*. Cambridge Monographs on Mathematical Physics. Cambridge University Press.

Bernstein, J. 2011. A memorandum that changed the world. *American Journal of Physics*, **79**(May), 440–446.

Berry, M. V. 1973. The statistical properties of echoes diffracted from rough surfaces. *Royal Society of London Philosophical Transactions Series A*, **273**(Feb.), 611–654.

Bertschinger, E. 1992. Large-scale structures and motions: Linear theory and statistics. Page 65 of: Martinez, V. J., Portilla, M., and Saez, D. (eds), *New Insights into the Universe*. Lecture Notes in Physics, Berlin Springer Verlag, Vol. 408.

Bethe, H. A. 1939. Energy production in stars. *Physical Review*, **55**(Mar.), 434–456.

Bethe, H. A., and Critchfield, C. L. 1938. The formation of deuterons by proton combination. *Physical Review*, **54**(Aug.), 248–254.

Betoule, M., Kessler, R. *et al.* 2014. Improved cosmological constraints from a joint analysis of the SDSS-II and SNLS supernova samples. *A&A*, **568**(Aug.), A22.

Bilicki, M., and Chodorowski, M. J. 2010. Influence of the local void on measurements of the clustering dipole. *MNRAS*, **406**(Aug.), 1358–1363.

Birrell, N. D., and Davies, P. C. W. 1982. *Quantum Fields in Curved Space*. Cambridge University Press.

Bičák, J., Katz, J., and Lynden-Bell, D. 2007. Cosmological perturbation theory, instantaneous gauges, and local inertial frames. *Phys. Rev. D*, **76**(*6*), 063501.

Blake, C., Kazin, E. A. *et al.* 2011. The WiggleZ Dark Energy Survey: mapping the distance–redshift relation with baryon acoustic oscillations. *MNRAS*, **418**(Dec.), 1707–1724.

Blake, C. *et al.* 2012. The WiggleZ Dark Energy Survey: joint measurements of the expansion and growth history at $z < 1$. *MNRAS*, **425**(Sept.), 405–414.

Blanchard, A. 1984. Angular fluctuations in the cosmological microwave background in a Universe with a cosmological constant. *A&A*, **132**(Mar.), 359.

Blau, M. 2014. *Lecture Notes on General Relativity*. Albert Einstein Centre for Fundamental Physics, Bern. [Online]. Available:. http://www.blau.itp.unibe.ch/newlecturesGR.pdf.

Bleem, L. E., Stalder, B. *et al.* 2015. Galaxy clusters discovered via the Sunyaev-Zel'dovich effect in the 2500-square-degree SPT-SZ Survey. *ApJs*, **216**(Feb.), 27.

Boardman, W. J. 1964. The radiative recombination coefficients of the hydrogen atom. *ApJs*, **9**(Aug.), 185–192.

Bodenner, J., and Will, C. M. 2003. Deflection of light to second order: A tool for illustrating principles of general relativity. *American Journal of Physics*, **71**(Aug.), 770–773.

Bond, J. R., Kofman, L., and Pogosyan, D. 1996. How filaments of galaxies are woven into the cosmic web. *Nature*, **380**(Apr.), 603–606.

Bondi, H. 1947. Spherically symmetrical models in general relativity. *MNRAS*, **107**, 410–425.

Bondi, H. 2009. *Cosmology*. 2nd edn. Dover Publications.

Bondi, H., and Gold, T. 1948. The steady-state theory of the expanding Universe. *MNRAS*, **108**, 252–270.

Bondi, H., Pirani, F. A. E., and Robinson, I. 1959. Gravitational waves in general relativity. III. Exact plane waves. *Proceedings of the Royal Society of London Series A*, **251**(June), 519–533.

Bonnor, W. B. 1954. The stability of cosmological models. *Zeitschrift fur Astrophysik*, **35**, 10.

Bonnor, W. B. 1957. Jeans' formula for gravitational instability. *MNRAS*, **117**, 104.

Born, M. 1909. Die Theorie des starren Elektrons in der Kinematik des Relativitätsprinzips. *Annalen der Physik*, **335**, 1–56.

Bortolot, V. J., Clauser, J. F., and Thaddeus, P. 1969. Upper limits to the intensity of background radiation at λ=1.32, 0.559, and 0.359 mm. *Physical Review Letters*, **22**(Feb.), 307–310.

Bosma, A. 1978. *The Distribution and Kinematics of Neutral Hydrogen in Spiral Galaxies of Various Morphological Types*. PhD thesis, Groningen Univ., (1978).

Bosma, A. 1981. 21-cm line studies of spiral galaxies. II. The distribution and kinematics of neutral hydrogen in spiral galaxies of various morphological types. *AJ*, **86**(Dec.), 1825–1846.

Boulliau, I. 1645. *Astronomia philolaica. Opus novum, in quo motus planetarum per novam ac veram hypothesim demonstrantur. The Inverse-Square Law of Attraction.* Simeonis Piget, Paris.

Boussinesq, J. 1877. Théorie des écoulements tourbillonnaries. *C.R. Acad. Sci Paris*, **23**.

Bousso, R. 2003. *Adventures in de Sitter Space.* Cambridge University Press, pp. 539–569.

Bower, R. G., Coles, P., Frenk, C. S., and White, S. D. M. 1993. Cooperative galaxy formation and large-scale structure. *ApJ*, **405**(Mar.), 403–412.

Box, T. C., and Roeder, R. C. 1984. The distribution of quasar emission-line redshifts. *A&A*, **134**(May), 234–239.

Branch, D., and Miller, D. L. 1993. Type IA supernovae as standard candles. *ApJl*, **405**(Mar.), L5–L8.

Brans, C., and Dicke, R. H. 1961. Mach's principle and a relativistic theory of gravitation. *Physical Review*, **124**(Nov.), 925–935.

Briggs, M. S. *et al.* 1999. Observations of GRB 990123 by the Compton Gamma Ray Observatory. *ApJ*, **524**(Oct.), 82–91.

Brinkmann, H.W. 1925. Einstein spaces which are mapped conformally on each other. *Mathematische Annalen*, **94**, 119–145.

Brout, R., Englert, F., and Gunzig, E. 1978. The creation of the Universe as a quantum phenomenon. *Annals of Physics*, **115**, 78–106.

Brown, H. 1949. A table of relative abundances of nuclear species. *Reviews of Modern Physics*, **21**(Oct.), 625–634.

Buchert, T. 2000. On average properties of inhomogeneous cosmologies. pp. 306–321 of: *Proceedings of 9th Workshop on General Relativity and Gravitation.*

Buchert, T. 2008. Dark energy from structure: a status report. *General Relativity and Gravitation*, **40**(Feb.), 467–527.

Buchert, T., and Ehlers, J. 1997. Averaging inhomogeneous Newtonian cosmologies. *A&A*, **320**(Apr.), 1–7.

Buddelmeijer, H. 2006. *Cosmotopology*. MPhil thesis, University of Groningen.

Bull, P., and Kamionkowski, M. 2013. What if Planck's Universe isn't flat? *Phys. Rev. D*, **87**(*8*), 081301.

Bunn, E. F., and Hogg, D. W. 2009. The kinematic origin of the cosmological redshift. *American Journal of Physics*, **77**(Aug.), 688–694.

Burbidge, E. M., Burbidge, G. R., Fowler, W. A., and Hoyle, F. 1957. Synthesis of the elements in stars. *Reviews of Modern Physics*, **29**, 547–650.

Burbidge, G. 1967. On the wavelengths of the absorption lines in quasi-stellar objects. *ApJ*, **147**(Feb.), 851.

Burbidge, G., and Napier, W. M. 2001. The distribution of redshifts in new samples of quasi-stellar objects. *AJ*, **121**(Jan.), 21–30.

Burbidge, G. R., and Burbidge, E. M. 1967. Limits to the distance of the quasi-stellar objects deduced from their absorption line spectra. *ApJl*, **148**(May), L107.

Burbidge, G. R., and O'dell, S. L. 1972. The distribution of redshifts of quasi-stellar objects and related emission-line objects. *ApJ*, **178**(Dec.), 583–606.

Burigana, C., Danese, L., and de Zotti, G. 1991. Formation and evolution of early distortions of the microwave background spectrum – A numerical study. *A&A*, **246**(June), 49–58.

Caderni, N., Fabbri, R., Melchiorri, B., Melchiorri, F., and Natale, V. 1978. Polarization of the microwave background radiation. II. An infrared survey of the sky. *Phys. Rev. D*, **17**(Apr.), 1908–1918.

Cai, Y.-C., Cole, S., Jenkins, A., and Frenk, C. 2009. Towards accurate modelling of the integrated Sachs–Wolfe effect: the nonlinear contribution. *MNRAS*, **396**(June), 772–778.

Calabretta, M. R., and Roukema, B. F. 2007. Mapping on the HEALPix grid. *MNRAS*, **381**(Oct.), 865–872.

Calder, L., and Lahav, O. 2008. Dark energy: back to Newton? *Astronomy and Geophysics*, **49**(*1*), 010000–1.

Caldwell, R. R., Kamionkowski, M., and Weinberg, N. N. 2003. Phantom energy: Dark energy with w < -1 causes a cosmic doomsday. *Physical Review Letters*, **91**(*7*), 071301.

Callan, C., Dicke, R. H., and Peebles, P. J. E. 1965. Cosmology and Newtonian mechanics. *American Journal of Physics*, **33**(Feb.), 105–108.

Campanelli, M., Lousto, C. O., Marronetti, P., and Zlochower, Y. 2006. Accurate evolutions of orbiting black-hole binaries without excision. *Physical Review Letters*, **96**(*11*), 111101.

Campbell, J.E. 1926. *A Course of Differential Geometry*. Clarendon Press.

Carroll, S. M., Press, W. H., and Turner, E. L. 1992. The cosmological constant. *ARA&A*, **30**, 499–542.

Carroll, S.M. 2003. *Spacetime and Geometry*. Addison Wesley.

Casella, G. 1989. An introduction to empirical Bayes data analysis. *Amer. Statistician*, **39**, 83–87.

Centrella, J., Baker, J. G., Kelly, B. J., and van Meter, J. R. 2010a. Black-hole binaries, gravitational waves, and numerical relativity. *Reviews of Modern Physics*, **82**(Oct.), 3069–3119.

Centrella, J., Baker, J. G., Kelly, B. J., and van Meter, J. R. 2010b. The final merger of black-hole binaries. *Annual Review of Nuclear and Particle Science*, **60**(Nov.), 75–100.

Chadwick, J. 1932. Possible existence of a neutron. *Nature*, **129**(Feb.), 312.

Challinor, A., and Lasenby, A. 1998. Relativistic corrections to the Sunyaev–Zeldovich effect. *ApJ*, **499**(May), 1–6.

Chan, K. L., and Jones, B. J. T. 1975a. Distortions of the 3 K background radiation spectrum – Observational constraints on the early thermal history of the universe. *ApJ*, **195**(Jan.), 1–11.

Chan, K. L., and Jones, B. J. T. 1975b. Distortions of the microwave background radiation spectrum in the submillimeter wavelength region. *ApJ*, **198**(June), 245–248.

Chan, K. L., and Jones, B. J. T. 1975c. The evolution of the cosmic radiation spectrum under the influence of turbulent heating. I – Theory. *ApJ*, **200**(Sept.), 454–460.

Chan, K. L., and Jones, B. J. T. 1975d. The evolution of the cosmic radiation spectrum under the influence of turbulent heating. II. Numerical calculation and application. *ApJ*, **200**(Sept.), 461–470.

Chandrasekhar, S. 2003. *An Introduction to the Study of Stellar Structure*. Dover Publications Inc.

Chandrasekhar, S., Gamow, G., and Tuve, M. A. 1938. The problem of stellar energy. *Nature*, **141**(May), 982.

Chang, K. 2006. *Black Holes Collide, and Gravity Quivers*. New York Times. [Online]. Available:
`http://www.nytimes.com/2006/05/02/science/space/02hole.html?`
`pagewanted=all&_r=0`.

Chluba, J., and Thomas, R. M. 2011. Towards a complete treatment of the cosmological recombination problem. *MNRAS*, **412**(Apr.), 748–764.

Chluba, J., Vasil, G. M., and Dursi, L. J. 2010. Recombinations to the Rydberg states of hydrogen and their effect during the cosmological recombination epoch. *MNRAS*, **407**(Sept.), 599–612.

Clifford, W.K. 1870. On the space-theory of matter. *Proc. Camb. Phil. Soc*, **2**, 157–158.

Clifton, T., Ferreira, P. G., Padilla, A., and Skordis, C. 2012. Modified gravity and cosmology. *Phys. Rep.*, **513**(Mar.), 1–189.

Cloutman, L.D. 1999. *Compressible Turbulence Transport Equations for Generalized Second Order Closure*. Informal Report UCRL-ID-134075. Lawrence Livermore National Laboratory.

Coc, A. 2013. Primordial nucleosynthesis. *Journal of Physics Conference Series*, **420**(*1*), 012136.

Cole, S., and Percival, W. J. *et al.* 2005. The 2dF Galaxy Redshift Survey: power-spectrum analysis of the final data set and cosmological implications. *MNRAS*, **362**(Sept.), 505–534.

Coleman, T. S., and Roos, M. 2003. Effective degrees of freedom during the radiation era. *Phys. Rev. D*, **68**(2), 027702.

Coles, P. 1996. *Einstein and the Birth of Big Science*. Postmodern Encounters. Totem Books.

Coles, P. 2001. Einstein, Eddington and the 1919 Eclipse. Page 21 of: Martínez, V. J., Trimble, V., and Pons-Bordería, M. J. (eds), *Historical Development of Modern Cosmology*. Astronomical Society of the Pacific Conference Series, Vol. 252.

Collins, H. 2004. *Gravity's Shadow: The Search for Gravitational Waves*. Univ. Chicago Press. (2nd. editiion).

Combrinck, L. 2013. *General Relativity and Space Geodesy*. Chap. 2, pp. 53–95, in Sciences of Geodesy – 11, ed. Guochang Xu, Springer.

Condon, J. J., Cotton, W. D., Fomalont, E. B., 2012. Resolving the radio source background: Deeper understanding through confusion. *ApJ*, **758**(Oct.), 23.

Conklin, E. K. 1969. Velocity of the Earth with respect to the cosmic background radiation. *Nature*, **222**(June), 971–972.

Conklin, E. K., and Bracewell, R. N. 1967. Limits on small scale variations in the cosmic background radiation. *Nature*, **216**(Nov.), 777–779.

Conley, A., Goldhaber, G., Wang, L. *et al.* 2006. Measurement of Ω_m, Ω_Λ from a blind analysis of type Ia supernovae with CMAGIC: Using color information to verify the acceleration of the Universe. *ApJ*, **644**(June), 1–20.

Conley, A., Guy, J., Sullivan, M. *et al.* 2011. Supernova constraints and systematic uncertainties from the first three years of the Supernova Legacy Survey. *ApJs*, **192**(Jan.), 1.

Corey, B. E., and Wilkinson, D. T. 1976. A measurement of the cosmic microwave background anisotropy at 19 GHz. *Bulletin of the American Astronomical Society*. **8**, 351.

Corry, L. 1999. From Mie's Electromagnetic Theory of Matter to Hilbert's Foundations of Physics. *Studies Hist. Phil. Modern Physics*, **30**, 159–183.

Corry, L., Renn, J., and J., Stachel. 1997. Belated decision in the Hilbert–Einstein priority dispute. *Science*, **278**, 1270–1273.

Couchman, H. M. P., and Carlberg, R. G. 1992. Large-scale structure in a low-bias universe. *ApJ*, **389**(Apr.), 453–463.

Courteau, S., and van den Bergh, S. 1999. The solar motion relative to the local group. *AJ*, **118**(July), 337–345.

Courtois, H. M., Pomarède, D., Tully, R. B., Hoffman, Y., and Courtois, D. 2013. Cosmography of the local Universe. *AJ*, **146**(Sept.), 69.

Cowan, Jr., C. L., Reines, F., Harrison, F. B., Kruse, H. W., and McGuire, A. D. 1956. Detection of the free neutrino: A confirmation. *Science*, **124**(July), 103–104.

Cox, D. R., and Hinkley, D. V. 1979. *Theoretical Statistics*. Chapman & Hall.

Cramér, H. 1946. *Mathematical Methods of Statistics*. Princeton Univ. Press.

Croom, S. M., Smith, R. J., Boyle, B. J. *et al.* 2004. The 2dF QSO Redshift Survey – XII. The spectroscopic catalogue and luminosity function. *MNRAS*, **349**(Apr.), 1397–1418.

Cunningham, E. 1908. Conformal representation and the transformation of Laplace's equation. *Proc. Lond. Math. Soc.*, vii. Communicated but not published.

Cunningham, E. 1914. *The Principle of Relativity*. Cambridge University Press.

Curtis, H. 1921. The scale of the Universe. *Bulletin of the National Research Council*, **2**, 194.

Curtis, H. D. 1920. Modern theories of the spiral nebulae. JRASC, **14**(Oct.), 317.

Cyburt, R. H., Fields, B. D., Olive, K. A., and Yeh, T.-H. 2016. Big bang nucleosynthesis: Present status. *Reviews of Modern Physics*, **88**(*1*), 015004.

Danby, G., Gaillard, J.-M., Goulianos, K. *et al.* 1962. Observation of high-energy neutrino reactions and the existence of two kinds of neutrinos. *Physical Review Letters*, **9**(July), 36–44.

Danese, L., and de Zotti, G. 1982. Double Compton process and the spectrum of the microwave background. *A&A*, **107**(Mar.), 39–42.

Das, S., Louis, T., Nolta, M. R. *et al.* 2014. The Atacama Cosmology Telescope: temperature and gravitational lensing power spectrum measurements from three seasons of data. *JCAP*, **4**(Apr.), 14.

Davidson, W. 1959. Steady-state cosmology treated according to general relativity. *MNRAS*, **119**, 309.

Davis, M., Efstathiou, G., Frenk, C. S., and White, S. D. M. 1992. The end of cold dark matter? *Nature*, **356**(Apr.), 489–494.

Davis, T. M., and Lineweaver, C. H. 2001. Superluminal recession velocities. pp. 348–351 of: Durrer, R., Garcia-Bellido, J., and Shaposhnikov, M. (eds), *Cosmology and Particle Physics*. American Institute of Physics Conference Series, Vol. 555.

Davis, T. M., and Lineweaver, C. H. 2004. Expanding confusion: Common misconceptions of cosmological horizons and the superluminal expansion of the Universe. *Publications of the Astronomical Society of Australia*, **21**, 97–109.

Davis, T. M., Mörtsell, E., Sollerman, J. *et al.* 2007. Scrutinizing exotic cosmological models using ESSENCE supernova data combined with other cosmological probes. *ApJ*, **666**(Sept.), 716–725.

de Bernardis, P. *et al.* 2000. A flat Universe from high-resolution maps of the cosmic microwave background radiation. *Nature*, **404**(Apr.), 955–959.

de Finetti, B. 1937. La prévision: ses lois logiques, ses sources subjectives. *Ann. Inst. Henri Poincaré*, **7**, 1–68.

de Finetti, B. 1974. *Theory of Probability (2 vols.)*. John Wiley and Sons.

de la Caille, N. 1755. Sur les Etoiles du Ciel Austral. *Histoire de l'Acad. Roy. des Sci. (Paris)*.

de Lapparent, V., Geller, M. J., and Huchra, J. P. 1986. A slice of the Universe. *ApJl*, **302**(Mar.), L1–L5.

de Martino, I., Atrio-Barandela, F., da Silva, A. *et al.* 2012. Measuring the redshift dependence of the cosmic microwave background monopole temperature with Planck data. *ApJ*, **757**(Oct.), 144.

de Sitter, W. 1917a. Einstein's theory of gravitation and its astronomical consequences. Third paper. *MNRAS*, **78**(Nov.), 3–28.

de Sitter, W. 1917b. On the relativity of inertia. Remarks concerning Einstein's latest hypothesis. *Koninklijke Nederlandse Akademie van Weteschappen Proceedings Series B Physical Sciences*, **19**, 1217–1225.

de Sitter, W. 1918. Further remarks on the solutions of the field-equations of the Einstein's theory of gravitation. *Koninklijke Nederlandse Akademie van Weteschappen Proceedings Series B Physical Sciences*, **20**, 1309–1312.

de Sitter, W. 1930. On the distances and radial velocities of extra-galactic nebulae, and the explanation of the latter by the relativity theory of inertia. *Proceedings of the National Academy of Science*, **16**(July), 474–488.

de Sitter, W. 1931. The expanding universe. *Scientia*, **49**, 1–10.

de Vaucouleurs, G., and de Vaucouleurs, A. 1964. *Reference Catalogue of Bright Galaxies*. University of Texas monographs in Astronomy, Austin.

de Vaucouleurs, G., de Vaucouleurs, A., and Corwin, Jr., H. G. 1976. *Second Reference Catalogue of Bright Galaxies. Containing Information on 4,364 Galaxies with References to Papers Published between 1964 and 1975.* University of Texas Press.

de Vaucouleurs, G., de Vaucouleurs, A., Corwin, Jr., H. G. *et al.* 1991. *Third Reference Catalogue of Bright Galaxies*, Vol. 1–3, Springer-Verlag.

Derlet, P. M., and Choy, T. C. 1996. Planck's radiation law: A many body theory perspective. *Australian Journal of Physics*, **49**, 589.

Dicke, R. H. 1963. Cosmology, Mach's principle and relativity. *American Journal of Physics*, **31**(July), 500–509.

Dicke, R. H., and Goldenberg, H. M. 1967. Solar oblateness and general relativity. *Physical Review Letters*, **18**(Feb.), 313–316.

Dicke, R. H., and Peebles, P. J. 1965. Gravitation and space science. *Space Sci. Rev.*, **4**(June), 419–460.

Dicke, R. H., Peebles, P. J. E., Roll, P. G., and Wilkinson, D. T. 1965. Cosmic black-body radiation. *ApJ*, **142**(July), 414–419.

Dirac, P. A. 1949. Forms of relativistic dynamics. *Reviews of Modern Physics*, **21**(July), 392–399.

Dirac, P. A. M. 1937. The cosmological constants. *Nature*, **139**(Feb.), 323.

Dirac, P. A. M. 1938. A new basis for cosmology. *Royal Society of London Proceedings Series A*, **165**(Apr.), 199–208.

Dodelson, S. 2003. *Modern Cosmology*. Academic Press.

Dolgov, A. D., Hansen, S. H., Pastor, S. *et al.* 2002. Cosmological bounds on neutrino degeneracy improved by flavor oscillations. *Nuclear Physics B*, **632**(June), 363–382.

DONUT Collaboration, Kodama, K., Ushida, N. *et al.* 2001. Observation of tau neutrino interactions. *Physics Letters B*, **504**(Apr.), 218–224.

Doob, J.L. 1949. Application of the theory of martingales. *Colloques Internationaux du CNRS*, 23–27.

Doroshkevich, A. G., and Novikov, I. D. 1964. Mean density of radiation in the metagalaxy and certain problems in relativistic cosmology. *Soviet Physics Doklady*, **9**(Aug.), 111.

Doroshkevich, A. G., and Shandarin, S. F. 1974. Galaxy formation in nonlinear gravitational-instability theory. *Soviet Ast.*, **18**(Aug.), 24.

Doroshkevich, A. G., Zel'dovich, Y. B., and Syunyaev, R. A. 1978. Fluctuations of the microwave background radiation in the adiabatic and entropic theories of galaxy formation. *Soviet Ast.*, **22**(Oct.), 523–528. Translation *Astronomicheskii Zhurnal*, **55**, (Sept.-Oct.) 1978, 913–921.

Doroshkevich, A. G., Kotok, E. V., Poliudov, A. N. *et al.* 1980. Two-dimensional simulation of the gravitational system dynamics and formation of the large-scale structure of the Universe. *MNRAS*, **192**(Aug.), 321–337.

Dreicer, H. 1964. Kinetic theory of an electron-photon gas. *Physics of Fluids*, **7**(May), 735–753.

Drever, R. W. P. 1977. Gravitational wave astronomy. *QJRAS*, **18**(Mar.), 9–27.

Droste, J. 1914. On the field of a single centre in Einstein's theory of gravitation. *Koninklijke Nederlandse Akademie van Wetenschappen Proceedings Series B Physical Sciences*, **17**, 998–1011.

Droste, J. 1917. The field of a single centre in Einstein's theory of gravitation, and the motion of a particle in that field. *Koninklijke Nederlandse Akademie van Wetenschappen Proceedings Series B Physical Sciences*, **19**, 197–215.

Duerbeck, H. W., and Seitter, W. C. 2001. *In Hubble's Shadow: Early Research on the Expansion of the Universe*. In *100 Years of Observational Astronomy and Astrophysics*, ed. Sterken, C. and Hearnshaw, J.B., Chapter 15, p. 231, VUB.

Dunham, Jr., T., and Adams, W. S. 1939. New interstellar lines in the ultra-violet spectrum. *Publications of the American Astronomical Society*, **9**. 5.

Dunkley, J., Spergel, D. N. *et al.* 2009. Five-year Wilkinson Microwave Anisotropy Probe (WMAP) observations: Bayesian estimation of cosmic microwave background polarization maps. *ApJ*, **701**(Aug.), 1804–1813.

Dünner, R., Hasselfield, M., Marriage, T. A., *et al.* 2013. The Atacama Cosmology Telescope: Data characterization and mapmaking. *ApJ*, **762**(Jan.), 10.

Dyson, F. J. 1969. The efficiency of energy release in gravitational collapse. *Comments on Astrophysics and Space Physics*, **1**(May), 75.

Dyson, F. W., Eddington, A. S., and Davidson, C. 1920. A determination of the deflection of light by the Sun's gravitational field, from observations made at the total eclipse of May 29, 1919. *Royal Society of London Philosophical Transactions Series A*, **220**, 291–333.

Eckart, C. 1940. The thermodynamics of irreversible processes. III. Relativistic theory of the simple fluid. *Physical Review*, **58**(Nov.), 919–924.

Eddington, A.S. 1923. *The Mathematical Theory of Relativity*. Reprinted 1960, 1965 edn. Cambridge Univerity Press.

Eddington, A.S. 1926. *The Internal Constitution of the Stars*. Cambridge Univerity Press.

Edwards, D. 1972. Exact expressions for the properties of the zero-pressure Friedmann models. *MNRAS*, **159**, 51.

Efron, B., and Morris, C. 1977. Stein's paradox in statistics. *Sci. Am.*, **236**, 119–127.

Efstathiou, G., Sutherland, W. J., and Maddox, S. J. 1990. The cosmological constant and cold dark matter. *Nature*, **348**(Dec.), 705–707.

Egan, C. A., and Lineweaver, C. H. 2010. A larger estimate of the entropy of the Universe. *ApJ*, **710**(Feb.), 1825–1834.

Ehlers, J. 1993. Contributions to the relativistic mechanics of continuous media. *General Relativity and Gravitation*, **25**(Dec.), 1225–1266. Translated by G.F.R. Ellis (with P.K.S. Dunsby).

Einasto, J. 2014. *Dark Matter and the Cosmic Web Story*. World Scientific Publishing Co.

Einasto, J., Einasto, M., Gottlöber, S. *et al.* 1997. A 120-Mpc periodicity in the three-dimensional distribution of galaxy superclusters. *Nature*, **385**(Jan.), 139–141.

Einstein, A. 1905. Zur Elektrodynamik bewegter Körper. *Annalen der Physik*, **322**, 891–921.

Einstein, A. 1914. Die formale Grundlage der allgemeinen Relativitätstheorie. *Sitzungsberichte der Königlich Preußischen Akademie der Wissenschaften (Berlin)*, 1030–1085. English translation:
http://einsteinpapers.press.princeton.edu/vol6-trans/42?ajax.

Einstein, A. 1915. Die Feldgleichungen der Gravitation. *Königlich Preussische Akademie der Wissenschaften*, 844–847. English translation:
http://einsteinpapers.press.princeton.edu/vol6-trans/129.

Einstein, A. 1915. Erklarung der Perihelionbewegung der Merkur aus der allgemeinen Relativitatstheorie. *Sitzungsber. preuss.Akad. Wiss.*, **47**, 831–839. English translation:
http://einsteinpapers.press.princeton.edu/vol6-trans/125?ajax.

Einstein, A. 1916a. Die Grundlage der allgemeinen Relativitätstheorie. *Annalen der Physik*, **354**, 769–822.

Einstein, A. 1916b. Näherungsweise Integration der Feldgleichungen der Gravitation. *Sitzungsberichte der Königlich Preußischen Akademie der Wissenschaften (Berlin)*, 688–696. English translation:
http://einsteinpapers.press.princeton.edu/vol6-trans/214?ajax.

Einstein, A. 1917a. Kosmologische Betrachtungen zur allgemeinen Relativitätstheorie. *Sitzungsberichte der Königlich Preußischen Akademie der Wissenschaften (Berlin)*, 142–152.

Einstein, A. 1917b. Zur Quantentheorie der Strahlung. *Physikalische Zeitschrift*, **18**, 121–128. English translation:
http://astro1.panet.utoledo.edu/~ljc/einstein_ab.pdf.

Einstein, A. 1918a. Prinzipielles zur allgemeinen Relativitätstheorie. *Annalen der Physik*, **55**, 241–244.

Einstein, A. 1918b. Über Gravitationswellen. *Sitzungsberichte der Königlich Preußischen Akademie der Wissenschaften (Berlin)*, 154–167. English translation:
http://einsteinpapers.press.princeton.edu/vol7-trans/25?ajax.

Einstein, A. 1919. Spielen Gravitationsfelder im Aufbau der materiellen Elementarteilchen eine wesentliche Rolle? *Sitzungsberichte der Königlich Preußischen Akademie der Wissenschaften (Berlin)*, 349–356. English translation:
http://einsteinpapers.press.princeton.edu/vol7-trans/96.

Einstein, A. 1922. Bemerkung zu arbeit von A. Friedman 'Uber die Krümmung des Raumes'. *Zeitschr. Phys.*, **11**, 326.

Einstein, A. 1923. Notiz zu Bemerkung zu arbeit von A. Friedman 'Uber die Krümmung des Raumes'. *Zeitschr. Phys.*, **16**, 228.

Einstein, A. 1931. Zum kosmologischen problem der allgemeinen Relativitätstheorie. *Sitzungsberichte der Königlich Preußischen Akademie der Wissenschaften (Berlin)*, 235 – 237.

Einstein, A., and de Sitter, W. 1932. On the relation between the expansion and the mean density of the Universe. *Proceedings of the National Academy of Science*, **18**(Mar.), 213–214. Contributions from the Mount Wilson Observatory, Vol. 3, pp.51–52.

Einstein, A., and Grossmann, M. 1913. Entwurf einer verallgemeinerten Relativitätstheorie und einer Theorie der Gravitation. *Zeitschrift für Mathematik und Physik*, **62**, 1– 38. Engish translation: http://www.pitt.edu/~jdnorton/teaching/GR&Grav_2007/pdf/Einstein_Entwurf_1913.pdf.

Einstein, A., and Straus, E. G. 1945. The influence of the expansion of space on the gravitation fields surrounding the individual stars. *Reviews of Modern Physics*, **17**(Apr.), 120–124.

Einstein, A., and Straus, E. G. 1946. Corrections and additional remarks to our paper: The influence of the expansion of space on the gravitation fields surrounding the individual stars. *Reviews of Modern Physics*, **18**(Jan.), 148–149.

Eisele, C., Nevsky, A. Y., and Schiller, S. 2009. Laboratory test of the isotropy of light propagation at the 10^{-17} level. *Physical Review Letters*, **103**(9), 090401.

Eisenhart, L.P. 1926. *Riemannian Geometry*. Princeton University Press.

Eisenstein, D., and White, M. 2004. Theoretical uncertainty in baryon oscillations. *Phys. Rev. D*, **70**(10), 103523.

Eisenstein, D. J., Zehavi, I. *et al.* 2005. Detection of the baryon acoustic peak in the large-scale correlation function of SDSS luminous red galaxies. *ApJ*, **633**(Nov.), 560–574.

Elfessi, A., and Reineke, D.M. 2001. A Bayesian look at classical estimation: The exponential distribution. *Journal of Statistics Education*, **9**. http://www.amstat.org/publications/jse/v9n1/elfessi.html

Ellis, G. F. R. 1971. Relativistic cosmology. 104–182 of: Sachs, R. K. (ed), *General Relativity and Cosmology*. Republished as a '*Golden Oldie*': Ellis (2009b).

Ellis, G. F. R. 2009a. Editorial note to: Pascual Jordan, Jürgen Ehlers and Wolfgang Kundt, Exact solutions of the field equations of the general theory of relativity. *General Relativity and Gravitation*, **41**(Sept.), 2179–2189.

Ellis, G. F. R. 2009b. Republication of: Relativistic cosmology. *General Relativity and Gravitation*, **41**(Mar.), 581–660.

Ellis, G. F. R. 2011. Inhomogeneity effects in cosmology. *Classical and Quantum Gravity*, **28**(*16*), 164001.

Ellis, G. F. R., and Rothman, T. 1993. Lost horizons. *American Journal of Physics*, **61**(Oct.), 883–893.

Ellis, G. F. R., and Rothman, T. 1995. Past light cone shape and refocusing in cosmology, A Response to Michael Rauch's 'Comments on "Lost Horizons" ' [*Am. J. Phys.* **63**, 87 (1995)]. *American Journal of Physics*, **63**(Jan.), 88–89.

Ellis, G. F. R., Maartens, R., and MacCallum, M. A. H. 2012. *Relativistic Cosmology.* Cambridge University Press.

Englert, F. 1999. Primordial inflation. *arXiv hep-th*, 9911185

Epstein, E. E. 1967. On the small-scale distribution at 3.4-mm wavelength of the reported 3 K background radiation. *ApJ*, **148**(June), L157.

Epstein, R.A. 1977. *The Theory of Gambling and Statistical Logic.* Academic Press.

Euler, L. 1757. Principes generaux du mouvement des fluides. *Mémoires de l'académie des sciences de Berlin*, **11**, 274–315.

Everitt, C.W.F. *et al.* 2011. Gravity probe B: Final results of a space experiment to test general relativity. Phys. Rev. Lett., **106**, 221101.

Ewen, H. I., and Purcell, E. M. 1951. Observation of a line in the galactic radio spectrum: Radiation from galactic hydrogen at 1,420 Mc./sec. *Nature*, **168**(Sept.), 356.

Exo-200 Collaboration. 2014. Search for Majorana neutrinos with the first two years of EXO-200 data. *Nature*, **510**(June), 229–234.

Fan, X. 2012. Observations of the first light and the epoch of reionization. *Research in Astronomy and Astrophysics*, **12**(Aug.), 865–890.

Faraoni, V. 2007. A common misconception about LIGO detectors of gravitational waves. *General Relativity and Gravitation*, **39**(May), 677–684.

Favre, A.J. 1965. *The Equations of a Compressible Turbulent Gas.* Annual Summary Report No.1. Institut de Mécanique Statistique de la Turbulence.

Fendt, W. A., Chluba, J., Rubiño-Martín, J. A., and Wandelt, B. D. 2009. RICO: A new approach for fast and accurate representation of the cosmological recombination history. *ApJs*, **181**(Apr.), 627–638.

Fermi, E., and Turkevich, A. 1949. Unpublished (classified document).

Ferris, T.A. 2003. *Coming of Age in the Milky Way.* Harper Perennial.

Feynman, R. P. 1965. *Feynman Lectures on Physics.* Addison Wesley.

Field, G. B., and Hitchcock, J. L. 1966a. Cosmic black-body radiation at λ=2.6 mm. *Physical Review Letters*, **16**(May), 817–818.

Field, G. B., and Hitchcock, J. L. 1966b. The radiation temperature of space at $\lambda = 2.6$ mm and the excitation of interstellar CN. *ApJ*, **146**(Oct.), 1.

Field, G. B., Herbig, G. H., and Hitchcock, J. 1966. Radiation temperature of space at $\lambda = 2.6$ mm. *AJ*, **71**(Feb.), 161–161.

Field, G. B., Rees, M. J., and Sciama, D. W. 1969. The astronomical significance of mass loss by gravitational radiation. *Comments on Astrophysics and Space Physics*, **1**(Nov.), 187.

Field, G. B., Arp, H. C., and Bahcall, J. N. 1973. *The Redshift Controversy.* Benjamin.

Filippenko, A. V. 1997. Optical spectra of supernovae. *ARA&A*, **35**, 309–355.

Fisher, R.A. 1922. On the mathematical foundations of theoretical statistics. *Phil. Transactions Roy. Soc.*, **222A**, 309–368.

Fisher., R.A. 1925. Theory of statistical estimation. *Proc. Camb. Phil. Soc.*, **22**, 700–725.

Fisher, R.A. 1925 (1st edition). *Statistical Methods for Research Workers.* Oliver and Boyd.

Fixsen, D. J. 2009. The temperature of the cosmic microwave background. *ApJ*, **707**(Dec.), 916–920.

Fixsen, D. J., Cheng, E. S., and Wilkinson, D. T. 1983. Large-scale anisotropy in the 2.7-K radiation with a balloon-borne maser radiometer at 24.5 GHz. *Physical Review Letters*, **50**(Feb.), 620–622.

Fixsen, D. J., Cheng, E. S., Cottingham, D. A. *et al.* 1994. Cosmic microwave background dipole spectrum measured by the COBE FIRAS instrument. *ApJ*, **420**(Jan.), 445–449.

Fixsen, D. J., Cheng, E. S., Gales, J. M. *et al.* 1996. The cosmic microwave background spectrum from the full COBE FIRAS data set. *ApJ*, **473**(Dec.), 576.

Flesch, E. W. 2015. The half million quasars (HMQ) catalogue. *Pub. Astron. Soc. Aus.*, **32**(Mar.), 10.

Ford, L. H. 1987. Cosmological-constant damping by unstable scalar fields. *Phys. Rev. D*, **35**(Apr.), 2339–2344.

Forger, M., and Römer, H. 2004. Currents and the energy-momentum tensor in classical field theory: a fresh look at an old problem. *Annals of Physics*, **309**(Feb.), 306–389.

Forward, R. L. 1978. Wideband laser-interferometer gravitational-radiation experiment. *Phys. Rev. D*, **17**(Jan.), 379–390.

Frankel, T. 2011. *The Geometry of Physics: An Introduction*. 3rd. edn. Cambridge University Press.

Freedman, W. L., and Madore, B. F. 2010. The Hubble constant. *ARA&A*, **48**(Sept.), 673–710.

Freedman, W. L., Madore, B. F., Gibson, B. K. *et al.* 2001. Final results from the Hubble Space Telescope key project to measure the Hubble constant. *ApJ*, **553**(May), 47–72.

Freedman, W. L., Madore, B. F., Scowcroft, V. *et al.* 2012. Carnegie Hubble Program: A mid-infrared calibration of the Hubble constant. *ApJ*, **758**(Oct.), 24.

Frieden, B. R. 1972. Restoring with maximum likelihood and maximum entropy. *Journal of the Optical Society of America (1917-1983)*, **62**(Apr.), 511.

Friedmann, A. 1922. On the curvature of space. *Zeitschrift f. Phys.*, **10**, 337. English translation in *General Relativity and Gravitation 1999*, **31**, 1991.

Friedmann, A. 1924. On the possibility of a world with constant negative curvature of space. *Zeitschrift f. Phys.*, **21**, 326. English translation in *General Relativity and Gravitation 1999*, **31**, 2001.

Frobenius, G. 1877. Ueber das Pfaffsche Problem. *Journal für die reine und angewandte Mathematik*, **82**, 230–315.

Fujita, T., Kawasaki, M., Tada, Y., and Takesako, T. 2013. A new algorithm for calculating the curvature perturbations in stochastic inflation. *JCAP*, **1312**(Dec.), 36.

Fujita, T., Kawasaki, M., and Tada, Y. 2014. Non-perturbative approach for curvature perturbations in stochastic δN formalism. *JCAP*, **10**(Oct.), 030.

Fukugita, M., and Peebles, P. J. E. 2004. The cosmic energy inventory. *ApJ*, **616**(Dec.), 643–668.

Futamase, T. 1988. Approximation scheme for constructing a clumpy Universe in general relativity. *Physical Review Letters*, **61**(Nov.), 2175–2178.

Futamase, T. 1989. An approximation scheme for constructing inhomogeneous Universes in general relativity. *MNRAS*, **237**(Mar.), 187–200.

Galama, T. J., Groot, P. J., van Paradijs, J. *et al.* 1997. GRB 970508. IAU Circ., **6655** (May), 1.

Gamow, G. 1928. Zur Quantentheorie des Atomkernes. *Zeitschrift fur Physik*, **51**(Mar.), 204–212.

Gamow, G. 1935. Nuclear transformations and the origin of the chemical elements. *Ohio J. Sci.*, **35**(5), 406–414.

Gamow, G. 1938. Nuclear energy sources and stellar evolution. *Physical Review*, **53**(Apr.), 595–604.

Gamow, G. 1946. Expanding universe and the origin of elements. *Physical Review*, **70**(Oct.), 572–573.

Gamow, G. 1948. The origin of elements and the separation of galaxies. *Physical Review*, **74**(Aug.), 505–506.

Gamow, G. 1949. On relativistic cosmogony. *Reviews of Modern Physics*, **21**(July), 367–373.

Gamow, G. 1952. *The Creation of the Universe.* Viking Press, reprinted Dover 2004.

Gamow, G. 1954a. On the steady-state theory of the universe. *AJ*, **59**(June), 200.

Gamow, G. 1954b. Possible mathematical relation between deoxyribonucleic acid and proteins. *Det Kong. Danske Viden. Selskab, Bio. Medd.*, **22**(3).

Gamow, G. 1956. The physics of the expanding Universe. *Vistas in Astronomy*, **2**, 1726–1732.

Gamow, G., and Critchfield, C.L. 1949. *Theory Of Atomic Nucleus And Nuclear Energy Sources.* Clarendon Press.

Gamow, G., and Teller, E. 1938. The rate of selective thermonuclear reactions. *Physical Review*, **53**(Apr.), 608–609.

Gamow, G., and Teller, E. 1939a. On the origin of great nebulae. *Physical Review*, **55**(Apr.), 654–657.

Gamow, G., and Teller, E. 1939b. The expanding Universe and the origin of the great nebulæ. *Nature*, **143**(Jan.), 116–117.

Gauss, C.F. 1827. *Allgemeine Flächentheorie (Disquisitiones generales circa superficies curvas).* Göttingen: Dieterich, translated into German 1828.

Gawiser, E., and Silk, J. 2000. The cosmic microwave background radiation. *Phys. Rep.*, **333**(Aug.), 245–267.

Gehrels, N. 1997. Use of νF_ν spectral energy distributions for multiwavelength astronomy. *Nuovo Cimento B Serie*, **112**(Jan.), 11–15.

Gelfand, A.E., and Smith, F.M. 1990. Sampling-based approaches to calculating marginal densities. *J. Amer. Stat. Assoc.*, **85**, 398–409.

Georgi, H., and Glashow, S. L. 1974. Unity of all elementary-particle forces. *Physical Review Letters*, **32**(Feb.), 438–441.

Gerber, P. 1898. Die Räumliche und zeitliche Ausbreitung der Gravitation. *Zeitschrift für Mathematik und Physik*, **43**, 93–104.

Gerber, P. 1917. Die Fortplfanzungsgeschwindigkeit der Gravitation. *Annalem der Physik*, **52**, 415–444.

Gibbons, G. W. 2003. Phantom matter and the cosmological constant. *arXiv hep-th*, 0302199

Gingerich, O. 1982. Obituary – Payne-Gaposchkin Cecilia. *QJRAS*, **23**(Mar.), 450.

Glover, S. C. O., Chluba, J., Furlanetto, S. R., Pritchard, J. R., and Savin, D. W. 2014. Chapter Three – Atomic, molecular, and optical physics in the early Universe: From recombination to reionization. *Advances in Atomic Molecular and Optical Physics*, **63**(Aug.), 135–270. Early version available at `http://user.astro.columbia.edu/~savin/papers/EarlyAMO121221.pdf` (may be volatile link).

Goldschmidt, V. M. 1938. *Geochemische Verteilungsgesetze der Elemente. IX Die Mengenverhältnisse der Elemente und der Atomarten*. Dybwad, Oslo.

Goldstein, H., Poole Jr., C.P., and Safko, J.L. 2001. *Classical Mechanics*. 3rd. edn. Addison-Wesley.

Goldstein, M. 2006. Subjective Bayesian analysis: Principles and practice. *Bayesian Analysis*, **1**(*3*), 403–420.

Golub, G. H., and van Loan, C. F. 1996. *Matrix Computations*. 3rd edn. Johns Hopkins University Press.

Gordon, C., Land, K., and Slosar, A. 2008. Determining the motion of the Solar system relative to the cosmic microwave background using Type Ia supernovae. *MNRAS*, **387**(June), 371–376.

Gorenstein, M. V., and Smoot, G. F. 1981. Large-angular-scale anisotropy in the cosmic background radiation. *ApJ*, **244**(Mar.), 361–381.

Górski, K. M., Hinshaw, G., Banday, A. J. *et al.* 1994. On determining the spectrum of primordial inhomogeneity from the COBE DMR sky maps: Results of two-year data analysis. *ApJ*, **430**(Aug.), L89–L92.

Górski, K. M., Hivon, E., and Wandelt, B.D. 1999. Analysis issues for large CMB data sets. Page 37 of: Banday, A. J., Sheth, R. K., and da Costa, L. N. (eds), *Evolution of Large Scale Structure : From Recombination to Garching*, MPA–ESO Cosmology Conference.

Górski, K. M., Hivon, E., Banday, A. J. *et al.* 2005. HEALPix: A framework for high-resolution discretization and fast analysis of data distributed on the sphere. *ApJ*, **622**(Apr.), 759–771.

Górski, K. M., Wandelt, B. D., Hansen, F. K., Hivon, E., and Banday, A. J. 2010. *The HEALPix Primer: version 2.15a*. First published 1999 for version 1.0. [Online]. Available: `http://healpix.jpl.nasa.gov/pdf/intro.pdf`.

Gotay, M.J., and Marsden, J.E. 1992. Stress–energy–momentum tensors and the Belinfante–rosenfeld formula. *Contemp. Math*, **132**, 367–392.

Gottlöber, S. 1994. Perturbations in the expanding universe. Page 415 of: Novello, M. (ed), *Cosmology and Gravitation* VII Brazilian School of Cosmology and Gravitation, Editions Frontières.

Gough, D. 2012. How oblate is the Sun? *Science*, **337**(Sept.), 1611.

Gould, R. J. 1984. The cross-section for double Compton scattering. *ApJ*, **285**(Oct.), 275–278.

Gradshteyn, I.S., and Ryzhik, I.M. 2007. *Table of Integrals, Series, and Products*. Academic Press; 7th edition.

Greason, M.R., Limon, M., Wollack, E., and the WMAP Science Working Group. 2012. *Wilkinson Microwave Anisotropy Probe (WMAP): Nine–Year Explanatory Supplement*. Greenbelt, MD: NASA/GSFC. Version 5. [Online]. Available: `http://`

lambda.gsfc.nasa.gov/product/map/dr5/pub_papers/nineyear/ supplement/WMAP_supplement.pdf.

Greenstein, J. L. 1963. Red-shift of the unusual radio source: 3C 48. *Nature*, **197**(Mar.), 1041–1042.

Griffiths, D. J., Proctor, T. C., and Schroeter, D. F. 2010. Abraham–Lorentz versus Landau–Lifshitz. *American Journal of Physics*, **78**(Apr.), 391–402.

Gull, S. F., and Daniell, G. J. 1978. Image reconstruction from incomplete and noisy data. *Nature*, **272**(Apr.), 686–690.

Gullstrand, A. 1922. Allgemeine Lösung des statischen Einkörperproblems in der Einsteinschen Gravitationstheorie. *Arkiv. Mat. Astron. Fys.*, **16**, 1–15.

Gunn, J. E., and Oke, J. B. 1975. Spectrophotometry of faint cluster galaxies and the Hubble diagram – an approach to cosmology. *ApJ*, **195**(Jan.), 255–268.

Gunn, J. E., and Peterson, B. A. 1965. On the density of neutral hydrogen in intergalactic space. *ApJ*, **142**(Nov.), 1633–1641.

Gunn, J. E., and Tinsley, B. M. 1975. An accelerating Universe. *Nature*, **257**(Oct.), 454–457.

Gurbatov, S. N., Saichev, A. I., and Shandarin, S. F. 1985a. A model for describing the development of the large-scale structure of the Universe. *Soviet Physics Doklady*, **30**(Nov.), 921.

Gurbatov, S. N., Saichev, A. I., and Shandarin, S. F. 1985b. Model description of the development of the large-scale structure of the Universe. *Akademiia Nauk SSSR Doklady*, **285**, 323–326.

Gurbatov, S. N., Saichev, A. I., and Shandarin, S. F. 1989. The large-scale structure of the Universe in the frame of the model equation of non-linear diffusion. *MNRAS*, **236**(Jan.), 385–402.

Gurbatov, S. N., Saichev, A. I., and Shandarin, S. F. 2012. Large-scale structure of the Universe. The Zel'dovich approximation and the adhesion model. *Physics Uspekhi*, **55**(Mar.), 223–249.

Gursky, H., Kellogg, E., Murray, S. *et al.* 1971. A strong X-ray source in the Coma Cluster observed by UHURU. *ApJ*, **167**(Aug.), L81.

Gursky, H., Solinger, A., Kellogg, E. M. *et al.* 1972. X-ray emission from rich clusters of galaxies. *ApJ*, **173**(May), L99.

Guth, A. H. 1981. Inflationary universe: A possible solution to the horizon and flatness problems. *Phys. Rev. D*, **23**(Jan.), 347–356.

Guth, A. H., and Nomura, Y. 2012. What can the observation of nonzero curvature tell us? *Phys. Rev. D*, **86**(2), 023534.

Guth, A. H., and Pi, S.-Y. 1982. Fluctuations in the new inflationary universe. *Physical Review Letters*, **49**(Oct.), 1110–1113.

Guy, J., and The SALT2 Collaboration. 2007. SALT2: using distant supernovae to improve the use of type Ia supernovae as distance indicators. *A&A*, **466**(Apr.), 11–21.

Hakkila, J., and Preece, R. D. 2011. Unification of pulses in long and short gamma-ray bursts: Evidence from pulse properties and their correlations. *ApJ*, **740**(Oct.), 104.

Halley, E. 1716. An account of several nebulae or lucid spots like clouds. *Phil. Transactions*, **29**, 390.

Hamuy, M., Phillips, M. M., Suntzeff, N. B. *et al.* 1996. The Hubble diagram of the Calan/Tololo type IA supernovae and the value of HO. *AJ*, **112**(Dec.), 2398.

Hannestad, S. 2006. Neutrinos in cosmology. *Progress in Particle and Nuclear Physics*, **57**(July), 309–323.

Hansen, L., Jorgensen, H. E., and Norgaard-Nielsen, H. U. 1987. Search for supernovae in distant clusters of galaxies. *The Messenger*, **47**(Mar.), 46–49.

Hansen, L., Jorgensen, H. E., Norgaard-Nielsen, H. U., Ellis, R. S., and Couch, W. J. 1989. A supernova at Z = 0.28 and the rate of distant supernovae. *A&A*, **211**(Feb.), L9–L11.

Harrison, E. 1991. Hubble spheres and particle horizons. *ApJ*, **383**(Dec.), 60–65.

Harrison, E. 1993. The redshift-distance and velocity-distance laws. *ApJ*, **403**(Jan.), 28–31.

Harrison, E. R. 1968a. Baryon inhomogeneity in the early Universe. *Physical Review*, **167**(Mar.), 1170–1175.

Harrison, E. R. 1968b. The early Universe. *Physics Today*, **21**, 31.

Harrison, E. R. 1970. Fluctuations at the threshold of classical cosmology. *Phys. Rev. D*, **1**(May), 2726–2730.

Harrison, E. R. 1995. Mining energy in an expanding Universe. *ApJ*, **446**(June), 63.

Harrison, E.R. 1989. *Darkness at Night: A Riddle of the Universe*. Harvard University Press.

Harrison, E. R. 2000. *Cosmology. The science of the Universe*. 2nd edn. Cambridge University Press.

Hasenöhrl, F. 1908. Zur Thermodynamik bewegter Systeme. *Sitzungsberichte der mathematisch-naturwissenschaftlichen Klasse der kaiserlichen Akademie der Wissenschaften, Wien*, **116**, 1391–1405. https://en.wikisource.org/wiki/Translation:On_the_Thermodynamics_of_MovingSystems.

Haugbølle, T., Hannestad, S., Thomsen, B. *et al.* 2007. The velocity field of the local universe from measurements of type Ia supernovae. *ApJ*, **661**(June), 650–659.

Hauser, M. G., and Peebles, P. J. E. 1973. Statistical analysis of catalogs of extragalactic objects. II. The Abell catalog of rich clusters. *ApJ*, **185**(Nov.), 757–786.

Hawking, S. W. 1966. Perturbations of an expanding Universe. *ApJ*, **145**(Aug.), 544.

Hawking, S. W. 1982. The development of irregularities in a single bubble inflationary Universe. *Physics Letters B*, **115**(Sept.), 295–297.

Hawking, S. W., and Ellis, G. F. R. 1975. *The Large-scale Structure of Space-Time*. 2nd edn. Cambridge University Press.

Hayashi, C. 1950. Proton-neutron concentration ratio in the expanding Universe at the stages preceding the formation of the elements. *Progress of Theoretical Physics*, **5**(Mar.), 224–235.

Hayashi, C., and Nishida, M. 1956. Formation of light nuclei in the expanding Universe. *Progress of Theoretical Physics*, **16**(Dec.), 613–624.

Hazard, C., Mackey, M. B., and Shimmins, A. J. 1963. Investigation of the radio source 3C 273 by the method of lunar occultations. *Nature*, **197**(Mar.), 1037–1039.

Heath, D. J. 1977. The growth of density perturbations in zero pressure Friedmann–Lemaître Universes. *MNRAS*, **179**(May), 351–358.

Heath, D. J. 1981. The growth of density perturbations in the Lemaître Universe. *Ap&SS*, **77**(June), 59–64.

Heath, D. J. 1989. Closed-form expressions for the rate of growth of perturbations in zero-pressure Friedmann–Lemaître Universes. *Ap&SS*, **154**(Apr.), 207–216.

Heath, D. J. 1991. Gravitational instability. *Ap&SS*, **175**(Jan.), 35–50.

Heckmann, O., and Schücking, E. 1955. Bemerkungen zur Newtonschen Kosmologie. I. Mit 3 Textabbildungen in 8 Einzeldarstellungen. *ZAp*, **38**, 95.

Heinrich, J. 2003a. *Benefits of Blind Analysis Techniques*. [Online]. Available: . `http://www-cdf.fnal.gov/physics/statistics/notes/cdf6576_blind.pdf`.

Heinrich, J. 2003b. Pitfalls of goodness-of-fit from likelihood. Page 52 of: Lyons, L., Mount, R., and Reitmeyer, R. (eds), *Statistical Problems in Particle Physics, Astrophysics, and Cosmology*. SLAC, Stanford, CA.

Heller, M. 1985. Friedman's cosmological views. *Acta Cosmologica*, **13**, 65.

Heller, M. 1989. Mach's predictions and relativistic cosmology. *Acta Cosmologica*, **16**, 47.

Heller, M. 2010. *Ultimate Explanations of the Universe*. Springer.

Henry, P. S. 1971. Isotropy of the 3 K background. *Nature*, **231**(June), 516–518.

Henry, R. C. 1999. Diffuse background radiation. *ApJl*, **516**(May), L49–L52.

Herschel, J. F. W. 1864. Catalogue of nebulae and clusters of stars. *Royal Society of London Philosophical Transactions Series I*, **154**, 1–137.

Herschel, W. 1789. Catalogue of a second thousand of new nebulae and clusters of stars; with a few introductory remarks on the construction of the heavens. *Royal Society of London Philosophical Transactions Series I*, **79**, 212–255.

Herschel, W. 1800. Catalogue of 500 new nebulae, nebulous stars, planetary nebulae, and clusters of stars; with remarks on the construction of the heavens. [Abstract]. *Royal Society of London Proceedings Series I*, **1**, 98–100.

Hertzsprung, E. 1913. Über die räumliche Verteilung der Veränderlichen vom δ Cephei-Typus. *Astronomische Nachrichten*, **196**(Nov.), 201.

Hewitt, A., and Burbidge, G. 1980. A revised optical catalog of quasi-stellar objects. *ApJs*, **43**(May), 57–158.

Hewitt, A., and Burbidge, G. 1993. A revised and updated catalog of quasi-stellar objects. *ApJs*, **87**(Aug.), 451–947.

Hilbert, D. 1915. Die Grundlagen der Physik. *Konigl. Gesell. d. Wiss. Göttingen, Nachr. Math.-Phys. Kl.*, 395–407.

Hinshaw, G., Larson, D. *et al.* 2013. Nine-year Wilkinson Microwave Anisotropy Probe (WMAP) observations: Cosmological parameter results. *ApJs*, **208**(Oct.), 19.

Hinshaw, G., Weiland, J. L. *et al.* 2009. Five-year Wilkinson Microwave Anisotropy Probe observations: Data processing, sky maps, and basic results. *ApJs*, **180**(Feb.), 225–245.

Hobson, M. P., Bridle, S. L., and Lahav, O. 2002. Combining cosmological data sets: hyperparameters and Bayesian evidence. *MNRAS*, **335**(Sept.), 377–388.

Hobson, M.P., Efstathiou, G.P., and Lasenby, A.N. 2006. *General Relativity: An Introduction for Physicists*. Cambridge University Press.

Hogg, D. W. 1999. Distance measures in cosmology. *arXiv astro-ph*, 9905116

Hoskin, M.A. 1976. *The 'Great Debate': What Really Happened. Journal for the History of Astronomy*, **7**, 169.

Hou, Z., Reichardt, C. L., Story, K. T. *et al.* 2012. Constraints on cosmology from the cosmic microwave background power spectrum of the 2500-square degree SPT-SZ survey. *arXiv astro-ph*, 1212.6267

Hoyle, F. 1946. The synthesis of the elements from hydrogen. *MNRAS*, **106**, 343.

Hoyle, F. 1947. On the formation of heavy elements in stars. *Proceedings of the Physical Society*, **59**(Nov.), 972–978.

Hoyle, F. 1948. A new model for the expanding Universe. *MNRAS*, **108**, 372–382.

Hoyle, F. 1950. Book reviews. *The Observatory*, **70**(Oct.), 194–197.

Hoyle, F. 1954. On nuclear reactions occurring in very hot stars.I. the synthesis of elements from carbon to nickel. *ApJS*, **1**(Sept.), 121.

Hoyle, F., and Narlikar, J. V. 1962. Mach's principle and the creation of matter. *Royal Society of London Proceedings Series A*, **270**(Nov.), 334–339.

Hoyle, F., and Narlikar, J. V. 1963. Mach's principle and the creation of matter. *Royal Society of London Proceedings Series A*, **273**(Apr.), 1–11.

Hoyle, F., and Narlikar, J. V. 1964. On the avoidance of singularities in C-field cosmology. *Royal Society of London Proceedings Series A*, **278**(Apr.), 465–478.

Hoyle, F., and Sandage, A. 1956. The second-order term in the redshift-magnitude relation. *PASP*, **68**(Aug.), 301.

Hoyle, F., and Tayler, R. J. 1964. The mystery of the cosmic helium abundance. *Nature*, **203**(Sept.), 1108–1110.

Hu, W., and Sugiyama, N. 1995. Anisotropies in the cosmic microwave background: an analytic approach. *ApJ*, **444**(May), 489–506.

Hu, W., and Sugiyama, N. 1996. Small-scale cosmological perturbations: an analytic approach. *ApJ*, **471**(Nov.), 542.

Hu, W., Hedman, M. M., and Zaldarriaga, M. 2003. Benchmark parameters for CMB polarization experiments. *Phys. Rev. D*, **67**(4), 043004.

Hubble, E. 1929. A relation between distance and radial velocity among extra-galactic nebulae. *Proceedings of the National Academy of Science*, **15**(Mar.), 168–173.

Hubble, E., and Humason, M. L. 1931. The velocity–distance relation among extra-galactic nebulae. *Contributions from the Mount Wilson Observatory / Carnegie Institution of Washington*, **427**, 1–38.

Hubble, E., and Tolman, R. C. 1935. Two methods of investigating the nature of the nebular redshift. *ApJ*, **82**(Nov.), 302.

Hubble, E. P. 1925. Cepheids in spiral nebulae. *Popular Astronomy*, **33**, 252–255.

Hubble, E. P. 1926. Extragalactic nebulae. *ApJ*, **64**(Dec.), 321–369.

Hubble, E. P. 1953. The law of red shifts (George Darwin Lecture). *MNRAS*, **113**, 658.

Hulse, R. A., and Taylor, J. H. 1975. Discovery of a pulsar in a binary system. *ApJ*, **195**(Jan.), L51–L53.

Humason, M. L. 1931. The large apparent velocities of extra-galactic nebulae. *Leaflet of the Astronomical Society of the Pacific*, **1**, 149.

Humason, M. L. 1936a. Is the Universe expanding? *Leaflet of the Astronomical Society of the Pacific*, **2**, 161.

Humason, M. L. 1936b. The apparent radial velocities of 100 extra-galactic nebulae. *ApJ*, **83**(Jan.), 10.

Humason, M. L., Mayall, N. U., and Sandage, A. R. 1956. Redshifts and magnitudes of extragalactic nebulae. *AJ*, **61**, 97–162.

Hung, C-L., Gurarie, V., and Chin, C. 2013. From cosmology to cold atoms: Observation of Sakharov oscillations in a quenched atomic superfluid. *Science*, **341**, 1213–1215.

Hunt, B.R. 1977. Bayesian methods in nonlinear digital image restoration. *IEEE Trans. Computers*, **C-26**, 219–229.

Hütsi, G. 2005. Acoustic oscillations in the SDSS luminous red galaxy sample power spectrum. *arXiv astro-ph*, 0507678

Hütsi, G. 2006a (May). *Cosmic sound: Measuring the Universe with baryonic acoustic oscillations*. PhD thesis, Max Planck Institut für Astrophysik. Available from https//edoc.ub.uni-muenchen.de/5400/.

Hütsi, G. 2006b. Power spectrum of the SDSS luminous red galaxies: constraints on cosmological parameters. *A&A*, **459**(Nov.), 375–389.

Huygens, C. 1690. *Traitè de la lumière*. Leyden.

IBM Archive – Exhibits. 2013 (Jan.). *Brunsviga Calculating Machine*. Exhibition of 19th century calculating machines at `http://www-03.ibm.com/ibm/history/exhibits/attic3/attic3_044.html`.

Icke, V., and van de Weygaert, R. 1987. Fragmenting the universe. *A&A*, **184**(Oct.), 16–32.

Isaacson, R. A. 1968a. Gravitational radiation in the limit of high frequency. I. The linear approximation and geometrical optics. *Physical Review*, **166**(Feb.), 1263–1271.

Isaacson, R. A. 1968b. Gravitational radiation in the limit of high frequency. II. Nonlinear terms and the effective stress tensor. *Physical Review*, **166**(Feb.), 1272–1279.

Israel, W., and Stewart, J. M. 1979. Transient relativistic thermodynamics and kinetic theory. *Annals of Physics*, **118**(Apr.), 341–372.

Israel, W., and Stewart, J. M. 1980. Progress in relativistic thermodynamics and electrodynamics of continuous media. Page 491 of: Held, A. (ed), *General Relativity and Gravitation II*, Vol. 2, Plenum Press.

Jackson, J.D. 1998. *Classical Electrodynamics*. 3rd. edn. John Wiley & Sons.

Jacobs, K. C. 1968. Spatially homogeneous and Euclidean cosmological models with shear. *ApJ*, **153**(Aug.), 661.

James, W., and Stein, C. 1961. Estimation with quadratic loss. *Proc. Fourth Berkeley Symp. on Math. Statist. and Prob., Vol. 1 (Univ. of Calif. Press, 1961)*, **1**, 361–379.

Janssen, M. 2007. What did Einstein know and when did he know it? A Besso memo dated August 1913. Pages 785–838 of: Janssen, M., Norton, J.D., Renn, J., Sauer, T., and Stachel, J. (eds), *The Genesis of General Relativity*. Boston Studies in the Philosophy of Science, Vol. 250.2. Springer.

Janssen, M., Norton, J.D., Renn, J., Sauer, T., and Stachel, J. 2007a. A commentary on the notes on gravity in the Zurich notebook. Pages 489–714 of: Janssen, M., Norton, J.D., Renn, J., Sauer, T., and Stachel, J. (eds), *The Genesis of General Relativity*. Boston Studies in the Philosophy of Science, Vol. 250.2. Springer.

Janssen, M., Norton, J.D., Renn, J., Sauer, T., and Stachel, J. 2007b. The Zurich notebook. Pages 313–487 of: Janssen, M., Norton, J.D., Renn, J., Sauer, T., and Stachel, J. (eds), *The Genesis of General Relativity*. Boston Studies in the Philosophy of Science, Vol. 250.1. Springer.

Jarrett, T. 2004. Large scale structure in the local Universe – The 2MASS galaxy catalog. *Publications of the Astronomical Society of Australia*, **21**, 396–403.

Jauncey, D. L. 1975. Radio surveys and source counts. *ARA&A*, **13**, 23–44.

Jaynes, E. T. 2003. *Probability Theory: The Logic of Science (Vol 1)*. Cambridge University Press.

Jeffreys, H. 1961. *Theory of Probability*. 3rd edn. Oxford Classic Texts in the Physical Sciences. Oxford University Press.

Jensen, W. B. 2007. The origin of the s, p, d, f orbital labels. *Journal of Chemical Education*, **84**(May), 757.

Jöeveer, M., Einasto, J., and Tago, M. 1977. The cell structure of the Universe. *Tartu Astrofüüs. Obs. Preprint, Nr. A-1, 45 p.*, **1**(July).

Johnstone Stoney, D. 1881. On the physical units of nature. *Phil. Mag. Ser. 5*, **11**(69), 381–390. Paper read before Royal Dublin Soc. Feb. 10, 1881.

Jones, B. J. T. 1973. Cosmic turbulence and the origin of galaxies. *ApJ*, **181**(Apr.), 269–294.

Jones, B. J. T. 1976. The origin of galaxies: A review of recent theoretical developments and their confrontation with observation. *Reviews of Modern Physics*, **48**(Jan.), 107–145.

Jones, B. J. T., and Mazure, A. 1996. Galaxy cluster luminosity functions and the distances to clusters. Page 197 of: Coles, P., Martinez, V., and Pons-Borderia, M.-J. (eds), *Mapping, Measuring, and Modelling the Universe*. Astronomical Society of the Pacific Conference Series, Vol. 94.

Jones, B. J. T., and Steigman, G. 1978. Distortions of the cosmic radiation spectrum in baryon-symmetric cosmologies. *MNRAS*, **183**(June), 585–594.

Jones, B. J. T., and Wyse, R. F. G. 1985. The ionisation of the primeval plasma at the time of recombination. *A&A*, **149**(Aug.), 144–150.

Jones, B. J. T., Martínez, V. J., Saar, E., and Trimble, V. 2004. Scaling laws in the distribution of galaxies. *Reviews of Modern Physics*, **76**(Oct.), 1211–1266.

Jones, J., and Jones, B. 1970. Physical sciences: Distribution of antimatter in the Universe. *Nature*, **227**(Aug.), 475–476.

Jones, W. C. *et al.* (BOOMERANG). 2006. A measurement of the angular power spectrum of the CMB temperature anisotropy from the 2003 Flight of BOOMERANG. *ApJ*, **647**(Aug.), 823–832.

Jonsson, R. M. 2005. Visualizing curved spacetime. *American Journal of Physics*, **73**(Mar.), 248–260.

Jordan, P., Ehlers, J., and Kundt, W. 2009. Republication of: Exact solutions of the field equations of the general theory of relativity. *General Relativity and Gravitation*, **41**(Sept.), 2191–2280.

Joseph, D. D., and Preziosi, L. 1989. Heat waves. *Reviews of Modern Physics*, **61**(Jan.), 41–73.

Kaluza, T. 1921. Zum Unitätsproblem in der Physik. *Sitzungsber. Preuss. Akad. Wiss. Berlin. (Math. Phys.)*, 966–972.

Kamionkowski, M., and Knox, L. 2003. Aspects of the cosmic microwave background dipole. *Phys. Rev. D*, **67**(6), 063001.

Kamionkowski, M., Kosowsky, A., and Stebbins, A. 1997. Statistics of cosmic microwave background polarization. *Phys. Rev. D*, **55**(June), 7368–7388.

Kasai, M. 1992. Construction of inhomogeneous Universes which are Friedmann–Lemaître–Robertson–Walker on average. *Physical Review Letters*, **69**(Oct.), 2330–2332.

Kasner, E. 1925a. An algebraic solution of the Einstein equations. *Trans. Am. Math. Soc.*, **27**, 101–105.

Kasner, E. 1925b. Solutions of the Einstein equations involving functions of only one variable. *Trans. Am. Math. Soc.*, **27**, 155–162.

Kass, R.E., and Raferty, A.E. 1995. Bayes factors. *J. Amer. Stat. Assoc.*, **90**, 773–795.

Kay, S. 1993. *Fundamentals of Statistical Signal Processing, Volume I: Estimation Theory*. Prentice Hall.

Kay, S. 1998. *Fundamentals of Statistical Signal Processing, Volume II: Detection Theory*. Prentice Hall.

Kay, S. 2013. *Fundamentals of Statistical Signal Processing, Volume III: Practical Algorithm Development*. Prentice Hall.

Kayser, R., Helbig, P., and Schramm, T. 1997. A general and practical method for calculating cosmological distances. *A&A*, **318**(Feb.), 680–686.

Kazanas, D. 1980. Dynamics of the universe and spontaneous symmetry breaking. *ApJ*, **241**(Oct.), L59–L63.

Keynes, J.M. 1921. *A Treatise on Probability*. London MacMillan & Co.

Kharbediya, L. I. 1976. Some exact solutions of the Friedmann equations with the cosmological term. *Soviet Astronomy*, **20**(Dec.), 647.

Kharbediya, L. I. 1983. Solutions to the Friedmann Equations with the lambda term for a dust radiation Universe. *Soviet Astronomy*, **27**(Aug.), 380.

Kholupenko, E. E., Ivanchik, A. V., Balashev, S. A., and Varshalovich, D. A. 2011. Advanced three-level approximation for numerical treatment of cosmological recombination. *MNRAS*, **417**(Nov.), 2417–2425.

Kiang, T. 1995. The Cosmological Redshift as Indicators of Distance and Velocity. *Irish Astronomical Journal*, **22**(July).

Kiang, T. 1997. Horizon and the question whether galaxies that recede faster than light are observable. *Chin. Astron. Astrophys.*, **21**(Feb.), 1–18.

Kiang, T. 2003. Can we observe galaxies that recede faster than light? A more clear-cut answer. *Chin. Astron. Astrophys.*, **27**(July), 247–251.

Kiang, T. 2004. Time, distance, velocity, redshift: A brief history of changes in basic physical concept. *Chin. Astron. Astrophys.*, **28**(July), 273–286.

Kinney, W. H. 2002. Cosmology, inflation, and the physics of nothing. In: Prosper, H.B., and Danilov, M. (eds), Lectures given at the *NATO Advanced Study Institute on Techniques and Concepts of High Energy Physics, St. Croix, USVI (2002)*. NATO Science Series Vol. 123. NATO.

Kipper, A. Y. 1952. Teoriya dvojnogo izlucheniya svetovyh kvantov dlya atoma vodoroda. The theory of duel rediation light rays of the hydrogen atom. *publications of the Tartu Astrofizica Observatory*, **32**, 63–93. In Russian. There are two articles in this one paper.

Kirshner, R. P. 2010. *Foundations of Supernova Cosmology*. In Ruiz-Lapuente, P. (ed), Dark Energy: Observational and Theoretical Approaches. Cambridge University Press. p.151.

Klein, O. 1926. Quantentheorie und fünfdimensionale Relativitätstheorie. *Zeits. f. Physik A*, 895–906.

Klypin, A. A., Sazhin, M. V., Strukov, I. A., and Skulachev, D. P. 1987. Limits on microwave background anisotropies – the Relikt experiment. *Soviet Astronomy Letters*, **13**(Apr.), 104.

Klypin, A. A., Strukov, I. A., and Skulachev, D. P. 1992. The Relikt missions – Results and prospects for detection of the microwave background anisotropy. *MNRAS*, **258**(Sept.), 71–81.

Knight, J. W., Sturrock, P. A., and Switzer, P. 1976. The distribution of redshifts of quasars and related objects. *ApJ*, **203**(Jan.), 286–290.

Kodama, H., and Sasaki, M. 1984. Cosmological perturbation theory. *Progress of Theoretical Physics Supplement*, **78**, 1.

Kofman, L., Pogosian, D., and Shandarin, S. 1990. Structure of the Universe in the two-dimensional model of adhesion. *MNRAS*, **242**(Jan.), 200–208.

Kogut, A., Spergel, D. N., Barnes, C. *et al.* 2003. First-year Wilkinson Microwave Anisotropy Probe (WMAP) observations: Temperature–polarization correlation. *ApJS*, **148**(Sept.), 161–173.

Kolb, E. W., and Turner, M. S. 1990. *The Early Universe*, Addison-Wesley.

Kolb, E.W. 2006. *Report of the Dark Energy Task Force*. HEPAP. `http://science.energy.gov/~/media/hep/pdf/files/pdfs/kolb_hepap_07_06.pdf`

Kolb, E.W. 2012. *Community Dark Energy Task Force Report*. [Online]. Available: `http://science.energy.gov/~/media/hep/hepap/pdf/20120827/Kolb_HEPAP_8_12_revised.pdf`.

Komar, A. 1956. Necessity of singularities in the solution of the field equations of general relativity. *Physical Review*, **104**(Oct.), 544–546.

Komatsu, E., Smith, K. M., and WMAP collaboration. 2010. Seven-year Wilkinson Microwave Anisotropy Probe (WMAP) observations: Cosmological interpretation. *arXiv astro-ph*, 1001.4538

Kompaneets, A.S. 1957. The establishment of thermal equilibrium between quanta and electrons. *Sov. Phys. – JETP*, **4**, 730.

Kottler, F. 1918. Über die physikalischen Grundlagen der Einsteinschen Gravitationstheorie. *Annalen der Physik*, **361**, 401–462.

Kovac, J. 2002. *The Degree Angular Scale Interferometer (DASI) in Antarctica has detected the polarization of the cosmic microwave background*. Univ. Chicago Press Release. [Online]. Available: `http://www-news.uchicago.edu/releases/photos/polarization/polmap_press.jpg`.

Kovac, J. M., Leitch, E. M., Pryke, C. *et al.* 2002. Detection of polarization in the cosmic microwave background using DASI. *Nature*, **420**(Dec.), 772–787.

Kowal, C. T. 1968. Absolute magnitudes of supernovae. *AJ*, **73**(Dec.), 1021–1024.

Kowalski, M., Rubin, D. *et al.* 2008. Improved cosmological constraints from new, old, and combined supernova data sets. *ApJ*, **686**(Oct.), 749–778.

Kragh, H. 2000. Max Planck: the reluctant revolutionary. *Physics World*, **13**, 31–35.

Kragh, H. 2008. The origin and earliest reception of Big-Bang cosmology. *Publications de l'Observatoire Astronomique de Beograd*, **85**(Oct.), 7–16.

Kragh, H., and Smith, R. W. 2003. Who discovered the expanding Universe? *History of Science*, **41**, 141–162.

Kragh, H.S. 2007. *Conceptions of Cosmos: From Myths to the Accelerating Universe: A History of Cosmology*. Oxford University Press.

Krasinski, A. 1997. *Inhomogeneous Cosmological Models*. Cambridge University Press.

Kristian, J., Sandage, A., and Westphal, J. A. 1978. The extension of the Hubble diagram. II – New redshifts and photometry of very distant galaxy clusters – First indication of a deviation of the Hubble diagram from a straight line. *ApJ*, **221**(Apr.), 383–394.

Kuhn, J. R., Bush, R., Emilio, M., and Scholl, I. F. 2012. The precise solar shape and its variability. *Science*, **337**(Sept.), 1638.

Lahav, O., and Liddle, A. R. 2014. The cosmological parameters 2014. *arXiv astro-ph*, 1401.1389

Lahav, O., and Suto, Y. 2004. Measuring our Universe from galaxy redshift surveys. *Living Reviews in Relativity*, **7**(July), 8.

Lambert, S. B., and Le Poncin-Lafitte, C. 2011. Improved determination of γ by VLBI. *A&A*, **529**(May), A70.

Landau, L.D., and Lifshitz, E.M. 1980. *The Classical Theory of Fields,*. 4th edn. Course of *Theoretical Physics Series*, Vol. 2. Butterworth–Heinemann.

Landau, L.D., and Lifshitz, E.M. 1987. *Fluid Mechanics*. 2nd edn. Butterworth–Heinemann Ltd.

Langlois, D. 2010 (Mar.). Inflation and cosmological perturbations. Pages 1–57 of: Wolschin, G. (ed), *Lecture Notes in Physics,* Springer Verlag, Vol. 800.

Larmor, J. 1897. On a dynamical theory of the electric and luminiferous medium, Part 3, Relations with material media. *Phil. Trans. Roy. Soc.*, **190**, 205–300.

Larmor, J. 2012. *Aether and Matter*. Forgotten Books.

Layzer, D. 1954. On the significance of Newtonian cosmology. *AJ*, **59**(Aug.), 268.

Layzer, D. 1964. The formation of stars and galaxies: Unified hypotheses. *ARA&A*, **2**, 341.

Layzer, D. 1971. Cosmogonic processes. Pages 151–233 of: *Brandeis Summer Institute 1968: Astrophysics and general relativity*, Vol. 2, pp. 151–233.

Layzer, D. 1975. *Galaxy Clustering: its Description and its Interpretation*. the University of Chicago Press. Page 665.

Layzer, D., and Burke, J. R. 1972. The weak-field approximation and the range of gravitational force. *General Relativity and Gravitation*, **3**(June), 121–121.

Leavitt, H. S. 1908. 1777 variables in the Magellanic Clouds. *Annals of Harvard College Observatory*, **60**, 87–108.

Leavitt, H. S., and Pickering, E. C. 1912. Periods of 25 variable stars in the small Magellanic Cloud. *Harvard College Observatory Circular*, **173**(Mar.), 1–3.

Lehmkuhl, D. 2011. Mass Energy Momentum: Only there because of spacetime? *Brit. J. Phil. Sci.*, **62**, 453–488.

Lemaître, G. 1925. Note on de Sitter's Universe. *M.I.T Journal Math.and Phys.*, **4**, 188–192.

Lemaître, G. 1927. Un Univers homogène de masse constante et de rayon croissant rendant compte de la vitesse radiale des nébuleuses extra-galactiques. *Annales de la Société Scientifique de Bruxelles*, **47**, 49–59.

Lemaître, G. 1931a. Expansion of the Universe, A homogeneous Universe of constant mass and increasing radius accounting for the radial velocity of extra-galactic nebulae. *MNRAS*, **91**(Mar.), 483–490.

Lemaître, G. 1931b. The expanding universe. *MNRAS*, **91**(Mar.), 490–501.

Lemaître, G. 1933. Schwarzschild's solution (actual title unknown). *Annales de la Societe Scientifique de Bruxelles, Ser. A*, **53**, 51–85.

Lemaître, G. 1958. Instability in the expanding Universe and its astronomical implications. *Ricerche Astronomiche*, **5**, 475.

Lemaître, G. 1961. Exchange of galaxies between clusters and field. *AJ*, **66**(Dec.), 603–606.

Letessier, J., and Rafelski, J. 2005. *Hadrons and Quark-Gluon Plasma*. Cambridge Monographs on Particle Physics, Nuclear Physics and Cosmology. Cambridge Univerity Press.

Lewis, H. R. 1973. Einstein's derivation of Planck's radiation law. *American Journal of Physics*, **41**(Jan.), 38–44.

Li, A. 2003. Cosmic needles versus cosmic microwave background radiation. *ApJ*, **584**(Feb.), 593–598.

Li, M.-H., Wang, P., Chang, Z., and Zhao, D. 2014. CosmoMC installation and running guidelines. *arXiv astro-ph*, 1409.1354

Liddle, A. R. 1989. Power-law inflation with exponential potentials. *Physics Letters B*, **220**(Apr.), 502–508.

Liddle, A. R., and Lyth, D. H. 1992. COBE, gravitational waves, inflation and extended inflation. *Physics Letters B*, **291**(Oct.), 391–398.

Lifshitz, E. M. 1946. On the gravitational stability of the expanding Universe. *J. Phys. (Moscow)*, **10**, 116.

LIGO Scientific Collaboration, and Virgo Collaboration. 2016. Observing gravitational-wave transient GW150914 with minimal assumptions. *arXiv astro-ph*, 1602.03843

Lilje, P. B., Yahil, A., and Jones, B. J. T. 1986. The tidal velocity field in the local supercluster. *ApJ*, **307**(Aug.), 91–96.

Limber, D. N. 1953. The analysis of counts of the extragalactic nebulae in terms of a fluctuating density field. *ApJ*, **117**(Jan.), 134.

Limber, D. N. 1954. The analysis of counts of the extragalactic nebulae in terms of a fluctuating density field. II. *ApJ*, **119**(May), 655.

Limber, D. N. 1957. Analysis of counts of the extragalactic nebulae in terms of a fluctuating density field. III. *ApJ*, **125**(Jan.), 9.

Linde, A. 2014. Inflationary cosmology after Planck 2013. *arXiv hep-th*, 1402.0526

Linde, A. D. 1982a. A new inflationary universe scenario: A possible solution of the horizon, flatness, homogeneity, isotropy and primordial monopole problems. *Physics Letters B*, **108**(Feb.), 389–393.

Linde, A. D. 1982b. Scalar field fluctuations in the expanding universe and the new inflationary universe scenario. *Physics Letters B*, **116**(Oct.), 335–339.

Linde, A. D. 1983. Chaotic inflation. *Physics Letters B*, **129**(Sept.), 177–181.

Linder, E. V. 1988a. Cosmological tests of generalized Friedmann models. *A&A*, **206**(Nov.), 175–189.

Linder, E. V. 1988b. Light propagation in generalized Friedmann Universes. *A&A*, **206**(Nov.), 190–198.

Linder, E. V. 2003. Exploring the expansion history of the Universe. *Physical Review Letters*, **90**(9), 091301.

Linder, E. V. 2005. Cosmic growth history and expansion history. *Phys. Rev. D*, **72**(4), 043529.

Linder, E. V. 2009. Extending the gravitational growth framework. *Phys. Rev. D*, **79**(6), 063519.

Linder, E. V., and Jenkins, A. 2003. Cosmic structure growth and dark energy. *MNRAS*, **346**(Dec.), 573–583.

Lineweaver, C. H. 1997. The CMB dipole: the most recent measurement and some history. Pages 69–75 of: Bouchet, F. R., Gispert, R., Guiderdoni, B., and Trân Thanh Vân, J. (eds), *Microwave Background Anisotropies*. Editions Frontières.

Lineweaver, C. H., and Davis, T. M. 2005. Misconceptions about the Big Bang. *Scientific American*, **292**(3), 030000–45.

Lineweaver, C. H., Tenorio, L., Smoot, G. F. *et al.* 1996. The dipole observed in the COBE DMR 4 year data. *ApJ*, **470**(Oct.), 38.

Livio, M. 2011. Lost in translation: Mystery of the missing text solved. *Nature*, **479**(Nov.), 171–173.

Longair, M. S. 1966. On the interpretation of radio source counts. *MNRAS*, **133**, 421.

Longair, M. S., and Einasto, J. 1978. *The Large Scale Structure of the Universe: symposium no. 79 held in Tallinn, Estonia, U.S.S.R., September 12-16, 1977*. Reidel.

Longair, M. S., and Sunyaev, R. A. 1969. The origin of the X-ray background. *Astrophys. Lett.*, **4**, 65.

Lorentz, H.-A. 1904. Electromagnetic phenomena in a system moving with any velocity smaller than that of light. *Proc. Acad. Amsterdam*, **6**, 809.

Lorentz, H.A., Einstein, A., Minkowski, H., and Weyl, H. 2000. *The Principle of Relativity*. Dover Publications Inc.

Lovelock, D. 1972. The four-dimensionality of space and the Einstein tensor. *Journal of Mathematical Physics*, **13**, 874–876.

Lubin, P., Villela, T., Epstein, G., and Smoot, G. 1985. A map of the cosmic background radiation at 3 millimeters. *ApJ*, **298**(Nov.), L1–L5.

Lubin, P. M., and Smoot, G. F. 1981. Polarization of the cosmic background radiation. *ApJ*, **245**(Apr.), 1–17.

Lucchin, F., and Matarrese, S. 1985. Power-law inflation. *Phys. Rev. D*, **32**(Sept.), 1316–1322.

Lundmark, K. 1924. The determination of the curvature of space-time in de Sitter's world. *MNRAS*, **84**(June), 747–770.

Lundmark, K. 1925. Nebulæ, The motions and the distances of spiral. *MNRAS*, **85**(June), 865.

Lundmark, K. 1956. On metagalactic distance-indicators. *Vistas in Astronomy*, **2**, 1607–1619.

Luzum, B. *et al.* 2011. The IAU 2009 system of astronomical constants: the report of the IAU working group on numerical standards for fundamental astronomy. *Celestial Mechanics and Dynamical Astronomy*, **110**(Aug.), 293–304.

Lynden-Bell, D., Katz, J., and Bicak, J. 1995. Mach's principle from the relativistic constraint equations. *MNRAS*, **272**(Jan.), 150–160.

Lyth, D. H., and Liddle, A. R. 2009. *The Primordial Density Perturbation*. Cambridge University Press.

Ma, C.-P., and Bertschinger, E. 1995. Cosmological perturbation theory in the synchronous and conformal Newtonian gauges. *ApJ*, **455**(Dec.), 7.

Maartens, R. 1996. Causal thermodynamics in relativity. *arXiv astro-ph*, 9609119

MacCallum, M. A. H. 2006 (June). Finding and using exact solutions of the Einstein equations. Pages 129–143 of: Mornas, L., and Diaz Alonso, J. (eds), *A Century of Relativity Physics: ERE 2005*. American Institute of Physics Conference Series, Vol. 841.

Mach, E., and McCormack, T. b. T. J. 2013. *The Science of Mechanics*. Cambridge University Press.

MacKay, D.J.C. 1992. Bayesian interpolation. *Neural Computation*, **4**, 415–447.

MacKay, D.J.C. 2003. *Information Theory, Inference and Learning Algorithms*. Cambridge University Press.

MacTavish, C. J. *et al.* 2006. Cosmological parameters from the 2003 flight of BOOMERANG. *ApJ*, **647**(Aug.), 799–812.

Maddox, S. J., Efstathiou, G., Sutherland, W. J., and Loveday, J. 1990. Galaxy correlations on large scales. *MNRAS*, **242**(Jan.), 43P–47P.

Maddox, S. J., Efstathiou, G., and Sutherland, W. J. 1996. The APM galaxy survey – III. An analysis of systematic errors in the angular correlation function and cosmological implications. *MNRAS*, **283**(Dec.), 1227–1263.

Madore, B. F., and Freedman, W. L. 1991. The Cepheid distance scale. *PASP*, **103**(Sept.), 933–957.

Madore, B. F., and Freedman, W. L. 1998. Calibration of the extragalactic distance scale. Page 263 of: Aparicio, A., Herrero, A., and Sánchez, F. (eds), *Stellar Astrophysics for the Local Group: VIII Canary Islands Winter School of Astrophysics*.

Malik, K. A., and Wands, D. 2009. Cosmological perturbations. *Phys. Rep.*, **475**(May), 1–51.

Mangano, G., Miele, G., Pastor, S. *et al.* 2005. Relic neutrino decoupling including flavour oscillations. *Nuclear Physics B*, **729**(Nov.), 221–234.

Margot, J.-L., and Giorgini, J. D. 2009 (May). Probing general relativity with radar astrometry in the inner solar system. Page 701 of: *IAU Symposium #261, American Astronomical Society*, Vol. 261.

Martin, J. 2015. The observational status of cosmic inflation after Planck. *arXiv astro-ph*, 1502.05733

Martin, J., and Musso, M. 2005. Stochastic quintessence. *Phys. Rev. D*, **71**(6), 063514.

Martin, J., Ringeval, C., and Vennin, V. 2014. Encyclopædia Inflationaris. *Physics of the Dark Universe*, **5**(Dec.), 75–235.

Martínez-González, E. 2009. Cosmic microwave background anisotropies: The power spectrum and beyond. Pages 79–120 of: Martínez, V. J., Saar, E., Martínez-González, E., and Pons-Bordería, M.-J. (eds), *Data Analysis in Cosmology*. Lecture Notes in Physics, Berlin Springer Verlag, Vol. 665.

Martínez-González, E., and Sanz, J. L. 1990. CMB anisotropies generated by cosmic voids and great attractors. *MNRAS*, **247**(Dec.), 473–478.

Martínez-González, E., Sanz, J. L., and Silk, J. 1990. Anisotropies in the microwave sky due to nonlinear structures. *ApJ*, **355**(May), L5–L9.

Mashhoon, B., and Wesson, P. 2007. An embedding for general relativity and its implications for new physics. *General Relativity and Gravitation*, **39**(Sept.), 1403–1412.

Massey, R., Rhodes, J., Ellis, R. *et al.* Dark matter maps reveal cosmic scaffolding. *Nature*, **445**(Jan.), 286–290.

Mather, J. C., and Boslough, J. 1996. *The Very First Light: the True Inside Story of the Scientific Journey Back to the Dawn of the Universe*. Basic Books.

Mather, J. C., and the COBE collaboration. 1990. A preliminary measurement of the cosmic microwave background spectrum by the Cosmic Background Explorer (COBE) satellite. *ApJ*, **354**(May), L37–L40.

Mathews, G. J., Kajino, T., and Shima, T. 2005. Big Bang nucleosynthesis with a new neutron lifetime. *Phys. Rev. D*, **71**(2), 021302.

Matsuda, T., Sato, H., and Takeda, H. 1971. Dissipation of primordial turbulence and thermal history of the Universe. *Progress of Theoretical Physics*, **46**(Aug.), 416–432.

Matthews, T. A., and Sandage, A. R. 1963. Optical identification of 3c 48, 3c 196, and 3c 286 with stellar objects. *ApJ*, **138**(July), 30.

Mattig, W. 1958. Über den Zusammenhang zwischen Rotverschiebung und scheinbarer Helligkeit. *Astronomische Nachrichten*, **284**(May), 109.

Maxwell, J. C. 1865. A dynamical theory of the electromagnetic field. *Phil. Trans. Roy. Soc. Lond.*, **155**, 459–512.

Maxwell, J.C. 1954. *Treatise on Electricity and Magnetism, Vol. 1*. 3rd. edn. Dover Publications.

McCrea, W. H. 1928. A note on Dr. P. A. Taylor's paper 'The equilibrium of the calcium chromosphere,'. *MNRAS*, **88**(June), 729.

McCrea, W. H. 1929. The hydrogen chromosphere. *MNRAS*, **89**(Mar.), 483.

McCrea, W. H. 1951. Relativity theory and the creation of matter. *Royal Society of London Proceedings Series A*, **206**(May), 562–575.

McCrea, W. H. 1964. Continual creation. *MNRAS*, **128**, 335.

McCrea, W. H. 1984. *The Influence of Radio Astronomy on Cosmology*, pp. 365–384 of Sullivan, W. T. (ed.) *The Early Years of Radio Astronomy – Reflections Fifty Years after Jansky's Discovery*. Cambridge University Press.

McCrea, W. H., and McVittie, G. C. 1930. Relativity: On the contraction of the Universe. *MNRAS*, **91**(Nov.), 128–133.

McCrea, W. H., and Milne, E. A. 1934. Newtonian Universes and the curvature of space. *The Quarterly Journal of Mathematics*, **5**, 73–80.

McCrea, W.H. 1950. The Steady State Theory of the expanding Universe. *Endeavour*, **9**, 3–10.

McKellar, A. 1940. Evidence for the molecular origin of some hitherto unidentified interstellar lines. *PASP*, **52**(June), 187.

McKellar, A. 1941. Molecular lines from the lowest states of diatomic molecules composed of atoms probably present in interstellar space. *Publications of the Dominion Astrophysical Observatory Victoria*, **7**(Mar.), 251–272.

McVittie, G. C. 1931. The problem of n bodies and the expansion of the Universe. *MNRAS*, **91**(Jan.), 274.

McVittie, G. C. 1932. Condensations in an expanding universe. *MNRAS*, **92**(Apr.), 500–518.

McVittie, G. C. 1933. The mass-particle in an expanding universe. *MNRAS*, **93**(Mar.), 325–339.

McVittie, G. C. 1965a. *General Relativity and Cosmology*. Chapman and Hall.

McVittie, G. C. 1965b. Some consequences of large redshifts. *ApJ*, **142**(Nov.), 1637.

Meerburg, P. D., Hložek, R., Hadzhiyska, B., and Meyers, J. 2015. Multiwavelength constraints on the inflationary consistency relation. *Phys. Rev. D*, **91**(10), 103505.

Messier, C. 1784. Catalogue des nebuleuses et des amas d'etoiles. *Connaissance des Temps*, 238–282.

Mészáros, P. 2006. Gamma-ray bursts. *Reports on Progress in Physics*, **69**(Aug.), 2259–2321.

Metzger, M. R., Djorgovski, S. G., Kulkarni, S. R. *et al.* 1997. Spectral constraints on the redshift of the optical counterpart to the γ-ray burst of 8 May 1997. *Nature*, **387**(June), 878–880.

Meyer, D. M., and Jura, M. 1984. The microwave background temperature at 2.64 and 1.32 millimeters. *ApJ*, **276**(Jan.), L1–L3.

Michie, R. W. 1969. On the growth of condensations in an expanding Universe. *Contributions from the Kitt Peak National Observatory*, **440**.

Milgrom, M. 1983. A modification of the Newtonian dynamics as a possible alternative to the hidden mass hypothesis. *ApJ*, **270**(July), 365–370.

Mills, B. Y., and Slee, O. B. 1957. A preliminary survey of radio sources in a limited region of the sky at the wavelength of 3.5 m. *Australian Journal of Physics*, **10**(Mar.), 162–194.

Mills, B. Y., Slee, O. B., and Hill, E. R. 1958. A catalogue of radio sources between declinations $+10°$ and $-20°$. *Australian Journal of Physics*, **11**(Sept.), 360.

Mills, B. Y., Slee, O. B., and Hill, E. R. 1960. A catalogue of radio sources between declinations $-20°$ and $-50°$. *Australian Journal of Physics*, **13**(Dec.), 676.

Mills, B. Y., Slee, O. B., and Hill, E. R. 1961. A catalogue of radio sources between declinations $-50°$ and $-80°$. *Australian Journal of Physics*, **14**(Dec.), 497.

Milne, E. A. 1933. Note on H. P. Robertson's paper on world-structure. *Zeitschrift fur Astrophysik*, **7**, 180.

Milne, E. A. 1935. *Relativity, Gravitation and World-Structure*. Oxford, The Clarendon press, 1935.

Minkowski, H. 1908. Die Grundgleichungen für die elektromagnetischen Vorgänge in bewegten Körpern. *Nachrichten der Königlichen Gesellschaft der Wissenschaften zu Göttinngen, Mathematisch-Physikalische Klasse*, 53–111. English title: The fundamental equations for electromagnetic processes in moving bodies.

Minkowski, H. 1909. Raum und Zeit. *Jahresberichte der Deutschen Mathematiker-Vereinigung*, **18**, 75–88. Engish title: Space and time. A lecture delivered before the Naturforscher Versammlung (Congress of Natural Philosophers) at Cologne – (21st September, 1908).

Minkowski, R. 1941. Spectra of supernovae. *PASP*, **53**(Aug.), 224.

Misner, C. W. 1968. The isotropy of the Universe. *ApJ*, **151**(Feb.), 431.

Misner, C. W. 1969. Mixmaster Universe. *Physical Review Letters*, **22**(May), 1071–1074.

Misner, C. W., and Sharp, D. H. 1965. Spherical gravitational collapse with energy transport by radiative diffusion. *Physics Letters*, **15**(Apr.), 279–281.

Misner, C. W., and Wheeler, J. A. 1957. Classical physics as geometry. *Annals of Physics*, **2**(Dec.), 525–603.

Misner, C. W., Thorne, K. S., and Wheeler, J. A. 1973. *Gravitation*. W.H. Freeman.

Mohr, P. J., Taylor, B. N., and Newell, D. B. 2012. CODATA recommended values of the fundamental physical constants: 2010. *Reviews of Modern Physics*, **84**(Oct.), 1527–1605.

Molina, R., Nunez, J., Cortijo, F. J., and Mateos, J. 2001. Image restoration in astronomy: a Bayesian perspective. *IEEE Signal Processing Magazine*, **18**(Mar.), 11–29.

Monthly Notices Obituary. 1854. Report of the Thirty-fourth Annual General Meeting of the Society. *MNRAS*, **14**(Feb.), 97.

Møller, C. 1967. Relativistic thermodynamics – A strange incident in the history of physics. *Kongl. Danske Vid. Selsk. mat-fys. Medd*, **36**, 1–27.

Mortlock, D. J., Warren, S. J. *et al.* 2011. A luminous quasar at a redshift of $z = 7.085$. *Nature*, **474**(June), 616–619.

Moschella, U. 2005. *The de Sitter and anti-de Sitter sightseeing tour*. [Online]. Available: `http://www.bourbaphy.fr/moschella.pdf`.

Moss, G. E., Miller, L. R., and Forward, R. L. 1971. Photon-noise-limited laser transducer for gravitational antenna. *Appl. Opt.*, **10**, 2495–2498.

Muller, C. A., and Oort, J. H. 1951. Observation of a line in the galactic radio spectrum: The interstellar hydrogen line at 1420 Mc/sec, and an estimate of galactic rotation. *Nature*, **168**(Sept.), 357–358.

Müller, I. 1967a. On the entropy inequality. *Arch. Ration. Mech. An.*, **26**(2), 118–141.

Müller, I. 1967b. Zum Paradoxon der Wärmeleitungstheorie. *Zeitschrift fur Physik*, **198**(Aug.), 329–344.

Muller, S., Beelen, A., Black, J. H. *et al.* 2013. A precise and accurate determination of the cosmic microwave background temperature at z = 0.89. *A&A*, **551**(Mar.), A109.

Nakahata, M. 2011. Astroparticle physics with solar neutrinos. *Proceeding of the Japan Academy, Series B*, **87**, 215–229.

Nanos, Jr., G. P. 1979. Polarization of the blackbody radiation at 3.2 centimeters. *ApJ*, **232**(Sept.), 341–347.

Narlikar, J.V. 2011. Machs principle. *Resonance*, **16**(4), 310–320.

NASA-LAMBDA Website. 2012. *LAMBDA – Data Products: CMB Experiments*. [Online]. Available: . `http://lambda.gsfc.nasa.gov/product/expt/`.

Neill, J. D., Hudson, M. J., and Conley, A. 2007. The peculiar velocities of local type Ia supernovae and their impact on cosmology. *ApJ*, **661**(June), L123–L126.

Netterfield, C. B. *et al.* 2002. A measurement by BOOMERANG of multiple peaks in the angular power spectrum of the cosmic microwave background. *ApJ*, **571**(June), 604–614.

Neumann, C. 1874. Über die den Kräften elecktrodynamischen Usprungs zuzuschreiben-den Elementargesetz. *Abh. Math.- Phys. Classe der Köngl. Sächsischen Gessellschaft de Wissenschaften*, **10**, 417–524.

Neumann, C. 1896. *Allgemeine Untersuchungen über das Newton'sche Princip der Fern-wirkungen mit besonderer Rücksicht auf die elektrische Wirkungen*. Leipzig: Teubner.

Newman, E., and Penrose, R. 1962. An approach to gravitational radiation by a method of spin coefficients. *Journal of Mathematical Physics*, **3**(May), 566–578.

Newman, E., and Penrose, R. 1963. Errata: an approach to gravitational radiation by a method of spin coefficients. *Journal of Mathematical Physics*, **4**(July), 998–998.

Newton, I. 1726. *The Principia: Mathematical Principles of Natural Philosophy (3rd edition)*. Royal Society of London.

Newton, I. 1999. *The Principia: Mathematical Principles of Natural Philosophy*. University of California Press.

Neyman, J. 1937. Outline of a theory of statistical estimation based on the classical theory of probability. *Phil. Trans. Roy Soc. A*, **236**(767), 333–380.

Neyman, J., and Scott, E.L. 1948. Consistent estimates based on partially consistent observations. *Econometrica*, **16**, 1–32.

Neyman, J., and Scott, E.L. 1960. Correction for bias introduced by a transformation of variables. *Ann. Math. Stat.*, **31**(3), 643–655.

Noether, E. 1918. Invariante Variationsprobleme. *Nachr. D. Knig. Gesellsch. D. Wiss. Zu Gttingen, Math-phys. Klasse*, 235–257.

Noonan, T. W. 1984. The gravitational contribution to the stress-energy tensor of a medium in general relativity. *General Relativity and Gravitation*, **16**(Nov.), 1103–1118.

Nordström, G. 1914. Über die Möglichkeit, das elektromagnetische Feld und das Gravita-tionsfeld zu vereinigen. *Physikalische Zeitschrift*, **15**, 504–506.

Norgaard-Nielsen, H. U., Hansen, L., Jorgensen, H. E., Aragon Salamanca, A., and Ellis, R. S. 1989. The discovery of a type IA supernova at a redshift of 0.31. *Nature*, **339**(June), 523–525.

North, J. D. 1965. *The Measure of the Universe. A History of Modern Cosmology*. Clarendon Press.

North, J.D. 2008. *Cosmos: An Illustrated History of Astronomy and Cosmology* . University Of Chicago Press.

Norton, J.D. 1999. The cosmological woes of Newtonian gravitation theory. Pages 271–322 of: Goenner, H., Renn, J., Ritter, J., and Sauer, T. (eds), *The Expanding Worlds of General Relativity: Einstein Studies*, Vol. 7. Birkhauser.

Novikov, I. D. 1964. On the evolution of a semiclosed world. *Soviet Ast.*, **7**(Feb.), 587.

Nussbaumer, H., and Bieri, L. 2011. Who discovered the expansion of the Universe? *The Observatory*, **131**(Dec.), 394–398.

O'Hagan, A., and Luce, B.R. 2003. *A Primer on Bayesian Statistics in Health Economics and Outcomes Research*. Tech. rept. Centre for Bayesian Statistics in Health Economics, University of Sheffield.

Ohanian, H. C., and Ruffini, R. 2013. *Gravitation and Spacetime*. 3rd. edn. Cambridge University Press.

Olive, D.J. 2012. *A Course in Statistical Theory*. David Olive.

Olive, K. A., and Particle Data Group. 2014. Review of particle physics. *Chinese Physics C*, **38**(9), 090001.

Olson, D. W. 1976. Density perturbations in cosmological models. *Phys. Rev. D*, **14**(July), 327–331.

Omnès, R. 1969. Possibility of matter–antimatter separation at high temperature. *Physical Review Letters*, **23**(July), 38–40.

Oort, J. H., Kerr, F. J., and Westerhout, G. 1958. The galactic system as a spiral nebula (Council Note). *MNRAS*, **118**, 379.

O'Raifeartaigh, C., and McCann, B. 2014. Einstein's cosmic model of 1931 revisited: an analysis and translation of a forgotten model of the Universe. *European Physical Journal H*, **39**(Feb.), 63–85.

O'Raifeartaigh, C., McCann, B., Nahm, W., and Mitton, S. 2014. Einstein's steady-state theory: an abandoned model of the cosmos. *European Physical Journal H*, **39**(Sept.), 353.

Osterbrock, D. E., and Rogerson, Jr., J. B. 1961. The helium and heavy-element content of gaseous-nebulae and the Sun. *PASP*, **73**(Apr.), 129.

Ostriker, J. P., and Tremaine, S. D. 1975. Another evolutionary correction to the luminosity of giant galaxies. *ApJ*, **202**(Dec.), L113–L117.

Overduin, J. M., and Wesson, P. S. 1997. Kaluza–Klein gravity. *Phys. Rep.*, **283**(Apr.), 303–378.

Overduin, J. M., and Wesson, P. S. 2004. Dark matter and background light. *Phys. Rep.*, **402**(Nov.), 267–406.

Padgett, M. 2014. A new twist on the Doppler shift. *Physics Today*, **67**(2), 58–59.

Padmanabhan , T. 2010. *Gravitation: Foundations and Frontiers*. Cambridge University Press.

Page, L. *et al.* 2003. First-year Wilkinson Microwave Anisotropy Probe (WMAP) observations: Interpretation of the TT and TE angular power spectrum peaks. *ApJs*, **148**(Sept.), 233–241.

Painlevé, P. 1921a. La gravitation dans la mécanique de Newton et dans la mécanique d'Einstein. *Comptes Rendus Acad. Sci. (Paris)*, **173**, 873–886.

Painlevé, P. 1921b. La mécanique classique et la thorie de la relativité. *Comptes Rendus Acad. Sci. (Paris)*, **173**, 677–680.

Pais, A. 1982. *Subtle is the Lord: The Science and the Life of Albert Einstein*. 1st edition, Oxford University Press.

Panofsky, W.K.H., and Phillips, M. 2005. *Classical Electricity and Magnetism: Second Edition*. Dover Publications.

Papoulis, A., and Pillai, S.U. 2002. *Probability, Random Variables and Stochastic Processes*. McGraw-Hill Higher Education. 4th edition.

Paranjape, A. 2008. Backreaction of cosmological perturbations in covariant macroscopic gravity. *Phys. Rev. D*, **78**(6), 063522.

Paranjape, A. 2009 (June). *The Averaging Problem in Cosmology*. PhD Thesis, 2009.

Paranjape, A., and Singh, T. P. 2008. Structure formation, backreaction and weak gravitational fields. *JCAP*, **3**(Mar.), 23.

Pâris, I., Petitjean, P. *et al.* 2014. The Sloan Digital Sky Survey quasar catalog: tenth data release. *A&A*, **563**(Mar.), A54.

Partridge, R. B. 1988. Review Article: The angular distribution of the cosmic background radiation. *Reports on Progress in Physics*, **51**(May), 647–706.

Partridge, R. B. 1991. Fluctuations in the cosmic microwave background: The first measurements. Pages 149–176 of: Blanchard, A., Celnikier, L., Lachieze-Rey, M., and Tran Thanh Van, J. (eds), *Physical Cosmology*. Editions Frontièrs.

Partridge, R. B. 1992. Recent observations of the cosmic microwave background and their cosmological implication. Page 97 of: de Carvalho, R. R. (ed), *Cosmology and Large-Scale Structure in the Universe*. Astronomical Society of the Pacific Conference Series, Vol. 24.

Partridge, R. B., and Wilkinson, D. T. 1967. Isotropy and homogeneity of the Universe from measurements of the cosmic microwave background. *Physical Review Letters*, **18**(Apr.), 557–559.

Partridge, R.B. 2007. *3K: The cosmic microwave background radiation*. Cambridge Astrophysics (Book 25). Cambridge University Press. First published 1995, reissued 2007.

Patton, C. M., and Wheeler, J. A. 1975. Is physics legislated by cosmogony. Pages 538–605 of: Isham, C. J., Penrose, R., and Sciama, D. W. (eds), *Quantum Gravity*. Oxford University Press.

Pawsey, J. L. 1951. Observation of a line in the galactic radio spectrum: The interstellar hydrogen line at 1420 Mc/sec, and an estimate of galactic rotation. *Nature*, **168**(Sept.), 358.

Payne, C. H. 1925a. Astrophysical data bearing on the relative abundance of the elements. *Proceedings of the National Academy of Science*, **11**(Mar.), 192–198.

Payne, C. H. 1925b. Stellar atmospheres; a contribution to the observational study of high temperature in the reversing layers of stars. PhD thesis, Radcliffe College.

Payne, W. T. 1952. Elementary spinor theory. *American Journal of Physics*, **20**(May), 253–262.

Payne, W. T. 1955. Spinor theory and relativity. I. *American Journal of Physics*, **23**(Nov.), 526–536.

Payne, W. T. 1959. Spinor theory and relativity. II. *American Journal of Physics*, **27**(May), 318–328.

Peacock, J. A. 1999. *Cosmological Physics*. Cambridge University Press.

Pearson, K. 1895. Contributions to the mathematical theory of evolution. II. Skew variation in homogeneous material. *Phil. Trans. Roy. Soc. A: Math., Phys. and Eng. Sci.*, **186**, 343–414.

Peebles, P. 1971. *Physical Cosmology*. Princeton Series in Physics. Princeton University Press.

Peebles, P. J. 1966a. Primeval helium abundance and the primeval fireball. *Physical Review Letters*, **16**(Mar.), 410–413.

Peebles, P. J., and Wilkinson, D. T. 1968. Comment on the anisotropy of the primeval fireball. *Physical Review*, **174**(Oct.), 2168–2168.

Peebles, P. J. E. 1965. The black-body radiation content of the Universe and the formation of galaxies. *ApJ*, **142**(Nov.), 1317.

Peebles, P. J. E. 1966b. Primordial helium abundance and the primordial fireball. II. *ApJ*, **146**(Nov.), 542.

Peebles, P. J. E. 1967. The gravitational instability of the Universe. *ApJ*, **147**(Mar.), 859.

Peebles, P. J. E. 1968. Recombination of the primeval plasma. *ApJ*, **153**(July), 1.

Peebles, P. J. E. 1973. Statistical analysis of catalogs of extragalactic objects. I. Theory. *ApJ*, **185**(Oct.), 413–440.

Peebles, P. J. E. 1993. *Principles of Physical Cosmology*. Princeton University Press.

Peebles, P. J. E. 2014. Discovery of the Hot Big Bang: What happened in 1948. *European Physical Journal H*, **39**(Apr.), 205–223.

Peebles, P. J. E., and Yu, J. T. 1970. Primeval adiabatic perturbation in an expanding Universe. *ApJ*, **162**(Dec.), 815.

Peebles, P.J.E. 1980. *Large-Scale Structure of the Universe*. Princeton Series in Physics. Princeton University Press.

Peebles, P.J.E., Page Jr., L.A., and Partridge, R.B. 2009. *Finding the Big Bang*. Cambridge University Press.

Penrose, R. 1960. A spinor approach to general relativity. *Annals of Physics*, **10**(June), 171–201.

Penzias, A. A., and Wilson, R. W. 1965. A measurement of excess antenna temperature at 4080 Mc/s. *ApJ*, **142**(July), 419–421.

Pequignot, D., Petitjean, P., and Boisson, C. 1991. Total and effective radiative recombination coefficients. *A&A*, **251**(Nov.), 680–688.

Percival, W. J. *et al.* 2001. The 2dF Galaxy Redshift Survey: the power spectrum and the matter content of the Universe. *MNRAS*, **327**(Nov.), 1297–1306.

Perl, M. L. *et al.* 1975. Evidence for anomalous lepton production in e^+e^- annihilation. *Physical Review Letters*, **35**(Dec.), 1489–1492.

Perley, D. A. *et al.* 2009. GRB 080503: Implications of a naked short gamma-ray burst dominated by extended emission. *ApJ*, **696**(May), 1871–1885.

Perlick, V. 2010. Gravitational lensing from a spacetime perspective. *arXiv gr-qc*, 1010.3416

Perlmutter, S. 2003. Supernovae, dark energy, and the accelerating Universe. *Physics Today*, **56**(4), 53–62.

Perlmutter, S. 2012. Nobel Lecture: Measuring the acceleration of the cosmic expansion using supernovae. *Reviews of Modern Physics*, **84**(July), 1127–1149.

Perlmutter, S. *et al.* 1998. Discovery of a supernova explosion at half the age of the universe. *Nature*, **391**(Jan.), 51.

Perlmutter, S. *et al.* (Supernova Cosmology Project). 1995. A supernova at $z = 0.458$ and implications for measuring the cosmological deceleration. *ApJ*, **440**(Feb.), L41–L44.

Perlmutter, S. *et al.* (Supernova Cosmology Project). 1997. Measurements of the cosmological parameters Ω and Λ from the first seven supernovae at $z \geq 0.35$. *ApJ*, **483**(July), 565–581.

Perlmutter, S. *et al.* (Supernova Cosmology Project). 1999. Measurements of omega and lambda from 42 high-redshift supernovae. *ApJ*, **517**(June), 565–586.

Petrosian, V., Bouvier, A., and Ryde, F. 2009. Gamma-ray bursts as cosmological tools. *arxiv astro-ph*, 0909.5051.

Phillips, M. M. 1993. The absolute magnitudes of Type IA supernovae. *ApJ*, **413**(Aug.), L105–L108.

Phillips, M. M., Lira, P., Suntzeff, N. B. *et al.* 1999. The reddening-free decline rate versus luminosity relationship for type IA supernovae. *AJ*, **118**(Oct.), 1766–1776.

Pina, R. K., and Puetter, R. C. 1993. Bayesian image reconstruction – The pixon and optimal image modeling. *PASP*, **105**(June), 630–637.

Pirani, F. A. 1957. Invariant formulation of gravitational radiation theory. *Physical Review*, **105**(Feb.), 1089–1099.

Planck, M. 1900a. Über eine Verbesserung der Wienschen Spektralgleichung. *Verhandl. Dtsch. phys. Ges.*, **2**, 202–204.

Planck, M. 1900b. Zur Theorie des Gesetzes der Energieverteilung im Normalspektrum. *Verhandl. Dtsch. phys. Ges.*, **2**, 237.

Planck, M. 1901. Ueber das Gesetz der Energieverteilung im Normalspectrum. *Annalen der Physik*, **309**, 553–563.

Planck, M. 1907. Zur Dynamik bewegter System. *Sitzungsberichte der Kniglich-Preussischen Akademie der Wissenschaften, Berlin*, **29**, 542–570.
https://archive.org/details/sitzungsbericht396klasgoog
https://en.wikisource.org/wiki/Translation:On_the_Dynamics_of_
Moving_Systems.

Planck Collaboration ES. 2013. *Planck Explanatory Supplement (Public Release 1.07)*. E.S.A., [Online]. Available: . http://wiki.cosmos.esa.int/planckpla/index.php/Main_Page.

Planck Collaboration I. 2013. Planck 2013 results. I. Overview of products and scientific results. *Astron. & Astrophys.*, **571**(Mar), A1.

Planck Collaboration I. 2015. Planck 2015 results. I. Overview of products and scientific results. *ArXiv astro-ph*, 1502.01582

Planck Collaboration VIII. 2015. Planck 2015 results. VIII. High frequency instrument data processing: Calibration and maps. *ArXiv astro-ph*, 1502.01587

Planck Collaboration XII. 2014. Planck 2013 results. XII. Diffuse component separation. *Astron. & Astrophys.*, **571**(A12), 31.

Planck Collaboration XIII. 2015. Planck 2015 results. XIII. Cosmological parameters. *arXiv astro-ph*, 1502.01589

Planck Collaboration XVI. 2013. Planck 2013 results: 16: Cosmological parameters. *Astron. & Astrophys.*, **571**.

Planck Team XX. 2015. Planck 2015 results. XX. Constraints on inflation. *arXiv astro-ph*, 1502.02114

Planck Collaboration XXII. 2014. Planck 2013 results. XXII. Constraints on inflation. *Astron. & Astrophys.*, **571**(Nov.), A22.

Platen, E., van de Weygaert, R., and Jones, B. J. T. 2007. A cosmic watershed: the WVF void detection technique. *MNRAS*, **380**(Sept.), 551–570.

Podolký, J. 1999. The structure of the extreme Schwarzschild–de Sitter space-time. *General Relativity and Gravitation*, **31**(Nov.), 1703.

Pogson, N. 1856. Magnitudes of thirty-six of the minor planets for the first day of each month of the year 1857. *MNRAS*, **17**(Nov.), 12–15.

Poincaré, H. 1905. Sur la dynamique de l'électron. *Comptes Rendus Acad. Sci.*, **23**, 140.

Poisson, E. 2007. *A Relativist's Toolkit: The Mathematics of Black-Hole Mechanics*. Cambridge University Press.

Pooley, G. G., and Ryle, M. 1968. The extension of the number-flux density relation for radio sources to very small flux densities. *MNRAS*, **139**, 515.

Popper, K.R. 2002. *The Logic of Scientific Discovery*. Routledge.

Porter, F. C. 2008. Testing consistency of two histograms. *arXiv data-an*, 0804.0380

Pound, R. V., and Rebka, G. A. 1960. Apparent weight of photons. *Physical Review Letters*, **4**(Apr.), 337–341.

Preskill, J. P. 1979. Cosmological production of superheavy magnetic monopoles. *Physical Review Letters*, **43**(Nov.), 1365–1368.

Press, W. H. 1980. Spontaneous production of the Zel'dovich spectrum of cosmological fluctuations. *Phys. Scr*, **21**, 702–707.

Press, W. H., and Thorne, K. S. 1972. Gravitational-wave astronomy. *ARA&A*, **10**, 335.

Press, W.H., Teukolsky, S., Vetterling, W.T., and Flannery, B.P. 2007. *Numerical Recipes in C++: The Art of Scientific Computing*. Cambridge University Press; 3rd. edition.

Pretorius, F. 2005. Evolution of binary black-hole spacetimes. *Physical Review Letters*, **95**(12), 121101.

Pretorius, F. 2006. Simulation of binary black hole spacetimes with a harmonic evolution scheme. *Classical and Quantum Gravity*, **23**(Aug.), S529–S552.

Proctor, R. A. 1869. Distribution of the nebulæ. *MNRAS*, **29**(June), 337.

Prokhorov, D. A. 2015. Upper bounds on matter-antimatter admixture from gamma-ray observations of colliding clusters of galaxies with the Fermi Large Area Telescope. *Phys. Rev. D*, **91**(8), 083002.

Pskovskii, I. P. 1977. Light curves, color curves, and expansion velocity of type I supernovae as functions of the rate of brightness decline. *Soviet Ast.*, **21**(Dec.), 675–682.

Pskovskii, Y. P. 1984. Photometric classification and basic parameters of type I supernovae. *Soviet Ast.*, **28**(Dec.), 658–664.

Rafelski, J., and Birrell, J. 2014. Traveling through the Universe: Back in time to the quark-gluon plasma era. *Journal of Physics Conference Series*, **509**(1), 012014.

Raine, D. J. 1975. Mach's principle in general relativity. *MNRAS*, **171**(June), 507–528.

Raine, D.W. 1981. *The Isotropic Universe*. Monographs on astronomical subjects. Adam Hilger.

Ramsey, F.P. 1931. Truth and probability. In *Mathematics and Other Logical Essays*, pp. 156–198. Kegan, Paul, Trench, Trubner & Co.

Rao, C.R. 1945. Information and the accuracy attainable in the estimation of statistical parameters. *Bull. Calcutta Math. Soc*, **37**, 81–89.

Räsänen, S. 2010. Applicability of the linearly perturbed FRW metric and Newtonian cosmology. *Phys. Rev. D*, **81**(10), 103512.

Raychaudhuri, A. 1955. Relativistic cosmology. I. *Physical Review*, **98**(May), 1123–1126.

Raychaudhuri, A. 1957. Singular state in relativistic cosmology. *Physical Review*, **106**(Apr.), 172–173.

Rees, M. J. 1968. Polarization and spectrum of the primeval radiation in an anisotropic Universe. *ApJ*, **153**(July), L1.

Rees, M. J., and Sciama, D. W. 1968. Large-scale density inhomogeneities in the Universe. *Nature*, **217**(Feb.), 511–516.

Reid, N. 2010. Likelihood inference. *Computational Statistics*, **2**(5), 517–525. (Not an open journal).

Reid, N. 2013. Aspects of likelihood inference. *Bernoulli*, **19**, 1404–1418.

Reines, F., and Cowan, C. L. 1953. Detection of the free neutrino. *Physical Review*, **92**(Nov.), 830–831.

Renn, J., and Schemmel, M. 2012. Theories of gravitation in the twilight of classical physics. Page 363 of: Lehner, C., Renn, J., and Schemmel, M. (eds), *Einstein and the Changing Worldviews of Physics*. Einstein Studies. Birkhäuser.

Richards, G. T., Strauss, M. A. *et al.* 2006. The Sloan Digital Sky Survey Quasar Survey: Quasar luminosity function from data release 3. *AJ*, **131**(June), 2766–2787.

Richardson, W. H. 1972. Bayesian-based iterative method of image restoration. *J. Optical Soc. America*, **62**, 55–59.

Riess, A. G. 2000. The case for an accelerating Universe from supernovae. *PASP*, **112**(Oct.), 1284–1299.

Riess, A. G., Filippenko, A. V. *et al.* (Hi-z team) 1998. Observational evidence from supernovae for an accelerating Universe and a cosmological constant. *AJ*, **116**(Sept.), 1009–1038.

Riess, A. G., Macri, L. *et al.* 2009. A redetermination of the Hubble constant with the Hubble Space Telescope from a differential distance ladder. *ApJ*, **699**(July), 539–563.

Riess, A. G., Macri, L. *et al.* 2011. A 3% solution: Determination of the Hubble constant with the Hubble Space Telescope and wide field camera 3. *ApJ*, **730**(Apr.), 119. This citation has been updated as per the errata *ApJ*. 2011, **730**, 119.

Riess, A. G., Strolger, L.-G. *et al.* (Hi-Z team) 2007. New Hubble Space Telescope discoveries of type Ia supernovae at $z \geq 1$: Narrowing constraints on the early behavior of dark energy. *ApJ*, **659**(Apr.), 98–121.

Riess, A. G., Press, W. H., and Kirshner, R. P. 1995. Using type IA supernova light curve shapes to measure the Hubble constant. *ApJ*, **438**(Jan.), L17–L20.

Riess, A. G., Press, W. H., and Kirshner, R. P. 1996. A precise distance indicator: Type IA supernova multicolor light-curve shapes. *ApJ*, **473**(Dec.), 88.

Rindler, W. 2006. *Relativity: Special, General, and Cosmological.* Oxford University Press.

Rindler, W., and Ishak, M. 2007. Contribution of the cosmological constant to the relativistic bending of light revisited. *Phys. Rev. D*, **76**(4), 043006.

Robert, C., and Casella, G. 2011. A short history of Markov Chain Monte Carlo: Subjective recollections from incomplete data. *Statistical Science*, **26**, 102–115.

Roberts, M. S. 1976. The rotation curves of galaxies. *Comments on Astrophysics*, **6**(July), 105.

Robertson, H. P. 1929. On the foundations of relativistic cosmology. *Proceedings of the National Academy of Science*, **15**(Nov.), 822–829.

Robertson, H. P. 1933. Relativistic cosmology. *Reviews of Modern Physics*, **5**(Jan.), 62–90.

Robertson, H. P. 1935. Kinematics and world-structure. *ApJ*, **82**(Nov.), 284.

Robertson, H. P. 1955. The theoretical aspects of the nebular redshift. *PASP*, **67**(Apr.), 82.

Robertson, H.P. 1927. Dynamical space-times which contain a conformal Euclidean 3-space. *Trans. Am. Math. Soc.*, **29**, 481–496.

Robertson, H.P. 1928. On relativistic cosmology. *Philosophical Magazine*, **5**, 835.

Robson, J. M. 1950. Radioactive decay of the neutron. *Physical Review*, **78**(May), 311–312.

Robson, J. M. 1951. The radioactive decay of the neutron. *Physical Review*, **83**(July), 349–358.

Roll, P. G., and Wilkinson, D. T. 1966. Cosmic background radiation at 3.2 cm: Support for cosmic black-body radiation. *Physical Review Letters*, **16**(Mar.), 405–407.

Romero, C., Tavakol, R., and Zalaletdinov, R. 1996. The embedding of general relativity in five dimensions. *General Relativity and Gravitation*, **28**(Mar.), 365–376.

Roos, M. 2003. *Introduction to Cosmology.* 3rd. edn. Wiley–Blackwell.

Rosenfeld, L. 1940. Sur le tenseur d'impulsion-energie. *Mem. Roy. Acad. Belg. Cl. Sci.*, **18**.

Roth, K. C., Meyer, D. M., and Hawkins, I. 1993. Interstellar cyanogen and the temperature of the cosmic microwave background radiation. *ApJ*, **413**(Aug.), L67–L71.

Rozelot, J.-P., and Damiani, C. 2011. History of solar oblateness measurements and interpretation. *European Physical Journal H*, **36**(Nov.), 407–436.

Rubin, D., Knop, R. A., Rykoff, E. *et al.* (Supernova Cosmology Project) 2013. Precision measurement of the most distant spectroscopically confirmed supernova Ia with the Hubble Space Telescope. *ApJ*, **763**(Jan.), 35.

Rubin, V. C. 1983. The rotation of spiral galaxies. *Science*, **220**(June), 1339–1344.

Rubin, V. C., Thonnard, N., and Ford, Jr., W. K. 1978. Extended rotation curves of high-luminosity spiral galaxies. IV – Systematic dynamical properties, SA through SC. *ApJ*, **225**(Nov.), L107–L111.

Rubin, V. C., Ford, W. K. J., and Thonnard, N. 1980. Rotational properties of 21 SC galaxies with a large range of luminosities and radii, from NGC 4605 $R = 4$kpc to UGC 2885 $R = 122$ kpc. *ApJ*, **238**(June), 471–487.

Ruhl, J. E. *et al.* 2003. Improved measurement of the angular power spectrum of temperature anisotropy in the cosmic microwave background from two new analyses of BOOMERANG observations. *ApJ*, **599**(Dec.), 786–805.Ruhl_etal_ NewAnalysisOfBoomerangData_0212229v2.pdf.

Ruiz-Lapuente, P. (ed). 2010. *Dark Energy: Observational and Theoretical Approaches.* Cambridge Univerity Press.

Rust, B. W. 1974a (June). The use of supernovae light curves for testing the expansion hypothesis. *Bulletin of the American Astronomical Society*, **6**, 309.

Rust, B. W. 1974b (Sept.). The velocity–distance relation for supernovae. *Bulletin of the American Astronomical Society*, **6**, 450.

Rust, B. W. 1974c (Dec.). Use of supernovae light curves for testing the expansion hypothesis and other cosmological relations. PhD thesis, Oak Ridge National Lab., TN.

Rust, B. W. 1975 (Mar.). The use of supernovae for determining the Hubble constant and estimating extragalactic distances. *Bulletin of the American Astronomical Society*, **7**, 236.

Ryden, B. 2003. *Introduction to Cosmology.* Addison-Wesley.

Ryle, M., and Scheuer, P. A. G. 1955. The spatial distribution and the nature of radio stars. *Royal Society of London Proceedings Series A*, **230**(July), 448–462.

Sachs, R. 1961. Gravitational waves in general relativity. VI. The outgoing radiation condition. *Royal Society of London Proceedings Series A*, **264**(Nov.), 309–338.

Sachs, R. K. 1967. *Gravitational Waves.* American Mathematical Society, pp. 129–140.

Sachs, R. K., and Wolfe, A. M. 1967. Perturbations of a cosmological model and angular variations of the microwave background. *ApJ*, **147**(Jan.), 73.

Sahni, V. 2002. The cosmological constant problem and quintessence. *Classical and Quantum Gravity*, **19**(July), 3435–3448.

Sakharov, A. D. 1966a. The initial stage of an expanding Universe and the appearance of a nonuniform distribution of matter. *Soviet Journal of Experimental and Theoretical Physics*, **22**(Jan.), 241. [Translation of 1965, *J. Expt. Theor. Phys. (U.S.S.R)*, **49**, 345].

Sandage, A. 1961a. The ability of the 200-inch telescope to discriminate between selected world models. *ApJ*, **133**(Mar.), 355.

Sandage, A. 1961b. *The Hubble Atlas of Galaxies.* Carnegie Institute of Washington.

Sandage, A. 1961c. The light travel time and the evolutionary correction to magnitudes of distant galaxies. *ApJ*, **134**(Nov.), 916.

Sandage, A. 1962. The change of redshift and apparent luminosity of galaxies due to the deceleration of selected expanding Universes. *ApJ*, **136**(Sept.), 319.

Sandage, A. 1968. A new determination of the Hubble constant from globular clusters in M87. *ApJ*, **152**(June), L149.

Sandage, A. 1975. *The Redshift*, p. 761. University of Chicago Press.

Sandage, A., and Tammann, G. A. 1981. *A Revised Shapley–Ames Catalog of Bright Galaxies.* Carnegie Institue of Washington.

Sandage, A., Sandage, M., and Kristian, J. (eds). 1975. *Stars and Stellar Systems IX: Galaxies and the Universe.* University of Chicago Press.

Sandage, A. R. 1970. Cosmology: a search for two numbers. *Physics Today*, **23**, 34–41.

Sandage, A. R. 1971. The age of the galaxies and globular clusters: Problems of finding the Hubble constant and deceleration parameter. In O'Connell, D. J. K. (ed), *Study Week on Nuclei of Galaxies*, p. 601.

Sandage, A. R. 1972. Distances to galaxies: the Hubble constant, the Friedmann time, and the edge of the world. *QJRAS*, **13**(Mar.), 282.

Sanders, R. H. 1986. Alternatives to dark matter. *MNRAS*, **223**(Dec.), 539–555.

Sanders, R. H. 2005. A tensor-vector-scalar framework for modified dynamics and cosmic dark matter. *MNRAS*, **363**(Oct.), 459–468.

Sanz, J. L. 1992. *The Microwave Sky*. ed. Sanchez, F., Collados, M. and Reboldo, R., p. 145.

Sanz, J. L. 1997. Elements of general relativity, cosmology and the cosmic microwave background. In Lineweaver, C. H., Bartlett, J. G., Blanchard, A., Signore, M., and Silk, J. (eds), *Cosmological Background Radiation*, pp. 33–65.

Sari, R., Piran, T., and Narayan, R. 1998. Spectra and light curves of gamma-ray burst afterglows. *ApJ*, **497**(Apr.), L17–L20.

Saro, A., Liu, J., Mohr, J. J. *et al.* 2014. Constraints on the CMB temperature evolution using multiband measurements of the Sunyaev-Zel'dovich effect with the South Pole Telescope. *MNRAS*, **440**(May), 2610–2615.

Sasaki, M., Nambu, Y., and Nakao, K.-I. 1988. Classical behavior of a scalar field in the inflationary universe. *Nuclear Physics B*, **308**(Oct.), 868–884.

Saslaw, W. C. 1999. *The Distribution of the Galaxies*. Cambridge University Press.

Sato, K. 1981a. Cosmological baryon-number domain structure and the first order phase transition of a vacuum. *Physics Letters B*, **99**(Feb.), 66–70.

Sato, K. 1981b. First-order phase transition of a vacuum and the expansion of the Universe. *MNRAS*, **195**(May), 467–479.

Sato, K. 1981c. The first order phase transition of a vacuum and baryon-number domain structure of the Universe. *Progress in Particle and Nuclear Physics*, **6**, 311–317.

Sauer, T. 2007. Einstein equations and Hilbert action: What is missing on page 8 of the proofs for Hilbert's first communication on the foundations of physics? Renn, J., Janssen, M., Norton, J. D. *et al.* (eds), *The Genesis of General Relativity*, p. 975.

Schaefer, B. E. 2007. The Hubble diagram to redshift >6 from 69 gamma-ray bursts. *ApJ*, **660**(May), 16–46.

Scheuer, P. A. G. 1957. A statistical method for analysing observations of faint radio stars. *Proceedings of the Cambridge Philosophical Society*, **53**, 764–773.

Schild, A. 1967. *Lectures on General Relativity Theory*. In *Lectures in Applied Mathematics*. Am. Math. Soc. ed. Ehlers, J. pp. 1–285.

Schmidt, B. P. *et al.* (Hi-Z team). 1998. The high-Z supernova search: Measuring cosmic deceleration and global curvature of the Universe using type IA supernovae. *ApJ*, **507**(Nov.), 46–63.

Schmidt, M. 1963. 3C 273 : A star-like object with large red-shift. *Nature*, **197**(Mar.), 1040.

Schmidt, M. 1965. Large redshifts of five quasi-stellar sources. *ApJ*, **141**(Apr.), 1295.

Schmidt, M. 1970. Space distribution and luminosity functions of quasars. *ApJ*, **162**(Nov.), 371.

Schulmann, R., Kox, A. J., Janssen, M., and Illy, J. 1998. *The Collected Papers of Albert Einstein. Vol.8: The Berlin Years: Correspondence 1914-1918*. Springer.

Schutz. 1980. *Geometrical Methods of Mathematical Physics*. Cambridge University Press.

Schwarzschild, K. 1916. On the gravitational field of a mass point according to Einstein's theory. *Abh. Konigl. Preuss. Akad. Wissenschaften Jahre 1906,92, Berlin,1907*, 189–196.

Sciama, D. W. 1953. On the origin of inertia. *MNRAS*, **113**, 34–42.

Sciama, D. W. 1967. Peculiar velocity of the Sun and the cosmic microwave background. *Physical Review Letters*, **18**(June), 1065–1067.

Sciama, D. W. 1972. Eppur Si Muove. *Comments on Astrophysics and Space Physics*, **4**(Mar.), 35.

Scott, D. 1991. Against the delta-ln(1 + z) of about 0.205 periodicity in quasar redshifts. *A&A*, **242**(Feb.), 1–12.

Scott, D., and Smoot, G.F. 2014. Cosmic microwave background. In Olive, K.A. *et al.* (eds), *Review of Particle Physics*, **C38**. Chin. Phys., p. 24.

Seeliger, H. 1895. Über das Newton'sche Gravitationsgesetz. *Astronomische Nachrichten*, **137**(Feb.), 129.

Serber, R. 1992. *The Los Alamos Primer: The First Lectures on How To Build an Atomic Bomb*. University of California Press. Originally published (classified) 1943.

Shakeshaft, J. R., Ryle, M., Baldwin, J. E., Elsmore, B., and Thomson, J. H. 1955. A survey of radio sources between declinations $-38°$ and $+83°$. *MmRAS*, **67**, 106.

Shapiro, I. I., Ash, M. E., and Smith, W. B. 1968. Icarus: Further confirmation of the relativistic perihelion precession. *Physical Review Letters*, **20**(June), 1517–1518.

Shapiro, I. I., Smith, W. B., Ash, M. E., and Herrick, S. 1971. General relativity and the orbit of Icarus. *AJ*, **76**(Sept.), 588.

Shapiro, S. S., Davis, J. L., Lebach, D. E., and Gregory, J. S. 2004. Measurement of the solar gravitational deflection of radio waves using geodetic very-long-baseline interferometry data, 1979–1999. *Physical Review Letters*, **92**(12), 121101.

Shapley, H. 1918a. Studies based on the colors and magnitudes in stellar clusters. VI. On the determination of the distances of globular clusters. *ApJ*, **48**(Sept.), 89–124.

Shapley, H. 1918b. Studies based on the colors and magnitudes in stellar clusters. VIII. The luminosities and distances of 139 Cepheid variables. *ApJ*, **48**(Dec.), 279–294.

Shapley, H. 1919. On the existence of external galaxies. JRASC, **13**(Dec.), 438.

Shapley, H. 1921. The scale of the Universe. *Bulletin of the National Research Council*, **2**, 171.

Shapley, H., and Ames, A. 1932. A survey of the external galaxies brighter than the thirteenth magnitude. *Annals of Harvard College Observatory*, **88**, 41–76.

Shaw, J. R., and Chluba, J. 2011. Precise cosmological parameter estimation using COSMOREC. *MNRAS*, **415**(Aug.), 1343–1354.

Sherwin, B. D., Dunkley, J., Das, S. *et al.* 2011. Evidence for dark energy from the cosmic microwave background alone using the Atacama Cosmology Telescope lensing measurements. *Physical Review Letters*, **107**(2), 021302.

Sievers, J. L., Hlozek, R. A., Nolta, M. R. *et al.* 2013. The Atacama Cosmology Telescope: cosmological parameters from three seasons of data. *J.C.A.P.*, **10**(Oct.), 60.

Silk, J. 1967. Fluctuations in the primordial fireball. *Nature*, **215**(Sept.), 1155–1156.

Silk, J. 1968. Cosmic black-body radiation and galaxy formation. *ApJ*, **151**(Feb.), 459.

Simak, C. 1950. *Cosmic Engineers*. Gnome Press.

Simon, J., Verde, L., and Jimenez, R. 2005. Constraints on the redshift dependence of the dark energy potential. *Phys. Rev. D*, **71**(12), 123001.

Singh, J. P., and Tiwari, R. K. 2008. Perfect fluid Bianchi Type-I cosmological models with time varying G and L. *Pramana*, **70**(Apr.), 565.

Skilling, J., and Bryan, R. K. 1984. Maximum entropy image reconstruction – general algorithm. *MNRAS*, **211**(Nov.), 111.

Skilling, J., Strong, A. W., and Bennett, K. 1979. Maximum-entropy image processing in gamma-ray astronomy. *MNRAS*, **187**(Apr.), 145–152.

Slee, O. B. 1959. Occultations of the Crab Nebula by the Solar Corona in June 1957 and 1958. *Australian Journal of Physics*, **12**(June), 134.

Slipher, V. M. 1915. Spectrographic observations of nebulae. *Popular Astronomy*, **23**(Jan.), 21–24.

Slipher, V. M. 1917. Nebulae. *Proceedings of the American Philosophical Society*, **56**, 403–409.

Smarr, L., and York, Jr., J. W. 1978. Kinematical conditions in the construction of spacetime. *Phys. Rev. D*, **17**(May), 2529–2551.

Smeenk, C., and Martin, C. 2005. Mie's theories of matter and gravitation. In *The Genesis of General Relativity*. Boston Studies in the Philosophy of Science, Vol. 250, pp. 1543–1553, Springer.

Smirnov, Y. N. 1965. Hydrogen and He4 formation in the pre-stellar Gamow Universe. *Soviet Astronomy*, **8**(June), 864.

Smolin, L. 2013. *Matters of Gravity*. American Scientist. [Online]. Available:. http://www.americanscientist.org/bookshelf/pub/matters-of-gravity.

Smoot, G., and Davidson, K. 1993. *Wrinkles in Time*. Avon Books.

Smoot, G. F., Gorenstein, M. V., and Muller, R. A. 1977. Detection of anisotropy in the cosmic blackbody radiation. *Physical Review Letters*, **39**(Oct.), 898–901.

Smoot, G. F., de Amici, G., Friedman, S. D. *et al.* 1985. Low-frequency measurements of the cosmic background radiation spectrum. *ApJ*, **291**(Apr.), L23–L27.

Smoot, G. F., Bennett, C. L., and Kogut, A. *et al.* 1991. Preliminary results from the COBE differential microwave radiometers – Large angular scale isotropy of the cosmic microwave background. *ApJ*, **371**(Apr.), L1–L5.

Smoot, G. F., Bennett, C. L., Kogut, A. *et al.* 1992. Structure in the COBE differential microwave radiometer first-year maps. *ApJ*, **396**(Sept.), L1–L5.

Snell, H., and Miller, L.C. 1948. On the radioactive decay of the neutron. *Physical Review*, **74**, 1217A–1218A. Abstract of talk presented at the American Physical Society Washington meeting, 29 April–1 May 1948.

Solomon, C., and Breckon, T. 2010. *Fundamentals of Digital Image Processing: A Practical Approach with Examples in Matlab*. Wiley-Blackwell.

Souradeep, T., and Sahni, V. 1992. Density perturbations, gravity waves and the cosmic microwave background. *Modern Physics Letters A*, **7**, 3541–3551.

Sparke, L. S., and Gallagher, III, J. S. 2007. *Galaxies in the Universe.* Cambridge University Press.

Spergel, D. N., Flauger, R., and Hložek, R. 2015. Planck data reconsidered. *Phys. Rev. D,* **91**(2), 023518.

Spitzer, Jr., L., and Greenstein, J. L. 1951. Continuous emission from planetary nebulae. *ApJ,* **114**(Nov.), 407.

Srianand, R., Petitjean, P., and Ledoux, C. 2000. The cosmic microwave background radiation temperature at a redshift of 2.34. *Nature,* **408**(Dec.), 931–935.

Starobinsky, A. A. 1979. Spectrum of relict gravitational radiation and the early state of the Universe. *Soviet Journal of Experimental and Theoretical Physics Letters,* **30**(Dec.), 682.

Starobinsky, A. A. 1982. Dynamics of phase transition in the new inflationary Universe scenario and generation of perturbations. *Physics Letters B,* **117**(Nov.), 175–178.

Starobinskii, A. A. 1983. The perturbation spectrum evolving from a nonsingular initially de sitter cosmology and the microwave background anisotropy. *Soviet Astronomy Letters,* **9**(June), 302–304.

Starobinskii, A. A. 1985. Cosmic background anisotropy induced by isotropic flat-spectrum gravitational-wave perturbations. *Soviet Astronomy Letters,* **11**(May), 133–136.

Starobinsky, A. A. 1986. Stochastic de Sitter (inflationary) stage in the early universe. In de Vega, H. J., and Sánchez, N. (eds), *Field Theory, Quantum Gravity and Strings.* Lecture Notes in Physics, Vol. 246, p. 107 Berlin Springer Verlag.

Steigman, G. 1976. Observational tests of antimatter cosmologies. *ARA&A,* **14**, 339–372.

Steigman, G. 2008. When clusters collide: constraints on antimatter on the largest scales. *JCAP,* **10**(Oct.), 001.

Steigman, G. 2012. Neutrinos and Big Bang nucleosynthesis. *Advances in High Energy Physics,* July, Article ID 268321.

Stein, C. 1956. Inadmissibility of the usual estimator for the mean of a multivariate normal distribution. In *Proc. Third Berkeley Symp. on Math. Statist. and Prob.,* Vol. 1, pp. 197–206. Univ. of Calif. Press.

Stelmach, J., Byrka, R., and Dabrowski, M. P. 1990. Large- and small-angle anisotropies of the microwave background in cosmological models with nonzero Λ term. *Phys. Rev. D,* **41**(Apr.), 2434–2443.

Stephani, H. 2008. *Relativity : an Introduction to Special and General Relativity.* 3rd. edn. Cambridge University Press.

Stephani, H., Kramer, D., MacCallum, M. A. H., Hoenselaers, C., and Herlt, E. 2009. *Exact Solutions of Einstein's Field Equations .* Cambridge University Press.

Stephenson, G, and Kilmister, C.W. 1962. *Special Relativity for Physicists.* Longmans. Reprinted by Dover Pub. 1987.

Stern, D., Jimenez, R., Verde, L., Kamionkowski, M., and Stanford, S. A. 2010. Cosmic chronometers: constraining the equation of state of dark energy. I: H(z) measurements. *JCAP,* **2**(Feb.), 8.

Stewart, I. 2008. *The Annotated Flatland: A Romance of Many Dimensions.* Basic Books.

Stewart, J. M. 1971. *Non-Equilibrium Relativistic Kinetic Theory.* Lecture Notes in Physics, Vol. 10. Springer-Verlag.

Stigler, S.M. 1983. Who discovered Bayes's theorem? *The American Statistician*, **37**, 290–296.

Story, K. T., Reichardt, C. L. *et al.* 2013. A measurement of the cosmic microwave background damping tail from the 2500-square-degree SPT-SZ survey. *ApJ*, **779** (Dec.), 86.

Strömgren, B. 1948. On the density distribution and chemical composition of the interstellar gas. *ApJ*, **108**(Sept.), 242.

Strukov, I. A., Skulachev, D. P., Boyarskiy, M. N., and Tkachev, A. N. 1987. Dipole component of RELICT radiation determined from Relikt experiment data. *JPRS Report Science Technology USSR Space*, **3**(Nov.), 59.

Strukov, I. A., Brukhanov, A. A., Skulachev, D. P., and Sazhin, M. V. 1992. Anisotropy of the microwave background radiation. *Soviet Astronomy Letters*, **18**(May), 153.

Sturrock, P. A. 1973. Evaluation of astrophysical hypotheses. *ApJ*, **182**(June), 569–580.

Sullivan, W. T. 1984. *The Early Years of Radio Astronomy – Reflections Fifty Years after Jansky's Discovery.* Cambridge University Press.

Sullivan, III, W. T. 1982. *Classics in Radio Astronomy. Studies in the History of Modern Science*, Vol. 10. Reidel.

Sung, R., and Coles, P. 2011. Temperature and polarization patterns in anisotropic cosmologies. *JCAP*, **6**(June), 36.

Sunyaev, R. A., and Chluba, J. 2009. Signals from the epoch of cosmological recombination (Karl Schwarzschild Award Lecture 2008). *Astronomische Nachrichten*, **330**(July), 657.

Sunyaev, R. A., and Zel'dovich, Ya. B. 1970a. Small-scale fluctuations of relic radiation. *Ap&SS*, **7**(Apr.), 3–19.

Sunyaev, R. A., and Zel'dovich, Ya. B. 1970b. The interaction of matter and radiation in the hot model of the Universe, II. *Ap&SS*, **7**(Apr.), 20–30.

Sunyaev, R. A., and Zel'dovich, Ya. B. 1972. The observations of relic radiation as a test of the nature of X-ray radiation from the clusters of galaxies. *Comments on Astrophysics and Space Physics*, **4**(Nov.), 173.

Sunyaev, R. A., and Zel'dovich, Ya. B. 1980. Microwave background radiation as a probe of the contemporary structure and history of the universe. *ARA&A*, **18**, 537–560.

Sussman, R. A. 2008. Quasi-local variables and inhomogeneous cosmological sources with spherical symmetry. Herrera-Aguilar, A., Murillo, F. S. G., Gómez, U. N., and Quiros, I. (eds), American Institute of Physics Conference Series, Vol. 1083, pp. 228–235.

Sussman, R. A. 2009. Quasilocal variables in spherical symmetry: Numerical applications to dark matter and dark energy sources. *Phys. Rev. D*, **79**(2), 025009.

Suzuki, N., Rubin, D., Lidman, C. *et al.* Supernova Cosmology Project. 2012. The Hubble Space Telescope cluster supernova survey. V. Improving the dark-energy constraints above $z > 1$ and building an early-type-hosted supernova sample. *ApJ*, **746**(Feb.), 85.

Szapudi, I., Prunet, S., Pogosyan, D., Szalay, A. S., and Bond, J. R. 2000. Fast CMB analyses via correlation functions. *arXiv astro-ph*, 0010256.

Szapudi, I., Prunet, S., Pogosyan, D., Szalay, A. S., and Bond, J. R. 2001. Fast cosmic microwave background analyses via correlation functions. *ApJ*, **548**(Feb.), L115–L118.

Takahara, F., and Sasaki, S. 1991. Effect of Rayleigh scattering on cosmic microwave background anisotropies. *Progress of Theoretical Physics*, **86**(Nov.), 1021–1030.

Tammann, G. A. 1979. Precise determination of the distances of galaxies. *NASA Conference Publication*. Vol. 2111, pp. 263–293.

Tanvir, N. R., Fox, D. B., Levan, A. J. *et al.* 2009. A γ-ray burst at a redshift of z~8.2. *Nature*, **461**(Oct.), 1254–1257.

Tawfik, A., and Harko, T. 2012. Quark-hadron phase transitions in the viscous early Universe. *Phys. Rev. D*, **85**(8), 084032.

Taylor, J. H., Hulse, R. A., Fowler, L. A., Gullahorn, G. E., and Rankin, J. M. 1976. Further observations of the binary pulsar PSR 1913+16. *ApJ*, **206**(May), L53–L58.

Taylor, J. H., Fowler, L. A., and McCulloch, P. M. 1979. Measurements of general relativistic effects in the binary pulsar PSR 1913+16. *Nature*, **277**(Feb.), 437–440.

Teanby, N. A. 2006. An icosahedron-based method for even binning of globally distributed remote sensing data. *Computers and Geosciences*, **32**(Nov.), 1442–1450.

Tegmark, M. 1996. An icosahedron-based method for pixelizing the celestial sphere. *ApJ*, **470**(Oct.), L81.

Terrell, J. 1977. The luminosity distance equation in Friedmann cosmology. *American Journal of Physics*, **45**(Sept.), 869–870.

Thaddeus, P., and Clauser, J. F. 1966. Cosmic microwave radiation at 2.63 mm from observations of interstellar CN. *Physical Review Letters*, **16**(May), 819–822.

Thomas, R. C., and Kantowski, R. 2000. Age-redshift relation for standard cosmology. *Phys. Rev. D*, **62**(10), 103507.

Tinsley, B. M. 1970. Possibility of a large evolutionary correction to the magnitude–redshift relation. *Ap&SS*, **6**(Mar.), 344–351.

Tinsley, B. M. 1972. Stellar evolution in elliptical galaxies. *ApJ*, **178**(Dec.), 319–336.

Tinsley, B. M. 1975 (Oct.). The evolution of galaxies and its significance for cosmology. In Bergman, P. G., Fenyves, E. J., and Motz, L. (eds), *Seventh Texas Symposium on Relativistic Astrophysics*. Annals of the New York Academy of Sciences, Vol. 262, pp. 436–448.

Tinsley, B. M. 1977a. Galaxy counts, color-redshift relations, and related quantities as probes of cosmology and galactic evolution. *ApJ*, **211**(Feb.), 621–637.

Tinsley, B. M. 1977b (Dec.). The cosmological constant and cosmological change. Papagiannis, M. D. (ed), *Eighth Texas Symposium on Relativistic Astrophysics*. Annals of the New York Academy of Sciences, Vol. 302, p. 423.

Tinsley, B. M. 1977c. The cosmological constant and cosmological change. *Physics Today*, **30**(June), 32–38.

Todorov, I. T. 2005. *Einstein and Hilbert: The Creation of General Relativity. arXiv astro-ph* 0504179

Tolman, R. C. 1931. Nonstatic model of Universe with reversible annihilation of matter. *Physical Review*, **38**(Aug.), 797–814.

Tolman, R. C. 1934a. Effect of inhomogeneity on cosmological models. *Proceedings of the National Academy of Science*, **20**(Mar.), 169–176.

Tolman, R. C. 1934b. *Relativity, Thermodynamics, and Cosmology*. Clarendon Press, Reprinted by Dover (1987).

Tolman, R. C. 1939. Static solutions of Einstein's field equations for spheres of fluid. *Physical Review*, **55**(Feb.), 364–373.

Tomita, K. 1969. Formation of gravitationally bound primordial gas clouds. II. *Progress of Theoretical Physics*, **42**(Oct.), 978–979.

Tomita, K., and Hayashi, C. 1963. The cosmical constant and the age of the Universe. *Progress of Theoretical Physics*, **30**(Nov.), 691–699.

Totsuji, H., and Kihara, T. 1969. The correlation function for the distribution of galaxies. *PASJ*, **21**, 221.

Trautman, A. 1964. *Foundations and Current Problems of General Relativity*. Prentice Hall, ed. Trautman, A., Pirani, F. A. E. and Bondi, H. pp. 1–248.

Trimble, V. 1975. The origin and abundances of the chemical elements. *Reviews of Modern Physics*, **47**(Oct.), 877–976.

Trimble, V. 1982. Supernovae. Part I: the events. *Reviews of Modern Physics*, **54**(Oct.), 1183–1224.

Trimble, V. 1983. Supernovae. Part II: the aftermath. *Reviews of Modern Physics*, **55**(Apr.), 511–563.

Trimble, V. 1995. The 1920 Shapley–Curtis discussion: Background, issues, and aftermath. *PASP*, **107**(Dec.), 1133.

Trombetti, T., and Burigana, C. 2012a. CMB polarization anisotropies from cosmological reionization: extension to B-modes. *arXiv astro-ph*, 1205.0463

Trombetti, T., and Burigana, C. 2012b. Imprints on CMB angular power spectrum modes from cosmological reionization. *J. Mod. Phys.*, **3**, 1918–1944.

Truesdell, C. 1983. *An Idiot's Fugitive Essays on Science*. Springer. pp. 97–132.

Truran, J. W., Hansen, C. J., and Cameron, A. G. W. 1965a. Helium in the interstellar medium. *AJ*, **70**(Mar.), 149.

Truran, J. W., Hansen, C. J., and Cameron, A. G. W. 1965b. The helium content of the galaxy. *Canadian Journal of Physics*, **43**, 1616.

Tryon, E. P. 1973. Is the Universe a vacuum fluctuation? *Nature*, **246**(Dec.), 396–397.

Tryon, E.P. 1984. What made the world? *New Scientist*, **101**, 14–16.

Tucker, W. 1975. *Radiation Processes in Astrophysics*. MIT Press.

Tully, R. B. 2007. In Our CMB motion: The role of the local void. In Metcalfe N. & Shanks T. (ed), *Cosmic Frontiers*. Astronomical Society of the Pacific Conference Series, Vol. 379, p. 240.

Tully, R. B., Shaya, E. J., Karachentsev, I. D. *et al.* 2008. Our peculiar motion away from the local void. *ApJ*, **676**(Mar.), 184–205.

Urey, H. C., Brickwedde, F. G., and Murphy, G. M. 1932. A hydrogen isotope of mass 2. *Physical Review*, **39**(Jan.), 164–165.

van de Hulst, H.C. 1945. Herkomst der radiogolven uit het wereldruim. *Ned. Tijdschrift vaar Natuurkunde*, **11**(Dec), 210.

van de Weygaert, M. A. M. 1991. Voids and the geometry of large scale structure. PhD thesis, University of Leiden.

van de Weygaert, R. 2002. Froth across the Universe. In Plionis, M., and Cotsakis, S. (eds), *Modern Theoretical and Observational Cosmology*. Astrophysics and Space Science Library, Vol. 276, p. 119. *arXiv astro-ph*, 0206427

van de Weygaert, R., and van Kampen, E. 1993. Voids in gravitational instability scenarios: Part One: Global density and velocity fields in an einstein–de-Sitter Universe. *MNRAS*, **263**(July), 481.

van den Bergh, S. 1975. *The Extragalactic Distance Scale*. The University of Chicago Press, ed. Sandage, A., Sandage, M. and Kristian, J., p. 509.

van Loan, C.F. 1987. *Computational Frameworks for the Fast Fourier Transform*. Frontiers in Applied Mathematics. Society for Industrial and Applied Mathematics.

Vennin, V., and Starobinsky, A. A. 2015. Correlation functions in stochastic inflation. *European Physical Journal C*, **75**(Sept.), 413.

Vessot, R. F. C., Levine, M. W., Mattison, E. M. *et al.* 1980. Test of relativistic gravitation with a space-borne hydrogen maser. *Physical Review Letters*, **45**(Dec.), 2081–2084.

Vilenkin, A. 1982. Creation of Universes from nothing. *Physics Letters B*, **117**(Nov.), 25–28.

von Laue, M. 1911. Zur Dynamik der Relativitätstheorie. *Annalen der Physik*, **35**, 524–542.

Waga, I. 1993. Decaying vacuum flat cosmological models – Expressions for some observable quantities and their properties. *ApJ*, **414**(Sept.), 436–448.

Wagoner, R. V., Fowler, W. A., and Hoyle, F. 1967. On the synthesis of elements at very high temperatures. *ApJ*, **148**(Apr.), 3.

Wald, R.M. 1984. *General Relativity*. Univ. Chicago Press.

Walke, H.J. 1935. Nuclear synthesis and stellar radiation. *Phil. Mag.*, **19**(126), 341–367.

Walker, A. G. 1935. On Riemannian spaces with spherical symmetry about a line, and the conditions for isotropy in genj relativity. *The Quarterly Journal of Mathematics*, **6**, 81–93.

Wall, J. V., and Cooke, D. J. 1975. Source counts at high spatial densities from pencil beam observations of background fluctuations. *MNRAS*, **171**(Apr.), 9–25.

Wang, L., and Steinhardt, P. J. 1998. Cluster abundance constraints for cosmological models with a time-varying, spatially inhomogeneous energy component with negative pressure. *ApJ*, **508**(Dec.), 483–490.

Ward, S. 1653. *In Ismaelis Bullialdi astronomiae philolaicae fundamenta inquisitio brevis*. Oxoniae, Leon Lichfield,.

Warwick, A. 2003. *Masters of Theory: Cambridge and the Rise of Mathematical Physics*. University of Chicago Press.

Waters, S. 1873. The distribution of clusters and nebulæ. *MNRAS*, **33**(June), 558.

Waterston, J.J. 1892. On the physics of media that are composed of free and perfectly elastic molecules in a state of motion. *Phil. Trans. Roy. Soc. A*, **183**, 1–79.

Webb, J. K., King, J. A., Murphy, M. T. *et al.* 2011. Indications of a spatial variation of the fine structure constant. *Physical Review Letters*, **107**(19), 191101.

Weber, J. 1958. *New Experiments in Gravitational Physics*.
3rd. Prize Essay Award of the Gravity Research Foundation
http://www.gravityresearchfoundation.org/pdf/awarded/1958/weber.pdf.

Weber, J. 1959. *Gravitational Waves.*
Ist Prize Essay Award of the Gravity Research Foundation.
`http://www.gravityresearchfoundation.org/pdf/awarded/1959/weber.`
`pdf` (Document barely legible).

Weber, J. 1960. Detection and generation of gravitational waves. *Physical Review*, **117**(Jan.), 306–313.

Weber, J. 1961. *General Relativity and Gravitational Waves.* Interscience (1961) and Dover (2013).

Weber, J. 1966. Observation of the thermal fluctuations of a gravitational-wave detector. *Physical Review Letters*, **17**(Dec.), 1228–1230.

Weber, J. 1967. Gravitational radiation. *Physical Review Letters*, **18**(Mar.), 498–501.

Weber, J. 1968. Gravitational-wave-detector events. *Physical Review Letters*, **20**(June), 1307–1308.

Weber, J. 1969. Evidence for discovery of gravitational radiation. *Physical Review Letters*, **22**(June), 1320–1324.

Weber, J. 1970. Anisotropy and polarization in the gravitational-radiation experiments. *Physical Review Letters*, **25**(July), 180–184.

Webster, A. S. 1974. The spectrum of the galactic non-thermal background radiational observations at 408, 610 and 1407 MHz. *MNRAS*, **166**(Feb.), 355–372.

Weinberg, S. 1972. *Gravitation and Cosmology: Principles and Applications of the General Theory of Relativity.* Wiley.

Weinberg, S. 1989. The cosmological constant problem. *Reviews of Modern Physics*, **61**(Jan.), 1–23.

Weinberg, S. 1993. *The First Three Minutes: A Modern View of the Origin of the Universe.* 3rd. edn. Basic Books.

Weinberg, S. 2008. *Cosmology.* Oxford University Press.

Weinstein, G. 2012a. *Albert Einstein's Methodology. arXiv hist-ph*, 1209.5181

Weinstein, G. 2012b. *From the Berlin 'Entwurf' Field Equations to the Einstein Tensor I: October 1914 until Beginning of November 1915. arXiv hist-ph*, 1201.5352

Weinstein, G. 2012c. *From the Berlin 'Entwurf' Field Equations to the Einstein Tensor II: November 1915 until March 1916. arXiv hist-ph*, 1201.5353

Weisberg, J. M., and Taylor, J. H. 2005 (July). The relativistic binary pulsar B1913+16: Thirty years of observations and analysis. In Rasio, F. A., and Stairs, I. H. (eds.), *Binary Radio Pulsars*. Astronomical Society of the Pacific Conference Series, Vol. 328, p. 25.

Weiss, R. 1972. Electronically coupled broadband gravitational antenna. *Quarterly Progress Report, Research Laboratory of Electronics (MIT) No. 105,* 54–76. Available from `http://www.hep.vanderbilt.edu/BTeV/test-DocDB/0009/` `000949/001/Weiss_1972.pdf`.

Westphal, J. A., Kristian, J., and Sandage, A. 1975. Absorption-line redshifts of galaxies in remote clusters obtained with a sky-subtraction spectrograph using an SIT television detector. *ApJ*, **197**(May), L95–L98.

Weyl, H. 2009. Republication of: On the general relativity theory. *General Relativity and Gravitation*, **41**(July), 1661–1666.

Weymann, R. 1965. Diffusion approximation for a photon gas interacting with a plasma via the Compton effect. *Physics of Fluids*, **8**(Nov.), 2112–2114.

Weymann, R. 1966. The energy spectrum of radiation in the expanding Universe. *ApJ*, **145**(Aug.), 560.

Will, C. M. 2006. *The Confrontation between General Relativity and Experiment*. Living Reviews in Relativity, **3**. [Online]. Available: . http://www.livingreviews.org/lrr-2006-3.

Will, C. M. 2009. The confrontation between general relativity and experiment. *Space Sci. Rev.*, **148**(Dec.), 3–13.

Will, C. M. 2014. *The Confrontation between General Relativity and Experiment*. Living Reviews in Relativity, **3**. [Online]. Available: . http://www.livingreviews.org/lrr-2014-4.

Will, C. M. 2015. Focus issue: Gravity probe B. In: *Classical and Quantum Gravity*, **32**.

Will, C. M., and Nordtvedt, Jr., K. 1972. Conservation laws and preferred frames in relativistic gravity. I. Preferred-frame theories and an extended PPN formalism. *ApJ*, **177**(Nov.), 757.

Wilson, O. C. 1939. Possible applications of supernovae to the study of the nebular red shifts. *ApJ*, **90**(Nov.), 634.

Wilson, R. N. 1990. 'Matching error' (spherical aberration) in the Hubble Space Telescope (HST) – Some technical comments. *The Messenger*, **61**(Sept.), 22–24.

Wilson, R. W. 1979. The cosmic microwave background radiation. *Reviews of Modern Physics*, **51**(July), 433–446.

Wiltshire, D. L. 2009. Average observational quantities in the timescape cosmology. *Phys. Rev. D*, **80**(12), 123512.

Winterberg, F. 2003. On 'Belated decision in the Hilbert–Einstein priority dispute', published by L. Corry, J. Renn, and J. Stachel. *Z. Naturforsch.*, **59a**, 715–719.

Wirtz, C. 1918. Über die Bewegungen der Nebelflecke. *Astronomische Nachrichten*, **206**, 109–116.

Wirtz, C. 1922. Notiz zur Radialbewegung der Spiralnebel. *Astronomische Nachrichten*, **216**(Sept.), 451.

Wirtz, C. 1924. de Sitters Kosmologie und die Radialbewegungen der Spiralnebel. *Astronomische Nachrichten*, **222**(July), 21.

Wong, W. Y., Moss, A., and Scott, D. 2008. How well do we understand cosmological recombination? *MNRAS*, **386**(May), 1023–1028.

Wood-Vasey, W. M. *et al.* 2007. Observational constraints on the nature of dark energy: First cosmological results from the ESSENCE supernova survey. *ApJ*, **666**(Sept.), 694–715.

Woody, D. P., and Richards, P. L. 1981. Near-millimeter spectrum of the microwave background. *ApJ*, **248**(Aug.), 18–37.

Wright, E. L., Smoot, G. F., Bennett, C. L., and Lubin, P. M. 1994a. Angular power spectrum of the microwave background anisotropy seen by the COBE differential microwave radiometer. *ApJ*, **436**(Dec.), 443–451.

Wright, E. L., Mather, J. C., Fixsen, D. J. *et al.* 1994b. Interpretation of the COBE FIRAS CMBR spectrum. *ApJ*, **420**(Jan.), 450–456.

Xu, L., Zhang, C., Chang, B., and Liu, H. 2007. Reconstruction of a deceleration parameter from the latest type ia supernovae gold dataset. *arXiv astro-ph*, 0701490

Xu, Y., Tegmark, M., and de Oliveira-Costa, A. 2002. CMB power spectrum at l=30–200 from QMASK. *Phys. Rev. D*, **65**(8), 083002.

Yue, A. T., Dewey, M. S., Gilliam, D. M. *et al.* 2013. Improved determination of the neutron lifetime. *Physical Review Letters*, **111**(22), 222501.

Yung, K.H. 1999. *Explaining the Stein Paradox*. Link may be volatile. [Online]. Available: `https://www.cs.nyu.edu/~roweis/csc2515-2006/readings/stein_paradox.pdf`.

Zaldarriaga, M., and Seljak, U. 1997. All-sky analysis of polarization in the microwave background. *Phys. Rev. D*, **55**(Feb.), 1830–1840.

Zangwill, A. 2013. *Modern Electrodynamics*. Cambridge University Press.

Zel'dovich, I. B., Einasto, J., and Shandarin, S. F. 1982. Giant voids in the universe. *Nature*, **300**(Dec.), 407–413.

Zel'dovich, Y. B. 1963. Pre-stellar state of matter. *Soviet Journal of Experimental and Theoretical Physics*, **16**, 1102.

Zel'dovich, Y. B. 1964. Special issue: the theory of the expanding Universe as originated by A. A. Fridman. *Soviet Physics Uspekhi*, **6**(Apr.), 475–494.

Zel'dovich, Y. B. 1972. A hypothesis, unifying the structure and the entropy of the Universe. *MNRAS*, **160**, 1P.

Zel'dovich, Y. B., and Khlopov, M. Y. 1978. On the concentration of relic magnetic monopoles in the Universe. *Physics Letters B*, **79**(Nov.), 239–241.

Zel'dovich, Y. B., and Raizer, Y. P. 1967. *Physics of shock waves and high-temperature hydrodynamic phenomena*. Academic Press.

Zel'dovich, Y. B., and Sunyaev, R. A. 1969. The interaction of matter and radiation in a hot-model Universe. *Ap&SS*, **4**(July), 301–316.

Zel'dovich, Y. B., Kurt, V. G., and Syunyaev, R. A. 1969. Recombination of hydrogen in the hot model of the Universe. *Soviet Journal of Experimental and Theoretical Physics*, **28**(Jan.), 146.

Zel'dovich, Y. B., Illarionov, A. F., and Syunyaev, R. A. 1972. The effect of energy release on the emission spectrum in a hot universe. *Soviet Journal of Experimental and Theoretical Physics*, **35**, 643.

Zel'dovich, Ya. B. 1970. Gravitational instability: An approximate theory for large density perturbations. *A&A*, **5**(Mar.), 84–89.

Zwicky, F. 1937. On the masses of nebulae and of clusters of nebulae. *ApJ*, **86**(Oct.), 217.

Index

Printed in the United States
by Baker & Taylor Publisher Services